CALCULUS
One and Several Variables

Sir Isaac Newton

CALCULUS:

One and Several Variables

with Analytic Geometry

Third Edition

S. L. Salas   **Einar Hille**

John Wiley & Sons
New York   Santa Barbara   Chichester   Brisbane   Toronto

This book was printed by Murray Printing and bound by Book Press.
It was set in Times Roman by Progressive Typographers
Christine Pines supervised production.
Picture research was done by Kathy Bendo.

Library of Congress Cataloging in Publication Data

Salas, Saturnino L
   Calculus.

   Includes index.
    I. Hille, Einar, 1894– joint author.
QA303.S17 1978   515   77-11630
ISBN 0-471-74983-4

Printed in the United States of America

10 9 8 7 6 5 4 3 2

Preface

Calculus: One and Several Variables is available both as a complete volume and in two parts. In the two-part version, Part 1 considers functions of one variable, analytic geometry, and sequences and series (Chapters 1 to 13 of the complete volume). Part 2 repeats the material on sequences and series and goes on to discuss functions of several variables and vector calculus (Chapters 12 to 19 of the complete volume).

Both in content and in spirit this edition is very similar to the previous ones. There are, however, some noticeable differences:

1. All the figures have been redrawn.
2. Chapters which were rather long in previous editions have been broken up, resulting in a marked increase in the number of chapters.
3. For the benefit of students who come weakly prepared, we have added a review section on lines and slowed the pace of our discussion of polar coordinates.
4. Although we continue to define the definite integral in terms of upper and lower sums, we also introduce Riemann sums and use them in several applications.
5. Indeterminate forms, formerly discussed after infinite series, now appear before the chapter on infinite series. This makes L'Hospital's rule available for radius of convergence arguments.
6. Line integrals, formerly appearing at the end of our chapter on functions of several variables, now form part of a new final chapter. In this new chapter we take up curl and divergence and give an elementary view of Green's Theorem, the Divergence Theorem, and Stokes's Theorem.

Here and there you may find other changes—a paragraph rewritten, an additional example, a new exercise—but most of these changes are minor and will probably pass unnoticed.

S. L. Salas
Haddam, Connecticut

Acknowledgments

We are particularly grateful to Wiley's Mathematics Editor, Gary W. Ostedt, whose judgment, imagination, good humor, and steadfast support sustained us throughout the revision.

Through the efforts of Mr. Ostedt we received valuable critiques from many distinguished colleagues. In this regard we wish to acknowledge our special debt to Professors John T. Anderson (Hamilton College), Harvey B. Keynes (University of Minnesota), Clifford Kottman (Simpson College), John W. Lee (Oregon State University), David Lovelock (University of Arizona), Stanley M. Lukawecki (Clemson University), Gordon D. Prichett (Hamilton College), and Donald R. Sherbert (University of Illinois at Urbana-Champaign).

Finally a word about an assistant who, to the point of exhaustion, read and reread and reread the manuscript, questioning this, questioning that, suggesting this, suggesting that. For all this, and more, we offer our thanks to Charles G. Salas.

S. L. S.

A Note From The Publisher

A *Student Supplement* is available for use either as a self-study guide or in conjunction with any calculus course based on this text. Preliminary editions of this supplement have been classroom tested since 1972.

The *Supplement* provides further explanations and perspectives on the material in the text. Additional examples are worked out in detail to help focus attention on the central concept(s) of each section.

The Greek Alphabet

A	α	alpha
B	β	beta
Γ	γ	gamma
Δ	δ	delta
E	ϵ	epsilon
Z	ζ	zeta
H	η	eta
Θ	θ	theta
I	ι	iota
K	κ	kappa
Λ	λ	lambda
M	μ	mu
N	ν	nu
Ξ	ξ	xi
O	o	omicron
Π	π	pi
P	ρ	rho
Σ	σ	sigma
T	τ	tau
Y	υ	upsilon
Φ	ϕ	phi
X	χ	chi
Ψ	ψ	psi
Ω	ω	omega

Gottfried Leibniz

Contents

Introduction

1.1 What is Calculus?

To a Roman in the days of the empire a "calculus" was a little pebble used in counting and in gambling. Centuries later the verb "calculare" came to mean "to compute," "to reckon," "to figure out." To the engineer and mathematician of today calculus is the branch of mathematics that takes in elementary algebra and geometry and adds one more ingredient: *the limit process*.

Calculus begins where elementary mathematics leaves off. It takes ideas from elementary mathematics and extends them to a much more general situation. Here are some examples. On the left-hand side you will find an idea from elementary mathematics; on the right, this same idea as enriched by calculus.

Elementary Mathematics	*Calculus*
slope of a line $y = mx + b$	slope of a curve $y = f(x)$

| tangent line to a circle | tangent line to a more general curve |

| average velocity, average acceleration | instantaneous velocity, instantaneous acceleration |

| distance moved under a constant velocity | distance moved under varying velocity |

| area of a region bounded by line segments | area of a region bounded by curves |

| sum of a finite collection of numbers $a_1 + a_2 + \cdots + a_n$ | sum on an infinite series $a_1 + a_2 + \cdots + a_n + \cdots$ |

| average of a finite collection of numbers | average value of a function on an interval |

| length of a line segment | length of a curve |

center of a circle

centroid of a region

volume of
a rectangular solid

volume of a solid
with a curved boundary

surface area of
a cylinder

surface area of
a more general solid

tangent plane to
a sphere

tangent plane to
a more general surface

work done by
a constant force

work done by
a varying force

mass of an object
of constant density

mass of an object
of varying density

center of a sphere

center of gravity of
a more general solid

It is fitting to say something about the discovery of calculus and about the men who made it. A certain amount of preliminary work was done during the first half of the seventeenth century, but the actual discovery can be credited to Sir Isaac Newton (1642–1727) and to Gottfried Wilhelm Leibniz (1646–1716), an Englishman and a German. Newton's discovery is one of the few good turns that the Great Plague did mankind. The plague forced the closing of Cambridge University in 1665 and young Isaac Newton of Trinity College returned to his home in Lincolnshire for eighteen months of meditation, out of which grew his *method of fluxions,* his *theory of gravitation* (*Philosophiae Naturalis Principia Mathematica,* 1687), and his *theory of light* (*Opticks,* 1704). The method of fluxions is what concerns us here. A treatise with this title was written by Newton in 1672, but it remained unpublished until 1736, nine years after his death. The new method (calculus to us) was first announced in 1687 but in vague general terms without symbolism, formulas, or applications. Newton himself seemed very reluctant to publish anything tangible about his discovery, and it is not surprising that the development on the Continent, in spite of a late start, soon overtook Newton and went beyond him.

Leibniz started his work in 1673, eight years after Newton. In 1675 he initiated the basic notation: dx and $\int$. His first publications appeared in 1684 and 1686. These made little stir in Germany, but the two brothers Bernoulli of Basel (Switzerland) took up the ideas and added profusely to them. From 1690 onward calculus grew rapidly. It reached approximately its present state in about one hundred years.

1.2   Notions and Formulas from Elementary Mathematics

The following outline is presented for review and easy reference.

I. Sets

the object x is in the set A (x is an element of A)	$x \in A$
the object x is not in the set A	$x \notin A$
containment	$A \subseteq B$
union	$A \cup B$
intersection	$A \cap B$
empty set	$\varnothing$
Cartesian products	$A \times B, \; A \times B \times C$

(These are the only notions from set theory that you will need for this book. If you are not familiar with them, see Appendix A.1 at the back of the book.)

II. Real Numbers

CLASSIFICATION

integers	$0, 1, -1, 2, -2, 3, -3$, etc.
rational numbers	$\dfrac{p}{q}$ where p and q are integers, $q \neq 0$.
irrational numbers	real numbers which are not rational; for example, $\sqrt{2}, \pi$.

ORDER PROPERTIES

(i) Either $a < b$, $b < a$, or $a = b$.
(ii) If $a < b$ and $b < c$, then $a < c$.
(iii) If $a < b$, then $a + c < b + c$ for all real numbers c.
(iv) If $a < b$ and $c > 0$, then $ac < bc$.
     If $a < b$ and $c < 0$, then $ac > bc$.

DENSITY

Between any two real numbers there is a rational number and an irrational number.

ABSOLUTE VALUE

$$\textit{definition} \quad |a| = \left\{ \begin{array}{ll} a, & \text{if } a \geq 0 \\ -a, & \text{if } a \leq 0 \end{array} \right\} .$$

$$\textit{other characterizations} \quad |a| = \max \{a, -a\}; \; |a| = \sqrt{a^2} .$$

geometric interpretation $|a|$ = distance between a and 0.

$|a - c|$ = distance between a and c.

properties   (i) $|a| = 0$ iff $a = 0$.†

(ii) $|-a| = |a|$.

(iii) $|ab| = |a||b|$.

(iv) $|a + b| \le |a| + |b|$.   (the triangle inequality)††

(v) $\big||a| - |b|\big| \le |a - b|$.   (a variant of the triangle inequality)

INTERVALS

We use $\{x: P\}$ to denote *the set of all x for which property P holds*. Suppose now that $a < b$. The *open interval* (a, b) is the set of all numbers between a and b:

$(a, b) = \{x: a < x < b\}$.

The *closed interval* $[a, b]$ is the open interval (a, b) together with the endpoints a and b:

$[a, b] = \{x: a \le x \le b\}$.

There are seven other types of intervals:

$(a, b] = \{x: a < x \le b\}$.

$[a, b) = \{x: a \le x < b\}$.

$(a, \infty) = \{x: a < x\}$.

$[a, \infty) = \{x: a \le x\}$.

$(-\infty, b) = \{x: x < b\}$.

$(-\infty, b] = \{x: x \le b\}$.

$(-\infty, \infty)$ = set of real numbers.

This interval notation is easy to remember: we use a bracket to include an endpoint; otherwise a parenthesis.

BOUNDEDNESS

A set S of real numbers is said to be

(i) *bounded above* iff there exists a real number M such that

$$x \le M \qquad \text{for all} \quad x \in S.$$

(M is called an *upper bound* for S.)

---

† By "iff" we mean "if and only if." This expression is used so often in mathematics that it's convenient to have an abbreviation for it.

†† The absolute value of the sum of two numbers cannot exceed the sum of their absolute values just as in a triangle the length of a side cannot exceed the sum of the lengths of the other two sides.

(ii) *bounded below* iff there exists a real number m such that

$$m \le x \quad \text{for all} \quad x \in S.$$

(m is called a *lower bound* for S.)

(iii) *bounded* iff it is bounded above and below.

Example. The intervals $(-\infty, 2]$ and $(-\infty, 2)$ are bounded above but not below.

Example. The set of positive integers is bounded below but not above.

Example. The intervals $[0, 1]$, $(0, 1)$, and $(0, 1]$ are bounded (both above and below).

III. Algebra and Geometry

GENERAL QUADRATIC FORMULA

The quadratic equation

$$ax^2 + bx + c = 0, \quad a \ne 0$$

has solutions

$$x = \frac{-b \pm \sqrt{b^2 - 4ac}}{2a}.$$

FACTORIALS

$$0! = 1$$
$$1! = 1$$
$$2! = 2 \cdot 1$$
$$3! = 3 \cdot 2 \cdot 1$$
$$.$$
$$.$$
$$.$$
$$n! = n(n - 1) \cdots (2)(1).$$

ELEMENTARY FIGURES

circle   area $= \pi r^2$; circumference $= 2\pi r$;
area of sector $= \frac{1}{2}r^2\theta$, where θ is the central angle measured in radians;
length of circular arc $= r\theta$, where θ is the central angle measured in radians.

right circular cylinder   volume $= \pi r^2 h$; lateral area $= 2\pi rh$;
total surface area $= 2\pi r(r + h)$.

$cone$ volume $= \frac{1}{3}\pi r^2 h$;

slant height $= \sqrt{r^2 + h^2}$;

lateral area $= \pi r \sqrt{r^2 + h^2}$;

total surface area $= \pi r(r + \sqrt{r^2 + h^2})$.

$sphere$ volume $= \frac{4}{3}\pi r^3$; surface area $= 4\pi r^2$.

POLYNOMIALS

A *polynomial* is a function of the form

$$P(x) = a_n x^n + a_{n-1} x^{n-1} + \cdots + a_1 x + a_0$$

where n is a positive integer. If $a_n \neq 0$, the polynomial is said to have *degree n*.
Recall the *factor theorem*: if P is a polynomial and c is a real number, then

$$P(c) = 0 \quad \text{iff} \quad x - c \text{ is a factor of } P(x).$$

IV. Analytic Geometry

RECTANGULAR COORDINATES AND THE DISTANCE FORMULA

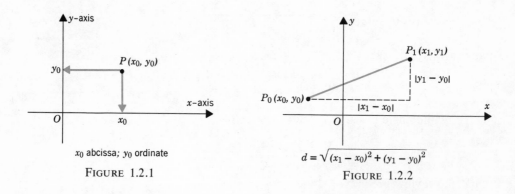

x_0 abcissa; y_0 ordinate

FIGURE 1.2.1

$d = \sqrt{(x_1 - x_0)^2 + (y_1 - y_0)^2}$

FIGURE 1.2.2

LINES

(i) Slope of a nonvertical line:

$$m = \frac{y_1 - y_0}{x_1 - x_0}$$

where $P_0(x_0, y_0)$ and $P_1(x_1, y_1)$ are any two points on the line;

$$m = \tan \theta$$

where θ, called *the inclination,* is the angle between the line and the x-axis.

$$\text{FIGURE } 1.2.3$$

general form	$Ax + By + C = 0$, with A and B not both 0.
vertical line	$x = a$.
slope-intercept form	$y = mx + b$.
point-slope form	$y - y_0 = m(x - x_0)$.
two-point form	$y - y_0 = \dfrac{y_1 - y_0}{x_1 - x_0}(x - x_0)$.
two-intercept form	$\dfrac{x}{a} + \dfrac{y}{b} = 1$.

(iii) Two nonvertical lines $l_1: y = m_1 x + b_1$ and $l_2: y = m_2 x + b_2$ are

$$\begin{array}{lll} \textit{parallel} & \text{iff} & m_1 = m_2; \\ \textit{perpendicular} & \text{iff} & m_1 m_2 = -1. \end{array}$$

If α is the angle between l_1 and l_2, then

$$\tan \alpha = \frac{m_1 - m_2}{1 + m_1 m_2}.$$

(Some review problems on lines are given in Section 1.3.)

EQUATIONS OF CONIC SECTIONS

circle of radius r centered at the origin	$x^2 + y^2 = r^2$.
circle of radius r centered at $P(x_0, y_0)$	$(x - x_0)^2 + (y - y_0)^2 = r^2$.
parabolas	$x^2 = 4cy, \quad y^2 = 4cx$.
ellipse	$\dfrac{x^2}{a^2} + \dfrac{y^2}{b^2} = 1$.
hyperbolas	$\dfrac{x^2}{a^2} - \dfrac{y^2}{b^2} = 1, \quad xy = c$.

(The conic sections are discussed extensively in Chapter 9.)

V. *Trigonometry*

ANGLE MEASUREMENT

$$\pi \cong 3.14159.†$$

$$2\pi \text{ radians} = 360 \text{ degrees} = \text{one complete revolution}.$$

$$\text{One radian} = \frac{180}{\pi} \text{ degrees} \cong 57.296 \text{ degrees}.$$

$$\text{One degree} = \frac{\pi}{180} \text{ radians} \cong 0.0175 \text{ radians}.$$

SECTORS

Figure 1.2.4 displays a sector in a circle of radius r. If θ is the central angle measured in radians, then

$$\text{the length of the subtended arc} = r\theta$$

and

$$\text{the area of the sector} = \tfrac{1}{2}r^2\theta.$$

SINE AND COSINE

Take θ in the interval $[0, 2\pi)$. In the unit circle (Figure 1.2.5) mark off the sector *OAP* with a central angle of θ radians:

$$\cos \theta = x\text{-coordinate of } P, \quad \sin \theta = y\text{-coordinate of } P.$$

These definitions are then extended periodically:

$$\cos \theta = \cos (\theta + 2n\pi), \quad \sin \theta = \sin (\theta + 2n\pi).$$

FIGURE 1.2.4

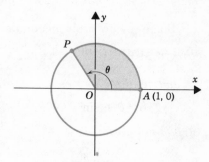

FIGURE 1.2.5

† We use $\cong$ to indicate approximate equality.

OTHER TRIGONOMETRIC FUNCTIONS

$$\tan \theta = \frac{\sin \theta}{\cos \theta}, \qquad \cot \theta = \frac{\cos \theta}{\sin \theta},$$

$$\sec \theta = \frac{1}{\cos \theta}, \quad \operatorname{cosec} \theta = \frac{1}{\sin \theta}.$$

IN TERMS OF A RIGHT TRIANGLE

For θ between 0 and $\frac{1}{2}\pi$,

$$\sin \theta = \frac{\text{opposite side}}{\text{hypotenuse}} \qquad \cos \theta = \frac{\text{adjacent side}}{\text{hypotenuse}}$$

$$\tan \theta = \frac{\text{opposite side}}{\text{adjacent side}} \qquad \cot \theta = \frac{\text{adjacent side}}{\text{opposite side}}$$

$$\sec \theta = \frac{\text{hypotenuse}}{\text{adjacent side}} \qquad \operatorname{cosec} \theta = \frac{\text{hypotenuse}}{\text{opposite side}}.$$

FIGURE 1.2.6

ODDNESS AND EVENNESS

cosine function is even $\cos (-\theta) = \cos \theta$. The graph is symmetric about the y-axis.

sine and tangent functions are odd $\sin (-\theta) = -\sin \theta$, $\tan (-\theta) = -\tan \theta$. The graphs are symmetric about the origin.

SOME IMPORTANT IDENTITIES

(i) *unit circle*

$$\cos^2 \theta + \sin^2 \theta = 1, \quad \tan^2 \theta + 1 = \sec^2 \theta, \quad \cot^2 \theta + 1 = \operatorname{cosec}^2 \theta.$$

(ii) *addition formulas*

$$\sin (\theta_1 + \theta_2) = \sin \theta_1 \cos \theta_2 + \cos \theta_1 \sin \theta_2,$$
$$\sin (\theta_1 - \theta_2) = \sin \theta_1 \cos \theta_2 - \cos \theta_1 \sin \theta_2,$$
$$\cos (\theta_1 + \theta_2) = \cos \theta_1 \cos \theta_2 - \sin \theta_1 \sin \theta_2,$$
$$\cos (\theta_1 - \theta_2) = \cos \theta_1 \cos \theta_2 + \sin \theta_1 \sin \theta_2;$$

$$\tan (\theta_1 + \theta_2) = \frac{\tan \theta_1 + \tan \theta_2}{1 - \tan \theta_1 \tan \theta_2}, \quad \tan (\theta_1 - \theta_2) = \frac{\tan \theta_1 - \tan \theta_2}{1 + \tan \theta_1 \tan \theta_2}.$$

(iii) *double-angle formulas*

$$\sin 2\theta = 2 \sin \theta \cos \theta, \quad \cos 2\theta = \cos^2 \theta - \sin^2 \theta.$$

(iv) *half-angle formulas*

$$\sin^2 \tfrac{1}{2}\theta = \tfrac{1}{2}(1 - \cos \theta), \quad \cos^2 \tfrac{1}{2}\theta = \tfrac{1}{2}(1 + \cos \theta).$$

FOR AN ARBITRARY TRIANGLE

$$area \quad \tfrac{1}{2}ab \sin C.$$

$$law \ of \ sines \quad \frac{\sin A}{a} = \frac{\sin B}{b} = \frac{\sin C}{c}.$$

$$law \ of \ cosines \quad a^2 = b^2 + c^2 - 2bc \cos A.$$

(You won't need much trigonometry until Chapter 7.)

VI. Induction

Axiom   Let S be a set of integers. If 1 is in S and if k is in S implies that $k + 1$ is in S, then all the positive integers are in S.

In general, we use induction when we want to prove that a certain proposition is true for all positive integers n. To do this, first we prove that the proposition is true for $n = 1$. Next we prove that, if it is true for $n = k$, then it is true for $n = k + 1$. It follows then from the axiom that the proposition is true for all positive integers n.

(We will use induction only sparingly. For a discussion of it see Appendix A.2.)

1.3   Some Problems on Lines

Problem.   Find the slope and y-intercept of each of the following lines:

$$l_1: 20x - 24y - 30 = 0, \quad l_2: 2x - 3 = 0, \quad l_3: 4y + 5 = 0.$$

SOLUTION.   The equation of l_1 can be written

$$y = \tfrac{5}{6}x - \tfrac{5}{4}.$$

This is in the form $y = mx + b$. The slope is $\tfrac{5}{6}$, and the y-intercept is $-\tfrac{5}{4}$.
The equation of l_2 can be written

$$x = \tfrac{3}{2}.$$

The line is vertical and the slope is not defined. Since the line does not cross the y-axis, it has no y-intercept.
The third equation can be written

$$y = -\tfrac{5}{4}.$$

The line is horizontal. Its slope is 0 and the y-intercept is $-\frac{5}{4}$. The three lines are drawn in Figure 1.3.1.　☐

$$y = \tfrac{5}{6}x - \tfrac{5}{4} \qquad\qquad x = \tfrac{3}{2} \qquad\qquad y = -\tfrac{5}{4}$$

FIGURE 1.3.1

Problem. Write an equation for the line l_2 that is parallel to

$$l_1: 3x - 5y + 8 = 0$$

and passes through the point $P(-3, 2)$.

SOLUTION. We can rewrite the equation of l_1 as

$$y = \tfrac{3}{5}x + \tfrac{8}{5}.$$

The slope of l_1 is $\frac{3}{5}$. The slope of l_2 must also be $\frac{3}{5}$.
　Since l_2 passes through $(-3, 2)$ with slope $\frac{3}{5}$, we can write its equation as

$$y - 2 = \tfrac{3}{5}(x + 3). \quad\square$$

Problem. Write an equation for the line that is perpendicular to

$$l_1: x - 4y + 8 = 0$$

and passes through the point $P(2, -4)$.

SOLUTION. The equation for l_1 can be written

$$y = \tfrac{1}{4}x + 2.$$

The slope of l_1 is $\frac{1}{4}$. The slope of l_2 is therefore -4. (*Remember:* for nonvertical perpendicular lines $m_1 m_2 = -1$.)
　Since l_2 passes through $(2, -4)$ with slope -4, we can write its equation as

$$y + 4 = -4(x - 2). \quad\square$$

Problem. Find the inclination of the line

$$3x - 4y + 2 = 0.$$

SOLUTION. In the slope-intercept form the equation becomes

$$y = \tfrac{3}{4}x + \tfrac{1}{2},$$

so that $m = \frac{3}{4}$. Here $\tan \theta = \frac{3}{4} = 0.7500$, and by Table 5 in the Appendix $\theta \cong 37°$. $\square$

Problem.   Find the point where the lines

$$l_1: 2x - 7y + 31 = 0 \quad \text{and} \quad l_2: x - 2y + 7 = 0$$

intersect, and determine the angle of intersection.

SOLUTION.   To find the point of intersection we solve the two equations simultaneously. As you can check, this gives $x = \frac{13}{3}$, $y = \frac{17}{3}$. The point of intersection is $P(\frac{13}{3}, \frac{17}{3})$.

To find the angle of intersection we use the formula

$$\tan \alpha = \frac{m_2 - m_1}{1 + m_1 m_2}.$$

Here, as you can verify, $m_1 = \frac{2}{7}$ and $m_2 = \frac{1}{2}$, so that

$$\tan \alpha = \frac{\frac{1}{2} - \frac{2}{7}}{1 + (\frac{1}{2})(\frac{2}{7})} = \frac{3}{16} = 0.1875.$$

By Table 5 in the Appendix, $\alpha \cong 11°$. $\square$

Exercises†

1. Find the slope of the line through the given points:
   *(a) $P_0(-1, 2)$, $P_1(3, 4)$.   (b) $P_0(4, 5)$, $P_1(6, -3)$.
   *(c) $P_0(3, 5)$, $P_1(6, 5)$.   (d) O, $P(x_0, y_0)$.
   *(e) $P(x_0, 0)$, $Q(0, y_0)$.   (f) $P(a, b)$, $Q(b, a)$.
2. Find the slope and the y-intercept of each of the following lines:
   *(a) $y = x + 4$.   (b) $y = 3x - 1$.
   *(c) $4x = 1$.   (d) $3y = x + 2$.
   *(e) $x + y + 1 = 0$.   (f) $4x - 2y + 6 = 0$.
   *(g) $7x + 4y + 4 = 0$.   (h) $7x - 4y + 4 = 0$.
   *(i) $2y + 5 = 0$.   (j) $\frac{1}{2}x + \frac{1}{3}y = \frac{1}{6}$.
3. Write an equation for the line with
   *(a) slope 5 and y-intercept 2.   (b) slope -5 and y-intercept 2.
   *(c) slope 5 and y-intercept -2.   (d) slope -5 and y-intercept -2.
4. Write an equation for the horizontal line 3 units
   *(a) above the x-axis.   (b) below the x-axis.
5. Write an equation for the vertical line 3 units
   (a) to the right of the y-axis.   *(b) to the left of the y-axis.
6. Find an equation for the line that passes through the point $P(2, 7)$ and is
   *(a) parallel to the x-axis.
   (b) parallel to the y-axis.

---

† The starred exercises have answers at the back of the book.

*(c) parallel to the line $x + y + 1 = 0$.

(d) perpendicular to the line $x + y + 1 = 0$.

(e) parallel to the line $3x - 2y + 6 = 0$.

*(f) perpendicular to the line $3x - 2y + 6 = 0$.

7. Write the equation $3y + 6x - 12 = 0$

   *(a) in the slope-intercept form.

   (b) in the point-slope form using the point $P(1, 2)$.

   *(c) in the two-intercept form.

8. Find the inclination of the line:

   *(a) $x - y + 2 = 0$.   (b) $6y + 5 = 0$.

   *(c) $2x - 3 = 0$.   (d) $2x - 3y + 6 = 0$.

9. Write an equation for the line with

   *(a) inclination 30°, y-intercept 2.   (b) inclination 60°, x-intercept 2.

   *(c) inclination 150°, x-intercept 3.   (d) inclination 135°, y-intercept 3.

10. Find the point where the lines intersect and determine the angle of intersection:

   *(a) $l_1: 2x - 3y - 3 = 0$, $l_2: 3x - 2y - 3 = 0$.

   (b) $l_1: 3x + 10y + 4 = 0$, $l_2: 7x - 10y - 1 = 0$.

   *(c) $l_1: 4x - y + 2 = 0$, $l_2: 19x + y = 0$.

   (d) $l_1: 5x - 6y + 1 = 0$, $l_2: 8x + 5y + 2 = 0$.

11. Find the angles of the triangle with the given vertices:

   *(a) $(-2, -5)$, $(0, 0)$, $(5, -2)$.   (b) $(1, -1)$, $(4, 0)$, $(3, 2)$.

   *(c) $(-4, -2)$, $(0, -1)$, $(1, 1)$.   (d) $(-4, 2)$, $(1, 0)$, $(1, 1)$.

1.4 Inequalities

To solve an inequality is to find the set of numbers that satisfy it. Inequalities play such an important role in calculus, that it's probably a good idea to refresh your memory on the subject. Here are some sample problems.

Problem.   Solve the inequality

$$3(1 - x) \leq 6.$$

SOLUTION

$$3(1 - x) \leq 6,$$
$$1 - x \leq 2, \qquad \text{(we divided by 3)}$$
$$-x \leq 1, \qquad \text{(subtracted 1)}$$
$$x \geq -1. \qquad \text{(multiplied by } -1)$$

As the solution we have the interval $[-1, \infty)$. $\square$

Problem.   Solve the inequality

$$2x^2 - 2x > 4.$$

SOLUTION

$$2x^2 - 2x > 4,$$
$$2x^2 - 2x - 4 > 0, \qquad \text{(we subtracted 4)}$$
$$x^2 - x - 2 > 0, \qquad \text{(divided by 2)}$$
$$(x + 1)(x - 2) > 0. \qquad \text{(factored)}$$

The product $(x + 1)(x - 2)$ is positive iff both factors are negative:

$$x + 1 < 0, \quad x - 2 < 0,$$

or both factors are positive:

$$x + 1 > 0, \quad x - 2 > 0.$$

The first situation arises when

$$x < -1$$

and the second situation arises when

$$x > 2.$$

The solution is

$$(-\infty, -1) \cup (2, \infty). \quad \square$$

Problem.   Solve the inequality

$$\tfrac{1}{5}(x^2 - 4x + 3) < 0.$$

SOLUTION

$$\tfrac{1}{5}(x^2 - 4x + 3) < 0,$$
$$x^2 - 4x + 3 < 0, \qquad \text{(we multiplied by 5)}$$
$$(x - 1)(x - 3) < 0. \qquad \text{(factored)}$$

The product $(x - 1)(x - 3)$ is negative iff the two factors have different signs. Since

$$x - 3 < x - 1,$$

it cannot happen that

$$x - 1 < 0 \quad \text{and} \quad x - 3 > 0. \qquad \text{(explain)}$$

The remaining possibility is that

$$x - 3 < 0 \quad \text{and} \quad x - 1 > 0.$$

Since $x - 3 < 0$ when

$$x < 3$$

and $x - 1 > 0$ when

$$x > 1,$$

both of these conditions hold only when

$$1 < x < 3.$$

The solution is the open interval $(1, 3)$. □

More generally, an expression of the form

$$(x - a_1)(x - a_2) \cdot \cdot \cdot (x - a_n)$$

is zero at $a_1, a_2, \ldots, a_n$. It is positive on those intervals where an even number of terms are negative, and it is negative on those intervals where an odd number of terms are negative.

As an example take the expression

$$(x + 2)(x - 1)(x - 3).$$

This product is zero at -2, 1, and 3. It is

negative on $(-\infty, -2)$,	(3 negative terms)
positive on $(-2, 1)$,	(2 negative terms)
negative on $(1, 3)$,	(1 negative term)
positive on $(3, \infty)$.	(0 negative terms)

Problem.   Solve the inequality

$$x^2 + 4x - 2 \leq 0.$$

SOLUTION.   To factor the quadratic we first complete the square:

$$x^2 + 4x - 2 \leq 0,$$
$$x^2 + 4x \leq 2,$$
$$x^2 + 4x + 4 \leq 2 + 4,$$
$$(x + 2)^2 \leq 6.$$

Now we subtract 6 from both sides:

$$(x + 2)^2 - 6 \leq 0$$

and factor the expression as the difference of two squares:

$$(x + 2 + \sqrt{6})(x + 2 - \sqrt{6}) \leq 0.$$

The product is 0 at $-2 - \sqrt{6}$ and at $-2 + \sqrt{6}$. It is negative between these two numbers. The solution is the closed interval $[-2 - \sqrt{6}, -2 + \sqrt{6}]$. □

Problem.   Solve the inequality

$$\frac{x + 2}{1 - x} > 1.$$

SOLUTION.   Since division by 0 is not defined, $1 - x$ cannot be 0 and therefore x

cannot be 1. To clear fractions we will have to multiply by $1 - x$. It is therefore important to keep track of the sign of $1 - x$.

CASE 1.   Let's assume first that

$$1 - x > 0, \quad \text{or equivalently, that} \quad x < 1.$$

In this situation,

$$\frac{x + 2}{1 - x} > 1$$

gives

$$x + 2 > 1 - x,$$
$$x > -1 - x,$$
$$2x > -1,$$
$$x > -\tfrac{1}{2}.$$

Of all the numbers less than 1, only those that are greater than $-\tfrac{1}{2}$ satisfy the initial inequality. This gives the open interval $(-\tfrac{1}{2}, 1)$.

FIGURE 1.4.1

CASE 2.   Let's assume now that

$$1 - x < 0, \quad \text{or equivalently, that} \quad x > 1.$$

In this situation,

$$\frac{x + 2}{1 - x} > 1$$

gives

$$x + 2 < 1 - x,$$
$$x < -1 - x,$$
$$2x < -1,$$
$$x < -\tfrac{1}{2}.$$

Of all the numbers greater than 1, only those that are less than $-\tfrac{1}{2}$ satisfy the initial inequality. There are no such numbers.

FIGURE 1.4.2

The interval $(-\frac{1}{2}, 1)$ that we obtained in the first case is therefore the total solution. □

Exercises

Solve the following inequalities.

*1. $2 + 3x < 5$.
2. $\frac{1}{2}(2x + 3) < 6$.

*3. $16x + 64 \le 16$.
4. $3x + 5 > \frac{1}{4}(x - 2)$.

*5. $3x - 2 \le 1 + 6x$.
6. $(x + 1)(x - 2) \le 0$.

*7. $x^2 - 1 < 0$.
8. $x^2 + x - 2 \le 0$.

*9. $4(x^2 - 3x + 2) > 0$.
10. $x^2 + 9x + 20 < 0$.

*11. $x(x - 1)(x - 2) > 0$.
12. $x(2x - 1)(3x - 5) \le 0$.

*13. $x^3 - 2x^2 + x \ge 0$.
14. $x^2 - 4x + 4 \le 0$.

*15. $\frac{1}{2}(1 + x) < \frac{1}{3}(1 - x)$.
16. $2 - x^2 \ge -4x$.

*17. $\frac{1}{2}(1 + x)^2 < \frac{1}{3}(1 - x)^2$.
18. $2x^2 + 9x + 6 \ge x + 2$.

*19. $1 - 3x^2 < \frac{1}{2}(2 - x^2)$.
20. $6x^2 + 4x + 4 \le 4(x - 1)^2$.

*21. $\dfrac{1}{x} < x$.
22. $x + \dfrac{1}{x} \ge 1$.

*23. $\dfrac{x}{x - 5} \ge 0$.
24. $\dfrac{x}{x + 5} < 0$.

*25. $\dfrac{x}{x - 5} > \dfrac{1}{4}$.
26. $\dfrac{1}{3x - 5} < 2$.

*27. $\dfrac{3x^2 - 1}{1 + x^2} \ge 0$.
28. $\dfrac{3x^2 + 1}{1 - x^2} < 0$.

29. Let a and b be nonnegative numbers. Show that

$$\text{if} \quad a^2 \le b^2 \quad \text{then} \quad a \le b.$$

HINT: $b^2 - a^2 = (b + a)(b - a)$.

30. Show that

$$\text{if} \; . \; 0 \le a \le b \quad \text{then} \quad \frac{a}{1 + a} \le \frac{b}{1 + b}.$$

31. Let a, b, c be nonnegative numbers. Show that

$$\text{if} \quad a \le b + c \quad \text{then} \quad \frac{a}{1 + a} \le \frac{b}{1 + b} + \frac{c}{1 + c}.$$

1.5 Inequalities and Absolute Value

Here we take up some inequalities that involve absolute values. With an eye toward Chapter 2 we introduce two Greek letters: δ (delta) and ϵ (epsilon). You can find the complete Greek alphabet in the front matter.

We begin with the inequality

$$|x| < \delta.$$

The easiest way to handle this inequality is to think of $|x|$ as the distance between 0 and x. It then becomes obvious that

(1.5.1) $\boxed{|x| < \delta \quad \text{iff} \quad -\delta < x < \delta.}$

The inequality

$$|x - c| < \delta$$

can be handled with similar ease. The number $|x - c|$ represents the distance between c and x. It follows that

(1.5.2) $\boxed{|x - c| < \delta \quad \text{iff} \quad c - \delta < x < c + \delta.}$

Somewhat more delicate is the inequality

$$0 < |x - c| < \delta.$$

Here we have $|x - c| < \delta$ with the additional requirement that x cannot be c. Consequently

(1.5.3) $\boxed{0 < |x - c| < \delta \quad \text{iff} \quad c - \delta < x < c \quad \text{or} \quad c < x < c + \delta.}$

Examples

1. $|x| < \frac{1}{2}$ iff $-\frac{1}{2} < x < \frac{1}{2}$.   SOLUTION: $(-\frac{1}{2}, \frac{1}{2})$.
2. $|x - 5| < 1$ iff $4 < x < 6$.   SOLUTION: $(4, 6)$.
3. $0 < |x - 5| < 1$ iff $4 < x < 5$ or $5 < x < 6$.

SOLUTION: $(4, 5) \cup (5, 6)$. □

One of the fundamental inequalities of calculus is the *triangle inequality*: for all real numbers a and b,

(1.5.4) $\boxed{|a + b| \le |a| + |b|.}$

PROOF.   The proof is easy if you think of $|x|$ as $\sqrt{x^2}$. Note first that

$$(a + b)^2 = a^2 + 2ab + b^2 \le |a|^2 + 2|a|\,|b| + |b|^2 = (|a| + |b|)^2.$$

Comparing the extremes of the inequality and taking square roots, we have

$$\sqrt{(a + b)^2} \le |a| + |b|. \quad \text{(Exercise 29, Section 1.4)}$$

The result follows from observing that

$$\sqrt{(a + b)^2} = |a + b|. \quad \square$$

Here is a variant of the triangle inequality that also comes up in calculus: for all real numbers a and b,

(1.5.5) $\boxed{\big||a| - |b|\big| \le |a - b|.}$

PROOF.   Once again we begin by squaring:

$$\big||a| - |b|\big|^2 = (|a| - |b|)^2 = |a|^2 - 2|a||b| + |b|^2$$
$$= a^2 - 2|a||b| + b^2$$
$$\leq a^2 - 2ab + b^2 \qquad \text{(Why?)}$$
$$= (a - b)^2.$$

Comparing the extremes, you can see that

$$\big||a| - |b|\big|^2 \leq (a - b)^2.$$

Now take square roots:

$$\big||a| - |b|\big| \leq \sqrt{(a - b)^2} = |a - b|. \quad \square$$

Problem. Solve the inequality

$$|3x - 4| < 2.$$

SOLUTION. Since

$$|3x - 4| = |3(x - \tfrac{4}{3})| = |3||x - \tfrac{4}{3}| = 3|x - \tfrac{4}{3}|,$$

the inequality can be rewritten

$$3|x - \tfrac{4}{3}| < 2.$$

This gives

$$|x - \tfrac{4}{3}| < \tfrac{2}{3},$$
$$\tfrac{4}{3} - \tfrac{2}{3} < x < \tfrac{4}{3} + \tfrac{2}{3},$$
$$\tfrac{2}{3} < x < 2.$$

The solution is the open interval $(\tfrac{2}{3}, 2)$. $\quad \square$

If you think of $|a|$ as the distance between a and 0, then obviously

(1.5.6) $\boxed{|a| > \epsilon \quad \text{iff} \quad a > \epsilon \quad \text{or} \quad a < -\epsilon.}$

Problem. Solve the inequality

$$|2x + 3| > 5.$$

SOLUTION. As you saw, in general

$$|a| > \epsilon \quad \text{iff} \quad a > \epsilon \quad \text{or} \quad a < -\epsilon.$$

So here

$$2x + 3 > 5 \quad \text{or} \quad 2x + 3 < -5.$$

The first possibility gives $2x > 2$ and thus

$$x > 1.$$

The second possibility gives $2x < -8$ and thus

$$x < -4.$$

The total solution is therefore the union

$$(-\infty, -4) \cup (1, \infty). \quad \square$$

Problem. Solve the inequality

$$|x^2 - 3| < 1.$$

SOLUTION. The distance between x^2 and 3 is less than 1 and therefore

$$3 - 1 < x^2 < 3 + 1,$$
$$2 < x^2 < 4,$$
$$\sqrt{2} < |x| < 2.$$

If $x > 0$, then

$$\sqrt{2} < x < 2.$$

If $x < 0$, then

$$\sqrt{2} < -x < 2$$

so that, multiplying by -1,

$$-2 < x < -\sqrt{2}. \qquad \text{(inequalities reversed)}$$

The solution is the union $(-2, -\sqrt{2}) \cup (\sqrt{2}, 2). \quad \square$

Problem. Solve the inequality

$$|x^2 - 9| > 2.$$

SOLUTION. By (1.5.6)

$$x^2 - 9 > 2 \quad \text{or} \quad x^2 - 9 < -2.$$

The first possibility gives

$$x^2 > 11,$$
$$|x| > \sqrt{11},$$
$$x > \sqrt{11} \quad \text{or} \quad x < -\sqrt{11}.$$

The second possibility gives

$$x^2 < 7,$$
$$|x| < \sqrt{7},$$
$$-\sqrt{7} < x < \sqrt{7}.$$

The total solution is the union

$$(-\infty, -\sqrt{11}) \cup (-\sqrt{7}, \sqrt{7}) \cup (\sqrt{11}, \infty). \quad \square$$

Exercises

Solve the following inequalities.

*1. $\lvert x \rvert < 2$.	2. $\lvert x \rvert \geq 1$.	*3. $\lvert x \rvert > 3$.
*4. $\lvert x - 1 \rvert < 1$.	5. $\lvert x - 2 \rvert < \frac{1}{2}$.	*6. $\lvert x - \frac{1}{2} \rvert < 2$.
*7. $\lvert x + 2 \rvert < \frac{1}{4}$.	8. $\lvert x - 3 \rvert < \frac{1}{4}$.	*9. $0 < \lvert x \rvert < 1$.
*10. $0 < \lvert x \rvert < \frac{1}{2}$.	11. $0 < \lvert x - 2 \rvert < \frac{1}{2}$.	*12. $0 < \lvert x - \frac{1}{2} \rvert < 2$.
*13. $0 < \lvert x - 3 \rvert < 8$.	14. $0 < \lvert x - 2 \rvert < 8$.	*15. $\lvert 2x - 3 \rvert < 1$.
*16. $\lvert 3x - 5 \rvert < 3$.	17. $\lvert 2x + 1 \rvert < \frac{1}{4}$.	*18. $\lvert 5x - 3 \rvert < \frac{1}{2}$.
*19. $\lvert 2x + 5 \rvert > 3$.	20. $\lvert 3x + 1 \rvert > 5$.	*21. $\lvert 5x - 1 \rvert > 9$.
*22. $\lvert 4x + 7 \rvert > 13$.	23. $\lvert x^2 - 4 \rvert < 5$.	*24. $\lvert x^2 - 2 \rvert < 1$.
*25. $\lvert x^2 - 1 \rvert < 1$.	26. $\lvert x^2 - 4 \rvert > 3$.	*27. $\lvert x^2 - 3 \rvert > 1$.
*28. $\lvert x^2 - 16 \rvert > 9$.	29. $\lvert x^2 - x + 1 \rvert > 1$.	*30. $\lvert x^2 - 4x + 1 \rvert > 1$.

31. Prove that for all real numbers a and b

$$\lvert a - b \rvert \leq \lvert a \rvert + \lvert b \rvert.$$

1.6 Functions

The fundamental processes of calculus (called *differentiation* and *integration*) are processes applied to functions. To understand these processes and to be able to carry them out, you have to be thoroughly familiar with functions. Here we review some of the basic notions.

Domain and Range

First we need a working definition for the word *function*. Let D be a set of real numbers. By a *function* on D we mean a rule which assigns a unique number to each number in D. The set D is called the *domain* of the function. The set of assignments that the function makes (the set of values that the function takes on) is called the *range* of the function.

Here are some examples. We begin with the squaring function

$$f(x) = x^2, \quad x \text{ real.}$$

The domain of f is explicitly given as the set of all real numbers. As x runs through all the real numbers, x^2 runs through all the nonnegative numbers. The range is therefore $[0, \infty)$. In abbreviated form we can write

$$\text{dom } (f) = (-\infty, \infty) \quad \text{and} \quad \text{ran } (f) = [0, \infty).$$

Now take the function

$$g(x) = \sqrt{x + 4}, \quad x \in [0, 5].$$

The domain of g is given as the closed interval $[0, 5]$. At 0, g takes on the value 2:

$$g(0) = \sqrt{0 + 4} = \sqrt{4} = 2.$$

At 5, g takes on the value 3:

$$g(5) = \sqrt{5 + 4} = \sqrt{9} = 3.$$

As x runs through all the numbers from 0 to 5, $g(x)$ runs through all the numbers from 2 to 3. The range of g is therefore the closed interval [2, 3]. In abbreviated form

$$\operatorname{dom}(g) = [0, 5], \quad \operatorname{ran}(g) = [2, 3].$$

The graph of f is the parabola depicted in Figure 1.6.1. The graph of g is the arc shown in Figure 1.6.2.

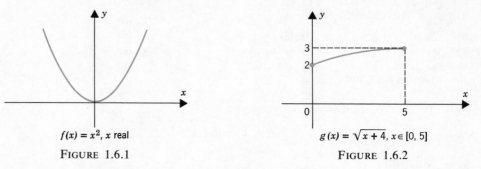

$f(x) = x^2$, x real

FIGURE 1.6.1

$g(x) = \sqrt{x+4}$, $x \in [0, 5]$

FIGURE 1.6.2

Sometimes the domain of a function is not explicitly given. Thus, for example, we might write

$$f(x) = x^3 \quad \text{or} \quad f(x) = \sqrt{x} \qquad \text{(Figure 1.6.3)}$$

without further explanation. In such cases take as the domain the maximal set of real numbers x for which $f(x)$ is itself a real number. For the cubing function take $\operatorname{dom}(f) = (-\infty, \infty)$ and for the square root function take $\operatorname{dom}(f) = [0, \infty)$.

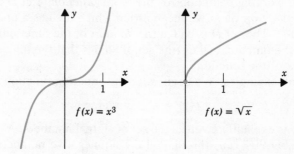

$f(x) = x^3$ $f(x) = \sqrt{x}$

FIGURE 1.6.3

Problem. Find the domain and range of

$$f(x) = \frac{1}{\sqrt{2-x}} + 5.$$

SOLUTION. First we look for the domain. To be able to form $\sqrt{2-x}$ we need $2 - x \geq 0$ and therefore $x \leq 2$. But at $x = 2$, $\sqrt{2-x} = 0$ and its reciprocal makes no sense. We must therefore restrict x to $x < 2$. The domain is $(-\infty, 2)$.

Now we look for the range. As x runs through $(-\infty, 2)$, $\sqrt{2-x}$ takes on all positive values and so does its reciprocal. The range of f is therefore $(5, \infty)$. □

Algebraic Combinations of Functions

Functions with a common domain can be added and subtracted:

$$(f + g)(x) = f(x) + g(x), \quad (f - g)(x) = f(x) - g(x).$$

They can be multiplied:

$$(fg)(x) = f(x)g(x).$$

If $g(x) \neq 0$, we can form the reciprocal:

$$\left(\frac{1}{g}\right)(x) = \frac{1}{g(x)},$$

and also the quotient:

$$\left(\frac{f}{g}\right)(x) = \frac{f(x)}{g(x)}.$$

With α and β real we can form scalar multiples:

$$(\alpha f)(x) = \alpha f(x)$$

and linear combinations:

$$(\alpha f + \beta g)(x) = \alpha f(x) + \beta g(x).$$

Example. If

$$f(x) = x^3 + x - 1 \quad \text{and} \quad g(x) = 3x^2 + 4,$$

then

$$(f + g)(x) = (x^3 + x - 1) + (3x^2 + 4) = x^3 + 3x^2 + x + 3,$$
$$(f - g)(x) = (x^3 + x - 1) - (3x^2 + 4) = x^3 - 3x^2 + x - 5,$$
$$\left(\frac{1}{g}\right)(x) = \frac{1}{3x^2 + 4},$$
$$\left(\frac{f}{g}\right)(x) = \frac{x^3 + x - 1}{3x^2 + 4},$$
$$(6f)(x) = 6(x^3 + x - 1),$$
$$(6f - 5g)(x) = 6(x^3 + x - 1) - 5(3x^2 + 4) = 6x^3 - 15x^2 + 6x - 26. \quad \square$$

Example. Now consider two functions defined piecewise:

$$f(x) = \begin{cases} 1 - x^2, & x \leq 0 \\ x, & x > 0 \end{cases} \quad \text{and} \quad g(x) = \begin{cases} -2x, & x < 1 \\ 1 - x, & x \geq 1 \end{cases}.$$

To find $f + g, f - g, fg$, we must fit the pieces together; namely, we must break up the domain of both functions in the same manner:

$$f(x) = \begin{cases} 1 - x^2, & x \leq 0 \\ x, & 0 < x < 1 \\ x, & x \geq 1 \end{cases}, \quad g(x) = \begin{cases} -2x, & x \leq 0 \\ -2x, & 0 < x < 1 \\ 1 - x, & x \geq 1 \end{cases}.$$

The rest is straightforward:

$$(f + g)(x) = f(x) + g(x) = \begin{cases} 1 - 2x - x^2, & x \le 0 \\ -x, & 0 < x < 1 \\ 1, & x \ge 1 \end{cases},$$

$$(f - g)(x) = f(x) - g(x) = \begin{cases} 1 + 2x - x^2, & x \le 0 \\ 3x, & 0 < x < 1 \\ -1 + 2x, & x \ge 1 \end{cases},$$

$$(fg)(x) = f(x)g(x) = \begin{cases} -2x + 2x^3, & x \le 0 \\ -2x^2, & 0 < x < 1 \\ x - x^2, & x \ge 1 \end{cases}. \quad \square$$

Exercises

Find the number(s), if any, where f takes on the value 1.

*1. $f(x) = \sqrt{1 + x}$.   2. $f(x) = |2 - x|$.

*3. $f(x) = x^2 + 4x + 5$.   4. $f(x) = 4 + 10x - x^2$.

Find the domain and the range.

*5. $f(x) = |x|$.   6. $g(x) = x^2 - 1$.

*7. $f(x) = 2x - 1$.   8. $g(x) = \sqrt{x} - 1$.

*9. $f(x) = \dfrac{1}{x^2}$.   10. $g(x) = \dfrac{1}{x}$.

*11. $f(x) = \sqrt{1 - x}$.   12. $g(x) = \sqrt{x - 1}$.

*13. $f(x) = \sqrt{1 - x} - 1$.   14. $g(x) = \sqrt{x - 1} - 1$.

*15. $f(x) = \dfrac{1}{\sqrt{1 - x}}$.   16. $g(x) = \dfrac{1}{\sqrt{4 - x^2}}$.

Sketch the graph.

17. $f(x) = 1$.   18. $f(x) = -1$.

19. $f(x) = 2x$.   20. $f(x) = 2x + 1$.

21. $f(x) = \frac{1}{2}x$.   22. $f(x) = -\frac{1}{2}x$.

23. $f(x) = \frac{1}{2}x + 2$.   24. $f(x) = -\frac{1}{2}x - 3$.

25. $f(x) = \sqrt{4 - x^2}$.   26. $f(x) = \sqrt{9 - x^2}$.

Sketch the graph and specify the domain and range.

*27. $f(x) = \begin{cases} -1, & x < 0 \\ 1, & x > 0 \end{cases}$.   28. $f(x) = \begin{cases} x^2, & x \le 0 \\ 1 - x, & x > 0 \end{cases}$.

*29. $f(x) = \begin{cases} 1 + x, & 0 \le x \le 1 \\ x, & 1 < x < 2 \\ \frac{1}{2}x + 1, & x \ge 2 \end{cases}$.   30. $f(x) = \begin{cases} x^2, & x < 0 \\ -1, & 0 < x < 2 \\ x, & x > 2 \end{cases}$.

*31. Given that

$$f(x) = \sqrt{x} - \frac{1}{\sqrt{x}} \quad \text{and} \quad g(x) = \sqrt{x} + \frac{2}{\sqrt{x}},$$

find (a) $6f + 3g$, (b) fg, (c) f/g.

32. Given

$$f(x) = \left\{ \begin{array}{ll} x, & x \le 0 \\ -1, & x > 0 \end{array} \right\}, \quad g(x) = \left\{ \begin{array}{ll} -x, & x < 1 \\ x^2, & x \ge 1 \end{array} \right\},$$

find (a) $f + g$, (b) $f - g$, (c) fg.

*33. Given

$$f(x) = \left\{ \begin{array}{ll} 1 - x, & x \le 1 \\ 2x - 1, & x > 1 \end{array} \right\}, \quad g(x) = \left\{ \begin{array}{ll} 0, & x < 2 \\ -1, & x \ge 2 \end{array} \right\},$$

find (a) $f + g$, (b) $f - g$, (c) fg.

Sketch the graphs of the following functions with f and g as in Figure 1.6.4.

34. $2f$.   *35. $\frac{1}{2}f$.   36. $-f$.

*37. $2g$.   38. $\frac{1}{2}g$.   *39. $-2g$.

40. $f + g$.   *41. $f - g$.   42. $f + 2g$.

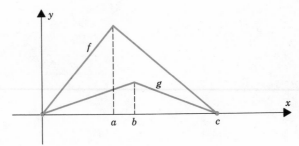

FIGURE 1.6.4

A function f is called *even* iff

$$f(-x) = f(x) \qquad \text{for all} \quad x \in \text{dom} (f).$$

It is called *odd* iff

$$f(-x) = -f(x) \qquad \text{for all} \quad x \in \text{dom} (f).$$

Test each of the functions below for evenness and oddness.

*43. $f(x) = x^3$.   44. $f(x) = x^2$.   *45. $g(x) = x(x - 1)$.

*46. $g(x) = x(x^2 + 1)$.   47. $f(x) = \dfrac{x^2}{1 - |x|}$.   *48. $f(x) = x + \dfrac{1}{x}$.

*49. What can you conclude about the product of two odd functions?

*50. What can you conclude about the product of two even functions?

*51. What can you conclude about the product of an odd function and an even function?

*52. What is the characteristic feature of the graph of an even function?

*53. What is the characteristic feature of the graph of an odd function?

Optional | 54. Show that every function defined for all real numbers can be written as the sum of an even function and an odd function.

1.7 The Composition of Functions

In the last section, you saw how to combine functions algebraically. There is another way to combine functions, called *composition*. To describe it, we begin with two functions f and g and a number x in the domain of g. By applying g to x, we get the number $g(x)$. If $g(x)$ is in the domain of f, then we can apply f to $g(x)$ and thereby obtain the number $f(g(x))$.

What is $f(g(x))$? It is the result of first applying g to x and then applying f to $g(x)$. The idea is illustrated in Figure 1.7.1.

If the range of g is completely contained in the domain of f (namely, if each value that g takes on is in the domain of f), then we can form $f(g(x))$ for *each x* in the domain of g and in this manner create a new function. This new function—it takes each x in the domain of g and assigns to it the value $f(g(x))$—is called the *composition* of f and g and is denoted by the symbol $f \circ g$. What is this function $f \circ g$? It is the function which results from first applying g and then f. (Figure 1.7.2)

FIGURE 1.7.1

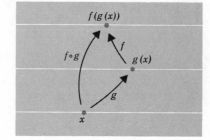

FIGURE 1.7.2

Definition 1.7.1 Composition

If the range of the function g is contained in the domain of the function f, then the *composition* $f \circ g$ (read "f circle g") is the function defined on the domain of g by setting

$$(f \circ g)(x) = f(g(x)).$$

Examples

1. Suppose that

$$g(x) = x^2 \qquad \text{(the squaring function)}$$

and

$$f(x) = x + 1. \qquad \text{(the function which adds 1)}$$

Then

$$(f \circ g)(x) = f(g(x)) = g(x) + 1 = x^2 + 1.$$

In other words, $f \circ g$ is the function which *first* squares and *then* adds 1.

2. If, on the other hand,

$$g(x) = x + 1 \qquad\qquad (g \text{ adds } 1)$$

and

$$f(x) = x^2, \qquad\qquad (f \text{ squares})$$

then

$$(f \circ g)(x) = f(g(x)) = [g(x)]^2 = (x + 1)^2.$$

Here $f \circ g$ is the function which *first* adds 1 and *then* squares.

3. If

$$g(x) = \frac{1}{x} + 1 \qquad (g \text{ adds } 1 \text{ to the reciprocal})$$

and

$$f(x) = x^2 - 5, \quad (f \text{ subtracts } 5 \text{ from the square})$$

then

$$(f \circ g)(x) = f(g(x)) = [g(x)]^2 - 5 = \left(\frac{1}{x} + 1\right)^2 - 5.$$

Here $f \circ g$ *first* adds 1 to the reciprocal and *then* subtracts 5 from the square.

4. If

$$f(x) = \frac{1}{x} + x^2 \quad \text{and} \quad g(x) = \frac{x^2 + 1}{x^4 + 1},$$

then

$$(f \circ g)(x) = f(g(x)) = \frac{1}{g(x)} + [g(x)]^2 = \frac{x^4 + 1}{x^2 + 1} + \left(\frac{x^2 + 1}{x^4 + 1}\right)^2. \quad \square$$

It is possible to form the composition of more than two functions. For example, the triple composition $f \circ g \circ h$ consists of first h, then g, and then f:

$$(f \circ g \circ h)(x) = f(g(h(x))).$$

Obviously we can go on in this manner with as many functions as we like.

Example.   If

$$f(x) = \frac{1}{x}, \quad g(x) = x + 2, \quad h(x) = (x^2 + 1)^2,$$

then

$$(f \circ g \circ h)(x) = f(g(h(x))) = \frac{1}{g(h(x))} = \frac{1}{h(x) + 2} = \frac{1}{(x^2 + 1)^2 + 2}. \quad \square$$

To apply some of the techniques of calculus you will need to be able to recognize composites. You will need to be able to start with a function, say $F(x) = (x + 1)^5$, and recognize how it is a composition.

Problem.   Find functions f and g such that $f \circ g = F$ is

$$F(x) = (x + 1)^5.$$

A SOLUTION.   The function consists of first adding 1 and then taking the fifth power. We can therefore set

$$g(x) = x + 1 \qquad\qquad \text{(adding 1)}$$

and

$$f(x) = x^5. \qquad\qquad \text{(taking the fifth power)}$$

As you can see,

$$(f \circ g)(x) = f(g(x)) = [g(x)]^5 = (x + 1)^5. \quad \square$$

Problem.   Find functions f and g such that $f \circ g = F$ if

$$F(x) = \frac{1}{x} - 6.$$

A SOLUTION. F takes the reciprocal and then subtracts 6. We can therefore set

$$g(x) = \frac{1}{x} \qquad\qquad \text{(taking the reciprocal)}$$

and

$$f(x) = x - 6. \qquad\qquad \text{(subtracting 6)}$$

As you can check,

$$(f \circ g)(x) = f(g(x)) = g(x) - 6 = \frac{1}{x} - 6. \quad \square$$

Problem.   Find three functions f, g, h such that $f \circ g \circ h = F$ if

$$F(x) = \frac{1}{|x| + 3}.$$

A SOLUTION. F takes the absolute value, adds 3, and then inverts. Let h take the absolute value:

$$\text{set} \quad h(x) = |x|.$$

Let g add 3:

$$\text{set} \quad g(x) = x + 3.$$

Let f do the inverting:

$$\text{set} \quad f(x) = \frac{1}{x}.$$

With this choice of f, g, h we have

$$(f \circ g \circ h)(x) = f(g(h(x))) = \frac{1}{g(h(x))} = \frac{1}{h(x) + 3} = \frac{1}{|x| + 3}. \quad \square$$

In the next problem we start with the same function

$$F(x) = \frac{1}{|x| + 3},$$

but this time we are asked to break it up as the composition of only two functions.

Problem. Find two functions f and g such that $f \circ g = F$ if

$$F(x) = \frac{1}{|x| + 3}.$$

SOME SOLUTIONS. F takes the absolute value, adds 3, and then inverts. We can let g do the first two things by setting

$$g(x) = |x| + 3$$

and then let f do the inverting by setting

$$f(x) = \frac{1}{x}.$$

Or, we could let g just take the absolute value,

$$g(x) = |x|,$$

and then have f add 3 and invert:

$$f(x) = \frac{1}{x + 3}.$$

There are still other possible choices for f and g. For example, we could set

$$g(x) = x \quad \text{and} \quad f(x) = \frac{1}{|x| + 3}$$

or

$$g(x) = \frac{1}{|x| + 3} \quad \text{and} \quad f(x) = x,$$

but these last choices of f and g don't advance us at all. They are much like factoring 12 into $12 \cdot 1$ or $1 \cdot 12$. $\square$

Exercises

Form the composition $f \circ g$:

*1. $f(x) = 2x + 5$, $g(x) = x^2$.

2. $f(x) = x^2$, $g(x) = 2x + 5$.

*3. $f(x) = \sqrt{x}$, $g(x) = x^2 + 5$.

4. $f(x) = x^2 + x$, $g(x) = \sqrt{x}$.

*5. $f(x) = \dfrac{1}{x}, \quad g(x) = \dfrac{1}{x}.$

6. $f(x) = x^2, \quad g(x) = \dfrac{1}{x - 1}.$

*7. $f(x) = x - 1, \quad g(x) = \dfrac{1}{x}.$

8. $f(x) = x^2 - 1, \quad g(x) = x(x - 1).$

*9. $f(x) = \dfrac{1}{\sqrt{x - 1}}, \quad g(x) = (x^2 + 2)^2.$

10. $f(x) = \dfrac{1}{x} - \dfrac{1}{x + 1}, \quad g(x) = \dfrac{1}{x^2}.$

Form the composition $f \circ g \circ h$:

*11. $f(x) = 4x, \quad g(x) = x - 1, \quad h(x) = x^2.$
12. $f(x) = x - 1, \quad g(x) = 4x, \quad h(x) = x^2.$
*13. $f(x) = x^2, \quad g(x) = x - 1, \quad h(x) = x^4.$
14. $f(x) = x - 1, \quad g(x) = x^2, \quad h(x) = 4x.$

*15. $f(x) = \dfrac{1}{x}, \quad g(x) = \dfrac{1}{2x + 1}, \quad h(x) = x^2.$

16. $f(x) = \dfrac{x + 1}{x}, \quad g(x) = \dfrac{1}{2x + 1}, \quad h(x) = x^2.$

17. Find f such that $f \circ g = F$ given that

*(a) $g(x) = \dfrac{1 - x}{1 + x}$ and $F(x) = \dfrac{1 + x}{1 - x}.$ $\qquad (x \neq \pm 1)$

(b) $g(x) = x^2$ and $F(x) = ax^2 + b.$

*(c) $g(x) = -x^2$ and $F(x) = \sqrt{a^2 - x^2}.$ $\qquad (-a \leq x \leq a)$

18. Find g such that $f \circ g = F$ given that

(a) $f(x) = x^2$ and $F(x) = \left(1 - \dfrac{1}{x^4}\right)^2.$

*(b) $f(x) = x + \dfrac{1}{x}$ and $F(x) = a^2 x^2 + \dfrac{1}{a^2 x^2}.$ $\qquad (x \neq 0)$

(c) $f(x) = x^2 + 1$ and $F(x) = (2x^3 - 1)^2 + 1.$

*19. Given that

$$f(x) = \begin{cases} 1 - x, & x \leq 0 \\ x^2, & x > 0 \end{cases} \quad \text{and} \quad g(x) = \begin{cases} -x, & x < 1 \\ 1 + x, & x \geq 1 \end{cases}$$

form (a) $f \circ g$ and (b) $g \circ f.$

20. Given that

$$f(x) = \begin{cases} 1 - 2x, & x < 1 \\ 1 + x, & x \geq 1 \end{cases} \quad \text{and} \quad g(x) = \begin{cases} x^2, & x < 0 \\ 1 - x, & x \geq 0 \end{cases}$$

form (a) $f \circ g$ and (b) $g \circ f.$

1.8 One-To-One Functions; Inverses

One-to-One Functions

A function can take on the same value at different points of its domain. Thus, for example, the squaring function takes on the same value at $-c$ as it does at c. In the case of

$$f(x) = (x - 3)(x - 5)$$

we have not only $f(3) = 0$ but also $f(5) = 0$. Functions for which this kind of repetition does not occur are called *one-to-one*.

Definition 1.8.1

A function is said to be *one-to-one* iff there are no two points at which it takes on the same value.

As examples take

$$f(x) = x^3 \quad \text{and} \quad g(x) = \sqrt{x}.$$

The cubing function is one-to-one because no two numbers have the same cube. The square root function is one-to-one because no two numbers have the same square root.

In the case of a one-to-one function no horizontal line intersects the graph more than once. (Figure 1.8.1) If a horizontal line intersects the graph more than once, then the function is not one-to-one. (Figure 1.8.2)

one–to–one function
FIGURE 1.8.1

function is not one–to–one
FIGURE 1.8.2

Inverses

I am thinking of a real number and its cube is 8. Can you figure out the number? Obviously yes. It must be 2.

Now I am thinking of another number and its square is 9. Can you figure out the number? No, not unless you are clairvoyant. It could be 3 but it could also be -3.

The difference between the two situations can be phrased this way: In the first case we are dealing with a one-to-one function—no two numbers have the same cube. So, knowing the cube, we can work back to the original number. In the second case we are dealing with a function which is not one-to-one—different numbers can have the same square. Knowing the square we still can't figure out what the original number was.

Throughout the rest of this section we shall be looking at functional values and tracing them back to where they came from. So as to be able to trace these values back, we shall restrict ourselves entirely to functions which are one-to-one.

We begin with a theorem about one-to-one functions.

Theorem 1.8.2

If f is a one-to-one function, then there is one and only one function g which is defined on the range of f and satisfies the equation

$$f(g(x)) = x \quad \text{for all } x \text{ in the range of } f.$$

PROOF.   The proof is easy. If x is in the range of f, then f must take on the value x at some number. Since f is one-to-one, there can be only one such number. We've called it $g(x)$.   □

The function that we've named g in the theorem is called the inverse of f and is usually denoted by f^{-1}.

Definition 1.8.3. Inverse Function

Let f be a one-to-one function. The *inverse* of f, denoted by f^{-1}, is the unique function which is defined on the range of f and satisfies the equation

$$f(f^{-1}(x)) = x \quad \text{for all } x \text{ in the range of } f.$$

Problem.   Find the inverse of f for

$$f(x) = x^3.$$

SOLUTION.   We begin by setting

$$f(f^{-1}(x)) = x.$$

Since f is the cubing function, we must have

$$f(f^{-1}(x)) = [f^{-1}(x)]^3$$

so that

$$[f^{-1}(x)]^3 = x.$$

Taking cube roots we have

$$f^{-1}(x) = x^{1/3}.$$

What we've found is that the cube root function is the inverse of the cubing function. As you can check,

$$f(f^{-1}(x)) = [f^{-1}(x)]^3 = [x^{1/3}]^3 = x. \quad \square$$

Problem.   Find the inverse of f for

$$f(x) = 3x - 5.$$

SOLUTION.   Set

$$f(f^{-1}(x)) = x.$$

With our choice of f this gives

$$3f^{-1}(x) - 5 = x,$$

which we can now solve for $f^{-1}(x)$:

$$3f^{-1}(x) = x + 5,$$
$$f^{-1}(x) = \tfrac{1}{3}x + \tfrac{5}{3}.$$

As you can check,

$$f(f^{-1}(x)) = 3f^{-1}(x) - 5 = 3(\tfrac{1}{3}x + \tfrac{5}{3}) - 5 = (x + 5) - 5 = x. \quad \square$$

Problem.   Find the inverse of f for

$$f(x) = (1 - x^3)^{1/5} + 2.$$

SOLUTION.   Once again we begin by setting

$$f(f^{-1}(x)) = x.$$

Since

$$f(f^{-1}(x)) = \{1 - [f^{-1}(x)]^3\}^{1/5} + 2,$$

we must have

$$\{1 - [f^{-1}(x)]^3\}^{1/5} + 2 = x.$$

We now solve for $f^{-1}(x)$:

$$\{1 - [f^{-1}(x)]^3\}^{1/5} = x - 2$$
$$1 - [f^{-1}(x)]^3 = (x - 2)^5$$
$$[f^{-1}(x)]^3 = 1 - (x - 2)^5$$
$$f^{-1}(x) = [1 - (x - 2)^5]^{1/3}. \quad \square$$

By definition f^{-1} satisfies the equation

(1.8.4) $\boxed{f(f^{-1}(x)) = x \quad \text{for all } x \text{ in the range of } f.}$

It is also true that

(1.8.5) $f^{-1}(f(x)) = x$ for all x in the domain of f.

PROOF.   We rewrite (1.8.4) replacing the letter x by the letter t:

(∗) $f(f^{-1}(t)) = t$ for all t in the range of f.

If x is in the domain of f, then $f(x)$ is in the range of f and therefore by (∗)

$$f(f^{-1}(f(x))) = f(x).$$ [let $t = f(x)$]

This tells us that f takes on the same value at $f^{-1}(f(x))$ as it does at x. With f one-to-one, this can only happen if

$$f^{-1}(f(x)) = x. \quad \square$$

FIGURE 1.8.3

FIGURE 1.8.4

Equation 1.8.5 tells us that f^{-1} "undoes" the work of f:

   if f takes x to $f(x)$,   then f^{-1} takes $f(x)$ back to x.   (Figure 1.8.3)

Equation 1.8.4 tells us that f "undoes" the work of f^{-1}:

   if f^{-1} takes x to $f^{-1}(x)$,   then f takes $f^{-1}(x)$ back to x.   (Figure 1.8.4)

The Graphs of f and f^{-1}

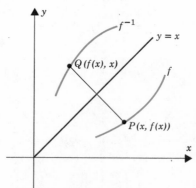

FIGURE 1.8.5

There is an important relation between the graph of a one-to-one function f and the graph of f^{-1}. (See Figure 1.8.5.) The graph of f consists of points of the form

$P(x, f(x))$. Since f^{-1} takes on the value x at $f(x)$, the graph of f^{-1} consists of points of the form $Q(f(x), x)$. Since, for each x, P and Q are symmetric with respect to the line $y = x$, the graph of f^{-1} is the graph f reflected in the line $y = x$.

Problem.   Given the graph of f in Figure 1.8.6, sketch the graph of f^{-1}.

FIGURE 1.8.6   FIGURE 1.8.7

SOLUTION.   First draw in the line $y = x$. Then reflect the graph of f in that line. See Figure 1.8.7.   □

Exercises

Determine whether or not the given function is one-to-one and, if so, find its inverse.

*1. $f(x) = 5x + 3$.   2. $f(x) = 3x + 5$.
*3. $f(x) = 4x - 7$.   4. $f(x) = 7x - 4$.
*5. $f(x) = 1 - x^2$.   6. $f(x) = x^5$.
*7. $f(x) = x^5 + 1$.   8. $f(x) = x^2 - 3x + 2$.
*9. $f(x) = 1 + 3x^3$.   10. $f(x) = x^3 - 1$.
*11. $f(x) = (1 - x)^3$.   12. $f(x) = (1 - x)^4$.
*13. $f(x) = (x + 1)^3 + 2$.   14. $f(x) = (4x - 1)^3$.
*15. $f(x) = x^{3/5}$.   16. $f(x) = 1 - (x - 2)^{1/3}$.
*17. $f(x) = (2 - 3x)^3$.   18. $f(x) = (2 - 3x^2)^3$.

*19. $f(x) = \dfrac{1}{x}$.   20. $f(x) = \dfrac{1}{1 - x}$.

*21. $f(x) = x + \dfrac{1}{x}$.   22. $f(x) = \dfrac{x}{|x|}$.

*23. $f(x) = \dfrac{1}{x^3 + 1}$.   24. $f(x) = \dfrac{1}{1 - x} - 1$.

*25. What is the relation between f and $(f^{-1})^{-1}$?

26. Sketch the graph of f^{-1} given the graph of f in (a) Figure 1.8.8. (b) Figure 1.8.9. (c) Figure 1.8.10.

FIGURE 1.8.8   FIGURE 1.8.9

FIGURE 1.8.10

Optional 27. (a) Show that the composition of two one-to-one functions is one-to-one.
(b) Express $(f \circ g)^{-1}$ in terms of f^{-1} and g^{-1}.

Limits and Continuity

2.1 The Idea of Limit

Chapter 1 was not calculus. It was just preliminary, just warm-up. Calculus begins with the idea of limit.

To introduce this idea we begin with a number l and a function f defined *near the number c but not necessarily defined at c itself.* A rough translation of

$$\lim_{x \to c} f(x) = l$$

(the limit of $f(x)$ as x tends to c is l)

might read

as *x approaches c, $f(x)$ approaches l*

or

for *x close* to c but *different* from c, $f(x)$ is *close* to l.

We illustrate the idea in Figure 2.1.1. The curve represents the graph of a function f. The number c appears on the x-axis, the limit l on the y-axis. As x approaches c along the x-axis, $f(x)$ approaches l along the y-axis.

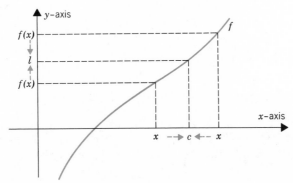

FIGURE 2.1.1

39

In taking the limit as x approaches c, it does not matter whether or not f is defined at c and, if so, how it is defined there. The only thing that matters is how f is defined *near* c. For example, in Figure 2.1.2 the graph of f is a broken curve defined peculiarly at c and yet

$$\lim_{x \to c} f(x) = l$$

FIGURE 2.1.2

because, as suggested in Figure 2.1.3,

as x approaches c, $f(x)$ approaches l.

FIGURE 2.1.3

Now consider the function

$$f(x) = \begin{cases} -1, & x < c \\ 1, & x > c \end{cases}. \qquad \text{(Figure 2.1.4)}$$

As x approaches c from the left, $f(x)$ is constantly -1. As x approaches c from the right, $f(x)$ is constantly 1. There is *no one number* which $f(x)$ approaches as x approaches c from both sides; namely, $\lim_{x \to c} f(x)$ does not exist.

FIGURE 2.1.4

Here are more functions for which $\lim\limits_{x \to c} f(x)$ does not exist:

$$f(x) = \frac{1}{x - c},$$

$$f(x) = \frac{1}{|x - c|},$$

$$f(x) = \begin{cases} 1, & x \text{ rational} \\ 0, & x \text{ irrational} \end{cases}.$$

We begin with

$$f(x) = \frac{1}{x - c}.$$

The graph is displayed in Figure 2.1.5. As x approaches c from the right, $f(x)$ becomes arbitrarily large—larger than any preassigned positive number. As x approaches c from the left, $f(x)$ becomes arbitrarily large negative—less than any preassigned negative number. Under these circumstances $f(x)$ can't possibly approach a fixed number.

 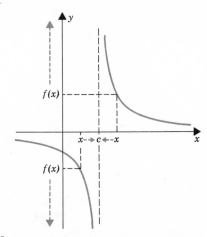

FIGURE 2.1.5

A slightly different situation occurs in the case of

$$f(x) = \frac{1}{|x - c|}.$$ (Figure 2.1.6)

Here, as x approaches c from either side, $f(x)$ becomes arbitrarily large. Becoming arbitrarily large, $f(x)$ cannot stay close to any fixed number l.

FIGURE 2.1.6

The final example,

$$f(x) = \begin{cases} 1, & x \text{ rational} \\ 0, & x \text{ irrational} \end{cases},$$ (Figure 2.1.7)

is somewhat exotic. As x approaches c, x passes through both rational and irrational numbers. As this happens, $f(x)$ jumps wildly back and forth between 0 and 1, and cannot stay close to any fixed number l.

FIGURE 2.1.7

If you've taken the attitude that all this is very imprecise, you're absolutely right. It is imprecise, but it need not remain so. One of the great triumphs of calculus has been its capacity to formulate limit statements with precision, but for this precision you'll have to wait until Section 2.2. Throughout the rest of this section we'll go on with examples in an informal, intuitive manner.

Example

$$\lim_{x \to 3} (2x + 5) = 11.$$

As x approaches 3, $2x$ approaches 6, and $2x + 5$ approaches 11. □

Example

$$\lim_{x \to 2} (x^2 - 1) = 3.$$

As x approaches 2, x^2 approaches 4, and $x^2 - 1$ approaches 3. □

Example

$$\lim_{x \to 3} \frac{1}{x - 1} = \frac{1}{2}.$$

As x approaches 3, $x - 1$ approaches 2, and the reciprocal of $x - 1$ approaches $\frac{1}{2}$. □

Example

$$\lim_{x \to 2} \frac{1}{x - 2} \quad \text{does not exist.}$$

It was argued before more generally that

$$\lim_{x \to c} \frac{1}{x - c} \quad \text{does not exist}$$

no matter what c is. (Figure 2.1.5) □

Before going on to the next examples, remember that, in taking the limit as x approaches a given number c, it does not matter whether or not the function is defined at the number c and, if so, how it is defined there. The only thing that matters is how the function is defined *near* the number c.

Example

$$\lim_{x \to 2} \frac{x^2 - 4x + 4}{x - 2} = 0.$$

At $x = 2$, the function is undefined: both numerator and denominator are 0. But that doesn't matter. For all $x \neq 2$, and therefore for all x near 2,

$$\frac{x^2 - 4x + 4}{x - 2} = x - 2.$$

It follows that

$$\lim_{x \to 2} \frac{x^2 - 4x + 4}{x - 2} = \lim_{x \to 2} (x - 2) = 0. □$$

Example

$$\lim_{x \to 2} \frac{x - 2}{x - 2} = 1.$$

At $x = 2$, the function is undefined: both numerator and denominator are 0. But, as we said before, that doesn't matter. For all $x \neq 2$, and therefore for all x near 2,

$$\frac{x - 2}{x - 2} = 1 \quad \text{so that} \quad \lim_{x \to 2} \frac{x - 2}{x - 2} = \lim_{x \to 2} 1 = 1. \quad \square$$

Example

$$\lim_{x \to 2} \frac{x - 2}{x^2 - 4x + 4} \quad \text{does not exist.}$$

This result does not follow from the fact that the function is not defined at 2 but rather from the fact that for all $x \neq 2$, and therefore for all x near 2,

$$\frac{x - 2}{x^2 - 4x + 4} = \frac{1}{x - 2}$$

and, as you saw before,

$$\lim_{x \to 2} \frac{1}{x - 2} \quad \text{does not exist.} \quad \square$$

Example

$$\text{If} \quad f(x) = \begin{cases} 3x - 1, & x \neq 2 \\ 45, & x = 2 \end{cases}, \quad \text{then} \quad \lim_{x \to 2} f(x) = 5.$$

It doesn't matter to us that $f(2) = 45$. For $x \neq 2$, and thus for all x near 2,

$$f(x) = 3x - 1 \quad \text{so that} \quad \lim_{x \to 2} f(x) = \lim_{x \to 2} (3x - 1) = 5. \quad \square$$

Example

$$\text{If} \quad f(x) = \begin{cases} -2x, & x < 1 \\ 2x, & x > 1 \end{cases}, \quad \text{then} \quad \lim_{x \to 1} f(x) \quad \text{does not exist.}$$

As x approaches 1 from the left, $f(x)$, being $-2x$, approaches -2. As x approaches 1 from the right, $f(x)$, being $2x$, approaches 2. There is no one number that $f(x)$ approaches as x approaches 2 from both sides. $\square$

Example

$$\text{If} \quad f(x) = \begin{cases} -2x, & x < 0 \\ 2x, & x > 0 \end{cases}, \quad \text{then} \quad \lim_{x \to 0} f(x) = 0.$$

As x approaches 0 from the left, $f(x)$, being $-2x$, approaches 0. As x approaches 0 from the right, $f(x)$, being $2x$, also approaches 0. As x approaches 0 from either side, $f(x)$ approaches 0. $\square$

Exercises

Decide on intuitive grounds whether or not the indicated limit exists. Evaluate the limits which do exist.

*1. $\lim\limits_{x \to 0} (2x - 1)$.

2. $\lim\limits_{x \to 1} (2 - 5x)$.

*3. $\lim\limits_{x \to -2} x^2$.

*4. $\lim\limits_{x \to 0} \dfrac{1}{|x|}$.

5. $\lim\limits_{x \to 1} \dfrac{3}{x + 1}$.

*6. $\lim\limits_{x \to 0} \dfrac{4}{x + 1}$.

*7. $\lim\limits_{x \to 2} \dfrac{1}{3x - 6}$.

8. $\lim\limits_{x \to 3} \dfrac{2x - 6}{x - 3}$.

*9. $\lim\limits_{x \to 3} \dfrac{x^2 - 6x + 9}{x - 3}$.

*10. $\lim\limits_{x \to 0} \left(x + \dfrac{1}{x} \right)$.

11. $\lim\limits_{x \to 1} \left(x + \dfrac{1}{x} \right)$.

*12. $\lim\limits_{x \to 0} \dfrac{2x - 5x^2}{x}$.

*13. $\lim\limits_{x \to 1} \dfrac{x^2 - 1}{x - 1}$.

14. $\lim\limits_{x \to 1} \dfrac{x^3 - 1}{x - 1}$.

*15. $\lim\limits_{x \to 1} \dfrac{x^2 + 1}{x^2 - 1}$.

16. $\lim\limits_{x \to 0} f(x)$ if $f(x) = \left\{ \begin{array}{ll} -x^2, & x < 0 \\ x^2, & x > 0 \end{array} \right]$.

*17. $\lim\limits_{x \to 0} f(x)$ if $f(x) = \left\{ \begin{array}{ll} x^2, & x < 0 \\ 1 + x, & x > 0 \end{array} \right]$.

18. $\lim\limits_{x \to 4} f(x)$ if $f(x) = \left\{ \begin{array}{ll} x^2, & x \neq 4 \\ 0, & x = 4 \end{array} \right]$.

*19. $\lim\limits_{x \to 0} f(x)$ if $f(x) = \left\{ \begin{array}{ll} 2, & x \text{ rational} \\ -2, & x \text{ irrational} \end{array} \right]$.

2.2   Definition of Limit

In Section 2.1 we tried to convey the idea of limit in an intuitive manner. Our remarks, however, were too vague to be called mathematics. In this section we shall be more precise.

Let f be a function and let c be a real number. We do not require that f be defined at c, but, to avoid complications at this point, we will assume that f is defined at least on a set of the form $(a, c) \cup (c, d)$. To say that

$$\lim_{x \to c} f(x) = l$$

is to say that the *difference between $f(x)$ and l can be made arbitrarily small (smaller than any preassigned positive number) simply by requiring that x be sufficiently close to c but different from c.*

Think of a positive number and call it ϵ (epsilon). If it is true that

$$\lim_{x \to c} f(x) = l,$$

then you can be sure that

$$|f(x) - l| < \epsilon$$

for all x sufficiently close to c but different from c; namely, you can be sure that there exists a positive number δ (delta) such that

$$\text{if} \quad 0 < |x - c| < \delta, \quad \text{then} \quad |f(x) - l| < \epsilon.$$

Definition 2.2.1   The Limit of a Function

$$\lim_{x \to c} f(x) = l \quad \text{iff} \quad \begin{cases} \text{for each } \epsilon > 0 \text{ there exists } \delta > 0 \text{ such that} \\ \text{if } \ 0 < |x - c| < \delta, \quad \text{then} \quad |f(x) - l| < \epsilon. \end{cases}$$

Figures 2.2.1 and 2.2.2 illustrate this definition.

FIGURE 2.2.1

FIGURE 2.2.2

In general, the choice of δ depends upon the previous choice of ϵ. We do not require that there exist a number δ which ''works'' for *all* ϵ but, rather, that for *each* ϵ there exist a δ which ''works'' for it.

In Figure 2.2.3, we give two choices of ϵ and for each we display a suitable δ. For a δ to be suitable, all the points within δ of c (with the possible exception of c itself) must be taken by the function to within ϵ of l. In part (b) of the figure we began with a smaller ϵ and had to use a smaller δ.

FIGURE 2.2.3

The δ of Figure 2.2.4 is too large for the given ϵ. In particular the points marked x_1 and x_2 in the figure are not taken by the function to within ϵ of l.

Below we apply the ϵ, δ definition of limit to a wide variety of functions. If you've never run across ϵ, δ arguments before, you may find them confusing at first. It usually takes a little while for the ϵ, δ idea to take hold.

FIGURE 2.2.4

Example.   Show that

$$\lim_{x \to 2} (2x - 1) = 3.$$ (Figure 2.2.5)

Finding a δ. Let $\epsilon > 0$. We seek a number $\delta > 0$ such that,

if $0 < |x - 2| < \delta$,   then $|(2x - 1) - 3| < \epsilon$.

What we have to do first is establish a connection between

$$|(2x - 1) - 3| \text{ and } |x - 2|.$$

The connection is simple:

$$|(2x - 1) - 3| = |2x - 4|$$

so that

(1) $$|(2x - 1) - 3| = 2|x - 2|.$$

To make $|(2x - 1) - 3|$ less than ϵ, we need only make $|x - 2|$ twice as small. This suggests that we choose $\delta = \frac{1}{2}\epsilon$.

$$\lim_{x \to 2} (2x - 1) = 3$$

FIGURE 2.2.5

Showing that the δ "works." If $0 < |x - 2| < \frac{1}{2}\epsilon$, then $2|x - 2| < \epsilon$ and, by (1), $|(2x - 1) - 3| < \epsilon.$† $\square$

Example.   Show that

$$\lim_{x \to -1} (2 - 3x) = 5.$$ (Figure 2.2.6)

Finding a δ. Let $\epsilon > 0$. We seek a number $\delta > 0$ such that,

$$\text{if}\quad 0 < |x - (-1)| < \delta,\quad \text{then}\quad |(2 - 3x) - 5| < \epsilon.$$

To find a connection between

$$|x - (-1)|\quad \text{and}\quad |(2 - 3x) - 5|,$$

we simplify both expressions:

$$|x - (-1)| = |x + 1|$$

and

$$|(2 - 3x) - 5| = |-3x - 3| = 3|-x - 1| = 3|x + 1|.$$

We can conclude that

(2) $$|(2 - 3x) - 5| = 3|x - (-1)|.$$

We can make the expression on the left less than ϵ by making $|x - (-1)|$ three times as small. This suggests that we set $\delta = \frac{1}{3}\epsilon$.

Showing that the δ *"works."* If $0 < |x - (-1)| < \frac{1}{3}\epsilon$, then $3|x - (-1)| < \epsilon$ and, by (2), $|(2 - 3x) - 5| < \epsilon.$ $\square$

$$\lim_{x \to -1} (2 - 3x) = 5$$

FIGURE 2.2.6

---

† Here we chose $\delta = \frac{1}{2}\epsilon$, but we could have chosen any $\delta \leq \frac{1}{2}\epsilon$. In general, if a certain δ "works", then any smaller δ will also work.

$$\lim_{x \to c} |x| = |c|$$

FIGURE 2.2.7

Example

(2.2.2) $$\boxed{\lim_{x \to c} |x| = |c|.}$$ (Figure 2.2.7)

PROOF.   Let $\epsilon > 0$. We seek $\delta > 0$ such that

$$\text{if}\quad 0 < |x - c| < \delta, \quad\text{then}\quad \big||x| - |c|\big| < \epsilon.$$

Since

$$\big||x| - |c|\big| \le |x - c|,$$ (1.5.5)

we can choose $\delta = \epsilon$; for

$$\text{if}\quad 0 < |x - c| < \epsilon, \quad\text{then}\quad \big||x| - |c|\big| < \epsilon. \quad \square$$

Example

(2.2.3) $$\boxed{\lim_{x \to c} x = c.}$$ (Figure 2.2.8)

PROOF.   Let $\epsilon > 0$. Here we must find $\delta > 0$ such that

$$\text{if}\quad 0 < |x - c| < \delta, \quad\text{then}\quad |x - c| < \epsilon.$$

Obviously we can choose $\delta = \epsilon$. $\square$

$$\lim_{x \to c} x = c$$

FIGURE 2.2.8

Example

(2.2.4) $$\lim_{x \to c} a = a.$$ (Figure 2.2.9)

PROOF.   Here we are dealing with the constant function

$$f(x) = a.$$

Let $\epsilon > 0$. We must find $\delta > 0$ such that

$$\text{if} \quad 0 < |x - c| < \delta, \quad \text{then} \quad |a - a| < \epsilon.$$

Since

$$|a - a| = 0,$$

we always have

$$|a - a| < \epsilon$$

no matter how δ is chosen; in short, any positive number will do for δ.   □

The remaining examples are more complicated. The ϵ, δ proofs that accompany the examples may be skipped at this stage.

$$\lim_{x \to c} a = a$$

FIGURE 2.2.9

$$\lim_{x \to 4} \sqrt{x} = 2$$

FIGURE 2.2.10

Example

$$\lim_{x \to 4} \sqrt{x} = 2.$$ (Figure 2.2.10)

Optional │ PROOF.   Let $\epsilon > 0$. We seek $\delta > 0$ such that

$$\text{if} \quad 0 < |x - 4| < \delta, \quad \text{then} \quad |\sqrt{x} - 2| < \epsilon.$$

First we want a relation between

$$|x - 4| \quad \text{and} \quad |\sqrt{x} - 2|.$$

Optional | To be able to form $\sqrt{x}$ at all we need $x \geq 0$. To insure this we must have $\delta \leq 4$. (Why?)

Remembering that we must have $\delta \leq 4$, let's go on. If $x \geq 0$, then we can form $\sqrt{x}$ and write

$$x - 4 = (\sqrt{x})^2 - 2^2 = (\sqrt{x} + 2)(\sqrt{x} - 2).$$

Taking absolute values, we have

$$|x - 4| = |\sqrt{x} + 2||\sqrt{x} - 2|.$$

Since

$$|\sqrt{x} + 2| > 1,$$

we have

$$|\sqrt{x} - 2| < |x - 4|.$$

This last inequality suggests that we simply set $\delta = \epsilon$. But remember now the previous requirement $\delta \leq 4$. We can meet all requirements by setting $\delta = $ minimum of 4 and ϵ.

Let's make sure that this choice of δ "works." We begin by assuming that

$$0 < |x - 4| < \delta.$$

Since $\delta \leq 4$, we have $x \geq 0$ and can write

$$|x - 4| = |\sqrt{x} + 2||\sqrt{x} - 2|.$$

Since

$$|\sqrt{x} + 2| > 1,$$

we can conclude that

$$|\sqrt{x} - 2| < |x - 4|.$$

Since

$$|x - 4| < \delta \quad \text{and} \quad \delta \leq \epsilon,$$

it does follow that

$$|\sqrt{x} - 2| < \epsilon. \quad \square$$

Example

If $h(x) = \begin{cases} -1, & x < 0 \\ 1, & x > 0 \end{cases}$, then $\displaystyle\lim_{x \to 0} h(x)$ does not exist. (Figure 2.2.11)

Optional | PROOF.   Suppose on the contrary that

$$\lim_{x \to 0} h(x) = l.$$

Optional | Then there must exist $\delta > 0$ such that

$$\text{if } \quad 0 < |x - 0| < \delta, \quad \text{ then } \quad |h(x) - l| < 1. \quad \text{ (take } \epsilon = 1)$$

Since

$$0 < |\tfrac{1}{2}\delta - 0| < \delta \quad \text{and} \quad h(\tfrac{1}{2}\delta) = 1,$$

we must have

$$|1 - l| < 1.$$

Since

$$0 < |-\tfrac{1}{2}\delta - 0| < \delta \quad \text{and} \quad h(-\tfrac{1}{2}\delta) = -1,$$

we must have

$$|-1 - l| < 1.$$

From $|1 - l| < 1$ we conclude that $l > 0$. From $|-1 - l| < 1$ we conclude that $l < 0$. Clearly no such number l exists. □

$$h(x) = \begin{cases} -1, x < 0 \\ 1, x > 0 \end{cases}; \quad \lim_{x \to 0} h(x) \text{ does not exist}$$

FIGURE 2.2.11

Example

$$\text{If } \quad g(x) = \begin{cases} x, & x \text{ rational} \\ 0, & x \text{ irrational} \end{cases}, \quad \text{ then } \quad \lim_{x \to 0} g(x) = 0.$$

Optional | **PROOF.** Let $\epsilon > 0$. We shall show that we can use ϵ itself as δ; namely, we shall show that

$$\text{if } \quad 0 < |x - 0| < \epsilon, \quad \text{ then } \quad |g(x) - 0| < \epsilon.$$

For suppose that

$$0 < |x - 0| < \epsilon.$$

If x is rational, then $g(x) = x$. If x is irrational, then $g(x) = 0$. In either case,

$$|g(x) - 0| < \epsilon. \quad \square$$

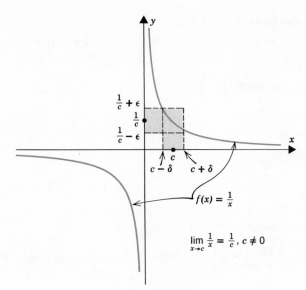

FIGURE 2.2.12

Example

(2.2.5)

$$\text{If} \quad c \neq 0, \quad \text{then} \quad \lim_{x \to c} \frac{1}{x} = \frac{1}{c} \cdot \qquad \text{(Figure 2.2.12)}$$

Optional PROOF.   Let $\epsilon > 0$. Our task is to show that there exists $\delta > 0$ such that

$$\text{if} \quad 0 < |x - c| < \delta, \quad \text{then} \quad \left| \frac{1}{x} - \frac{1}{c} \right| < \epsilon.$$

Note that, for $x \neq 0$,

$$\left| \frac{1}{x} - \frac{1}{c} \right| = \left| \frac{x - c}{xc} \right| = \frac{|x - c|}{|x| \, |c|} \cdot$$

Let δ be any positive number satisfying

$$\delta < \tfrac{1}{2}|c| \quad \text{and} \quad \delta < \tfrac{1}{2}|c|^2 \epsilon.$$

(Why this choice of δ will be clear in a moment.) Now if

$$0 < |x - c| < \delta,$$

then

$$|x| > \tfrac{1}{2}|c| \quad \text{and} \quad |x - c| < \tfrac{1}{2}|c|^2 \epsilon. \qquad \text{(explain)}$$

It follows then that

$$\frac{1}{|x|} < \frac{1}{\tfrac{1}{2}|c|} \quad \text{and} \quad \frac{|x - c|}{\tfrac{1}{2}|c| \, |c|} < \epsilon$$

Optional | and therefore that

$$\frac{|x - c|}{|x|\,|c|} < \epsilon.$$

Since, for such x,

$$\left|\frac{1}{x} - \frac{1}{c}\right| = \frac{|x - c|}{|x|\,|c|},$$

we have

$$\left|\frac{1}{x} - \frac{1}{c}\right| < \epsilon. \quad \square$$

Example   (*The Dirichlet function*)

If $f(x) = \begin{cases} 1, & x \text{ rational} \\ 0, & x \text{ irrational} \end{cases}$,   then   for no number c does $\lim\limits_{x \to c} f(x)$ exist.

PROOF.   Suppose, on the contrary, that

$$\lim_{x \to c} f(x) = l \quad \text{for some particular } c.$$

Taking $\epsilon = \frac{1}{2}$, there must exist $\delta > 0$ such that,

$$\text{if } 0 < |x - c| < \delta, \quad \text{then } |f(x) - l| < \tfrac{1}{2}.$$

Let x_1 be a rational number satisfying

$$0 < |x_1 - c| < \delta$$

and x_2 an irrational number satisfying

$$0 < |x_2 - c| < \delta.$$

(That such numbers exist follows from the fact that every interval contains both rational and irrational numbers.) Now

$$f(x_1) = 1 \quad \text{and} \quad f(x_2) = 0.$$

Thus we must have both

$$|1 - l| < \tfrac{1}{2} \quad \text{and} \quad |0 - l| < \tfrac{1}{2}.$$

From the first inequality we conclude that $l > \frac{1}{2}$. From the second, we conclude that $l < \frac{1}{2}$. Clearly no such number l exists. $\square$

Exercises

Decide in the manner of Section 2.1 whether or not the indicated limit exists. Evaluate the limits which do exist.

*1. $\lim\limits_{x \to 1} \dfrac{x}{x + 1}$.

2. $\lim\limits_{x \to 0} \dfrac{x^2(1 + x)}{2x}$.

*3. $\lim\limits_{x \to 0} \dfrac{x(1 + x)}{2x^2}$.

*4. $\lim\limits_{x\to 4} \dfrac{x}{\sqrt{x}+1}$ ·

5. $\lim\limits_{x\to 4} \dfrac{x}{\sqrt{x}+1}$ ·

*6. $\lim\limits_{x\to -1} \dfrac{1-x}{x+1}$ ·

*7. $\lim\limits_{x\to 1} \dfrac{x^4-1}{x-1}$ ·

8. $\lim\limits_{x\to 0} \dfrac{x}{|x|}$ ·

*9. $\lim\limits_{x\to 2} \dfrac{x}{|x|}$ ·

*10. Referring to Figure 2.2.13, which of the δ's displayed "works" for the given ϵ?

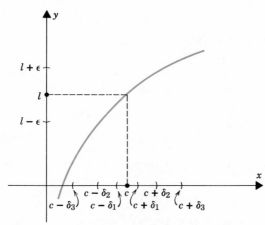

FIGURE 2.2.13

For the following limits, find the largest δ which "works" for a given arbitrary ϵ.

11. $\lim\limits_{x\to 1} 2x = 2$.

*12. $\lim\limits_{x\to 4} 5x = 20$.

13. $\lim\limits_{x\to 2} \tfrac{1}{2}x = 1$.

*14. $\lim\limits_{x\to 2} \tfrac{1}{5}x = \tfrac{2}{5}$.

Give an ϵ, δ proof for the following limits.

15. $\lim\limits_{x\to 4} (2x - 5) = 3$.

*16. $\lim\limits_{x\to 2} (3x - 1) = 5$.

17. $\lim\limits_{x\to 3} (6x - 7) = 11$.

*18. $\lim\limits_{x\to 0} (2 - 5x) = 2$.

19. $\lim\limits_{x\to 2} |1 - 3x| = 5$.

*20. $\lim\limits_{x\to 2} |x - 2| = 0$.

Optional Give an ϵ, δ proof for the following limits.

21. $\lim\limits_{x\to 1} \sqrt{x} = 1$.

*22. $\lim\limits_{x\to 9} \sqrt{x} = 3$.

23. $\lim\limits_{x\to 2} x^2 = 4$.

*24. $\lim\limits_{x\to 5} \sqrt{x - 1} = 2$.

2.3   Some Limit Theorems

As you saw in the last section, it can be rather tedious to apply the ϵ, δ limit test to individual functions. By proving some general theorems about limits we can avoid some of this repetitive work. The theorems themselves of course (at least the first ones) will have to be proved by ϵ, δ methods.

We begin by showing that, if a limit exists, it is unique.

Theorem 2.3.1   The Uniqueness of a Limit

If

$$\lim_{x \to c} f(x) = l \quad \text{and} \quad \lim_{x \to c} f(x) = m,$$

then

$$l = m.$$

PROOF.   Take $\epsilon > 0$ and assume that

$$\lim_{x \to c} f(x) = l \quad \text{and} \quad \lim_{x \to c} f(x) = m.$$

We will show that $l = m$ by showing that $|l - m| = 0$. Observe first that $\frac{1}{2}\epsilon > 0$. From the first limit statement, there exists $\delta_1 > 0$ such that

$$\text{if} \quad 0 < |x - c| < \delta_1, \quad \text{then} \quad |f(x) - l| < \tfrac{1}{2}\epsilon.$$

From the second limit statement, there exists $\delta_2 > 0$ such that

$$\text{if} \quad 0 < |x - c| < \delta_2, \quad \text{then} \quad |f(x) - m| < \tfrac{1}{2}\epsilon.$$

Fox x_1 satisfying

$$0 < |x_1 - c| < \text{minimum of } \delta_1 \text{ and } \delta_2,$$

we have both

$$|f(x_1) - l| < \tfrac{1}{2}\epsilon \quad \text{and} \quad |f(x_1) - m| < \tfrac{1}{2}\epsilon.$$

Rewriting

$$|f(x_1) - l| \quad \text{as} \quad |l - f(x_1)|,$$

we have

$$(*) \qquad |l - f(x_1)| < \tfrac{1}{2}\epsilon \quad \text{and} \quad |f(x_1) - m| < \tfrac{1}{2}\epsilon.$$

It follows that

$$|l - m| = |[l - f(x_1)] + [f(x_1) - m]| \underset{\text{by the triangle inequality}}{\leq} |l - f(x_1)| + |f(x_1) - m|$$
$$\underset{\text{by } (*)}{<} \tfrac{1}{2}\epsilon + \tfrac{1}{2}\epsilon = \epsilon.$$

In short we have found that

$$|l - m| < \epsilon.$$

But since ϵ is arbitrary, $|l - m|$ is a nonnegative number which is strictly less than *every* positive number. Think about it, and you'll see that $|l - m|$ must be zero. ☐

Theorem 2.3.2

If

$$\lim_{x \to c} f(x) = l \quad \text{and} \quad \lim_{x \to c} g(x) = m,$$

then

(i) $$\lim_{x \to c} [f(x) + g(x)] = l + m,$$

(ii) $$\lim_{x \to c} [\alpha f(x)] = \alpha l \qquad \text{for each real } \alpha,$$

(iii) $$\lim_{x \to c} [f(x) g(x)] = lm.$$

PROOF.   Let $\epsilon > 0$. To prove (i) we must show that there exists $\delta > 0$ such that

$$\text{if} \quad 0 < |x - c| < \delta, \qquad \text{then} \quad |[f(x) - g(x)] - [l + m]| < \epsilon.$$

Note that

(*)
$$\begin{aligned}|[f(x) + g(x)] - [l + m]| &= |[f(x) - l] + [g(x) - m]| \\ &\leq |f(x) - l| + |g(x) - m|.\end{aligned}$$

We will make

$$|[f(x) + g(x)] - [l + m]|$$

less than ϵ by making

$$|f(x) - l| \quad \text{and} \quad |g(x) - m|$$

each less than $\frac{1}{2}\epsilon$. Since $\epsilon > 0$, we know that $\frac{1}{2}\epsilon > 0$. Since

$$\lim_{x \to c} f(x) = l \quad \text{and} \quad \lim_{x \to c} g(x) = m,$$

we know that there exist positive numbers δ_1 and δ_2 such that

$$\text{if} \quad 0 < |x - c| < \delta_1, \qquad \text{then} \quad |f(x) - l| < \tfrac{1}{2}\epsilon$$

and

$$\text{if} \quad 0 < |x - c| < \delta_2, \qquad \text{then} \quad |g(x) - m| < \tfrac{1}{2}\epsilon.$$

Now we set

$$\delta = \text{minimum of } \delta_1 \text{ and } \delta_2$$

and note that,

if $0 < |x - c| < \delta$, then $|f(x) - l| < \frac{1}{2}\epsilon$ and $|g(x) - m| < \frac{1}{2}\epsilon$,

and thus by ($*$)

$$|[f(x) + g(x)] - [l + m]| < \epsilon.$$

In summary, by setting $\delta = \min\{\delta_1, \delta_2\}$, we found that,

if $0 < |x - c| < \delta$, then $|[f(x) + g(x)] - [l + m]| < \epsilon$.

Thus (i) is proved.

To prove (ii), we consider two cases: $\alpha \neq 0$ and $\alpha = 0$. If $\alpha \neq 0$, then $\epsilon/|\alpha| > 0$ and, since

$$\lim_{x \to c} f(x) = l,$$

we know that there exists $\delta > 0$ such that,

if $0 < |x - c| < \delta$, then $|f(x) - l| < \dfrac{\epsilon}{|\alpha|}.$

From the last inequality we obtain

$$|\alpha||f(x) - l| < \epsilon \quad \text{and thus} \quad |\alpha f(x) - \alpha l| < \epsilon.$$

The case $\alpha = 0$ was treated before in (2.2.4). For a proof of (iii), see the supplement at the end of this section. $\square$

The results of Theorem 2.3.2 are easily extended to any finite number of functions; namely, if

$$\lim_{x \to c} f_1(x) = l_1, \quad \lim_{x \to c} f_2(x) = l_2, \quad \ldots, \quad \lim_{x \to c} f_n(x) = l_n$$

then

(2.3.3) $\quad \lim\limits_{x \to c} [\alpha_1 f_1(x) + \alpha_2 f_2(x) + \cdots + \alpha_n f_n(x)] = \alpha_1 l_1 + \alpha_2 l_2 + \cdots + \alpha_n l_n$

and

(2.3.4) $\quad \lim\limits_{x \to c} [f_1(x)f_2(x) \cdots f_n(x)] = l_1 l_2 \cdots l_n.$

It follows that for every polynomial $P(x) = a_n x^n + \cdots + a_1 x + a_0$

(2.3.5) $\qquad \boxed{\lim\limits_{x \to c} P(x) = P(c).}$

PROOF. We already know that

$$\lim_{x \to c} x = c.$$

Applying (2.3.4) we have

$$\lim_{x \to c} x^k = c^k \qquad \text{for each positive integer } k.$$

We also know that

$$\lim_{x \to c} a_0 = a_0.$$

If follows now from (2.3.3) that

$$\lim_{x \to c} (a_n x^n + \cdots + a_1 x + a_0) = a_n c^n + \cdots + a_1 c + a_0;$$

that is,

$$\lim_{x \to c} P(x) = P(c). \quad \square$$

Examples

$$\lim_{x \to 1} (5x^2 - 12x + 2) = 5(1)^2 - 12(1) + 2 = -5,$$

$$\lim_{x \to 0} (14x^5 - 7x^2 + 2x + 8) = 14(0)^5 - 7(0)^2 + 2(0) + 8 = 8,$$

$$\lim_{x \to -1} (2x^3 + x^2 - 2x - 3) = 2(-1)^3 + (-1)^2 - 2(-1) - 3 = -2. \quad \square$$

We come now to reciprocals and quotients.

Theorem 2.3.6

$$\text{If} \quad \lim_{x \to c} g(x) = m \quad \text{and} \quad m \neq 0, \quad \text{then} \quad \lim_{x \to c} \frac{1}{g(x)} = \frac{1}{m}.$$

PROOF.   See the supplement to this section. $\square$

Examples

$$\lim_{x \to 4} \frac{1}{x^2} = \frac{1}{16}, \quad \lim_{x \to 2} \frac{1}{x^3 - 1} = \frac{1}{7}, \quad \lim_{x \to -3} \frac{1}{|x|} = \frac{1}{|-3|} = \frac{1}{3}. \quad \square$$

Once you know that reciprocals present no trouble, quotients become easy to handle.

Theorem 2.3.7

If

$$\lim_{x \to c} f(x) = l \quad \text{and} \quad \lim_{x \to c} g(x) = m \quad \text{with} \quad m \neq 0,$$

then

$$\lim_{x \to c} \frac{f(x)}{g(x)} = \frac{l}{m}.$$

PROOF. The key here is to observe that the quotient can be written as a product

$$\frac{f(x)}{g(x)} = f(x)\,\frac{1}{g(x)}.$$

With

$$\lim_{x \to c} f(x) = l \quad \text{and} \quad \lim_{x \to c} \frac{1}{g(x)} = \frac{1}{m},$$

the product rule [part (iii) of Theorem 2.3.2] gives

$$\lim_{x \to c} \frac{f(x)}{g(x)} = l\,\frac{1}{m} = \frac{l}{m}. \quad \square$$

As an immediate consequence of this theorem on quotients you can see that, if P and Q are polynomials and $Q(c) \neq 0$, then

(2.3.8)
$$\boxed{\lim_{x \to c} \frac{P(x)}{Q(x)} = \frac{P(c)}{Q(c)}.}$$

Examples

$$\lim_{x \to 2} \frac{3x - 5}{x^2 + 1} = \frac{6 - 5}{4 + 1} = \frac{1}{5}, \quad \lim_{x \to 3} \frac{x^3 - 3x^2}{1 - x^2} = \frac{27 - 27}{1 - 9} = 0. \quad \square$$

There is no point looking for a limit that does not exist. The last theorem of this section gives a condition under which a quotient does not have a limit.

Theorem 2.3.9

If

$$\lim_{x \to c} f(x) = l \quad \text{with } l \neq 0 \qquad \text{and} \qquad \lim_{x \to c} g(x) = 0,$$

then

$$\lim_{x \to c} \frac{f(x)}{g(x)} \quad \text{does not exist.}$$

PROOF. Suppose on the contrary that there exists a real number L such that

$$\lim_{x \to c} \frac{f(x)}{g(x)} = L.$$

Then

$$l = \lim_{x \to c} f(x) = \lim_{x \to c} \left[g(x) \cdot \frac{f(x)}{g(x)} \right] = \lim_{x \to c} g(x) \cdot \lim_{x \to c} \frac{f(x)}{g(x)} = 0 \cdot L = 0,$$

which contradicts our assumption that $l \neq 0$. $\quad \square$

Examples.   From Theorem 2.3.9 you can see that

$$\lim_{x\to 1}\frac{x^2}{x-1},\qquad \lim_{x\to 2}\frac{3x-7}{x^2-4},\qquad \text{and}\qquad \lim_{x\to 0}\frac{5}{x}$$

all fail to exist.   □

Problem.   Evaluate the limits that exist:

(a) $\displaystyle\lim_{x\to 3}\frac{x^2-x-6}{x-3}$ (b) $\displaystyle\lim_{x\to 4}\frac{(x^2-3x-4)^2}{x-4}$ (c) $\displaystyle\lim_{x\to -1}\frac{x+1}{(2x^2+7x+5)^2}.$

SOLUTION.   In each case both numerator and denominator tend to zero, and so we have to be careful.

(a) First we factor the numerator:

$$\frac{x^2-x-6}{x-3}=\frac{(x+2)(x-3)}{x-3}.$$

For $x\ne 3$,

$$\frac{x^2-x-6}{x-3}=x+2.$$

Thus

$$\lim_{x\to 3}\frac{x^2-x-6}{x-3}=\lim_{x\to 3}(x+2)=5.$$

(b) Note that

$$\frac{(x^2-3x-4)^2}{x-4}=\frac{[(x+1)(x-4)]^2}{x-4}=\frac{(x+1)^2(x-4)^2}{x-4}$$

so that, for $x\ne 4$,

$$\frac{(x^2-3x-4)^2}{x-4}=(x+1)^2(x-4).$$

It follows then that

$$\lim_{x\to 4}\frac{(x^2-3x-4)^2}{x-4}=\lim_{x\to 4}(x+1)^2(x-4)=0.$$

(c) Since

$$\frac{x+1}{(2x^2+7x+5)^2}=\frac{x+1}{[(2x+5)(x+1)]^2}=\frac{x+1}{(2x+5)^2(x+1)^2},$$

you can see that, for $x\ne -1$,

$$\frac{x+1}{(2x^2+7x+5)^2}=\frac{1}{(2x+5)^2(x+1)}.$$

By Theorem 2.3.9,

$$\lim_{x\to -1}\frac{1}{(2x+5)^2(x+1)}\quad \text{does not exist,}$$

and therefore

$$\lim_{x \to -1} \frac{x + 1}{(2x^2 + 7x + 5)^2} \quad \text{does not exist either.} \quad \Box$$

Exercises

1. Evaluate the limits that exist.

*(a) $\displaystyle\lim_{x \to 2} \frac{x^2 + x + 1}{x^2 + 2x}$.

(b) $\displaystyle\lim_{x \to 2} \frac{x - 2}{x^2 - 4}$.

*(c) $\displaystyle\lim_{x \to 2} \frac{x^2 - 4}{x - 2}$.

(d) $\displaystyle\lim_{x \to -1} (\sqrt{2}x - 6x + 1)$.

*(e) $\displaystyle\lim_{x \to 0} \left(x - \frac{4}{x} \right)$.

(f) $\displaystyle\lim_{x \to 1} \frac{x - x^2}{(3x - 1)(x^4 - 2)}$.

*(g) $\displaystyle\lim_{x \to 5} \frac{2 - x^2}{4x}$.

(h) $\displaystyle\lim_{x \to 1} \frac{x - x^2}{1 - x}$.

*(i) $\displaystyle\lim_{x \to -1} \frac{x^2 + 1}{3x^5 + 4}$.

(j) $\displaystyle\lim_{x \to 0} \frac{x^2 + 1}{x}$.

*(k) $\displaystyle\lim_{x \to 0} \frac{x^2}{x^2 + 1}$.

(l) $\displaystyle\lim_{x \to 2} \frac{3x}{x + 4}$.

*(m) $\displaystyle\lim_{x \to -2} \frac{(x^2 - x - 6)^2}{x + 2}$.

(n) $\displaystyle\lim_{x \to -2} \frac{(x^2 - x - 6)^2}{(x + 2)^2}$.

*(o) $\displaystyle\lim_{x \to -2} \frac{x^2 - x - 6}{(x + 2)^2}$.

(p) $\displaystyle\lim_{x \to 1} \frac{x^2 - x - 6}{(x + 2)^2}$.

*(q) $\displaystyle\lim_{x \to 2} \frac{x^2 + 4}{x^2 + 2x + 1}$.

(r) $\displaystyle\lim_{x \to 2} \frac{x}{x^2 - 4}$.

*(s) $\displaystyle\lim_{x \to -3} \frac{5x + 15}{x^3 + 3x^2 - x - 3}$.

(t) $\displaystyle\lim_{x \to 0} \frac{x^4}{3x^3 + 2x^2 + x}$.

*(u) $\displaystyle\lim_{x \to -4} \left(\frac{3x}{x + 4} + \frac{8}{x + 4} \right)$.

(v) $\displaystyle\lim_{x \to -4} \left(\frac{2x}{x + 4} + \frac{8}{x + 4} \right)$.

2. Evaluate the limits that exist.

*(a) $\displaystyle\lim_{t \to 1} \frac{t + 1}{t^2 - 1}$.

(b) $\displaystyle\lim_{t \to 1} \frac{t - 1}{t^2 - 1}$.

*(c) $\displaystyle\lim_{t \to 0} \frac{t + 1/t}{t - 1/t}$.

(d) $\displaystyle\lim_{t \to 0} \frac{t + a/t}{t + b/t}$.

*(e) $\displaystyle\lim_{t \to -2} \frac{|t|}{t - 2}$.

(f) $\displaystyle\lim_{t \to -1} \frac{1 + |t|}{1 - |t|}$.

*(g) $\displaystyle\lim_{t \to -1} \frac{t^2 + 6t + 5}{t^2 + 3t + 2}$.

(h) $\displaystyle\lim_{t \to 1} \frac{t^2 - 2t + 1}{t^3 - 3t + 3t - 1}$.

3. Given that

$$\lim_{x \to c} f(x) = 2, \quad \lim_{x \to c} g(x) = -1, \quad \text{and} \quad \lim_{x \to c} h(x) = 0,$$

evaluate the limits that exist.

*(a) $\lim\limits_{x \to c} [f(x) - g(x)]$.

(b) $\lim\limits_{x \to c} [f(x)]^2$.

(c) $\lim\limits_{x \to c} \dfrac{f(x)}{g(x)}$.

*(d) $\lim\limits_{x \to c} \dfrac{h(x)}{f(x)}$.

*(e) $\lim\limits_{x \to c} \dfrac{f(x)}{h(x)}$.

(f) $\lim\limits_{x \to c} \dfrac{1}{f(x) - g(x)}$.

4. Given that $f(x) = x^3$, evaluate the limits that exist.

*(a) $\lim\limits_{x \to 3} \dfrac{f(x) - f(3)}{x - 3}$.

(b) $\lim\limits_{x \to 3} \dfrac{f(x) - f(2)}{x - 3}$.

*(c) $\lim\limits_{x \to 3} \dfrac{f(x) - f(3)}{x - 2}$.

(d) $\lim\limits_{x \to 1} \dfrac{f(x) - f(1)}{x - 1}$.

5. Given that

$$\lim_{x \to c} g(x) = 0 \quad \text{and} \quad f(x)g(x) = 1 \quad \text{for all real } x,$$

prove that $\lim\limits_{x \to c} f(x)$ does not exist.

*6. Give an example to show that the existence of

$$\lim_{x \to c} [f(x) + g(x)]$$

does not imply the existence of

$$\lim_{x \to c} f(x) \quad \text{or} \quad \lim_{x \to c} g(x).$$

7. Give an example to show that the existence of

$$\lim_{x \to c} [f(x)g(x)]$$

does not imply the existence of

$$\lim_{x \to c} f(x) \quad \text{or} \quad \lim_{x \to c} g(x).$$

Optional | **Supplement to Section 2.3**

PROOF OF THEOREM 2.3.2 (iii)

$$\begin{aligned}
|f(x)g(x) - lm| &\leq |f(x)g(x) - f(x)m| + |f(x)m - lm| \\
&= |f(x)||g(x) - m| + |m||f(x) - l| \\
&\leq |f(x)||g(x) - m| + (1 + |m|)|f(x) - l|.
\end{aligned}$$

Let $\epsilon > 0$. Since

$$\lim_{x \to c} f(x) = l \quad \text{and} \quad \lim_{x \to c} g(x) = m,$$

we know

(1) that there exists $\delta_1 > 0$ such that, if $0 < |x - c| < \delta_1$, then

$$|f(x) - l| < 1 \quad \text{and thus} \quad |f(x)| < 1 + |l|;$$

(2) that there exists $\delta_2 > 0$ such that

$$\text{if} \quad 0 < |x - c| < \delta_2, \quad \text{then} \quad |g(x) - m| < \left(\frac{\frac{1}{2}\epsilon}{1 + |l|}\right);$$

(3) that there exists $\delta_3 > 0$ such that

$$\text{if} \quad 0 < |x - c| < \delta_3, \quad \text{then} \quad |f(x) - l| < \left(\frac{\frac{1}{2}\epsilon}{1 + |m|}\right).$$

We can now set $\delta = \min\{\delta_1, \delta_2, \delta_3\}$ and observe that,

if $\qquad\qquad\qquad 0 < |x - c| < \delta,$

then

$$|f(x)g(x) - lm| < (1 + |l|)\left(\frac{\frac{1}{2}\epsilon}{1 + |l|}\right) + (1 + |m|)\left(\frac{\frac{1}{2}\epsilon}{1 + |m|}\right) = \epsilon. \quad \square$$

PROOF OF THEOREM 2.3.6.   For $g(x) \neq 0$,

$$\left|\frac{1}{g(x)} - \frac{1}{m}\right| = \frac{|g(x) - m|}{|m|\,|g(x)|}.$$

Choose $\delta_1 > 0$ such that

$$\text{if} \quad 0 < |x - c| < \delta_1, \quad \text{then} \quad |g(x) - m| < \frac{|m|}{2}.$$

For such x,

$$|g(x)| > \frac{|m|}{2}, \qquad \frac{1}{|g(x)|} < \frac{2}{|m|}$$

and thus

$$\left|\frac{1}{g(x)} - \frac{1}{m}\right| \le \frac{2}{|m|^2}\,|g(x) - m|.$$

Now let $\epsilon > 0$ and choose $\delta_2 > 0$ such that

$$\text{if} \quad 0 < |x - c| < \delta_2, \quad \text{then} \quad |g(x) - m| < \frac{|m|^2}{2}\,\epsilon.$$

Setting $\delta = \min\{\delta_1, \delta_2\}$, we find that

$$\text{if} \quad 0 < |x - c| < \delta, \quad \text{then} \quad \left|\frac{1}{g(x)} - \frac{1}{m}\right| < \epsilon. \quad \square$$

2.4 More on Limits

As we have indicated before, whether or not $\lim_{x \to c} f(x)$ exists and, if so, what it is, depends only on the behavior of f for x close to c but different from c. If, for such x, f is trapped between two functions both of which tend to the same limit l

(see Figure 2.4.1), then $f(x)$ must also tend to l. This idea is the substance of what we call the "pinching theorem."

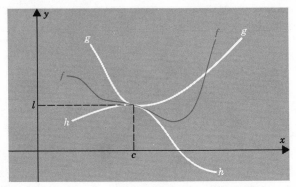

FIGURE 2.4.1

Theorem 2.4.1   The Pinching Theorem

Let $p > 0$. Suppose that, for all x such that $0 < |x - c| < p$,

$$h(x) \le f(x) \le g(x).$$

If

$$\lim_{x \to c} h(x) = l \quad \text{and} \quad \lim_{x \to c} g(x) = l,$$

then

$$\lim_{x \to c} f(x) = l.$$

PROOF.   Let $\epsilon > 0$. Let $p > 0$ be such that

$$\text{if} \quad 0 < |x - c| < p, \quad \text{then} \quad h(x) \le f(x) \le g(x).$$

Choose $\delta_1 > 0$ such that

$$\text{if} \quad 0 < |x - c| < \delta_1, \quad \text{then} \quad l - \epsilon < h(x) < l + \epsilon.$$

Choose $\delta_2 > 0$ such that

$$\text{if} \quad 0 < |x - c| < \delta_2, \quad \text{then} \quad l - \epsilon < g(x) < l + \epsilon.$$

Let $\delta = \min \{p, \delta_1, \delta_2\}$. For x satisfying $0 < |x - c| < \delta$, we have

$$l - \epsilon < h(x) \le f(x) \le g(x) < l + \epsilon$$

and thus

$$|f(x) - l| < \epsilon. \quad \square$$

Example.   Suppose that

$$-x^2 \le f(x) \le x^2 \qquad \text{if } 0 < |x| < 1.$$

Since

$$\lim_{x \to 0} - x^2 = 0 \quad \text{and} \quad \lim_{x \to 0} x^2 = 0,$$

we can conclude from the pinching theorem that

$$\lim_{x \to 0} f(x) = 0. \quad \square$$

Example. Suppose that

$$x^2 + 12x - 9 \le f(x) \le 5x^2 \qquad \text{for all } x \ne \tfrac{3}{2}.$$

As you can check,

$$\lim_{x \to 3/2} (x^2 + 12x - 9) = \tfrac{45}{4} \quad \text{and} \quad \lim_{x \to 3/2} 5x^2 = \tfrac{45}{4}.$$

It follows from the pinching theorem that

$$\lim_{x \to 3/2} f(x) = \tfrac{45}{4}. \quad \square$$

Example

$$\text{If} \quad f(x) = \begin{cases} x^2, & x \text{ rational} \\ x^4, & x \text{ irrational} \end{cases}, \qquad \text{then} \quad \lim_{x \to 0} f(x) = 0.$$

To verify this, note that, when $|x| < 1$,

$$0 \le f(x) \le x^2. \qquad\qquad \text{(explain)}$$

Since

$$\lim_{x \to 0} x^2 = 0 \quad \text{(and, of course, } \lim_{x \to 0} 0 = 0),$$

the pinching theorem guarantees that

$$\lim_{x \to 0} f(x) = 0. \quad \square$$

There are many different ways of formulating the same limit statement. Sometimes one formulation is more convenient, sometimes another. In any case, it is useful to recognize that the following are equivalent:

(2.4.2)

$$\begin{array}{ll} \text{(i) } \lim_{x \to c} f(x) = l. & \text{(ii) } \lim_{x \to c} (f(x) - l) = 0. \\[2mm] \text{(iii) } \lim_{x \to c} |f(x) - l| = 0. & \text{(iv) } \lim_{h \to 0} f(c + h) = l. \end{array}$$

The equivalence of these four statements is obvious. It is, however, a good exercise in ϵ, δ technique to show that (i) is equivalent to (iv).

Example. For $f(x) = x^3$, we have

$$\lim_{x \to 2} x^3 = 8, \qquad \lim_{x \to 2} (x^3 - 8) = 0,$$

$$\lim_{x \to 2} |x^3 - 8| = 0, \qquad \lim_{h \to 0} (2 + h)^3 = 8. \quad \square$$

While we are on the subject of equivalent statements, note that

$$\lim_{x \to c} f(x) = l \quad \text{and} \quad \lim_{x \to c} |f(x)| = |l|$$

are *not* equivalent. While

$$\lim_{x \to c} f(x) = l \quad \text{implies} \quad \lim_{x \to c} f(x) = |l|, \qquad \text{(Exercise 4)}$$

the converse is false. There are instances in which

$$\lim_{x \to c} |f(x)| \text{ exists} \quad \text{but} \quad \lim_{x \to c} f(x) \text{ does not.} \qquad \text{(Exercise 5)}$$

Exercises

1. Find the following limits.

   *(a) $\lim_{x \to 0} f(x)$ given that $0 \le f(x) \le |x|$ if $0 < |x| < 1$.

   (b) $\lim_{x \to 0} f(x)$ given that $2 - |x| \le f(x) \le 2 + |x|$ for $x \ne 0$.

   *(c) $\lim_{x \to 3} f(x)$ given that $1 \le f(x) \le (x - 3)^2 + 1$ for $x \ne 3$.

   (d) $\lim_{x \to 2} f(x)$ given that $-(x - 2)^2 \le f(x) \le 0$ if $0 < |x - 2| < 1$.

   *(e) $\lim_{x \to 5} f(x)$ given that $x + 5 \le f(x) \le x^2 - 9x + 30$ for $x \ne 5$.

2. Find the limits that exist.

   *(a) $\lim_{h \to 0} (3 + h)^2$. (b) $\lim_{h \to 0} (2 - h)^2$. *(c) $\lim_{x \to 1} 2|x - 1|$.

   *(d) $\lim_{x \to 2} |x^2 - 4|$. (e) $\lim_{h \to 0} h\left(1 + \dfrac{1}{h}\right)$. *(f) $\lim_{h \to 0} h\left(1 - \dfrac{1}{h}\right)$.

   *(g) $\lim_{h \to 0} h^2\left(1 + \dfrac{1}{h}\right)$. (h) $\lim_{h \to 0} h\left(1 + \dfrac{1}{h^2}\right)$. *(i) $\lim_{h \to 0} \dfrac{1 - 1/h}{1 + 1/h}$.

   *(j) $\lim_{h \to 0} \dfrac{1 - 1/h^2}{1 + 1/h^2}$. (k) $\lim_{h \to 0} \dfrac{1 - 1/h^2}{1 - 1/h}$. *(l) $\lim_{h \to 0} \dfrac{1 + 1/h}{1 + 1/h^2}$.

   *(m) $\lim_{h \to 1} \dfrac{h}{|h - 1|}$. (n) $\lim_{h \to 1} \dfrac{|h - 1|}{h}$. *(o) $\lim_{x \to c} \dfrac{x - c}{|x - c|}$.

3. Give the four limit statements displayed in 2.4.2, taking

   *(a) $f(x) = \dfrac{1}{x - 1}$, $c = 3$. (b) $f(x) = \dfrac{x}{x^2 + 2}$, $c = 1$.

4. Show that

   $$\text{if} \quad \lim_{x \to c} f(x) = l \quad \text{then} \quad \lim_{x \to c} |f(x)| = |l|.$$

*5. Given that

   $$f(x) = \begin{cases} -1, & x < 0 \\ 1, & x > 0 \end{cases},$$

   find (a) $\lim_{x \to 0} f(x)$ and (b) $\lim_{x \to 0} |f(x)|$.

6. Show that

$$\lim_{x \to c} \sqrt{x} = \sqrt{c} \quad \text{for each } c > 0.$$

HINT: If x and c are positive, then

$$0 \le |\sqrt{x} - \sqrt{c}| = \frac{|x - c|}{\sqrt{x} + \sqrt{c}} \le \frac{1}{\sqrt{c}} |x - c|.$$

7. Find the following limits.

*(a) $\lim_{x \to 0} f(x)$ given that $\left| \dfrac{f(x)}{x} \right| \le M$ for $x \ne 0$.

(b) $\lim_{x \to c} f(x)$ given that $\left| \dfrac{f(x) - f(c)}{x - c} \right| \le M$ for $x \ne c$.

8. (a) Verify that

$$\max \{f(x), g(x)\} = \tfrac{1}{2}\{[f(x) + g(x)] + |f(x) - g(x)|\}.$$

*(b) Find a similar expression for min $\{f(x), g(x)\}$.

9. Let

$$h(x) = \min \{f(x), g(x)\} \quad \text{and} \quad H(x) = \max \{f(x), g(x)\}.$$

Show that

if $\lim_{x \to c} f(x) = l$ and $\lim_{x \to c} g(x) = l$ then $\lim_{x \to c} h(x) = l$ and $\lim_{x \to c} H(x) = l$.

HINT: Use Exercise 8.

Optional | 10. Give an ϵ, δ proof that statement (i) of 2.4.2 is equivalent to statement (iv).

11. Show that,

if $f(x) \le g(x)$ for all real x,   then $\lim_{x \to c} f(x) \le \lim_{x \to c} g(x)$

provided both limits exist.

2.5   One-Sided Limits

Whether or not $\lim_{x \to c} f(x)$ exists, and, if so, what it is, depends on the behavior of f on both sides of c. The situation is simpler in the case of one-sided limits: the *left-hand* limit, $\lim_{x \uparrow c} f(x)$, and the *right-hand* limit, $\lim_{x \downarrow c} f(x)$.†

You can think of the left-hand limit

$$\lim_{x \uparrow c} f(x)$$

as the number that $f(x)$ approaches as x comes into c from the left, and the

---

† The left-hand limit is sometimes written $\lim_{x \to c^-} f(x)$ and the right-hand limit, $\lim_{x \to c^+} f(x)$.

right-hand limit

$$\lim_{x \downarrow c} f(x)$$

as the number that $f(x)$ approaches as x comes into c from the right. See Figure 2.5.1.

FIGURE 2.5.1

Here are the appropriate ϵ, δ definitions:

Definition 2.5.1 Left-Hand Limit

Let f be a function defined at least on an interval of the form (a, c).

$$\lim_{x \uparrow c} f(x) = l \quad \text{iff} \quad \begin{cases} \text{for each } \epsilon > 0 \text{ there exists } \delta > 0 \text{ such that} \\ \text{if } \quad c - \delta < x < c, \quad \text{then} \quad |f(x) - l| < \epsilon. \end{cases}$$

Definition 2.5.2 Right-Hand Limit

Let f be a function defined at least on an interval of the form (c, d).

$$\lim_{x \downarrow c} f(x) = l \quad \text{iff} \quad \begin{cases} \text{for each } \epsilon > 0 \text{ there exists } \delta > 0 \text{ such that} \\ \text{if } \quad c < x < c + \delta, \quad \text{then} \quad |f(x) - l| < \epsilon. \end{cases}$$

Example.   If

$$h(x) = \begin{cases} -1, & x < 0 \\ 1, & x > 0 \end{cases},$$

then

$$\lim_{x \uparrow 0} h(x) = -1 \quad \text{and} \quad \lim_{x \downarrow 0} h(x) = 1$$

although, as you saw before, $\lim_{x \to 0} h(x)$ does not exist. $\square$

Example.   Since the square-root function is not defined on both sides of 0, we cannot consider

$$\lim_{x \to 0} \sqrt{x}.$$

However, it is true that

$$\lim_{x \downarrow 0} \sqrt{x} = 0. \qquad\qquad \text{(Figure 2.5.2)}$$

For a formal proof, begin with $\epsilon > 0$ and take ϵ^2 as δ:

$$\text{if} \quad 0 < x < \epsilon^2 \quad \text{then} \quad |\sqrt{x} - 0| = \sqrt{x} < \epsilon. \quad \square$$

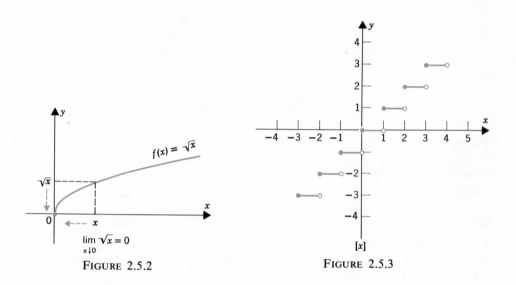

FIGURE 2.5.2   FIGURE 2.5.3

Example.   An interesting function to look at when dealing with one-sided limits is the *greatest-integer function*:

(2.5.3) $[x] = \text{greatest integer} \le x.$

For the graph we refer to Figure 2.5.3. At 0 the function is 0 and it remains 0 throughout the interval $[0, 1)$. At 1 the function jumps up to 1 and it remains 1 throughout the interval $[1, 2)$. At 2 the function jumps up to 2 and so on. It is easy to see that for each integer n,

$$\lim_{x \uparrow n} [x] = n - 1 \quad \text{and} \quad \lim_{x \downarrow n} [x] = n.$$

For example,

$$\lim_{x \uparrow 2} [x] = 1 \quad \text{and} \quad \lim_{x \downarrow 2} [x] = 2.$$

If c is a number which is not an integer, then c lies between consecutive integers n

and $n + 1$. For such c, we have both

$$\lim_{x \uparrow c} [x] = n \quad \text{and} \quad \lim_{x \downarrow c} [x] = n.$$

For example,

$$\lim_{x \uparrow 3/2} [x] = 1 \quad \text{and} \quad \lim_{x \downarrow 3/2} [x] = 1. \quad \square$$

Results like those proved in Sections 2.3 and 2.4 for two-sided limits also hold for one-sided limits. Thus, for instance, if

$$\lim_{x \uparrow c} f(x) = l \quad \text{and} \quad \lim_{x \uparrow c} g(x) = m,$$

then

$$\lim_{x \uparrow c} [f(x) + g(x)] = l + m, \quad \lim_{x \uparrow c} [\alpha f(x)] = \alpha l, \quad \lim_{x \uparrow c} [f(x)g(x)] = lm.$$

Moreover, if $m \neq 0$, then

$$\lim_{x \uparrow c} \frac{1}{g(x)} = \frac{1}{m} \quad \text{and} \quad \lim_{x \uparrow c} \frac{f(x)}{g(x)} = \frac{l}{m}.$$

Similar results hold for right-hand limits. Proofs can be obtained by mimicking the proofs given before.

One-sided limits give us a simple way of determining whether or not a (two-sided) limit exists:

(2.5.4)　　$\boxed{\lim_{x \to c} f(x) = l \quad \text{iff} \quad \lim_{x \uparrow c} f(x) = l \quad \text{and} \quad \lim_{x \downarrow c} f(x) = l.}$

The proof is left as an exercise. It is easy because any δ that "works" for the limit will work for the two one-sided limits and any δ that "works" for the two one-sided limits will work for the limit.

We illustrate the effectiveness of (2.5.4) by some examples.

Example. For

$$h(x) = \begin{cases} -1, & x < 0 \\ 1, & x > 0 \end{cases},$$

we have

$$\lim_{x \uparrow 0} h(x) = -1 \quad \text{and} \quad \lim_{x \downarrow 0} h(x) = 1.$$

Since the two one-sided limits are not equal,

$$\lim_{x \to 0} h(x) \quad \text{does not exist.} \quad \square$$

Example. With

$$g(x) = \begin{cases} x^2, & x < 0 \\ \sqrt{x}, & x > 0 \end{cases},$$

we have both

$$\lim_{x \uparrow 0} g(x) = 0 \quad \text{and} \quad \lim_{x \downarrow 0} g(x) = 0.$$

Consequently,

$$\lim_{x \to 0} g(x) = 0. \quad \square$$

Example. Set

$$k(x) = \begin{cases} 1, & \text{negative rational } x \\ 0, & \text{negative irrational } x \\ x, & \text{positive } x \end{cases}.$$

Here

$$\lim_{x \uparrow 0} k(x) \quad \text{does not exist}$$

and therefore

$$\lim_{x \to 0} k(x) \quad \text{does not exist either.} \quad \square$$

Exercises

1. (a) Sketch the graph of

$$f(x) = \begin{cases} x - 1, & x < 0 \\ 1 + x, & x > 0 \end{cases}.$$

(b) Evaluate the limits that exist.

    *(i) $\lim_{x \uparrow 0} f(x)$.   (ii) $\lim_{x \downarrow 0} f(x)$.   *(iii) $\lim_{x \to 0} f(x)$.

2. (a) Sketch the graph of

$$g(x) = \begin{cases} x^2, & x < 1 \\ x, & 1 < x < 4 \\ 4 - x, & x > 4 \end{cases}.$$

(b) Evaluate the limits that exist.

    *(i) $\lim_{x \uparrow 1} g(x)$.   (ii) $\lim_{x \downarrow 1} g(x)$.   *(iii) $\lim_{x \to 1} g(x)$.

    (iv) $\lim_{x \uparrow 2} g(x)$.   *(v) $\lim_{x \downarrow 2} g(x)$.   (vi) $\lim_{x \to 2} g(x)$.

    *(vii) $\lim_{x \uparrow 4} g(x)$.   (viii) $\lim_{x \downarrow 4} g(x)$.   *(ix) $\lim_{x \to 4} g(x)$.

3. Evaluate the limits that exist.

    *(a) $\lim_{x \uparrow 2} \dfrac{x^2}{x + 2}$.   (b) $\lim_{x \downarrow 2} \dfrac{x^2}{x + 2}$.   *(c) $\lim_{x \uparrow 2} \dfrac{x - 2}{|x - 2|}$.

    *(d) $\lim_{x \downarrow 2} \dfrac{x - 2}{|x - 2|}$.   (e) $\lim_{x \to 1} \dfrac{x^3 - 1}{|x^3 - 1|}$.   *(f) $\lim_{x \downarrow 0} \dfrac{1 + \sqrt{x}}{1 - \sqrt{x}}$.

*(g) $\lim_{x \uparrow 1} \sqrt{|x| - x}$.

(h) $\lim_{x \downarrow 1} \sqrt{|x| - x}$.

*(i) $\lim_{x \uparrow 1} \sqrt{x - [x]}$.

*(j) $\lim_{x \downarrow 1} \sqrt{x - [x]}$.

(k) $\lim_{x \downarrow 0} \dfrac{[x] - x}{x}$.

*(l) $\lim_{x \uparrow 1} (x^2 + 2)^{[x]}$.

*(m) $\lim_{x \uparrow 2} \left(\dfrac{1}{x - 2} - \dfrac{1}{|x - 2|} \right)$.

(n) $\lim_{x \downarrow 2} \left(\dfrac{1}{x - 2} - \dfrac{1}{|x - 2|} \right)$.

4. (a) Sketch the graph of

$$f(x) = \begin{cases} -1, & x < -1 \\ x^3, & -1 \le x \le 1 \\ 1 - x, & 1 < x < 2 \\ 3 - x^2, & x \ge 2 \end{cases}.$$

(b) Evaluate the limits that exist.

*(i) $\lim_{x \uparrow -1} f(x)$.

(ii) $\lim_{x \downarrow -1} f(x)$.

*(iii) $\lim_{x \to -1} f(x)$.

(iv) $\lim_{x \uparrow 1} f(x)$.

*(v) $\lim_{x \downarrow 1} f(x)$.

(vi) $\lim_{x \to 1} f(x)$.

*(vii) $\lim_{x \uparrow 2} f(x)$.

(viii) $\lim_{x \downarrow 2} f(x)$.

*(ix) $\lim_{x \to 2} f(x)$.

5. (a) Sketch the graph of $f(x) = [2x]$.

(b) Evaluate the limits that exist.

*(i) $\lim_{x \uparrow 1/2} [2x]$.

*(ii) $\lim_{x \downarrow 1/2} [2x]$.

*(iii) $\lim_{x \uparrow 3/4} [2x]$.

*(iv) $\lim_{x \downarrow 3/4} [2x]$.

*(v) $\lim_{x \to 1/2} [2x]$.

*(vi) $\lim_{x \to 3/4} [2x]$.

6. Give an ϵ, δ proof of (2.5.4).

Optional 7. Given an ϵ, δ proof that

(a) $\lim_{x \uparrow c} f(x) = l$ iff $\lim_{h \to 0} f(c - |h|) = l$.

(b) $\lim_{x \downarrow c} f(x) = l$ iff $\lim_{h \to 0} f(c + |h|) = l$.

8. State the pinching theorem (2.4.1) (a) for left-hand limits. (b) for right-hand limits.

9. Evaluate the limits that exist.

*(a) $\lim_{x \to 1} [2x](x - 1)$.

(b) $\lim_{x \to 0} [x][x + 1]$.

*(c) $\lim_{x \to 0} x[1/x]$.

(d) $\lim_{x \to 0} x^3[1/x]$.

HINT: Use one-sided limits. For (c) and (d) note that $t - 1 \le [t] \le t$ and use the appropriate pinching theorem.

2.6 Continuity

In ordinary language, to say that a certain process is "continuous" is to say that it goes on without interruption and without abrupt changes. In mathematics the word "continuous" has much the same meaning.

The idea of continuity is so important to calculus that we will discuss it with some care. First we introduce the idea of continuity at a point (or number) c, and then we talk about continuity on an interval.

Continuity at a Point

Let f be a function defined at least on an open interval $(c - \delta, c + \delta)$.

Definition 2.6.1

The function f is said to be *continuous* at c iff $\lim\limits_{x \to c} f(x) = f(c)$.

If the domain of f contains an interval $(c - \delta, c + \delta)$, then f can fail to be continuous at c only for one of two reasons: either $f(x)$ does not have a limit as x tends to c (in which case c is called an *essential discontinuity*), or it does have a limit, but the limit is not $f(c)$. In this case c is called a *removable discontinuity;* the discontinuity can be removed by redefining f at c.

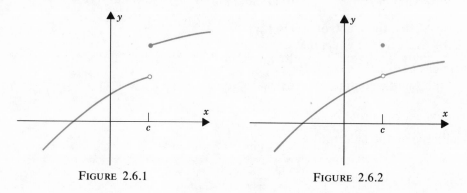

FIGURE 2.6.1   FIGURE 2.6.2

The function depicted in Figure 2.6.1 is discontinuous at c because it does not have a limit at c. The discontinuity is essential.

The function depicted in Figure 2.6.2 does have a limit at c. It is discontinuous at c only because its limit at c is not its value at c. The discontinuity is removable; it can be removed by lowering the dot into place.

FIGURE 2.6.3

In Figure 2.6.3, we have tried to suggest the Dirichlet function

$$f(x) = \left[\begin{cases} 1, & x \text{ rational} \\ 0, & x \text{ irrational} \end{cases}\right].$$

At no point c does f have a limit. It is therefore everywhere discontinuous and every point is an essential discontinuity.

Most of the functions that you have encountered so far are continuous at each point of their domains. In particular, this is true for polynomials,

$$\lim_{x \to c} P(x) = P(c),$$

for rational functions, by which we mean quotients of polynomials,

$$\lim_{x \to c} \frac{P(x)}{Q(x)} = \frac{P(c)}{Q(c)} \quad \text{if} \quad Q(c) \neq 0,$$

and for the absolute-value function,

$$\lim_{x \to c} |x| = |c|.$$

In one of the exercises you showed (or at least were asked to show) that

$$\lim_{x \to c} \sqrt{x} = \sqrt{c} \quad \text{for each } c > 0.$$

This makes the square-root function continuous at each positive number. What happens at 0, we discuss later.

With f and g continuous at c, we have

$$\lim_{x \to c} f(x) = f(c), \quad \lim_{x \to c} g(x) = g(c)$$

and thus, by the limit theorems,

$$\lim_{x \to c} [f(x) + g(x)] = f(c) + g(c)$$
$$\lim_{x \to c} [\alpha f(x)] = \alpha f(c)$$
$$\lim_{x \to c} [f(x)g(x)] = f(c)g(c)$$

and, if $g(c) \neq 0$,

$$\lim_{x \to c} \frac{f(x)}{g(x)} = \frac{f(c)}{g(c)}.$$

This tells us that, *if the functions f and g are continuous at c, then $f + g$, αf, and fg are continuous at c; moreover, if $g(c) \neq 0$, then the quotient f/g is also continuous at c.* These results can obviously be extended to any finite number of functions.

Our next topic is the continuity of composite functions. Before getting into this, however, let's take a look at continuity in terms of ϵ, δ. A direct translation of

$$\lim_{x \to c} f(x) = f(c)$$

into ϵ, δ terms reads like this: for each $\epsilon > 0$ there exists $\delta > 0$ such that

$$\text{if} \quad 0 < |x - c| < \delta \quad \text{then} \quad |f(x) - f(c)| < \epsilon.$$

Here the restriction $0 < |x - c|$ is unnecessary. We can allow $|x - c| = 0$ because then $x = c$, $f(x) = f(c)$, and thus $|f(x) - f(c)| = 0$. Being 0, $|f(x) - f(c)|$ is certainly less than ϵ.

In summary, an ϵ, δ characterization of continuity at c reads as follows:

$$(2.6.2) \quad \boxed{f \text{ is continuous at } c \quad \text{iff} \quad \begin{cases} \text{for each } \epsilon > 0 \text{ there exists } \delta > 0 \text{ such that} \\ \text{if } |x - c| < \delta \quad \text{then} \quad |f(x) - f(c)| < \epsilon. \end{cases}}$$

In simple intuitive language

$$\boxed{f \text{ is continuous at } c \quad \text{iff} \quad \text{for } x \text{ close to } c, f(x) \text{ is close to } f(c).}$$

We are now ready to take up the continuity of composites. Remember the defining formula: $(f \circ g)(x) = f(g(x))$.

Theorem 2.6.3

If g is continuous at c and f is continuous at $g(c)$, then the composition $f \circ g$ is continuous at c.

The idea here is simple: with g continuous at c, we know that

$$\text{for } x \text{ close to } c, \quad g(x) \text{ is close to } g(c);$$

from the continuity of f at $g(c)$, we know that

$$\text{with } g(x) \text{ close to } g(c), \quad f(g(x)) \text{ is close to } f(g(c)).$$

In summary,

$$\text{with } x \text{ close to } c, \quad f(g(x)) \text{ is close to } f(g(c)). \quad \square$$

The argument we just gave is too vague to be a proof. Here in contrast is a proof. We begin with $\epsilon > 0$. We must show that there exists some number $\delta > 0$ such that

$$\text{if } |x - c| < \delta, \quad \text{then} \quad |f(g(x)) - f(g(c))| < \epsilon.$$

In the first place, we observe that since f is continuous at $g(c)$ there does exist a number $\delta_1 > 0$ such that

$$\text{if } |t - g(c)| < \delta_1, \quad \text{then} \quad |f(t) - f(g(c))| < \epsilon.$$

With $\delta_1 > 0$, we know from the continuity of g at c that there exists a number $\delta > 0$ such that

$$\text{if } |x - c| < \delta, \quad \text{then} \quad |g(x) - g(c)| < \delta_1.$$

Putting these two statements together, we see that

$$\text{if } |x - c| < \delta, \quad \text{then} \quad |g(x) - g(c)| < \delta_1$$

and thus

$$|f(g(x)) - f(g(c))| < \epsilon. \quad \square$$

This proof is illustrated in Figure 2.6.4. The numbers within δ of c are taken by g to within δ_1 of $g(c)$ and then by f to within ϵ of $f(g(c))$.

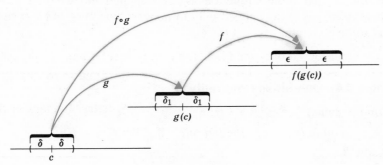

FIGURE 2.6.4

It's time to look at some examples.

Example.   The function

$$F(x) = \sqrt{x^2 + 1}$$

is everywhere continuous. To see this, note that $F = f \circ g$, where

$$f(x) = \sqrt{x} \quad \text{and} \quad g(x) = x^2 + 1.$$

Now take c as an arbitrary real number. The function g is continuous at c (it is everywhere continuous), and since $g(c) > 0$, f is continuous at $g(c)$. Thus, by the theorem, F is continuous at c. $\square$

Example.   The function

$$F(x) = \sqrt{\frac{x^2 + 1}{(x - 8)^3}}$$

is continuous at all numbers greater than 8. To see this, note that $F = f \circ g$, where

$$g(x) = \frac{x^2 + 1}{(x - 8)^3} \quad \text{and} \quad f(x) = \sqrt{x}.$$

Now take $c > 8$. The function g is continuous at c, and since $g(c)$ is positive, f is continuous at $g(c)$. By the theorem F is continuous at c. $\square$

The continuity of composites holds for any finite number of functions. The only requirement is that each function be continuous *where it is applied.*

Example.   The function

$$F(x) = \frac{1}{5 - \sqrt{x^2 + 16}}$$

is continuous everywhere except at $x = \pm 3$, where it is not defined. To see this, note that $F = f \circ g \circ k \circ h$, where

$$f(x) = \frac{1}{x}, \quad g(x) = 5 - x, \quad k(x) = \sqrt{x}, \quad h(x) = x^2 + 16$$

and observe that each of these functions is being applied only where it is continuous. In particular, f is being applied only to nonzero numbers and k is being applied only to positive numbers. $\square$

Just as we considered one-sided limits, we can consider one-sided continuity.

Definition 2.6.4   One-Sided Continuity

A function f is called

continuous from the left at c iff $\lim\limits_{x \uparrow c} f(x) = f(c)$.

It is called

continuous from the right at c iff $\lim\limits_{x \downarrow c} f(x) = f(c)$.†

It is clear from (2.5.4) that a function f is continuous at c iff it is continuous there from both sides.

At each integer the greatest-integer function

$$f(x) = [x] \hspace{3cm} \text{(Figure 2.6.5)}$$

is continuous from the right but discontinuous from the left. This situation is reversed in the case of

$$g(x) = [x] = \text{greatest integer less than } x.$$

For the graph, see Figure 2.6.6. At each integer, g is continuous from the left but discontinuous from the right.

FIGURE 2.6.5   FIGURE 2.6.6

---

† Continuity *from the left* (*from the right*) is sometimes called continuity *from below* (*from above*).

For more examples of one-sided continuity we refer to Figures 2.6.7 and 2.6.8. In Figure 2.6.7 we have continuity from the right at 0. In Figure 2.6.8 we have continuity from the left at 0.

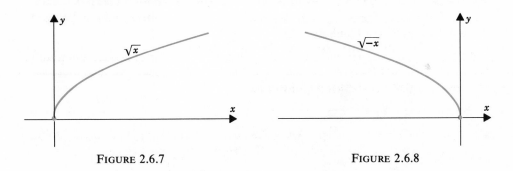

FIGURE 2.6.7   FIGURE 2.6.8

Continuity on [a, b]

From a function defined on a closed interval $[a, b]$ the most continuity that we can possibly expect is

1. continuity at each point c of the open interval (a, b),
2. continuity from the right at a, and
3. continuity from the left at b.

Any function f that fulfills these requirements is called *continuous on* $[a, b]$.
As an example we take the function

$$f(x) = \sqrt{1 - x^2}.$$

Its graph is the semicircle displayed in Figure 2.6.9. The function is continuous on $[-1, 1]$ because it is continuous at each number c in $(-1, 1)$, continuous from the right at -1, and continuous from the left at 1.

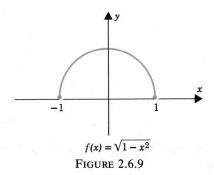

$f(x) = \sqrt{1 - x^2}$

FIGURE 2.6.9

Functions which are continuous on a closed interval $[a, b]$ have special properties that, in general, discontinuous functions do not have. We close this section by
emphasizing two of these properties.

In the first place, a function which is continuous on an interval does not "skip"
any values, and thus its graph is an "unbroken curve." There are no "holes" in it
and no "gaps." This is the idea behind the *intermediate-value theorem*.

Theorem 2.6.5   The Intermediate-Value Theorem

If f is continuous on $[a, b]$ and C is a number between $f(a)$ and $f(b)$, then there is at
least one number c between a and b such that $f(c) = C$.

We illustrate the theorem in Figure 2.6.10. What can happen in the discontinuous case is illustrated in Figure 2.6.11. There the number C has been "skipped."

FIGURE 2.6.10   FIGURE 2.6.11

We come now to the maximum-minimum property of continuous functions.

Theorem 2.6.6   The Maximum-Minimum Theorem

If f is continuous on $[a, b]$, then f takes on both a maximum value M and a minimum value m on $[a, b]$.

The situation is illustrated in Figure 2.6.12. The maximum value M is taken
on at the point marked x_2, and the minimum value m is taken on at the point
marked x_1.

FIGURE 2.6.12

It is interesting to note that in the maximum-minimum theorem the full hypothesis is needed. If we drop the continuity requirement, the conclusion does not follow. As an example, we can take the function

$$f(x) = \begin{cases} \frac{1}{2}(a + b), & x = a \\ x, & a < x < b \\ \frac{1}{2}(a + b), & x = b \end{cases}.$$

The graph of this function is pictured in Figure 2.6.13. The function is defined on $[a, b]$, but it takes on neither a maximum nor a minimum value. If, instead of dropping the continuity requirement, we drop the requirement that the interval be of the form $[a, b]$, then again the result fails. For example, in the case of

$$g(x) = x \quad \text{for } x \in (a, b),$$

g is continuous at each point of (a, b) but again there is no maximum value and no minimum value. See Figure 2.6.14. (Proofs of the intermediate-value theorem and the maximum-minimum theorem can be found in Appendix B.)

FIGURE 2.6.13

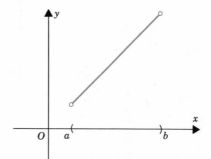

FIGURE 2.6.14

Exercises

Determine whether or not the function is continuous at the indicated point. If not, determine whether the discontinuity is essential or removable.

*1. $f(x) = x^3 - 5x + 1; \quad x = 2.$

2. $g(x) = \sqrt{(x - 1)^2 + 5}; \quad x = 1.$

*3. $f(x) = [x]; \quad x = 0.$

4. $g(x) = (x - 3)[x]; \quad x = 3.$

*5. $f(x) = \begin{cases} x^2 + 4, & x < 2 \\ x^3, & x \geq 2 \end{cases}; \quad x = 2.$

6. $g(x) = \begin{cases} x^2 + 5, & x < 2 \\ x^3, & x \geq 2 \end{cases}; \quad x = 2.$

*7. $f(x) = \begin{cases} x^2 + 4, & x < 2 \\ 5, & x = 2 \\ x^3, & x > 2 \end{cases}; \quad x = 2.$

8. $g(x) = \begin{cases} x^2 + 5, & x < 2 \\ 10, & x = 2 \\ 1 + x^3, & x > 2 \end{cases}$; $x = 2$.

*9. $f(x) = \begin{cases} -1, & x < 0 \\ 0, & x = 0 \\ 1, & x > 0 \end{cases}$; $x = 0$.

10. $g(x) = \begin{cases} 1 - x, & x < 1 \\ 1, & x = 1 \\ x^2 - 1, & x > 1 \end{cases}$; $x = 1$.

*11. $f(x) = \begin{cases} \dfrac{x^2 - 1}{x + 1}, & x \neq -1 \\ -2, & x = -1 \end{cases}$; $x = -1$.

12. $g(x) = \begin{cases} \dfrac{1}{x + 1}, & x \neq -1 \\ 0, & x = -1 \end{cases}$; $x = -1$.

*13. $f(x) = \begin{cases} -x^2, & x < 0 \\ 1 - \sqrt{x}, & x \geq 0 \end{cases}$; $x = 0$.

14. $g(x) = \dfrac{x(x + 1)(x + 2)}{\sqrt{(x - 1)(x - 2)}}$; $x = -2$.

Sketch the graph and classify the discontinuities (if any).

*15. $f(x) = |x - 1|$.

16. $g(x) = |x^2 - 1|$.

*17. $f(x) = \begin{cases} x - 1, & x < 1 \\ 0, & x = 1 \\ x^2, & x > 1 \end{cases}$.

18. $g(x) = \begin{cases} 2x - 1, & x < 1 \\ 0, & x = 1 \\ x^2, & x > 1 \end{cases}$.

*19. $f(x) = \max\{x, x^2\}$.

20. $g(x) = \min\{x, x^2\}$.

*21. $f(x) = \begin{cases} -1, & x < -1 \\ x^3, & -1 \leq x \leq 1 \\ 1, & x > 1 \end{cases}$.

22. $g(x) = \begin{cases} 1, & x \leq -2 \\ \frac{1}{2}x, & -2 < x < 4 \\ \sqrt{x}, & x \geq 4 \end{cases}$.

*23. $f(x) = \begin{cases} 1, & x \leq 0 \\ x^2, & 0 < x < 1 \\ 1, & 1 \leq x < 2 \\ x, & x \geq 2 \end{cases}$.

24. $g(x) = \begin{cases} -x^2, & x < -1 \\ 3, & x = -1 \\ 2 - x, & -1 < x \leq 1 \\ x^2, & x > 1 \end{cases}$.

Each of the functions f below is defined everywhere except at $x = 1$. Where possible, define $f(1)$ so that f becomes continuous at 1.

*25. $f(x) = \dfrac{x^2 - 1}{x - 1}$.

26. $f(x) = \dfrac{1}{x - 1}$.

*27. $f(x) = \dfrac{x - 1}{|x - 1|}$.

28. $f(x) = \dfrac{(x - 1)^2}{|x - 1|}$.

29. Define a function which is everywhere continuous from
    (a) the left but discontinuous from the right at $x = \frac{1}{2}$.
    (b) the right but discontinuous from the left at $x = \frac{1}{2}$.

30. Define a function which is continuous on the open interval $(0, 1)$ and has (a) no maximum value there, (b) a maximum value there.

*31. At what points (if any) are the following functions continuous?

  (a) $f(x) = \begin{cases} 1, & x \text{ rational} \\ 0, & x \text{ irrational} \end{cases}$.   (b) $g(x) = \begin{cases} x, & x \text{ rational} \\ 0, & x \text{ irrational} \end{cases}$.

32. (Important) Prove that

$$f \text{ is continuous at } c \quad \text{iff} \quad \lim_{h \to 0} f(c + h) = f(c).$$

33. (Important) Let f and g be continuous at c. Prove that if
   (a) $f(c) > 0$, then there exists $\delta > 0$ such that

$$f(x) > 0 \qquad \text{for all } x \in (c - \delta, c + \delta).$$

   (b) $f(c) < 0$, then there exists $\delta > 0$ such that

$$f(x) < 0 \qquad \text{for all } x \in (c - \delta, c + \delta).$$

   (c) $f(c) < g(c)$, then there exists $\delta > 0$ such that

$$f(x) < g(x) \qquad \text{for all } x \in (c - \delta, c + \delta).$$

Optional 34. The intermediate-value theorem can be used to prove that each polynomial equation of odd degree

$$x^n + a_{n-1}x^{n-1} + \cdots + a_1 x + a_0 = 0 \qquad \text{with } n \text{ odd}$$

has at least one real root. Prove that each of the equations below has at least one real root.
   (a) $x^3 - x + 5 = 0$.   (b) $x^5 + 2x^2 + x - 100 = 0$.

35. (*Fixed point property.*) Show that, if f is continuous and $0 \le f(x) \le 1$ for all $x \in [0, 1]$, then there exists at least one point c in $[0, 1]$ such that $f(c) = c$. HINT: Apply the intermediate-value theorem to $g(x) = x - f(x)$.

36. Let n be a positive integer.
   (a) Prove that, if $0 \le a < b$, then $a^n < b^n$.
   (b) Prove that every nonnegative real number x has a unique nonnegative nth root $x^{1/n}$. HINT: The existence of $x^{1/n}$ can be seen by applying the intermediate-value theorem to the function $f(t) = t^n$ for $t \ge 0$. The uniqueness follows from part (a).

2.7 Additional Exercises

Evaluate the limits that exist.

*1. $\lim_{x \uparrow 3} \dfrac{x - 3}{|x - 3|}$.

2. $\lim_{x \downarrow 3} \dfrac{x - 3}{|x - 3|}$.

*3. $\lim_{x \downarrow 3} \dfrac{\sqrt{x - 3}}{|x - 3|}$.

*4. $\lim_{x \to 3} \dfrac{(x - 3)^2}{x + 3}$.

5. $\lim_{x \to 3} \dfrac{x^2 - 9}{x^2 - 5x + 6}$.

*6. $\lim_{x \to 3} \dfrac{x^2 - 3}{x - 3}$.

*7. $\lim\limits_{x\to 0} \left(\dfrac{1}{x} - \dfrac{1-x}{x} \right)$.

8. $\lim\limits_{x\to 0} \left(\dfrac{1}{x} - \dfrac{x}{1-x} \right)$.

*9. $\lim\limits_{x\to -1} \dfrac{x^2 - x - 2}{x + 1}$.

*10. $\lim\limits_{x\uparrow 2} (x - [x])$.

11. $\lim\limits_{x\downarrow 2} (x - [x])$.

*12. $\lim\limits_{x\to 1/2} (x - [x])$.

*13. $\lim\limits_{x\uparrow 0} \dfrac{[x]}{x}$.

14. $\lim\limits_{x\downarrow 0} \dfrac{[x]}{x}$.

*15. $\lim\limits_{x\to 1/2} \dfrac{[x]}{x}$.

*16. $\lim\limits_{x\uparrow 0} \dfrac{\sqrt{x} - 1}{x - 1}$.

17. $\lim\limits_{x\downarrow 0} \dfrac{\sqrt{x} - 1}{x - 1}$.

*18. $\lim\limits_{x\uparrow 1} \dfrac{\sqrt{x} - 1}{x - 1}$.

*19. $\lim\limits_{x\downarrow 3} \dfrac{x^2 - 2x - 3}{\sqrt{x} - 3}$.

20. $\lim\limits_{x\downarrow 3} \dfrac{\sqrt{x^2 - 2x - 3}}{x - 3}$.

*21. $\lim\limits_{x\downarrow 0} \left(\dfrac{1}{x} \right)^{[x]}$.

*22. $\lim\limits_{x\to 1} \dfrac{1 - 1/x^2}{1 - x}$.

23. $\lim\limits_{x\to 1} \dfrac{1 - 1/x}{1 - x^2}$.

*24. $\lim\limits_{x\to 1} \dfrac{1 + 1/x}{1 - x^2}$.

*25. $\lim\limits_{x\to -1} \dfrac{1 + 1/x^2}{1 - x}$.

26. $\lim\limits_{x\to 1} \dfrac{1 - 1/x}{1 - 1/x^2}$.

*27. $\lim\limits_{x\to -1} \dfrac{1 + 1/x}{1 + 1/x^2}$.

Give an ϵ, δ proof for the following limits.

*28. $\lim\limits_{x\to 2} (5x - 4) = 6$.

29. $\lim\limits_{x\to 3} (1 - 2x) = -5$.

*30. $\lim\limits_{x\to -4} |2x + 5| = 3$.

31. $\lim\limits_{x\to 3} |1 - 2x| = 5$.

32. (a) Sketch the graph of

$$
f(x) = \begin{cases}
-2x, & x < -1 \\
0, & x = -1 \\
x^2 + 1, & -1 < x \le 0 \\
\sqrt{x}, & 0 < x \le 1 \\
x^2 - 1, & x > 1
\end{cases}.
$$

(b) Evaluate the limits that exist.

*(i) $\lim\limits_{x\uparrow -1} f(x)$.

(ii) $\lim\limits_{x\downarrow -1} f(x)$.

*(iii) $\lim\limits_{x\to -1} f(x)$.

*(iv) $\lim\limits_{x\uparrow 0} f(x)$.

(v) $\lim\limits_{x\downarrow 0} f(x)$.

*(vi) $\lim\limits_{x\to 0} f(x)$.

*(vii) $\lim\limits_{x\uparrow 1} f(x)$.

(viii) $\lim\limits_{x\downarrow 1} f(x)$.

*(ix) $\lim\limits_{x\to 1} f(x)$.

(c) *(i) At what points is f discontinuous from the left?

(ii) At what points is f discontinuous from the right?

*(iii) Where are the removable discontinuities of f?

(iv) Where are the essential discontinuities of f?

*33. Find $\lim\limits_{x\to c} f(x)$ given that $|f(x)| \le M|x - c|$ for all $x \ne c$.

34. Given that $\lim\limits_{x\to 0} [f(x)/x]$ exists, prove that $\lim\limits_{x\to 0} f(x) = 0$.

35. Given an ϵ, δ proof for the following limits:

(a) $\lim_{x \to 9} \sqrt{x - 5} = 2$.   (b) $\lim_{x \to 2} \sqrt{x^2 - 4} = 0$.

36. Prove that there exists a real number x_0 such that

$$x_0^5 - 4x_0 + 1 = 7.21.$$

37. Given that f and g are continuous and $f(0) < g(0) < g(1) < f(1)$, prove that, for some point c between 0 and 1, $f(c) = g(c)$.

Differentiation

3

3.1 The Derivative

Introduction: The Tangent to a Curve

We begin with a function f, and on the graph we choose a point $(x, f(x))$. See Figure 3.1.1. What line, if any, should be called tangent to the graph at that point?

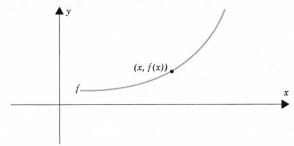

FIGURE 3.1.1

To answer this question, we choose a small number $h \neq 0$ and on the graph mark the point $(x + h, f(x + h))$. Now we draw the secant line that passes through these two points. The situation is pictured in Figure 3.1.2, first with $h > 0$ and then with $h < 0$.

FIGURE 3.1.2

As h tends to zero from the right (see the figure), the secant line tends to a limiting position, and it tends to the same limiting position as h tends to zero from the left. The line at this limiting position is what we call "the tangent to the graph at the point $(x, f(x))$."

Since the approximating secants have slope of the form

$$(*) \qquad\qquad \frac{f(x + h) - f(x)}{h}, \qquad\qquad \text{(check this out)}$$

you can expect the tangent, the limiting position of these secants, to have slope

$$(**) \qquad\qquad \lim_{h \to 0} \frac{f(x + h) - f(x)}{h}.$$

While $(*)$ measures the steepness of the line which passes through $(x, f(x))$ and $(x + h, f(x + h))$, $(**)$ measures the steepness of the graph at $(x, f(x))$ and is called the "slope of the graph" at that point.

Derivatives and Differentiation

In the introduction we spoke informally about tangent lines and gave a geometric interpretation to limits of the form

$$\lim_{h \to 0} \frac{f(x + h) - f(x)}{h}.$$

Here we begin the systematic study of such limits, what mathematicians call the theory of *differentiation*.

Definition 3.1.1

A function f is said to be *differentiable* at x iff

$$\lim_{h \to 0} \frac{f(x + h) - f(x)}{h} \qquad \text{exists.}$$

If this limit exists, it is called the *derivative* of f at x and is denoted by $f'(x)$.†

We interpret

$$f'(x) = \lim_{h \to 0} \frac{f(x + h) - f(x)}{h}$$

as the *slope of the graph at the point* $(x, f(x))$. The line through $(x, f(x))$ which has this slope is called the *tangent line at* $(x, f(x))$.

It is time we looked at some examples.

---

† This prime notation goes back to the French mathematician Joseph Louis Lagrange (1736–1813). Another will be introduced later.

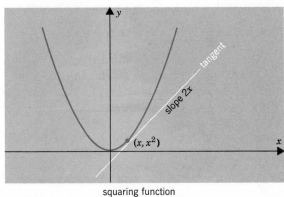

squaring function

FIGURE 3.1.3

Example.   We begin with the squaring function

$$f(x) = x^2. \qquad \text{(Figure 3.1.3)}$$

To find $f'(x)$ we form the *difference quotient*

$$\frac{f(x + h) - f(x)}{h} = \frac{(x + h)^2 - x^2}{h}.$$

Since

$$\frac{(x + h)^2 - x^2}{h} = \frac{x^2 + 2xh + h^2 - x^2}{h} = \frac{2xh + h^2}{h} = 2x + h,$$

we have

$$\frac{f(x + h) - f(x)}{h} = 2x + h.$$

Obviously then

$$f'(x) = \lim_{h \to 0} \frac{f(x + h) - f(x)}{h} = 2x. \quad \square$$

Example.   In the case of a linear function

$$f(x) = mx + b$$

we have

$$f'(x) = m.$$

In other words, the slope is constantly m. To verify this, note that

$$\frac{f(x + h) - f(x)}{h} = \frac{[m(x + h) + b] - [mx + b]}{h} = \frac{mh}{h} = m.$$

It follows that

$$f'(x) = \lim_{h \to 0} \frac{f(x + h) - f(x)}{h} = m. \quad \square$$

The derivative

$$f'(x) = \lim_{h \to 0} \frac{f(x + h) - f(x)}{h}$$

is a two-sided limit. Therefore it cannot be taken at an endpoint of the domain. In our next example we will be dealing with the square-root function. Although this function is defined for all $x \geq 0$, you can expect a derivative only for $x > 0$.

Example.   The square-root function

$$f(x) = \sqrt{x}, \quad x \geq 0 \qquad \text{(Figure 3.1.4)}$$

has derivative

$$f'(x) = \frac{1}{2\sqrt{x}}, \quad \text{for } x > 0.$$

To verify this, we begin with $x > 0$ and form the difference quotient

$$\frac{f(x + h) - f(x)}{h} = \frac{\sqrt{x + h} - \sqrt{x}}{h}.$$

To remove the radicals from the numerator we multiply both the numerator and the denominator by $\sqrt{x + h} + \sqrt{x}$. This gives

$$\frac{f(x + h) - f(x)}{h} = \left(\frac{\sqrt{x + h} - \sqrt{x}}{h} \right)\left(\frac{\sqrt{x + h} + \sqrt{x}}{\sqrt{x + h} + \sqrt{x}} \right) = \frac{1}{\sqrt{x + h} + \sqrt{x}}$$

and thus

$$f'(x) = \lim_{h \to 0} \frac{f(x + h) - f(x)}{h} = \frac{1}{2\sqrt{x}}. \quad \square$$

square root function

FIGURE 3.1.4

To *differentiate* a function is to find its derivative.

Example.   Here we differentiate the function

$$f(x) = \sqrt{1 - x^2}. \qquad \text{(Figure 3.1.5)}$$

Its graph is the upper half of the unit circle. The domain of f is the closed interval $[-1, 1]$, but we can expect differentiability at most on the open interval $(-1, 1)$. For x in $(-1, 1)$ we form the difference quotient

$$\frac{f(x-h) - f(x)}{h} = \frac{\sqrt{1-(x+h)^2} - \sqrt{1-x^2}}{h}.$$

Once again we remove the radicals from the numerator, this time by multiplying both numerator and denominator by $\sqrt{1-(x+h)^2} + \sqrt{1-x^2}$. After simplification these calculations yield

$$\frac{f(x+h) - f(x)}{h} = \frac{-2x-h}{\sqrt{1-(x+h)^2} + \sqrt{1-x^2}}$$

and thus

$$f'(x) = \lim_{h \to 0} \frac{f(x+h) - f(x)}{h} = -\frac{x}{\sqrt{1-x^2}}. \quad \square$$

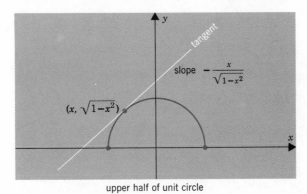

upper half of unit circle

FIGURE 3.1.5

Evaluating Derivatives

Problem.   Find $f'(-2)$ if

$$f(x) = 1 - x^2.$$

SOLUTION.   We can first find $f'(x)$ in general:

$$f'(x) = \lim_{h \to 0} \frac{f(x+h) - f(x)}{h}$$

$$= \lim_{h \to 0} \frac{[1-(x+h)^2] - [1-x^2]}{h} = \lim_{h \to 0} \frac{-2xh - h^2}{h} = \lim_{h \to 0} -2x - h = -2x$$

and then substitute -2 for x:

$$f'(-2) = -2(-2) = 4.$$

We can also evaluate $f'(-2)$ more directly:

$$f'(-2) = \lim_{h \to 0} \frac{f(-2+h) - f(-2)}{h}$$

$$= \lim_{h \to 0} \frac{[1-(-2+h)^2]-[1-(-2)^2]}{h} = \lim_{h \to 0} \frac{4h-h^2}{h} = \lim_{h \to 0} 4-h = 4. \quad \square$$

Problem. Find $f'(0)$ if

$$f(x) = \begin{cases} 3x^2 + 1, & x \le 0 \\ x^3 + 1, & 0 < x < 1 \end{cases}.$$

SOLUTION. We will evaluate

$$f'(0) = \lim_{h \to 0} \frac{f(0+h) - f(0)}{h}$$

by taking the corresponding one-sided limits:

$$\lim_{h \uparrow 0} \frac{f(0+h) - f(0)}{h} = \lim_{h \uparrow 0} \frac{(3h^2+1)-1}{h} = \lim_{h \uparrow 0} 3h = 0;$$

$$\lim_{h \downarrow 0} \frac{f(0+h) - f(0)}{h} = \lim_{h \downarrow 0} \frac{(h^3+1)-1}{h} = \lim_{h \downarrow 0} h^2 = 0.$$

It follows now that

$$f'(0) = \lim_{h \to 0} \frac{f(0+h) - f(0)}{h} = 0. \quad \square$$

Differentiability and Continuity

A function can be continuous at some number x without being differentiable there. For example, the absolute-value function

$$f(x) = |x|$$

is continuous at 0 (it is everywhere continuous), but it is not differentiable at 0:

$$\frac{f(0+h) - f(0)}{h} = \frac{|0+h| - |0|}{h} = \frac{|h|}{h} = \begin{cases} -1, & h < 0 \\ 1, & h > 0 \end{cases},$$

so that

$$\lim_{h \uparrow 0} \frac{f(0+h) - f(0)}{h} = -1, \quad \lim_{h \downarrow 0} \frac{f(0+h) - f(0)}{h} = 1$$

and

$$\lim_{h \to 0} \frac{f(0+h) - f(0)}{h} \quad \text{does not exist.} \quad \square$$

The failure of the absolute-value function to be differentiable at 0 is reflected by its graph, displayed in Figure 3.1.6. At (0, 0) the graph suddenly changes direction and there is no tangent line at that point.

no derivative at 0

FIGURE 3.1.6

You can see a similar sudden change of direction in the graph of

$$f(x) = \begin{cases} x^2, & x \le 1 \\ x, & x > 1 \end{cases}$$ (Figure 3.1.7)

at the point (1, 1). Once again f is everywhere continuous (you can verify that), but it is not differentiable at 1:

$$\lim_{h \uparrow 1} \frac{f(1 + h) - f(1)}{h} = \lim_{h \uparrow 1} \frac{(1 + h)^2 - 1}{h} = \lim_{h \uparrow 1} \frac{h^2 + 2h}{h} = \lim_{h \uparrow 1} h + 2 = 2;$$

$$\lim_{h \downarrow 1} \frac{f(1 + h) - f(1)}{h} = \lim_{h \downarrow 1} \frac{(1 + h) - 1}{h} = \lim_{h \downarrow 1} 1 = 1.$$

Since these one-sided limits are different, the two-sided limit

$$f'(1) = \lim_{h \to 1} \frac{f(1 + h) - f(1)}{h} \quad \text{cannot exist.} \quad \square$$

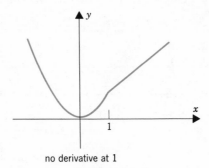

no derivative at 1

FIGURE 3.1.7

Although not every continuous function is differentiable, every differentiable function is continuous. In particular,

(3.1.2)   if f is differentiable at x,   then f is continuous at x.

PROOF.   For $h \neq 0$ and $x + h$ in the domain of f,

$$f(x + h) - f(x) = \frac{f(x + h) - f(x)}{h} \cdot h.$$

With f differentiable at x,

$$\lim_{h \to 0} \frac{f(x + h) - f(x)}{h} = f'(x).$$

Since $\lim\limits_{h \to 0} h = 0$,

$$\lim_{h \to 0} [f(x + h) - f(x)] = \left[\lim_{h \to 0} \frac{f(x + h) - f(x)}{h} \right] \cdot \left[\lim_{h \to 0} h \right] = f'(x) \cdot 0 = 0.$$

It follows that

$$\lim_{h \to 0} f(x + h) = f(x) \qquad \text{(explain)}$$

and thus (by Exercise 32, Section 2.6) that f is continuous at x. □

Exercises

Differentiate each of the following functions by forming a suitable difference quotient and taking the limit as h tends to 0.

*1. $f(x) = \alpha$.   2. $f(x) = 3x$.   *3. $f(x) = 3x + 5$.

4. $f(x) = mx + b$.   *5. $f(x) = 1/x$.   6. $f(x) = 1/x^2$.

*7. $f(x) = \frac{1}{2}x^2$.   8. $f(x) = -x^2$.   *9. $f(x) = 2x^3$.

10. $f(x) = x^4$.   *11. $f(x) = (x - 1)^2$.   12. $f(x) = (2x - 3)^2$.

*13. $f(x) = 1/\sqrt{x}$.   14. $f(x) = \sqrt{x - 1}$.   *15. $f(x) = 1/\sqrt{x - 1}$.

For each of the functions below find $f'(2)$ by forming the difference quotient

$$\frac{f(2 + h) - f(2)}{h}$$

and taking the limit as $h \to 0$.

16. $f(x) = x^2$. $\left[\dfrac{f(2 + h) - f(2)}{h} = \dfrac{(2 + h)^2 - 2^2}{h} = 4 + h, \text{ etc.} \right]$

*17. $f(x) = \dfrac{1}{x + 1}$.   18. $f(x) = \dfrac{1}{2x}$.

*19. $f(x) = \sqrt{x + 2}$.   20. $f(x) = \dfrac{1}{\sqrt{x + 2}}$.

Draw the graph of each of the following functions and indicate where it is not differentiable.

21. $f(x) = |2x - 5|$.   22. $f(x) = |x^2 - 4|$.

23. $f(x) = \sqrt{|x|}$.   24. $f(x) = |x^3 - 1|$.

Find $f'(c)$ if it exists.

*25. $f(x) = \begin{cases} 4x, & x < 1 \\ 2x^2 + 2, & x \geq 1 \end{cases}$; $c = 1$.   26. $f(x) = \begin{cases} 3x^2, & x \leq 1 \\ 2x^3 + 1, & x > 1 \end{cases}$; $c = 1$.

*27. $f(x) = \begin{cases} x + 1, & c \le -1 \\ (x + 1)^2, & c > -1 \end{cases}$; $c = -1$.   28. $f(x) = \begin{cases} -\frac{1}{2}x^2, & x < 3 \\ -3x, & x \ge 3 \end{cases}$; $c = 3$.

*29. $f(x) = [x]$; $c = 0$.   30. $f(x) = \frac{1}{3}x^3 + \frac{1}{2}x^2 + x + 1$; $c = 1$.

3.2   Some Differentiation Formulas

Calculating the derivative of

$$f(x) = (x^3 - 2)(4x + 1) \quad \text{or} \quad f(x) = \frac{6x^2 - 1}{x^4 + 5x + 1}$$

by forming the proper difference quotient

$$\frac{f(x + h) - f(x)}{h}$$

and then taking the limit as h tends to 0 is somewhat laborious. Here we derive some general formulas which make such calculations quite simple.

For future reference we first point out that the constant functions have derivative 0:

(3.2.1)

$$\boxed{\text{if} \quad f(x) = \alpha, \qquad \text{then} \quad f'(x) = 0}$$

and the identity function has derivative 1:

(3.2.2)

$$\boxed{\text{if} \quad f(x) = x, \qquad \text{then} \quad f'(x) = 1.}$$

PROOF.   For $f(x) = \alpha$,

$$f'(x) = \lim_{h \to 0} \frac{f(x + h) - f(x)}{h} = \lim_{h \to 0} \frac{\alpha - \alpha}{h} = \lim_{h \to 0} 0 = 0.$$

For $f(x) = x$,

$$f'(x) = \lim_{h \to 0} \frac{f(x + h) - f(x)}{h} = \lim_{h \to 0} \frac{(x + h) - x}{h} = \lim_{h \to 0} \frac{h}{h} = \lim_{h \to 0} 1 = 1. \quad \Box$$

Definition 3.2.3   Derivatives of Sums and Scalar Multiples

Let α be a real number. If f and g are differentiable at x, then $f + g$ and αf are differentiable at x. Moreover,

$$(f + g)'(x) = f'(x) + g'(x) \quad \text{and} \quad (\alpha f)'(x) = \alpha f'(x).$$

PROOF.   To verify the first formula note that

$$\frac{(f + g)(x + h) - (f + g)(x)}{h} = \frac{[f(x + h) + g(x + h)] - [f(x) + g(x)]}{h}$$

$$= \frac{f(x + h) - f(x)}{h} + \frac{g(x + h) - g(x)}{h}.$$

By definition,

$$\lim_{h \to 0} \frac{f(x + h) - f(x)}{h} = f'(x), \quad \lim_{h \to 0} \frac{g(x + h) - g(x)}{h} = g'(x).$$

Thus

$$\lim_{h \to 0} \frac{(f + g)(x + h) - (f + g)(x)}{h} = f'(x) + g'(x),$$

which means that

$$(f + g)'(x) = f'(x) + g'(x).$$

To verify the second formula we must show that

$$\lim_{h \to 0} \frac{(\alpha f)(x + h) - (\alpha f)(x)}{h} = \alpha f'(x).$$

This follows directly from the fact that

$$\frac{(\alpha f)(x + h) - (\alpha f)(x)}{h} = \alpha \left[\frac{f(x + h) - f(x)}{h} \right]. \quad \square$$

Theorem 3.2.4   The Product Rule

If f and g are differentiable at x, then so is their product, and

$$(fg)'(x) = f(x)g'(x) + g(x)f'(x).$$

PROOF.   We form the difference quotient

$$\frac{(fg)(x + h) - (fg)(x)}{h} = \frac{f(x + h)g(x + h) - f(x)g(x)}{h}$$

and rewrite it as

$$f(x + h) \left[\frac{g(x + h) - g(x)}{h} \right] + g(x) \left[\frac{f(x + h) - f(x)}{h} \right].$$

(Here we've added and subtracted $f(x + h)g(x)$ in the numerator and then regrouped the terms so as to display difference quotients for f and g.) Since f is differentiable at x, we know that f is continuous at x (3.1.2) and thus

$$\lim_{h \to 0} f(x + h) = f(x).$$

Since

$$\lim_{h \to 0} \frac{g(x + h) - g(x)}{h} = g'(x) \quad \text{and} \quad \lim_{h \to 0} \frac{f(x + h) - f(x)}{h} = f'(x),$$

we obtain

$$\lim_{h \to 0} \frac{(fg)(x + h) - (fg)(x)}{h} = f(x)g'(x) + g(x)f'(x). \quad \square$$

Using the product rule it is not hard to prove that

(3.2.5)

> for each positive integer n
>
> $p(x) = x^n$ has derivative $p'(x) = nx^{n-1}$.

In particular

$$p(x) = x \quad \text{has derivative} \quad p'(x) = 1 \cdot x^0 = 1,$$
$$p(x) = x^2 \quad \text{has derivative} \quad p'(x) = 2x,$$
$$p(x) = x^3 \quad \text{has derivative} \quad p'(x) = 3x^2,$$
$$p(x) = x^4 \quad \text{has derivative} \quad p'(x) = 4x^3,$$

and so on.

PROOF OF (3.2.5).   We proceed by induction on n. If $n = 1$, then we have the identity function

$$p(x) = x,$$

which we know satisfies

$$p'(x) = 1 = 1 \cdot x^0.$$

This means that the formula holds for $n = 1$.
   We suppose now that the result holds for $n = k$ and go on to show that it holds for $n = k + 1$. We let

$$p(x) = x^{k+1}$$

and note that

$$p(x) = x \cdot x^k.$$

Applying the product rule (3.2.4) and our inductive hypothesis, we obtain

$$p'(x) = x \cdot kx^{k-1} + 1 \cdot x^k = (k + 1)x^k.$$

This shows that the formula holds for $k + 1$. $\square$

   The formula for differentiating polynomials follows from (3.2.3) and (3.2.5):

(3.2.6)

> if $\quad$ $P(x) = a_n x^n + \cdots + a_2 x^2 + a_1 x + a_0,$
>
> then $\quad$ $P'(x) = na_n x^{n-1} + \cdots + 2a_2 x + a_1.$

For example,

$$P(x) = 12x^3 - 6x - 2 \quad \text{has derivative} \quad P'(x) = 36x^2 - 6$$

and

$$P(x) = \tfrac{1}{4}x^4 - 2x^2 + x + 5 \quad \text{has derivative} \quad P'(x) = x^3 - 4x + 1.$$

Problem.   Differentiate

$$F(x) = (x^3 - 2)(4x + 1)$$

and find $F'(-1)$.

SOLUTION.   We have a product $F(x) = f(x)g(x)$ with

$$f(x) = x^3 - 2 \quad \text{and} \quad g(x) = 4x + 1.$$

The product rule gives

$$\begin{aligned}
F'(x) &= f(x)g'(x) + g(x)f'(x) \\
&= (x^3 - 2) \cdot 4 + (4x + 1)(3x^2) \\
&= 4x^3 - 8 + 12x^3 + 3x^2 = 16x^3 + 3x^2 - 8.
\end{aligned}$$

Setting $x = -1$ we have

$$F'(-1) = 16(-1)^3 + 3(-1)^2 - 8 = -16 + 3 - 8 = -21. \quad \square$$

Problem.   Differentiate

$$F(x) = (ax + b)(cx + d).$$

SOLUTION.   We have a product $F(x) = f(x)g(x)$ with

$$f(x) = ax + b \quad \text{and} \quad g(x) = cx + d.$$

Again we use the product rule

$$F'(x) = f(x)g'(x) + g(x)f'(x).$$

In this case

$$F'(x) = (ax + b)c + (cx + d)a = 2acx + bc + ad.$$

We can also do the problem without using the product rule by first carrying out the multiplication:

$$F(x) = acx^2 + bcx + adx + bd$$

and then differentiating:

$$F'(x) = 2acx + bc + ad.$$

The result is the same. $\square$

   We come now to reciprocals.

Theorem 3.2.7   The Reciprocal Rule

If g is differentiable at x and $g(x) \neq 0$, then $1/g$ is differentiable at x and

$$\left(\frac{1}{g}\right)'(x) = -\frac{g'(x)}{[g(x)]^2}.$$

PROOF.   Since g is differentiable at x, g is continuous at x. (3.1.2) Since $g(x) \neq 0$, we know that $1/g$ is continuous at x, and thus that

$$\lim_{h \to 0} \frac{1}{g(x+h)} = \frac{1}{g(x)} \cdot$$

For h different from zero and sufficiently small, $g(x+h) \neq 0$ and

$$\frac{1}{h} \left[\frac{1}{g(x+h)} - \frac{1}{g(x)} \right] = - \left[\frac{g(x+h) - g(x)}{h} \right] \frac{1}{g(x+h)} \frac{1}{g(x)} \cdot$$

Taking the limit as h tends to zero, we see that the right-hand side (and thus the left) tends to

$$-\frac{g'(x)}{[g(x)]^2} \cdot \quad \square$$

From this last result it is easy to see that the formula for the derivative of a power, x^n, also applies to negative powers; namely,

(3.2.8)

for each negative integer n

$p(x) = x^n$ has derivative $p'(x) = nx^{n-1}$.

This formula holds except of course at $x = 0$ where no negative power is even defined. In particular

$$p(x) = x^{-1} \quad \text{has derivative} \quad p'(x) = (-1)x^{-2} = -x^{-2},$$
$$p(x) = x^{-2} \quad \text{has derivative} \quad p'(x) = -2x^{-3},$$
$$p(x) = x^{-3} \quad \text{has derivative} \quad p'(x) = -3x^{-4},$$

and so on.

PROOF OF (3.2.8).   Note that

$$p(x) = \frac{1}{g(x)} \quad \text{where} \quad g(x) = x^{-n} \quad \text{and} \quad -n \text{ is a positive integer.}$$

The rule for reciprocals gives

$$p'(x) = -\frac{g'(x)}{[g(x)]^2} = -\frac{(-nx^{-n-1})}{x^{-2n}} = nx^{n-1}. \quad \square$$

Problem.   Differentiate

$$f(x) = 5x^2 - \frac{6}{x}$$

and find $f'(\tfrac{1}{2})$.

SOLUTION

$$f(x) = 5x^2 - 6x^{-1}$$

so that

$$f'(x) = 10x + 6x^{-2},$$

which, if you don't like negative exponents, you can rewrite as

$$f'(x) = 10x + \frac{6}{x^2}.$$

Setting $x = \frac{1}{2}$ we have

$$f'(\tfrac{1}{2}) = 10(\tfrac{1}{2}) + \frac{6}{(\tfrac{1}{2})^2} = 5 + 24 = 29. \quad \square$$

Problem.   Differentiate

$$f(x) = \frac{1}{ax^2 + bx + c}.$$

SOLUTION.   Here we have a reciprocal $f(x) = 1/g(x)$ with

$$g(x) = ax^2 + bx + c.$$

The reciprocal rule (3.2.7) gives

$$f'(x) = -\frac{g'(x)}{[g(x)]^2} = -\frac{2ax + b}{[ax^2 + bx + c]^2}. \quad \square$$

Finally we come to quotients in general.

Theorem 3.2.9   The Quotient Rule

If f and g are differentiable at x and $g(x) \neq 0$, then the quotient f/g is differentiable at x and

$$\left(\frac{f}{g}\right)'(x) = \frac{g(x)f'(x) - f(x)g'(x)}{[g(x)]^2}.$$

The proof is left to you. HINT: $f/g = f \cdot 1/g. \quad \square$

From the quotient rule you can see that all rational functions (quotients of polynomials) are differentiable wherever they are defined.

Problem.   Differentiate

$$F(x) = \frac{ax + b}{cx + d}.$$

SOLUTION.   We are dealing with a quotient $F(x) = f(x)/g(x)$. The quotient rule,

$$F'(x) = \frac{g(x)f'(x) - f(x)g'(x)}{[g(x)]^2},$$

gives

$$F'(x) = \frac{(cx + d) \cdot a - (ax + b) \cdot c}{(cx + d)^2} = \frac{ad - bc}{(cx + d)^2}. \quad \square$$

Problem. Differentiate

$$F(x) = \frac{6x^2 - 1}{x^4 + 5x + 1}.$$

SOLUTION. Again we are dealing with a quotient $F(x) = f(x)/g(x)$. The quotient rule gives

$$F'(x) = \frac{(x^4 + 5x + 1)(12x) - (6x^2 - 1)(4x^3 + 5)}{(x^4 + 5x + 1)^2}. \quad \square$$

Problem. Find $f'(0)$, $f'(1)$, and $f'(2)$ for the function

$$f(x) = \frac{5x}{1 + x}.$$

SOLUTION. First we find a general expression for $f'(x)$, and then we evaluate it at 0, 1, and 2. Using the quotient rule, we get

$$f'(x) = \frac{(1 + x)5 - 5x(1)}{(1 + x)^2} = \frac{5}{(1 + x)^2}.$$

This gives

$$f'(0) = \frac{5}{(1 + 0)^2} = 5, \quad f'(1) = \frac{5}{(1 + 1)^2} = \frac{5}{4}, \quad f'(2) = \frac{5}{(1 + 2)^2} = \frac{5}{9}. \quad \square$$

Problem. Find $f'(-1)$ for

$$f(x) = \frac{x^2}{ax^2 + b}.$$

SOLUTION. We first find $f'(x)$ in general by the quotient rule:

$$f'(x) = \frac{(ax^2 + b)2x - x^2(2ax)}{(ax^2 + b)^2} = \frac{2bx}{(ax^2 + b)^2}.$$

Now we evaluate f' at -1:

$$f'(-1) = -\frac{2b}{(a + b)^2}. \quad \square$$

Exercises

Differentiate the following functions.

*1. $F(x) = 1 - x$.

2. $F(x) = 2(1 + x)$.

*3. $F(x) = 11x^5 - 6x^3 + 8$.

4. $F(x) = \frac{x^4}{4} - \frac{x^3}{3} + \frac{x^2}{2} - \frac{x}{1}$.

*5. $F(x) = ax^2 + bx + c$.

6. $F(x) = \frac{3}{x^2}$.

*7. $F(x) = -\dfrac{1}{x^2}.$

8. $F(x) = \dfrac{(x^2 + 2)}{x^3}.$

*9. $F(x) = \dfrac{ax - b}{cx - d}.$

10. $F(x) = x - \dfrac{1}{x}.$

*11. $G(x) = \dfrac{x^3}{1 - x}.$

12. $G(x) = \dfrac{x^2 - 1}{2x + 3}.$

*13. $G(x) = (x^2 - 1)(x - 3).$

14. $G(x) = \dfrac{7x^4 + 11}{x + 1}.$

*15. $G(x) = (x - 1)(x - 2).$

16. $G(x) = \dfrac{2x^2 + 1}{x + 2}.$

*17. $G(x) = \dfrac{6 - 1/x}{x - 2}.$

18. $G(x) = \dfrac{1 + x^4}{x^2}.$

*19. $G(x) = (9x^8 - 8x^9)\left(x + \dfrac{1}{x}\right).$

20. $G(x) = \left(1 + \dfrac{1}{x}\right)\left(1 + \dfrac{1}{x^2}\right).$

Find $f'(0)$ and $f'(1)$.

*21. $f(x) = \dfrac{1}{x - 2}.$

22. $f(x) = x^2(x + 1).$

*23. $f(x) = \dfrac{1 - x^2}{1 + x^2}.$

24. $f(x) = \dfrac{2x^2 + x + 1}{x^2 + 2x + 1}.$

*25. $f(x) = \dfrac{ax + b}{cx + d}.$

26. $f(x) = \dfrac{ax^2 + bx + c}{cx^2 + bx + a}.$

Given that $h(0) = 3$ and $h'(0) = 2$, find $f'(0)$.

*27. $f(x) = xh(x).$

28. $f(x) = 3x^2 h(x) - 5x.$

*29. $f(x) = h(x) - \dfrac{1}{h(x)}.$

30. $f(x) = h(x) + \dfrac{x}{h(x)}.$

31. Prove the validity of the quotient rule.

32. Verify that, if f, g, h are differentiable, then

$$(fgh)'(x) = f'(x)g(x)h(x) + f(x)g'(x)h(x) + f(x)g(x)h'(x).$$

33. Use the product rule (3.2.4) to show that, if f is differentiable, then

$$g(x) = [f(x)]^2 \quad \text{has derivative} \quad g'(x) = 2f(x)f'(x).$$

Optional 34. Show that, if f is differentiable, then

$$g(x) = [f(x)]^n \quad \text{has derivative} \quad g'(x) = n[f(x)]^{n-1}f'(x)$$

for each nonzero integer n.

3.3 The d/dx Notation

So far we have indicated the derivative by a prime, but there are other notations which are widely used, particularly in science and engineering. The most popular of these is the double-d notation of Leibniz. In the Leibniz notation the derivative of a function y is indicated by writing

$$\frac{dy}{dx}, \quad \frac{dy}{dt}, \quad \text{or} \quad \frac{dy}{dz}, \quad \text{etc.},$$

depending upon whether the letter x, t, or z, etc., is being used for the elements of the domain of y. For instance, if y is initially defined by

$$y(x) = x^3,$$

then the Leibniz notation gives

$$\frac{dy}{dx}(x) = 3x^2.$$

Usually writers drop the x on the left and simply write

$$y = x^3 \quad \text{and} \quad \frac{dy}{dx} = 3x^2.$$

The symbols

$$\frac{d}{dx}, \quad \frac{d}{dt}, \quad \frac{d}{dz}, \quad \text{etc.},$$

are also used as prefixes before expressions to be differentiated. For example,

$$\frac{d}{dx}(x^3 - 4x) = 3x^2 - 4, \quad \frac{d}{dt}(t^2 + 3t + 1) = 2t + 3, \quad \frac{d}{dz}(z^5 - 1) = 5z^4.$$

In the Leibniz notation the differentiation formulas look like this:

I. $$\frac{d}{dx}[f(x) + g(x)] = \frac{d}{dx}[f(x)] + \frac{d}{dx}[g(x)].$$

II. $$\frac{d}{dx}[\alpha f(x)] = \alpha \frac{d}{dx}[f(x)].$$

III. $$\frac{d}{dx}[f(x)g(x)] = f(x)\frac{d}{dx}[g(x)] + g(x)\frac{d}{dx}[f(x)].$$

IV. $$\frac{d}{dx}\left[\frac{1}{g(x)}\right] = -\frac{1}{[g(x)]^2}\frac{d}{dx}[g(x)].$$

V. $$\frac{d}{dx}\left[\frac{f(x)}{g(x)}\right] = \frac{g(x)\frac{d}{dx}[f(x)] - f(x)\frac{d}{dx}[g(x)]}{[g(x)]^2}.$$

Often the functions f and g are replaced by u and v and the x is left out altogether. Formulas I to V then look like this:

I. $$\frac{d}{dx}(u + v) = \frac{du}{dx} + \frac{dv}{dx}.$$

II. $$\frac{d}{dx}(\alpha u) = \alpha \frac{du}{dx}.$$

III. $$\frac{d}{dx}(uv) = u\frac{d}{dx} + v\frac{du}{dx}.$$

IV. $$\frac{d}{dx}\left(\frac{1}{v}\right) = -\frac{1}{v^2}\frac{dv}{dx}.$$

V.
$$\frac{d}{dx}\left(\frac{u}{v}\right) = \frac{v\,\dfrac{du}{dx} - u\,\dfrac{dv}{dx}}{v^2}.$$

Probably the only way to develop a feeling for this new notation is to use it. Below we work out some practice problems.

Problem.   Find

$$\frac{dy}{dx} \quad \text{if} \quad y = \frac{x-1}{x+2}.$$

SOLUTION.   Here we use the quotient rule (V):

$$\frac{dy}{dx} = \frac{(x+2)\dfrac{d}{dx}(x-1) - (x-1)\dfrac{d}{dx}(x+2)}{(x+2)^2}$$

$$= \frac{(x+2)-(x-1)}{(x+2)^2} = \frac{3}{(x+2)^2}. \quad \square$$

Problem.   Find

$$\frac{dy}{dx} \quad \text{if} \quad y = (x^3+1)(3x^5+2x-1).$$

SOLUTION.   Here we can use the product rule (III):

$$\frac{dy}{dx} = (x^3+1)\frac{d}{dx}(3x^5+2x-1) + (3x^5+2x-1)\frac{d}{dx}(x^3+1)$$

$$= (x^3+1)(15x^4+2) + (3x^5+2x-1)(3x^2)$$

$$= (15x^7+15x^4+2x^3+2) + (9x^7+6x^3-3x^2)$$

$$= 24x^7+15x^4+8x^3-3x^2+2.$$

We could have achieved the same result by carrying out the initial multiplication:

$$y = (x^3+1)(3x^5+2x-1)$$

$$= (3x^8+2x^4-x^3) + (3x^5+2x-1)$$

$$= 3x^8+3x^5+2x^4-x^3+2x-1$$

and then differentiating:

$$\frac{dy}{dx} = 24x^7+15x^4+8x^3-3x^2+2. \quad \square$$

Problem.   Find

$$\frac{d}{dt}\left(t^3 - \frac{t}{t^2-1}\right).$$

SOLUTION

$$\frac{d}{dt}\left(t^3 - \frac{t}{t^2 - 1}\right) = \frac{d}{dt}(t^3) - \frac{d}{dt}\left(\frac{t}{t^2 - 1}\right)$$

$$= 3t^2 - \left[\frac{(t^2 - 1)(1) - t(2t)}{(t^2 - 1)^2}\right] = 3t^2 + \frac{t^2 + 1}{(t^2 - 1)^2} \cdot \quad \square$$

Problem.   Find

$$\frac{du}{dx} \quad \text{if} \quad u = \frac{ax + b}{cx + d} - \frac{cx + d}{ax + b}.$$

SOLUTION

$$\frac{du}{dx} = \frac{d}{dx}\left(\frac{ax + b}{cx + d}\right) - \frac{d}{dx}\left(\frac{cx + d}{ax + b}\right)$$

$$= \frac{(cx + d)a - (ax + b)c}{(cx + d)^2} - \frac{(ax + b)c - (cx + d)a}{(ax + b)^2}$$

$$= \frac{ad - bc}{(cx + d)^2} - \frac{bc - ad}{(ax + b)^2} = (ad - bc)\left[\frac{1}{(cx + d)^2} + \frac{1}{(ax + b)^2}\right]. \quad \square$$

Problem.   Find

$$\frac{dy}{dx} \quad \text{if} \quad y = x(x + 1)(x + 2).$$

SOLUTION.   You can think of y as

$$[x(x + 1)](x + 2) \quad \text{or as} \quad x[(x + 1)(x + 2)].$$

From the first point of view,

$$\frac{dy}{dx} = [x(x + 1)](1) + (x + 2)\frac{d}{dx}[x(x + 1)]$$

$$= x(x + 1) + (x + 2)[x(1) + (x + 1)(1)]$$

(∗) $$= x(x + 1) + (x + 2)(2x + 1).$$

From the second point of view,

$$\frac{dy}{dx} = x\frac{d}{dx}[(x + 1)(x + 2)] + (x + 1)(x + 2)(1)$$

$$= x[(x + 1)(1) + (x + 2)(1)] + (x + 1)(x + 2)$$

(∗∗) $$= x(2x + 3) + (x + 1)(x + 2).$$

Both (∗) and (∗∗) can be multiplied out to give

$$\frac{dy}{dx} = 3x^2 + 6x + 2.$$

This same result can be obtained by first carrying out the entire multiplication and then differentiating:

$$y = x(x + 1)(x + 2) = x(x^2 + 3x + 2) = x^3 + 3x^2 + 2x$$

so that

$$\frac{dy}{dx} = 3x^2 + 6x + 2. \quad \square$$

Problem.   Evaluate dy/dx at $x = 0$ and $x = 1$ if

$$y = \frac{x^2}{x^2 - 4}.$$

SOLUTION

$$\frac{dy}{dx} = \frac{(x^2 - 4)2x - x^2(2x)}{(x^2 - 4)^2} = -\frac{8x}{(x^2 - 4)^2}.$$

At $x = 0$, $dy/dx = 0$; at $x = 1$, $dy/dx = -\frac{8}{9}$. $\square$

Exercises

Find dy/dx.

*1. $y = 3x^4 - x^2 + 1$.

2. $y = x^2 + 2x^{-4}$.

*3. $y = x - \dfrac{1}{x}$.

4. $y = \dfrac{2x}{1 - x}$.

*5. $y = \dfrac{x}{1 + x^2}$.

6. $y = x(x - 2)(x + 1)$.

*7. $y = \dfrac{x^2}{1 - x}$.

8. $y = \left(\dfrac{x}{1 + x}\right)\left(\dfrac{2 - x}{3}\right)$.

*9. $y = \dfrac{x^3 + 1}{x^3 - 1}$.

10. $y = \dfrac{x^2}{(1 + x)^2}$.

Find the indicated derivative.

*11. $\dfrac{d}{dx}(2x - 5)$.

12. $\dfrac{d}{dx}(5x + 2)$.

*13. $\dfrac{d}{dx}(7x^3 - 1)$.

14. $\dfrac{d}{dx}[(1 - x)^2(1 + x)]$.

*15. $\dfrac{d}{dt}\left(\dfrac{t^2 + 1}{t^2 - 1}\right)$.

16. $\dfrac{d}{dt}\left(\dfrac{2t^3 + 1}{t^4}\right)$.

*17. $\dfrac{d}{dt}\left(\dfrac{t^4}{2t^3 - 1}\right)$.

18. $\dfrac{d}{dt}\left[\dfrac{t}{(1 + t)^2}\right]$.

*19. $\dfrac{d}{du}\left(\dfrac{2u}{1 - 2u}\right)$.

20. $\dfrac{d}{du}\left(\dfrac{u^2}{u^3 + 1}\right)$.

*21. $\dfrac{d}{du}\left(\dfrac{u}{u - 1} - \dfrac{u}{u + 1}\right)$.

22. $\dfrac{d}{du}[u^2(1 - u^2)(1 - u^3)]$.

*23. $\dfrac{d}{dx}\left(\dfrac{x^2}{1-x^2}-\dfrac{1-x^2}{x^2}\right).$

24. $\dfrac{d}{dx}\left(\dfrac{3x^4+2x+1}{x^4+x-1}\right).$

*25. $\dfrac{d}{dx}\left(\dfrac{x^3+x^2+x+1}{x^3-x^2+x-1}\right).$

26. $\dfrac{d}{dx}\left(\dfrac{x^3+x^2+x-1}{x^3-x^2+x+1}\right).$

Evaluate dy/dx at $x = 2$.

*27. $y = (x+1)(x+2)(x+3).$

28. $y = (x+1)(x^2+2)(x^3+3).$

*29. $y = \dfrac{(x-1)(x-2)}{(x+2)}.$

30. $y = \dfrac{(x^2+1)(x^2-2)}{x^2+2}.$

3.4 The Derivative as a Rate of Change

In the case of a linear function

$$y = mx + b$$

the graph is a straight line and the slope m gives *the rate of change of y with respect to x*:

$$m = \frac{\text{change in } y}{\text{change in } x}. \qquad\qquad \text{(Figure 3.4.1)}$$

In other words, on a line of slope m, y changes m times as fast as x.

$$m = \frac{\text{change in } y}{\text{change in } x}$$

FIGURE 3.4.1

In the more general case of a differentiable function

$$y = f(x)$$

the graph is a curve. The slope

$$\frac{dy}{dx} = f'(x)$$

still gives *the rate of change of y with respect to x,* but this rate can vary from point to point. At x_1 (see Figure 3.4.2) the rate of change is $f'(x_1)$; at x_2, the rate of change is $f'(x_2)$; at x_3, the rate of change is $f'(x_3)$.

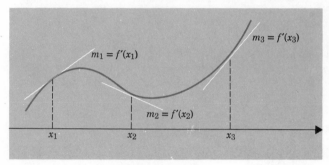

FIGURE 3.4.2

The derivative as a rate of change is one of the fundamental ideas of the calculus. Keep it in mind whenever you see a derivative.

Problem.   Find the rate of change of the area of a circle with respect to its radius r. Evaluate this rate of change when $r = 2$.

SOLUTION.   Since $A = \pi r^2$, the rate of change of A with respect to r is given by

$$\frac{dA}{dr} = 2\pi r,$$

which, by the way, is the circumference. When $r = 2$, the rate of change is 4π. This means that, when the radius is 2, the area is changing 4π times as fast as the radius.   □

Problem.   Find the rate of change of the area of an equilateral triangle with respect to the length of a side. Evaluate this rate of change when the side has length $\sqrt{3}$.

SOLUTION.   If the side has length s, the area is given by the formula

$$A = \tfrac{1}{4}\sqrt{3}\, s^2. \qquad\qquad \text{(check this out)}$$

The rate of change is therefore

$$\frac{dA}{ds} = \tfrac{1}{2}\sqrt{3}\, s.$$

When $s = \sqrt{3}$, the rate of change is $\tfrac{3}{2}$. In other words, when the side has length $\sqrt{3}$, the area is changing $\tfrac{3}{2}$ times as fast as the length of the side.   □

Problem.   The volume of a circular cylinder is given by the formula $V = \pi r^2 h$ where r is the radius and h is the height. Find the rate of change of h with respect to r if V is to remain constant as r increases.

SOLUTION.   Applying the product rule to

$$V = \pi r^2 h$$

we get

$$\frac{dV}{dr} = \pi r^2 \frac{dh}{dr} + h \frac{d}{dr}(\pi r^2) = \pi r^2 \frac{dh}{dr} + 2\pi r h.$$

Since V is to remain constant, $dV/dr = 0$, and therefore

$$\pi r^2 \frac{dh}{dr} + 2\pi r h = 0.$$

Dividing through by πr, we get

$$r \frac{dh}{dr} + 2h = 0$$

which gives

$$\frac{dh}{dr} = -\frac{2h}{r}.$$

For the volume to remain constant, h must *decrease* $2h/r$ times as fast as r increases.   □

REMARK.   More sophisticated rates of change problems are presented later, particularly in Section 4.8 where we discuss rates of change per unit time.

Exercises

*1. Find the rate of change of the area of a square with respect to the length s of one of its sides. Evaluate this rate of change when $s = 4$.

 2. Find the rate of change of the volume of a cube with respect to the length s of one of its sides. Evaluate this rate of change when $s = 4$.

*3. Find the rate of change of the area of a square with respect to the length z of one of its diagonals. Evaluate this rate of change when $z = 4$.

 4. Find the rate of change of $y = 1/x$ with respect to x when $x = -1$.

*5. Find the rate of change of

$$y = \frac{1}{x(x + 1)}$$

with respect to x when $x = 2$.

 6. Find the values of x for which the rate of change of $y = x^3 - 12x^2 + 45x - 1$ is zero.

 7. Find the rate of change of the volume of a ball with respect to its radius, given that $V = \frac{4}{3}\pi r^3$.

*8. Find the rate of change of the surface area of a ball with respect to its radius given that $A = 4\pi r^2$. What is this rate of change when $r = r_0$? How must r_0 be chosen so that the rate of change is 1?

 9. Find x_0 given that the rate of change of $y = 2x^2 + x - 1$ with respect to x is 4 when $x = x_0$.

*10. The dimensions of a rectangle are changing in such a way that the area of the rectangle remains constant. Find the rate of change of the height h with respect to the base b.

11. The area of a sector in a circle is given by the formula $A = \frac{1}{2}r^2\theta$ where r is the radius and θ is the central angle measured in radians.

   *(a) Find the rate of change of A with respect to θ if r remains constant.

   (b) Find the rate of change of A with respect to r if θ remains constant.

   *(c) Find the rate of change of θ with respect to r if A remains constant.

12. The total surface area of a right circular cylinder is given by the formula $A = 2\pi r(r + h)$ where r is the radius and h is the height.

   (a) Find the rate of change of A with respect to h if r remains constant.

   *(b) Find the rate of change of A with respect to r if h remains constant.

   (c) Find the rate of change of h with respect to r if A remains constant.

13. For what value of x is the rate of change of

$$y = ax^2 + bx + c \quad \text{with respect to } x$$

the same as the rate of change of

$$z = bx^2 + ax + c \quad \text{with respect to } x?$$

Assume $a \neq b$.

*14. Find the rate of change of the product $f(x)g(x)h(x)$ with respect to x when $x = 1$ given that

$$f(1) = 0, \quad g(1) = 2, \quad h(1) = -2,$$
$$f'(1) = 1, \quad g'(1) = -1, \quad h'(1) = 0.$$

Optional

15. Given that $A = \frac{1}{2}r^2\theta$, find the rate of change of r with respect to θ if A remains constant.

*16. Given that $A = 2\pi r(r + h)$ find the rate of change of r with respect to h if A remains constant.

3.5 The Chain Rule

Here we take up the differentiation of composite functions. Until we get to Theorem 3.5.7, our approach is completely intuitive—no real definitions, no proofs, just informal discussion ("arm waving," if you like). Our purpose is to give you some experience with the standard computational procedures and some insight into why these procedures work. Theorem 3.5.7 puts it all on a sound footing.

Assume for the moment that the composition of differentiable functions is differentiable. Knowing the derivative of f and the derivative of g, how can we find the derivative of $f \circ g$? That is, how can we find

$$\frac{d}{dx}[f(g(x))]?$$

The key to answering this question lies in the following two ideas:

I. For any differentiable function h, $h(t)$ changes $h'(t)$ times as fast as t. [This comes from $h'(t)$ being the rate of change of $h(t)$ with respect to t.]

II. If A changes m times as fast as B and B changes n times as fast as C, then A changes mn times as fast as C.

Once you accept I and II, the computation of $\dfrac{d}{dx}[f(g(x))]$ is easy. By I,

$$f(g(x)) \text{ changes } f'(g(x)) \text{ times as fast as } g(x)$$

and

$$g(x) \text{ changes } g'(x) \text{ times as fast as } x.$$

It follows by II that

$$f(g(x)) \text{ changes } f'(g(x))g'(x) \text{ times as fast as } x.$$

This makes

$$f'(g(x))g'(x) \text{ the rate of change of } f(g(x)) \text{ with respect to } x,$$

which means that

(3.5.1)

$$\boxed{\frac{d}{dx}[f(g(x))] = f'(g(x))g'(x).\dagger}$$

Formula (3.5.1) is called the *chain rule*. The name is particularly appropriate when three or more functions are involved. For the composition of three differentiable functions we have

$$\frac{d}{dx}[f(g(h(x)))] = f'(g(h(x)))g'(h(x))h'(x).$$

If four functions are involved, then a new link is added to the chain. The pattern is obvious.

An an application of the chain rule we show how to differentiate integral powers of a differentiable function. The formula is this:

(3.5.2)

$$\boxed{\frac{d}{dx}[g(x)]^n = n[g(x)]^{n-1}g'(x).}$$

To derive this formula we set

$$f(x) = x^n$$

so that

$$\frac{d}{dx}[g(x)]^n \quad \text{becomes} \quad \frac{d}{dx}[f(g(x))].$$

$\dagger$ If you wish to use the prime notation throughout, then the left-hand side becomes $(f \circ g)'(x)$ and the formula reads

$$\boxed{(f \circ g)'(x) = f'(g(x))g'(x).}$$

The chain rule now gives

(1) $$\frac{d}{dx}[g(x)]^n = f'(g(x))g'(x).$$

Since

$$f(x) = x^n \quad \text{has derivative} \quad f'(x) = nx^{n-1},$$

we have

$$f'(g(x)) = n[g(x)]^{n-1}.$$

Substituting this into (1), we get the desired formula. $\square$

Formula (3.5.2) can also be written

(3.5.3) $$\boxed{\frac{d}{dx}[g(x)]^n = n[g(x)]^{n-1}\frac{d}{dx}[g(x)].}$$

Example

$$\frac{d}{dx}(x^2 + 1)^3 = 3(x^2 + 1)^2\frac{d}{dx}(x^2 + 1) = 3(x^2 + 1)^2 2x = 6x(x^2 + 1)^2. \quad \square$$

Example

$$\frac{d}{dx}\left(x + \frac{1}{x}\right)^{-3} = -3\left(x + \frac{1}{x}\right)^{-4}\frac{d}{dx}\left(x + \frac{1}{x}\right) = -3\left(x + \frac{1}{x}\right)^{-4}\left(1 - \frac{1}{x^2}\right). \quad \square$$

Example

$$\frac{d}{dx}[1 - (2 + 3x)^2]^3 = 3[1 - (2 + 3x)^2]^2\frac{d}{dx}[1 - (2 + 3x)^2].$$

Since

$$\frac{d}{dx}[1 - (2 + 3x)^2] = -2(2 + 3x)\frac{d}{dx}(3x) = -6(2 + 3x),$$

we have

$$\frac{d}{dx}[1 - (2 + 3x)^2]^3 = -18(2 + 3x)[1 - (2 + 3x)^2]^2. \quad \square$$

Another Formulation of the Chain Rule

If y is expressed in terms of u [$y = f(u)$], and u is expressed in terms of x [$u = g(x)$], then y can be expressed in terms of x:

$$y = f(u) = f(g(x)).$$

In this situation

$$\frac{d}{dx}[f(g(x))] = \frac{dy}{dx}, \qquad f'(g(x)) = f'(u) = \frac{dy}{du}, \qquad g'(x) = \frac{du}{dx}.$$

Formula 3.5.1 can then be written

(3.5.4)
$$\frac{dy}{dx} = \frac{dy}{du}\frac{du}{dx}.$$

This formulation of the chain rule is particularly easy to understand. It simply points out that

with y changing $\dfrac{dy}{du}$ times as fast as u

and u changing $\dfrac{du}{dx}$ times as fast as x,

y must change $\dfrac{dy}{du}\dfrac{du}{dx}$ times as fast as x.

Problem.   Find dy/dx given that

$$y = 3u^2 + 2 \quad \text{and} \quad u = \frac{1}{x - 1}.$$

SOLUTION

$$\frac{dy}{du} = 6u = \frac{6}{x - 1}, \qquad \frac{du}{dx} = -\frac{1}{(x - 1)^2}$$

so that

$$\frac{dy}{dx} = \frac{dy}{du}\frac{du}{dx} = \frac{6}{x - 1}\left[-\frac{1}{(x - 1)^2}\right] = -\frac{6}{(x - 1)^3}. \quad \Box$$

ALTERNATIVE SOLUTION.   We can achieve the same result by first expressing y in terms of x and then differentiating. With

$$y = 3u^2 + 2 \quad \text{and} \quad u = \frac{1}{x - 1}$$

we have

$$y = \frac{3}{(x - 1)^2} + 2$$

so that

$$\frac{dy}{dx} = -\frac{6}{(x - 1)^3}. \quad \Box$$

Problem.   Find dy/dx for $x = 2$ given that

$$y = \frac{u - 1}{u + 1} \quad \text{and} \quad u = x^2.$$

SOLUTION

$$\frac{dy}{du} = \frac{(u + 1)1 - (u - 1)1}{(u + 1)^2} = \frac{2}{(u + 1)^2} = \frac{2}{(x^2 + 1)^2} \quad \text{and} \quad \frac{du}{dx} = 2x$$

so that

$$\frac{dy}{dx} = \frac{dy}{du}\frac{du}{dx} = \frac{2}{(x^2 + 1)^2}\,(2x) = \frac{4x}{(x^2 + 1)^2}.$$

At $x = 2$,

$$\frac{dy}{dx} = \frac{4(2)}{(2^2 + 1)^2} = \frac{8}{25}. \quad \square$$

ALTERNATIVE SOLUTION.   With

$$y = \frac{u - 1}{u + 1} \quad \text{and} \quad u = x^2$$

we have

$$y = \frac{x^2 - 1}{x^2 + 1}$$

so that

$$\frac{dy}{dx} = \frac{(x^2 + 1)2x - (x^2 - 1)2x}{(x^2 + 1)^2} = \frac{4x}{(x^2 + 1)^2}.$$

At $x = 2$,

$$\frac{dy}{dx} = \frac{4(2)}{(2^2 + 1)^2} = \frac{8}{25}. \quad \square$$

The formula

$$\frac{dy}{dx} = \frac{dy}{du}\frac{du}{dx}$$

can easily be extended to more variables. For example, if x itself depends on s, then we have

(3.5.5)
$$\boxed{\frac{dy}{ds} = \frac{dy}{du}\frac{du}{dx}\frac{dx}{ds}.}$$

If, in addition, s depends on t, then

(3.5.6)
$$\boxed{\frac{dy}{dt} = \frac{dy}{du}\frac{du}{dx}\frac{dx}{ds}\frac{ds}{dt}}$$

and so on. Each new dependence adds a new link to the chain.

A More Rigorous Approach

So far our approach to the differentiation of composite functions has been entirely intuitive. As we said at the beginning, our purpose was to give you some familiarity with the standard formulas and some insight into why these formulas "work."

It is time to be more careful. How do we know that the composition of differentiable functions is differentiable? What assumptions do we need? Precisely under what circumstances does

$$(f \circ g)'(x) = f'(g(x))g'(x)?$$

The following theorem takes care of these loose ends.

Theorem 3.5.7   The Chain Rule Theorem

If g is differentiable at x and f is differentiable at $g(x)$, then the composition $f \circ g$ is differentiable at x and

$$(f \circ g)'(x) = f'(g(x))g'(x).$$

For a proof see the supplement at the end of this section. $\square$

Exercises

Differentiate.

*1. $f(x) = (1 - 2x)^{-1}$.

2. $f(x) = (1 + 2x)^5$.

*3. $f(x) = (x^5 - x^{10})^{20}$.

4. $f(x) = \left(x^2 + \dfrac{1}{x^2}\right)^3$.

*5. $f(x) = \left(x - \dfrac{1}{x}\right)^4$.

6. $f(x) = \left(x + \dfrac{1}{x}\right)^3$.

*7. $f(x) = (x - x^3 - x^5)^4$.

*9. $f(t) = (t^2 - 1)^{100}$.

8. $f(t) = \left(\dfrac{1}{1 + t}\right)^4$.

*11. $f(t) = (t^{-1} + t^{-2})^4$.

10. $f(t) = (c^2 + t^2)^3$.

*13. $f(x) = \left(\dfrac{3x}{x^2 + 1}\right)^4$.

12. $f(x) = \left(\dfrac{ax + b}{cx + d}\right)^3$.

*15. $f(x) = (x^4 + x^2 + x)^2$.

14. $f(x) = [(2x + 1)^2 + (x + 1)^2]^3$.

*17. $f(x) = \left(\dfrac{x^3}{3} + \dfrac{x^2}{2} + \dfrac{x}{1}\right)^{-1}$.

16. $f(x) = (x^2 + 2x + 1)^3$.

18. $f(x) = \left(\dfrac{x^2 + 2}{x^2 + 1}\right)^5$.

*19. $f(x) = \left(\dfrac{1}{x + 2} - \dfrac{1}{x - 2}\right)^3$.

20. $f(x) = [(6x + x^5)^{-1} + x]^2$.

Find dy/dx.

*21. $y = \dfrac{1}{1 + u^2}$, $u = 2x + 1$.

22. $y = u + \dfrac{1}{u}$, $u = (3x + 1)^4$.

*23. $y = \dfrac{2u}{1 - 4u}$, $u = (5x^2 + 1)^4$.

24. $y = u^3 - u + 1$, $u = \dfrac{1 - x}{1 + x}$.

Find dy/dt.

*25. $y = \dfrac{1 - 7u}{1 + u^2}$, $u = 1 + x^2$, $x = 2t - 5$.

26. $y = 1 + u^2, \quad u = \dfrac{1 - 7x}{1 + x^2}, \quad x = 5t + 2.$

Given that

$$f(0) = 1, \quad f'(0) = 2, \quad f(1) = 0, \quad f'(1) = 1, \quad f(2) = 1, \quad f'(2) = 1,$$
$$g(0) = 2, \quad g'(0) = 1, \quad g(1) = 1, \quad g'(1) = 0, \quad g(2) = 2, \quad g'(2) = 1,$$
$$h(0) = 1, \quad h'(0) = 2, \quad h(1) = 2, \quad h'(1) = 1, \quad h(2) = 0, \quad h'(2) = 2,$$

evaluate the following.

*27. $(f \circ g)'(0).$ 28. $(f \circ g)'(1).$ *29. $(f \circ g)'(2).$
30. $(g \circ f)'(0).$ *31. $(g \circ f)'(1).$ 32. $(g \circ f)'(2).$
*33. $(f \circ h)'(0).$ 34. $(h \circ f)'(1).$ *35. $(h \circ f)'(0).$
36. $(f \circ h \circ g)'(1).$ *37. $(g \circ f \circ h)'(2).$ 38. $(g \circ h \circ f)'(0).$

Find the following derivatives.

*39. $\dfrac{d}{dx}[f(x^2 + 1)].$ 40. $\dfrac{d}{dx}\left[f\left(\dfrac{x-1}{x+1}\right)\right].$

*41. $\dfrac{d}{dx}[[f(x)]^2 + 1].$ 42. $\dfrac{d}{dx}\left[\dfrac{f(x)-1}{f(x)+1}\right].$

Determine the values of x for which
    (a) $f'(x) = 0.$ (b) $f'(x) > 0.$ (c) $f'(x) < 0.$

*43. $f(x) = (1 + x^2)^{-2}.$ 44. $f(x) = (1 - x^2)^2.$
*45. $f(x) = x(1 + x^2)^{-1}.$ 46. $f(x) = x(1 - x^2)^3.$

47. Let f be a differentiable function. Show that
    (a) if f is even, then f' is odd.   (b) if f is odd, then f' is even.
*48. Exercise 15 of Section 3.4 if previously omitted.
49. Exercise 16 of Section 3.4 if previously omitted.

Differentiate.

*50. $f(x) = [(7x - x^{-1})^{-2} + 3x^2]^{-1}$ 51. $f(x) = [(x^2 - x^{-2})^3 - x]^5.$
*52. $f(x) = [(x + x^{-1})^2 - (x^2 + x^{-2})^{-1}]^3.$ 53. $f(x) = [(x^{-1} + 2x^{-2})^3 + 3x^{-3}]^4.$

Optional | **Supplement to Section 3.5**

To prove Theorem 3.5.7 it is convenient to use a slightly different formulation of derivative.

Theorem 3.5.8

The function f is differentiable at x iff

$$\lim_{t \to x} \frac{f(t) - f(x)}{t - x} \quad \text{exists.}$$

If this limit exists, it is $f'(x)$.

Optional | Proof. For each t in the domain of f, $t \neq x$, define

$$G(t) = \frac{f(t) - f(x)}{t - x}.$$

The theorem follows from noting that f is differentiable at x iff

$$\lim_{h \to 0} G(x + h) \quad \text{exists,}$$

and recalling that

$$\lim_{h \to 0} G(x + h) = l \quad \text{iff} \quad \lim_{t \to x} G(t) = l.\dagger \quad \square$$

Proof of theorem 3.5.7. By Theorem 3.5.8 it is enough to show that

$$\lim_{t \to x} \frac{f(g(t)) - f(g(x))}{t - x} = f'(g(x))g'(x).$$

We begin by defining an auxiliary function F on the domain of f by setting

$$F(y) = \begin{cases} \dfrac{f(y) - f(g(x))}{y - g(x)}, & y \neq g(x) \\ f'(g(x)), & y = g(x) \end{cases}.$$

F is continuous at $g(x)$ since

$$\lim_{y \to g(x)} F(y) = \lim_{y \to g(x)} \frac{f(y) - f(g(x))}{y - g(x)},$$

and the right-hand side is, by Theorem 3.5.8, $f'(g(x))$, which is the value of F at $g(x)$. For $t \neq x$,

$$(1) \qquad \frac{f(g(t)) - f(g(x))}{t - x} = F(g(t))\left[\frac{g(t) - g(x)}{t - x}\right].$$

To see this we note that, if $g(t) = g(x)$, then both sides are 0. If $g(t) \neq g(x)$, then

$$F(g(t)) = \frac{f(g(t)) - f(g(x))}{g(t) - g(x)},$$

so that again we have equality.

Since g, being differentiable at x, is continuous at x and since F is continuous at $g(x)$, we know that the composition $F \circ g$ is continuous at x. Thus

$$\lim_{t \to x} F(g(t)) = F(g(x)) = f'(g(x)).$$

---

† Equation (2.4.2).

Optional | This, together with Equation (1), gives

$$\lim_{t \to x} \frac{f(g(t)) - f(g(x))}{t - x} = f'(g(x))g'(x). \quad \square$$

3.6　Derivatives of Higher Order

If a function f is differentiable, then we can form a new function f'. If f' is itself differentiable, then we can form its derivative, called the *second derivative of f*, and denoted by f''. So long as we have differentiability, we can continue in this manner, forming f''', etc. The prime notation is not used beyond the third order. For the fourth derivative we write $f^{(4)}$ and more generally, for the nth derivative, $f^{(n)}$.

If $f(x) = x^5$, then

$$f'(x) = 5x^4, \quad f''(x) = 20x^3, \quad f'''(x) = 60x^2, \quad f^{(4)}(x) = 120x, \quad f^{(5)}(x) = 120.$$

All higher derivatives are identically zero. As a variant of this notation you can write $y = x^5$ and then

$$y' = 5x^4, \quad y'' = 20x^3, \quad y''' = 60x^2, \quad \text{etc.}$$

Since each polynomial P has a derivative P' which is in turn a polynomial and each rational function Q has derivative Q' which is in turn a rational function, polynomials and rational functions have derivatives of all orders. In the case of a polynomial of degree n, derivatives of order greater than n are all identically zero. (Explain.)

In the Leibniz notation the derivatives of higher order are written

$$\frac{d^2y}{dx^2} = \frac{d}{dx}\left(\frac{dy}{dx}\right), \frac{d^3y}{dx^3} = \frac{d}{dx}\left(\frac{d^2y}{dx^2}\right), \ldots, \frac{d^ny}{dx^n} = \frac{d}{dx}\left(\frac{d^{n-1}y}{dx^{n-1}}\right), \ldots$$

or

$$\frac{d^2}{dx^2}[f(x)] = \frac{d}{dx}\left[\frac{d}{dx}[f(x)]\right], \frac{d^3}{dx^3}[f(x)] = \frac{d}{dx}\left[\frac{d^2}{dx^2}[f(x)]\right], \ldots,$$

$$\frac{d^n}{dx^n}[f(x)] = \frac{d}{dx}\left[\frac{d^{n-1}}{dx^{n-1}}[f(x)]\right], \ldots$$

Below we work out some examples.

Example.　We begin with

$$y = x^{-1}.$$

In the Leibniz notation

$$\frac{dy}{dx} = -x^{-2}, \quad \frac{d^2y}{dx^2} = 2x^{-3}, \quad \frac{d^3y}{dx^3} = -6x^{-4}, \quad \frac{d^4y}{dx^4} = 24x^{-5}$$

and, more generally,

$$\frac{d^ny}{dx^n} = (-1)^n n! x^{-n-1}.$$

In the prime notation

$$y' = -x^{-2}, \quad y'' = 2x^{-3}, \quad y''' = -6x^{-4}, \quad y^{(4)} = 24x^{-5}$$

and

$$y^{(n)} = (-1)^n n! x^{-n-1}. \quad \square$$

Example

$$\frac{d}{dx}\left(\frac{x}{1+x}\right) = \frac{(1+x)(1) - x(1)}{(1+x)^2} = \frac{1}{(1+x)^2}$$

so that

$$\frac{d^2}{dx^2}\left(\frac{x}{1+x}\right) = \frac{d}{dx}\left[\frac{1}{(1+x)^2}\right] = -\frac{2}{(1+x)^3}$$

and

$$\frac{d^3}{dx^3}\left(\frac{x}{1+x}\right) = \frac{d}{dx}\left[-\frac{2}{(1+x)^3}\right] = \frac{6}{(1+x)^4}. \quad \square$$

Example.   If

$$f(x) = \frac{3x^2}{1+x^2}$$

then

$$f'(x) = \frac{(1+x^2)6x - 3x^2(2x)}{(1+x^2)^2} = \frac{6x}{(1+x^2)^2}$$

and

$$f''(x) = \frac{(1+x^2)^2 \cdot 6x(2)(1+x^2)2x}{(1+x^2)^4} = \frac{(1+x^2)(6-18x^2)}{(1+x^2)^4} = \frac{6(1-3x^2)}{(1+x^2)^3}. \quad \square$$

Exercises

Find the second derivative.

*1. $f(x) = 7x^3 - 6x^5$.

2. $f(x) = 2x^5 - 6x^4 + 2x - 1$.

*3. $f(x) = \dfrac{a+bx}{a-bx}$.

4. $f(x) = x^2 - \dfrac{1}{x^2}$.

*5. $y = ax^3 + bx^{-3}$.

6. $y = (a^2 - b^2x^2)^3$.

*7. $y = \dfrac{x}{x^2+a^2}$.

8. $y = \dfrac{x^2}{x^2+a^2}$.

Find d^3y/dx^3.

*9. $y = \frac{1}{3}x^3 + \frac{1}{2}x^2 + x + 1$.

10. $y = (1+5x)^2$.

*11. $y = (1+2x)^3$.

12. $y = (ax+b)^{-1}$.

Find the indicated derivative.

*13. $\dfrac{d^2}{dx^2}\left(\dfrac{ax+b}{cx+d}\right)$.

14. $\dfrac{d^2}{dx^2}\left(\dfrac{ax^2-b}{cx^2-d}\right)$.

*15. $\dfrac{d^5}{dx^5}(ax^4 + bx^3 + cx^2 + dx + e)$.   16. $\dfrac{d^4}{dx^4}[(1 - 2x)^4]$.

*17. $\dfrac{d}{dx}\left[x \dfrac{d^2}{dx^2}\left(\dfrac{1}{1 + x}\right)\right]$.   18. $\dfrac{d}{dx}\left[\dfrac{1}{x} \dfrac{d^2}{dx^2}\left(\dfrac{1}{1 + x}\right)\right]$.

19. Show that in general

$$(fg)''(x) \neq f(x)g''(x) + f''(x)g(x).$$

20. Verify the identity

$$f(x)g''(x) - f''(x)g(x) = \frac{d}{dx}[f(x)g'(x) - f'(x)g(x)].$$

Determine the values of x for which

   (a) $f''(x) = 0$.   (b) $f''(x) > 0$.   (c) $f''(x) < 0$.

*21. $f(x) = x^3$.   22. $f(x) = x^4$.
*23. $f(x) = x^4 + 2x^3 - 12x^2 + 1$.   24. $f(x) = x^4 + 3x^3 - 6x^2 - x$.

25. Prove by induction that

$$\text{if}\quad y = x^{-1}\quad \text{then}\quad \frac{d^n y}{dx^n} = (-1)^n n! x^{-n-1}.$$

Find a formula for the nth derivative.

*26. $y = (a + bx)^n$.   27. $y = \dfrac{a}{bx + c}$.

3.7 Differentiating Inverses; Fractional Exponents

Differentiating Inverses

If f is a one-to-one function, then f has an inverse f^{-1}. This you already know. Suppose now that f is differentiable. Is f^{-1} necessarily differentiable? Yes, if f' does not take on the value 0. You'll find a proof of this in Appendix B.3. Right now we assume this result and go on to describe how to calculate the derivative of f^{-1}.

To simplify notation we set $f^{-1} = g$. Then

$$f(g(x)) = x\quad \text{for all } x \text{ in the range of } f.$$

Differentiation gives

$$\frac{d}{dx}[f(g(x)] = 1,$$

$$f'(g(x))g'(x) = 1,\qquad \text{(we are assuming that } f \text{ and } g \text{ are differentiable)}$$

$$g'(x) = \frac{1}{f'(g(x))}.\qquad \text{(we are assuming that } f' \text{ is never 0)}$$

Substituting f^{-1} back in for g, we have

(3.7.1) $$\boxed{(f^{-1})'(x) = \frac{1}{f'(f^{-1}(x))}.}$$

This is the standard formula for the derivative of an inverse. □

nth Roots

For $x > 0$, x^n is one-to-one and its derivative, nx^{n-1}, does not take on the value 0. The inverse $x^{1/n}$ is therefore differentiable and we can use the method just described to find its derivative.

$$(x^{1/n})^n = x,$$

$$\frac{d}{dx}\,[(x^{1/n})]^n = 1,$$

$$n(x^{1/n})^{n-1}\frac{d}{dx}\,[x^{1/n}] = 1,$$

$$\frac{d}{dx}\,[x^{1/n}] = \frac{1}{n}\,(x^{1/n})^{1-n}.$$

Since $(x^{1/n})^{1-n} = x^{(1/n)-1}$, we have the differentiation formula

(3.7.2)
$$\boxed{\frac{d}{dx}\,[x^{1/n}] = \frac{1}{n}\,x^{(1/n)-1}.}$$

We derived this formula for $x > 0$. If n is odd, the formula also holds for $x < 0$. Same argument. □

For suitable values of x,

$$\frac{d}{dx}\,(x^{1/2}) = \frac{1}{2}\,x^{-1/2}, \qquad \frac{d}{dx}\,(x^{1/3}) = \frac{1}{3}\,x^{-2/3}, \qquad \frac{d}{dx}\,(x^{1/4}) = \frac{1}{4}\,x^{-3/4}, \qquad \text{etc.}$$

Rational Powers

We come now to rational exponents p/q. We take q as positive and absorb the sign in p. You can see then that

$$\frac{d}{dx}\,[x^{p/q}] = \frac{d}{dx}\,[(x^{1/q})^p] = p(x^{1/q})^{p-1}\frac{d}{dx}\,[x^{1/q}] = p(x^{1/q})^{p-1}\frac{1}{q}\,x^{(1/q)-1} = \frac{p}{q}\,x^{(p/q)-1}.$$

power formula (3.5.3)⏐   root formula (3.7.2)

In short

(3.7.3)
$$\boxed{\frac{d}{dx}\,[x^{p/q}] = \frac{p}{q}\,x^{(p/q)-1}.}$$

Here are some simple examples:

$$\frac{d}{dx}\,(x^{2/3}) = \frac{2}{3}\,x^{-1/3}, \qquad \frac{d}{dx}\,(x^{5/2}) = \frac{5}{2}\,x^{3/2}, \qquad \frac{d}{dx}\,(x^{-7/9}) = -\frac{7}{9}\,x^{-16/9}. \quad \Box$$

It's easy to go from (3.7.3) to

(3.7.4)
$$\frac{d}{dx}[g(x)]^{p/q} = \frac{p}{q}[g(x)]^{(p/q)-1}\frac{d}{dx}[g(x)].$$

Set $f(x) = x^{p/q}$ and apply the chain rule to $f(g(x)) = [g(x)]^{p/q}$. The details are left as an exercise. $\square$

The question of the applicability of (3.7.4) remains. If q is odd, then (3.7.4) is valid for all $g(x) \neq 0$. If q is even, then (3.7.4) holds only for $g(x) > 0$. In both instances, of course, g must be differentiable. It's time for some examples.

Example

$$\frac{d}{dx}[(x^2+1)^{1/5}] = \frac{1}{5}(x^2+1)^{-4/5}\cdot 2x = \frac{2}{5}x(x^2+1)^{-4/5}. \quad \square$$

Example

$$\frac{d}{dx}[\sqrt[3]{3x^2+7}] = \frac{d}{dx}[(3x^2+7)^{1/3}]$$

$$= \frac{1}{3}(3x^2+7)^{-2/3}\cdot 6x = \frac{2x}{(\sqrt[3]{3x^2+7})^2}. \quad \square$$

Example

$$\frac{d}{dx}\left[\sqrt{\frac{x^2}{1+x^2}}\right] = \frac{d}{dx}\left[\left(\frac{x^2}{1+x^2}\right)^{1/2}\right]$$

$$= \frac{1}{2}\left(\frac{x^2}{1+x^2}\right)^{-1/2}\frac{d}{dx}\left(\frac{x^2}{1+x^2}\right)$$

$$= \frac{1}{2}\left(\frac{x^2}{1+x^2}\right)^{-1/2}\frac{(1+x^2)2x - x^2(2x)}{(1+x^2)^2}$$

$$= \frac{1}{2}\left(\frac{1+x^2}{x^2}\right)^{1/2}\frac{2x}{(1+x^2)^2} = \frac{x}{(1+x^2)^2}\sqrt{\frac{1+x^2}{x^2}}. \quad \square$$

For some geometric reassurance of the validity of Formula 3.7.1 we refer you to Figure 3.7.1. The graphs of f and f^{-1} are reflections of one another in the line $y = x$. The tangent lines l_1 and l_2 are also reflections of one another. From the figure

$$(f^{-1})'(x) = \text{slope of } l_1 = \frac{f^{-1}(x) - b}{x - b},$$

$$f'(f^{-1}(x)) = \text{slope of } l_2 = \frac{x - b}{f^{-1}(x) - b};$$

namely, $(f^{-1})'(x)$ and $f'(f^{-1}(x))$ are indeed reciprocals. $\square$

Optional | In Leibniz's notation formula (3.7.1) reads simply

(3.7.5)
$$\frac{dx}{dy} = \frac{1}{dy/dx}.$$

Optional | You can see how this formula evolves by beginning with

$$y = f(x), \qquad x = f^{-1}(y).$$

The derivatives are then

$$\frac{dy}{dx} = f'(x), \qquad \frac{dx}{dy} = (f^{-1})'(y).$$

Differentiation of the identity

$$y = f(f^{-1}(y))$$

yields

$$1 = f'(f^{-1}(y))(f^{-1})'(y) = f'(x)(f^{-1})'(y) = \frac{dy}{dx}\frac{dx}{dy}$$

so that

$$\frac{dx}{dy} = \frac{1}{dy/dx} \cdot \quad \square$$

One last remark. You can get an intuitive grasp of (3.7.5) by looking at the matter this way: since y changes dy/dx times as fast as x, x must change $1/(dy/dx)$ times as fast as y, and consequently we must have $dx/dy = 1/(dy/dx)$. This is hardly a proof, but the idea is right.

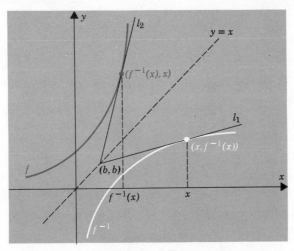

FIGURE 3.7.1

Exercises

Find dy/dx.

*1. $y = (x^3 + 1)^{1/2}$.   2. $y = (x + 1)^{1/3}$.

*3. $y = x\sqrt{x^2 + 1}$.   4. $y = x^2\sqrt{x^2 + 1}$.

*5. $y = \sqrt[4]{2x^2 + 1}$.   6. $y = (x + 1)^{1/3}(x + 2)^{2/3}$.

*7. $y = \sqrt{2 - x^2}\,\sqrt{3 - x^2}$.   8. $y = \sqrt{(x^4 - x + 1)^3}$.

Compute.

*9. $\dfrac{d}{dx}\left(\sqrt{x} + \dfrac{1}{\sqrt{x}}\right)$.   10. $\dfrac{d}{dx}\left(\sqrt{\dfrac{3x + 1}{2x + 5}}\right)$.   *11. $\dfrac{d}{dx}\left(\dfrac{x}{\sqrt{x^2 + 1}}\right)$.

*12. $\dfrac{d}{dx}\left(\dfrac{\sqrt{x^2 + 1}}{x}\right)$.   13. $\dfrac{d}{dx}\left(\sqrt[3]{x} + \dfrac{1}{\sqrt[3]{x}}\right)$.   *14. $\dfrac{d}{dx}\left(\sqrt{\dfrac{ax + b}{cx + d}}\right)$.

15. Show the general form of the graph of $f(x) = x^{1/n}$ (a) for n even. (b) for n odd.

Find the second derivative.

*16. $y = \sqrt{a^2 + x^2}$.   17. $y = \sqrt{a + bx}$.   *18. $y = x\sqrt{a^2 - x^2}$.

19. Our geometric argument for the validity of Formula 3.7.1 collapses if the tangent marked l_1 in Figure 3.7.1 does not intersect the line $y = x$. Fill in this loophole.

Compute.

*20. $\dfrac{d}{dx}\,[f(\sqrt{x} + 1)]$.   21. $\dfrac{d}{dx}\,[\sqrt{[f(x)]^2 + 1}]$.

*22. $\dfrac{d}{dx}\,[\sqrt{f(x^2 + 1)}]$.   23. $\dfrac{d}{dx}\,[f(\sqrt{x^2 + 1})]$.

Find a formula for $(f^{-1})'(x)$ given that f is one-to-one and its derivative satisfies the indicated equation.

*24. $f'(x) = f(x)$.   25. $f'(x) = 1 + [f(x)]^2$.   *26. $f'(x) = \sqrt{1 - [f(x)]^2}$.

27. In economics the *elasticity* of demand is given by the formula

$$\epsilon = \frac{P}{Q}\left|\frac{dQ}{dP}\right|,$$

where P is price and Q quantity. The demand is said to be

$$\begin{cases} \text{inelastic} & \text{where } \epsilon < 1 \\ \text{unitary} & \text{where } \epsilon = 1 \\ \text{elastic} & \text{where } \epsilon > 1 \end{cases}.$$

Describe the elasticity of each of the following demand curves. Keep in mind that Q and P must remain positive.

*(a) $Q = \dfrac{1}{P^2}$.   (b) $Q = \dfrac{1}{\sqrt{P}}$.   *(c) $Q = \dfrac{1}{P}$.

(d) $Q = 200 - \tfrac{1}{5}P$.   *(e) $Q = 200 - 5P$.   (f) $Q = \sqrt{300 - P}$.

3.8 Additional Practice in Differentiation

Differentiate.

*1. $y = x^{2/3} - a^{2/3}$.

2. $y = 2x^{3/4} + 4x^{-1/4}$.

*3. $y = \dfrac{a + bx + cx^2}{x}$.

*4. $y = \dfrac{\sqrt{x}}{2} - \dfrac{2}{\sqrt{x}}$.

5. $y = \sqrt{ax} + \dfrac{a}{\sqrt{ax}}$.

*6. $s = \dfrac{a + bt + ct^2}{\sqrt{t}}$.

*7. $r = \sqrt{1 - 2\theta}$.

8. $f(t) = (2 - 3t^2)^3$.

*9. $f(x) = \dfrac{1}{\sqrt{a^2 - x^2}}$.

*10. $y = \left(a - \dfrac{b}{x}\right)^2$.

11. $y = \left(a + \dfrac{b}{x^2}\right)^3$.

*12. $y = x\sqrt{a + bx}$.

*13. $s = t\sqrt{a^2 + t^2}$.

14. $y = \dfrac{a - x}{a + x}$.

*15. $y = \dfrac{a^2 + x^2}{a^2 - x^2}$.

*16. $y = \dfrac{\sqrt{a^2 + x^2}}{x}$.

17. $y = \dfrac{2 - x}{1 + 2x^2}$.

*18. $y = \dfrac{x}{\sqrt{a^2 - x^2}}$.

*19. $y = \dfrac{x}{\sqrt{a - bx}}$.

20. $r = \theta^2\sqrt{3 - 4\theta}$.

*21. $y = \sqrt{\dfrac{1 - cx}{1 + cx}}$.

*22. $f(x) = x\sqrt[3]{2 + 3x}$.

23. $y = \sqrt{\dfrac{a^2 + x^2}{a^2 - x^2}}$.

*24. $s = \sqrt[3]{\dfrac{2 + 3t}{2 - 3t}}$.

*25. $r = \dfrac{\sqrt[3]{a + b\theta}}{\theta}$.

26. $s = \sqrt{2t - \dfrac{1}{t^2}}$.

*27. $y = \dfrac{b}{a}\sqrt{a^2 - x^2}$.

*28. $y = (a^{3/5} - x^{3/5})^{5/3}$.

29. $y = (a^{2/3} - x^{2/3})^{3/2}$.

*30. $y = (x + 2)^2\sqrt{x^2 + 2}$.

Evaluate dy/dx at the given x.

*31. $y = (x^2 - x)^3$, $x = 3$.

32. $y = (4 - x^2)^3$, $x = 3$.

*33. $y = \sqrt[3]{x} + \sqrt{x}$, $x = 64$.

34. $y = x\sqrt{3 + 2x}$, $x = 3$.

*35. $y = (2x)^{1/3} + (2x)^{2/3}$, $x = 4$.

36. $y = \sqrt{9 + 4x^2}$, $x = 2$.

*37. $y = \dfrac{1}{\sqrt{25 - x^2}}$, $x = 3$.

38. $y = \dfrac{x^2 + 2}{2 - x^2}$, $x = 2$.

*39. $y = \dfrac{\sqrt{16 + 3x}}{x}$, $x = 3$.

40. $y = \dfrac{\sqrt{5 - 2x}}{2x + 1}$, $x = \frac{1}{2}$.

*41. $y = x\sqrt{8 - x^2}$, $x = 2$.

42. $y = \sqrt{\dfrac{4x + 1}{5x - 1}}$, $x = 2$.

*43. $y = x^2\sqrt{1 + x^3}$, $x = 2$.

44. $y = \sqrt{\dfrac{x^2 - 5}{10 - x^2}}$, $x = 3$.

3.9 Tangent Lines and Normal Lines

We begin with a differentiable function f and choose a point (x_0, y_0) on the graph. The tangent line through this point has slope $f'(x_0)$. To get an equation for this tangent line, we use the point-slope formula

$$y - y_0 = m(x - x_0).$$

In this case $m = f'(x_0)$ and the equation becomes

(3.9.1)
$$y - y_0 = f'(x_0)(x - x_0).$$

The line through (x_0, y_0) which is perpendicular to the tangent is called the *normal line*. Since the slope of the tangent is $f'(x_0)$, the slope of the normal is $-1/f'(x_0)$, provided of course that $f'(x_0) \neq 0$. (Remember: for nonvertical perpendicular lines, $m_1 m_2 = -1$.) As an equation for this normal line we have

(3.9.2)
$$y - y_0 = -\frac{1}{f'(x_0)}(x - x_0).$$

Some tangents and normals are pictured in Figure 3.9.1.

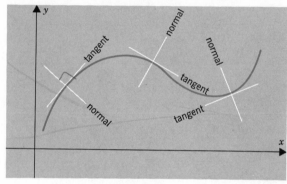

FIGURE 3.9.1

Problem.   Find equations for the tangent and normal to the graph of

$$f(x) = \frac{2x + 1}{3 - x}$$

at the point $(2, 5)$.

SOLUTION

$$f'(x) = \frac{(3 - x)2 - (2x + 1)(-1)}{(3 - x)^2} = \frac{7}{(3 - x)^2}.$$

At the point $(2, 5)$ the slope is $f'(2) = 7$. As an equation for the tangent we have

$$y - 5 = 7(x - 2).$$

The equation for the normal can be written

$$y - 5 = -\tfrac{1}{7}(x - 2). \quad \square$$

Problem.   Find an equation for the tangent to the circle

$$x^2 + y^2 = 25$$

at the point $(3, 4)$ and show that the normal line passes through the origin.

SOLUTION. The point (3, 4) lies on the upper semicircle

$$y = \sqrt{25 - x^2}.$$

Here

$$\frac{dy}{dx} = \frac{d}{dx}[(25 - x^2)^{1/2}] = \tfrac{1}{2}(25 - x^2)^{-1/2}(-2x) = -\frac{x}{\sqrt{25 - x^2}}.$$

At the point (3, 4) the tangent has slope

$$-\frac{3}{\sqrt{25 - 9}} = -\frac{3}{4}.$$

As an equation for the tangent we have

$$y - 4 = -\tfrac{3}{4}(x - 3).$$

The equation for the normal can be written

$$y - 4 = \tfrac{4}{3}(x - 3).$$

Since this equation is satisfied by setting $x = 0$ and $y = 0$, the normal line does pass through the origin.† □

If $f'(x_0) = 0$, equation 3.9.2 does not apply. If $f'(x_0) = 0$, the tangent line at (x_0, y_0) is horizontal (equation $y = y_0$) and the normal line is vertical (equation $x = x_0$).

In Figure 3.9.2 you can see several instances of a horizontal tangent. In each case the point of tangency is the origin, the tangent line is the x-axis ($y = 0$), and the normal line is the y-axis ($x = 0$).

FIGURE 3.9.2

A Note on Vertical Tangents

It's also possible for the graph of a function to have a vertical tangent. This cannot occur at a point of differentiability but only at a point where the derivative becomes infinite. Take, for example, the cube root function $f(x) = x^{1/3}$. The deriva-

---

† More generally, for any point on any circle the normal line passes through the center of the circle.

tive becomes infinite at 0:

$$\lim_{h \to 0} \frac{f(0 + h) - f(0)}{h} = \lim_{h \to 0} \frac{h^{1/3} - 0}{h} = \lim_{h \to 0} \frac{1}{h^{2/3}} \quad \text{is infinite.}$$

At the origin the function has a vertical tangent ($x = 0$) and a horizontal normal ($y = 0$). See Figure 3.9.3.

Vertical tangents are mentioned here only in passing. They are discussed in Section 11.5.

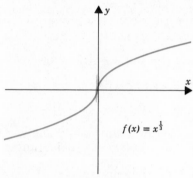

FIGURE 3.9.3

Exercises

1. Find equations for the tangent and normal at the given point.

   *(a) $f(x) = x^3 - x$; $(-2, -6)$.   (b) $f(x) = x + \dfrac{1}{x}$; $(5, \frac{26}{5})$.

   *(c) $f(x) = \dfrac{1}{x - 1}$; $(\frac{1}{2}, -2)$.   (d) $f(x) = \dfrac{x}{1 + x^2}$; $(1, \frac{1}{2})$.

2. Find the points where the tangent is parallel to the line $y = x$.
   *(a) $y = \sqrt{x}$.   (b) $y = (x - 1)^3$.   *(c) $y = x(5 - x)$.

3. Find the points where the tangent to the curve
   *(a) $y = -x^2 - 6$ is parallel to the line $y = 4x - 1$.
   (b) $xy = 1$ is perpendicular to the line $y = x$.
   *(c) $x + y^2 = 1$ is parallel to the line $x + 2y = 0$.
   (d) $x + y^2 = 1$ is perpendicular to the line $x + 2y = 0$.

4. Find the points on the curve $y = \frac{2}{3}x^{3/2}$ where the inclination of the tangent line is *(a) 45°. (b) 60°. *(c) 30°.

5. Find an equation for the tangent to the curve $y = x^3$ which passes through the point $(0, 2)$.

6. Find an equation for the line tangent to the graph of f^{-1} at the point (b, a) given that the slope of the graph of f at (a, b) is m.

*7. Find the area of the triangle formed by the x-axis and the lines tangent and normal to the curve $y = 6x - x^2$ at the point $(5, 5)$.

8. Find the area of the triangle formed by the y-axis and the lines tangent and normal to the curve $y^2 = 9 - x$ at the point $(5, 2)$.

*9. Find an equation for the normal to the parabola $y = 5x + x^2$ which makes an angle of 45° with the x-axis.

10. Find equations for the tangents to the circle $x^2 + y^2 = 58$ which are parallel to the line $3x - 7y - 19 = 0$.

*11. Determine the coefficients A, B, C so that the curve $y = Ax^2 + Bx + C$ will pass through the point $(1, 3)$ and be tangent to the line $4x + y = 8$ at the point $(2, 0)$.

12. Determine A, B, C, D so that the curve $y = Ax^3 + Bx^2 + Cx + D$ will be tangent to the line $y = 3x - 3$ at the point $(1, 0)$ and tangent to the line $y = 18x - 27$ at the point $(2, 9)$.

*13. Isocost lines and indifference curves are studied in economics. For what value of C is the isocost line

$$Ax + By = 1 \qquad (A > 0, B > 0)$$

tangent to the indifference curve

$$y = \frac{1}{x} + C, \qquad x > 0?$$

Find the point of tangency (called the equilibrium point).

14. Let

$$P = f(Q) \qquad (P = \text{price}, \quad Q = \text{output})$$

be the supply function of Figure 3.9.4. Compare angles θ and ϕ at points where the elasticity

$$\epsilon = \frac{f(Q)}{Q|f'(Q)|}$$

(a) is 1.   (b) is less than 1.   (c) is greater than 1.

FIGURE 3.9.4

3.10   Implicit Differentiation; The Angle between Two Curves

Suppose you know that y is a function of x which satisfies the equation

$$3x^3y - 4y - 2x + 1 = 0.$$

One way to find dy/dx is to solve first for y:

$$(3x^3 - 4)y - 2x + 1 = 0,$$

$$(3x^3 - 4)y = 2x - 1,$$

$$y = \frac{2x - 1}{3x^3 - 4},$$

and then differentiate:

$$\frac{dy}{dx} = \frac{(3x^3 - 4)2 - (2x - 1)(9x^2)}{(3x^3 - 4)^2}$$

$$= \frac{6x^3 - 8 - 18x^3 + 9x^2}{(3x^3 - 4)^2}$$

(1)
$$= -\frac{12x^3 - 9x^2 + 8}{(3x^3 - 4)^2}.$$

It is also possible to find dy/dx without first solving the equation for y. The technique is called *implicit differentiation*.

Let's return to the equation

$$3x^3y - 4y - 2x + 1 = 0.$$

Differentiating both sides of this equation (remembering that y is a function of x), we have

$$\left(3x^3 \frac{dy}{dx} + 9x^2y\right) - 4\frac{dy}{dx} - 2 = 0,$$

$$(3x^3 - 4)\frac{dy}{dx} = 2 - 9x^2y,$$

(2)
$$\frac{dy}{dx} = \frac{2 - 9x^2y}{3x^3 - 4}.$$

The answer looks different from what we obtained before because this time y appears on the right-hand side. This is generally no disadvantage. To satisfy yourself that the two answers are really the same all you have to do is substitute

$$y = \frac{2x - 1}{3x^3 - 4}$$

in (2). This gives

$$\frac{dy}{dx} = \frac{2 - 9x^2\left(\dfrac{2x - 1}{3x^3 - 4}\right)}{3x^3 - 4} = \frac{6x^3 - 8 - 18x^3 + 9x^2}{(3x^3 - 4)^2} = -\frac{12x^3 - 9x^2 + 8}{(3x^3 - 4)^2}.$$

This is the answer (1) we obtained before. □

Implicit differentiation is particularly useful where it is inconvenient (or impossible) to first solve the given equation for y.

Problem.   Find dy/dx given that

$$4x^2 + 2xy - xy^3 = 0.$$

SOLUTION.   The relation

$$4x^2 + 2xy - xy^3 = 0$$

gives

$$\frac{d}{dx}(4x^2 + 2xy - xy^3) = 0,$$

$$8x + \left(2x\frac{dy}{dx} + 2y\right) - \left(x3y^2\frac{dy}{dx} + y^3\right) = 0,$$

$$8x + 2y - y^3 + x(2 - 3y^2)\frac{dy}{dx} = 0,$$

$$\frac{dy}{dx} = \frac{y^3 - 2y - 8x}{x(2 - 3y^2)}. \quad \square$$

Problem.   Find the slope of the curve

$$x^3 - 3xy^2 + y^3 = 1 \qquad \text{at } (2, -1).$$

SOLUTION.   Differentiation gives

$$3x^2 - 3x\left(2y\frac{dy}{dx}\right) - 3y^2 + 3y^2\frac{dy}{dx} = 0,$$

$$3x^2 - 6xy\frac{dy}{dx} - 3y^2 + 3y^2\frac{dy}{dx} = 0.$$

At $x = 2$ and $y = -1$, the equation becomes

$$12 + 12\frac{dy}{dx} - 3 + 3\frac{dy}{dx} = 0$$

$$15\frac{dy}{dx} = -9$$

$$\frac{dy}{dx} = -\frac{3}{5}.$$

The slope is $-\frac{3}{5}$. $\square$

The angle between two curves is the angle between their tangents at the point of intersection. If the slopes are m_1 and m_2, then the angle of intersection can be obtained from the formula

$$\tan \theta = \frac{m_1 - m_2}{1 + m_1 m_2}.$$

Problem.   Find the angles at which the circles

$$C_1: (x - 1)^2 + y^2 = 10 \quad \text{and} \quad C_2: x^2 + (y - 2)^2 = 5$$

intersect.

SOLUTION.   Solve the two equations simultaneously, and you'll find that $(2, 3)$

and $(-2, 1)$ are the points of intersection. Differentiating the first equation, we have

$$2(x - 1) + 2y \frac{dy}{dx} = 0 \quad \text{and therefore} \quad m_1 = \frac{dy}{dx} = \frac{1 - x}{y}.$$

Differentiating the second equation, we have

$$2x + 2(y - 2) \frac{dy}{dx} = 0 \quad \text{and therefore} \quad m_2 = \frac{dy}{dx} = \frac{x}{2 - y}.$$

At $(2, 3)$,

$$m_1 = \frac{1 - 2}{3} = \frac{1}{3}, \qquad m_2 = \frac{2}{2 - 3} = -2,$$

so that

$$\tan \theta = \frac{m_1 - m_2}{1 + m_1 m_2} = \frac{-\frac{1}{3} + 2}{1 + \frac{2}{3}} = 1 \quad \text{and} \quad \theta = \tfrac{1}{4}\pi \text{ radians } (45°).$$

By symmetry (the configuration is symmetric about the line that joins the centers of the two circles) the angle of intersection at $(-2, 1)$ is also $\tfrac{1}{4}\pi$ radians (45°).   □

We can also find higher derivatives by implicit differentiation.

Problem. Find d^2y/dx^2 given that

$$b^2x^2 - a^2y^2 = a^2b^2.$$

SOLUTION. Differentiation with respect to x gives

$$2b^2x - 2a^2y \frac{dy}{dx} = 0$$

and therefore

(1) $$b^2x - a^2y \frac{dy}{dx} = 0.$$

Differentiating again we have

(2) $$b^2 - a^2y \frac{d^2y}{dx^2} - a^2 \left(\frac{dy}{dx}\right)^2 = 0.$$

From (1) we know that

$$\frac{dy}{dx} = \frac{b^2x}{a^2y}.$$

Substituting this in (2) we have

$$b^2 - a^2y \frac{d^2y}{dx^2} - a^2 \left(\frac{b^2x}{a^2y}\right)^2 = 0$$

which, as you can check, gives

$$\frac{d^2y}{dx^2} = \frac{b^2(a^2y^2 - b^2x^2)}{a^4y^3}.$$

We could leave the answer as it stands, but, since we know that

$$b^2x^2 - a^2y^2 = a^2b^2, \qquad \text{(our initial equation)}$$

we can write our answer more neatly as

$$\frac{d^2y}{dx^2} = -\frac{b^4}{a^2y^3}. \quad \square$$

Exercises

Find dy/dx in terms of x and y by implicit differentiation.

*1. $x^2 + y^2 = r^2$.
2. $x^3 + y^3 - 3axy = 0$.
*3. $b^2x^2 + a^2y^2 = a^2b^2$.
4. $\sqrt{x} + \sqrt{y} = \sqrt{a}$.
*5. $x^{2/3} + y^{2/3} = a^{2/3}$.
6. $y^2 = 4cx$.
*7. $x^4 + 4x^3y + y^4 = 1$.
8. $(2y)^{1/2} + (3y)^{1/3} = x$.
*9. $x + 2xy + y = 1$.
10. $x^2 + axy + y^2 = b^2$.

Find equations for the tangent and normal at the indicated point.

*11. $2x + 3y = 5$; $(-2, 3)$.
12. $9x^2 + 4y^2 = 72$; $(2, 3)$.
*13. $x^2 + xy + 2y^2 = 28$; $(-2, -3)$.
14. $x^3 - axy + 3ay^2 = 3a^3$; (a, a).

*15. At what angles do the parabolas $y^2 = 2px + p^2$ and $y^2 = p^2 - 2px$ intersect?
16. At what angle does the line $y = 2x$ cut the curve $x^2 - xy + 2y^2 = 28$?
17. Show that the hyperbola $x^2 - y^2 = 5$ and the ellipse $4x^2 + 9y^2 = 72$ intersect at right angles.

Find d^2y/dx^2 in terms of x and y.

*18. $y^2 + 2xy = 16$.
19. $ax^2 + 2xy + by^2 = 1$.

Express d^2y/dx^2 in terms of y alone.

20. $x^2 + y^2 = r^2$.
*21. $y^2 = 4ax$.
22. $\sqrt{x} + \sqrt{y} = 1$.
*23. $b^2x^2 + a^2y^2 = a^2b^2$.

Evaluate dy/dx and d^2y/dx^2 at the given point.

*24. $x^2 - 4y^2 = 9$; $(5, 2)$.
25. $x^2 + 4xy + y^2 + 3 = 0$; $(2, -1)$.

*26. Find equations for the tangents to the ellipse $4x^2 + y^2 = 72$ that pass through the point $(4, 4)$.
27. Find equations for the normals to the hyperbola $4x^2 - y^2 = 36$ that are parallel to the line $2x + 5y - 4 = 0$.

$$\frac{x^2}{b^2} + \frac{y^2}{A^2} = 1$$

The Mean-Value Theorem and Applications

4.1 The Mean-Value Theorem

The main object of this section is to state and prove a result which is known as *the mean-value theorem*. At first glance the theorem may appear innocuous. It is certainly easy to understand. Its remarkable feature is its immense usefulness.

Theorem 4.1.1   The Mean-Value Theorem†

If f is differentiable on (a, b) and continuous on $[a, b]$, then there is at least one number c in (a, b) at which

$$f'(c) = \frac{f(b) - f(a)}{b - a}.$$

The number

$$\frac{f(b) - f(a)}{b - a}$$

is the slope of the line l which passes through the points $(a, f(a))$ and $(b, f(b))$. To say that there is at least one number c at which

$$f'(c) = \frac{f(b) - f(a)}{b - a}$$

is to say that the graph of f has at least one point $(c, f(c))$ at which the tangent is parallel to l. See Figure 4.1.1.

---

† This result was first stated by Joseph Louis Lagrange (1736–1813) who, born in Turin, spent twenty years at the court of Frederick the Great.

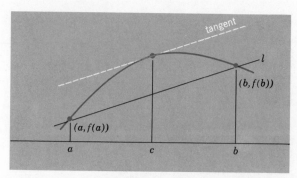

FIGURE 4.1.1

To prove the mean-value theorem we need first some other results.

Theorem 4.1.2

Let f be differentiable at x_0. If $f'(x_0) > 0$, then

$$f(x_0 - h) < f(x_0) < f(x_0 + h)$$

for all positive h sufficiently small. If $f'(x_0) < 0$, then

$$f(x_0 + h) < f(x_0) < f(x_0 - h)$$

for all positive h sufficiently small.

PROOF.   We take the case $f'(x_0) > 0$ and leave the other case to you. By the definition of derivative,

$$\lim_{k \to 0} \frac{f(x_0 + k) - f(x_0)}{k} = f'(x_0).$$

With $f'(x_0) > 0$ we can use $f'(x_0)$ itself as ϵ and conclude that there exists $\delta > 0$ such that

$$\text{if} \quad 0 < |k| < \delta, \quad \text{then} \quad \left| \frac{f(x_0 + k) - f(x_0)}{k} - f'(x_0) \right| < f'(x_0).$$

For such k we have

$$\frac{f(x_0 + k) - f(x_0)}{k} > 0. \tag{why?}$$

If now $0 < h < \delta$, then

$$\frac{f(x_0 + h) - f(x_0)}{h} > 0 \quad \text{and} \quad \frac{f(x_0 - h) - f(x_0)}{-h} > 0,$$

and consequently

$$f(x_0) < f(x_0 + h) \quad \text{and} \quad f(x_0 - h) < f(x_0). \quad \square$$

Next we prove a special case of the mean-value theorem, known as Rolle's theorem (after the French mathematician Michel Rolle who first announced the result in 1691). In Rolle's theorem we make the additional assumption that $f(a)$ and $f(b)$ are both 0. (See Figure 4.1.2.) In this case the line joining $(a, f(a))$ to $(b, f(b))$ is horizontal. (It is the x-axis.) The conclusion is that there is a point $(c, f(c))$ at which the tangent is horizontal.

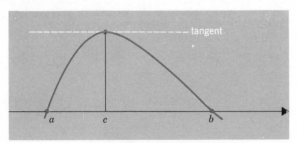

FIGURE 4.1.2

Theorem 4.1.3   Rolle's Theorem

Let f be differentiable on (a, b) and continuous on $[a, b]$. If $f(a)$ and $f(b)$ are both 0, then there is at least one number c in (a, b) at which

$$f'(c) = 0.$$

PROOF.   If f is constantly 0 on $[a, b]$, the result is obvious. If f is not constantly 0 on $[a, b]$, then either f takes on some positive values or some negative ones. We assume the former and leave the other case to you.

Since f is continuous on $[a, b]$, at some point c of $[a, b]$, f must take on a maximum value. (Theorem 2.6.6.) This maximum value, $f(c)$, must be positive. Since $f(a)$ and $f(b)$ are both 0, c cannot be a and it cannot be b. This means that c must lie in the open interval (a, b) and therefore $f'(c)$ exists. Now $f'(c)$ cannot be greater than 0 and it cannot be less than 0 because each of these conditions would imply that f takes on values greater than $f(c)$. (This follows from Theorem 4.1.2.) We conclude therefore that $f'(c) = 0$.   □

We are now ready to give a proof of the mean-value theorem. To do this, we create a function g which satisfies the conditions of Rolle's theorem and is so related to f that the conclusion $g'(c) = 0$ leads to the conclusion

$$f'(c) = \frac{f(b) - f(a)}{b - a}.$$

It is not hard to see that

$$g(x) = f(x) - \left[\frac{f(b) - f(a)}{b - a}(x - a) + f(a)\right]$$

is exactly such a function. Geometrically $g(x)$ is represented in Figure 4.1.3. The line that passes through $(a, f(a))$ and $(b, f(b))$ has the equation

$$y = \frac{f(b) - f(a)}{b - a}(x - a) + f(a).$$

[This is not hard to verify. The slope is right, and, when $x = a, y = f(a)$.] The difference

$$g(x) = f(x) - \left[\frac{f(b) - f(a)}{b - a}(x - a) + f(a)\right]$$

is simply the vertical separation between the graph of f and the line in question.

FIGURE 4.1.3

If f is differentiable on (a, b) and continuous on $[a, b]$, then so is g. As you can check, $g(a)$ and $g(b)$ are both 0. Therefore by Rolle's theorem there is at least one point c in (a, b) at which $g'(c) = 0$. Since in general

$$g'(x) = f'(x) - \frac{f(b) - f(a)}{b - a},$$

in particular

$$g'(c) = f'(c) - \frac{f(b) - f(a)}{b - a}.$$

With $g'(c) = 0$, we have

$$f'(c) = \frac{f(b) - f(a)}{b - a}. \quad \square$$

Exercises

After checking that the function satisfies the conditions of the mean-value theorem on the indicated interval, find the admissible values of c.

*1. $f(x) = x^2$; $[1, 2]$.   2. $f(x) = 3\sqrt{x} - 4x$; $[1, 4]$.
*3. $f(x) = x^3$; $[1, 3]$.   4. $f(x) = x^{2/3}$; $[1, 8]$.
*5. $f(x) = \sqrt{1 - x^2}$; $[0, 1]$.   6. $f(x) = x^3 - 3x$; $[-1, 1]$.

*7. Determine whether the function $f(x) = (1 - x^2)^{1/2}(3 + x^2)^{-1}$ satisfies the conditions of Rolle's theorem on the interval $[-1, 1]$. If so, find the admissible values of c.

8. Set $f(x) = x^{-1}$, $a = -1$, $b = 1$. Verify that there is no number c such that

$$f'(c) = \frac{f(b) - f(a)}{b - a}.$$

Explain how this does not violate the mean-value theorem.

*9. Graph the function $f(x) = |2x - 1| - 3$ and compute its derivative. Verify that $f(-1) = 0 = f(2)$ and yet $f'(x)$ is never 0. Explain how this does not violate Rolle's theorem.

10. Sketch the graph of

$$f(x) = \begin{cases} 2 + x^3, & x \leq 1 \\ 3x, & x > 1 \end{cases}$$

and compute its derivative. Determine whether f satisfies the conditions of the mean-value theorem on the interval $[-1, 2]$ and, if so, find all admissible values of c.

11. Let P be a nonconstant polynomial

$$P(x) = a_n x^n + \cdots + a_1 x + a_0.$$

Show that between any two consecutive roots of the equation $P'(x) = 0$ there is at most one root of the equation $P(x) = 0$.

*12. Show that the equation $6x^4 - 7x + 1 = 0$ does not have more than two real roots. (Use Rolle's theorem.)

13. Show that the equation $6x^5 + 13x + 1 = 0$ has exactly one real root. (Use Rolle's theorem and the intermediate-value theorem.)

*14. Show that the equation $x^3 + 9x^2 + 33x - 8 = 0$ has exactly one real root.

15. Let f be twice differentiable. Show that, if the equation $f(x) = 0$ has n real roots, then the equation $f'(x) = 0$ has at least $n - 1$ real roots and the equation $f''(x) = 0$ has at least $n - 2$ real roots.

16. Show that the equation $x^3 + ax + b = 0$ has exactly one real root if $a \geq 0$ and at most one real root between $-\frac{1}{3}\sqrt{3}|a|$ and $\frac{1}{3}\sqrt{3}|a|$ if $a < 0$.

17. Given that $|f'(x)| \leq 1$ for all real numbers x, show that

$$|f(x_1) - f(x_2)| \leq |x_1 - x_2| \qquad \text{for all real numbers } x_1 \text{ and } x_2.$$

4.2 Increasing and Decreasing Functions

It is often useful to know on what intervals a given function is "increasing" and on what intervals it is "decreasing." If the function in question is differentiable, we can answer such questions by looking at the sign of the derivative. As we shall see (and as one would expect from the interpretation of the derivative as the slope of the graph), a function is

(a) "increasing" on any interval on which its derivative is positive,
(b) "decreasing" on any interval on which its derivative is negative, and
(c) constant on any interval on which its derivative is identically 0.

Before attempting to prove these results, we will explain exactly what we mean by a function "increasing" or "decreasing" on an interval.

Definition 4.2.1

A function f is said to be

(i) *increasing* on the interval I iff for every two numbers x_1, x_2 in I

$$x_1 < x_2 \quad \text{implies} \quad f(x_1) < f(x_2).$$

(ii) *decreasing* on the interval I iff for every two numbers x_1, x_2 in I

$$x_1 < x_2 \quad \text{implies} \quad f(x_1) > f(x_2).$$

Preliminary Examples

1. The squaring function

$$f(x) = x^2 \qquad \text{(Figure 4.2.1)}$$

is decreasing on $(-\infty, 0]$ and increasing on $[0, \infty)$.

2. The function

$$f(x) = \begin{cases} 1, & x < 0 \\ x, & x \geq 0 \end{cases}, \qquad \text{(Figure 4.2.2)}$$

is constant on $(-\infty, 0)$ and increasing on $[0, \infty)$.

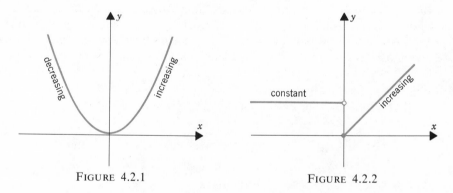

FIGURE 4.2.1   FIGURE 4.2.2

3. The cubing function

$$f(x) = x^3 \qquad \text{(Figure 4.2.3)}$$

is everywhere increasing.

4. In the case of the Dirichlet function

$$g(x) = \begin{cases} 1, & x \text{ rational} \\ 0, & x \text{ irrational} \end{cases}, \qquad \text{(Figure 4.2.4)}$$

there is no interval on which the function increases and no interval on which the function decreases. On every interval it jumps back and forth between 0 and 1 an infinite number of times. □

FIGURE 4.2.3   FIGURE 4.2.4

Now we start proving some theorems. As you will see, all of them are consequences of the mean-value theorem.

Theorem 4.2.2

Let f be differentiable on the open interval I.

(i) If $f'(x) > 0$ for all x in I, then f is increasing on I.
(ii) If $f'(x) < 0$ for all x in I, then f is decreasing on I.
(iii) If $f'(x) = 0$ for all x in I, then f is constant on I.

PROOF.   We take x_1 and x_2 in I with $x_1 < x_2$. Since f is differentiable on (x_1, x_2) and continuous on $[x_1, x_2]$, we see from the mean-value theorem that there is a number c in (x_1, x_2) such that

$$f'(c) = \frac{f(x_2) - f(x_1)}{x_2 - x_1}.$$

In (i), we have

$$\frac{f(x_2) - f(x_1)}{x_2 - x_1} > 0 \quad \text{and thus} \quad f(x_1) < f(x_2).$$

In (ii), we have

$$\frac{f(x_2) - f(x_1)}{x_2 - x_1} < 0 \quad \text{and thus } f(x_1) > f(x_2).$$

In (iii), we have

$$\frac{f(x_2) - f(x_1)}{x_2 - x_1} = 0 \quad \text{and thus} \quad f(x_1) = f(x_2). \quad \square$$

The theorem we just proved is useful, but it has some deficiencies. For example, in the case of the squaring function

$$f(x) = x^2$$

the derivative

$$f'(x) = 2x$$

is negative for x in $(-\infty, 0)$, zero at $x = 0$, and positive for x in $(0, \infty)$. Theorem 4.2.2 assures us that

f decreases on $(-\infty, 0)$ and increases on $(0, \infty)$,

but actually stronger results are true:

f decreases on $(-\infty, 0]$ and increases on $[0, \infty)$.

To get these stronger results we need a theorem that works for closed intervals.
   To extend Theorem 4.2.2 so that it works for a closed interval I, the only thing we need is continuity at the endpoint(s).

Theorem 4.2.3

Let f be continuous on a closed interval I and differentiable on the interior of I.

  (i) If $f'(x) > 0$ for all x in the interior of I, then f is increasing on all of I.
 (ii) If $f'(x) < 0$ for all x in the interior of I, then f is decreasing on all of I.
(iii) If $f'(x) = 0$ for all x in the interior of I, then f is constant on all of I.

A proof of this result is not hard to construct. As you can verify, a word-for-word copy of our proof of Theorem 4.2.2 also works here.   □
   It is time to give more examples.

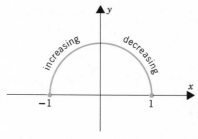

FIGURE 4.2.5

Example.   The function

$$f(x) = \sqrt{1 - x^2}$$

has as its graph the upper semicircle of Figure 4.2.5. It is therefore obvious that f increases on $[-1, 0]$ and decreases on $[0, 1]$. With Theorem 4.2.3 we can get this information without reference to the figure.
   That f increases on $[-1, 0]$ can be seen from the fact that the derivative

$$f'(x) = -\frac{x}{\sqrt{1-x^2}}$$

is positive for x in $(-1, 0)$ and f itself is continuous on $[-1, 0]$. That f decreases on $[0, 1]$ can be seen from the fact that $f'(x)$ is negative for x in $(0, 1)$ and f is continuous on $[0, 1]$. □

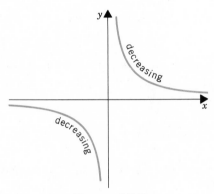

FIGURE 4.2.6

Example.   The function

$$f(x) = \frac{1}{x}$$

has derivative

$$f'(x) = -\frac{1}{x^2}.$$

Since $f'(x) < 0$ for all $x \neq 0$ and f is not defined at 0, f decreases both on $(-\infty, 0)$ and on $(0, \infty)$. (Figure 4.2.6) As x approaches 0, the graph becomes steeper; it flattens out as x tends away from 0. □

Example.   If

$$g(x) = 4x^5 - 15x^4 - 20x^3 + 110x^2 - 120x + 40,$$

then g is everywhere differentiable and

$$\begin{aligned}
g'(x) &= 20x^4 - 60x^3 - 60x^2 + 220x - 120 \\
&= 20(x^4 - 3x^3 - 3x^2 + 11x - 6) \\
&= 20(x + 2)(x - 1)^2(x - 3).
\end{aligned}$$

The derivative g' takes on the value 0 at -2, at 1, and at 3. As you can verify by the methods of Section 1.4,

$$g'(x) \text{ is } \left\{\begin{array}{llll} \text{positive,} & \text{for} & x \in (-\infty, -2) \\ \text{negative,} & \text{for} & x \in (-2, 1) \\ \text{negative,} & \text{for} & x \in (1, 3) \\ \text{positive,} & \text{for} & x \in (3, \infty) \end{array}\right].$$

Since g' is positive on $(-\infty, -2)$, we can conclude that g increases on $(-\infty, -2]$. Since g' is negative on $(-2, 1)$ and on $(1, 3)$, we can conclude that g decreases both on $[-2, 1]$ and on $[1, 3]$. This tells us that g actually decreases on $[-2, 3]$. (Explain.) Finally, since g' is positive on $(3, \infty)$, g must increase on $[3, \infty)$. $\square$

If two differentiable functions differ by a constant,

$$f(x) = g(x) + C,$$

then their derivatives are obviously equal:

$$f'(x) = g'(x).$$

The converse is also true. In fact, we have the following theorem.

Theorem 4.2.4

I. If $f'(x) = g'(x)$ for all x in an open interval I, then f and g differ by a constant on I.

II. If $f'(x) = g'(x)$ for all x in the interior of a closed interval I and f and g are continuous at the endpoint(s), then f and g differ by a constant on I.

PROOF.   Set $h = f - g$. For the first assertion apply (iii) of Theorem 4.2.2 to h. For the second assertion apply (iii) of Theorem 4.2.3 to h. We leave the details as an exercise. $\square$

For an illustration of this theorem see Figure 4.2.7. At points with the same first coordinate the slopes are equal, and thus the curves have the same steepness. The separation between the curves remains constant.

FIGURE 4.2.7

From what you have learned about differentiating rational powers you can see that for rational $n \neq -1$

(4.2.5)
$$\boxed{\frac{d}{dx}\left(\frac{1}{n + 1} x^{n+1}\right) = x^n.}$$

Thus, for example,

$$\frac{d}{dx}(x) = x^0 = 1, \quad \frac{d}{dx}(\tfrac{1}{2}x^2) = x, \quad \frac{d}{dx}(\tfrac{1}{3}x^3) = x^2, \quad \text{etc.}$$

$$\frac{d}{dx}(-x^{-1}) = x^{-2}, \quad \frac{d}{dx}(-\tfrac{1}{2}x^{-2}) = x^{-3}, \quad \frac{d}{dx}(-\tfrac{1}{3}x^{-3}) = x^{-4}, \quad \text{etc.}$$

Problem.   Find a general expression for $f(x)$ given that

$$f'(x) = 4x^2 + 5x \quad \text{for all real } x \quad \text{and} \quad f(0) = 1.$$

SOLUTION

$$\frac{d}{dx}(\tfrac{1}{3}x^3) = x^2 \qquad\qquad \frac{d}{dx}(\tfrac{1}{2}x^2) = x$$

$$\frac{d}{dx}(\tfrac{4}{3}x^3) = 4x^2 \qquad\qquad \frac{d}{dx}(\tfrac{5}{2}x^2) = 5x$$

$$\frac{d}{dx}(\tfrac{4}{3}x^3 + \tfrac{5}{2}x^2) = 4x^2 + 5x.$$

By Theorem 4.2.4 there exists a constant C such that

$$f(x) = \tfrac{4}{3}x^3 + \tfrac{5}{2}x^2 + C.$$

On the one hand

$$f(0) = C, \qquad\qquad\qquad\qquad \text{(set } x = 0\text{)}$$

and on the other

$$f(0) = 1. \qquad\qquad\qquad\qquad \text{(given)}$$

This forces $C = 1$ and

$$f(x) = \tfrac{4}{3}x^3 + \tfrac{5}{2}x^2 + 1. \quad \square$$

Problem.   Find a general expression for $f(x)$ on $(0, \infty)$ given that

$$f'(x) = 3x^4 - x^{-3} \quad \text{on } (0, \infty) \quad \text{and} \quad f(1) = \tfrac{1}{2}.$$

SOLUTION

$$\frac{d}{dx}(\tfrac{1}{5}x^5) = x^4 \qquad\qquad \frac{d}{dx}(-\tfrac{1}{2}x^{-2}) = x^{-3}$$

$$\frac{d}{dx}(\tfrac{3}{5}x^5) = 3x^4 \qquad\qquad \frac{d}{dx}(\tfrac{1}{2}x^{-2}) = -x^{-3}$$

$$\frac{d}{dx}(\tfrac{3}{5}x^5 + \tfrac{1}{2}x^{-2}) = 3x^4 - x^{-3}.$$

By Theorem 4.2.4 there exists a constant C such that

$$f(x) = \tfrac{3}{5}x^5 + \tfrac{1}{2}x^{-2} + C.$$

Since $f(1) = \tfrac{1}{2}$, we must have $\tfrac{1}{2} = \tfrac{3}{5} + \tfrac{1}{2} + C$. Thus $C = -\tfrac{3}{5}$ and

$$f(x) = \tfrac{3}{5}x^5 + \tfrac{1}{2}x^{-2} - \tfrac{3}{5}. \quad \square$$

Exercises

Find the intervals on which f is increasing and those on which it is decreasing, if $f(x)$ is as follows.

*1. $x^3 - 3x + 2$.

2. $x^3 - 3x^2 + 6$.

*3. $x + \dfrac{1}{x}$.

4. $x^3(1 + x)$.

*5. $x(x + 1)(x + 2)$.

6. $(x + 1)^4$.

*7. $\dfrac{1}{|x - 2|}$.

8. $\dfrac{x}{1 + x^2}$.

*9. $\dfrac{x^2 + 1}{x^2 - 1}$.

10. $\dfrac{x^2}{x^2 + 1}$.

*11. $|x^2 - 5|$.

12. $x^2(1 + x)^2$.

*13. $\dfrac{x - 1}{x + 1}$.

14. $x^2 + \dfrac{16}{x^2}$.

*15. $\left(\dfrac{1 - \sqrt{x}}{1 + \sqrt{x}}\right)^7$.

16. $\sqrt{\dfrac{2 + x}{1 + x}}$.

*17. $\sqrt{\dfrac{1 + x^2}{2 + x^2}}$.

18. $|x + 1|\,|x - 2|$.

Find a general expression for $f(x)$ given the following.

*19. $f'(x) = x^2 - 1$ for all real x and $f(0) = 1$.

20. $f'(x) = x^2 - 1$ for all real x and $f(1) = 2$.

*21. $f'(x) = 2x - 5$ for all real x and $f(2) = 4$.

22. $f'(x) = 5x^4 + 4x^3 + 3x^2 + 2x + 1$ for all real x and $f(0) = 5$.

*23. $f'(x) = 4x^{-3}$ for $x > 0$ and $f(1) = 0$.

24. $f'(x) = x^{1/3} - x^{1/2}$ for $x > 0$ and $f(0) = 1$.

*25. $f'(x) = x^{-5} - 5x^{-1/5}$ for $x > 0$ and $f(1) = 0$.

26. Given that $f'(x) > g'(x)$ for all real x and $f(0) = g(0)$, compare $f(x)$ with $g(x)$ on $(-\infty, 0)$ and on $(0, \infty)$. Justify your answers.

27. Prove Theorem 4.2.4.

4.3 Maxima and Minima

Closely related to the material just discussed is the problem of determining the maximum and minimum values of functions. There is a practical aspect to this matter. In many problems of economics, engineering, and physics it is important to find out how large or how small a certain quantity may become. If the problem admits a mathematical formulation, it is often reducible to the problem of finding the extreme values of some function.

In this section we consider functions which are defined on *an open interval* or on *the union of open intervals*. We begin with a definition.

(4.3.1)

A function f is said to have a *local* (or *relative*) *maximum* at c iff

$$f(c) \geq f(x) \qquad \text{for all } x \text{ sufficiently close to } c.$$

It is said to have a *local* (or *relative*) *minimum* at c iff

$$f(c) \leq f(x) \qquad \text{for all } x \text{ sufficiently close to } c.$$

We illustrate these notions in Figure 4.3.1 and Figure 4.3.2. A careful look at the figures suggests that local maxima and minima occur only at points where the tangent is horizontal [$f'(c) = 0$] or where there is no tangent line [$f'(c)$ does not exist]. This is indeed the case.

FIGURE 4.3.1

FIGURE 4.3.2

Theorem 4.3.2

If f has a local maximum or minimum at c, then

$$\text{either} \quad f'(c) = 0 \quad \text{or} \quad f'(c) \text{ does not exist.}$$

PROOF.   If $f'(c) > 0$ or $f'(c) < 0$, then, by Theorem 4.1.2, there must be numbers x_1 and x_2 which are arbitrarily close to c and yet satisfy

$$f(x_1) < f(c) < f(x_2).$$

This makes it impossible for a local maximum or minimum to occur at c.   □

In view of this result, when searching for the local maxima and minima of a function f, the only points we need to test are *those points c in the domain of f at which $f'(c) = 0$ or $f'(c)$ does not exist.* Such points are called *critical points.*

We illustrate the technique for finding local maxima and minima by some examples. In each example the first step will be to find the critical points. We begin with very simple cases.

Example. For

$$f(x) = 3 - x^2 \qquad\qquad \text{(Figure 4.3.3)}$$

the derivative

$$f'(x) = -2x$$

exists everywhere. Since $f'(x) = 0$ only at $x = 0$, 0 is the only critical point. The number $f(0) = 3$ is obviously a local maximum. □

Example. In the case of

$$f(x) = |x + 1| + 2 \qquad\qquad \text{(Figure 4.3.4)}$$

differentiation gives

$$f'(x) = \left\{ \begin{array}{ll} -1, & x < -1 \\ 1, & x > -1 \end{array} \right].$$

This derivative is never 0. It fails to exist only at -1. The number -1 is the only critical point. The value $f(-1) = 2$ is a local minimum. □

local maximum = 3

local minimum = 2

$f(x) = 3 - x^2$ $f(x) = |x + 1| + 2$

FIGURE 4.3.3   FIGURE 4.3.4

Example. In the case of

$$f(x) = \frac{1}{x - 1} \qquad\qquad \text{(Figure 4.3.5)}$$

the derivative

$$f'(x) = -\frac{1}{(x - 1)^2}$$

exists throughout the domain of f and is never 0. There are therefore no critical points and no local extreme values. □

$$f(x) = \frac{1}{x-1}$$

FIGURE 4.3.5

CAUTION.   The fact that c is a critical point of f does not guarantee that $f(c)$ is a local extreme value. This is illustrated in the next two examples.

Example.   In the case of the cubing function

$$f(x) = x^3 \qquad \text{(Figure 4.3.6)}$$

the derivative $f'(x) = 3x^2$ is 0 at 0, but $f(0) = 0$ is not a local extreme value. The cubing function is everywhere increasing.   □

Example.   The function

$$f(x) = \begin{cases} 2x, & x < 1 \\ \frac{1}{2}x + \frac{3}{2}, & x \geq 1 \end{cases}, \qquad \text{(Figure 4.3.7)}$$

is everywhere increasing. Although 1 is a critical point [$f'(1)$ does not exist], $f(1) = 2$ is not a local extreme value.   □

$f(x) = x^3$

FIGURE 4.3.6

$$f(x) = \begin{cases} 2x, x < 1 \\ \frac{1}{2}x + \frac{3}{2}, x \geq 1 \end{cases}$$

FIGURE 4.3.7

The following test is often useful in determining the behavior of a function at a critical point.

Test 1. (*The first-derivative test*)   Suppose that c is a critical point for f and that f is continuous at c. If there exists an interval $(c - \delta, c + \delta)$ such that $f'(x) > 0$ for all x in $(c - \delta, c)$ and $f'(x) < 0$ for all x in $(c, c + \delta)$, then $f(c)$ is a local maximum. If both inequalities are reversed, then $f(c)$ is a local minimum.

PROOF.   The proof is easy. In the first case f increases on $(c - \delta, c]$ and decreases on $[c, c + \delta)$. This makes $f(c)$ a local maximum. (Figures 4.3.8 and 4.3.9) In the second case f decreases on $(c - \delta, c]$ and increases on $[c, c + \delta)$. This makes $f(c)$ a local minimum. (Figures 4.3.10 and 4.3.11) $\square$

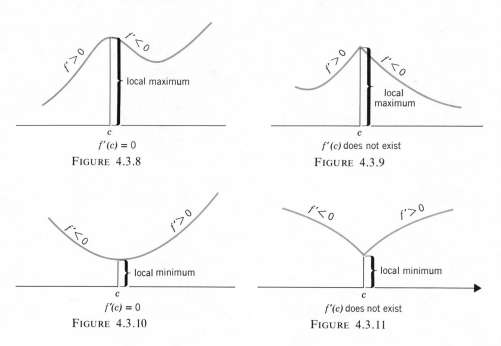

FIGURE 4.3.8   FIGURE 4.3.9

FIGURE 4.3.10   FIGURE 4.3.11

Example.   In the case of

$$f(x) = |x^2 - 1| \qquad \text{(Figure 4.3.12)}$$

the derivative is given by

$$f'(x) = \begin{cases} 2x, & x < -1 \\ -2x, & -1 < x < 1 \\ 2x, & x > 1 \end{cases} \qquad \text{(check this out)}$$

The critical points are $0, -1,$ and 1. At $0, f'(x) = 0$. At -1 and $1, f'(x)$ does not exist. To analyze the behavior of f at the critical points we use Test 1. Since

$$f'(x) \text{ is } \begin{cases} \text{negative,} & \text{for} & x < -1 \\ \text{positive,} & \text{for} & -1 < x < 0 \\ \text{negative,} & \text{for} & 0 < x < 1 \\ \text{positive,} & \text{for} & x > 1 \end{cases},$$

you can see that $f(-1) = 0$ is a local minimum, $f(0) = 1$ is a local maximum, and $f(1) = 0$ is a local minimum. □

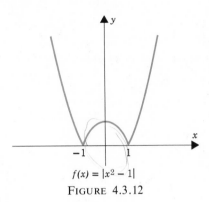

$$f(x) = |x^2 - 1|$$
FIGURE 4.3.12

Example.   Here we begin with

$$f(x) = (x - 2)(x - 1)^4.$$

Differentiating we obtain

$$f'(x) = (x - 2)4(x - 1)^3 + (x - 1)^4$$
$$= (x - 1)^3(4x - 8 + x - 1)$$
$$= (x - 1)^3(5x - 9).$$

The critical points are 1 and $\frac{9}{5}$. Since

$$f'(x) \text{ is } \begin{cases} \text{positive,} & \text{for} & x < 1 \\ \text{negative,} & \text{for} & 1 < x < \frac{9}{5} \\ \text{positive,} & \text{for} & x > \frac{9}{5} \end{cases},$$

we see that $f(1) = 0$ is a local maximum and $f(\frac{9}{5}) = -4^4/5^5$ is a local minimum. □

In certain cases it may be difficult to determine the sign of the first derivative both to the left of c and to the right of c. If f is twice differentiable at c, the following test may be easier to apply.

Test 2. (*The second-derivative test*)   Suppose that $f'(c) = 0$.
  If $f''(c) > 0$,   then $f(c)$ is a local minimum value.
  If $f''(c) < 0$,   then $f(c)$ is a local maximum value.

PROOF.   We handle the case $f''(c) > 0$. The other is left as an exercise. Since f'' is the derivative of f', we see from Theorem 4.1.2 that there exists $\delta > 0$ such that, if

$$c - \delta < x_1 < c < x_2 < c + \delta,$$

then

$$f'(x_1) < f'(c) < f'(x_2).$$

Since $f'(c) = 0$, we have

$$f'(x) < 0 \quad \text{for } x \text{ in } (c - \delta, c) \quad \text{and} \quad f'(x) > 0 \quad \text{for } x \text{ in } (c, c + \delta).$$

By Test 1 this shows that $f(c)$ is a local minimum. $\square$

Now we illustrate the second-derivative test.

Example.   For

$$f(x) = x^3 - x$$

we have

$$f'(x) = 3x^2 - 1.$$

The critical points are $-\frac{1}{3}\sqrt{3}$ and $\frac{1}{3}\sqrt{3}$. At each of these points the derivative is 0. As a second derivative we have

$$f''(x) = 6x.$$

Since $f''(-\frac{1}{3}\sqrt{3}) < 0$ and $f''(\frac{1}{3}\sqrt{3}) > 0$, we can conclude from the second-derivative test that $f(-\frac{1}{3}\sqrt{3}) = \frac{2}{9}\sqrt{3}$ is a local maximum and $f(\frac{1}{3}\sqrt{3}) = -\frac{2}{9}\sqrt{3}$ is a local minimum. $\square$

We can apply the first-derivative test even at points where the function is not differentiable, provided that it is continuous there. On the other hand, the second-derivative test can only be applied at points where the function is twice differentiable and then only if the second derivative is different from zero. This is a drawback. Take, for example, the functions

$$f(x) = x^{4/3} \quad \text{and} \quad g(x) = x^4.$$

In the first case we have $f'(x) = \frac{4}{3}x^{1/3}$ so that

$$f'(0) = 0, \quad f'(x) < 0 \quad \text{for } x < 0, \quad f'(x) > 0 \quad \text{for } x > 0.$$

By the first-derivative test, $f(0) = 0$ is a local minimum. We cannot get this information from the second-derivative test because $f''(x) = \frac{4}{9}x^{-2/3}$ does not exist at $x = 0$. In the case of $g(x) = x^4$, we have $g'(x) = 4x^3$ so that

$$g'(0) = 0, \quad g'(x) < 0 \quad \text{for } x < 0, \quad g'(x) > 0 \quad \text{for } x > 0.$$

By the first-derivative test, $g(0) = 0$ is a local minimum. But here again the second-derivative test is of no avail: $g''(x) = 12x^2$ is 0 at $x = 0$. $\square$

Exercises

Find the critical points and the local extreme values.

*1. $x^3 + 3x - 2$.   2. $2x^4 - 4x^2 + 6$.   *3. $x + \dfrac{1}{x}$.

4. $x^2(1 - x)$.   *5. $x(x + 1)(x + 2)$.   6. $(1 - x)^2(1 + x)$.

*7. $\dfrac{1}{|x-2|}$.

8. $\dfrac{1+x}{1-x}$.

*9. $\dfrac{2-3x}{2+x}$.

10. $\dfrac{2}{x(x+1)}$.

*11. $|x^2-16|$.

12. $x^3(1-x)^2$.

*13. $\left(\dfrac{x-2}{x+2}\right)^3$.

14. $(1-2x)(x-1)^3$.

*15. $(1-x)(1+x)^3$.

16. $\dfrac{x^2}{1+x}$.

*17. $\dfrac{|x|}{1+|x|}$.

18. $(3x-5)^3$.

*19. $|x-1||x+2|$.

20. $x\sqrt[3]{1-x}$.

*21. $-\dfrac{x^3}{x+1}$.

22. $\dfrac{1}{x+1}-\dfrac{1}{x-2}$.

*23. $\dfrac{1}{x+1}-\dfrac{1}{x+2}$.

24. $|x-3|+|2x+1|$.

25. Prove the validity of the second-derivative test in the case that $f''(c) < 0$.

Optional | 26. Find the critical points and the local extreme values of the polynomial $P(x) = x^4 - 8x^3 + 22x^2 - 24x + 4$. Then show that the equation $P(x) = 0$ has exactly two real roots, both positive.

4.4 More on Maxima and Minima

Endpoint Maxima and Minima

For continuous functions defined on an open interval or on the union of open intervals the critical points are those at which the derivative is 0 or the derivative does not exist. For continuous functions defined on a closed interval

$$[a, b], \quad [a, \infty), \quad \text{or} \quad (-\infty, b]$$

the *endpoints* of the domain (a and b in the case of $[a, b]$, a in the case of $[a, \infty)$, and b in the case of $(-\infty, b]$) are also called *critical points*.

Endpoints can give rise to what are called *endpoint maxima* and *endpoint minima*. See Figures 4.4.1, 4.4.2, and 4.4.3.

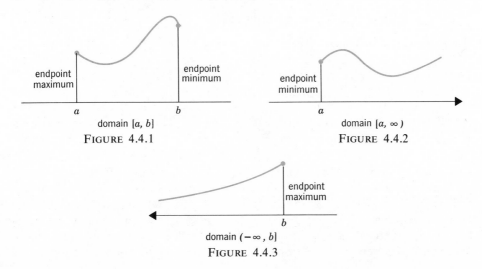

endpoint maximum

endpoint minimum

domain $[a, b]$

FIGURE 4.4.1

endpoint minimum

domain $[a, \infty)$

FIGURE 4.4.2

endpoint maximum

domain $(-\infty, b]$

FIGURE 4.4.3

The figures render the general idea so obvious that a formal definition seems unnecessary. Nevertheless, here it is.

Definition 4.4.1   Endpoint Extremes

If c is an endpoint of the domain of f, then f is said to have an *endpoint maximum* at c iff

$$f(c) \geq f(x) \qquad \text{for all } x \text{ in the domain of } f \text{ which are sufficiently close to } c.$$

It is said to have an *endpoint minimum* at c iff

$$f(c) \leq f(x) \qquad \text{for all } x \text{ in the domain of } f \text{ which are sufficiently close to } c.$$

Endpoints are usually tested by examining the sign of the derivative at nearby points. Suppose, for example, that a is the left endpoint. If $f'(x) < 0$ for x close to a, then f decreases on an interval of the form $[a, a + \epsilon]$ and therefore $f(a)$ must be an endpoint maximum. (Figure 4.4.1) On the other hand, if $f'(x) > 0$ for x close to a, then f increases on an interval of the form $[a, a + \epsilon]$ and therefore $f(a)$ must be an endpoint minimum. (Figure 4.4.2) Similar reasoning can be applied to right endpoints.

Absolute Maxima and Minima

Whether or not a function f has a local or endpoint extreme at some point depends entirely on the behavior of f for x close to that point. Absolute extreme values, which we define below, depend on the behavior of the function on its entire domain.

We begin with a number d in the domain of f. Here d can be an interior point or an endpoint.

Definition 4.4.2   Absolute Extreme Values

$f(d)$ is called *the absolute maximum* (*value*) of f iff

$$f(d) \geq f(x) \qquad \text{for all } x \text{ in the domain of } f;$$

$f(d)$ is called *the absolute minimum* (*value*) of f iff

$$f(d) \leq f(x) \qquad \text{for all } x \text{ in the domain of } f.$$

A function can be continuous on an interval (or even differentiable there) without taking on any absolute extreme values. All we can say in general is that, if f takes on an absolute extreme value, then it does so at a critical point. If, however, the domain is a bounded closed interval $[a, b]$, then the continuity of the function guarantees the existence of both an absolute maximum and an absolute minimum. (Theorem 2.6.6) In this case the most practical way of determining the absolute extremes is to gather together the local extremes and the endpoint extremes. The largest of these numbers is obviously the absolute maximum, and the smallest is the absolute minimum.

Summary

We summarize here the central ideas of these last two sections by outlining a step-by-step procedure for finding all the extreme values (local, endpoint, and absolute) of a function f.

1. Find the critical points. These are the interior points c at which $f'(c) = 0$ or $f'(c)$ does not exist, and the endpoints of the domain.
2. Test each interior critical point for local extremes:
   (a) by applying Test 1 (the first-derivative test),
   (b) or, if $f''(c)$ exists and is not 0, by applying Test 2 (the second-derivative test).
3. Test each endpoint for endpoint extremes. (By inspection, or if necessary, by checking the sign of f' at nearby points.)
4. Determine whether any of the local or endpoint extremes are absolute extremes.

Problem.   Find the critical points and classify the extreme values of

$$f(x) = \frac{\sqrt{x} - 1}{\sqrt{x} + 1}.$$

SOLUTION.   The domain here is $[0, \infty)$. First we check the interior points $x \in (0, \infty)$. On $(0, \infty)$ the function f is differentiable and

$$f'(x) = \frac{(\sqrt{x} + 1)\dfrac{1}{2\sqrt{x}} - (\sqrt{x} - 1)\dfrac{1}{2\sqrt{x}}}{(\sqrt{x} + 1)^2} = \frac{1}{\sqrt{x}(\sqrt{x} + 1)^2}.$$

Since the derivative $f'(x)$ is never 0, there are no interior critical points.

The number 0 is a critical point because it is an endpoint. Since $f'(x) > 0$ for all x in $(0, \infty)$ and f is continuous on $[0, \infty)$, f must be increasing on $[0, \infty)$. This makes $f(0) = -1$ an endpoint minimum. It is also the absolute minimum. You can see the graph in Figure 4.4.4.   □

FIGURE 4.4.4

Problem.   Find the critical points and classify the extreme values of

$$f(x) = x^2 - 2|x| + 2, \quad x \in [-\tfrac{1}{2}, \tfrac{3}{2}].$$

SOLUTION.   First we check the interior points $x \in (-\tfrac{1}{2}, \tfrac{3}{2})$. On the open interval $(-\tfrac{1}{2}, \tfrac{3}{2})$ the function is differentiable except at 0:

$$f'(x) = \begin{cases} 2x + 2, & -\tfrac{1}{2} < x < 0 \\ 2x - 2, & 0 < x < \tfrac{3}{2} \end{cases}.$$ (check this out)

This makes 0 a critical point. Since $f'(x) = 0$ at $x = 1$, 1 is a critical point. The endpoints $-\tfrac{1}{2}$ and $\tfrac{3}{2}$ are also critical points.

A little to the left of $0, f' > 0$ and a little to the right of $0, f' < 0$. By Test 1, this means that $f(0) = 2$ is a local maximum.

A little to the left of $1, f' < 0$ and a little to the right of $1, f' > 0$. This makes $f(1) = 1$ a local minimum. (Again by Test 1.)

Now we check the endpoints. Since $f' > 0$ a little to the right of $-\tfrac{1}{2}$,

$$f(-\tfrac{1}{2}) = (-\tfrac{1}{2})^2 - 2|-\tfrac{1}{2}| + 2 = \tfrac{1}{4} - 1 + 2 = 1\tfrac{1}{4}$$

is an endpoint minimum. (Explain.) Since $f' > 0$ a little to the left of $\tfrac{3}{2}$,

$$f(\tfrac{3}{2}) = (\tfrac{3}{2})^2 - 2|\tfrac{3}{2}| + 2 = \tfrac{9}{4} - 3 + 2 = 1\tfrac{1}{4}$$

is an endpoint maximum. (Explain.)

In short, the critical points are $-\tfrac{1}{2}$, 0, 1, and $\tfrac{3}{2}$.

$\quad f(-\tfrac{1}{2}) = 1\tfrac{1}{4}$ is an endpoint minimum.

$\quad f(0) = 2$ is a local maximum; it is also the absolute maximum.

$\quad f(1) = 1$ is a local minimum; it is also the absolute minimum.

$\quad f(\tfrac{3}{2}) = 1\tfrac{1}{4}$ is an endpoint maximum.

For the graph see Figure 4.4.5.   ☐

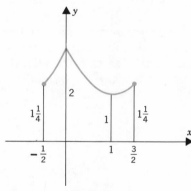

FIGURE 4.4.5

Problem.   Find the critical points and classify the extreme values of

$$f(x) = (1 - x^2)^{1/2} + \tfrac{1}{2}x.$$

SOLUTION. The domain is the closed interval $[-1, 1]$. The endpoints -1 and 1 are critical points. To check the interior points $x \in (-1, 1)$, we differentiate. This gives

$$f'(x) = -x(1 - x^2)^{-1/2} + \tfrac{1}{2}.$$

Setting $f'(x) = 0$, we get

$$2x = (1 - x^2)^{1/2}$$

and, squaring both sides,

$$4x^2 = 1 - x^2, \quad 5x^2 = 1, \quad x = \pm \tfrac{1}{5}\sqrt{5}.$$

Of these two numbers only $\tfrac{1}{5}\sqrt{5}$ is a root of the equation $f'(x) = 0$. The other, $-\tfrac{1}{5}\sqrt{5}$, is an extraneous root that we introduced when we did the squaring. In short there are three critical points to examine: -1, $\tfrac{1}{5}\sqrt{5}$, and 1. Since $f' > 0$ a little to the right of -1, $f(-1) = -\tfrac{1}{2}$ is an endpoint minimum. To test $f(\tfrac{1}{5}\sqrt{5})$ we apply the second-derivative test:

$$f''(x) = (-x)(-\tfrac{1}{2})(1 - x^2)^{-3/2}(-2x) - (1 - x^2)^{-1/2}$$

$$= (1 - x^2)^{-3/2}[-x^2 - (1 - x^2)] = -(1 - x^2)^{-3/2}.$$

Obviously $f''(\tfrac{1}{5}\sqrt{5}) < 0$, and therefore

$$f(\tfrac{1}{5}\sqrt{5}) = \sqrt{1 - \tfrac{1}{5}} + \tfrac{1}{10}\sqrt{5} = \tfrac{2}{5}\sqrt{5} + \tfrac{1}{10}\sqrt{5} = \tfrac{1}{2}\sqrt{5}$$

is a local maximum value. Since $f' < 0$ a little to the left of 1, $f(1) = \tfrac{1}{2}$ is an endpoint minimum.

To summarize, the critical points are -1, $\tfrac{1}{5}\sqrt{5}$, and 1.

$f(-1) = -\tfrac{1}{2}$ is an endpoint minimum; it is also the absolute minimum.

$f(\tfrac{1}{5}\sqrt{5}) = \tfrac{1}{2}\sqrt{5}$ is a local maximum; it is also the absolute maximum.

$f(1) = \tfrac{1}{2}$ is an endpoint minimum. $\square$

Exercises

Find the critical points and classify the extreme values.

*1. $f(x) = \sqrt{x + 2}$.

2. $f(x) = (x - 1)(x - 2)$.

*3. $f(x) = x^2 - 4x + 1$, $x \in [0, 3]$.

4. $f(x) = 2x^2 + 5x - 1$, $x \in [-2, 0]$.

*5. $f(x) = x^2 + \dfrac{1}{x}$.

6. $f(x) = x + \dfrac{1}{x^2}$.

*7. $f(x) = x^2 + \dfrac{1}{x}$, $x \in [\tfrac{1}{10}, 2]$.

8. $f(x) = x + \dfrac{1}{x^2}$, $x \in (-1, 0)$.

*9. $f(x) = (x - 1)(x - 2)$, $x \in [0, 2]$.

10. $f(x) = (x - 1)^2(x - 2)^2$, $x \in [0, 4]$.

*11. $f(x) = \dfrac{1 - 3\sqrt{x}}{3 - \sqrt{x}}$.

12. $f(x) = \dfrac{x^2}{1 + x^2}$, $x \in [-1, 2]$.

*13. $f(x) = (x - \sqrt{x})^2$

14. $f(x) = x\sqrt{4 - x^2}$.

*15. $f(x) = x\sqrt{3 - x}$.

16. $f(x) = \sqrt{x} - \dfrac{1}{\sqrt{x}}$.

*17. $f(x) = 1 - \sqrt[3]{x - 1}$.

18. $f(x) = (4x - 1)^{1/3}(2x - 1)^{2/3}$.

*19. $f(x) = \begin{cases} -2x, & x < 1 \\ x - 3, & 1 \le x \le 4 \\ 5 - x, & x > 4 \end{cases}$.

20. $f(x) = \begin{cases} -x^2, & 0 \le x < 1 \\ -2x, & 1 < x < 2 \\ -\frac{1}{2}x^2, & 2 \le x \le 3 \end{cases}$.

4.5 Some Max-Min Problems

The techniques of the last two sections can be brought to bear on a wide variety of applied max-min problems. The *key step* in solving such problems is to express the quantity to be maximized or minimized as a differentiable function of one variable. All that remains then is to differentiate and analyze the results. Here are some examples.

Problem. A window in the shape of a rectangle capped by a semicircle is to be surrounded by p inches of metal border. Find the radius of the semicircular part if the total area of the window is to be a maximum.

FIGURE 4.5.1

SOLUTION. As in Figure 4.5.1 we let r be the radius of the semicircular part and let $h(r)$ be the height of the rectangular part. The total area of the window is then given by the function

(*) $$A(r) = 2rh(r) + \tfrac{1}{2}\pi r^2, \quad r > 0.$$

We need $h(r)$ explicity in terms of r. Since the total perimeter of the window is p, we have

$$p = 2r + 2h(r) + \pi r \quad \text{and thus} \quad h(r) = \tfrac{1}{2}(p - 2r - \pi r).$$

Substituting this expression for $h(r)$ in (*) we have

$$A(r) = 2r[\tfrac{1}{2}(p - 2r - \pi r)] + \tfrac{1}{2}\pi r^2,$$

which simplifies to

$$A(r) = pr - 2r^2 - \tfrac{1}{2}\pi r^2. \qquad \text{(key step completed)}$$

Our problem now is to find the value of r in $(0, \infty)$ that maximizes this function A. Differentiation gives

$$A'(r) = p - 4r - \pi r = p - r(4 + \pi).$$

Setting $A'(r) = 0$, we have $r = p/(4 + \pi)$. Since $A'(r) > 0$ for $0 < r < p/(4 + \pi)$ and $A'(r) < 0$ for $r > p/(4 + \pi)$, $r = p/(4 + \pi)$ does give rise to the desired maximum. $\square$

Problem. The sum of two nonnegative numbers is 3. Find the two numbers if the square of twice the first diminished by twice the square of the second is a minimum.

SOLUTION. Denote the first number by x; the second number becomes $3 - x$. We want the minimum value of

$$f(x) = (2x)^2 - 2(3 - x)^2 \quad \text{for } 0 \le x \le 3.$$

(Key step completed.) Since

$$f'(x) = 8x + 4(3 - x) = 4x + 12 = 4(x + 3),$$

we have $f'(x) > 0$ on $(0, 3)$. Since f is continuous on $[0, 3]$, f increases on $[0, 3]$. The minimum value of f occurs at $x = 0$. The first number must be 0 and the second 3. $\square$

Problem. Find the dimensions of the base of the rectangular box of greatest volume that can be constructed from 200 square inches of cardboard if the base is to be three times as long as it is wide.

SOLUTION. Call the dimensions of the base x and $3x$, and call the height $h(x)$.

FIGURE 4.5.2

The volume is then

$$V(x) = 3x^2 h(x).$$

The conditions of the problem require that $0 < x < \frac{10}{3}\sqrt{3}$.† To obtain $h(x)$ in

---

† Since x is the width, x must be positive. Also we must have $6x^2 < 200$ and thus $x^2 < \frac{100}{3}$, for otherwise we would run out of cardboard.

terms of x, we use the fact that the total surface area is 200 square inches. This gives

$$200 = 6x^2 + 6xh(x) + 2xh(x) = 6x^2 + 8xh(x).$$

and thus

$$h(x) = \frac{200 - 6x^2}{8x} = \frac{100 - 3x^2}{4x}.$$

Substitution of this last expression for $h(x)$ into the volume formula gives

$$V(x) = 3x^2 \left(\frac{100 - 3x^2}{4x} \right),$$

which simplifies to

$$V(x) = \tfrac{3}{4}(100x - 3x^3). \qquad \text{(key step completed)}$$

Our problem now is to find the value of x between 0 and $\frac{10}{3}\sqrt{3}$ which maximizes this function V. Differentiation yields

$$V'(x) = \tfrac{3}{4}(100 - 9x^2).$$

Setting $V'(x) = 0$, we get

$$100 - 9x^2 = 0 \quad \text{and thus} \quad x = \pm \tfrac{10}{3}.$$

Of these two answers the only one that fits the conditions of the problem is $x = \frac{10}{3}$. Since

$$V'(x) \text{ is } \begin{cases} \text{positive,} & \text{for} \quad 0 < x < \tfrac{10}{3} \\ \text{zero,} & \text{for} \quad x = \tfrac{10}{3} \\ \text{negative,} & \text{for} \quad \tfrac{10}{3} < x < \tfrac{10}{3}\sqrt{3} \end{cases},$$

you can see that V increases on $(0, \frac{10}{3}]$ and decreases on $[\frac{10}{3}, \frac{10}{3}\sqrt{3})$. Thus $x = \frac{10}{3}$ does give rise to the desired maximum. The rectangle must be $\frac{10}{3}$ inches wide and 10 inches long. $\square$

Problem.   A manufacturing plant has a capacity of 25 articles per week. Experience has shown that n articles per week can be sold at a price of p dollars each, where $p = 110 - 2n$, and the cost of producing n articles is $(600 + 10n + n^2)$ dollars. How many articles should be made each week to give the largest profit?

SOLUTION.   The profit (P dollars) on the sale of n articles is

$$P = np - (600 + 10n + n^2),$$

which simplifies to

$$P = 100n - 600 - 3n^2.$$

In this problem n must be an integer, and thus it makes no sense to differentiate P

with respect to n. The formula shows that P is negative if n is less than 8. By direct calculation we construct Table 4.5.1. The table shows that the largest profit is obtained when 17 articles per week are manufactured.

TABLE 4.5.1

n	P	n	P	n	P
8	8	14	212	20	200
9	57	15	225	21	117
10	100	16	232	22	148
11	137	17	233	23	113
12	168	18	228	24	72
13	193	19	217	25	25

We can avoid such detailed computation by considering the function

$$f(x) = 100x - 600 - 3x^2, \qquad 0 \le x \le 25. \qquad \text{(key step)}$$

This function is differentiable with respect to x, and for integral values of x it agrees with P. Differentiation of f gives

$$f'(x) = 100 - 6x.$$

Obviously $f'(x) = 0$ at $x = \frac{100}{6} = 16\frac{2}{3}$. Since $f'(x) > 0$ on $(0, 16\frac{2}{3})$, f increases on $[0, 16\frac{2}{3}]$. Since $f'(x) < 0$ on $(16\frac{2}{3}, 25)$, f decreases on $[16\frac{2}{3}, 25]$. The largest value of f corresponding to an integral value of x will therefore occur at $x = 16$ or $x = 17$. Direct calculation of $f(16)$ and $f(17)$ shows that the choice of $x = 17$ is correct. $\square$

Exercises

*1. Find the two positive numbers whose sum is 40 and whose product is a maximum.

2. Find the dimensions of the rectangle of perimeter p that has the largest area.

*3. A rectangular garden 200 square feet in area is to be fenced off against rabbits. Find the dimensions that will require the least amount of fencing if one side of the garden is already protected by a barn.

4. Find the largest possible area for a rectangle with base on the x-axis and upper vertices on the curve $y = 4 - x^2$.

*5. Find the largest possible area for a rectangle inscribed in a circle of radius r.

6. Let Q denote output, R revenue, C cost, and P profit. Given that $P = R - C$, what is the relation between marginal revenue dR/dQ and marginal cost dC/dQ when profits are at a maximum?

*7. Find the coordinates of P which maximize the area of the rectangle in Figure 4.5.3.

FIGURE 4.5.3   FIGURE 4.5.4

8. A triangle is formed by the coordinate axes and a line through the point (a, b) as in Figure 4.5.4. Determine the slope of this line if the area of the triangle is to be (a) a minimum; (b) a maximum.

*9. A manufacturer finds that the total cost of producing Q tons is $aQ^2 + bQ + c$ dollars and the price at which each ton can be sold is $\beta - \alpha Q$. Assuming that a, b, c, α, β are all positive, what is the output for maximum profit?

10. Davis Rent-A-TV derives an average net profit of $15 per customer if it services 1000 customers or less. If it services over 1000 customers, then the average profit decreases 1¢ for each customer above that number. How many customers give the maximum net profit?

*11. The cross section of a beam is in the form of a rectangle of length l and width w. (Figure 4.5.5.) Assuming that the strength of the beam varies directly with $w^2 l$, what are the dimensions of the strongest beam that can be sawed from a round log of diameter d?

FIGURE 4.5.5

12. Let x and y be two positive numbers such that $x + y = 100$, and let n be a positive integer. Show that by setting $x = y$ (a) $x^n + y^n$ is minimized; (b) $x^n y^n$ is maximized.

*13. Find the dimensions of the isosceles triangle with perimeter p which has the largest area.

14. Let $P = f(Q)$ be a demand curve. (P = price, Q = output) Show that at an

output which maximizes total revenue (price × output) the elasticity

$$\epsilon = \frac{f(Q)}{Q|f'(Q)|}$$

is 1.

*15. The base of a triangle is on the x-axis, one side lies along the line $y = 3x$, and the third side passes through the point $(1, 1)$. What is the slope of the third side if the area of the triangle is to be a minimum?

16. Find the point(s) on the parabola $y = \frac{1}{8}x^2$ closest to the point $(0, 6)$.

*17. Find the point(s) on the parabola $x = y^2$ closest to the point $(0, 3)$.

18. What are the dimensions of the base of the rectangular box of greatest volume that can be constructed from 100 square inches of cardboard if the base is to be twice as long as it is wide?
(a) Assume that the box has a top. (b) Assume that the box has no top.

*19. From a rectangular piece of cardboard of dimensions $a \times b$ four congruent squares are to be cut out, one at each corner. See Figure 4.5.6. The remaining crosslike piece is then to be folded into an open box. What size squares should be cut out if the volume of the resulting box is to be a maximum?

FIGURE 4.5.6

20. A lighting fixtures manufacturer finds that he can sell x standing lamps per week at p dollars each, where $5x = 375 - 3p$. The cost of production is $(500 + 15x + \frac{1}{5}x^2)$ dollars. Show that maximum profit is obtained when production is about 30 units per week.

21. In Exercise 20 suppose that the relation between x and p is given by the equation $x = 100 - 4\sqrt{5p}$. Show that for maximum profit the manufacturer should produce only about 25 units per week.

22. In Exercise 20 suppose that the relation between x and p is $x^2 = 2500 - 20p$. What production produces maximum profit in this instance?

*23. Let ABC be a triangle with vertices $A = (-a, 0), B = (0, 2a), C = (a, 0)$. Let P be a point on the line segment which joins B to the origin. Find the position of P which minimizes the sum of the distances between P and the vertices.

24. Exercise 23 with $A = (-2a, 0), B = (0, a), C = (2a, 0)$.

25. Conical paper cups are usually made so that the depth is $\sqrt{2}$ times the radius of the rim. Show that this design requires the least amount of paper per unit volume.

*26. A string of length l is to be cut into two pieces, one piece to form an equilateral triangle and the other to form a circle. How should the string be cut so as to (a) maximize the sum of the two areas? (b) minimize the sum of the two areas?

27. Figure 4.5.7 shows a cylinder inscribed in a right circular cone of height h and base radius r. Find the dimensions of the cylinder *(a) if its volume is to be a maximum; (b) if the area of its curved surface is to be a maximum.

FIGURE 4.5.7

28. Find the maximum area for a parallelogram inscribed in a triangle ABC in such a way that one vertex coincides with A while the others fall one on each side of the triangle.

*29. Find the dimensions of the isosceles triangle of least area that can circumscribe a circle of radius r.

30. What is the maximum area possible for a triangle inscribed in a circle of radius r?

31. Figure 4.5.8 shows a cylinder inscribed in a sphere of radius R. Find the dimensions of the cylinder (a) if its volume is to be a maximum; *(b) if the area of its curved surface is to be a maximum.

FIGURE 4.5.8   FIGURE 4.5.9

32. A right circular cone is inscribed in a sphere of radius R as in Figure 4.5.9. Find the dimensions of the cone if its volume is to be a maximum.

*33. What is the greatest possible volume for a right circular cone of slant height a?

Optional   *34. The cost of erecting an office building is $1,000,000 for the first story, $1,100,000 for the second, $1,200,000 for the third, and so on. Other expenses (lot, basement, etc.) are $5,000,000. Assume that the net annual income is $200,000 per story. How many stories will provide the greatest rate of return on investment?

35. Draw the parabola $y = x^2$. On the parabola mark a point $P \neq 0$. Through P draw the normal line. This normal line intersects the parabola at another point Q. Show that the distance between P and Q is minimal when $P = (\pm \frac{1}{2}\sqrt{2}, \frac{1}{2})$.

36. *The problem of Viviani.*† Two parallel lines are cut by a given line *AB*. See Figure 4.5.10. From the point *C* we draw a line to the point marked *Q*. How should *Q* be chosen so that the sum of the areas of the two triangles is a minimum?

FIGURE 4.5.10

4.6   Concavity and Points of Inflection

If you look at the graph of the cubing function (Figure 4.6.1), you will see that for *x* in $(-\infty, 0]$ the graph is "curving down" and for *x* in $[0, \infty)$ the graph is "curving up." In the language of calculus we say that the graph is *concave down* on $(-\infty, 0]$ and *concave up* on $[0, \infty)$.

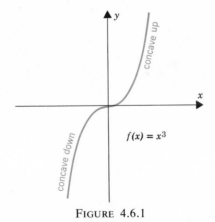

FIGURE 4.6.1

Definition 4.6.1   Concavity

The graph of a differentiable function *f* is said to be *concave up* on the interval *I* iff *f'* is increasing on *I*; it is said to be *concave down* on *I* iff *f'* is decreasing on *I*.

---

† Vicenzo Viviani (1622–1703), a pupil of Galileo.

If f'' exists, it provides a good test for concavity. If $f''(x) > 0$ for all x in I, then f' is increasing there, so that the graph of f is concave up on I. If $f''(x) < 0$ for all x in I, then f' is decreasing there, so that the graph of f is concave down on I.

Of special interest in curve sketching are the points which separate arcs of opposite concavity. These are called *points of inflection.*

Definition 4.6.2 Point of Inflection

A point $(c, f(c))$ at which there is a tangent line is called *a point of inflection* iff there exists $\delta > 0$ such that the graph of f is concave in one sense on $(c - \delta, c)$ and concave in the opposite sense on $(c, c + \delta)$.

It's not hard to see that, if $(c, f(c))$ is a point of inflection and $f'(c)$ exists, then the derivative f' has either a local maximum at c (Figure 4.6.2) or a local minimum at c (Figure 4.6.3). It follows therefore that

(4.6.3)

> if $(c, f(c))$ is a point of inflection, then
> either $f''(c) = 0$ or $f''(c)$ does not exist.

The proof is left to you as an exercise. $\square$

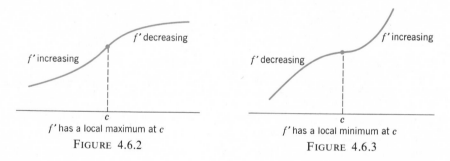

f' increasing f' decreasing f' decreasing f' increasing

c c

f' has a local maximum at c f' has a local minimum at c

FIGURE 4.6.2   FIGURE 4.6.3

Example.   For

$$f(x) = x^3 + \tfrac{1}{2}x^2 - 2x + 1$$

we have

$$f'(x) = 3x^2 + x - 2 \quad \text{and} \quad f''(x) = 6x + 1.$$

Since

$$f''(x) \text{ is } \begin{cases} \text{negative,} & \text{for } x < -\tfrac{1}{6} \\ 0, & \text{for } x = -\tfrac{1}{6} \\ \text{positive,} & \text{for } x > -\tfrac{1}{6} \end{cases},$$

the graph of f is concave down on $(-\infty, -\tfrac{1}{6}]$ and concave up on $[-\tfrac{1}{6}, \infty)$. The point $(-\tfrac{1}{6}, f(-\tfrac{1}{6}))$ is a point of inflection. $\square$

Example. For

$$f(x) = 3x^{5/3} - x$$

we have

$$f'(x) = 5x^{2/3} - 1 \quad \text{and} \quad f''(x) = \tfrac{10}{3}x^{-1/3}.$$

The second derivative fails to exist at 0. Since

$$f''(x) \text{ is } \begin{cases} \text{negative,} & \text{for} \quad x < 0 \\ \text{positive,} & \text{for} \quad x > 0 \end{cases},$$

the point $(0, f(0)) = (0, 0)$ is a point of inflection. The graph is concave down on $(-\infty, 0]$ and concave up on $[0, \infty)$. $\square$

Exercises

Describe the concavity of the graph and find the points of inflection (if any).

*1. $\dfrac{1}{x}$.

2. $x + \dfrac{1}{x}$.

*3. $x^3 - 3x + 2$.

4. $2x^2 - 5x + 2$.

*5. $\tfrac{1}{4}x^4 - \tfrac{1}{2}x^2$.

6. $x^3(1 - x)$.

*7. $\dfrac{x}{x^2 - 1}$.

8. $\dfrac{x + 2}{x - 2}$.

*9. $(1 - x)^2(1 + x)^2$.

10. $\dfrac{6x}{x^2 + 1}$.

*11. $\dfrac{1 - \sqrt{x}}{1 + \sqrt{x}}$.

12. $(x - 3)^{1/5}$.

*13. Find d given that $(d, f(d))$ is a point of inflection of

$$f(x) = (x - a)(x - b)(x - c).$$

14. Find c given that the graph of $f(x) = cx^2 + x^{-2}$ has an inflection point at $(1, f(1))$.

*15. Find a and b given that the graph of $f(x) = ax^3 + bx^2$ passes through $(-1, 1)$ and has an inflection point when $x = \tfrac{1}{3}$.

16. Determine A and B so that the curve $y = Ax^{1/2} + Bx^{-1/2}$ will have a point of inflection at $(1, 4)$.

17. Prove that, if $(c, f(c))$ is a point of inflection, then

$$\text{either} \quad f''(c) = 0 \quad \text{or} \quad f''(c) \text{ does not exist.}$$

4.7 Some Curve Sketching

During the course of the last few sections you have seen how to determine the extreme values of a function and how to find where a function is increasing and where it is decreasing. Now you have seen how to determine the concavity of the graph and how to find the points of inflection. By putting all this information together you can get a good idea of what the graph of a given function looks like.

Example. In the case of

$$f(x) = \tfrac{1}{4}(x^3 - \tfrac{3}{2}x^2 - 6x + 2),$$

$f(0) = \frac{1}{2}$ and therefore the graph intersects the y-axis at the point $(0, \frac{1}{2})$.† Differentiation gives

$$f'(x) = \tfrac{1}{4}(3x^2 - 3x - 6) = \tfrac{3}{4}(x + 1)(x - 2)$$

and

$$f''(x) = \tfrac{1}{4}(6x - 3) = \tfrac{3}{4}(2x - 1).$$

The first derivative is 0 at $x = -1$ and $x = 2$. These are the only critical points. Since $f''(-1) < 0$,

$$f(-1) = \tfrac{1}{4}[(-1)^3 - \tfrac{3}{2}(-1)^2 - 6(-1) + 2] = \tfrac{1}{4}[-1 - \tfrac{3}{2} + 6 + 2] = 1\tfrac{3}{8}$$

is a local maximum. Since $f''(2) > 0$,

$$f(2) = \tfrac{1}{4}[(2)^3 - \tfrac{3}{2}(2)^2 - 6(2) + 2] = -2$$

is a local minimum. Since

$$f''(x) \text{ is } \begin{cases} \text{negative,} & \text{for} \quad x < \frac{1}{2} \\ \qquad 0, & \text{for} \quad x = \frac{1}{2} \\ \text{positive,} & \text{for} \quad x > \frac{1}{2} \end{cases},$$

the point $(\frac{1}{2}, f(\frac{1}{2})) = (\frac{1}{2}, -\frac{5}{16})$ is the only point of inflection. The graph is concave down on $(-\infty, \frac{1}{2}]$ and concave up on $[\frac{1}{2}, \infty)$.

If you now put together all this information and plot a few points, you will not find it hard to come up with a graph that looks like the one displayed in Figure 4.7.1. □

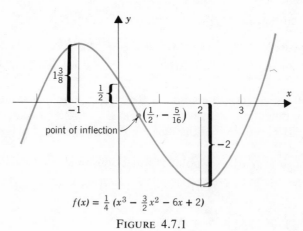

$$f(x) = \tfrac{1}{4}(x^3 - \tfrac{3}{2}x^2 - 6x + 2)$$

FIGURE 4.7.1

Example.   The function

$$g(x) = \frac{1}{1 + x^2}$$

---

† It would be nice to find out exactly where the graph intersects the x-axis, but this would require that we solve the equation $x^3 - \frac{3}{2}x^2 - 6x + 2 = 0$.

remains positive throughout. Since $g(-x) = g(x)$, the function is even and its graph is symmetric with respect to the y-axis. When $x = 0$, $g(x) = 1$. This is clearly the maximum value of g. As x tends away from zero,

$$1 + x^2 \quad \text{increases without bound}$$

and therefore

$$g(x) = \frac{1}{1 + x^2} \quad \text{decreases toward 0.}$$

So far we have not used calculus in this example, but we need it now to determine the concavity of the graph. Differentiation gives

$$g'(x) = -\frac{2x}{(1 + x^2)^2} \quad \text{and} \quad g''(x) = \frac{2(3x^2 - 1)}{(1 + x^2)^3}.$$

Note that

$$g''(x) \text{ is} \begin{cases} \text{positive,} & \text{for} & x < -\tfrac{1}{3}\sqrt{3} \\ \text{0,} & \text{for} & x = -\tfrac{1}{3}\sqrt{3} \\ \text{negative,} & \text{for} & -\tfrac{1}{3}\sqrt{3} < x < \tfrac{1}{3}\sqrt{3} \\ \text{0,} & \text{for} & x = \tfrac{1}{3}\sqrt{3} \\ \text{positive,} & \text{for} & x > \tfrac{1}{3}\sqrt{3} \end{cases}.$$

This shows that the graph is concave down on $[-\tfrac{1}{3}\sqrt{3}, \tfrac{1}{3}\sqrt{3}]$ and concave up on $(-\infty, -\tfrac{1}{3}\sqrt{3}]$ and $[\tfrac{1}{3}\sqrt{3}, \infty)$. The points $(-\tfrac{1}{3}\sqrt{3}, \tfrac{3}{4})$ and $(\tfrac{1}{3}\sqrt{3}, \tfrac{3}{4})$ are points of inflection.

The graph appears in Figure 4.7.2. □

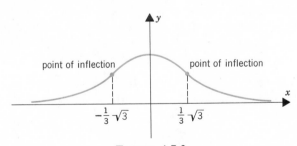

FIGURE 4.7.2

Example.   As our final example we take the function

$$f(x) = 3(x^{5/3} - x^{4/3}).$$

We can rewrite this as

$$f(x) = 3x^{4/3}(x^{1/3} - 1).$$

The function is zero at $x = 0$ and at $x = 1$. The graph crosses the x-axis at the origin and at the point $(1, 0)$.

Differentiation gives

$$f'(x) = 5x^{2/3} - 4x^{1/3} = x^{1/3}(5x^{1/3} - 4),$$
$$f''(x) = \tfrac{10}{3}x^{-1/3} - \tfrac{4}{3}x^{-2/3} = \tfrac{2}{3}x^{-2/3}(5x^{1/3} - 2).$$

The first derivative is zero at $x = 0$ and at $x = 4^3/5^3 = \frac{64}{125}$. These are the only critical points. Note that

$$f'(x) \text{ is } \begin{cases} \text{positive,} & \text{for} & x < 0 \\ 0, & \text{for} & x = 0 \\ \text{negative,} & \text{for} & 0 < x < \frac{64}{125} \\ 0, & \text{for} & x = \frac{64}{125} \\ \text{positive,} & \text{for} & x > \frac{64}{125} \end{cases}.$$

This shows that

$$f(0) = 0 \quad \text{is a local maximum}$$

and

$$f(\tfrac{65}{125}) = -\tfrac{768}{3125} \cong -\tfrac{1}{4} \quad \text{is a local minimum.}$$

The second derivative does not exist at $x = 0$. It is zero at $x = 2^3/5^3 = \frac{8}{125}$. Since

$$f''(x) \text{ is } \begin{cases} \text{negative,} & \text{for} & x < 0 \\ \text{nonexistent,} & \text{for} & x = 0 \\ \text{negative,} & \text{for} & 0 < x < \frac{8}{125} \\ 0, & \text{for} & x = \frac{8}{125} \\ \text{positive,} & \text{for} & x > \frac{8}{125} \end{cases},$$

the graph is concave down on all of $(-\infty, \frac{8}{125}]$ (explain) and concave up on $[\frac{8}{125}, \infty)$. The point

$$(\tfrac{8}{125}, f(\tfrac{8}{125})) \cong (0.06, -0.04)$$

is the only point of inflection. For our graph this is very close to the origin.

Before sketching the graph we gather a little more information. The easiest numbers to work with are the perfect cubes:

at $x = -\frac{64}{125}$, $f(x)$ is about -2.2 and the slope is 6.4;

at $x = -\frac{8}{27}$, $f(x)$ is about -0.1 and the slope is about 5;

at $x = 1$, the slope is 1;

at $x = \frac{125}{64}$, $f(x)$ is about 1.8 and the slope is about 3.

Use this information and you will come up with a curve much like the one displayed in Figure 4.7.3. $\square$

REMARK. So far we have paid little attention to questions of asymptotic behavior. Asymptotes, infinite limits, limits as $x \to \pm\infty$ are discussed in Chapter 12.

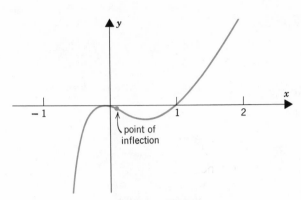

FIGURE 4.7.3

Exercises

For each of the following functions (i) find the critical points, (ii) find and classify the extreme values (local, endpoint, absolute), (iii) indicate where the function is increasing and where it is decreasing, (iv) indicate the concavity of the graph, (v) specify the points of inflection, and then (vi) sketch the graph.

*1. $f(x) = (x - 2)^2$.

2. $f(x) = 1 - (x - 2)^2$.

*3. $f(x) = x^3 - 2x^2 + x + 1$.

4. $f(x) = x^3 - 9x^2 + 24x - 7$.

*5. $f(x) = x^3 + 6x^2, \quad x \in [-4, 4]$.

6. $f(x) = x^4 - 8x^2, \quad x \in [0, \infty)$.

*7. $f(x) = \frac{2}{3}x^3 - \frac{1}{2}x^2 - 10x - 1$.

8. $f(x) = 2 + (x - 4)^{1/3}, \quad x \in [4, \infty)$.

*9. $f(x) = 3x^4 - 4x^3 + 1$.

10. $f(x) = x(x^2 + 4)^2$.

*11. $f(x) = 2\sqrt{x} - x, \quad x \in [0, 4]$.

12. $f(x) = \frac{1}{4}x - \sqrt{x}, \quad x \in [0, 9]$.

*13. $f(x) = 2 + (x + 1)^{6/5}$.

14. $f(x) = 2 + (x + 1)^{7/5}$.

*15. $f(x) = 3x^5 + 5x^3$.

*16. $f(x) = 3x^4 + 4x^3$.

17. $f(x) = 1 + (x - 2)^{4/3}$.

*18. $f(x) = 1 + (x - 2)^{5/3}$.

*19. $f(x) = x^2(1 + x)^2$.

20. $f(x) = x^2(1 + x)^3$.

*21. $f(x) = x\sqrt{1 - x}$.

*22. $f(x) = \sqrt{x - x^2}$.

23. $f(x) = \dfrac{x}{3x - 1}$.

*24. $f(x) = \dfrac{2x}{4x - 3}$.

*25. $f(x) = \dfrac{x^2}{3x + 1}$.

26. $f(x) = \dfrac{2x^2}{x + 1}$.

*27. $f(x) = \dfrac{x}{(3x + 1)^2}$.

*28. $f(x) = \dfrac{2x}{(x + 1)^2}$.

29. $f(x) = \dfrac{2x}{x^2 + 2}$.

*30. $f(x) = \dfrac{2x^2}{x^2 + 2}$.

4.8 Rates of Change per Unit Time

You've already seen that the derivative of a function gives its rate of change. If the domain of f is a time interval, then f' gives the rate of change of f *per unit time.* While the difference quotient

$$\frac{f(t + h) - f(t)}{h}$$

gives an average rate of change, *the average rate of change of f from time t to*

$t + h$, the derivative

$$f'(t) = \lim_{h \to 0} \frac{f(t + h) - f(t)}{h}$$

gives an instantaneous rate of change, *the rate of change of f at time t.*

FIGURE 4.8.1

Example.   A spherical balloon (Figure 4.8.1) is expanding under the influence of solar radiation. If its radius is increasing at the rate of 2 inches per minute, at what rate is the volume increasing when the radius is 5 inches?

SOLUTION.   We can express the volume at time t in terms of the radius at time t by the formula

$$V(t) = \tfrac{4}{3}\pi[r(t)]^3.$$

Differentiation gives

$$V'(t) = 4\pi[r(t)]^2 r'(t).$$

If at time t_0 the radius is 5 inches, then

$$V'(t_0) = 4\pi(25)2 = 200\pi.$$

This means that the volume is increasing at the rate of 200π cubic inches per minute.   □

Example.   A particle moves along its circular orbit $x^2 + y^2 = 1$. (Figure 4.8.2) As it passes through the point $(\tfrac{1}{2}, \tfrac{1}{2}\sqrt{3})$, its y-coordinate decreases at the rate of 3 units per second. At what rate is the x-coordinate changing?

SOLUTION.   At each time t

$$[x(t)]^2 + [y(t)]^2 = 1.$$

By differentiation

$$2x(t)x'(t) + 2y(t)y'(t) = 0 \quad \text{and thus} \quad x(t)x'(t) + y(t)y'(t) = 0.$$

If the particle passes through the point $(\tfrac{1}{2}, \tfrac{1}{2}\sqrt{3})$ when $t = t_0$, then

$$\tfrac{1}{2}x'(t_0) + \tfrac{1}{2}\sqrt{3}(-3) = 0 \quad \text{and consequently} \quad x'(t_0) = 3\sqrt{3}.$$

As the particle passes through the point $(\tfrac{1}{2}, \tfrac{1}{2}\sqrt{3})$, the x-coordinate increases at the rate of $3\sqrt{3}$ units per second.   □

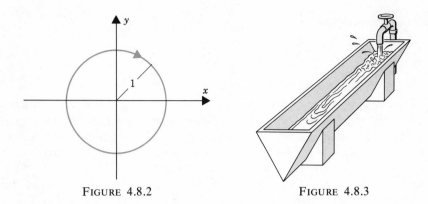

FIGURE 4.8.2   FIGURE 4.8.3

Example. A water trough with vertical cross section in the shape of an equilateral triangle is being filled at the rate of 4 cubic feet per minute. (Figure 4.8.3) Given that the trough is 12 feet long, how fast is the water level rising when the water reaches a depth of $1\frac{1}{2}$ feet?

SOLUTION. When the water is x feet deep, the area of a vertical cross section of water is $\frac{1}{3}\sqrt{3}x^2$. Since the trough is 12 feet long, the volume of the water is then $4\sqrt{3}x^2$. If we note the depth of the water at time t by $x(t)$, then for the volume at time t we have

$$V(t) = 4\sqrt{3}[x(t)]^2.$$

Differentiation gives

$$V'(t) = 8\sqrt{3}x(t)x'(t).$$

If the water is $1\frac{1}{2}$ feet deep when $t = t_0$, then

$$4 = 8\sqrt{3}(1\tfrac{1}{2})x'(t_0) \quad \text{and thus} \quad x'(t_0) = \tfrac{1}{9}\sqrt{3}.$$

When the depth is $1\frac{1}{2}$ feet, the water is rising at the rate of $\frac{1}{9}\sqrt{3}$ feet per minute. $\square$

Velocity and Acceleration

Suppose now that an object is moving along a straight line and that for each time t during a certain time interval the object has position (coordinate) $x(t)$. If $x'(t)$ exists, then $x'(t)$ gives the rate of change of position at time t. This rate of change of position is called the *velocity* of the object at time t; in symbols,

(4.8.1) $$\boxed{v(t) = x'(t).}$$

If the velocity function is itself differentiable, then its rate of change at time t is called the *acceleration* at time t; in symbols,

(4.8.2) $$\boxed{a(t) = v'(t) = x''(t).}$$

In the Leibniz notation,

(4.8.3)
$$v = \frac{dx}{dt} \quad \text{and} \quad a = \frac{dv}{dt} = \frac{d^2x}{dt^2}.$$

A positive velocity indicates an increase in coordinate and thus a motion in the positive direction. A negative velocity indicates a decrease in coordinate and thus a motion in the negative direction. When the direction of motion is immaterial, we speak of *speed*. This is by definition the absolute value of the velocity:

(4.8.4)
$$\text{speed at time } t = |v(t)|.$$

A word about units is in order. If distance is measured in feet and time in seconds, then velocity and speed are given in feet per second and acceleration is given in feet per second per second.

We can illustrate these ideas by considering the motion of an object which falls freely under the influence of gravity. If we neglect air resistance, then the height of the object at time t is given by the equation

(4.8.5)
$$x(t) = \tfrac{1}{2}gt^2 + v_0 t + x_0.$$

Here v_0 represents the velocity at time $t = 0$ (called the *initial velocity*), and x_0 represents the height at time $t = 0$ (called the *initial position*). The constant g, called the gravitational constant, is the acceleration due to gravity. If we measure time in seconds and distances in feet, then g is approximately -32 feet per second per second. (Why is it negative?) In making numerical calculations we shall take g as -32 feet per second per second. Doing so we obtain

$$x(t) = -16t^2 + v_0 t + x_0.$$

Problem.   A stone is dropped from a height of 1600 feet. How long does it take to reach the ground and what is its speed at impact?

SOLUTION.   Here $x_0 = 1600$ and $v_0 = 0$. Consequently

$$x(t) = -16t^2 + 1600.$$

To find t at the moment of impact we set $x(t) = 0$. This gives

$$-16t^2 + 1600 = 0, \quad t^2 = 100, \quad t = \pm 10.$$

We can disregard the negative answer because the stone had not even been dropped before $t = 0$. We conclude that it takes 10 seconds for the stone to reach the ground.

The velocity at impact is the velocity when $t = 10$. Since

$$v(t) = x'(t) = -32t.$$

we have

$$v(10) = -320.$$

The speed at impact is therefore 320 feet per second.   □

Problem.   A stone is projected from ground level vertically upward with an initial velocity of 72 feet per second.

(a) How many seconds later does it attain its maximum height?
(b) What is this maximum height?
(c) How fast is it traveling when it reaches a height of 32 feet?
   (i) going up?   (ii) going down?

SOLUTION.   The basic equation is again

$$x(t) = -16t^2 + v_0 t + x_0.$$

Here $x_0 = 0$ (it starts at ground level) and $v_0 = 72$ (the initial velocity is 72 feet per second). The equation of motion is therefore

$$x(t) = -16t^2 + 72t.$$

Differentiation gives

$$v(t) = x'(t) = -32t + 72.$$

The maximum height is obviously reached when the velocity is 0. This occurs when $t = \frac{72}{32} = \frac{9}{4}$. To get the maximum height attained we need only evaluate $x(t)$ at $t = \frac{9}{4}$. Since $x(\frac{9}{4}) = 81$, we see that the maximum height attained is 81 feet.
    To answer part (c) we must find those numbers t for which $x(t) = 32$. Since

$$x(t) = -16t^2 + 72t,$$

we are led to the quadratic equation

$$16t^2 - 72t + 32 = 0.$$

This quadratic has two solutions, $t = \frac{1}{2}$ and $t = 4$. For (i) we have $v(\frac{1}{2}) = 56$; and for (ii), $v(4) = -56$. The speed in each case is 56 feet per second. □

Exercises

  *1. A point moves along the straight line $x + 2y = 2$. Find: (a) the rate of change of the y-coordinate, given that the x-coordinate increases 4 units per second; (b) the rate of change of the x-coordinate, given that the y-coordinate decreases 2 units per second.

   2. A rubbish heap in the shape of a cube is being compacted. Given that the volume decreases at the rate of 2 cubic meters per minute, find the rate of change of an edge of the cube when the volume is 27 cubic meters. What is the rate of change of the surface area of the cube?

  *3. A particle is moving in the circular orbit $x^2 + y^2 = 25$. As it passes through the point $(3, 4)$, its y-coordinate is decreasing at the rate of 2 units per second. How is the x-coordinate changing?

   4. At a certain instant the dimensions of a rectangle are a and b. These dimensions are changing at the rates m, n, respectively. Find the rate at which the area is changing.

*5. The side of an equilateral triangle is α centimeters long and is increasing at the rate of k centimeters per minute. How fast is the area increasing?

6. A boat is held by a bow line which is wound about a windlass 6 feet higher than the bow of the boat. If the boat is drifting away at the rate of 8 feet per minute, how fast is the line unwinding when the bow is 30 feet from the windlass?

*7. Two boats are racing with constant speed toward a finish marker, boat A sailing from the south at 13 mph and boat B approaching from the east. When equidistant from the marker the boats are 16 miles apart and the distance between them is decreasing at the rate of 17 mph. Which boat will win the race?

8. A ladder 13 feet long is leaning against a wall. If the base of the ladder is being pulled away from the wall at the rate of 0.5 feet per second, how fast is the top of the ladder dropping when the base is 5 feet from the wall?

*9. A tank contains 1000 cubic feet of natural gas at a pressure of 5 pounds per square inch. If the pressure is decreasing at the rate of 0.05 pounds per square inch per hour, find the rate of increase of the volume. (Assume Boyle's law: *pressure* × *volume* = *constant*.)

10. In the special theory of relativity the mass of a particle moving at velocity v is

$$m\left(1 - \frac{v^2}{c^2}\right)^{-1/2},$$

where m is the mass at rest and c is the speed of light. At what rate is the mass changing when the particle's velocity is $\frac{1}{2}c$ and the rate of change of the velocity is $0.01c$ per second?

*11. Taking

$$x(t) = -t\sqrt{t + 1}, \quad t \geq 0$$

as the equation of motion, find
(a) the velocity when $t = 1$,
(b) the acceleration when $t = 1$,
(c) the speed when $t = 1$,
(d) the rate of change of the speed when $t = 1$.

12. An object moves in a straight line for 10 seconds with

$$x(t) = 25t^2 - \tfrac{5}{3}t^3, \quad t \in [0, 10]$$

as the equation of motion. Here t is measured in seconds and x in meters. (a) How far does the object go? (b) What is its maximum velocity? (c) How far has the object gone when it reaches its maximum velocity?

*13. The volume of a spherical balloon is increasing at the constant rate of 8 cubic feet per minute. How fast is the radius increasing when the radius is 10 feet? How fast is the surface area increasing at this instant?

14. The adiabatic law for the expansion of air is $PV^{1.4} = C$. If at a given time the volume is observed to be 10 cubic feet and the pressure is 50 pounds per square inch, at what rate is the pressure changing if the volume is decreasing 1 cubic foot per second?

*15. The shadow cast by a man standing 3 feet from a lamp post is 4 feet long. If the man is 6 feet tall and walks away from the lamp post at a speed of 400 feet per minute, at what rate will his shadow lengthen?

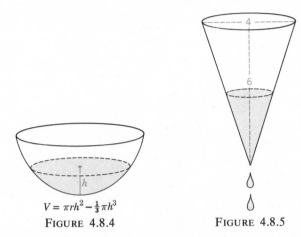

$$V = \pi r h^2 - \tfrac{1}{3}\pi h^3$$
FIGURE 4.8.4   FIGURE 4.8.5

16. Water flows from a faucet into a hemispherical basin of diameter 14 inches at the rate of 2 cubic inches per second. (Figure 4.8.4) How fast is the water rising (a) when the water is halfway to the top? (b) just as it runs over? (The volume of a spherical segment is $\pi r h^2 - \tfrac{1}{3}\pi h^3$ where r is the radius of the sphere and h is the depth of the segment.)

*17. A conical paper cup 4 inches across the top and 6 inches deep is leaking water at the rate of 1 cubic inch per minute. (Figure 4.8.5) At what rate is the water level dropping (a) when the water is 3 inches deep? (b) when the cup is half full?

18. A stone is thrown upward from ground level with an initial speed of 32 feet per second. (a) How many seconds later will it hit the ground? (b) What will be the maximum height attained? (c) With what initial speed should it be thrown if it is to reach a maximum height of 36 feet?

*19. To estimate the height of a bridge a man drops a stone into the water below. How high is the bridge (a) if the stone hits the water 3 seconds later? (b) if the man hears the splash 3 seconds later? (Use 1080 feet per second as the speed of sound.)

20. A falling stone is observed to be at a height of 100 feet. Two seconds later it is observed to be at a height of 16 feet. (a) From what height was it dropped? (b) If it was thrown down with an initial speed of 5 feet per second, from what height was it thrown? (c) If it was thrown upward with an initial speed of 10 feet per second, from what height was it thrown?

4.9 The Little-$o(h)$ Idea; Differentials

In calculus we are often called upon to make comparisons between small quantities. In this connection there is a useful notation which we now introduce.

If h is small, then πh^2 is small compared to h, in the sense that the ratio of πh^2 to h tends to 0 with h:

$$\lim_{h \to 0} \frac{\pi h^2}{h} = 0.$$

If h is small, then $h^{5/3}\sqrt{1-h}$ is small compared to h, in the sense that the ratio of $h^{5/3}\sqrt{1-h}$ to h tends to 0 with h:

$$\lim_{h \to 0} \frac{h^{5/3}\sqrt{1-h}}{h} = 0.$$

Expressions $g(h)$, such as πh^2 and $h^{5/3}\sqrt{1-h}$, that tend to 0 with h when divided by h are called *little-o(h)*. More briefly,

(4.9.1)

$$g(h) \text{ is } little\text{-}o(h) \quad \text{iff} \quad \lim_{h \to 0} \frac{g(h)}{h} = 0.$$

To indicate that $g(h)$ is *little-o(h)* we write

$$g(h) = o(h).$$

As you can check,

$$h^3 = o(h) \quad \text{and} \quad \frac{h^2}{h-1} = o(h),$$

but

$$h^{1/3} \neq o(h) \quad \text{and} \quad \frac{h}{h-1} \neq o(h).$$

We can use the *little-o(h)* notation to advantage in connection with differentiation. If f is differentiable at x, then

$$\lim_{h \to 0} \frac{f(x+h) - f(x)}{h} = f'(x).$$

Since obviously

$$f'(x) = \lim_{h \to 0} \frac{f'(x)h}{h},$$

we have

$$\lim_{h \to 0} \frac{[f(x+h) - f(x)] - f'(x)h}{h} = 0.$$

This says that

$$[f(x+h) - f(x)] - f'(x)h = o(h)$$

and therefore

(4.9.2)

$$f(x+h) - f(x) = f'(x)h + o(h).$$

Definition 4.9.3

The difference $f(x + h) - f(x)$ is called the *increment* (of f from x to $x + h$) and is denoted by Δf:

$$\Delta f = f(x + h) - f(x).$$

The product $f'(x)h$ is called the *differential* (at x with increment h) and is denoted by df:

$$df = f'(x)h.$$

Equation (4.9.2) says that, for small h, Δf and df are approximately equal:

$$\boxed{\Delta f \cong df.}$$

This approximate equality is illustrated in Figure 4.9.1. While Δf gives the precise change in f from x to $x + h$, df gives what the change in f would have been had the graph of f proceeded from x to $x + h$ with constant slope $f'(x)$. The difference between Δf and df is the vertical separation between the graph of f and the tangent line as measured at $x + h$. The less the graph curves, the more accurate the differential estimate becomes.

FIGURE 4.9.1

To illustrate the use of differentials we begin with a square of side x. The area is

$$f(x) = x^2.$$

An increase h in the length of each side produces a change in area

$$\Delta f = f(x + h) - f(x) = (x + h)^2 - x^2 = (x^2 + 2xh + h^2) - x^2 = 2xh + h^2.$$

FIGURE 4.9.2

As an estimate for this change we can use the differential

$$df = f'(x)h = 2xh.$$ (Figure 4.9.2)

The error of our estimate—the difference between the actual change and the estimated change—is the difference

$$\Delta f - df = h^2.$$

Problem.   Given that

$$f(x) = x^{2/5},$$

estimate the change in f
(a) if x is increased from 32 to 34.   (b) if x is decreased from 1 to $\frac{9}{10}$.

SOLUTION.   Since

$$f'(x) = \tfrac{2}{5}(1/x)^{3/5},$$

we have

$$df = f'(x)h = \tfrac{2}{5}(1/x)^{3/5}h.$$

For part (a) we set $x = 32$ and $h = 2$. The differential then becomes

$$\tfrac{2}{5}(\tfrac{1}{32})^{3/5}2 = \tfrac{1}{10} = 0.10.$$

A change in x from 32 to 34 increases the value of f by approximately 0.10.
   For part (b) we set $x = 1$ and $h = -\frac{1}{10}$. The differential then becomes

$$\tfrac{2}{5}(\tfrac{1}{1})^{3/5}(-\tfrac{1}{10}) = -\tfrac{2}{50} = -\tfrac{1}{25} = -0.04.$$

A change in x from 1 to $\frac{9}{10}$ decreases the value of f by approximately 0.04.   □

Problem.   Use a differential to estimate $\sqrt{104}$.

SOLUTION.   We know $\sqrt{100}$. What we need is an estimate for the increase of

$$f(x) = \sqrt{x}$$

from 100 to 104. Differentiation gives

$$f'(x) = \frac{1}{2\sqrt{x}}.$$

so that in general

$$df = f'(x)h = \frac{h}{2\sqrt{x}}.$$

With $x = 100$ and $h = 4$, df becomes

$$\frac{4}{2\sqrt{100}} = \frac{1}{5} = 0.2.$$

A change in x from 100 to 104 increases the value of the square root by approximately 0.2. It follows that

$$\sqrt{104} \cong \sqrt{100} + 0.2 = 10 + 0.2 = 10.2.$$

As you can check, $(10.2)^2 = 104.04$, so that we are not far off.　□

Optional | Let's go back to the little-$o(h)$ idea. We have used

$$df = f'(x)h$$

to approximate

$$\Delta f = f(x + h) - f(x).$$

The expression $f'(x)h$ is linear in h:

it is of the form $A + Bh$　　　　$[A = 0, \quad B = f'(x)]$

and it is an $o(h)$ approximation to $f(x + h) - f(x)$:

$$f(x + h) - f(x) = f'(x)h + o(h).$$

We now assert that $f'(x)h$ is the best linear approximation to $f(x + h) - f(x)$, the best in the sense that it is the only linear approximation whose difference from $f(x + h) - f(x)$ is $o(h)$.

Theorem 4.9.4

If f is differentiable at x and

$$f(x + h) - f(x) = A + Bh + o(h),$$

then

$$A = 0 \quad \text{and} \quad B = f'(x).$$

PROOF.　Assume that

$$f(x + h) - f(x) = A + Bh + o(h).$$

Taking the limit of both sides as h tends to zero, we get 0 on the left and on the right A. (Explain.) Thus we have

$$f(x + h) - f(x) = Bh + o(h).$$

Optional | Dividing by h we have

$$\frac{f(x + h) - f(x)}{h} = B + \frac{o(h)}{h}.$$

Once again we take the limit of both sides as h tends to 0. The left-hand side tends to $f'(x)$, and the right-hand side tends to B. (Explain.) It follows that $B = f'(x)$. □

Exercises

1. Use a differential to estimate the change in the volume of a cube caused by an increase h in the length of each side. Interpret geometrically the error of your estimate $\Delta V - dV$.
2. Use a differential to estimate the area of a ring of inner radius r and width h. What is the exact area?

Estimate the following by differentials.

*3. $\sqrt[3]{1010}$.   4. $\sqrt{125}$.   *5. $\sqrt[4]{15}$.   6. $\dfrac{1}{\sqrt{24}}$.

*7. $\sqrt[5]{30}$.   8. $(26)^{2/3}$.   *9. $(33)^{3/5}$.   10. $(33)^{-1/5}$.

11. Show that

$$\text{if} \quad g(h) = o(h) \qquad \text{then} \quad \lim_{h \to 0} g(h) = 0.$$

12. Estimate $f(5.4)$ given that $f(5) = 1$ and $f'(x) = \sqrt[3]{x^2 + 2}$.
*13. Find the approximate volume of a thin cylindrical sheet with open ends if the inner radius is r, the height is h, and the thickness is t.
14. The diameter of a steel ball is measured to be 16 centimeters, with a maximum error of 0.3 centimeters. Estimate by differentials the maximum error in (a) the surface area when calculated by the formula $S = 4\pi r^2$; *(b) the volume when calculated by the formula $V = \frac{4}{3}\pi r^3$.
*15. A box is to be constructed in the form of a cube to hold 1000 cubic feet. Use a differential to estimate how accurately the inner edge must be made so that the volume will be correct to within 3 cubic feet.
16. Use differentials to estimate the values of x for which

   *(a) $\sqrt{x + 1} - \sqrt{x} < 0.01$.   (b) $\sqrt[4]{x + 1} - \sqrt[4]{x} < 0.002$.
*17. The time of one vibration of a pendulum is given by the formula

$$t = \pi \sqrt{\frac{l}{g}},$$

where t is measured in seconds, $g = 32.2$, and l is the length of the pendulum measured in feet. Taking $\pi = 3.14$, and given that a pendulum of length 3.26 feet vibrates once a second, find the approximate change in t if the pendulum is lengthened 0.01 feet.

18. When a metal cube is heated, each edge increases $\frac{1}{10}\%$ per degree increase in temperature. Show by differentials that the surface area increases about $\frac{2}{10}\%$ per degree and the volume increases about $\frac{3}{10}\%$ per degree.

*19. Estimate by a differential how precisely the diameter of a circle must be measured for the area to be correct within 1%.

20. Estimate by differentials how precisely x must be determined if
*(a) x^n is to be accurate within 1%; (b) $x^{1/n}$ is to be accurate within 1%.

4.10   Additional Exercises

*1. The rate of change of f at x_0 is twice its rate of change at 1. Find x_0 if
(a) $f(x) = x^2$.   (b) $f(x) = 2x^3$.   (c) $f(x) = \sqrt{x}$.

2. An object moves on a line according to the law $x = \sqrt{t+1}$. (a) Show that the acceleration is negative and proportional to the cube of the velocity. (b) Use differentials to find numerical estimates for the position, velocity, and acceleration of the object at time $t = 17$. Base your estimate on $t = 15$.

*3. A rectangular banner has a red border and a white center. The width of the border at top and bottom is 8 inches and along the sides it is 6 inches. The total area is 27 square feet. What should be the dimensions of the banner if the area of the white center is to be a maximum?

*4. Determine the coefficients A, B, C so that the curve $y = Ax^2 + Bx + C$ will pass through the point $(1, 3)$ and be tangent to the line $x - y + 1 = 0$ at the point $(2, 3)$.

5. Determine the coefficients A, B, C, D so that the curve $y = Ax^3 + Bx^2 + Cx + D$ will be tangent to the line $y = 5x - 4$ at the point $(1, 1)$ and tangent to the line $y = 9x$ at the point $(-1, -9)$.

*6. Determine the coefficients A, B so that the function $y = Ax^{-1/2} + Bx^{1/2}$ has a minimum value of 6 at $x = 9$.

7. Find a relation between A, B, C given that the function $y = Ax^3 + 3Bx^2 + 3Cx$ has a critical point in common with its derivative.

Test the following functions for extreme values.

*8. $f(x) = \dfrac{x^2 + x + 4}{x^2 + 2x + 4}$.   9. $f(x) = \dfrac{1}{x} + \dfrac{1}{1-x}$.

*10. $f(x) = \dfrac{(x-1)(2-x)}{x^2}$.   11. $f(x) = \dfrac{x^2}{x^2 + a^2}$.

*12. If three sides of a trapezoid are each 10 inches long, how long must the fourth side be if the area is to be a maximum?

13. A rectangular box with a square base and an open top is to be made. Find the volume of the largest box that can be made from 1200 square feet of material.

*14. A point moves along the parabola $y^2 = 12x$ in such a way that its x-coordinate increases uniformly at the rate of 2 inches per second. At what point do the x-coordinate and the y-coordinate increase at the same rate?

15. A circular metal plate is expanded by heat. Given that the radius grows at the rate of 0.02 inches per second, at what rate is the area of the plate increasing when the radius is 2 inches?

*16. A light is hung 12 feet directly above a straight horizontal path on which a boy 5 feet tall is walking. How fast is the boy's shadow lengthening if he is walking away from the light at the rate of 168 feet per minute?

*17. One ship was sailing south at 6 miles per hour; another east at 8 miles per hour. At 4 p.m. the second crossed the track of the first, where the first had been 2 hours before. (a) How was the distance between the ships changing at 3 p.m.? (b) How at 5 p.m.?

*18. A railroad track crosses a highway at an angle of 60°. A locomotive is 500 feet from the intersection and moving away from it at the rate of 60 miles per hour. An automobile is 500 feet from the intersection and moving toward it at the rate of 30 miles per hour. What is the rate of change of the distance between them?   HINT: Use the law of cosines.

19. A rectangular box with square base and top is to be made to contain 1250 cubic feet. The material for the base costs 35 cents per square foot, for the top 15 cents per square foot, and for the sides 20 cents per square foot. Find the dimensions that will minimize the cost of the box.

*20. The period (P seconds) of a complete oscillation of a pendulum of length l inches is given by the formula $P = 0.32\sqrt{l}$. (a) Find the rate of change of the period with respect to the length when $l = 9$ inches. (b) Use a differential to estimate the change in P caused by an increase in l from 9 to 9.2 inches.

*21. A stone dropped from a balloon which was rising at the rate of 5 feet per second reached the ground in 8 seconds. How high was the balloon when the stone was dropped?

22. In Exercise 21, if the balloon had been falling at the rate of 5 feet per second, how long would the stone have taken to reach the ground?

*23. A ball thrown upward from the ground reaches a height of 24 feet in 1 second. How high will the ball go?

24. An isosceles trapezoid is inscribed in a circle of radius r. (Figure 4.10.1) Given that one base has length $2r$, find the length of the other base if the area is a maximum.

FIGURE 4.10.1   FIGURE 4.10.2

25. Find the dimensions of the rectangular solid of maximum volume that can be cut from a solid sphere of radius r. (Figure 4.10.2)

*26. Find the base and altitude of the isosceles triangle of minimum area that circumscribes the ellipse $b^2x^2 + a^2y^2 = a^2b^2$ if the base is parallel to the x-axis.

27. A triangle is formed in the first quadrant by the coordinate axes and a tangent to the ellipse $b^2x^2 + a^2y^2 = a^2b^2$. Find the point of tangency if the triangle is to have minimum area.

*28. The diameter and height of a right circular cylinder are found at a certain instant to be 10 and 20 inches respectively. If the diameter is increasing at the rate of 1 inch per minute, what change in the height will keep the volume constant?

*29. The total cost of producing Q units per week is

$$C(Q) = \tfrac{1}{3}Q^3 - 20Q^2 + 600Q + 1000 \quad \text{dollars.}$$

Given that the total revenues are

$$R(Q) = 420Q - 2Q^2 \quad \text{dollars,}$$

find the output which maximizes profit.

30. A steel plant is capable of producing Q_1 tons per day of low-grade steel and Q_2 tons per day of high-grade steel, where

$$Q_2 = \frac{40 - 5Q_1}{10 - Q_1}.$$

If the market price of low-grade steel is half that of high-grade steel, show that about $5\tfrac{1}{2}$ tons of low-grade steel should be produced per day for maximum receipts.

31. The total cost of producing Q articles per week is $aQ^2 + bQ + c$ dollars, and the price at which each can be sold is $p = \beta - \alpha Q^2$. Show that the output for maximum profit is

$$Q = \frac{\sqrt{a^2 + 3\alpha(\beta - b)} - a}{3\alpha}. \quad \text{(Take } a, b, c, \alpha, \beta \text{ as positive.)}$$

*32. In a certain industry an output of Q units can be obtained at a total cost of

$$C(Q) = aQ^2 + bQ + c \quad \text{dollars,}$$

bringing in total revenues of

$$R(Q) = \beta Q - \alpha Q^2 \quad \text{dollars.}$$

The government decides to impose an excise tax on the product. What tax rate (dollars per unit) will maximize the government's revenues from this tax? (Take a, b, c, α, β as positive.)

*33. Find the absolute maximum value of $y = x(r^2 + x^2)^{-3/2}$.

34. Given that $\overline{PQ}$ is the longest or shortest line segment that can be drawn from $P(a, b)$ to the differentiable curve $y = f(x)$, show that $\overline{PQ}$ is perpendicular to the tangent to the curve at Q.

*35. What point on the curve $y = x^{3/2}$ is closest to $P(\tfrac{1}{2}, 0)$?

*36. The equation of the path of a ball is $y = mx - \tfrac{1}{800}(m^2 + 1)x^2$, where the origin is taken as the point from which the ball is thrown and m is the slope of the curve at the origin. For what value of m will the ball strike (a) at the greatest distance along the same horizontal level? (b) at the greatest height on a vertical wall 300 feet away?

*37. A horizontal trough 12 feet long has a vertical cross section in the shape of a trapezoid, the bottom being 3 feet wide and the sides inclined to the vertical at an angle whose sine is $\tfrac{4}{5}$. Water is being poured into it at the rate of 10 cubic feet per minute. How fast is the water level rising when the water is 2 feet deep?

38. In Exercise 37, at what rate is the water being drawn from the trough if the level is falling 0.1 feet per minute when the water is 3 feet deep?

*39. A point P moves along the parabola $y = x^2$ so that its x-coordinate increases at the constant rate of k units per second. The projection of P on the x-axis is M. At what rate is the area of triangle OMP changing when P is at the point where $x = a$?

*40. If $y = 4x - x^3$ and x is increasing steadily at the rate of $\frac{1}{3}$ unit per second, how fast is the slope of the graph changing at the instant when $x = 2$?

41. The sum of the surface areas of a sphere and a cube being given, show that the sum of the volumes will be least when the diameter of the sphere is equal to the edge of the cube. When will the sum of the volumes be greatest?

*42. A miner wishes to dig a tunnel from a point A to a point B 200 feet below and 600 feet to the east of A. Below the level of A it is bedrock, and above A is soft earth. If the cost of tunneling through earth is $5 per linear foot and through rock $13, find the minimum cost of a tunnel.

*43. The distance between two sources of heat A and B, with intensities a and b, respectively, is s. The intensity of heat at a point P between A and B is given by the formula

$$I = \frac{a}{x^2} + \frac{b}{(s - x)^2},$$

where x is the distance between P and A. For what position P will the temperature be lowest?

44. Let $P(x_0, y_0)$ be a point in the first quadrant. Draw a line through P which cuts the positive x-axis at $A(a, 0)$ and the positive y-axis at $B(0, b)$. Find a and b in each of the following cases:
(a) when the area of $\triangle OAB$ is a minimum.
(b) when the length of $\overline{AB}$ is a minimum.
(c) when $a + b$ is a minimum.
(d) when the perpendicular distance from O to $\overline{AB}$ is a maximum.

Integration

5

5.1 Motivation: An Area Problem; A Speed-Distance Problem

We were led to the notion of derivative by considering a tangent problem. By considering two new problems, we will be led to the notion of integral.

An Area Problem

In Figure 5.1.1 we display a region Ω which is bounded above by the graph of a continuous nonnegative function f, bounded below by the x-axis, bounded to the left by $x = a$, and bounded to the right by $x = b$.

FIGURE 5.1.1

What number, if any, should be called the area of Ω? This is the question we want to resolve.

To do this, we begin by splitting up the interval $[a, b]$ into a finite number of non-overlapping intervals

$$[x_0, x_1], [x_1, x_2], \ldots, [x_{n-1}, x_n] \quad \text{with} \quad a = x_0 < x_1 < \cdots < x_n = b.$$

This breaks up the region Ω into n subregions:

$$\Omega_1, \Omega_2, \ldots, \Omega_n.$$ (Figure 5.1.2)

FIGURE 5.1.2

We can estimate the total area of Ω by estimating the area of each subregion Ω_j and adding up the results. Let's denote by M_j the maximum value of f on $[x_{j-1}, x_j]$ and by m_j the minimum value. Consider now the rectangles r_j and R_j of Figure 5.1.3.

FIGURE 5.1.3

Since

$$r_j \subseteq \Omega_j \subseteq R_j,$$

we must have

$$\text{area of } r_j \le \text{area of } \Omega_j \le \text{area of } R_j.$$

Taking the area of rectangles to be given by length times width, we obtain

$$m_j(x_j - x_{j-1}) \le \text{area of } \Omega_j \le M_j(x_j - x_{j-1}).$$

Setting $x_j - x_{j-1} = \Delta x_j$ we have

$$m_j \Delta x_j \le \text{area of } \Omega_j \le M_j \Delta x_j.$$

This is presumably true for $j = 1, j = 2, \ldots$ and so on. Adding up these results, we get on the one hand

(5.1.1) $$m_1 \Delta x_1 + m_2 \Delta x_2 + \cdots + m_n \Delta x_n \le \text{area of } \Omega$$

and on the other hand

(5.1.2)   area of $\Omega \leq M_1 \Delta x_1 + M_2 \Delta x_2 + \cdots + M_n \Delta x_n.$

A sum of the form

$$m_1 \Delta x_1 + m_2 \Delta x_2 + \cdots + m_n \Delta x_n \qquad \text{(Figure 5.1.4)}$$

is called a *lower sum* for f. A sum of the form

$$M_1 \Delta x_1 + M_2 \Delta x_2 + \cdots + M_n \Delta x_n \qquad \text{(Figure 5.1.5)}$$

is called an *upper sum* for f.

area of shaded region is a lower
sum for f

FIGURE 5.1.4

area of shaded region is an upper
sum for f

FIGURE 5.1.5

Inequalities (5.1.1) and (5.1.2) together tell us that for a number to be a candidate for the title of area of Ω it must be greater than or equal to every lower sum for f and less than or equal to every upper sum for f. By an argument that we omit, it can be proven that with f continuous on $[a, b]$ there is one and only one such number. This number we call the area of Ω. $\square$

Later we'll return to the subject of area. At this point we turn to a speed-distance problem. As you'll see, this new problem can be solved by the same technique that we just applied to the area problem.

A Speed-Distance Problem

If an object moves at a constant speed for a given period of time, then the total distance traveled is given by the familiar formula

$$\text{distance} = (\text{speed}) \times (\text{time}).$$

Suppose now that during the course of the motion the speed does not remain constant but instead varies continuously. How can the total distance traveled be computed then?

To answer this question, we suppose that the motion begins at time a, ends at time b, and that during the time interval $[a, b]$ the speed varies continuously.

As in the case of the area problem we begin by breaking up the interval $[a, b]$ into a finite number of non-overlapping intervals

$$[t_0, t_1], [t_1, t_2], \ldots, [t_{n-1}, t_n] \quad \text{with} \quad a = t_0 < t_1 < \cdots < t_n = b.$$

On each subinterval $[t_{j-1}, t_j]$ the object attains a certain maximum speed M_j and a certain minimum speed m_j. (How do we know this?) If throughout the time interval $[t_{j-1}, t_j]$ the object were to move constantly at its minimum speed, m_j, then it would cover a distance of $m_j \Delta t_j$ units. If instead it were to move constantly at its maximum speed, M_j, then it would cover a distance of $M_j \Delta t_j$ units. As it is, the actual distance traveled, call it s_j, must lie somewhere in between; namely, we must have

$$m_j \Delta t_j \leq s_j \leq M_j \Delta t_j.$$

The total distance traveled during the time interval $[a, b]$, call it s, must be the sum of the distances traveled during the subintervals $[t_{j-1}, t_j]$. In other words we must have

$$s = s_1 + s_2 + \cdots + s_n.$$

Since

$$m_1 \Delta t_1 \leq s_1 \leq M_1 \Delta t_1$$
$$m_2 \Delta t_2 \leq s_2 \leq M_2 \Delta t_2$$
$$\cdot$$
$$\cdot$$
$$\cdot$$
$$m_n \Delta t_n \leq s_n \leq M_n \Delta t_n,$$

it follows by the addition of these inequalities that

$$m_1 \Delta t_1 + m_2 \Delta t_2 + \cdots + m_n \Delta t_n \leq s \leq M_1 \Delta t_1 + M_2 \Delta t_2 + \cdots + M_n \Delta t_n.$$

A sum of the form

$$m_1 \Delta t_1 + m_2 \Delta t_2 + \cdots + m_n \Delta t_n$$

is called a *lower sum* for the speed function. A sum of the form

$$M_1 \Delta t_1 + M_2 \Delta t_2 + \cdots + M_n \Delta t_n$$

is called an *upper sum* for the speed function. The inequality we just obtained for *s* tells us that *s* must be greater than or equal to every lower sum for the speed function and less than or equal to every upper sum for the speed function. As in the case of the area problem, it turns out that there is one and only one such number, and this is the total distance traveled. □

5.2 The Definite Integral of a Continuous Function

The procedure applied to the two problems of the last section is called integration, and the numbers so obtained are called definite integrals. Our purpose here is to establish these notions more precisely.

(5.2.1)
> By a *partition* of $[a, b]$ we mean a finite subset of $[a, b]$ which contains the points a and b.

It is convenient to index the elements of a partition according to their natural order. Thus if we write

$$P = \{x_0, x_1, \ldots, x_n\} \text{ is a partition of } [a, b],$$

you can conclude that

$$a = x_0 < x_1 < \cdots < x_n = b.$$

Example. The sets

$$\{0, 1\}, \quad \{0, \tfrac{1}{2}, 1\}, \quad \{0, \tfrac{1}{4}, \tfrac{1}{2}, 1\}, \quad \text{and} \quad \{0, \tfrac{1}{4}, \tfrac{1}{3}, \tfrac{1}{2}, \tfrac{5}{8}, 1\}$$

are all partitions of the interval $[0, 1]$. □

If $P = \{x_0, x_1, \ldots, x_n\}$ is a partition of $[a, b]$, then P breaks up $[a, b]$ into a finite number of non-overlapping intervals

$[x_0, x_1], [x_1, x_2], \ldots, [x_{n-1}, x_n]$ of lengths $\Delta x_1, \Delta x_2, \ldots, \Delta x_n$, respectively.

Suppose now that f is continuous on $[a, b]$. Then on each such interval $[x_{j-1}, x_j]$ the function f takes on a maximum value, M_j, and a minimum value, m_j.

(5.2.2)
> The number
> $$U_f(P) = M_1 \Delta x_1 + M_2 \Delta x_2 + \cdots + M_n \Delta x_n$$
> is called the *P upper sum* for f, and the number
> $$L_f(P) = m_1 \Delta x_1 + m_2 \Delta x_2 + \cdots + m_n \Delta x_n$$
> is called the *P lower sum* for f.

Example. If $[a, b] = [0, 1]$, $P = \{0, \tfrac{1}{4}, \tfrac{1}{2}, 1\}$,   and $f(x) = x^2$,   then
$$U_f(P) = \tfrac{1}{16}(\tfrac{1}{4}) + \tfrac{1}{4}(\tfrac{1}{4}) + 1(\tfrac{1}{2}) = \tfrac{37}{64} \quad \text{and} \quad L_f(P) = 0(\tfrac{1}{4}) + \tfrac{1}{16}(\tfrac{1}{4}) + \tfrac{1}{4}(\tfrac{1}{2}) = \tfrac{9}{64}. \quad \square$$

Example. If $[a, b] = [-1, 0]$, $P = \{-1, -\frac{1}{4}, 0\}$, and $f(x) = -(x + 1)$, then

$$U_f(P) = (0)(\tfrac{3}{4}) + (-\tfrac{3}{4})(\tfrac{1}{4}) = -\tfrac{3}{16} \quad \text{and} \quad L_f(P) = (-\tfrac{3}{4})(\tfrac{3}{4}) + (-1)(\tfrac{1}{4}) = -\tfrac{13}{16}. \quad \square$$

By an argument that we omit here (it appears in Appendix B.4) it can be proved that, with f continuous on $[a, b]$, there is one and only one number I which satisfies the inequality

$$L_f(P) \leq I \leq U_f(P) \quad \text{for } \textit{all} \text{ partitions } P \text{ of } [a, b].$$

This is the number we want.

Definition 5.2.3   The Definite Integral

The unique number I which satisfies the inequality

$$L_f(P) \leq I \leq U_f(P) \quad \text{for all partitions } P \text{ of } [a, b]$$

is called the *definite integral* (or more simply the *integral*) of f from a to b and is denoted by

$$\int_a^b f(x)\, dx.\dagger$$

The symbol $\int$ dates back to Leibniz and is called an *integral sign*. It is really an elongated S—as in *Sum*. The numbers a and b are called *the limits of integration*, and we'll speak of *integrating* a function from a to b.$\dagger\dagger$ The function being integrated is called the *integrand*.

In the expression

$$\int_a^b f(x)\, dx$$

the letter x is a "dummy variable"; in other words, it may be replaced by any other letter not already engaged. Thus, for example, there is no difference between

$$\int_a^b f(x)\, dx, \quad \int_a^b f(t)\, dt, \quad \text{and} \quad \int_a^b f(z)\, dz.$$

All of these denote the definite integral of f from a to b.

$\dagger$ This is not the only notation. Some mathematicians omit the dx and write simply

$$\int_a^b f.$$

We will keep the dx. As we go on, you'll see that it does serve a useful purpose.

$\dagger\dagger$ The word "limit" in this setting has no connection with the limits discussed in Chapter 2.

Section 5.1 gives two immediate applications of the definite integral:

I. If f is nonnegative on $[a, b]$, then

$$A = \int_a^b f(x)\, dx$$

gives the area below the graph of f. See Figure 5.1.1.

II. If $|v(t)|$ is the speed of an object at time t, then

$$s = \int_a^b |v(t)|\, dt$$

gives the distance traveled from time a to time b.

We'll come back to these ideas later. Right now, some simple computations.

Example. If $f(x) = \alpha$ for all x in $[a, b]$, then

$$\int_a^b f(x)\, dx = \alpha(b - a).$$

To see this, take $P = \{x_0, x_1, \ldots, x_n\}$ as an arbitrary partition of $[a, b]$. Since f is constantly α on $[a, b]$, it is constantly α on each subinterval $[x_{j-1}, x_j]$. Thus M_j and m_j are both α. It follows that

$$U_f(P) = \alpha \Delta x_1 + \alpha \Delta x_2 + \cdots + \alpha \Delta x_n$$
$$= \alpha(\Delta x_1 + \Delta x_2 + \cdots + \Delta x_n) = \alpha(b - a)$$

and

$$L_f(P) = \alpha \Delta x_1 + \alpha \Delta x_2 + \cdots + \alpha \Delta x_n$$
$$= \alpha(\Delta x_1 + \Delta x_2 + \cdots + \Delta x_n) = \alpha(b - a).$$

Obviously then

$$L_f(P) \leq \alpha(b - a) \leq U_f(P).$$

Since this inequality holds for all partitions P of $[a, b]$, we can conclude that

$$\int_a^b f(x)\, dx = \alpha(b - a). \quad \square$$

This last result is usually written

(5.2.4)
$$\boxed{\int_a^b \alpha\, dx = \alpha(b - a).}$$

Thus for instance

$$\int_{-1}^1 4\, dx = 4(2) = 8 \quad \text{and} \quad \int_4^{10} -2\, dx = -2(6) = -12.$$

If $\alpha > 0$, the region below the graph is a rectangle of height α erected on the interval $[a, b]$. (Figure 5.2.1) The integral gives the area of this rectangle.

FIGURE 5.2.1

Example

(5.2.5)

$$\int_a^b x \, dx = \tfrac{1}{2}(b^2 - a^2).$$

To see this, take $P = \{x_0, x_1, \ldots, x_n\}$ as an arbitrary partition of $[a, b]$. On each subinterval $[x_{j-1}, x_j]$ the function $f(x) = x$ has a maximum $M_j = x_j$ and a minimum $m_j = x_{j-1}$. It follows that

$$U_f(P) = x_1 \Delta x_1 + x_2 \Delta x_2 + \cdots + x_n \Delta x_n$$
$$= x_1(x_1 - x_0) + x_2(x_2 - x_1) + \cdots + x_n(x_n - x_{n-1})$$

and

$$L_f(P) = x_0 \Delta x_1 + x_1 \Delta x_2 + \cdots + x_{n-1} \Delta x_n$$
$$= x_0(x_1 - x_0) + x_1(x_2 - x_1) + \cdots + x_{n-1}(x_n - x_{n-1}).$$

For each index j

$$x_{j-1} \leq \tfrac{1}{2}(x_j + x_{j-1}) \leq x_j,$$

and therefore

$$x_{j-1}(x_j - x_{j-1}) \leq \tfrac{1}{2}(x_j^2 - x_{j-1}^2) \leq x_j(x_j - x_{j-1}).$$

Thus

$$L_f(P) \leq \tfrac{1}{2}(x_1^2 - x_0^2) + \tfrac{1}{2}(x_2^2 - x_1^2) + \cdots + \tfrac{1}{2}(x_n^2 - x_{n-1}^2) \leq U_f(P).$$

The middle collapses to

$$\tfrac{1}{2}(x_n^2 - x_0^2) = \tfrac{1}{2}(b^2 - a^2),$$

and consequently

$$L_f(P) \leq \tfrac{1}{2}(b^2 - a^2) \leq U_f(P).$$

Since P was chosen arbitrarily, we can conclude that this inequality holds for all partitions P of $[a, b]$. It follows therefore that

$$\int_a^b x \, dx = \tfrac{1}{2}(b^2 - a^2). \quad \square$$

If the interval $[a, b]$ lies to the right of the origin, then the region below the graph of

$$f(x) = x, \qquad x \in [a, b]$$

is the trapezoid of Figure 5.2.2. The integral

$$\int_a^b x \, dx$$

gives the area of this trapezoid.

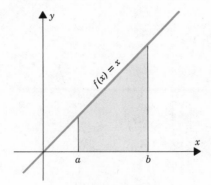

FIGURE 5.2.2

Example

$$\int_0^1 x^2 \, dx = \tfrac{1}{3}.$$

This time take $P = \{x_0, x_1, \dots, x_n\}$ as a partition of $[0, 1]$. On each subinterval $[x_{j-1}, x_j]$ the function $f(x) = x^2$ has a maximum $M_j = x_j^2$ and a minimum $m_j = x_{j-1}^2$. It follows that

$$U_f(P) = x_1^2 \Delta x_1 + \cdots + x_n^2 \Delta x_n$$
$$= x_1^2(x_1 - x_0) + \cdots + x_n^2(x_n - x_{n-1})$$

and

$$L_f(P) = x_0^2 \Delta x_1 + \cdots + x_{n-1}^2 \Delta x_n$$
$$= x_0^2(x_1 - x_0) + \cdots + x_{n-1}^2(x_n - x_{n-1}).$$

For each index j,

$$x_{j-1}^2 \le \tfrac{1}{3}(x_{j-1}^2 + x_{j-1}x_j + x_j^2) \le x_j^2. \qquad \text{(check this out)}$$

If we now multiply this inequality by $x_j - x_{j-1}$, the middle term reduces to

$$\tfrac{1}{3}(x_j^3 - x_{j-1}^3), \qquad \text{(check this out)}$$

and consequently we get

$$x_{j-1}^2(x_j - x_{j-1}) \leq \tfrac{1}{3}(x_j^3 - x_{j-1}^3) \leq x_j^2(x_j - x_{j-1}).$$

Adding up the terms on the left, we get $L_f(P)$. Adding up the middle terms, we get a collapsing sum which reduces to $\tfrac{1}{3}$:

$$\tfrac{1}{3}(x_1^3 - x_0^3 + x_2^3 - x_1^3 + \cdots + x_n^3 - x_{n-1}^3) = \tfrac{1}{3}(x_n^3 - x_0^3) = \tfrac{1}{3}(1^3 - 0^3) = \tfrac{1}{3}.$$

Adding up the terms on the right, we get $U_f(P)$. It follows therefore that

$$L_f(P) \leq \tfrac{1}{3} \leq U_f(P).$$

Since P was chosen arbitrarily, we can conclude that this inequality holds for all P. It follows therefore that

$$\int_0^1 x^2 \, dx = \tfrac{1}{3}. \quad \square$$

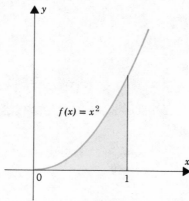

area of the shaded region: $\int_0^1 x^2 \, dx = \tfrac{1}{3}$

FIGURE 5.2.3

Exercises

Find $L_f(P)$ and $U_f(P)$ for each of the following:

*1. $f(x) = 2x$, $x \in [0, 1]$; $P = \{0, \tfrac{1}{4}, \tfrac{1}{2}, 1\}$.

2. $f(x) = 1 - x$, $x \in [0, 2]$; $P = \{0, \tfrac{1}{3}, \tfrac{3}{4}, 1, 2\}$.

*3. $f(x) = x^2$, $x \in [-1, 0]$; $P = \{-1, -\tfrac{1}{2}, -\tfrac{1}{4}, 0\}$.

4. $f(x) = 1 - x^2$, $x \in [0, 1]$; $P = \{0, \tfrac{1}{4}, \tfrac{1}{2}, 1\}$.

*5. $f(x) = 1 + x^3$, $x \in [0, 1]$; $P = \{0, \tfrac{1}{2}, 1\}$.

6. $f(x) = \sqrt{x}$, $x \in [0, 1]$; $P = \{0, \tfrac{1}{25}, \tfrac{4}{25}, \tfrac{9}{25}, \tfrac{16}{25}, 1\}$.

*7. $f(x) = |x|$, $x \in [-1, 1]$; $P = \{-1, -\tfrac{1}{2}, 0, \tfrac{1}{4}, 1\}$.

8. $f(x) = |x|$, $x \in [-1, 1]$; $P = \{-1, -\tfrac{1}{2}, -\tfrac{1}{4}, \tfrac{1}{2}, 1\}$.

*9. $f(x) = x^2$, $x \in [-1, 1]$; $P = \{-1, -\tfrac{1}{4}, \tfrac{1}{4}, \tfrac{1}{2}, 1\}$.

10. $f(x) = x^2$, $x \in [-1, 1]$; $P = \{-1, -\tfrac{3}{4}, -\tfrac{1}{4}, \tfrac{1}{4}, \tfrac{1}{2}, 1\}$.

11. Explain why each of the following statements must be false. Take P as a partition of $[-1, 1]$.

   (a) $L_f(P) = 3$ and $U_f(P) = 2$.

   (b) $L_f(P) = 3$, $U_f(P) = 6$,   and $\displaystyle\int_{-1}^{1} f(x)\, dx = 2$.

   (c) $L_f(P) = 3$, $U_f(P) = 6$,   and $\displaystyle\int_{-1}^{1} f(x)\, dx = 10$.

12. (a) Given that $P = \{x_0, x_1, \ldots, x_n\}$ is an arbitrary partition of $[a, b]$, find $L_f(P)$ and $U_f(P)$ if $f(x) = 1 + 2x$.

   (b) Use your answers to part (a) to evaluate

$$\int_a^b (1 + 2x)\, dx.$$

*13. (a) Given that $P = \{x_0, x_1, \ldots, x_n\}$ is an arbitrary partition of $[a, b]$, find $L_f(P)$ and $U_f(P)$ if $f(x) = -3x$.

   (b) Use your answers to part (a) to evaluate

$$\int_a^b -3x\, dx.$$

Optional | 14. Evaluate

$$\int_0^1 x^3\, dx$$

   by the methods of this section.

5.3   The function $F(x) = \int_a^x f(t)\, dt$

The evaluation of

$$\int_a^b f(x)\, dx$$

by its definition as the unique number I satisfying

$$L_f(P) \leq I \leq U_f(P) \qquad \text{for all partitions } P \text{ of } [a, b]$$

is at best a laborious process. Try for example to evaluate

$$\int_2^5 \left(x^3 + x^{5/2} - \frac{2x}{1 - x^2} \right) dx \quad \text{or} \quad \int_{-1/2}^{1/4} \frac{x}{1 - x^2}\, dx$$

in this manner. Fortunately there is another way that we can evaluate such integrals. This other way doesn't work for all functions, but it does work for many of the elementary functions. This other way of evaluating definite integrals is described in Section 5.4 in what is called *the fundamental theorem of integral calculus*. Its validity depends on a connection between differentiation and integration described in Theorem 5.3.5. To prove Theorem 5.3.5 we first need some other results.

Theorem 5.3.1

Let P and Q be partitions of the interval $[a, b]$. If $P \subseteq Q$, then

$$L_f(P) \leq L_f(Q) \quad \text{and} \quad U_f(Q) \leq U_f(P).$$

This result is easy to see. By adding points to a partition we tend to make the sub-intervals $[x_{j-1}, x_j]$ smaller. This tends to make the minima, m_j, larger and the maxima, M_j, smaller. Thus the lower sums are made bigger, and the upper sums are made smaller. This idea is illustrated in Figures 5.3.1 and 5.3.2.

as points are added to a partition, the lower sums tend to get bigger

FIGURE 5.3.1

as points are added to a partition, the upper sums tend to get smaller

FIGURE 5.3.2

The next theorem says that the integral is additive on intervals.

Theorem 5.3.2

$$\text{If} \quad a < c < b, \quad \text{then} \quad \int_a^c f(t)\, dt + \int_c^b f(t)\, dt = \int_a^b f(t)\, dt.$$

For nonnegative functions this theorem is easily understood in terms of area. See Figure 5.3.3. The area of part I is given by

$$\int_a^c f(t)\, dt.$$

The area of part II is given by

$$\int_c^b f(t)\,dt.$$

The area of the entire region is given by

$$\int_a^b f(t)\,dt.$$

The theorem says that

the area of part I + the area of part II = the area of the entire region.

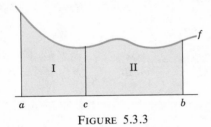

FIGURE 5.3.3

Theorem 5.3.2 can also be viewed in terms of speed and distance. With $f(t) \geq 0$, we can interpret $f(t)$ as the speed of an object at time t. With this interpretation,

$$\int_a^c f(t)\,dt = \text{distance traveled from time } a \text{ to time } c,$$

$$\int_c^b f(t)\,dt = \text{distance traveled from time } c \text{ to time } b.,$$

$$\int_a^b f(t)\,dt = \text{distance traveled from time } a \text{ to time } b.$$

It is not surprising that the sum of the first two distances should equal the third.

The fact that the additivity theorem is so easy to understand does not relieve us of the necessity to prove it. Here is a proof.

PROOF OF THEOREM 5.3.2.   To prove the theorem we need only show that for each partition P of $[a, b]$

$$L_f(P) \leq \int_a^c f(t)\,dt + \int_c^b f(t)\,dt \leq U_f(P). \qquad \text{(why?)}$$

We begin with

$$P = \{x_0, x_1, \ldots, x_n\}$$

as an arbitrary partition of $[a, b]$. Since the partition $Q = P \cup \{c\}$ contains P, we know from the first theorem that

(1) $L_f(P) \leq L_f(Q) \quad \text{and} \quad U_f(Q) \leq U_f(P).$

The sets

$$Q_1 = Q \cap [a, c] \quad \text{and} \quad Q_2 = Q \cap [c, b]$$

are partitions of $[a, c]$ and $[c, b]$ respectively and clearly satisfy

$$L_f(Q_1) + L_f(Q_2) = L_f(Q) \quad \text{and} \quad U_f(Q_1) + U_f(Q_2) = U_f(Q).$$

Since

$$L_f(Q_1) \leq \int_a^c f(t)\, dt \leq U_f(Q_1) \quad \text{and} \quad L_f(Q_2) \leq \int_c^b f(t)\, dt \leq U_f(Q_2),$$

we have

$$L_f(Q_1) + L_f(Q_2) \leq \int_a^c f(t)\, dt + \int_c^b f(t)\, dt \leq U_f(Q_1) + U_f(Q_2).$$

It follows that

$$L_f(Q) \leq \int_a^c f(t)\, dt + \int_a^b f(t)\, dt \leq U_f(Q)$$

and thus by (1) that

$$L_f(P) \leq \int_a^c f(t)\, dt + \int_c^b f(t)\, dt \leq U_f(P). \quad \square$$

Until now we have considered the definite integral only over an interval $[a, b]$; namely, in writing

$$\int_a^b f(t)\, dt$$

we have assumed that a was less than b. We can also integrate in the other direction; by definition

(5.3.3)
$$\boxed{\int_b^a f(t)\, dt = -\int_a^b f(t)\, dt.}$$

Moreover, the integral from any point c to itself is defined to be zero:

(5.3.4)
$$\boxed{\int_c^c f(t)\, dt = 0.}$$

With these two extra conventions it's not hard to show that, if f is continuous on an interval I, then the additivity condition

$$\int_a^c f(t)\, dt + \int_c^b f(t)\, dt = \int_a^b f(t)\, dt$$

holds for all choices of a, b, c from I, no matter what the order is.

We are now ready to establish the basic link that exists between differentiation and integration. To describe this link, we begin with a function f that is continuous on a closed interval $[a, b]$. For each x in $[a, b]$ the integral

$$\int_a^x f(t)\, dt$$

is a number, and consequently we can define a function F on $[a, b]$ by setting

$$F(x) = \int_a^x f(t)\, dt.$$

Theorem 5.3.5

If f is continuous on $[a, b]$, the function F defined on $[a, b]$ by setting

$$F(x) = \int_a^x f(t)\, dt$$

is continuous on $[a, b]$, differentiable on (a, b), and has derivative

$$F'(x) = f(x) \qquad \text{for all } x \text{ in } (a, b).$$

PROOF.   We begin with x in the half-open interval $[a, b)$ and show that

$$\lim_{h \downarrow 0} \frac{F(x + h) - F(x)}{h} = f(x).$$

(For a pictorial outline of the proof in the case that f is nonnegative see Figure 5.3.4.)

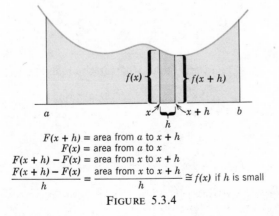

$F(x + h)$ = area from a to $x + h$
$F(x)$ = area from a to x
$F(x + h) - F(x)$ = area from x to $x + h$
$\dfrac{F(x + h) - F(x)}{h} = \dfrac{\text{area from } x \text{ to } x + h}{h} \cong f(x)$ if h is small

FIGURE 5.3.4

If $x < x + h \leq b$, then

$$F(x + h) - F(x) = \int_a^{x+h} f(t)\, dt - \int_a^x f(t)\, dt.$$

It follows that

$$F(x + h) - F(x) = \int_x^{x+h} f(t)\, dt. \qquad \text{(justify this step)}$$

Now we set

$$M_h = \text{maximum value of } f \text{ on } [x, x + h]$$

and

$$m_h = \text{minimum value of } f \text{ on } [x, x + h].$$

Since

$$M_h[(x + h) - x] = M_h \cdot h$$

is an upper sum for f on $[x, x + h]$ and

$$m_h[(x + h) - x] = m_h \cdot h$$

is a lower sum for f on $[x, x + h]$, we know that

$$m_h \cdot h \le \int_x^{x+h} f(t)\, dt \le M_h \cdot h$$

and thus that

$$m_h \le \frac{F(x + h) - F(x)}{h} \le M_h.$$

Since f is continuous on $[x, x + h]$, it follows that

(1) $$\lim_{h \downarrow 0} m_h = f(x) = \lim_{h \downarrow 0} M_h$$

and thus that

(2) $$\lim_{h \downarrow 0} \frac{F(x + h) - F(x)}{h} = f(x).$$

In a similar manner you can verify that, if x is in the half-open interval $(a, b]$, then

(3) $$\lim_{h \uparrow 0} \frac{F(x + h) - F(x)}{h} = f(x).$$

If x is in the open interval (a, b), then both (2) and (3) hold, and therefore

$$F'(x) = \lim_{h \to 0} \frac{F(x + h) - F(x)}{h} = f(x).$$

This proves the differentiability condition. Applying (2) to $x = a$, we have

$$\lim_{h \downarrow 0} \frac{F(a + h) - F(a)}{h} = f(a).$$

This implies that

$$\lim_{h \downarrow 0} F(a + h) - F(a) = 0$$

and thus that

$$\lim_{h \downarrow 0} F(a + h) = F(a).$$

This shows that F is continuous from the right at a. That F is continuous from the left at b can be shown by applying (3) to $x = b$. $\square$

Exercises

*1. Given that

$$\int_0^1 f(x)\, dx = 6, \quad \int_0^2 f(x)\, dx = 4, \quad \int_2^5 f(x)\, dx = 1,$$

find each of the following:

(a) $\displaystyle\int_0^5 f(x)\, dx.$ (b) $\displaystyle\int_1^2 f(x)\, dx.$ (c) $\displaystyle\int_1^5 f(x)\, dx.$

(d) $\displaystyle\int_0^0 f(x)\, dx.$ (e) $\displaystyle\int_2^0 f(x)\, dx.$ (f) $\displaystyle\int_5^1 f(x)\, dx.$

2. Explain why each of the following statements must be false.
   *(a) $U_f(P_1) = 4$ for the partition $P_1 = \{0, 1, \frac{3}{2}, 2\}$ and
   $U_f(P_2) = 5$ for the partition $P_2 = \{0, \frac{1}{4}, 1, \frac{3}{2}, 2\}$.
   (b) $L_f(P_1) = 5$ for the partition $P_1 = \{0, 1, \frac{3}{2}, 2\}$ and
   $L_f(P_2) = 4$ for the partition $P_2 = \{0, \frac{1}{4}, 1, \frac{3}{2}, 2\}$.

3. For $x > -1$ set $F(x) = \displaystyle\int_0^x t\sqrt{t + 1}\, dt.$

   (a) Find $F(0)$.   *(b) Find $F'(x)$.   (c) Find $F'(2)$.
   *(d) Express $F(2)$ as an integral of $t\sqrt{t + 1}$.
   (e) Express $-F(x)$ as an integral of $t\sqrt{t + 1}$.

4. Use upper and lower sums to show that

$$0.5 < \int_1^2 \frac{dx}{x} < 1.$$

Optional   5. Show that, if f is continuous on the interval I, then

$$\int_a^c f(t)\, dt + \int_c^b f(t)\, dt = \int_a^b f(t)\, dt$$

for *every* choice of a, b, c in I.

6. Show the validity of step (1) in the proof of Theorem 5.3.5.
7. Complete the proof of Theorem 5.3.5 by showing that

$$\lim_{h \uparrow 0} \frac{F(x + h) - F(x)}{h} = f(x) \qquad\qquad \text{for } x \in (a, b].$$

8. Extend Theorem 5.3.5 by showing that, if f is continuous in $[a, b]$ and c is *any* point in $[a, b]$, then

$$F(x) = \int_c^x f(t)\, dt$$

is continuous on $[a, b]$, differentiable on (a, b), and satisfies

$$F'(x) = f(x) \qquad \text{for all } x \text{ in } (a, b).$$

HINT: $\displaystyle \int_c^x f(t)\, dt = \int_c^a f(t)\, dt + \int_a^x f(t)\, dt.$

9. Set

$$F(x) = \int_0^x \frac{dt}{t^2 + 9}$$

and find the following: *(a) $F'(-1)$. (b) $F'(0)$. *(c) $F'(\tfrac{1}{2})$.

10. (*The First Mean-Value Theorem for Integrals*.)† Show that, if f is continuous on $[a, b]$, then there is at least one number c in (a, b) such that

$$\int_a^b f(x)\, dx = f(c)(b - a).$$

HINT: Apply the mean-value theorem to the function

$$F(x) = \int_a^x f(t)\, dt \quad \text{on } [a, b].$$

5.4   The Fundamental Theorem of Integral Calculus

Definition 5.4.1   Antiderivative

A function G is called an *antiderivative* for f on $[a, b]$ iff

(i) G is continuous on $[a, b]$ and   (ii) $G'(x) = f(x)$ for all $x \in (a, b)$.

Theorem 5.3.5 says that, if f is continuous on $[a, b]$, then

$$F(x) = \int_a^x f(t)\, dt$$

is an antiderivative for f on $[a, b]$. This gives us a prescription for constructing antiderivatives. It tells us that we can construct an antiderivative for f by integrating f.

The so-called "fundamental theorem" goes the other way. It gives us a prescription, not for finding antiderivatives, but for evaluating integrals. It tells us that we can evaluate

$$\int_a^b f(t)\, dt$$

by finding an antiderivative for f.

---

† This theorem will play a major role in Chapter 10.

Theorem 5.4.2 **The Fundamental Theorem of Integral Calculus**

Let f be continuous on $[a, b]$. If G is an antiderivative for f on $[a, b]$, then

$$\int_a^b f(t)\, dt = G(b) - G(a).$$

PROOF. From Theorem 5.3.5 we know that the function

$$F(x) = \int_a^x f(t)\, dt$$

is an antiderivative for f on $[a, b]$. If G is also an antiderivative for f on $[a, b]$, then we have both F and G continuous on $[a, b]$ and satisfying $F'(x) = G'(x)$ for all x in (a, b). From Theorem 4.2.4 we can conclude that there exists a constant C such that

$$F(x) = G(x) + C \qquad \text{for all } x \text{ in } [a, b].$$

Since $F(a) = 0$, we must have

$$G(a) + C = 0 \quad \text{and thus} \quad C = -G(a).$$

This means that

$$F(x) = G(x) - G(a) \qquad \text{for all } x \text{ in } [a, b].$$

In particular,

$$F(b) = G(b) - G(a).$$

Since

$$F(b) = \int_a^b f(t)\, dt,$$

the theorem is proved. □

Applying the Fundamental Theorem

Problem. Evaluate

$$\int_1^4 x\, dx.$$

SOLUTION. As an antiderivative we can use the function

$$G(x) = \tfrac{1}{2}x^2. \qquad\qquad \text{(check this out)}$$

By the fundamental theorem

$$\int_1^4 x\, dx = G(4) - G(1) = \tfrac{1}{2}(4)^2 - \tfrac{1}{2}(1)^2 = 7\tfrac{1}{2}. \quad □$$

Problem.   Evaluate

$$\int_0^1 8x\,dx.$$

SOLUTION.   Here we can use the antiderivative $G(x) = 4x^2$:

$$\int_0^1 8x\,dx = G(1) - G(0) = 4(1)^2 - 4(0)^2 = 4. \quad \square$$

Expressions of the form $G(b) - G(a)$ are conveniently abbreviated

$$\left[G(x) \right]_a^b.$$

In this notation

$$\int_1^4 x\,dx = \left[\tfrac{1}{2}x^2 \right]_1^4 = \tfrac{1}{2}(4)^2 - \tfrac{1}{2}(1)^2 = 7\tfrac{1}{2};$$

$$\int_0^1 8x\,dx = \left[4x^2 \right]_0^1 = 4(1)^2 - 4(0)^2 = 4. \quad \square$$

From your study of differentiation you know that for all positive integers n

$$\frac{d}{dx}\left(\frac{1}{n+1}x^{n+1} \right) = x^n.$$

It follows then that

(5.4.3) $$\int_a^b x^n\,dx = \left[\frac{1}{n+1}x^{n+1} \right]_a^b = \frac{1}{n+1}(b^{n+1} - a^{n+1}).$$

Thus for example

$$\int_a^b x^2\,dx = \left[\tfrac{1}{3}x^3 \right]_a^b = \tfrac{1}{3}(b^3 - a^3),$$

$$\int_a^b x^3\,dx = \left[\tfrac{1}{4}x^4 \right]_a^b = \tfrac{1}{4}(b^4 - a^4),$$

$$\int_a^b x^4\,dx = \left[\tfrac{1}{5}x^5 \right]_a^b = \tfrac{1}{5}(b^5 - a^5), \text{ etc. } \quad \square$$

Problem.   Evaluate

$$\int_0^1 (2x - 6x^4 + 5)\,dx.$$

SOLUTION.   As an antiderivative we use $G(x) = x^2 - \tfrac{6}{5}x^5 + 5x$:

$$\int_0^1 (2x - 6x^4 + 5)\,dx = \left[x^2 - \tfrac{6}{5}x^5 + 5x \right]_0^1 = 1 - \tfrac{6}{5} + 5 = 4\tfrac{4}{5}. \quad \square$$

Problem.   Evaluate

$$\int_{-1}^{1} (x - 1)(x + 2)\, dx.$$

SOLUTION.   First we carry out the indicated multiplication:

$$(x - 1)(x + 2) = x^2 + x - 2.$$

As an antiderivative we use $G(x) = \frac{1}{3}x^3 + \frac{1}{2}x^2 - 2x$:

$$\int_{-1}^{1} (x - 1)(x + 2)\, dx = \left[\tfrac{1}{3}x^3 + \tfrac{1}{2}x^2 - 2x\right]_{-1}^{1} = -\tfrac{10}{3}. \quad \square$$

Formula 5.4.3 holds not only for positive integer exponents but for all rational exponents other than -1 provided that the integrand, the function being integrated, is continuous over the interval of integration.

$$\int_{1}^{2} \frac{dx}{x^2} = \int_{1}^{2} x^{-2}\, dx = \left[-x^{-1}\right]_{1}^{2} = \left[-\frac{1}{x}\right]_{1}^{2} = \left(-\tfrac{1}{2}\right) - \left(-1\right) = \tfrac{1}{2}.$$

$$\int_{4}^{9} \frac{dt}{\sqrt{t}} = \int_{4}^{9} t^{-1/2}\, dt = \left[2t^{1/2}\right]_{4}^{9} = 2(3) - 2(2) = 2.$$

$$\int_{0}^{1} t^{5/3}\, dt = \left[\tfrac{3}{8}t^{8/3}\right]_{0}^{1} = \tfrac{3}{8}(1)^{8/3} - \tfrac{3}{8}(0)^{8/3} = \tfrac{3}{8}. \quad \square$$

Finally here are some slightly more complicated examples. The essential step in each case is the determination of a suitable antiderivative. Check each computation in detail.

$$\int_{1}^{2} \frac{x^4 + 1}{x^2}\, dx = \int_{1}^{2} (x^2 + x^{-2})\, dx = \left[\tfrac{1}{3}x^3 - x^{-1}\right]_{1}^{2} = \tfrac{17}{6}.$$

$$\int_{1}^{5} \sqrt{x - 1}\, dx = \int_{1}^{5} (x - 1)^{1/2}\, dx = \left[\tfrac{2}{3}(x - 1)^{3/2}\right]_{1}^{5} = \tfrac{16}{3}.$$

$$\int_{0}^{1} (4 - \sqrt{x})^2\, dx = \int_{0}^{1} (16 - 8\sqrt{x} + x)\, dx = \left[16x - \tfrac{16}{3}x^{3/2} + \tfrac{1}{2}x^2\right]_{0}^{1} = \tfrac{67}{6}.$$

$$\int_{1}^{2} -\frac{dt}{(t + 2)^2} = \int_{1}^{2} -(t + 2)^{-2}\, dt = \left[(t + 2)^{-1}\right]_{1}^{2} = -\tfrac{1}{12}. \quad \square$$

Exercises

Evaluate the following integrals by applying the fundamental theorem.

*1. $\int_{0}^{1} (2x - 3)\, dx.$ 2. $\int_{0}^{1} (3x + 2)\, dx.$ *3. $\int_{-1}^{0} 5x^4\, dx.$

*4. $\int_{1}^{2} (2x + x^2)\, dx.$ 5. $\int_{1}^{4} \sqrt{x}\, dx.$ *6. $\int_{0}^{4} \sqrt{x}\, dx.$

*7. $\int_{1}^{5} 2\sqrt{x - 1}\, dx.$ 8. $\int_{1}^{2} \left(\frac{3}{x^3} + 5x\right) dx.$ *9. $\int_{-2}^{0} (x + 1)(x - 2)\, dx.$

*10. $\int_{2}^{0} \frac{dx}{(x + 1)^2}.$ 11. $\int_{0}^{1} (x^{3/2} - x^{1/2})\, dx.$ *12. $\int_{0}^{1} (x^{3/4} - 2x^{1/2})\, dx.$

*13. $\displaystyle\int_0^1 (x + 1)^{17}\, dx.$ 14. $\displaystyle\int_0^a (a^2x - x^3)\, dx.$ *15. $\displaystyle\int_0^a (\sqrt{a} - \sqrt{x})^2\, dx.$

*16. $\displaystyle\int_{-1}^1 (x - 2)^2\, dx.$ 17. $\displaystyle\int_1^2 \frac{6 - t}{t^3}\, dt.$ *18. $\displaystyle\int_1^2 \frac{2 - t}{t^3}\, dt.$

*19. $\displaystyle\int_0^1 x^2(x - 1)\, dx.$ 20. $\displaystyle\int_1^3 \left(x^2 - \frac{1}{x^2}\right) dx.$ *21. $\displaystyle\int_1^2 2x(x^2 + 1)\, dx.$

22. Compare

$$\frac{d}{dx}\left[\int_a^x f(t)\, dt\right] \quad \text{with} \quad \int_a^x \frac{d}{dt}\, [f(t)]\, dt.$$

5.5 Some Area Problems; Some Problems of Motion

Problem.   Find the area of the region bounded above by the curve $y = 4 - x^2$ and below by the x-axis.

SOLUTION.   The curve intersects the x-axis at $x = -2$ and $x = 2$. The region in question is depicted in Figure 5.5.1.
   The area is given by the integral

$$\int_{-2}^2 (4 - x^2)\, dx = \left[4x - \tfrac{1}{3}x^3\right]_{-2}^2 = \tfrac{32}{3}. \quad \square$$

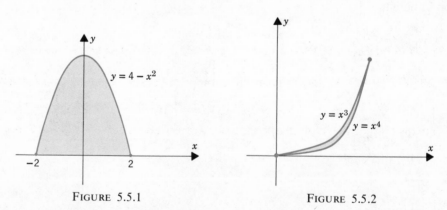

FIGURE 5.5.1   FIGURE 5.5.2

Problem.   Find the area of the crescent depicted in Figure 5.5.2.

SOLUTION.   The two curves meet at $x = 0$ and at $x = 1$:

$$x^4 = x^3 \quad \text{iff} \quad x^4 - x^3 = 0 \quad \text{iff} \quad x(x^3 - 1) = 0.$$

We can obtain the area of the crescent by finding the area below $y = x^3$ and subtracting from it the area below $y = x^4$. This gives

$$A = \int_0^1 x^3\, dx - \int_0^1 x^4\, dx = \left[\tfrac{1}{4}x^4\right]_0^1 - \left[\tfrac{1}{5}x^5\right]_0^1 = \tfrac{1}{4} - \tfrac{1}{5} = \tfrac{1}{20}. \quad \square$$

By the fundamental theorem

$$f(b) - f(a) = \int_a^b f'(x)\,dx$$

and thus

(5.5.1) $$f(b) = f(a) + \int_a^b f'(x)\,dx.$$

Problem.   Find $f(2)$ given that $f(0) = 1$ and $f'(x) = x^3 + 1$.

SOLUTION.   Here

$$f(2) = f(0) + \int_0^2 f'(x)\,dx$$

so that

$$f(2) = 1 + \int_0^2 (x^3 + 1)\,dx = 1 + \left[\tfrac{1}{4}x^4 + x\right]_0^2 = 1 + 6 = 7. \quad \square$$

Problem.   An object moves along a coordinate line with velocity

$$v(t) = (1 - t)(2 - t) \quad \text{units per second.}$$

Its initial position (its position at time $t = 0$) is 2 units to the right of the origin.

(a) Find the position of the object 4 seconds later.
(b) Find the total distance traveled by the object during those first 4 seconds.

SOLUTION.   (a) Let $x(t)$ be the position (coordinate) of the object at time t. Then

$$x'(t) = v(t).$$

This means that the position function is an antiderivative for the velocity function. From (5.5.1)

$$x(4) = x(0) + \int_0^4 v(t)\,dt.$$

Since the initial position of the object is 2 units to the right of the origin, we have $x(0) = 2$. Since

$$v(t) = (1 - t)(2 - t) = 2 - 3t + t^2,$$

we see that

$$\int_0^4 v(t)\,dt = \int_0^4 (2 - 3t + t^2)\,dt = \left[2t - \tfrac{3}{2}t^2 + \tfrac{1}{3}t^3\right]_0^4 = 5\tfrac{1}{3}.$$

Having found that

$$x(0) = 2 \quad \text{and} \quad \int_0^4 v(t)\,dt = 5\tfrac{1}{3}$$

we have

$$x(4) = x(0) + \int_0^4 v(t)\, dt = 2 + 5\tfrac{1}{3} = 7\tfrac{1}{3}.$$

At the end of 4 seconds the object is $7\tfrac{1}{3}$ units to the right of the origin.

(b) The total distance traveled by the object during the first 4 seconds is given by the integral

$$s = \int_0^4 |v(t)|\, dt.$$

In this instance

$$s = \int_0^4 |(1 - t)(2 - t)|\, dt.$$

To evaluate this integral we must first remove the absolute-value sign. As you can verify

$$|(1 - t)(2 - t)| = \left\{ \begin{array}{ll} 2 - 3t + t^2, & t \le 1 \\ -2 + 3t - t^2, & 1 \le t \le 2 \\ 2 - 3t + t^2, & 2 \le t \end{array} \right].$$

It follows now that

$$\begin{aligned} s &= \int_0^4 |(1 - t)(2 - t)|\, dt \\ &= \int_0^1 (2 - 3t + t^2)\, dt + \int_1^2 (-2 + 3t - t^2)\, dt + \int_2^4 (2 - 3t + t^2)\, dt \\ &= \left[2t - \tfrac{3}{2}t^2 + \tfrac{1}{3}t^3 \right]_0^1 + \left[-2t + \tfrac{3}{2}t^2 - \tfrac{1}{3}t^3 \right]_1^2 + \left[2t - \tfrac{3}{2}t^2 + \tfrac{1}{3}t^3 \right]_2^4 = 5\tfrac{2}{3}. \end{aligned}$$

The total distance traveled is $5\tfrac{2}{3}$ units.   □

Exercises

*1. In the motion we just discussed an object started out 2 units to the right of the origin. After traveling a distance of $5\tfrac{2}{3}$ units it ended up only $7\tfrac{1}{3}$ units to the right of the origin. Explain how this is possible.

In Exercises 2–7 find the area below the graph.

2. $f(x) = 2 + x^3$, $x \in [0, 1]$.   *3. $f(x) = \dfrac{1}{(x + 1)^2}$, $x \in [0, 2]$.

4. $f(x) = \sqrt{x + 1}$, $x \in [3, 8]$.   *5. $f(x) = \dfrac{1}{2\sqrt{x + 1}}$, $x \in [0, 8]$.

6. $f(x) = (2x^2 + 1)^2$, $x \in [0, 1]$.   *7. $f(x) = x^2(3 + x)$, $x \in [-3, 0]$.

Sketch the region bounded above by (a) and below by (b) and determine its area.

*8. (a) $y = \sqrt{x}$; (b) $y = x^2$.   9. (a) $y = 5 - x^2$; (b) $y = 3 - x$.

*10. (a) $y = 8 - x^2$; (b) $y = x^2$.   11. (a) $y = 6x - x^2$; (b) $y = 0$.

*12. (a) $y = 8$;   (b) $y = x^2 + 2x$.   13. (a) $y = \sqrt{x}$;   (b) $y = \frac{1}{4}x$.

*14. Find $f(1)$ given that $f(0) = 0$ and $f'(x) = ax^2 + bx + c$.

15. Find $f(4)$ given that $f(9) = 1$ and $f'(x) = (x - 9)^3$.

*16. An object moves along a coordinate line with velocity $v(t) = t(1 - t)$ units per second. Its initial position (its position at time $t = 0$) is 2 units to the left of the origin. (a) Find the position of the object 10 seconds later. (b) Find the total distance traveled by the object during the first 10 seconds.

17. An object moves along a coordinate line with acceleration

$$a(t) = \frac{1}{\sqrt{t + 1}} \text{ units per second per second.}$$

(a) Find its velocity at time t_0 given that its initial velocity is 1 unit per second.

(b) Find the position of the object at time t_0 given that its initial velocity is 1 unit per second and its initial position is the origin.

*18. A point moves along the plane in such a manner that its x-coordinate is changing at the rate of $(t^2 + 1)^2$ units per second and its y-coordinate is changing at the rate of $\sqrt{t}$ units per second. Find the position of the point one second after the motion starts if the initial position is $x = x_0$, $y = y_0$.

19. An automobile with varying velocity $v(t)$ moves in a fixed direction for 5 minutes and covers a distance of 4 miles. What theorem would you invoke to argue that for at least one instant the speedometer must have read 48 miles per hour?

5.6 The Linearity of the Integral

Here are some simple properties of the integral that are often used in computation.

I. Constants may be factored through the integral sign:

(5.6.1)
$$\int_a^b \alpha f(x)\, dx = \alpha \int_a^b f(x)\, dx.$$

Examples

$$\int_0^{10} \tfrac{3}{7}(x - 5)\, dx = \tfrac{3}{7}\int_0^{10} (x - 5)\, dx = \tfrac{3}{7}\left[\tfrac{1}{2}(x - 5)^2\right]_0^{10} = 0.$$

$$\int_0^1 \frac{8}{(x + 1)^3}\, dx = 8\int_0^1 \frac{dx}{(x + 1)^3} = 8\left[-\tfrac{1}{2}(x + 1)^{-2}\right]_0^1 = 8(\tfrac{3}{8}) = 3. \quad \square$$

II. The integral of a sum is the sum of the integrals:

(5.6.2)
$$\int_a^b [f(x) + g(x)]\, dx = \int_a^b f(x)\, dx + \int_a^b g(x)\, dx.$$

Example

$$\int_1^2 \left[(x - 1)^2 + \frac{1}{(x + 2)^2} \right] dx = \int_1^2 (x - 1)^2 \, dx + \int_1^2 \frac{dx}{(x + 2)^2}$$

$$= \left[\tfrac{1}{3}(x - 1)^3 \right]_1^2 + \left[-(x + 2)^{-1} \right]_1^2$$

$$= \tfrac{1}{3} - \tfrac{1}{4} + \tfrac{1}{3} = \tfrac{5}{12}. \quad \square$$

Combining I and II, we get the linearity property.

III. The integral of a linear combination is the linear combination of the integrals:

(5.6.3)
$$\int_a^b [\alpha f(x) + \beta g(x)] \, dx = \alpha \int_a^b f(x) \, dx + \beta \int_a^b g(x) \, dx.$$

Example

$$\int_2^{5/2} \left[\frac{3}{(x - 1)^2} - 4x \right] dx = 3 \int_2^{5/2} \frac{dx}{(x - 1)^2} - 2 \int_2^{5/2} 2x \, dx$$

$$= 3 \left[-(x - 1)^{-1} \right]_2^{5/2} - 2 \left[x^2 \right]_2^{5/2}$$

$$= 3(-\tfrac{2}{3} + 1) - 2(\tfrac{25}{4} - 4) = -\tfrac{7}{2}. \quad \square$$

Properties I and II are just particular instances of property III. To prove III, take f and g as continuous, and choose for these functions antiderivatives F and G respectively. The details are left to you.

As a special case of (5.6.1) we have

(5.6.4)
$$\int_a^b - f(x) \, dx = - \int_a^b f(x) \, dx.$$ (take $\alpha = -1$)

If, as in Figure 5.6.1, $f(x) \le 0$ for all $x \in [a, b]$, then $-f(x) \ge 0$ for all $x \in [a, b]$.

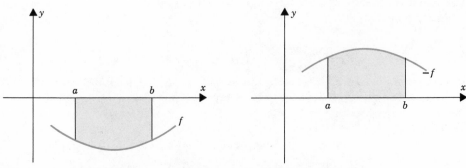

FIGURE 5.6.1

Since the area A between the x-axis and the graph of f is obviously the same as the area between the x-axis and the graph of $-f$, we have

$$A = \int_a^b -f(x)\,dx = -\int_a^b f(x)\,dx$$

and therefore

$$\int_a^b f(x)\,dx = -A.$$

The integral of a negative function is thus the negative of the area between the x-axis and the graph of the function.

Example. Figure 5.6.2 shows a continuous function f which takes on both positive and negative values. Here

$$\int_a^d f(x)\,dx = \int_a^b f(x)\,dx + \int_b^c f(x)\,dx + \int_c^d f(x)\,dx = 2 - \tfrac{3}{4} + 1 = 2\tfrac{1}{4}. \quad \square$$

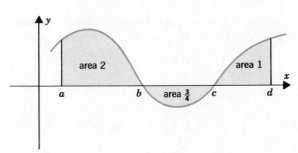

FIGURE 5.6.2

5.7 Indefinite Integrals

If F is an antiderivative for f on $[a, b]$, then we have

$$\int_a^b f(x)\,dx = \Big[F(x)\Big]_a^b.$$

The result is obviously unchanged if we replace $F(x)$ by $F(x) + C$, where C is an arbitrary constant:

$$\int_a^b f(x)\,dx = \Big[F(x) + C\Big]_a^b.$$

If we have no particular interest in the interval $[a, b]$ but wish instead to emphasize that F is an antiderivative for f on *some* interval, then we can omit the a and the b and simply write

$$\int f(x)\,dx = F(x) + C.$$

Antiderivatives expressed in this manner are called *indefinite integrals.*

Examples

$$\int x^2 \, dx = \tfrac{1}{3}x^3 + C, \quad \int \sqrt{s} \, ds = \tfrac{2}{3}s^{3/2} + C,$$

$$\int \frac{4}{\sqrt{t+2}} \, dt = 8\sqrt{t+2} + C. \quad \square$$

The following problem will serve to illustrate the notation further.

Problem. Find the general law of motion of an object which moves in a straight line with constant acceleration g.

SOLUTION. The acceleration is given by the equation

$$a(t) = g.$$

The velocity function is an indefinite integral of the acceleration:

$$v(t) = \int a(t) \, dt = \int g \, dt = gt + C.$$

The constant C is not determined by the acceleration but is arbitrary. Physically C represents the initial velocity v_0. We replace C by v_0 and write

$$v(t) = gt + v_0.$$

The position function is an indefinite integral of the velocity function:

$$x(t) = \int v(t) \, dt = \int (gt + v_0) \, dt = \tfrac{1}{2}gt^2 + v_0 t + K.$$

The constant K (we avoid C because we used it before) is not determined by the velocity function but is arbitrary. Physically it represents the initial position x_0. We therefore write

$$x(t) = \tfrac{1}{2}gt^2 + v_0 t + x_0.$$

This is the general law of motion. It is an equation you saw before in connection with free-falling bodies. $\square$

Exercises

*1. $\displaystyle \int \frac{dx}{x^4}.$

2. $\displaystyle \int (x-1)^2 \, dx.$

*3. $\displaystyle \int (ax+b) \, dx.$

4. $\displaystyle \int (ax^2+b) \, dx.$

*5. $\displaystyle \int \frac{dx}{\sqrt{1+x}}.$

6. $\displaystyle \int \left(\frac{x^3+1}{x^5} \right) dx.$

*7. $\displaystyle \int \left(\sqrt{x} - \frac{1}{\sqrt{x}} \right) dx.$

8. $\displaystyle \int \left(\frac{x^3-1}{x^2} \right) dx.$

*9. $\int (t - a)(t - b)\, dt.$ 10. $\int (t^2 - a)(t^2 - b)\, dt.$

*11. $\int \dfrac{g'(x)}{[g(x)]^2}\, dx.$ 12. $\int \dfrac{g(x)f'(x) - g'(x)f(x)}{[g(x)]^2}\, dx.$

*13. Compare

$$\frac{d}{dx}\left[\int f(x)\, dx\right] \quad \text{with} \quad \int \frac{d}{dx}[f(x)]\, dx.$$

14. Find the general law of motion of an object which moves in a straight line with acceleration *(a) $a(t) = 2A + 6Bt.$ (b) $a(t) = 45t^{1/2} + t^{-1/2}.$

15. Show that, if the acceleration of a particle is proportional to the square of the velocity, $a(t) = -k[v(t)]^2$, then $[v(t)]^{-1} = v_0^{-1} + kt$ where v_0 is the initial velocity.

5.8   The u-Substitution; Change of Variables

Integrals of the form

$$\int f(g(x))g'(x)\, dx$$

can be calculated by computing

$$\int f(u)\, du \quad \text{and then setting} \quad u = g(x).$$

For suppose that F is an antiderivative for f. Then

$$\int f(g(x))g'(x)\, dx = \int F'(g(x))g'(x)\, dx = F(g(x)) + C$$

by the chain rule

and

$$\int f(u)\, du = F(u) + C = F(g(x)) + C. \quad \square$$

by setting $u = g(x)$

You can transform

$$\int f(g(x))g'(x)\, dx \quad \text{into} \quad \int f(u)\, du$$

by setting

$$u = g(x); \qquad du = g'(x)\, dx.\dagger$$

---

† Think of $du = g'(x)\, dx$ as a "formal differential," writing dx for h.

Problem. Find

$$\int \frac{dx}{(3 + 2x)^2}.$$

SOLUTION. Set

$$u = 3 + 2x; \quad du = 2\, dx, \quad \tfrac{1}{2}du = dx.$$

Then

$$\int \frac{dx}{(3 + 2x)^2} = \frac{1}{2} \int \frac{du}{u^2} = -\frac{1}{2u} + C = -\frac{1}{2(3 + 2x)} + C. \quad \square$$

Problem. Find

$$\int x\sqrt{a^2 + b^2x^2}\, dx.$$

SOLUTION. Set

$$u = a^2 + b^2x^2; \quad du = 2b^2x\, dx, \quad \frac{du}{2b^2} = x\, dx.$$

$$\int x\sqrt{a^2 + b^2x^2}\, dx = \frac{1}{2b^2} \int \sqrt{u}\, du$$

$$= \frac{1}{2b^2}\left(\frac{2}{3}\right) u^{3/2} + C = \frac{1}{3b^2}(a^2 + b^2x^2)^{3/2} + C. \quad \square$$

For definite integrals we have the following change of variables formula:

(5.8.1)
$$\boxed{\int_a^b f(g(x))g'(x)\, dx = \int_{g(a)}^{g(b)} f(u)\, du.}$$

The formula holds provided that f and g' are both continuous. More precisely, g' must be continuous on an interval that joins a and b, and f must be continuous on the set of values taken on by g.

PROOF. Let F be an antiderivative for f. Then $F' = f$ and

$$\int_a^b f(g(x))g'(x)\, dx = \int_a^b F'(g(x))g'(x)\, dx$$

$$= \Big[F(g(x))\Big]_a^b = F(g(b)) - F(g(a)) = \int_{g(a)}^{g(b)} f(u)\, du. \quad \square$$

Problem. Evaluate

$$\int_0^1 (x^2 - 1)^4 2x\, dx.$$

SOLUTION. Set

$$u = x^2 - 1; \quad du = 2x\, dx.$$

At $x = 0$, $u = -1$. At $x = 1$, $u = 0$.

$$\int_0^1 (x^2 - 1)^4 2x \, dx = \int_{-1}^0 u^4 \, du = \left[\tfrac{1}{5}u^5\right]_{-1}^0 = \tfrac{1}{5}. \quad \square$$

Problem.   Evaluate

$$\int_1^2 \frac{10x^2}{(x^3 + 1)^2} \, dx.$$

SOLUTION.   Set

$$u = x^3 + 1; \qquad du = 3x^2 dx, \quad \tfrac{10}{3} \, du = 10x^2 \, dx.$$

At $x = 1$, $u = 2$. At $x = 2$, $u = 9$.

$$\int_1^2 \frac{10x^2}{(x^3 + 1)^2} \, dx = \frac{10}{3} \int_2^9 \frac{du}{u^2} = \frac{10}{3}\left[-\frac{1}{u}\right]_2^9 = \frac{35}{27}. \quad \square$$

Problem.   Evaluate

$$\frac{1}{8}\int_0^2 \frac{x}{\sqrt{4x^2 + 9}} \, dx.$$

SOLUTION.   Set

$$u = 4x^2 + 9; \qquad du = 8x \, dx, \quad \tfrac{1}{8} \, du = x \, dx.$$

At $x = 0$, $u = 9$. At $x = 2$, $u = 25$.

$$\int_0^2 \frac{x}{\sqrt{4x^2 + 9}} \, dx = \frac{1}{8} \int_9^{25} \frac{du}{\sqrt{u}} = \frac{1}{4}\left[\sqrt{u}\right]_9^{25} = \frac{1}{2}. \quad \square$$

Exercises

Work out the following integrals by a *u*-substitution.

*1. $\displaystyle\int \frac{dx}{(2 - 3x)^2}.$

2. $\displaystyle\int \frac{dx}{\sqrt{2x + 1}}.$

*3. $\displaystyle\int \sqrt{2x + 1} \, dx.$

4. $\displaystyle\int \sqrt{ax + b} \, dx.$

*5. $\displaystyle\int (ax + b)^{3/4} \, dx.$

6. $\displaystyle\int (2ax + b)(ax^2 + bx + c)^4 \, dx.$

*7. $\displaystyle\int \frac{x}{(4x^2 + 9)^2} \, dx.$

8. $\displaystyle\int \frac{3x}{(x^2 + 1)^2} \, dx.$

*9. $\displaystyle\int x^2(5x^3 + 9)^4 \, dx.$

10. $\displaystyle\int t(1 + t^2)^3 \, dt.$

*11. $\int x^2(1 + x^3)^{1/4}\, dx$.

12. $\int x^{n-1}\sqrt{a + bx^n}\, dx$.

*13. $\int \dfrac{s}{(1 + s^2)^3}\, ds$.

14. $\int \dfrac{2x}{\sqrt[3]{6 - 5x^2}}\, dx$.

*15. $\int \dfrac{x}{\sqrt{x^2 + 1}}\, dx$.

16. $\int \dfrac{3ax^2 - 2bx}{\sqrt{ax^3 - bx^2}}\, dx$.

*17. $\int \dfrac{b^3 x^3}{\sqrt{1 - a^4 x^4}}\, dx$.

18. $\int \dfrac{x^{n-1}}{\sqrt{a + bx^n}}\, dx$.

Evaluate the following integrals by a u-substitution.

*19. $\displaystyle\int_0^1 x(x^2 + 1)^3\, dx$.

20. $\displaystyle\int_{-1}^0 3x^2(4 + 2x^3)^2\, dx$.

*21. $\displaystyle\int_0^1 5x(1 + x^2)^4\, dx$.

22. $\displaystyle\int_1^2 (6 - x)^{-3}\, dx$.

*23. $\displaystyle\int_{-1}^1 \dfrac{x}{(1 + x^2)^4}\, dx$.

24. $\displaystyle\int_0^3 \dfrac{r}{\sqrt{r^2 + 16}}\, dr$.

*25. $\displaystyle\int_0^a x\sqrt{a^2 - x^2}\, dx$.

26. $\displaystyle\int_0^a x\sqrt{a^2 + x^2}\, dx$.

Work out the following integrals by a u-substitution.

*27. $\int x\sqrt{x + 1}\, dx$. [set $u = x + 1$]

28. $\int 2x\sqrt{x - 1}\, dx$.

*29. $\int x\sqrt{2x - 1}\, dx$.

30. $\int x^2\sqrt{x + 1}\, dx$.

5.9 Some Further Properties of the Definite Integral

In this section we feature some important general properties of the integral. The proofs are left mostly to you. You can assume throughout that the functions involved are continuous and that $a < b$.

　I. The integral of a nonnegative function is nonnegative:

(5.9.1)　　if $f(x) \geq 0$ for all $x \in [a, b]$, then $\displaystyle\int_a^b f(x)\, dx \geq 0$.

The integral of a positive function is positive:

(5.9.2)　　if $f(x) > 0$ for all $x \in [a, b]$, then $\displaystyle\int_a^b f(x)\, dx > 0$.

The next property is an immediate consequence of property I and the linearity property.

　II. The integral is order-preserving:

(5.9.3) $\quad$ if $\quad f(x) \leq g(x) \quad$ for all $\ x \in [a, b], \qquad$ then $\ \displaystyle\int_a^b f(x)\, dx \leq \int_a^b g(x)\, dx$

and

(5.9.4) $\quad$ if $\quad f(x) < g(x) \quad$ for all $\ x \in [a, b], \qquad$ then $\ \displaystyle\int_a^b f(x)\, dx < \int_a^b g(x)\, dx.$

III. Just as the absolute value of a sum is less than or equal to the sum of the absolute values,

$$|x_1 + x_2 + \cdots + x_n| \leq |x_1| + |x_2| + \cdots + |x_n|,$$

the absolute value of an integral is less than or equal to the integral of the absolute value:

(5.9.5) $\qquad\qquad \displaystyle\left| \int_a^b f(x)\, dx \right| \leq \int_a^b |f(x)|\, dx.$

HINT FOR PROOF OF III: Show that

$$\int_a^b f(x)\, dx \quad \text{and} \quad - \int_a^b f(x)\, dx$$

are both less than or equal to

$$\int_a^b |f(x)|\, dx.$$

The next property is one with which you are already familiar.

IV. If m is the minimum value of f on $[a, b]$ and M is the maximum, then

(5.9.6) $\qquad\qquad m(b - a) \leq \displaystyle\int_a^b f(x)\, dx \leq M(b - a).$

We come now to a generalization of Theorem 5.3.5. Its importance will become apparent in Chapter 6.

V. If g is differentiable (and f is continuous), then

(5.9.7) $\qquad\qquad \dfrac{d}{dx}\left(\displaystyle\int_a^{g(x)} f(t)\, dt \right) = f(g(x))g'(x).$

PROOF.   Set

$$H(x) = \int_a^{g(x)} f(t)\, dt$$

and recognize that H is the composition of differentiable functions:

$$H(x) = F(g(x)) \qquad \text{with} \qquad F(x) = \int_a^x f(t)\, dt.$$

The chain rule gives

$$H'(x) = F'(g(x))g'(x).$$

Theorem 5.3.5 gives

$$F'(x) = f(x).$$

It therefore follows that

$$H'(x) = f(g(x))g'(x). \quad \square$$

Problem.   Find

$$\frac{d}{dx} \left(\int_0^{x^3} \frac{dt}{1+t} \right).$$

SOLUTION.   At this stage you'd probably be hard put to carry out the integration. But here that doesn't matter. By Property V you know that

$$\frac{d}{dx} \left(\int_0^{x^3} \frac{dt}{1+t} \right) = \frac{1}{1+x^3} \, 3x^2 = \frac{3x^2}{1+x^3}$$

without carrying out the integration. $\square$

Exercises

Assume: f and g continuous, $a < b$, and

$$\int_a^b f(x) \, dx > \int_a^b g(x) \, dx.$$

Answer questions 1–6 giving supporting reasons.

*1. Does it necessarily follow that

$$\int_a^b [f(x) - g(x)] \, dx > 0?$$

*2. Does it necessarily follow that

$$f(x) > g(x) \qquad \text{for all } x \in [a, b]?$$

*3. Does it necessarily follow that

$$f(x) > g(x) \qquad \text{for at least some } x \in [a, b]?$$

*4. Does it necessarily follow that

$$\left| \int_a^b f(x) \, dx \right| > \left| \int_a^b g(x) \, dx \right| ?$$

*5. Does it necessarily follow that

$$\int_a^b |f(x)| \, dx > \int_a^b g(x) \, dx?$$

*6. Does it necessarily follow that

$$\int_a^b |f(x)|\, dx > \int_a^b |g(x)|\, dx?$$

Assume: f continuous, $a < b$, and

$$\int_a^b f(x)\, dx = 0.$$

Answer questions 7–15 giving supporting reasons.

7. Does it necessarily follow that $f(x) = 0$ for all $x \in [a, b]$?
8. Does it necessarily follow that $f(x) = 0$ for at least some $x \in [a, b]$?
9. Does it necessarily follow that

$$\left| \int_a^b f(x)\, dx \right| = 0?$$

10. Does it necessarily follow that

$$\int_a^b |f(x)|\, dx = 0?$$

11. Must all upper sums $U_f(P)$ be nonnegative?
12. Must all upper sums $U_f(P)$ be positive?
13. Can a lower sum, $L_f(P)$, be positive?
14. Does it necessarily follow that

$$\int_a^b [f(x)]^2\, dx = 0?$$

15. Does it necessarily follow that

$$\int_a^b [f(x) + 1]\, dx = b - a?$$

Compute

*16. $\dfrac{d}{dx}\left(\displaystyle\int_1^{x^2} \dfrac{dt}{t} \right).$

*17. $\dfrac{d}{dx}\left(\displaystyle\int_0^{1+x^2} \dfrac{dt}{\sqrt{2t + 5}} \right).$

18. $\dfrac{d}{dx}\left(\displaystyle\int_0^{x^3} \dfrac{dt}{\sqrt{1 + t^2}} \right).$

19. $\dfrac{d}{dx}\left(\displaystyle\int_x^a f(t)\, dt \right).$

20. Show that

$$\frac{d}{dx}\left(\int_{g_1(x)}^{g_2(x)} f(t)\, dt \right) = f(g_2(x))g_2'(x) - f(g_1(x))g_1'(x).$$

Here g_1 and g_2 are assumed to be differentiable and f continuous. HINT: Take a number a from the domain of f. Express the integral as the difference of two integrals each with lower limit a.

Compute using Exercise 20.

*21. $\dfrac{d}{dx}\left(\displaystyle\int_x^{x^2} \dfrac{dt}{t} \right).$

22. $\dfrac{d}{dx}\left(\displaystyle\int_{1-x}^{1+x} \dfrac{t - 1}{t}\, dt \right).$

23. Prove Property I: (a) by considering lower sums $L_f(P)$; (b) by using an anti-derivative.
24. Prove Property II.
25. Prove Property III.

Optional 26. Prove that, if f is continuous on $[a, b]$ and

$$\int_a^b |f(x)|\, dx = 0,$$

then

$$f(x) = 0 \qquad \text{for all } x \text{ in } [a, b].$$

5.10 More on Area

We begin with two functions f and g both continuous on an interval $[a, b]$. If f and g are both positive and if $f(x) \geq g(x)$ for all x in $[a, b]$ (see Figure 5.10.1), then the area between the graphs is given by the integral

$$\int_a^b [f(x) - g(x)]\, dx.$$

This you already know.

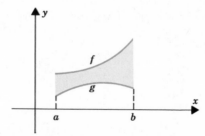

FIGURE 5.10.1

What we want now is a formula which applies also to the region depicted in Figure 5.10.2. Here neither f nor g remains positive. Moreover, neither function remains greater than the other.

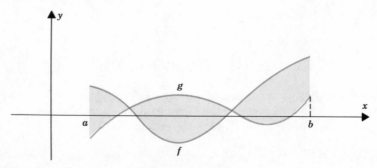

FIGURE 5.10.2

One way to proceed is to raise the entire region a fixed number of units so that both functions become positive. (See Figure 5.10.3.) The new boundaries are now of the form

$$F(x) = f(x) + C \quad \text{and} \quad G(x) = g(x) + C.$$

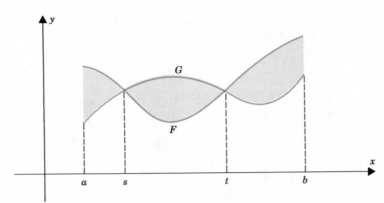

FIGURE 5.10.3

The area of the first part is

$$\int_a^s [F(x) - G(x)] \, dx = \int_a^s [f(x) - g(x)] \, dx = \int_a^s |f(x) - g(x)| \, dx;$$

the area of the second part is

$$\int_s^t [G(x) - F(x)] \, dx = \int_s^t [g(x) - f(x)] \, dx = \int_s^t |f(x) - g(x)| \, dx;$$

the area of the third part is

$$\int_t^b [F(x) - G(x)] \, dx = \int_t^b [f(x) - g(x)] \, dx = \int_t^b |f(x) - g(x)| \, dx.$$

The total area is therefore the sum

$$\int_a^s |f(x) - g(x)| \, dx + \int_s^t |f(x) - g(x)| \, dx + \int_t^b |f(x) - g(x)| \, dx.$$

By the additivity of the integral, this sum is simply

$$\int_a^b |f(x) - g(x)| \, dx.$$

In short,

(5.10.1) $\qquad \boxed{\text{the area between the graphs} = \int_a^b |f(x) - g(x)| \, dx.}$

The expression $|f(x) - g(x)|$ represents the vertical separation between the two graphs at x. We can find the area between the two graphs by integrating this vertical separation from $x = a$ to $x = b$.

Problem.   Find the area between

$$y = x^2 \quad \text{and} \quad y = x^{1/3}, \qquad \text{taking } x \in [-1, 1].$$

SOLUTION.   The region in question is displayed in Figure 5.10.4. The vertical separation is $|x^2 - x^{1/3}|$, and the area is given by the integral

$$\int_{-1}^{1} |x^2 - x^{1/3}| \, dx.$$

To evaluate this integral note that $x^2 - x^{1/3}$ is positive between $x = -1$ and $x = 0$ and negative between $x = 0$ and $x = 1$. Thus

$$\text{area} = \int_{-1}^{0} (x^2 - x^{1/3}) \, dx + \int_{0}^{1} (x^{1/3} - x^2) \, dx$$

$$= \left[\tfrac{1}{3}x^3 - \tfrac{3}{4}x^{4/3} \right]_{-1}^{0} + \left[\tfrac{3}{4}x^{4/3} - \tfrac{1}{3}x^3 \right]_{0}^{1} = (\tfrac{1}{3} + \tfrac{3}{4}) + (\tfrac{3}{4} - \tfrac{1}{3}) = \tfrac{3}{2}. \quad \square$$

FIGURE 5.10.4

In Figures 5.10.5 and 5.10.6 we display regions in which the boundary curves are not functions of x but functions of y instead. In such cases the area is given by the integral

(5.10.2)
$$\int_{c}^{d} |f(y) - g(y)| \, dy.$$

The expression $|f(y) - g(y)|$ represents the horizontal separation between the two curves at y. We can find the area between the two curves by integrating this horizontal separation from $y = c$ to $y = d$.

FIGURE 5.10.5   FIGURE 5.10.6

Problem.   Find the area between the two parabolas

$$x = y^2 \quad \text{and} \quad x = 3 - 2y^2.$$

SOLUTION.   The region is displayed in Figure 5.10.7. The two parabolas intersect at $y = -1$ and $y = 1$. For each y between -1 and 1 the horizontal separation between the two curves is

$$|(3 - 2y^2) - y^2| = |3 - 3y^2| = 3 - 3y^2.$$

We find the area of the region by integrating from $y = -1$ to $y = 1$:

$$A = \int_{-1}^{1} (3 - 3y^2)\, dy = 4. \quad \square$$

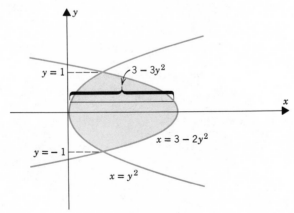

FIGURE 5.10.7

Problem.   Find the area between

$$x = \tfrac{1}{8}y^3 \quad \text{and} \quad x = \tfrac{1}{4}y^3 + \tfrac{1}{8}y^2 - \tfrac{1}{4}y, \qquad \text{taking } y \in [-2, 1].$$

SOLUTION. To see where these curves intersect, we solve the two equations simultaneously:

$$\tfrac{1}{8}y^3 = \tfrac{1}{4}y^3 + \tfrac{1}{8}y^2 - \tfrac{1}{4}y$$
$$0 = \tfrac{1}{8}y^3 + \tfrac{1}{8}y^2 - \tfrac{1}{4}y$$
$$0 = y^3 + y^2 - 2y = y(y^2 + y - 2) = y(y - 1)(y + 2).$$

The two curves intersect at

$$y = -2, \quad y = 0, \quad y = 1.$$

The region between the two curves is displayed in Figure 5.10.8.

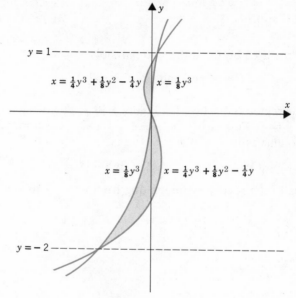

FIGURE 5.10.8

To find the area, we integrate the horizontal separation

$$\left| (\tfrac{1}{4}y^3 + \tfrac{1}{8}y^2 - \tfrac{1}{4}y) - (\tfrac{1}{8}y^3) \right| = \left| \tfrac{1}{8}y^3 + \tfrac{1}{8}y^2 - \tfrac{1}{4}y \right|$$

from $y = -2$ to $y = 1$. Note that

$$\tfrac{1}{8}y^3 + \tfrac{1}{8}y^2 - \tfrac{1}{4}y = \tfrac{1}{8}(y^3 + y^2 - 2y) = \tfrac{1}{8}y(y - 1)(y + 2)$$

is positive between $y = -2$ and $y = 0$, and is negative between $y = 0$ and $y = 1$. It follows that

$$A = \int_{-2}^{1} \left| \tfrac{1}{8}y^3 + \tfrac{1}{8}y^2 - \tfrac{1}{4}y \right| \, dy$$
$$= \int_{-2}^{0} (\tfrac{1}{8}y^3 + \tfrac{1}{8}y^2 - \tfrac{1}{4}y) \, dy + \int_{0}^{1} -(\tfrac{1}{8}y^3 + \tfrac{1}{8}y^2 - \tfrac{1}{4}y) \, dy$$
$$= \tfrac{1}{3} + \tfrac{5}{96} = \tfrac{37}{96}. \quad \square$$

You can sketch a curve $x = f(y)$ by first sketching $y = f(x)$ and then reflecting in the line $y = x$. (Figure 5.10.9)

FIGURE 5.10.9

Exercises

Sketch the region bounded by the curves and find its area.

*1. $y = x^2$, $y = 2x + 3$.
*3. $y^2 - 27x = 0$, $x + y = 0$.
*5. $y^2 = 2x$, $x - y = 4$.
*7. $a^2x = a^2y - y^2$, $4x - y = 0$.
*9. $4y = x^4 - x^2 - 4$, $y = 2$.

2. $x - y^2 + 3 = 0$, $x - 2y = 0$.
4. $x^3 - 10y^2 = 0$, $x - y = 0$.
6. $3y - 3x + x^2 = 0$, $x - 2y = 0$.
8. $x + y^2 - 4 = 0$, $x + y = 2$.
10. $x + y - y^3 = 0$, $x - y + y^2 = 0$.

*11. Find the area bounded by the parabola $y = 6 + 4x - x^2$ and the chord joining $(-2, -6)$ to $(4, 6)$.

12. Find the area bounded by the semicubical parabola $y^3 = x^2$ and the chord joining $(-1, 1)$ to $(8, 4)$.

*13. Find the area between

$$x = y^3 + 4y^2 - 4y + 1 \quad \text{and} \quad x = 4y^2 + 1, \qquad \text{taking } y \in [-1, 2].$$

14. Find the area between

$$x = y^3 - 3y^2 + 2y + 2 \quad \text{and} \quad x = 2y^2 - 4y + 2, \qquad \text{taking } y \in [0, 3].$$

5.11 Additional Exercises

Work out the following integrals.

*1. $\int (\sqrt{x - a} - \sqrt{x - b})\, dx$.

2. $\int ax\sqrt{1 + bx^2}\, dx$.

*3. $\int x^{-1/3}(x^{2/3} - 1)^2\, dx$.

4. $\int t^2(1 + t^3)^{10}\, dt$.

*5. $\int (1 + 2\sqrt{x})^2\, dx$.

6. $\int \frac{1}{\sqrt{x}} (1 + 2\sqrt{x})^5\, dx$.

*7. $\displaystyle\int \frac{(a + b\sqrt{x + 1})^2}{\sqrt{x + 1}}\, dx.$

8. $\displaystyle\int x\sqrt{x}(1 + x^2\sqrt{x})^2\, dx.$

*9. $\displaystyle\int \frac{g'(x)}{[g(x)]^3}\, dx.$

10. $\displaystyle\int \frac{g(x)g'(x)}{\sqrt{1 + [g(x)]^2}}\, dx.$

Find the area below the graph.

*11. $y = x\sqrt{2x^2 + 1}, \quad x \in [0, 2].$

12. $y = \dfrac{x}{(2x^2 + 1)^2}, \quad x \in [0, 2].$

*13. Find the area of the portion of the first quadrant that is bounded above by $y = 2x$ and below by $y = x\sqrt{3x^2 + 1}$.

14. At every point of the curve the slope is given by the equation

$$\frac{dy}{dx} = x\sqrt{x^2 + 1}.$$

Find an equation for the curve given that it passes through the point $(0, 1)$.

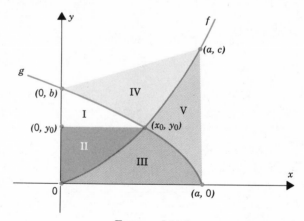

FIGURE 5.11.1

*15. Find a formula for the area of each region depicted in Figure 5.11.1:
    (a) I.   (b) II.   (c) III.   (d) IV.   (e) V.

Sketch the region bounded by the curves and find its area.

*16. $y^2 = 2x, \quad x^2 = 3y.$

17. $y^2 = 6x, \quad x^2 + 4y = 0.$

*18. $y = 4 - x^2, \quad y = x + 2.$

19. $y = 4 - x^2, \quad x + y + 2 = 0.$

*20. $y^2 = x, \quad y = x^3.$

21. $x + y = 2y^2, \quad y = x^3.$

*22. $4y = x^2 - x^4, \quad x + y + 1 = 0.$

23. $x^2y = a^2$, the x-axis, $x = a$, $x = 2a$.

Assume that f and g are continuous, that $a < b$, and that

$$\int_a^b f(x)\, dx > \int_a^b g(x)\, dx.$$

Which of the following statements necessarily hold for all partitions P of $[a, b]$? Support your answers.

*24. $L_g(P) < U_f(P)$.   25. $L_g(P) < L_f(P)$.   *26. $L_g(P) < \int_a^b f(x)\,dx$.

*27. $U_g(P) < U_f(P)$.   28. $U_f(P) > \int_a^b g(x)\,dx$.   *29. $U_g(P) < \int_a^b f(x)\,dx$.

A curve which passes through the point $(4, \frac{1}{3})$ has varying slope

$$\frac{dy}{dx} = -\frac{1}{2\sqrt{x}(1 + \sqrt{x})^2}.$$

30. Find an equation for the curve.
31. Sketch the curve.
32. Estimate the area under the curve between $x = 0$ and $x = 1$ by means of the partition $P = \{0, \frac{1}{25}, \frac{4}{25}, \frac{9}{25}, \frac{16}{25}, 1\}$.

Carry out the differentiation.

*33. $\dfrac{d}{dx}\left(\displaystyle\int_0^x \dfrac{dt}{1 + t^2}\right)$.   34. $\dfrac{d}{dx}\left(\displaystyle\int_0^{x^2} \dfrac{dt}{1 + t^2}\right)$.   *35. $\dfrac{d}{dx}\left(\displaystyle\int_x^{x^2} \dfrac{dt}{1 + t^2}\right)$.

36. Where does the function

$$f(x) = \int_0^x \frac{t - 1}{1 + t^2}\,dt, \quad x \in [0, 2]$$

take on its minimum value?

Optional | 37. Show that

$$\int_{a+c}^{b+c} f(x - c)\,dx = \int_a^b f(x)\,dx.$$

38. Show that for $c \neq 0$

$$\frac{1}{c}\int_{ac}^{bc} f\left(\frac{x}{c}\right) dx = \int_a^b f(x)\,dx.$$

39. Show that, if f is even, then

$$\int_{-a}^a f(x)\,dx = 2\int_0^a f(x)\,dx.$$

40. Show that, if f is odd, then

$$\int_{-a}^a f(x)\,dx = 0.$$

5.12   The Integral as the Limit of Riemann Sums

Let f be some function continuous on a closed interval $[a, b]$. With our approach to integration the definite integral

$$\int_a^b f(x)\,dx$$

is the unique number which satisfies the inequality

$$L_f(P) \leq \int_a^b f(x)\, dx \leq U_f(P)$$

for all partitions P of $[a, b]$. This method of obtaining the definite integral by *squeezing* with upper and lower sums is called the *Darboux method*.[†]

There is another way to obtain the integral which is frequently used. Take a partition $P = \{x_0, x_1, \ldots, x_n\}$ of $[a, b]$. P breaks up $[a, b]$ into n subintervals

$$[x_0, x_1], [x_1, x_2], \ldots, [x_{n-1}, x_n]$$

of lengths

$$\Delta x_1, \Delta x_2, \ldots, \Delta x_n.$$

Define $\|P\|$, the *norm* of P, by setting

$$\|P\| = \max \Delta x_i \qquad i = 1, 2, \ldots, n.$$

Now pick a point x_1^* from $[x_0, x_1]$ and form the product $f(x_1^*)\Delta x_1$; pick a point x_2^* from $[x_1, x_2]$ and form the product $f(x_2^*)\Delta x_2$; go on in this manner until you have formed the products

$$f(x_1^*)\Delta x_1, f(x_2^*)\Delta x_2, \ldots, f(x_n^*)\Delta x_n.$$

The sum of these products

$$S^*(P) = f(x_1^*)\Delta x_1 + f(x_2^*)\Delta x_2 + \cdots + f(x_n^*)\Delta x_n$$

is called a *Riemann sum*.[††]

The definite integral can be viewed as the *limit* of such Riemann sums in the following sense: given any $\epsilon > 0$, there exists a $\delta > 0$ such that

$$\text{if} \quad \|P\| < \delta \qquad \text{then} \quad \left| S^*(P) - \int_a^b f(x)\, dx \right| < \epsilon,$$

no matter how the x_i^* are chosen within $[x_{i-1}, x_i]$.

In symbols we write

(5.12.1)
$$\int_a^b f(x)\, dx = \lim_{\|P\| \mapsto 0} [f(x_1^*)\Delta x_1 + f(x_2^*)\Delta x_2 + \cdots + f(x_n^*)\Delta x_n].$$

A proof of this assertion appears in Appendix B.5. Figure 5.12.1 illustrates the idea. Here the base interval is broken up into 8 subintervals. The point x_1^* is chosen from $[x_0, x_1]$, x_2^* from $[x_1, x_2]$, and so on. While the integral represents the area under the curve, the Riemann sum represents the sum of the areas of the shaded rectangles. The difference between the two can be made as small as we wish (less than ϵ) simply by making the maximum length of the base subintervals sufficiently small—that is, by making $\|P\|$ sufficiently small.

In numerical computations the base interval $[a, b]$ is usually broken up into n

---

[†] After the French mathematician J. G. Darboux (1842–1917).
[††] After the German mathematician G. F. B. Riemann (1826–1866).

FIGURE 5.12.1

subintervals each of length $(b - a)/n$. The Riemann sums then take the form

$$S_n^* = \frac{b - a}{n} [f(x_1^*) + f(x_2^*) + \cdots + f(x_n^*)].$$

We'll carry out some numerical computations in Section 8.8.

Exercises

Choose $\epsilon > 0$. Break up the interval $[0, 1]$ into n subintervals each of length $1/n$. Take $x_1^* = 1/n, \ x_2^* = 2/n, \ \ldots, \ x_n^* = n/n$.

*1. (a) Determine the Riemann sum S_n^* for

$$\int_0^1 x \, dx.$$

(b) Show that

$$\left| S_n^* - \int_0^1 x \, dx \right| < \epsilon \qquad\qquad \text{if } n > 1/\epsilon.$$

HINT: $1 + 2 + \cdots + n = \frac{1}{2} n(n + 1)$.

2. (a) Determine the Riemann sum S_n^* for

$$\int_0^1 x^2 \, dx.$$

(b) Show that

$$\left| S_n^* - \int_0^1 x^2 \, dx \right| < \epsilon \qquad\qquad \text{if } n > 1/\epsilon.$$

HINT: $1^2 + 2^2 + \cdots + n^2 = \frac{1}{6} n(n + 1)(2n + 1)$.

*3. (a) Determine the Riemann sum S_n^* for

$$\int_0^1 x^3 \, dx.$$

(b) Show that

$$\left| S_n^* - \int_0^1 x^3 \, dx \right| < \epsilon \qquad\qquad \text{if } n > 1/\epsilon.$$

HINT: $1^3 + 2^3 + \cdots + n^3 = (1 + 2 + \cdots + n)^2$.

The Logarithm and Exponential Functions

6.1 In Search of a Notion of Logarithm

In elementary mathematics the logarithm to a base $B > 0$ is defined by setting

$$C = \log_B A \quad \text{iff} \quad B^C = A.$$

Generally the base 10 is chosen and the defining relation becomes

$$C = \log_{10} A \quad \text{iff} \quad 10^C = A.$$

The basic properties of $\log_{10}$ can then be summarized as follows:

(a) $\quad \log_{10} XY = \log_{10} X + \log_{10} Y.$ $\hspace{2cm}$ $(X > 0, Y > 0)$
(b) $\quad \log_{10} 1 = 0.$
(c) $\quad \log_{10} (1/Y) = -\log_{10} Y.$ $\hspace{3cm}$ $(Y > 0)$
(d) $\quad \log_{10} (X/Y) = \log_{10} X - \log_{10} Y.$ $\hspace{1cm}$ $(X > 0, Y > 0)$
(e) $\quad \log_{10} X^Y = Y \log_{10} X.$ $\hspace{3cm}$ $(X > 0)$
(f) $\quad \log_{10} 10 = 1.$

This elementary notion of logarithm is inadequate for calculus. It is unclear: what is meant by 10^C if C is irrational? It does not lend itself well to the methods of calculus: how would you differentiate $Y = \log_{10} X$ knowing only that $10^Y = X$?

Here we take an entirely different approach to logarithms. Instead of trying to tamper with the elementary definition, we discard it altogether. From our point of view the fundamental property of logarithms is that they transform multiplication into addition:

$$\text{the log of a product} = \text{the sum of the logs.}$$

Taking this as the central idea we are led to a general notion of logarithm which encompasses the elementary notion, lends itself well to the methods of calculus, and leads us naturally to a choice of base that simplifies many calculations.

Definition 6.1.1

A *logarithm* function is a nonconstant differentiable function f defined on the set of positive numbers such that for all $x > 0$ and $y > 0$

$$f(xy) = f(x) + f(y).$$

Let's assume for the time being that such logarithm functions exist, and let's see what we can find out about them. In the first place, if f is such a function, then

$$f(1) = f(1 \cdot 1) = f(1) + f(1) = 2f(1)$$

and so

$$f(1) = 0.$$

Taking $y > 0$, we have

$$0 = f(1) = f(y \cdot 1/y) = f(y) + f(1/y)$$

and therefore

$$f(1/y) = -f(y).$$

Taking $x > 0$ and $y > 0$, we have

$$f(x/y) = f(x \cdot 1/y) = f(x) + f(1/y),$$

which, in view of the previous result, means that

$$f(x/y) = f(x) - f(y).$$

We are now ready to look for the derivative. (Remember, we are *assuming* that f is differentiable.) We begin by forming the difference quotient

$$\frac{f(x + h) - f(x)}{h}.$$

From what we have discovered about f,

$$f(x + h) - f(x) = f\left(\frac{x + h}{x}\right) = f(1 + h/x),$$

and therefore

$$\frac{f(x + h) - f(x)}{h} = \frac{f(1 + h/x)}{h}.$$

Remembering that $f(1) = 0$ and multiplying the denominator by x/x, we have

$$\frac{f(x + h) - f(x)}{h} = \frac{1}{x}\left[\frac{f(1 + h/x) - f(1)}{h/x}\right].$$

As h tends to 0, the left side tends to $f'(x)$ and, while $1/x$ remains fixed, the bracketed expression tends to $f'(1)$. In short,

(6.1.2)

$$f'(x) = \frac{1}{x}f'(1).$$

We have now established that, if f is a logarithm, then

$$f(1) = 0 \quad \text{and} \quad f'(x) = \frac{1}{x} f'(1).$$

We can't have $f'(1) = 0$ for that would make f constant. (Explain.) The most natural alternative is to set $f'(1) = 1.\dagger$ The derivative then becomes $1/x$.

This function, which takes on the value 0 at 1 and has derivative $1/x$ for $x > 0$, must, by the fundamental theorem of integral calculus, take the form

$$\int_1^x \frac{dt}{t} . \qquad\qquad \text{(check this out)}$$

6.2 The Logarithm Function, Part I

Definition 6.2.1

The function

$$L(x) = \int_1^x \frac{dt}{t}, \qquad x > 0$$

is called the (*natural*) *logarithm function*.

Here are some obvious properties of L:

(1) L is defined on $(0, \infty)$ with derivative

$$L'(x) = \frac{1}{x} \qquad \text{for all } x > 0.$$

L' is positive on $(0, \infty)$, and therefore L is an increasing function.

(2) L is continuous on $(0, \infty)$ since it is there differentiable.

(3) For $x > 1$, $L(x)$ gives the area of the region shaded in Figure 6.2.1.

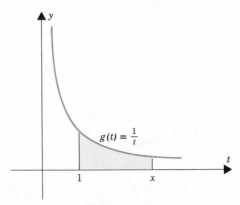

FIGURE 6.2.1

$\dagger$ This, as you'll see later, is tantamount to a choice of base.

(4) $L(x)$ is negative if $0 < x < 1$, zero at $x = 1$, positive if $x > 1$.

The following result is fundamental.

Theorem 6.2.2

If x and y are positive, then

$$L(xy) = L(x) + L(y).$$

PROOF

$$L(xy) = \int_1^{xy} \frac{dt}{t} = \int_1^x \frac{dt}{t} + \int_x^{xy} \frac{dt}{t}.$$

Since

$$L(x) = \int_1^x \frac{dt}{t},$$

we need only show that

$$L(y) = \int_x^{xy} \frac{dt}{t}.$$

To do this, set

$$xu = t; \qquad x\,du = dt.$$

At $t = x$, $u = 1$ and at $t = xy$, $u = y$. Thus

$$\int_x^{xy} \frac{dt}{t} = \int_1^y \frac{x\,du}{xu} = \int_1^y \frac{du}{u} = L(y). \quad \square$$

The next result follows from the multiplicative property $L(xy) = L(x) + L(y)$.

Theorem 6.2.3

If x is positive and p/q is rational, then

$$L(x^{p/q}) = \frac{p}{q} L(x).$$

PROOF.   First we show that for all nonnegative integers n

$$L(x^n) = nL(x).$$

We proceed by induction on n. If n is 0 or 1, the result is trivial. Let's suppose that the result holds for $n = k$. Now

$$L(x^{k+1}) = L(x^k \cdot x) = L(x^k) + L(x),$$

the second equality holding because of L's special multiplicative property. By our inductive assumption

$$L(x^k) = kL(x).$$

Thus

$$L(x^{k+1}) = kL(x) + L(x) = (k + 1)L(x),$$

which means that the result holds for $n = k + 1$. This shows that

$$L(x^n) = nL(x) \quad \text{for all nonnegative integers } n.$$

Since

$$0 = L(1) = L(x \cdot x^{-1}) = L(x) + L(x^{-1}),$$

you can see that

$$L(x^{-1}) = -L(x).$$

If n is a positive integer, then

$$L(x^{-n}) = L([x^{-1}]^n) = nL(x^{-1}) = -nL(x),$$

so that the formula holds for all negative integers.

Now take q as a positive integer. Since

$$L(x) = L([x^{1/q}]^q) = qL(x^{1/q}),$$

we have

$$L(x^{1/q}) = \frac{1}{q} L(x).$$

Finally take p/q as an arbitrary rational number. We can assume that q is positive and absorb the sign in p. Since p is an integer, we have

$$L(x^{p/q}) = L([x^{1/q}]^p) = pL(x^{1/q}).$$

Since we just proved that

$$L(x^{1/q}) = \frac{1}{q} L(x),$$

we can conclude that

$$L(x^{p/q}) = \frac{p}{q} L(x). \quad \square$$

The domain of L is $(0, \infty)$. What is the range of L?

Theorem 6.2.4

The range of L is $(-\infty, \infty)$.

PROOF.   Since L is continuous on $(0, \infty)$, we know from the intermediate-value theorem that it "skips" no values. Thus, its range is an interval. To show that the interval is $(-\infty, \infty)$, we need only show that it is unbounded above and unbounded below. We can do this by taking M as an arbitrary positive number and showing that L takes on values greater than M and values less than $-M$.

Since

$$L(2) = \int_1^2 \frac{dt}{t}$$

is positive (explain), we know that some positive multiple of $L(2)$ must be greater than M; namely, we know that there exists a positive integer n such that

$$nL(2) > M.$$

Multiplying this equation by -1 we have

$$-nL(2) < -M.$$

Since

$$nL(2) = L(2^n) \quad \text{and} \quad -nL(2) = L(2^{-n}),$$

we have

$$L(2^n) > M \quad \text{and} \quad L(2^{-n}) < -M.$$

This proves the unboundedness. $\square$

The Number e

Since the range of L is $(-\infty, \infty)$ and L is an increasing function, we know that L takes on every value and does so only once. In particular, there is one and only one number at which the function L takes on the value 1. *This unique number is denoted by the letter e.*†

Since

(6.2.5) $\boxed{L(e) = 1}$

it follows from Theorem 6.2.3 that

(6.2.6) $\boxed{L(e^{p/q}) = \dfrac{p}{q} \quad \text{for all rational numbers } \dfrac{p}{q}.}$

Because of this relation, we call L *the logarithm to the base e* and sometimes write

$$L(x) = \log_e x.$$

---

† After Leonard Euler (1707–1783), considered by many the greatest mathematician of the eighteenth century.

Usually we drop the subscript and write

(6.2.7) $$\boxed{L(x) = \log x.}^{\dagger}$$

The Graph of the Logarithm Function

You know that the logarithm function

$$\log x = \int_1^x \frac{dt}{t}$$

has domain $(0, \infty)$, range $(-\infty, \infty)$, and derivative

$$\frac{d}{dx}(\log x) = \frac{1}{x} > 0.$$

For small x the derivative is large (near 0 the curve is steep); for large x the derivative is small (far out the curve flattens out). At $x = 1$ the logarithm is 0 and its derivative $1/x$ is 1. (The graph crosses the x-axis at the point $(1, 0)$, and the tangent at that point is parallel to the line $y = x$.) The second derivative

$$\frac{d^2}{dx^2}(\log x) = -\frac{1}{x^2}$$

is negative on $(0, \infty)$. (The graph is concave down throughout.) For a picture of the logarithm curve see Figure 6.2.2.

$y = \log x$

FIGURE 6.2.2

$\dagger$ Another common notation for $L(x)$ is $\ln x$. We will stay with $\log x$. Logarithms to bases other than e will be taken up later [they arise by other choices of $f'(1)$], but by far the most important logarithm in calculus is the logarithm to the base e. So much so, that when we speak of the logarithm of a number x and don't specify the base, you can be sure that we are talking about the *natural logarithm*

$$\log x = \int_1^x \frac{dt}{t}.$$

Problem.   Estimate log 2.

SOLUTION

$$\log 2 = \int_1^2 \frac{dt}{t} \, .$$

We can estimate the integral by upper and lower sums. Using the partition

$$P = \{1 = \tfrac{10}{10}, \tfrac{11}{10}, \tfrac{12}{10}, \tfrac{13}{10}, \tfrac{14}{10}, \tfrac{15}{10}, \tfrac{16}{10}, \tfrac{17}{10}, \tfrac{18}{10}, \tfrac{19}{10}, \tfrac{20}{10} = 2\},$$

we have

$$L_f(P) = \tfrac{1}{10}(\tfrac{10}{11} + \tfrac{10}{12} + \tfrac{10}{13} + \tfrac{10}{14} + \tfrac{10}{15} + \tfrac{10}{16} + \tfrac{10}{17} + \tfrac{10}{18} + \tfrac{10}{19} + \tfrac{10}{20})$$

$$= \tfrac{1}{11} + \tfrac{1}{12} + \tfrac{1}{13} + \tfrac{1}{14} + \tfrac{1}{15} + \tfrac{1}{16} + \tfrac{1}{17} + \tfrac{1}{18} + \tfrac{1}{19} + \tfrac{1}{20}$$

and

$$U_f(P) = \tfrac{1}{10}(\tfrac{10}{10} + \tfrac{10}{11} + \tfrac{10}{12} + \tfrac{10}{13} + \tfrac{10}{14} + \tfrac{10}{15} + \tfrac{10}{16} + \tfrac{10}{17} + \tfrac{10}{18} + \tfrac{10}{19})$$

$$= \tfrac{1}{10} + \tfrac{1}{11} + \tfrac{1}{12} + \tfrac{1}{13} + \tfrac{1}{14} + \tfrac{1}{15} + \tfrac{1}{16} + \tfrac{1}{17} + \tfrac{1}{18} + \tfrac{1}{19}.$$

Here a hand calculator can come in handy. After determining that

$$0.100 = \tfrac{1}{10} = 0.100$$
$$0.090 < \tfrac{1}{11} < 0.091$$
$$0.083 < \tfrac{1}{12} < 0.084$$
$$0.076 < \tfrac{1}{13} < 0.077$$
$$0.071 < \tfrac{1}{14} < 0.072$$
$$0.066 < \tfrac{1}{15} < 0.067$$
$$0.062 < \tfrac{1}{16} < 0.063$$
$$0.058 < \tfrac{1}{17} < 0.059$$
$$0.055 < \tfrac{1}{18} < 0.056$$
$$0.052 < \tfrac{1}{19} < 0.053$$
$$0.050 = \tfrac{1}{20} = 0.050$$

you can see after a little addition that

$$0.663 < L_f(P) < \log 2 < U_f(P) < 0.722.$$

The average of these two estimates is

$$\tfrac{1}{2}(0.663 + 0.722) = 0.6925.$$

We are not far off. Three-place tables carry the estimate 0.693.   □

   Table 6.2.1 gives the natural logarithms of the integers 2 through 10 rounded off to the nearest hundredth.†

---

† A more extended table appears at the end of the book, but throughout this chapter all numerical calculations will be based on Table 6.2.1.

TABLE 6.2.1

n	$\log n$	n	$\log n$
1	0.00	6	1.79
2	0.69	7	1.95
3	1.10	8	2.08
4	1.39	9	2.20
5	1.61	10	2.30

Problem.   Use Table 6.2.1 to estimate the following logarithms:

(a) log 0.2.   (b) log 0.25.   (c) log 2.4.   (d) log 90.

SOLUTION

(a) $\log 0.2 = \log \frac{1}{5} = -\log 5 \cong -1.61$.

(b) $\log 0.25 = \log (0.5)^2 = 2 \log \frac{1}{2} = -2 \log 2 \cong -0.138$.

(c) $\log 2.4 = \log [(3)(0.8)] = \log 3 + \log 8 - \log 10 \cong 0.88$.

(d) $\log 90 = \log [(9)(10)] = \log 9 + \log 10 \cong 4.50$.   □

Problem.   Use Table 6.2.1 to estimate e.

SOLUTION.   So far all we know is that $\log e = 1$. From the table you can see that

$$3 \log 3 - \log 10 \cong 1.$$

The expression on the left can be written

$$\log 3^3 - \log 10 = \log 27 - \log 10 = \log \left(\tfrac{27}{10}\right) = \log 2.7.$$

With $\log 2.7 \cong 1$ we have $e \cong 2.7$. (Eight-place tables give the estimate $e \cong 2.71828182$. For most purposes we take $e \cong 2.72$.)   □

Exercises

Use Table 6.2.1 to estimate the following natural logarithms.

*1. log 20.   2. log 16.   *3. log 1.6.
*4. $\log 3^4$.   5. log 0.1.   *6. log 2.5.
*7. log 7.2.   8. $\log \sqrt{630}$.   *9. $\log \sqrt{2}$.

10. Interpret the equation

$$\log n = \log mn - \log m$$

in terms of area under the curve $y = 1/x$. Draw a figure.

*11. Determine the area under the graph of $y = 1/t$ from $t = x$ to $t = 1$ if $0 < x < 1$.

12. Estimate

$$\log 1.5 = \int_1^{1.5} \frac{dt}{t}$$

using the approximation

$$\tfrac{1}{2}[L_f(P) + U_f(P)] \quad \text{with} \quad P = \{1 = \tfrac{8}{8}, \tfrac{9}{8}, \tfrac{10}{8}, \tfrac{11}{8}, \tfrac{12}{8} = 1.5\}.$$

13. Taking $\log 5 \cong 1.61$, use differentials to estimate
    *(a) $\log 5.2$.   (b) $\log 4.8$.   *(c) $\log 5.5$.
14. Taking $\log 10 \cong 2.30$, use differentials to estimate
    (a) $\log 10.3$.   *(b) $\log 9.6$.   (c) $\log 11$.

Solve the following equations for x.

*15. $\log x = 2$.   16. $\log x = -1$.
17. $(2 - \log x) \log x = 0$.   *18. $\tfrac{1}{2}\log x = \log (2x - 1)$.
*19. $\log [(2x + 1)(x + 2)] = 2 \log (x + 2)$.   20. $2 \log (x + 2) - \tfrac{1}{2}\log x^4 = 1$.

6.3   The Logarithm Function, Part II

Differentiation and Graphing

Problem.   Find the domain of f and find $f'(x)$ given that

$$f(x) = (\log x^2)^3.$$

SOLUTION.   Since the logarithm function is defined only for positive numbers, we must have $x^2 > 0$. The domain of f consists of all $x \neq 0$. Before differentiating note that

$$f(x) = (\log x^2)^3 = (2 \log x)^3 = 8(\log x)^3.$$

It follows that

$$f'(x) = 24(\log x)^2 \frac{d}{dx} (\log x) = 24(\log x)^2 \frac{1}{x} = \frac{24}{x} (\log x)^2. \quad \square$$

Problem.   Find the domain of f and find $f'(x)$ if

$$f(x) = \log (x\sqrt{1 + 3x}).$$

SOLUTION.   For x to be in the domain of f, we must have $x\sqrt{1 + 3x} > 0$, and consequently, $x > 0$. The domain is the set of positive numbers.

Before differentiating f, we make use of the special properties of the logarithm:

$$f(x) = \log (x\sqrt{1 + 3x}) = \log x + \tfrac{1}{2}\log(1 + 3x).$$

From this we have

$$f'(x) = \frac{1}{x} + \frac{1}{2}\left(\frac{3}{1 + 3x}\right). \quad \square$$

Problem.   Sketch the graph of

$$f(x) = \log |x|.$$

SOLUTION.   This is an even function, $f(-x) = f(x)$, defined for all numbers $x \neq 0$. The graph has two branches:

$$y = \log (-x), \quad x < 0 \quad \text{and} \quad y = \log x, \quad x > 0.$$

Each branch is the mirror image of the other. (Figure 6.3.1)   □

FIGURE 6.3.1

Problem.   Show that

(6.3.1)
$$\frac{d}{dx} (\log |x|) = \frac{1}{x}$$

for all $x \neq 0$.

SOLUTION.   For $x > 0$,

$$\frac{d}{dx} (\log |x|) = \frac{d}{dx} (\log x) = \frac{1}{x}.$$

For $x < 0$,

$$\frac{d}{dx} (\log |x|) = \frac{d}{dx} [\log (-x)] = \frac{1}{-x} \frac{d}{dx} (-x) = \left(\frac{1}{-x} \right) (-1) = \frac{1}{x}. \quad \square$$

Problem.   Show that, if g is differentiable and $g(x) \neq 0$, then

(6.3.2)
$$\frac{d}{dx} (\log |g(x)|) = \frac{g'(x)}{g(x)}.$$

SOLUTION.   Note that

$$\log |g(x)| = f(g(x)), \quad \text{where} \quad f(x) = \log |x|.$$

By the chain rule

$$\frac{d}{dx}\left(\log|g(x)|\right) = f'(g(x))g'(x) = \frac{g'(x)}{g(x)} \cdot \quad \square$$

Here are two examples:

$$\frac{d}{dx}\left(\log|1 - x^3|\right) = \frac{-3x^2}{1 - x^3} = \frac{3x^2}{x^3 - 1} \cdot$$

$$\frac{d}{dx}\left(\log\left|\frac{x - 1}{x - 2}\right|\right) = \frac{d}{dx}\left(\log|x - 1|\right) - \frac{d}{dx}\left(\log|x - 2|\right) = \frac{1}{x - 1} - \frac{1}{x - 2} \cdot \quad \square$$

Problem.   Let

$$f(x) = \log\left(\frac{x^4}{x - 1}\right) \cdot$$

(1) Specify the domain of f. (2) On what intervals is f increasing? Decreasing? (3) Find the extreme values of f. (4) Determine the concavity of the graph and find the points of inflection. (5) Sketch the graph.

SOLUTION.   Since the logarithm function is defined only for positive numbers, the domain of f consists of all numbers greater than 1.

Making use of the special properties of the logarithm we get

$$f(x) = \log x^4 - \log(x - 1) = 4\log x - \log(x - 1).$$

This gives

$$f'(x) = \frac{4}{x} - \frac{1}{x - 1} = \frac{3x - 4}{x(x - 1)}$$

and

$$f''(x) = -\frac{4}{x^2} + \frac{1}{(x - 1)^2} = -\frac{(x - 2)(3x - 2)}{x^2(x - 1)^2} \cdot$$

Since the domain of f is $(1, \infty)$, we need concern ourselves only with $x > 1$. It is easy to see that

$$f'(x) \text{ is } \begin{cases} \text{negative,} & \text{if} \quad 1 < x < \frac{4}{3} \\ 0, & \text{if} \qquad x = \frac{4}{3} \\ \text{positive,} & \text{if} \qquad x > \frac{4}{3} \end{cases} \cdot$$

This means that f decreases on $(1, \frac{4}{3}]$ and increases on $[\frac{4}{3}, \infty)$. By the first derivative test

$$f(\tfrac{4}{3}) = 4\log 4 - 3\log 3 \cong 2.26 \qquad \text{(Table 6.2.1)}$$

is a local and absolute minimum. There are no other extreme values.

Since

$$f''(x) \text{ is } \begin{cases} \text{positive,} & \text{if} \quad 1 < x < 2 \\ 0, & \text{if} \qquad x = 2 \\ \text{negative,} & \text{if} \qquad x > 2 \end{cases},$$

the graph is concave up on $(1, 2]$ and concave down on $[2, \infty)$. The point

$$(2, f(2)) = (2, 4 \log 2) \cong (2, 2.76)$$

is the only point of inflection.

Before sketching the graph note that the derivative

$$f'(x) = \frac{4}{x} - \frac{1}{x - 1}$$

is very large negative for x close to 1 and very close to 0 for x large. This means that the graph is very steep for x close to 1 and very flat for x large. The graph appears in Figure 6.3.2. $\square$

FIGURE 6.3.2

Integration

The integral counterpart of (6.3.1) takes the form

(6.3.3)
$$\int \frac{dx}{x} = \log |x| + C.$$

The integral counterpart of (6.3.2) can be written

(6.3.4)
$$\int \frac{g'(x)}{g(x)} \, dx = \log |g(x)| + C,$$

provided of course that $g(x) \neq 0$. We can also derive (6.3.4) from (6.3.3) by the usual u-substitution: set

$$u = g(x); \qquad du = g'(x) \, dx.$$

$$\int \frac{g'(x)}{g(x)} \, dx = \int \frac{du}{u} = \log |u| + C = \log |g(x)| + C. \quad \square$$

Problem.   Find

$$\int \frac{8x}{x^2 - 1}\, dx.$$

SOLUTION.   Set

$$u = x^2 - 1;\qquad du = 2x\, dx,\quad 4\, du = 8x\, dx.$$

$$\int \frac{8x}{x^2 - 1}\, dx = 4 \int \frac{du}{u} = 4 \log |u| + C = 4 \log |x^2 - 1| + C. \quad \Box$$

Problem.   Find

$$\int \frac{x^2}{1 - 4x^3}\, dx.$$

SOLUTION.   Set

$$u = 1 - 4x^3;\qquad du = -12x^2\, dx,\quad -\tfrac{1}{12}du = x^2\, dx.$$

$$\int \frac{x^2}{1 - 4x^3}\, dx = -\frac{1}{12}\int \frac{du}{u} = -\frac{1}{12}\log |u| + C = -\frac{1}{12}\log |1 - 4x^3| + C. \quad \Box$$

Problem.   Find

$$\int \frac{\log x}{x}\, dx.$$

SOLUTION.   Set

$$u = \log x;\qquad du = \frac{1}{x}dx.$$

$$\int \frac{\log x}{x}\, dx = \int u\, du = \tfrac{1}{2}u^2 + C = \tfrac{1}{2}(\log x)^2 + C. \quad \Box$$

Problem.   Evaluate

$$\int_1^2 \frac{6x^2 - 2}{x^3 - x + 1}\, dx.$$

SOLUTION.   Set

$$u = x^3 - x + 1;\qquad du = (3x^2 - 1)\, dx,\quad 2\, du = (6x^2 - 2)\, dx$$

At $x = 1$, $u = 1$; at $x = 2$, $u = 7$.

$$\int_1^2 \frac{6x^2 - 2}{x^3 - x + 1}\, dx = 2 \int_1^7 \frac{du}{u} = 2 \left[\log |u| \right]_1^7 = 2\,(\log 7 - \log 1) = 2 \log 7. \quad \Box$$

Logarithmic Differentiation

To differentiate a complicated product

$$g(x) = g_1(x)g_2(x) \cdots g_n(x)$$

you can first form

$$\log|g(x)| = \log|g_1(x)| + \log|g_2(x)| + \cdots + \log|g_n(x)|$$

and then differentiate

$$\frac{g'(x)}{g(x)} = \frac{g_1'(x)}{g_1(x)} + \frac{g_2'(x)}{g_2(x)} + \cdots + \frac{g_n'(x)}{g_n(x)}.$$

This process is called *logarithmic differentiation.*

Problem.   Use logarithmic differentiation to find $g'(x)$ for

$$g(x) = x(x-1)(x-2)(x-3).$$

SOLUTION

$$\log|g(x)| = \log|x| + \log|x-1| + \log|x-2| + \log|x-3|,$$

$$\frac{g'(x)}{g(x)} = \frac{1}{x} + \frac{1}{x-1} + \frac{1}{x-2} + \frac{1}{x-3},$$

$$g'(x) = x(x-1)(x-2)(x-3)\left(\frac{1}{x} + \frac{1}{x-1} + \frac{1}{x-2} + \frac{1}{x-3}\right). \quad \square$$

You can also use logarithmic differentiation when quotients are involved.

Problem.   Use logarithmic differentiation to find $g'(x)$ for

$$g(x) = \frac{(x^2+1)^3(2x-5)^2}{(x^2+5)^2}.$$

What is $g'(1)$?

SOLUTION

$$\log|g(x)| = 3\log|x^2+1| + 2\log|2x-5| - 2\log|x^2+5|,$$

$$\frac{g'(x)}{g(x)} = \frac{6x}{x^2+1} + \frac{4}{2x-5} - \frac{4x}{x^2+5},$$

$$g'(x) = \frac{(x^2+1)^3(2x-5)^2}{(x^2+5)^2}\left(\frac{6x}{x^2+1} + \frac{4}{2x-5} - \frac{4x}{x^2+5}\right),$$

$$g'(1) = \frac{2^3(-3)^2}{6^2}\left(\frac{6}{2} + \frac{4}{(-3)} - \frac{4}{6}\right) = 2. \quad \square$$

Exercises

Find the domain and compute the derivative.

*1. $f(x) = \log 4x$.

2. $f(x) = \log (2x + 1)$.

*3. $f(x) = \log (x^3 + 1)$.

4. $f(x) = \log [(x + 1)^3]$.

*5. $f(x) = \log \sqrt{1 + x^2}$.

6. $f(x) = (\log x)^3$.

*7. $f(x) = \log |x^4 - 1|$.

8. $f(x) = \log (\log x)$.

*9. $f(x) = \dfrac{1}{\log x}$.

10. $f(x) = \log \left|\dfrac{x + 2}{x^3 - 1}\right|$.

*11. $f(x) = x \log x$.

12. $f(x) = \log \sqrt[4]{x^2 + 1}$.

*13. $f(x) = \dfrac{\log (x + 1)}{x + 1}$.

14. $f(x) = \log \sqrt{\dfrac{1 - x}{2 - x}}$.

Work out the following integrals.

*15. $\displaystyle\int \dfrac{dx}{x + 1}$.

16. $\displaystyle\int \dfrac{dx}{3 - x}$.

*17. $\displaystyle\int \dfrac{x}{3 - x^2}\, dx$.

18. $\displaystyle\int \dfrac{x + 1}{x^2}\, dx$.

*19. $\displaystyle\int \dfrac{x}{(3 - x^2)^2}\, dx$.

20. $\displaystyle\int \dfrac{\log (x + a)}{x + a}\, dx$.

*21. $\displaystyle\int \left(\dfrac{1}{x - a} - \dfrac{1}{x - b}\right) dx$.

22. $\displaystyle\int \dfrac{x^2}{2x^3 - 1}\, dx$.

*23. $\displaystyle\int x \left(\dfrac{1}{x^2 - a^2} - \dfrac{1}{x^2 - b^2}\right) dx$.

24. $\displaystyle\int \dfrac{2x - 1}{x(x - 1)}\, dx$.

*25. $\displaystyle\int \dfrac{\sqrt{x}}{1 + x\sqrt{x}}\, dx$.

26. $\displaystyle\int \dfrac{dx}{x (\log x)^2}$.

Evaluate.

*27. $\displaystyle\int_1^e \dfrac{dx}{x}$.

28. $\displaystyle\int_1^{e^2} \dfrac{dx}{x}$.

*29. $\displaystyle\int_e^{e^2} \dfrac{dx}{x}$.

30. $\displaystyle\int_3^4 \dfrac{dx}{1 - x}$.

*31. $\displaystyle\int_4^5 \dfrac{x}{x^2 - 1}\, dx$.

32. $\displaystyle\int_1^e \dfrac{\log x}{x}\, dx$.

*33. $\displaystyle\int_0^1 \dfrac{\log (x + 1)}{x + 1}\, dx$.

34. $\displaystyle\int_0^1 \left(\dfrac{1}{x + 1} - \dfrac{1}{x + 2}\right) dx$.

Find the derivative by logarithmic differentiation.

*35. $g(x) = (x^2 + 1)^2 (x - 1)^5 x^3$.

36. $g(x) = (x^2 + a^2)^2 (x^3 + b^3)^3$.

*37. $g(x) = \dfrac{x^4 (x - 1)}{x + 2}$.

38. $g(x) = \dfrac{(1 + x)(2 + x)x}{4 + x}$.

*39. $g(x) = \sqrt{\dfrac{(x - 1)(x - 2)}{(x - 3)(x - 4)}}$.

40. $g(x) = \dfrac{\sqrt{x^2 + 1} - x}{\sqrt{x^2 + 1} + x}$.

41. Find the area of that portion of the first quadrant which lies between
    *(a) $x + 4y - 5 = 0$ and $xy = 1$.　　(b) $x + y - 3 = 0$ and $xy = 2$.
42. A particle moves along a line with acceleration

$$a(t) = -\frac{1}{(t+1)^2} \quad \text{feet per second per second.}$$

Find the distance traveled by the particle during the time interval $[0, 4]$, given that the initial velocity $v(0)$ is: *(a) 1 foot per second. (b) 2 feet per second.

43. Find a formula for the nth derivative:

(a) $\dfrac{d^n}{dx^n} (\log x)$.　　　　　　(b) $\dfrac{d^n}{dx^n} [\log (1 - x)]$.

44. (a) Verify that

$$\log x = \int_1^x \frac{dt}{t} < \int_1^x \frac{dt}{\sqrt{t}} \quad \text{for } x > 1.$$

(b) Use (a) to show that

$$0 < \log x < 2(\sqrt{x} - 1) \quad \text{for } x > 1.$$

(c) Use (b) to show that

$$2x\left(1 - \frac{1}{\sqrt{x}}\right) < x \log x < 0 \quad \text{for } 0 < x < 1.$$

(d) Use (c) to show that

$$\lim_{x \downarrow 0} (x \log x) = 0.$$

45. For each of the functions below, (i) find the domain, (ii) find the intervals where it is increasing and the intervals where it is decreasing, (iii) find the extreme values, (iv) determine the concavity of the graph and find the points of inflection, and finally, (v) sketch the graph.
    *(a) $f(x) = \log 2x$.　　　　　　(b) $f(x) = x - \log x$.
    *(c) $f(x) = \log (4 - x)$.　　　　(d) $f(x) = \log (4 - x^2)$.
    *(e) $f(x) = x \log x$.　　　　　　(f) $f(x) = \log (8x - x^2)$.

    *(g) $f(x) = \log \left(\dfrac{x}{1 + x^2}\right)$.　　(h) $f(x) = \log \left(\dfrac{x^3}{x - 1}\right)$.

6.4　The Exponential Function

Rational powers of e already have an established meaning: by $e^{p/q}$ we mean the qth root of e raised to the pth power. But what is meant by $e^{\sqrt{2}}$ or e^π?

Earlier we proved that each rational power $e^{p/q}$ had logarithm p/q:

(6.4.1) 　　　　　　　　　　$$\log e^{p/q} = \frac{p}{q}.$$

Inspired by this relation, we make the following definition:

Definition 6.4.2

If z is irrational, then by e^z we mean the unique number which has logarithm z:

$$\log e^z = z.$$

What is $e^{\sqrt{2}}$? It is the unique number which has logarithm $\sqrt{2}$. What is e^{π}? It is the unique number which has logarithm π. Note that e^x now has meaning for every real value of x.

Definition 6.4.3

The function

$$E(x) = e^x \qquad \text{for all real } x$$

is called the *exponential function*.

Some properties of the exponential function follow.

  (1) Combining 6.4.1 and 6.4.2 we have

(6.4.4)
$$\boxed{\log e^x = x \qquad \text{for all real } x.}$$

Writing

$$L(x) = \log x \quad \text{and} \quad E(x) = e^x$$

we have

$$L(E(x)) = x \qquad \text{for all real } x.$$

This says that the *exponential function is the inverse of the logarithm function*.
  (2) The graph of the exponential function appears in Figure 6.4.1. It can be obtained from the graph of the logarithm by reflection in the line $y = x$.

FIGURE 6.4.1

TABLE 6.4.1

t	e^t	t	e^t
0.1	1.11	1.1	3.00
0.2	1.22	1.2	3.32
0.3	1.35	1.3	3.67
0.4	1.49	1.4	4.06
0.5	1.65	1.5	4.48
0.6	1.82	1.6	4.95
0.7	2.01	1.7	5.47
0.8	2.23	1.8	6.05
0.9	2.46	1.9	6.68
1.0	2.72	2.0	7.39

(3) Since the graph of the logarithm remains to the right of the y-axis, the graph of the exponential function remains above the x-axis; namely,

(6.4.5)
$$e^x > 0 \qquad \text{for all real } x.$$

(4) Since the graph of the logarithm crosses the x-axis at $(1, 0)$, the graph of the exponential function crosses the y-axis at $(0, 1)$:

$$\log 1 = 0 \quad \text{gives} \quad e^0 = 1.$$

(5) That the exponential function and logarithm function are inverses can also be expressed by writing

(6.4.6)
$$e^{\log x} = x \qquad \text{for all } x > 0.$$

You can verify this equation by observing that both sides have the same logarithm:

$$\log (e^{\log x}) = \log x \log e = \log x. \quad \square$$

You know that for rational exponents

$$e^{(p/q + r/s)} = e^{p/q} \cdot e^{r/s}.$$

This fundamental property holds for all exponents, including the irrational ones.

Theorem 6.4.7

$$e^{x+y} = e^x \cdot e^y \qquad \text{for all real } x \text{ and } y.$$

PROOF

$$\log (e^x \cdot e^y) = \log e^x + \log e^y = x + y = \log e^{x+y}.$$

Since the logarithm is one-to-one, we must have

$$e^{x+y} = e^x \cdot e^y. \quad \square$$

We leave it to you to show that

(6.4.8)
$$e^{-y} = \frac{1}{e^y} \qquad \text{and} \qquad e^{x-y} = \frac{e^x}{e^y}.$$

Table 6.4.1 gives some values of the exponential function rounded off to the nearest hundredth.†

Problem. Use Table 6.4.1 to estimate the following powers of e:
(a) $e^{-0.2}$.   (b) $e^{2.4}$.   (c) $e^{3.1}$.

---

† More extended tables appear at the end of the book, but throughout this chapter all numerical calculations will be based on Table 6.4.1.

SOLUTION.   The idea, of course, is to use the laws of exponents.

(a) $e^{-0.2} = \dfrac{1}{e^{0.2}} \cong \dfrac{1}{1.22} \cong 0.82.$

(b) $e^{2.4} = e^{2+0.4} = (e^2)(e^{0.4}) \cong (7.39)(1.49) \cong 11.01.$
(c) $e^{3.1} = e^{1.7+1.4} = (e^{1.7})(e^{1.4}) \cong (5.47)(4.06) \cong 22.21.$ $\square$

We come now to one of the most important results in all of calculus. It is marvelously simple.

Theorem 6.4.9

The exponential function is its own derivative:

$$\frac{d}{dx}(e^x) = e^x.$$

PROOF.   The logarithm function is differentiable, and its derivative is never 0. It follows from our discussion in Section 3.7 that its inverse, the exponential function, is also differentiable. Knowing this, we can show that

$$\frac{d}{dx}(e^x) = e^x$$

by differentiating the identity

$$\log e^x = x.$$

On the left-hand side, the chain rule gives

$$\frac{d}{dx}(\log e^x) = \frac{1}{e^x}\frac{d}{dx}(e^x).$$

On the right-hand side, the derivative is 1. Equating the derivatives, we get

$$\frac{1}{e^x}\frac{d}{dx}(e^x) = 1 \quad \text{and thus} \quad \frac{d}{dx}(e^x) = e^x. \quad \square$$

You will frequently run across expressions of the form $e^{f(x)}$. If f is differentiable, the chain rule gives

(6.4.10) $\boxed{\dfrac{d}{dx}[e^{f(x)}] = e^{f(x)}f'(x).}$ †

---

† If you have trouble seeing this, write

$$e^{f(x)} = E(f(x)).$$

Then

$$\frac{d}{dx}[e^{f(x)}] = \frac{d}{dx}[E(f(x))] = E'(f(x))f'(x).$$

Since the exponential function is its own derivative,

$$E'(f(x)) = E(f(x)) = e^{f(x)}.$$

The result now follows. $\square$

Examples

$$\frac{d}{dx}(e^{kx}) = e^{kx} \cdot k = ke^{kx}.$$

$$\frac{d}{dx}(e^{\sqrt{x}}) = e^{\sqrt{x}}\frac{d}{dx}(\sqrt{x}) = e^{\sqrt{x}}\left(\frac{1}{2\sqrt{x}}\right) = \frac{1}{2\sqrt{x}}e^{\sqrt{x}}.$$

$$\frac{d}{dx}(e^{-x^2}) = e^{-x^2}\frac{d}{dx}(-x^2) = e^{-x^2}(-2x) = -2xe^{-x^2}. \quad \square$$

The relation

$$\frac{d}{dx}(e^x) = e^x \quad \text{and its corollary} \quad \frac{d}{dx}(e^{kx}) = ke^{kx}$$

have important applications to engineering, physics, chemistry, biology, and economics. We will discuss some of these applications in Section 6.6.

Problem. Let

$$f(x) = xe^{-x} \qquad \text{for all real } x.$$

(a) On what intervals is f increasing? Decreasing? (b) Find the extreme values of f. (c) Determine the concavity of the graph and find the points of inflection. (d) Sketch the graph.

SOLUTION. We have

$$f(x) = xe^{-x},$$
$$f'(x) = xe^{-x}(-1) + e^{-x} = (1-x)e^{-x},$$
$$f''(x) = (1-x)e^{-x}(-1) - e^{-x} = (x-2)e^{-x}.$$

Since

$$f'(x) \text{ is } \begin{cases} \text{positive,} & \text{for} \quad x < 1 \\ 0, & \text{at} \quad x = 1 \\ \text{negative,} & \text{for} \quad x > 1 \end{cases},$$

f increases on $(-\infty, 1]$ and decreases on $[1, \infty)$. The number

$$f(1) = \frac{1}{e} \cong \frac{1}{2.72} \cong 0.37$$

is a local and absolute maximum. There are no other extreme values. Since

$$f''(x) \text{ is } \begin{cases} \text{negative,} & \text{for} \quad x < 2 \\ 0, & \text{at} \quad x = 2 \\ \text{positive,} & \text{for} \quad x > 2 \end{cases},$$

the graph is concave down on $(-\infty, 2]$ and concave up on $[2, \infty)$. The point

$$(2, f(2)) = (2, 2e^{-2}) \cong \left(2, \frac{2}{(2.72)^2}\right) \cong (2, 0.27)$$

is the only point of inflection. A sketch of the graph is given in Figure 6.4.2. $\quad \square$

FIGURE 6.4.2   FIGURE 6.4.3

Problem.   Let

$$f(x) = xe^x \qquad \text{for all } x \in (-1, 1).$$

(a) On what intervals is f increasing? Decreasing? (b) Find the extreme values of f. (c) Determine the concavity of the graph and find the points of inflection. (d) Sketch the graph.

SOLUTION.   Here

$$f(x) = xe^{x^2},$$
$$f'(x) = xe^{x^2}(2x) + e^{x^2} = (1 + 2x^2)e^{x^2},$$
$$f''(x) = (1 + 2x^2)e^{x^2}(2x) + 4xe^{x^2} = 2x(2x^2 + 3)e^{x^2}.$$

Since $f'(x) > 0$ for all $x \in (-1, 1)$, f increases on $(-1, 1)$ and there are no extreme values. Since

$$f''(x) \text{ is } \begin{cases} \text{negative,} & \text{for } x \in (-1, 0) \\ 0, & \text{at } x = 0 \\ \text{positive,} & \text{for } x \in (0, 1) \end{cases},$$

the graph is concave down on $(-1, 0]$, concave up on $[0, 1)$. The point $(0, f(0)) = (0, 0)$ is the only point of inflection. Since $f(-x) = -f(x)$, f is an odd function, and thus its graph (Figure 6.4.3) is symmetric with respect to the origin.   □

The integral counterpart of (6.4.9) can be written

(6.4.11) $$\int e^x \, dx = e^x + C.$$

Problem. Find

$$\int 6e^{3x}\, dx.$$

SOLUTION. Set

$$u = e^{3x}; \quad du = 3e^{3x}\, dx, \quad 2du = 6e^{3x}\, dx.$$

$$\int 6e^{3x}\, dx = 2 \int du = 2u + C = 2e^{3x} + C.$$

If you recognize at the very beginning that

$$3e^{3x} = \frac{d}{dx}(e^{3x}),$$

then you can dispense with the u-substitution and write

$$\int 6e^{3x}\, dx = 2 \int 3e^{3x}\, dx = 2e^{3x} + C. \quad \square$$

Problem. Find

$$\int \frac{e^{\sqrt{x}}}{\sqrt{x}}\, dx.$$

SOLUTION. Set

$$u = e^{\sqrt{x}}; \quad du = e^{\sqrt{x}} \frac{1}{2\sqrt{x}}\, dx, \quad 2du = \frac{e^{\sqrt{x}}}{\sqrt{x}}\, dx.$$

$$\int \frac{e^{\sqrt{x}}}{\sqrt{x}}\, dx = 2 \int du = 2u + C = 2e^{\sqrt{x}} + C.$$

If you recognize immediately that

$$\frac{1}{2}\left(\frac{e^{\sqrt{x}}}{\sqrt{x}}\right) = \frac{d}{dx}(e^{\sqrt{x}}),$$

then you can dispense with the u-substitution and integrate directly:

$$\int \frac{e^{\sqrt{x}}}{\sqrt{x}}\, dx = 2 \int \frac{1}{2}\left(\frac{e^{\sqrt{x}}}{\sqrt{x}}\right) dx = 2e^{\sqrt{x}} + C. \quad \square$$

REMARK. The u-substitution simplifies many calculations, but there is no reason to use it where you find that you can carry out the integration more quickly without it.

Problem. Find

$$\int \frac{e^{3x}}{e^{3x} + 1}\, dx.$$

SOLUTION. Set

$$u = e^{3x} + 1; \qquad du = 3e^{3x}\, dx, \quad \tfrac{1}{3}du = e^{3x}\, dx.$$

$$\int \frac{e^{3x}}{e^{3x} + 1}\, dx = \frac{1}{3} \int \frac{du}{u} = \frac{1}{3} \log |u| + C = \frac{1}{3} \log (e^{3x} + 1) + C. \quad \square$$

Problem.   Evaluate

$$\int_0^{\log 2} e^x\, dx.$$

SOLUTION

$$\int_0^{\log 2} e^x\, dx = \left[e^x \right]_0^{\log 2} = e^{\log 2} - e^0 = 2 - 1 = 1.$$

The u-substitution is obviously unnecessary here, but if you prefer to use it you certainly can: set

$$u = e^x; \qquad du = e^x\, dx.$$

$$x = 0, \quad u = 1; \qquad x = \log 2, \quad u = 2.$$

$$\int_0^{\log 2} e^x\, dx = \int_1^2 du = \left[u \right]_1^2 = 2 - 1 = 1. \quad \square$$

Problem.   Evaluate

$$\int_0^1 e^x(e^x + 1)^{1/5}\, dx.$$

SOLUTION.   Set

$$u = e^x + 1; \qquad du = e^x\, dx.$$

$$x = 0, \quad u = 2; \qquad x = 1, \quad u = e + 1.$$

$$\int_0^1 e^x(e^x + 1)^{1/5}\, dx = \int_2^{e+1} u^{1/5}\, du = \left[\tfrac{5}{6}u^{6/5} \right]_2^{e+1} = \tfrac{5}{6}[(e + 1)^{6/5} - 2^{6/5}] + C. \quad \square$$

Exercises

Differentiate the following functions.

*1. $y = e^{-2x}$.

2. $y = 3e^{2x+1}$.

*3. $y = e^{x^2-1}$.

*4. $y = 2e^{-4x}$.

5. $y = e^x \log x$.

*6. $y = x^2 e^x$.

*7. $y = x^{-1}e^{-x}$.

8. $y = e^{\sqrt{x}+1}$.

*9. $y = \tfrac{1}{2}(e^x + e^{-x})$.

*10. $y = \tfrac{1}{2}(e^x - e^{-x})$.

11. $y = e^{\sqrt{x}} \log \sqrt{x}$.

*12. $y = (1 - e^{4x})^2$.

*13. $y = (e^x + e^{-x})^2$.

14. $y = (3 - 2e^{-x})^3$.

*15. $y = (e^{x^2} + 1)^2$.

*16. $y = (e^{2x} - e^{-2x})^2$.

17. $y = e^{\sqrt{1-x^2}}$.

*18. $y = x^2 e^x - xe^{x^2}$.

*19. $y = \dfrac{e^x - 1}{e^x + 1}$.

20. $y = \dfrac{e^{2x} - 1}{e^{2x} + 1}$.

*21. $y = \dfrac{e^{ax} - e^{bx}}{e^{ax} + e^{bx}}$.

Work out the following indefinite integrals.

*22. $\int e^{2x}\,dx.$ 23. $\int e^{-2x}\,dx.$ *24. $\int e^{kx}\,dx.$

*25. $\int e^{ax+b}\,dx.$ 26. $\int xe^{x^2}\,dx.$ *27. $\int xe^{-x^2}\,dx.$

*28. $\int \dfrac{e^{1/x}}{x^2}\,dx.$ 29. $\int \dfrac{e^{2\sqrt{x}}}{\sqrt{x}}\,dx.$ *30. $\int (e^{-x}-1)^2\,dx.$

*31. $\int \dfrac{4}{\sqrt{e^x}}\,dx.$ 32. $\int \dfrac{e^x}{e^x+1}\,dx.$ *33. $\int \dfrac{e^x}{\sqrt{e^x+1}}\,dx.$

*34. $\int \dfrac{2e^x}{\sqrt[3]{e^x+1}}\,dx.$ 35. $\int \dfrac{e^{2x}}{2e^{2x}+3}\,dx.$ *36. $\int \dfrac{xe^{ax^2}}{e^{ax^2}+1}\,dx.$

Evaluate the following definite integrals.

*37. $\int_0^1 e^x\,dx.$ 38. $\int_0^1 e^{-kx}\,dx.$ *39. $\int_0^{\log \pi} e^{-6x}\,dx.$

*40. $\int_0^1 xe^{-x^2}\,dx.$ 41. $\int_0^1 \dfrac{e^x+e^{-x}}{2}\,dx.$ *42. $\int_0^1 \dfrac{4-e^x}{e^x}\,dx.$

*43. $\int_0^1 \dfrac{e^x}{4-e^x}\,dx.$ 44. $\int_0^1 x(e^{x^2}+2)\,dx.$ *45. $\int_1^2 (2-e^{-x})^2\,dx.$

Estimate by using Table 6.4.1.

*46. $e^{-0.4}.$ 47. $e^{2.6}.$ *48. $e^{2.8}.$ 49. $e^{-2.1}.$

Estimate by using differentials.

*50. $e^{2.03}.$ $(e^2 \cong 7.39)$ 51. $e^{-0.15}.$ $(e^0 = 1)$
*52. $e^{3.15}.$ $(e^3 \cong 20.09)$ 53. $e^{2a+h}.$ $(e^a = A)$

54. A particle moves on a coordinate line so that its position at time t is given by $x(t) = Ae^{ct} + Be^{-ct}$. Show that the acceleration of the particle is proportional to its position.

Sketch the region between the curves and compute the area.

55. $y = e^x, \quad x \in [-1, 1]; \qquad y = e^{-x}, \quad x \in [-1, 1].$
*56. $y = e^{ax}, \quad x \in [-a, a]; \qquad y = e^{-ax}, \quad x \in [-a, a].$

57. We refer here to Figure 6.4.4 on p. 258. (a) Find the points of tangency, marked A and B. (b) Find the area of region I. (c) Find the area of region II.

For each of the functions below, (i) find the domain, (ii) find the intervals where it is increasing and the intervals where it is decreasing, (iii) find the extreme values, (iv) determine the concavity and find the points of inflection, and finally, (v) sketch the graph.

*58. $f(x) = e^{x^2}.$ 59. $f(x) = e^{-x^2}.$
*60. $f(x) = \frac{1}{2}(e^x + e^{-x}).$ 61. $f(x) = \frac{1}{2}(e^x - e^{-x}).$
*62. $f(x) = xe^x.$ 63. $f(x) = (1 - x)e^x.$
*64. $f(x) = e^{(1/x)^2}.$ 65. $f(x) = x^2 e^{-x}.$

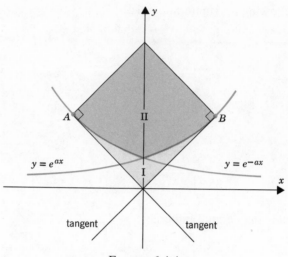

FIGURE 6.4.4

Optional **66.** Prove that for all $x > 0$ and all positive integers n

$$e^x > 1 + x + \frac{x^2}{2!} + \frac{x^3}{3!} + \cdots + \frac{x^n}{n!},$$

where $n!$, read "n factorial," is shorthand for

$$n(n-1)(n-2) \cdots 2 \cdot 1.$$

HINT: $e^x = 1 + \displaystyle\int_0^x e^t \, dt > 1 + \int_0^x dt = 1 + x,$

$$e^x = 1 + \int_0^x e^t \, dt > 1 + \int_0^x (1+t) \, dt = 1 + x + \frac{x^2}{2}, \quad \text{etc.}$$

67. Prove that, if n is a positive integer, then

$$e^x > x^n \qquad \text{for all } x \text{ sufficiently large.}$$

HINT: Use Exercise 66.

6.5 Arbitrary Powers; Other Bases; Estimating e

The Function x^r

The elementary notion of exponent applies only to rational numbers. Expressions such as

$$x^5, \quad x^{1/3}, \quad x^{-4/7}$$

make sense, but so far we have attached no meaning to expressions such as

$$x^{\sqrt{2}}, \quad x^\pi, \quad x^{-\sqrt{3}}.$$

The extension of our sense of exponent to allow for irrational exponents is conveniently done by making use of the logarithm and exponential functions. The heart of the matter is to observe that for $x > 0$ and p/q rational

$$x^{p/q} = e^{(p/q)\log x}.$$

(To verify this, take the log of both sides.) The right-hand side of this equation already makes sense for irrational numbers z; namely,

$$e^{z \log x}$$

already makes sense. To define x^z, we can simply set

$$x^z = e^{z \log x}.$$

Definition 6.5.1

If z is irrational and x is positive, we define the number x^z by setting

$$x^z = e^{z \log x}.$$

Having made this definition, we now have

(6.5.2)
$$\boxed{x^r = e^{r \log x} \qquad \text{for all real numbers } r.}$$

For example,

$$10^{\sqrt{2}} = e^{\sqrt{2} \log 10}, \quad 2^{\pi} = e^{\pi \log 2}, \quad 7^{-\sqrt{3}} = e^{-\sqrt{3} \log 7}. \quad \square$$

With this extended sense of exponent the usual laws of exponents

(6.5.3)
$$\boxed{x^{r+s} = x^r x^s, \qquad x^{r-s} = \frac{x^r}{x^s}, \qquad (x^r)^s = x^{rs}}$$

still hold:

$$x^{r+s} = e^{(r+s)\log x} = e^{r \log x} \cdot e^{s \log x} = x^r x^s,$$

$$x^{r-s} = e^{(r-s)\log x} = e^{r \log x} \cdot e^{-s \log x} = \frac{e^{r \log x}}{e^{s \log x}} = \frac{x^r}{x^s},$$

$$(x^r)^s = e^{s \log x^r} = e^{rs \log x} = x^{rs}. \quad \square$$

The differentiation of arbitrary powers has a familiar look:

(6.5.4)
$$\boxed{\frac{d}{dx}(x^r) = rx^{r-1}.}$$

PROOF

$$\frac{d}{dx}(x^r) = \frac{d}{dx}(e^{r \log x}) = e^{r \log x}\,\frac{d}{dx}(r \log x) = x^r \frac{r}{x} = rx^{r-1}.$$

You can also set

$$f(x) = x^r$$

and use logarithmic differentiation:

$$\log f(x) = r \log x$$

$$\frac{f'(x)}{f(x)} = \frac{r}{x}$$

$$f'(x) = \frac{rf(x)}{x} = \frac{rx^r}{x} = rx^{r-1}. \quad \square$$

Thus

$$\frac{d}{dx}(x^{\sqrt{2}}) = \sqrt{2}\, x^{\sqrt{2}-1}, \qquad \frac{d}{dx}(x^\pi) = \pi x^{\pi-1}.$$

If g is a positive differentiable function, then

(6.5.5)
$$\boxed{\frac{d}{dx}[g(x)]^r = r[g(x)]^{r-1}g'(x).}$$

For example,

$$\frac{d}{dx}[(x^2 + 5)^{\sqrt{3}}] = \sqrt{3}\,(x^2 + 5)^{\sqrt{3}-1}(2x) = 2\sqrt{3}\,x(x^2 + 5)^{\sqrt{3}-1}.$$

The proof is left to you as an exercise. $\square$

Problem.   Find

$$\int \frac{x^3}{(2x^4 + 1)^\pi}\, dx.$$

SOLUTION.   Set

$$u = 2x^4 + 1; \qquad du = 8x^3\, dx, \quad \tfrac{1}{8}du = x^3\, dx.$$

$$\int \frac{x^3}{(2x^4 + 1)^\pi}\, dx = \frac{1}{8} \int \frac{du}{u^\pi} = \frac{1}{8}\left(\frac{u^{1-\pi}}{1 - \pi}\right) + C = \frac{(2x^4 + 1)^{1-\pi}}{8(1 - \pi)} + C. \quad \square$$

Problem.   Find

$$\frac{d}{dx}(x^x).$$

SOLUTION.   One way to do this is to observe that

$$x^x = e^{x \log x}$$

and then differentiate:

$$\frac{d}{dx}(x^x) = \frac{d}{dx}(e^{x \log x}) = e^{x \log x}\left(x \cdot \frac{1}{x} + \log x\right) = x^x(1 + \log x).$$

Another way to do this problem is to set

$$f(x) = x^x$$

and use logarithmic differentiation:

$$\log f(x) = x \log x$$

$$\frac{f'(x)}{f(x)} = x \cdot \frac{1}{x} + \log x = 1 + \log x$$

$$f'(x) = f(x)(1 + \log x) = x^x(1 + \log x). \quad \square$$

The Function p^x

To form the function

$$f(x) = x^r$$

we take a positive variable x and raise it to a constant power r. To form the function

$$g(x) = p^x$$

we take a positive constant p and raise it to a variable power x.

The high status enjoyed by Euler's number e comes from the fact that

$$\frac{d}{dx}(e^x) = e^x.$$

For other bases the derivative is a little more complicated:

(6.5.6)
$$\boxed{\frac{d}{dx}(p^x) = p^x \log p.}$$

PROOF

$$\frac{d}{dx}(p^x) = \frac{d}{dx}\left[e^{x \log p}\right] = e^{x \log p} \log p = p^x \log p. \quad \square$$

The chain rule gives

(6.5.7)
$$\boxed{\frac{d}{dx}\left[p^{g(x)}\right] = p^{g(x)} g'(x) \log p.}$$

Examples

$$\frac{d}{dx}(e^x) = e^x \quad \text{but} \quad \frac{d}{dx}(2^x) = 2^x \log 2.$$

$$\frac{d}{dx}(e^{3x^2}) = e^{3x^2}(6x) \quad \text{but} \quad \frac{d}{dx}(2^{3x^2}) = 2^{3x^2}(6x)(\log 2). \quad \square$$

Problem.   Find

$$\int 2^x \, dx.$$

SOLUTION.   Set

$$u = 2^x; \quad du = 2^x \log 2 \, dx, \quad \frac{du}{\log 2} = 2^x \, dx.$$

$$\int 2^x \, dx = \frac{1}{\log 2} \int du = \frac{1}{\log 2} u + C = \frac{1}{\log 2} 2^x + C. \quad \square$$

Problem.   Evaluate

$$\int_1^2 3^{-x} \, dx.$$

SOLUTION.   Set

$$u = 3^{-x}; \quad du = -3^{-x} \log 3 \, dx, \quad -\frac{du}{\log 3} = 3^{-x} \, dx.$$

$$x = 1, \quad u = \tfrac{1}{3}; \quad x = 2, \quad u = \tfrac{1}{9}.$$

$$\int_1^2 3^{-x} \, dx = -\frac{1}{\log 3} \int_{1/3}^{1/9} du = -\frac{1}{\log 3} \left(\frac{1}{9} - \frac{1}{3} \right) = \frac{2}{9 \log 3}. \quad \square$$

The Function $\log_p x$

If p is positive, then

$$\log p^t = t \log p.$$

If in addition p is different from 1, then $\log p \neq 0$, and therefore

$$\frac{\log p^t}{\log p} = t.$$

This means that the function

$$f(x) = \frac{\log x}{\log p}$$

satisfies the relation

$$f(p^t) = t.$$

In view of this we call

$$\frac{\log x}{\log p}$$

the logarithm of x to the base p and write

(6.5.8)
$$\boxed{\log_p x = \frac{\log x}{\log p}.}$$

As examples, we have

$$\log_2 32 = \frac{\log 32}{\log 2} = \frac{\log 2^5}{\log 2} = \frac{5 \log 2}{\log 2} = 5$$

and

$$\log_{100} \left(\tfrac{1}{10}\right) = \frac{\log \left(\tfrac{1}{10}\right)}{\log 100} = \frac{\log 10^{-1}}{\log 10^2} = \frac{-\log 10}{2 \log 10} = -\frac{1}{2}. \quad \square$$

We can obtain the same results more quickly by using the fact that

(6.5.9) $$\boxed{\log_p p^t = t.}$$

Thus

$$\log_2 32 = \log_2 2^5 = 5 \quad \text{and} \quad \log_{100} \left(\tfrac{1}{10}\right) = \log_{100} (100^{-1/2}) = -\tfrac{1}{2}. \quad \square$$

By differentiating

$$\log_p x = \frac{\log x}{\log p}$$

we obtain the formula

(6.5.10) $$\boxed{\frac{d}{dx} (\log_p x) = \frac{1}{x \log p}.}^{\dagger}$$

When p is e, the factor $\log p$ in the denominator becomes 1, and the formula is simply

$$\frac{d}{dx} (\log_e x) = \frac{1}{x}.$$

We view the logarithm to the base e, $\log = \log_e$, as the *"natural logarithm"* because it is the logarithm with the simplest derivative.

For completeness note that, if g is a positive differentiable function, then

(6.5.11) $$\boxed{\frac{d}{dx} [\log_p g(x)] = \frac{g'(x)}{g(x) \log p}.}$$

This follows directly from (6.5.9) and the chain rule.

---

† The function $f(x) = \log_p x$ satisfies

$$f'(x) = \frac{1}{x \log p}, \quad f'(1) = \frac{1}{\log p}.$$

This means that in general

$$f'(x) = \frac{1}{x} f'(1).$$

We predicted this from general considerations in Section 6.1. (See Formula 6.1.2.)

Examples

$$\frac{d}{dx}(\log_5 x) = \frac{1}{x \log 5}, \qquad \frac{d}{dx}[\log_2(3x^2 + 1)] = \frac{6x}{(3x^2 + 1)\log 2}. \quad \square$$

Estimating the Number *e*

We have defined the number *e* as the number which satisfies

$$1 = \int_1^e \frac{dt}{t}.$$

It is time to discuss how to obtain a numerical estimate for *e*.

Since *e* is irrational (Exercise 30, Section 13.6), we cannot hope to express *e* as a terminating decimal. (Nor as a repeating decimal.) What we can do is describe a simple technique by which one can compute the value of *e* to any desired degree of accuracy.

Theorem 6.5.12

For each positive integer n,

$$\left(1 + \frac{1}{n}\right)^n \leq e \leq \left(1 + \frac{1}{n}\right)^{n+1}.$$

PROOF.   Refer throughout to Figure 6.5.1.

$$\log\left(1 + \frac{1}{n}\right) = \int_1^{1+1/n} \frac{dt}{t} \leq \int_1^{1+1/n} 1\, dt = \frac{1}{n}.$$

since $\frac{1}{t} \leq 1$ throughout the interval of integration

$$\log\left(1 + \frac{1}{n}\right) = \int_1^{1+1/n} \frac{dt}{t} \geq \int_1^{1+1/n} \frac{dt}{1 + 1/n} = \frac{1}{1 + 1/n} \cdot \frac{1}{n} = \frac{1}{n + 1}.$$

since $\frac{1}{t} \geq \frac{1}{1 + 1/n}$ throughout the interval of integration

We have thus shown that

$$\frac{1}{n + 1} \leq \log\left(1 + \frac{1}{n}\right) \leq \frac{1}{n}.$$

From the inequality on the right

$$1 + \frac{1}{n} \leq e^{1/n} \quad \text{and thus} \quad \left(1 + \frac{1}{n}\right)^n \leq e.$$

From the inequality on the left

$$e^{1/(n+1)} \leq 1 + \frac{1}{n} \quad \text{and thus} \quad e \leq \left(1 + \frac{1}{n}\right)^{n+1}. \quad \square$$

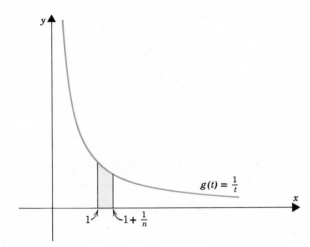

FIGURE 6.5.1

It is easy to see how to estimate e by using Theorem 6.5.12. Since

$$(1 + \tfrac{1}{1})^1 = 2 \quad \text{and} \quad (1 + \tfrac{1}{1})^2 = 4,$$

we have

$$2 \leq e \leq 4.$$

Since

$$(1 + \tfrac{1}{2})^2 = \tfrac{9}{4} = 2.25 \quad \text{and} \quad (1 + \tfrac{1}{2})^3 = \tfrac{27}{8} = 3.375,$$

we have

$$2.25 \leq e \leq 3.375.$$

With a little more patience you can verify that

$$2.59 \leq (1 + \tfrac{1}{10})^{10} \quad \text{and} \quad (1 + \tfrac{1}{10})^{11} \leq 2.86$$

and thus arrive at

$$2.59 \leq e \leq 2.86.$$

Lengthier calculations show that

$$2.70 \leq (1 + \tfrac{1}{100})^{100} \quad \text{and} \quad (1 + \tfrac{1}{100})^{101} \leq 2.74$$

and thus

$$2.70 \leq e \leq 2.74.$$

Still lengthier calculations show that

$$2.71 \le (1 + \tfrac{1}{200})^{200} \quad \text{and} \quad (1 + \tfrac{1}{200})^{201} \le 2.73,$$

and thus

$$2.71 \le e \le 2.73.$$

A look at an eight-place table reveals the estimate

$$2.71828182.$$

For our purposes the estimate 2.72 is sufficient.

The inequality

$$\left(1 + \frac{1}{n}\right)^{n} \le e \le \left(1 + \frac{1}{n}\right)^{n+1}$$

is rather elegant, but, as you saw, it does not provide a very efficient way of estimating e. (We have to go to $n = 200$ to show that $2.71 \le e \le 2.73$.) We shall have a much more efficient way of estimating e when we have infinite series at our disposal.

Exercises

1. Find.
   *(a) $\log_2 64$.   (b) $\log_2 \frac{1}{64}$.   *(c) $\log_{64} \frac{1}{2}$.
   *(d) $\log_{10} 0.01$.   (e) $\log_5 1$.   *(f) $\log_5 0.2$.
   *(g) $\log_5 125$.   (h) $\log_2 4^3$.   *(i) $\log_{32} 8$.
   *(j) $\log_{100} 10^{-4/5}$.   (k) $\log_{10} 100^{-4/5}$.   *(l) $\log_9 \sqrt{3}$.

2. Show that
   (a) $\log_p xy = \log_p x + \log_p y$,   (b) $\log_p \dfrac{x}{y} = \log_p x - \log_p y$,
   (c) $\log_p x^y = y \log_p x$.

3. Find those numbers x, if any, for which
   *(a) $10^x = e^x$.   (b) $\log_5 x = 0.04$.   *(c) $\log_x 2 = \log_3 x$.

   *(d) $\log_x 10 = \log_4 100$.   (e) $\log_2 x = \displaystyle\int_2^x \frac{dt}{t}$.   *(f) $\log_x 10 = \log_2 (\tfrac{1}{10})$.

4. *(a) Estimate $\log a$ given that $e^{t_1} < a < e^{t_2}$.
   *(b) Estimate e^b given that $\log x_1 < b < \log x_2$.

5. Show that, if a, b, c are positive, then

   $$\log_a c = \log_a b \, \log_b c$$

   provided that a and b are both different from 1.

6. Work out the following integrals.

   *(a) $\displaystyle\int 3^x \, dx$.   (b) $\displaystyle\int x^3 \, dx$.   *(c) $\displaystyle\int 2^{-x} \, dx$.

   *(d) $\displaystyle\int x 10^{x^2} \, dx$.   (e) $\displaystyle\int x 10^{-x^2} \, dx$.   *(f) $\displaystyle\int (2^x + 2^{-x}) \, dx$.

   *(g) $\displaystyle\int \frac{dx}{x \log 5}$.   (h) $\displaystyle\int \frac{\log_5 x}{x} \, dx$.   *(i) $\displaystyle\int \frac{\log_3 \sqrt{x} - 1}{x} \, dx$.

7. Find $f'(e)$.
   *(a) $f(x) = \log_3 x$.
   *(c) $f(x) = \log (\log x)$.
   (b) $f(x) = x \log_3 x$.
   (d) $f(x) = \log_3 (\log_2 x)$.

8. Derive the formula for $f'(x)$ by logarithmic differentiation.
   (a) $f(x) = p^x$.　　　(b) $f(x) = p^{g(x)}$.

9. Find by logarithmic differentiation.

   *(a) $\dfrac{d}{dx} [(x + 1)^x]$.　　(b) $\dfrac{d}{dx} [(\log x)^x]$.　　　*(c) $\dfrac{d}{dx} [(x^2 + 2)^{\log x}]$.

   *(d) $\dfrac{d}{dx} \left[\left(\dfrac{1}{x} \right)^x \right]$.　　(e) $\dfrac{d}{dx} [(\log x)^{x^2+2}]$.　　　*(f) $\dfrac{d}{dx} [(\log x)^{\log x}]$.

10. Sketch figures in which you compare the following pairs of graphs.
    (a) $f(x) = e^x$　and　$g(x) = 2^x$.　　(b) $f(x) = e^x$　and　$g(x) = 3^x$.
    (c) $f(x) = e^x$　and　$g(x) = e^{-x}$.　　(d) $f(x) = 2^x$　and　$g(x) = 2^{-x}$.
    (e) $f(x) = 3^x$　and　$g(x) = 3^{-x}$.　　(f) $f(x) = \log x$　and　$g(x) = \log_2 x$.
    (g) $f(x) = 2^x$　and　$g(x) = \log_2 x$.　　(h) $f(x) = 10^x$　and　$g(x) = \log_{10} x$.

11. For each of the functions below (i) specify the domain, (ii) find the intervals where the function is increasing and those where it is decreasing, (iii) find the extreme values.
    *(a) $f(x) = 10^{1-x^2}$.　　　　　　*(b) $f(x) = 10^{1/(1-x^2)}$.
    *(c) $f(x) = 10^{\sqrt{1-x^2}}$.　　　　　(d) $f(x) = \log_{10} \sqrt{1 - x^2}$.

12. Evaluate.

    *(a) $\displaystyle\int_0^1 (1 + x)^{2\pi} \, dx$.　　(b) $\displaystyle\int_1^2 2^{-x} \, dx$.　　　*(c) $\displaystyle\int_0^1 4^x \, dx$.

    *(d) $\displaystyle\int_0^1 b a^{2x} \, dx$.　　(e) $\displaystyle\int_0^1 x 10^{1+x^2} \, dx$.　　*(f) $\displaystyle\int_1^4 \dfrac{dx}{x \log 2}$.

    *(g) $\displaystyle\int_{10}^{100} \dfrac{dx}{x \log_{10} x}$.　　(h) $\displaystyle\int_0^2 p^{x/2} \, dx$.　　*(i) $\displaystyle\int_0^1 \dfrac{5p^{\sqrt{x+1}}}{\sqrt{x + 1}} \, dx$.

13. Prove that, if g is positive and differentiable, then for each real exponent r

    $$\frac{d}{dx} [g(x)]^r = r[g(x)]^{r-1} g'(x)$$

    (a) using the chain rule and (6.5.4).　　　(b) by logarithmic differentiation.

14. Use Table 6.2.1 to estimate the following logarithms:
    *(a) $\log_{10} 4$.　　(b) $\log_{10} 7$.　　*(c) $\log_{10} 12$.　　(d) $\log_{10} 45$.

Optional　| 15. Prove that

$$\lim_{h \to 0} (1 + h)^{1/h} = e.$$

HINT:　At $x = 1$ the logarithm function has derivative 1:

$$\lim_{h \to 0} \frac{\log (1 + h) - \log 1}{h} = 1.$$

6.6   Exponential Growth and Decline

An exponential of the form

$$f(x) = Ce^{kx}$$

has the property that its derivative $f'(x)$ is proportional at each point to $f(x)$:

$$f'(x) = Cke^{kx} = kCe^{kx} = kf(x).$$

Moreover it is the only such function:

Theorem 6.6.1

If

$$f'(x) = kf(x) \qquad \text{for all } x \text{ in some interval } I,$$

then f is of the form

$$f(x) = Ce^{kx} \qquad \text{for all } x \text{ in } I.$$

PROOF

$$f'(x) = kf(x),$$
$$f'(x) - kf(x) = 0,$$
$$e^{-kx}f'(x) - ke^{-kx}f(x) = 0, \qquad \text{(we multiplied by } e^{-kx}\text{)}$$
$$\frac{d}{dx}[e^{-kx}f(x)] = 0,$$
$$e^{-kx}f(x) = C,$$
$$f(x) = Ce^{kx}. \quad \square$$

The constant C is the value of f at 0:

$$f(0) = Ce^0 = C.$$

This is usually called the *initial value of f*.

Problem.   A function everywhere defined has the property that at each point of its graph the slope is twice the y-coordinate. Given that the graph passes through the point $(0, \sqrt{2})$, find the function.

SOLUTION

$$f'(x) = 2f(x) \qquad \text{for all real } x.$$

This means that f is of the form

$$f(x) = Ce^{2x}.$$

Since the graph passes through the point $(0, \sqrt{2})$, we know that $f(0) = \sqrt{2} = C$. The function must be

$$f(x) = \sqrt{2}e^{2x}. \quad \square$$

Problem.　The number of bacteria present in a given culture increases at a rate proportional to the number present. When first observed, the culture contained n_0 bacteria, an hour later, n_1.

(a) Find the number present t hours after the observations began.
(b) How long did it take for the number of bacteria to double?

SOLUTION.　(a) Let $n(t)$ = the number of bacteria present at time t. The basic equation is of the form

$$n'(t) = kn(t).$$

From this it follows that

$$n(t) = Ce^{kt}.$$

Since initially (at time 0) there were n_0 bacteria, we know that $C = n_0$. The equation is therefore

$$n(t) = n_0e^{kt}.$$

Since $n(1) = n_1$, we have in turn

$$n_1 = n_0e^k, \quad n_1/n_0 = e^k, \quad (n_1/n_0)^t = e^{kt}.$$

The general law of growth is

$$n(t) = n_0\left(\frac{n_1}{n_0}\right)^t.$$

(b) To find out how long it took for the number of bacteria to double we set

$$2n_0 = n_0\left(\frac{n_1}{n_0}\right)^t$$

and solve for t:

$$2 = \left(\frac{n_1}{n_0}\right)^t, \quad \log 2 = t(\log n_1 - \log n_0), \quad t = \frac{\log 2}{\log n_1 - \log n_0} \text{ hours.} \quad \square$$

Problem.　The rate of decay of radioactive material is proportional to the amount of such material present.

(a) Find the amount of material present t years later given that the initial amount is A_0 pounds and it takes 5 years for a third of the material to decay.
(b) How long does it take for half of the material to decay?

SOLUTION.　(a) Let $A(t)$ be the amount of radioactive material present at time t. Since the rate of decay is proportional to the amount present, we know that

$$A'(t) = kA(t) \quad \text{and thus} \quad A(t) = Ce^{kt}.$$

Since A_0 is the initial amount, we have $C = A_0$. The general equation is thus of the form

(1) $$A(t) = A_0e^{kt}.$$

At the end of 5 years, one-third of A_0 has decayed and therefore two-thirds of A_0 remains:

$$A(5) = \tfrac{2}{3}A_0.$$

We can use this relation to eliminate k from (1):

$$A(5) = A_0 e^{5k} = \tfrac{2}{3}A_0, \quad e^{5k} = \tfrac{2}{3}, \quad e^k = (\tfrac{2}{3})^{1/5},$$

and, consequently, (1) becomes

$$A(t) = A_0(\tfrac{2}{3})^{t/5}.$$

(b) Here we must find the value of t for which

$$A_0(\tfrac{2}{3})^{t/5} = \tfrac{1}{2}A_0.$$

Dividing by A_0 we have

$$(\tfrac{2}{3})^{t/5} = \tfrac{1}{2}$$

so that

$$(\tfrac{2}{3})^t = (\tfrac{1}{2})^5, \quad t \log \tfrac{2}{3} = 5 \log \tfrac{1}{2}, \quad t(\log 2 - \log 3) = -5 \log 2,$$

and

$$t = \frac{5 \log 2}{\log 3 - \log 2} \text{ years.}$$

With $\log 2 \cong 0.69$ and $\log 3 \cong 1.10$, you'll find that $t \cong 8.41$ years. It takes about 8 years and 5 months for half of the material to decay. □

Compound Interest

Consider money invested at interest rate r. If the accumulated interest is credited once a year, then the interest is said to be compounded annually; if twice a year, then semiannually; if four times a year, then quarterly. Many savings banks are now compounding interest daily.

The idea can be pursued further. Interest can be credited every hour, every second, every half-second, and so on. In the limiting case, interest is credited instantaneously. Economists call this *continuous compounding*.

The economists' formula for continuous compounding is a simple exponential:

(6.6.2)
$$\boxed{A(t) = A_0 e^{rt}.}$$

Here t is measured in years,

$$A(t) = \text{the principal in dollars at time } t,$$
$$A_0 = A(0) = \text{the initial investment,}$$
$$r = \text{the annual interest rate.}$$

The rate r is called the *nominal* interest rate. Compounding makes the effective rate higher.

A DERIVATION OF THE COMPOUND INTEREST FORMULA. We take h as a small time increment and observe that

$$A(t + h) - A(t) = \text{interest earned from time } t \text{ to time } t + h.$$

Had the principal remained $A(t)$ from time t to time $t + h$, the interest earned during this time period would have been

$$rhA(t).$$

Had the principal been $A(t + h)$ throughout the time interval, the interest earned would have been

$$rhA(t + h).$$

The actual interest earned must be somewhere in between:

$$rhA(t) \leq A(t + h) - A(t) \leq rhA(t + h).$$

Dividing by h, we get

$$rA(t) \leq \frac{A(t + h) - A(t)}{h} \leq rA(t + h).$$

If A varies continuously, then, as h tends to zero, $rA(t + h)$ tends to $rA(t)$ and (by the pinching theorem) the difference quotient in the middle must also tend to $rA(t)$:

$$\lim_{h \to 0} \frac{A(t + h) - A(t)}{h} = rA(t).$$

This says that

$$A'(t) = rA(t),$$

from which it follows that

$$A(t) = Ce^{rt}.$$

With A_0 as the initial investment, we have $C = A_0$, and therefore

$$A(t) = A_0 e^{rt}. \quad \square$$

Problem. Find the amount of interest drawn by \$100 compounded continuously at 6% for 5 years.

SOLUTION

$$A(5) = 100e^{(0.06)5} = 100e^{0.30} \cong 135. \qquad \text{(Table 6.4.1)}$$

The interest drawn is approximately \$35. $\square$

Problem. A sum of money is drawing interest at the rate of 10% compounded continuously. What is the effective interest rate?

SOLUTION. At the end of one year each dollar grows to

$$e^{0.10} \cong 1.11.$$

The interest accrued is approximately 11¢. The effective interest rate is about 11%. □

Problem. How long does it take a sum of money to double at 5% compounded continuously?

SOLUTION. In general

$$A(t) = A_0 e^{0.05t}.$$

We set

$$2A_0 = A_0 e^{0.05t}$$

and solve for t:

$$2 = e^{0.05t}, \quad \log 2 = 0.05t = \tfrac{1}{20}t, \quad t = 20 \log 2 \cong 13.8. \text{ (Table 6.2.1)}$$

It takes about 13 years, 9 months, and 18 days. □

A dollar today is better than a dollar tomorrow because during that time we could be earning interest.

Problem. What is the present value of $1000 forty months from now? Assume continuous compounding at 6%.

SOLUTION. We begin with

$$A(t) = A_0 e^{0.06t}$$

and solve for A_0:

$$A_0 = A(t)e^{-0.06t}.$$

Since forty months is $\tfrac{10}{3}$ years, $t = \tfrac{10}{3}$ and

$$A_0 = A(\tfrac{10}{3})e^{(-0.06)(10/3)} = 1000e^{-0.20} \cong 820.$$

The present value of $1000 forty months from now is about $820. □

Problem. A producer of brandy finds that the value of his inventory increases with time (the brandy improves with age) according to the formula

$$V(t) = V_0 e^{\sqrt{t}/2}. \qquad\qquad (t \text{ measured in years})$$

How long should he keep the brandy in storage? Neglect storage costs and assume continuous compounding at 8%.

SOLUTION. Future dollars should be discounted by a factor of $e^{-0.08t}$. What we want to maximize is not $V(t)$ but the product

$$f(t) = V(t)e^{-0.08t} = V_0 e^{(\sqrt{t}/2)-0.08t}.$$

Differentiation gives

$$f'(t) = V_0 e^{(\sqrt{t}/2) - 0.08t} \left(\frac{1}{4\sqrt{t}} - 0.08 \right).$$

Setting $f'(t) = 0$, we find that

$$\frac{1}{4\sqrt{t}} = 0.08, \quad \sqrt{t} = \frac{1}{0.32}, \quad t = \left(\frac{1}{0.32} \right)^2 \cong 9.76.$$

He should store it for about nine years and nine months.   □

Exercises

1. A positive function everywhere defined has the property that at each point of its graph the slope is three times the y-coordinate. Find the function given that the graph passes through the point *(a) (0, 2). (b) (1, 1). *(c) (2, e).

2. Find the amount of interest drawn by $500 compounded continuously for 10 years *(a) at 5%. (b) at 6%. *(c) at 7%.

3. What is the present value of a $1000 bond that matures 5 years from now? Assume continuous compounding *(a) at 4%. (b) at 6%. *(c) at 8%.

4. How long does it take for a sum of money to double when compounded continuously *(a) at 4%? (b) at 6%? *(c) at 8%?

5. At what rate r of continuous compounding does a sum of money increase by a factor of e (a) in one year? (b) in n years?

6. At what rate r of continuous compounding does a sum of money *(a) triple in 20 years? (b) double in 10 years?

*7. Water is run into a tank to dilute a saline solution. The volume V of the mixture is kept constant. Given that s, the amount of salt in the tank, varies with respect to x, the amount of water which has been run through, according to the formula

$$\frac{ds}{dx} = -\frac{s}{V},$$

find the amount of water which must be used to wash down 50% of the salt. Take V as 10,000 gallons.

8. (*Newton's law of cooling*) If the temperature of an object is x degrees above that of the surrounding air, the time rate of decrease of x is proportional to x. Given that x is at first 80 degrees and one minute later 70 degrees, what will it be after two minutes! In how many minutes will it decrease 20 degrees!

*9. Atmospheric pressure p varies with altitude h according to the law

$$\frac{dp}{dh} = kp, \quad \text{where } k \text{ is a constant.}$$

Given that p is 15 pounds per square inch at sea level and 10 pounds per square inch at 10,000 feet, find p at (a) 5000 feet. (b) 15,000 feet.

10. Given that

$$f'(t) = k[2 - f(t)] \quad \text{and} \quad f(0) = 0$$

show that the constant k satisfies the equation

$$k = \frac{1}{t} \log \left[\frac{2}{2 - f(t)} \right].$$

*11. In the inversion of raw sugar, the time rate of change varies directly with the amount of raw sugar remaining. If after 10 hours, 1000 pounds of raw sugar have been reduced to 800 pounds, how much raw sugar will remain after 20 hours?

12. The rate of decay of a radioactive substance is proportional to the amount of the substance present.
  (a) How long does it take for half of the substance to decay
     *(i) if it takes 4 years for a quarter of the substance to decay?
      (ii) if it takes 3 years for a quarter of the substance to decay?
  (b) A year ago there were 4 grams of the substance. Now there are 3 grams.
     *(i) How much was there 2 years ago?
      (ii) How much was there 10 years ago?
     *(iii) How much will remain 10 years from now?

13. A radioactive substance weighed n grams at time $t = 0$. Today, five years later, the substance weighs m grams. How much will it weigh (a) 5 years from now? (b) 10 years from now?

*14. A lumber company finds from experience that the value of its standing timber increases with time according to the formula

$$V(t) = V_0 (\tfrac{3}{2})^{\sqrt{t}}.$$

Here t is measured in years and V_0 is the value at planting time. How long should the company wait before cutting the timber? Neglect costs and assume continuous compounding at 5%.

15. Population tends to grow with time at a rate roughly proportional to the population present. According to the Bureau of the Census, the population of the United States in 1960 was approximately 179 million and in 1970, 205 million.
  (a) Use this information to estimate the population of 1940. (The actual figure was about 132 million.)
  (b) Predict the populations for 1990 and for the year 2000.
  (c) Estimate how long it takes for the population to double.

6.7 Integration by Parts

We begin with the differentiation formula

$$f(x)g'(x) + f'(x)g(x) = (f \cdot g)'(x).$$

Integrating both sides, we get

$$\int f(x)g'(x)\, dx + \int f'(x)g(x)\, dx = \int (f \cdot g)'(x)\, dx.$$

Since

$$\int (f \cdot g)'(x)\, dx = f(x)g(x) + C,$$

we have

$$\int f(x)g'(x)\, dx + \int f'(x)g(x)\, dx = f(x)g(x) + C$$

and therefore

$$\int f(x)g'(x)\, dx = f(x)g(x) - \int f'(x)g(x)\, dx + C.$$

Since the computation of

$$\int f'(x)g(x)\, dx$$

will yield its own arbitrary constant, there is no reason to keep the constant C. We therefore drop it and write

(6.7.1)
$$\boxed{\int f(x)g'(x)\, dx = f(x)g(x) - \int f'(x)g(x)\, dx.}$$

This formula, called the formula for *integration by parts,* enables us to find

$$\int f(x)g'(x)\, dx$$

by computing

$$\int f'(x)g(x)\, dx$$

instead. It is, of course, of practical use only if the second integral is easier to compute than the first.

In practice we usually set

$$u = f(x), \qquad dv = g'(x)\, dx,$$
$$du = f'(x)\, dx, \qquad v = g(x).$$

The formula for integration by parts then becomes

(6.7.2)
$$\boxed{\int u\, dv = uv - \int v\, du.}$$

Success with this formula depends on choosing u and dv so that

$$\int v\, du \quad \text{becomes easier to calculate than} \quad \int u\, dv.$$

Problem 1.   Calculate

$$\int xe^x \, dx.$$

Solution.   Set

$$u = x, \qquad dv = e^x \, dx,$$
$$du = dx, \qquad v = e^x.$$

$$\int xe^x \, dx = \int u \, dv = uv - \int v \, du = xe^x - \int e^x \, dx = xe^x - e^x + C.$$

Our choice of u and dv worked out very well. Had we set

$$u = e^x, \qquad dv = x \, dx,$$
$$du = e^x, \qquad v = \tfrac{1}{2}x^2,$$

integration by parts would have given

$$\int xe^x \, dx = \int u \, dv = uv - \int v \, du = \tfrac{1}{2}x^2 e^x - \tfrac{1}{2} \int x^2 e^x \, dx,$$

an integral more complicated than the one we started with. A propitious choice of u and dv is crucial.   □

Problem 2.   Calculate

$$\int \log x \, dx.$$

Solution.   Set

$$u = \log x, \qquad dv = dx,$$
$$du = \frac{1}{x} \, dx, \qquad v = x.$$

$$\int \log x \, dx = \int u \, dv = uv - \int v \, du$$

$$= x \log x - \int x \left(\frac{1}{x}\right) dx = x \log x - \int dx$$

$$= x \log x - x + C.$$ □

Problem 3.   Calculate

$$\int \frac{xe^x}{(x + 1)^2} \, dx.$$

Solution.   Again the challenge is to figure out how to choose u and dv. After several false starts we found that the following decomposition works out well:

$$u = xe^x, \qquad\qquad\qquad dv = \frac{1}{(x + 1)^2} \, dx,$$

$$du = (xe^x + e^x) \, dx = (x + 1)e^x \, dx, \qquad v = -\frac{1}{x + 1}.$$

$$\int \frac{xe^x}{(x + 1)^2} \, dx = \int u \, dv = uv - \int v \, du$$

$$= -\frac{xe^x}{x + 1} + \int e^x \, dx$$

$$= -\frac{xe^x}{x + 1} + e^x + C = \frac{e^x}{x + 1} + C. \quad \square$$

Problem 4.   Calculate

$$\int x^5 e^{-x^3} \, dx.$$

SOLUTION.   Set

$$u = -\tfrac{1}{3}x^3, \qquad dv = -3x^2 e^{-x^3} \, dx,$$
$$du = -x^2, \qquad v = e^{-x^3}.$$

$$\int x^5 e^{-x^3} \, dx = uv - \int v \, du = -\tfrac{1}{3}x^3 e^{-x^3} - \int -x^2 e^{-x^3} \, dx.$$

Since

$$\int -x^2 e^{-x^3} \, dx = \tfrac{1}{3}e^{-x^3} + C, \qquad\qquad \text{(check this out)}$$

we have

$$\int x^5 e^{-x^3} \, dx = -\tfrac{1}{3}x^3 e^{-x^3} - \tfrac{1}{3}e^{-x^3} + C.$$

This same integral can be calculated somewhat more simply by setting

$$u = -x^3; \qquad du = -3x^2 \, dx, \quad \tfrac{1}{3}du = -x^2 \, dx.$$

Then

$$\int x^5 e^{-x^3} \, dx = \tfrac{1}{3} \int u e^u \, du = \tfrac{1}{3}(ue^u - e^u) + C = -\tfrac{1}{3}x^3 e^{-x^3} - \tfrac{1}{3}e^{-x^3} + C. \quad \square$$

by Problem 1

To calculate some integrals you may have to integrate by parts more than once.

Problem 5.   Evaluate

$$\int_0^1 x^2 e^x \, dx.$$

SOLUTION.   First we calculate the indefinite integral

$$\int x^2 e^x \, dx.$$

We set

$$u = x^2, \qquad dv = e^x \, dx,$$
$$du = 2x \, dx, \qquad v = e^x.$$

$$\int x^2 e^x \, dx = \int u \, dv = uv - \int v \, du = x^2 e^x - \int 2x e^x \, dx.$$

We now compute the integral on the right again by parts. This time we set

$$u = 2x, \qquad dv = e^x \, dx,$$
$$du = 2 \, dx, \qquad v = e^x.$$

$$\int 2x e^x \, dx = \int u \, dv = uv - \int v \, du = 2x e^x - \int 2e^x \, dx = 2x e^x - 2e^x + C.$$

This together with our earlier calculations gives

$$\int x^2 e^x \, dx = x^2 e^x - 2x e^x + 2e^x + C.$$

It follows now that

$$\int_0^1 x^2 e^x \, dx = \left[x^2 e^x - 2x e^x + 2e^x \right]_0^1 = (e - 2e + 2e) - 2 = e - 2. \quad \square$$

Exercises

Work out the following integrals and check your answers by differentiation.

*1. $\displaystyle\int x e^{-x} \, dx.$ 2. $\displaystyle\int x \log x \, dx.$ *3. $\displaystyle\int x \log (x + 1) \, dx.$

*4. $\displaystyle\int x 2^x \, dx.$ 5. $\displaystyle\int x^2 \log x \, dx.$ *6. $\displaystyle\int \log (-x) \, dx.$

*7. $\displaystyle\int \sqrt{x} \log x \, dx.$ 8. $\displaystyle\int x^2 e^{-x^3} \, dx.$ *9. $\displaystyle\int x^3 e^{-x^2} \, dx.$

*10. $\displaystyle\int x^2 (e^x - 1) \, dx.$ 11. $\displaystyle\int \frac{\log (x + 1)}{\sqrt{x + 1}} \, dx.$ *12. $\displaystyle\int \frac{x^2}{\sqrt{1 - x}} \, dx.$

*13. $\displaystyle\int x \sqrt{x + 1} \, dx.$ 14. $\displaystyle\int x \log x^2 \, dx.$ *15. $\displaystyle\int (\log x)^2 \, dx.$

*16. $\displaystyle\int x^2 e^{-x} \, dx.$ 17. $\displaystyle\int x(x + 5)^{14} \, dx.$ *18. $\displaystyle\int x(x + 5)^{-14} \, dx.$

*19. $\displaystyle\int x^2 (x + 1)^9 \, dx.$ 20. $\displaystyle\int x^2 (2x - 1)^{-7} \, dx.$ *21. $\displaystyle\int_e^{e^2} \frac{dx}{x(\log x)^3}.$

*22. $\displaystyle\int_0^1 \frac{(e^x + 2x)^2}{2} \, dx.$ 23. $\displaystyle\int_1^e x^n \log x \, dx.$ *24. $\displaystyle\int_0^1 (2^x + x^2)^2 \, dx.$

Optional | **6.8   The Equation $y'(x) + P(x)y(x) = Q(x)$**

Differential equations (equations involving functions and their derivatives) arise so frequently in scientific problems that their study constitutes an important branch of mathematics.

The exponential function, being fixed under differentiation, that is, being endowed with the property

$$\frac{d}{dx}(e^x) = e^x,$$

plays an interesting role in the theory of differential equations.

In the case of linear differential equations

$$y^{(n)}(x) + P_1(x)y^{(n-1)}(x) + \cdots + P_{n-1}(x)y'(x) + P_n(x)y(x) = Q(x),$$

the exponential function plays a dominant role. In this section we consider the first-order case (where no derivative higher than the first appears):

(1) $$y'(x) + P(x)y(x) = Q(x).$$

We assume that the coefficients P and Q are both continuous. To solve Equation (1) we begin by multiplying both sides by $e^{\int P(x)\,dx}$:

$$e^{\int P(x)\,dx}\, y'(x) + e^{\int P(x)\,dx}\, P(x)y(x) = e^{\int P(x)\,dx}\, Q(x).$$

Once we've done this, the left-hand side becomes

$$\frac{d}{dx}\left[e^{\int P(x)\,dx}\, y(x)\right]$$

and we can rewrite the equation as

$$\frac{d}{dx}\left[e^{\int P(x)\,dx}\, y(x)\right] = e^{\int P(x)\,dx}\, Q(x).$$

We can now integrate both sides and get

$$e^{\int P(x)\,dx}\, y(x) = \int e^{\int P(x)\,dx}\, Q(x)\,dx + C.$$

To get $y(x)$ by itself, we multiply the equation by $e^{-\int P(x)dx}$:

$$y(x) = e^{-\int P(x)\,dx}\left[\int e^{\int P(x)\,dx}\, Q(x)\,dx + C\right].$$

This is called the *general solution* of Equation (1). Different choices of C give different *particular solutions*.

Example. To solve the differential equation

$$y'(x) + 2y(x) = e^x$$

we multiply through by

$$e^{2\int dx} = e^{2x}. \qquad \text{(we don't carry the constant here)}$$

Optional | This gives

$$e^{2x}y'(x) + e^{2x}2y(x) = e^{3x},$$

$$\frac{d}{dx}[e^{2x}y(x)] = e^{3x},$$

$$e^{2x}y(x) = \tfrac{1}{3}e^{3x} + C,$$

$$y(x) = \tfrac{1}{3}e^{x} + Ce^{-2x}.$$

This is the general solution. Different choices of C give different particular solutions. Setting $C = 0$, we get

$$y(x) = \tfrac{1}{3}e^{x}.$$

Setting $C = 1$, we get

$$y(x) = \tfrac{1}{3}e^{x} + e^{-2x}.$$

If we want the solution which takes on the value 1 at 0, then we set

$$y(0) = \tfrac{1}{3} + C = 1.$$

This forces $C = \tfrac{2}{3}$ and

$$y(x) = \tfrac{1}{3}e^{x} + \tfrac{2}{3}e^{-2x}.$$

More generally, if we want the solution which takes on the value y_0 at x_0, then we set

$$y(x_0) = \tfrac{1}{3}e^{x_0} + Ce^{-2x_0} = y_0.$$

This forces

$$C = (y_0 - \tfrac{1}{3}e^{x_0})e^{2x_0}$$

and

$$y(x) = \tfrac{1}{3}e^{x} + (y_0 - \tfrac{1}{3}e^{x_0})e^{2(x_0 - x)}. \quad \square$$

Example. To solve the differential equation

$$xy'(x) - 2y(x) = 2$$

we first divide the equation by x so that the leading coefficient becomes 1:

$$y'(x) - \frac{2}{x}y(x) = \frac{2}{x}.$$

Then we multiply through by

$$e^{\int -2x^{-1}\,dx} = e^{-2\log x} = e^{\log x^{-2}} = x^{-2}.$$

This gives

$$x^{-2}y'(x) - 2x^{-3}y(x) = 2x^{-3},$$

$$\frac{d}{dx}[x^{-2}y(x)] = 2x^{-3},$$

Optional

$$x^{-2}y(x) = -x^{-2} + C,$$
$$y(x) = Cx^2 - 1.$$

This is the general solution. □

Problem. Water from a polluted reservoir (pollution level p_0 grams per liter) is constantly being drawn off at the rate of n liters per hour and replaced by less polluted water (pollution level p_1 grams per liter). Given that the capacity of the reservoir is M liters, how long will it take to reduce the pollution level to p grams per liter?

SOLUTION. Let $A(t)$ be the total number of grams of pollution in the reservoir at time t. We want to find the time t at which

$$\frac{A(t)}{M} = p.$$

At each time t, pollutants leave the reservoir at the rate of

$$n\frac{A(t)}{M} \quad \text{grams per hour}$$

and enter the reservoir at the rate of

$$np_1 \quad \text{grams per hour.}$$

It follows that

$$A'(t) = -n\frac{A(t)}{M} + np_1$$

and

$$A'(t) + \frac{n}{M}A(t) = np_1.$$

To solve the equation for $A(t)$ we multiply through by $e^{\int (n/M)\,dt} = e^{nt/M}$. This gives

$$e^{nt/M}A'(t) + e^{nt/M}\frac{n}{M}A(t) = np_1e^{nt/M},$$

$$\frac{d}{dt}\left[e^{nt/M}A(t)\right] = np_1e^{nt/M},$$

$$e^{nt/M}A(t) = Mp_1e^{nt/M} + C,$$

$$A(t) = Mp_1 + Ce^{-nt/M},$$

$$\frac{A(t)}{M} = p_1 + \frac{C}{M}e^{-nt/M}.$$

The constant C is determined by the initial pollution level p_0:

$$p_0 = \frac{A(0)}{M} = p_1 + \frac{C}{M} \quad \text{so that} \quad C = M(p_0 - p_1).$$

Optional | We therefore have

$$\frac{A(t)}{M} = p_1 + (p_0 - p_1)e^{-nt/M}.$$

This equation gives the pollution level at each time t. To find the time at which the pollution level has been reduced to p, we set

$$p_1 + (p_0 - p_1)e^{-nt/M} = p$$

and solve for t:

$$e^{-nt/M} = \frac{p - p_1}{p_0 - p_1}, \quad -\frac{n}{M} t = \log \frac{p - p_1}{p_0 - p_1}, \quad t = \frac{M}{n} \log \frac{p_0 - p_1}{p - p_1}. \quad \Box$$

Problem.   In the problem just considered, how long would it take to reduce the pollution by a factor of 50% if the reservoir contained 50,000,000 liters, the water were replaced at the rate of 25,000 liters per hour, and the replacement water were completely pollution free?

SOLUTION.   In general

$$t = \frac{M}{n} \log \frac{p_0 - p_1}{p - p_1}.$$

Here

$$M = 50{,}000{,}000, \quad n = 25{,}000, \quad p_1 = 0, \quad p = \tfrac{1}{2}p_0,$$

so that

$$t = 2000 \log 2 \cong 2000(0.69) = 1380.$$

It would take about 1380 hours. $\Box$

Exercises

Find the general solution.

*1. $y'(x) - 2y(x) = 1$.   2. $xy'(x) - 2y(x) = -x$.
*3. $2y'(x) + 5y(x) = 2$.   4. $y'(x) - y(x) = -2e^{-x}$.
*5. $y'(x) - 2y(x) = 1 - 2x$.   6. $xy'(x) - 2y(x) = -3x$.
*7. $xy'(x) - 3y(x) = -2nx$.   8. $y'(x) + y(x) = 2 + 2x$.

*9. $y'(x) + \dfrac{2}{x + 1} y(x) = (x + 1)^{5/2}$.   10. $y'(x) + y(x) = \dfrac{1}{1 + e^x}$.

*11. $xy'(x) + y(x) = (1 + x)e^x$.   12. $y'(x) - y(x) = e^x$.

Find the particular solution determined by the given side condition.

*13. $y'(x) + y(x) = x$, $y(0) = 1$.
 14. $y'(x) - y(x) = e^{2x}$, $y(1) = 1$.

*15. $y'(x) + y(x) = \dfrac{1}{1 + e^x}$, $y(0) = e$.

Optional   16. $y'(x) + y(x) = \dfrac{1}{1 + 2e^x}$, $y(0) = e$.

*17. $xy'(x) - 2y(x) = x^3 e^x$, $y(1) = 0$.

18. $x^2 y'(x) + 2xy(x) = 1$, $y(1) = 2$.

*19. $y'(x) + \dfrac{2}{x} y(x) = e^{-x}$, $y(1) = -1$.

20. $y'(x) - \dfrac{2}{x + 1} y(x) = (x + 1)^3$, $y(0) = 1$.

*21. A 200-liter tank initially full of water develops a leak at the bottom. Given that 20% of the water leaks out in the first 5 minutes, find the amount of water left in the tank t minutes after the leak develops

   (a) if the water drains off at a rate that is proportional to the amount of water present;

   (b) if the water drains off at a rate that is proportional to the product of the time elapsed and the amount of water present.

22. An object falling in air is subject not only to gravitational force but also to air resistance. Find $v(t)$, the velocity of the object at time t, given that

$$v'(t) + Kv(t) = 32 \quad \text{and} \quad v(0) = 0. \qquad (\text{take } K > 0)$$

Show that $v(t)$ cannot exceed $32/K$. ($32/K$ is called the *terminal velocity*.)

*23. At a certain moment a 100-gallon mixing tank is full of brine containing 0.25 pounds of salt per gallon. Find the amount of salt present t minutes later if the brine is being continuously drawn off at the rate of 3 gallons per minute and replaced by brine containing 0.2 pounds of salt per gallon.

24. The current i in an electric circuit varies with time according to the formula

$$L \frac{di}{dt} + Ri = E.$$

Take E (the voltage), L (the inductance), and R (the resistance) as constants. Measure the current in amperes and t in seconds, and suppose that initially the current is 0.

   (a) Find a formula for the current at each subsequent time t.

   (b) What upper limit does the current approach as t increases?

   (c) In how many seconds will the current reach 90% of its upper limit?

6.9   Additional Exercises

*1. Find the minimum value of $y = ae^{kx} + be^{-kx}$, $a > 0$, $b > 0$.

2. Show that, if $y = \frac{1}{2}a(e^{x/a} + e^{-x/a})$, then $y'' = y/a^2$.

*3. Find the points of inflection of the graph of $y = e^{-x^2}$.

4. A rectangle has one side on the x-axis and the two upper vertices on the graph of $y = e^{-x^2}$. Where should the upper vertices be placed so as to maximize the area of the rectangle?

*5. A telegraph cable consists of a core of copper wires with a covering made of nonconducting material. If x denotes the ratio of the radius of the core to the thickness of the covering, it is known that the speed of signaling is proportional to $x^2 \log (1/x)$. For what value of x is the speed a maximum?

6. A rectangle has two sides along the coordinate axes and one vertex at a point P which moves along the curve $y = e^x$ in such a way that y increases at the rate of $\frac{1}{2}$ unit per minute. How fast is the area of the rectangle increasing when $y = 3$?

*7. Find

$$\lim_{h \to 0} \frac{1}{h} (e^h - 1).$$

8. Show that $y = Ae^{\alpha x} + Be^{-\alpha x}$ satisfies the differential equation

$$y'' - \alpha^2 y = 0.$$

9. Find a function y which satisfies the differential equation of Exercise 8 and the initial conditions below.
  *(a) $y(0) = 0, \quad y'(0) = \alpha$.   (b) $y(0) = \alpha, \quad y'(0) = 0$.
  *(c) $y(0) = 1, \quad y'(0) = \alpha$.   (d) $y(0) = \alpha^2, \quad y(1) = \alpha$.

10. Find the extreme values and points of inflection of

$$y = \frac{x}{\log x}.$$

*11. Find the area of the region bounded by the curve $xy = a^2$, the x-axis, and the vertical lines $x = a, x = 2a$.

12. A boat moving in still water is subject to a retardation proportional to its velocity. Show that the velocity t seconds after the power is shut off is given by the formula $v = ce^{-kt}$, where c is the velocity at the instant the power is shut off.

*13. At a certain instant a boat drifting in still water has a velocity of 4 miles per hour. One minute later the velocity is 2 miles per hour. Find the distance moved.

14. An object falls from rest under the action of its weight and a small resistance which varies with velocity:

$$x(0) = 0, \quad v(0) = 0, \quad a(t) = g - kv(t).$$

Show that the following relations hold.

(a) $v(t) = \frac{g}{k} (1 - e^{-kt})$.   (b) $x(t) = \frac{g}{k^2} (kt + e^{-kt} - 1)$.

(c) $kx(t) + v(t) + \frac{g}{k} \log \left(1 - \frac{kv(t)}{g}\right) = 0$.

15. Show that for all positive integers $m < n$

$$\log \frac{n+1}{m} < \frac{1}{m} + \frac{1}{m+1} + \cdots + \frac{1}{n} < \log \frac{n}{m-1}.$$

HINT:

$$\frac{1}{k+1} < \int_k^{k+1} \frac{dx}{x} < \frac{1}{k}.$$

The Trigonometric and Hyperbolic Functions

7.1 Differentiating the Trigonometric Functions

An outline review of trigonometry appears in Chapter 1. Here we single out for attention some trigonometric identities.

(1) Unit Circle

$$\cos^2 \theta + \sin^2 \theta = 1, \quad \tan^2 \theta + 1 = \sec^2 \theta, \quad \cot^2 \theta + 1 = \operatorname{cosec}^2 \theta.$$

(2) Addition Formulas

$$\sin (\theta_1 + \theta_2) = \sin \theta_1 \cos \theta_2 + \cos \theta_1 \sin \theta_2,$$
$$\sin (\theta_1 - \theta_2) = \sin \theta_1 \cos \theta_2 - \cos \theta_1 \sin \theta_2,$$
$$\cos (\theta_1 + \theta_2) = \cos \theta_1 \cos \theta_2 - \sin \theta_1 \sin \theta_2,$$
$$\cos (\theta_1 - \theta_2) = \cos \theta_1 \cos \theta_2 + \sin \theta_1 \sin \theta_2,$$
$$\tan (\theta_1 + \theta_2) = \frac{\tan \theta_1 + \tan \theta_2}{1 - \tan \theta_1 \tan \theta_2}, \qquad \tan(\theta_1 - \theta_2) = \frac{\tan \theta_1 - \tan \theta_2}{1 + \tan \theta_1 \tan \theta_2}.$$

(3) Double-Angle Formulas

$$\sin 2\theta = 2 \sin \theta \cos \theta, \quad \cos 2\theta = \cos^2 \theta - \sin^2 \theta.$$

(4) Half-Angle Formulas

$$\sin^2 \tfrac{1}{2}\theta = \tfrac{1}{2}(1 - \cos \theta), \quad \cos^2 \tfrac{1}{2}\theta = \tfrac{1}{2}(1 + \cos \theta).$$

Just as the calculus of logarithms is simplified by the use of the base e, the calculus of the trigonometric functions is simplified by the use of radian measure. Throughout our discussion we will use radian measure and refer to degrees only in passing.

FIGURE 7.1.1

FIGURE 7.1.2

FIGURE 7.1.3

We begin our discussion with the sine function. To differentiate the sine function we will form the difference quotient

$$\frac{\sin (x + h) - \sin x}{h}$$

and take its limit as h tends to 0. Since

$$\sin (x + h) = \sin x \cos h + \cos x \sin h,$$

we obtain for $h \neq 0$

$$\frac{\sin (x + h) - \sin x}{h} = \cos x \frac{\sin h}{h} - \sin x \frac{1 - \cos h}{h}.$$

We are thus led to consider the behavior of

$$\frac{\sin h}{h} \quad \text{and} \quad \frac{1 - \cos h}{h}$$

as h tends to 0. In the exercises you are asked to show that

$$|\sin h| \leq |h|$$

and then that

$$\lim_{h \to 0} \sin h = 0 \quad \text{and} \quad \lim_{h \to 0} \cos h = 1.$$

Assuming these results, we can prove that

(7.1.1) $$\boxed{\lim_{h \to 0} \frac{\sin h}{h} = 1 \quad \text{and} \quad \lim_{h \to 0} \frac{1 - \cos h}{h} = 0.}$$

PROOF.   First we prove that

$$\lim_{h \to 0} \frac{\sin h}{h} = 1.$$

To follow the argument see Figure 7.1.4.

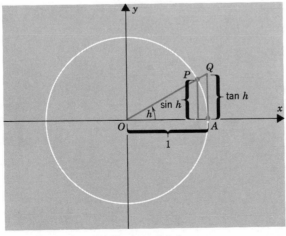

FIGURE 7.1.4

For small $h > 0$

$$\text{area of triangle } OAP = \frac{1}{2} \sin h,$$

$$\text{area of sector} = \frac{1}{2} h,$$

$$\text{area of triangle } OAQ = \frac{1}{2} \tan h = \frac{1}{2} \frac{\sin h}{\cos h}.$$

Since

$$\text{triangle } OAP \subseteq \text{sector} \subseteq \text{triangle } OAQ$$

(and these are all proper containments), we must have

$$\frac{1}{2} \sin h < \frac{1}{2} h < \frac{1}{2} \frac{\sin h}{\cos h},$$

$$\sin h < h < \frac{\sin h}{\cos h},$$

$$\frac{\cos h}{\sin h} < \frac{1}{h} < \frac{1}{\sin h},$$

$$\cos h < \frac{\sin h}{h} < 1.$$

This last inequality was derived for $h > 0$, but, since

$$\cos(-h) = \cos h \quad \text{and} \quad \frac{\sin(-h)}{-h} = \frac{-\sin h}{-h} = \frac{\sin h}{h},$$

the inequality also holds for $h < 0$.
  With

$$\cos h < \frac{\sin h}{h} < 1 \quad \text{together with} \quad \lim_{h \to 0} \cos h = 1,$$

we can conclude from the pinching theorem that

$$\lim_{h \to 0} \frac{\sin h}{h} = 1.$$

  Our proof that

$$\lim_{h \to 0} \frac{1 - \cos h}{h} = 0$$

can be outlined as follows:

$$\left| \frac{1 - \cos h}{h} \right| \underset{\substack{\smile \\ \text{since } |\sin h| < |h|}}{\leq} \left| \frac{1 - \cos h}{\sin h} \right| \underset{\substack{\nearrow \\ \text{since } 1 - \cos^2 h = \sin^2 h}}{=} \left| \frac{\sin h}{1 + \cos h} \right| \to \frac{0}{2} = 0.$$

The details are not hard to fill in. $\square$

From

$$\frac{\sin (x + h) - \sin x}{h} = \cos x \, \frac{\sin h}{h} - \sin x \, \frac{1 - \cos h}{h}$$

together with

$$\lim_{h \to 0} \frac{\sin h}{h} = 1 \quad \text{and} \quad \lim_{h \to 0} \frac{1 - \cos h}{h} = 0,$$

it follows that

(7.1.2)
$$\boxed{\frac{d}{dx} (\sin x) = \cos x.}$$

To find the derivative of the cosine function note that $\cos x = \sin (x + \frac{1}{2}\pi)$. It follows then that

$$\frac{d}{dx} (\cos x) = \frac{d}{dx} [\sin (x + \tfrac{1}{2}\pi)] = \cos (x + \tfrac{1}{2}\pi) = -\sin x.$$

In short

(7.1.3)
$$\boxed{\frac{d}{dx} (\cos x) = -\sin x.}$$

Problem. Differentiate

$$f(x) = \sin (a^2 + b^2x^2) \quad \text{and} \quad g(x) = \cos (a^2 + b^2x^2).$$

SOLUTION. These are composite functions, and therefore we must use the chain rule:

$$f'(x) = \cos (a^2 + b^2x^2) \frac{d}{dx} (a^2 + b^2x^2) = 2b^2x \cos (a^2 + b^2x^2),$$

$$g'(x) = -\sin (a^2 + b^2x^2) \frac{d}{dx} (a^2 + b^2x^2) = -2b^2x \sin (a^2 + b^2x^2). \quad \square$$

We come now to the tangent function. Since

$$\tan x = \frac{\sin x}{\cos x},$$

we have

$$\frac{d}{dx} (\tan x) = \frac{\cos x \, \dfrac{d}{dx} (\sin x) - \sin x \, \dfrac{d}{dx} (\cos x)}{\cos^2 x}$$

$$= \frac{\cos^2 x + \sin^2 x}{\cos^2 x} = \frac{1}{\cos^2 x} = \sec^2 x.$$

The derivative of the tangent function is the secant squared:

(7.1.4)
$$\frac{d}{dx}(\tan x) = \sec^2 x.$$

The derivatives of the other trigonometric functions are as follows:

(7.1.5)
$$\frac{d}{dx}(\cot x) = -\csc^2 x,$$
$$\frac{d}{dx}(\sec x) = \sec x \tan x,$$
$$\frac{d}{dx}(\csc x) = -\csc x \cot x.$$

The verification of these formulas is left as an exercise. $\square$

Problem. Find

$$\frac{d}{dx}[\tan(\sqrt{1-x^2})].$$

SOLUTION

$$\frac{d}{dx}[\tan(\sqrt{1-x^2})] = \sec^2(\sqrt{1-x^2}) \cdot \frac{d}{dx}(\sqrt{1-x^2})$$

$$= \sec^2(\sqrt{1-x^2})\frac{-x}{\sqrt{1-x^2}} = -\frac{x\sec^2(\sqrt{1-x^2})}{\sqrt{1-x^2}}. \quad \square$$

Problem. Find

$$\frac{d}{dx}[(\cos ax)(\sin ax)].$$

SOLUTION

$$\frac{d}{dx}[(\cos ax)(\sin ax)] = \cos ax \frac{d}{dx}(\sin ax) + \sin ax \frac{d}{dx}(\cos ax)$$

$$= (\cos ax)(a\cos ax) + (\sin ax)(-a\sin ax)$$

$$= a(\cos^2 ax - \sin^2 ax) = a\cos 2ax. \quad \square$$

Problem. Find

$$\frac{d}{dx}(\sqrt{x}\sec\sqrt{x}).$$

SOLUTION

$$\frac{d}{dx}(\sqrt{x}\ \sec \sqrt{x}) = \sqrt{x}\frac{d}{dx}(\sec \sqrt{x}) + (\sec \sqrt{x})\frac{d}{dx}(\sqrt{x})$$

$$= \sqrt{x}\ (\sec \sqrt{x})(\tan \sqrt{x})\left(\frac{1}{2\sqrt{x}}\right) + (\sec \sqrt{x})\left(\frac{1}{2\sqrt{x}}\right)$$

$$= \frac{\sec \sqrt{x}}{2\sqrt{x}}[\sqrt{x}\ \tan \sqrt{x} + 1]. \quad \square$$

Problem. Find

$$\frac{d}{dx}(\log |\cos x|).$$

SOLUTION

$$\frac{d}{dx}(\log |\cos x|) = \frac{1}{\cos x}\frac{d}{dx}(\cos x) = \frac{1}{\cos x}(-\sin x) = -\tan x. \quad \square$$

Problem. Find

$$\frac{d}{dx}(x^{\sin x}).$$

SOLUTION

$$\frac{d}{dx}(x^{\sin x}) = \frac{d}{dx}(e^{\sin x \log x})$$

$$= e^{\sin x \log x}\frac{d}{dx}(\sin x \log x)$$

$$= x^{\sin x}\left[\sin x \frac{d}{dx}(\log x) + \log x \frac{d}{dx}(\sin x)\right]$$

$$= x^{\sin x}\left[\sin x \left(\frac{1}{x}\right) + \log x \cos x\right].$$

We can also set $g(x) = x^{\sin x}$ and use logarithmic differentiation:

$$\log g(x) = \sin x \log x,$$

$$\frac{g'(x)}{g(x)} = \sin x \left(\frac{1}{x}\right) + \cos x \log x,$$

$$g'(x) = g(x)\left[\sin x \left(\frac{1}{x}\right) + \cos x \log x\right]$$

$$= x^{\sin x}\left[\sin x \left(\frac{1}{x}\right) + \cos x \log x\right]. \quad \square$$

Problem. Given that

$$\log y = \sin (x + y)$$

find dy/dx by implicit differentiation.

SOLUTION

$$\frac{1}{y}\frac{dy}{dx} = \cos{(x + y)}\left[1 + \frac{dy}{dx}\right]$$

$$\frac{dy}{dx} = y\cos{(x + y)}\left[1 + \frac{dy}{dx}\right]$$

$$[1 - y\cos{(x + y)}]\frac{dy}{dx} = y\cos{(x + y)}$$

$$\frac{dy}{dx} = \frac{y\cos{(x + y)}}{1 - y\cos{(x + y)}}. \quad \square$$

TABLE 7.1.1†

x	$\sin x$	$\cos x$	$\tan x$
0	0.000	1.000	0.000
0.1	0.100	0.995	0.100
0.2	0.199	0.980	0.203
0.3	0.296	0.955	0.309
0.4	0.389	0.921	0.423
0.5	0.479	0.878	0.546
0.6	0.565	0.825	0.684
0.7	0.644	0.765	0.842
0.8	0.717	0.697	1.030
0.9	0.783	0.622	1.260
1.0	0.841	0.540	1.557
1.1	0.891	0.454	1.965
1.2	0.932	0.362	2.572
1.3	0.964	0.267	3.602
1.4	0.985	0.170	5.798
1.5	0.997	0.071	14.101
1.57	0.999+	0.001	1255.770

$\pi \cong 3.14159$

Problem.   Use a differential to estimate sin 0.43.

SOLUTION.   From Table 7.1.1

$$\sin 0.40 \cong 0.389 \quad \text{and} \quad \cos 0.40 \cong 0.921.$$

What we need is an estimate for the increase of

$$f(x) = \sin x$$

from 0.40 to 0.43. Differentiation gives

$$f'(x) = \cos x$$

---

† More extended tables appear at the end of the book.

so that in general

$$df = f'(x)h = (\cos x)h.$$

With $x = 0.40$ and $h = 0.03$, df becomes

$$(\cos 0.40)(0.03) \cong (0.92)(0.03) \cong 0.028.$$

A change from 0.40 to 0.43 increases the sine by approximately 0.028:

$$\sin 0.43 \cong \sin 0.40 + 0.028 \cong 0.389 + 0.028 = 0.417. \quad \square$$

TABLE 7.1.2

x		$\sin x$	$\cos x$	$\tan x$	
0	[0°]	0	1	0	
$\frac{1}{6}\pi$	[30°]	$\frac{1}{2}$	$\frac{1}{2}\sqrt{3}$	$\frac{1}{3}\sqrt{3}$	
$\frac{1}{4}\pi$	[45°]	$\frac{1}{2}\sqrt{2}$	$\frac{1}{2}\sqrt{2}$	1	$1° \cong 0.0175$ radians
$\frac{1}{3}\pi$	[60°]	$\frac{1}{2}\sqrt{3}$	$\frac{1}{2}$	$\sqrt{3}$	
$\frac{1}{2}\pi$	[90°]	1	0	——	
$\frac{2}{3}\pi$	[120°]	$\frac{1}{2}\sqrt{3}$	$-\frac{1}{2}$	$-\sqrt{3}$	$\sqrt{2} \cong 1.414$
$\frac{3}{4}\pi$	[135°]	$\frac{1}{2}\sqrt{2}$	$-\frac{1}{2}\sqrt{2}$	-1	$\sqrt{3} \cong 1.732$
$\frac{5}{6}\pi$	[150°]	$\frac{1}{2}$	$-\frac{1}{2}\sqrt{3}$	$-\frac{1}{3}\sqrt{3}$	
π	[180°]	0	-1	0	

Problem. Given that $1° \cong 0.0175$ radians, use a differential to estimate $\tan 46°$.

SOLUTION. Table 7.1.2 gives

$$\tan 45° = 1, \quad \sec 45° = \frac{1}{\cos 45°} = \sqrt{2}.$$

Note that

$$45° = \tfrac{1}{4}\pi \text{ radians} \quad \text{and} \quad 46° \cong \tfrac{1}{4}\pi + 0.0175 \text{ radians}.$$

For $f(x) = \tan x$

$$df = f'(x)h = (\sec^2 x)h.$$

It follows that

$$\tan 46° \cong \tan (\tfrac{1}{4}\pi + 0.0175) \cong \tan \tfrac{1}{4}\pi + (\sec^2 \tfrac{1}{4}\pi)(0.0175)$$
$$= 1 + 2(0.0175) = 1.035. \quad \square$$

Problem. Let

$$f(x) = \sin^2 x, \quad x \in [0, \pi].$$

(a) On what intervals is f increasing? Decreasing? (b) Find the extreme values of f.
(c) Determine the concavity of the graph and find the points of inflection.
(d) Sketch the graph.

SOLUTION.   Here

$$f(x) = \sin^2 x,$$

$$f'(x) = 2 \sin x \, \frac{d}{dx} (\sin x) = 2 \sin x \cos x = \sin 2x,$$

$$f''(x) = 2 \cos 2x.$$

Obviously f is positive between 0 and π, and it is 0 at the endpoints. This makes 0 an endpoint minimum and also the absolute minimum. Since

$$f'(x) \text{ is } \begin{cases} \text{positive,} & \text{for} \quad 0 < x < \tfrac{1}{2}\pi \\ 0, & \text{at} \quad\quad\quad x = \tfrac{1}{2}\pi \\ \text{negative,} & \text{for} \quad \tfrac{1}{2}\pi < x < \pi \end{cases},$$

f increases on $[0, \tfrac{1}{2}\pi]$ and decreases on $[\tfrac{1}{2}\pi, \pi]$. The number $f(\tfrac{1}{2}\pi) = 1$ is a local and absolute maximum. There are no other extreme values. Since

$$f''(x) \text{ is } \begin{cases} \text{positive,} & \text{for} \quad 0 < x < \tfrac{1}{4}\pi \\ 0, & \text{at} \quad\quad\quad x = \tfrac{1}{4}\pi \\ \text{negative,} & \text{for} \quad \tfrac{1}{4}\pi < x < \tfrac{3}{4}\pi \\ 0, & \text{at} \quad\quad\quad x = \tfrac{3}{4}\pi \\ \text{positive,} & \text{for} \quad \tfrac{3}{4}\pi < x < \pi \end{cases},$$

the graph is concave up on $[0, \tfrac{1}{4}\pi]$, concave down on $[\tfrac{1}{4}\pi, \tfrac{3}{4}\pi]$, and concave up again on $[\tfrac{3}{4}\pi, \pi]$. There are two points of inflection: $(\tfrac{1}{4}\pi, f(\tfrac{1}{4}\pi)) = (\tfrac{1}{4}\pi, \tfrac{1}{2})$ and $(\tfrac{3}{4}\pi, f(\tfrac{3}{4}\pi)) = (\tfrac{3}{4}\pi, \tfrac{1}{2})$. The graph is given in Figure 7.1.5. $\square$

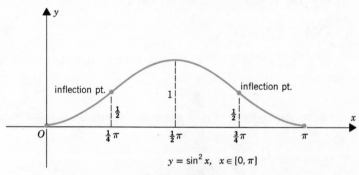

$$y = \sin^2 x, \quad x \in [0, \pi]$$

FIGURE 7.1.5

Problem.   Sketch the graph of

$$y = e^{-\frac{1}{4}x} \sin \tfrac{1}{2}\pi x, \qquad x \geq 0.$$

SOLUTION.   Since

$$-1 \leq \sin \tfrac{1}{2}\pi x \leq 1,$$

the graph will stay entirely between the curves

$$y = -e^{-\frac{1}{4}x} \quad \text{and} \quad y = e^{-\frac{1}{4}x}.$$

These are called *boundary curves*. The graph will touch the upper boundary when

$$\sin \tfrac{1}{2}\pi x = 1;$$

namely, at $x = 1, 5, 9$, etc. It will touch the lower boundary when

$$\sin \tfrac{1}{2}\pi x = -1;$$

namely, at $x = 3, 7, 11$, etc. It will cross the x-axis when

$$\sin \tfrac{1}{2}\pi x = 0;$$

namely, at $x = 0, 2, 4$, etc. The graph appears in Figure 7.1.6. The oscillations of the sine function have been *damped* by the boundary curves.   □

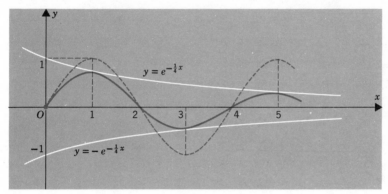

FIGURE 7.1.6

Problem.   Find

$$\lim_{x \to 0} \frac{\sin 4x}{x}.$$

SOLUTION.   At the beginning of this section we proved that

$$\lim_{h \to 0} \frac{\sin h}{h} = 1.$$

It follows now that

$$\lim_{x \to 0} \frac{\sin 4x}{4x} = 1 \quad \text{and} \quad \lim_{x \to 0} \frac{\sin 4x}{x} = \lim_{x \to 0} 4 \left(\frac{\sin 4x}{4x} \right) = 4. \quad \square$$

Our treatment of the trigonometric functions has been based entirely on radian measure. When degrees are used, the derivatives of the trigonometric functions contain the extra factor $\frac{1}{180}\pi \cong 0.0175$.

Problem.   Find

$$\frac{d}{dx} (\sin x°).$$

SOLUTION.   Since $x° = \frac{1}{180}\pi x$ radians, we have

$$\frac{d}{dx}(\sin x°) = \frac{d}{dx}(\sin \tfrac{1}{180}\pi x) = \tfrac{1}{180}\pi \cos \tfrac{1}{180}\pi x = \tfrac{1}{180}\pi \cos x°. \quad \square$$

This extra factor $\frac{1}{180}\pi$ is a disadvantage, particularly in problems where it occurs repeatedly. It tends to discourage the use of degree measure in theoretical work.

Exercises

Differentiate.

*1. $y = \sin 3x$.

2. $y = \sin x \cos x$.

*3. $y = x \tan x$.

*4. $y = \sin x + \cos x$.

5. $y = \tan x^2$.

*6. $y = \sin \sqrt{x}$.

*7. $y = \cos \pi x \sin \pi x$.

8. $y = \sec x \tan x$.

*9. $y = x^2 \sin \pi x$.

*10. $y = \frac{1}{2}\sin^2 x$.

11. $y = \log|\sin x|$.

*12. $y = \tan^3 2x$.

*13. $y = e^x \sin x$.

14. $y = e^x \cos x$.

*15. $y = \sqrt{\cos x^2}$.

*16. $y = (\cos \pi x)^x$.

17. $y = \sec ax$.

*18. $y = e^{-kx} \sin kx$.

*19. $y = 10^x \tan \pi x$.

20. $y = x^{2\pi} \cos 2\pi x$.

*21. $y = \log \sqrt{\tan x}$.

*22. $y = e^x \sin \frac{1}{2}\pi x$.

23. $y = x \cot x$.

*24. $y = \tan^4 \frac{1}{4}\pi x$.

*25. $y = e^{\sin x}$.

26. $y = e^{\sin x} \cos x$.

*27. $y = \sec^2 x \tan^2 x$.

*28. $y = \sin nx \sin^n x$.

29. $y = \cos^2(x - a)$.

*30. $y = \log \sqrt{\cos 2x}$.

*31. $\rho = \tan^2 \theta$.

32. $\rho = 2^\theta \cos \pi\theta$.

*33. $\rho = \operatorname{cosec}^2 3\theta$.

Find the second derivative.

*34. $y = \sin kx$.

35. $y = \cos kx$.

*36. $y = x \cos x$.

*37. $s = e^{2t} \cos t$.

38. $s = e^{-t} \sin t$.

*39. $s = e^{-t} \sin 2t$.

Find dy/dx by implicit differentiation.

*40. $y = \cos(x - y)$.

41. $e^y = \sin(x + y)$.

*42. $\cos y = \log(x + y)$.

43. $\tan y = 3x^2 + \tan(x + y)$.

*44. The sine curve intersects the cosine curve at $x = \frac{1}{4}\pi$. What is the angle of intersection?

45. Find a formula for the nth derivative: (a) $y = \sin x$. (b) $y = \cos x$.

46. Use differentials and Table 7.1.1 to estimate the following.
    *(a) $\cos 0.52$.   (b) $\sin 0.73$.   *(c) $\cos 1.32$.

47. Use differentials and Table 7.1.2 to estimate the following.
    *(a) $\sin 62°$.   (b) $\cos 57°$.   *(c) $\tan 33°$.

Find the limits which exist.

*48. $\lim\limits_{x\to 0} \dfrac{\sin 3x}{x}$.

49. $\lim\limits_{x\to 0} \dfrac{\sin x}{3x}$.

*50. $\lim\limits_{x\to 0} \dfrac{\sin 3x}{2x}$.

*51. $\lim\limits_{x\to 0} \dfrac{\sin x^2}{x}$.

52. $\lim\limits_{x\to 0} \dfrac{\sin x^2}{x^2}$.

*53. $\lim\limits_{x\to 0} \dfrac{\sin x}{x^2}$.

*54. $\lim\limits_{x\to 0} \dfrac{\sin^2 2x}{x^2}$.

55. $\lim\limits_{x\to 0} \dfrac{\tan 3x}{2x}$.

*56. $\lim\limits_{x\to 0} \dfrac{\tan 3x^2}{2x^2}$.

*57. $\lim\limits_{x\to 0} \dfrac{\tan^2 3x}{2x^2}$.

58. $\lim\limits_{x\to 0} \dfrac{2x^2 + x}{\sin x}$.

*59. $\lim\limits_{x\downarrow 0} \left(x^2 \log \dfrac{x}{\sin x}\right)$.

60. Using Figure 7.1.6 as a model, sketch the graph of

$$y = e^{-\frac{1}{4}x} \cos \tfrac{1}{2}\pi x, \qquad x \geq 0.$$

61. Prove that for all real x and y
    (a) $|\cos x - \cos y| \leq |x - y|$.   (b) $|\sin x - \sin y| \leq |x - y|$.
    HINT: Apply the mean-value theorem.
62. Show that

$$|\sin h| \leq |h|. \qquad \text{(begin with Figure 7.1.4)}$$

63. Show that

$$\lim_{h \to 0} \sin h = 0. \qquad \text{(use Exercise 62)}$$

64. Show that

$$\lim_{h \to 0} \cos h = 1. \qquad (\sin^2 \tfrac{1}{2}h = \tfrac{1}{2}[1 - \cos h])$$

For each of the functions below, (a) find the intervals where it is increasing and the intervals where it is decreasing, (b) find the extreme values, (c) determine the concavity of the graph and find the points of inflection, and finally, (d) sketch the graph.

*65. $f(x) = x + \sin x, \quad 0 \leq x \leq 2\pi$.
 66. $f(x) = x - \sin x, \quad 0 \leq x \leq 2\pi$.
*67. $f(x) = e^x \cos x, \quad 0 \leq x \leq 2\pi$.
 68. $f(x) = e^x(\cos x + \sin x), \quad 0 \leq x \leq 2\pi$.
*69. $f(x) = \cos^2 x, \quad 0 \leq x \leq \pi$.
Optional | 70. $f(x) = x \tan x, \quad -\tfrac{1}{2}\pi < x < \tfrac{1}{2}\pi$.

7.2 Integrating the Trigonometric Functions

In the last section you saw that each trigonometric function is differentiable on its domain and that

$$
\begin{array}{ll}
\dfrac{d}{dx}(\sin x) = \cos x. & \dfrac{d}{dx}(\cos x) = -\sin x. \\[2mm]
\dfrac{d}{dx}(\tan x) = \sec^2 x. & \dfrac{d}{dx}(\cot x) = -\operatorname{cosec}^2 x. \\[2mm]
\dfrac{d}{dx}(\sec x) = \sec x \tan x. & \dfrac{d}{dx}(\operatorname{cosec} x) = -\operatorname{cosec} x \cot x.
\end{array}
$$

(7.2.1)

Here we integrate the trigonometric functions. As we shall show in a moment, the following formulas are valid:

(i) $\displaystyle \int \sin x \, dx = -\cos x + C.$

(ii) $\displaystyle \int \cos x \, dx = \sin x + C.$

(7.2.2)

$$\text{(iii)} \int \tan x \, dx = \log |\sec x| + C.$$

$$\text{(iv)} \int \cot x \, dx = \log |\sin x| + C.$$

$$\text{(v)} \int \sec x \, dx = \log |\sec x + \tan x| + C.$$

$$\text{(vi)} \int \cosec x \, dx = \log |\cosec x - \cot x| + C.$$

DERIVATION OF FORMULAS (i)–(vi)

(i) $\int \sin x \, dx = - \int - \sin x \, dx = - \int \dfrac{d}{dx} (\cos x) \, dx = - \cos x + C. \quad \square$

(ii) $\int \cos x \, dx = \int \dfrac{d}{dx} (\sin x) \, dx = \sin x + C. \quad \square$

The remaining formulas can all be derived by writing the integrals in the form

$$\int \frac{du}{u} = \log |u| + C.$$

(iii) $\displaystyle\int \tan x \, dx = \int \dfrac{\sin x}{\cos x} \, dx$ [set $u = \cos x$; $du = - \sin x \, dx$, $- du = \sin x \, dx$]

$$= - \int \frac{du}{u} = - \log |u| + C$$

$$= - \log |\cos x| + C = \log \frac{1}{|\cos x|} + C = \log |\sec x| + C. \quad \square$$

(iv) $\displaystyle\int \cot x \, dx = \int \dfrac{\cos x}{\sin x} \, dx$ \hspace{2em} [set $u = \sin x$; $du = \cos x \, dx$]

$$= \int \frac{du}{u} = \log |u| + C = \log |\sin x| + C. \quad \square$$

(v) $\displaystyle\int \sec x \, dx \overset{\dagger}{=} \int \sec x \dfrac{\sec x + \tan x}{\sec x + \tan x} \, dx = \int \dfrac{\sec x \tan x + \sec^2 x}{\sec x + \tan x} \, dx.$

[set $u = \sec x + \tan x$; $du = (\sec x \tan x + \sec^2 x) \, dx$]

$$\int \sec x \, dx = \int \frac{du}{u} = \log |u| + C = \log |\sec x + \tan x| + C. \quad \square$$

The derivation of formula (vi) is left to you as an exercise.

Problem. Find

$$\int \sin x \cos x \, dx.$$

SOLUTION. Set

$$u = \sin x; \quad du = \cos x \, dx.$$

$$\int \sin x \cos x \, dx = \int u \, du = \tfrac{1}{2} u^2 + C = \tfrac{1}{2} \sin^2 x + C. \quad \square$$

_____

† Only experience prompts us to multiply numerator and denominator by $\sec x + \tan x$.

Problem. Find

$$\int \sec^3 x \tan x \, dx.$$

SOLUTION. Set

$$u = \sec x; \quad du = \sec x \tan x \, dx.$$

$$\int \sec^3 x \tan x \, dx = \int \sec^2 x \, (\sec x \tan x) \, dx$$

$$= \int u^2 \, du = \tfrac{1}{3} u^3 + C = \tfrac{1}{3} \sec^3 x + C. \quad \square$$

Problem. Find

$$\int x \cos \pi x^2 \, dx.$$

SOLUTION. Set

$$u = \pi x^2; \quad du = 2\pi x \, dx, \quad \frac{1}{2\pi} \, du = x \, dx.$$

$$\int x \cos \pi x^2 \, dx = \frac{1}{2\pi} \int \cos u \, du = \frac{1}{2\pi} \sin u + C = \frac{1}{2\pi} \sin \pi x^2 + C. \quad \square$$

Problem. Find

$$\int \frac{\sec^2 x}{1 - \tan x} \, dx.$$

SOLUTION. Set

$$u = 1 - \tan x; \quad du = -\sec^2 x \, dx, \quad -du = \sec^2 x \, dx.$$

$$\int \frac{\sec^2 x}{1 - \tan x} \, dx = -\int \frac{du}{u} = -\log |u| + C = -\log |1 - \tan x| + C. \quad \square$$

FIGURE 7.2.1

Problem. Find the area between

$$y = \cos x \quad \text{and} \quad y = \sin x, \quad x \in [0, 2\pi].$$

SOLUTION.   The region in question is displayed in Figure 7.2.1. The area is given by the integral

$$\int_0^{2\pi} |\cos x - \sin x| \, dx.$$

To evaluate this integral note that

$$\cos x - \sin x \quad \text{is} \quad \begin{cases} \geq 0, & \text{for} \quad 0 \leq x \leq \frac{1}{4}\pi \\ \leq 0, & \text{for} \quad \frac{1}{4}\pi \leq x \leq \frac{5}{4}\pi \\ \geq 0, & \text{for} \quad \frac{5}{4}\pi \leq x \leq 2\pi \end{cases}$$

Consequently

$$\text{area} = \int_0^{\frac{1}{4}\pi} (\cos x - \sin x) \, dx + \int_{\frac{1}{4}\pi}^{\frac{5}{4}\pi} (\sin x - \cos x) \, dx + \int_{\frac{5}{4}\pi}^{2\pi} (\cos x - \sin x) \, dx$$

$$= \Big[\sin x + \cos x \Big]_0^{\frac{1}{4}\pi} + \Big[-\cos x - \sin x \Big]_{\frac{1}{4}\pi}^{\frac{5}{4}\pi} + \Big[\sin x + \cos x \Big]_{\frac{5}{4}\pi}^{2\pi}$$

$$= (\sqrt{2} - 1) + 2\sqrt{2} + (1 + \sqrt{2}) = 4\sqrt{2}. \quad \square$$

Problem.   Find

$$\int x \sin x \, dx.$$

SOLUTION.   What causes difficulty here is the factor x. We can eliminate the x by integrating by parts. Set

$$u = x, \qquad dv = \sin x \, dx,$$
$$du = dx, \qquad v = -\cos x.$$

$$\int x \sin x \, dx = \int u \, dv = uv - \int v \, du$$

$$= -x \cos x + \int \cos x \, dx = -x \cos x + \sin x + C. \quad \square$$

Problem.   Find

$$\int e^x \cos x \, dx.$$

SOLUTION.   Here we integrate by parts twice. First we set

$$u = e^x, \qquad dv = \cos x \, dx,$$
$$du = e^x \, dx, \qquad v = \sin x.$$

This gives

$$(1) \quad \int e^x \cos x \, dx = \int u \, dv = uv - \int v \, du = e^x \sin x - \int e^x \sin x \, dx.$$

Now we work with the integral on the right. Set

$$u = e^x, \qquad dv = \sin x \, dx,$$
$$du = e^x \, dx, \qquad v = -\cos x.$$

This gives

(2) $\quad \int e^x \sin x \, dx = \int u \, dv = uv - \int v \, du = -e^x \cos x + \int e^x \cos x \, dx.$

Substituting (2) in (1), we get

$$\int e^x \cos x \, dx = e^x \sin x + e^x \cos x - \int e^x \cos x \, dx,$$

$$2 \int e^x \cos x \, dx = e^x (\sin x + \cos x),$$

$$\int e^x \cos x \, dx = \tfrac{1}{2} e^x (\sin x + \cos x).$$

Since this integral is an indefinite integral, we add an arbitrary constant C:

$$\int e^x \cos x \, dx = \tfrac{1}{2} e^x (\sin x + \cos x) + C. \quad \square$$

Exercises

Work out the following indefinite integrals.

*1. $\displaystyle \int \cos (3x - 1) \, dx.$

2. $\displaystyle \int \cos \tfrac{1}{2} \pi x \, dx.$

*3. $\displaystyle \int 3 \sin^2 x \cos x \, dx.$

*4. $\displaystyle \int \sec 2x \tan 2x \, dx.$

5. $\displaystyle \int \cos^4 x \sin x \, dx.$

*6. $\displaystyle \int \sqrt{1 + \cos x} \, \sin x \, dx.$

*7. $\displaystyle \int e^{-\sin x} \cos x \, dx.$

8. $\displaystyle \int x \tan x^2 \, dx.$

*9. $\displaystyle \int x^{-1/2} \sin x^{1/2} \, dx.$

*10. $\displaystyle \int \frac{\cos x}{\sqrt{1 - \sin x}} \, dx.$

11. $\displaystyle \int \frac{\sin x}{\sqrt{1 + \cos x}} \, dx.$

*12. $\displaystyle \int \frac{\sin x}{1 + \cos x} \, dx.$

*13. $\displaystyle \int \cos^2 \pi x \sin \pi x \, dx.$

14. $\displaystyle \int \tan \pi x \, dx.$

*15. $\displaystyle \int \cot \pi x \, dx.$

*16. $\displaystyle \int \frac{\cos x}{1 + 2 \sin x} \, dx.$

17. $\displaystyle \int \frac{dx}{\cos^2 x}.$

*18. $\displaystyle \int \frac{\cos 2x}{2 - \sin 2x} \, dx.$

*19. $\displaystyle \int \sec^2 x \tan x \, dx.$

20. $\displaystyle \int (1 + \tan^2 x) \sec^2 x \, dx.$

*21. $\displaystyle \int \sqrt{1 + \tan x} \, \sec^2 x \, dx.$

*22. $\displaystyle \int x^2 \sin (a^3 x^3 + b^3) \, dx.$

Calculate these integrals using the half-angle formulas.

*23. $\displaystyle \int \cos^2 \tfrac{1}{2} x \, dx.$

24. $\displaystyle \int \sin^2 \tfrac{1}{2} x \, dx.$

*25. $\int \cos^2 \pi x \, dx$.   26. $\int \sin^2 \pi x \, dx$.

*27. Find the area between $y = \sin 2x$ and $y = \sin x$ for $x \in [0, \pi]$.

28. Find the area between $y = \sin^2 x$ and $y = \cos^2 x$ for $x \in [0, 2\pi]$.

Integrate by parts.

*29. $\int x \cos x \, dx$.   30. $\int x \cos (x + \pi) \, dx$.

*31. $\int e^x \sin x \, dx$.   32. $\int \sec^3 x \, dx$.

33. Derive formula (vi):

$$\int \operatorname{cosec} x \, dx = \log |\operatorname{cosec} x - \cot x| + C.$$

7.3 The Inverse Trigonometric Functions

The Arc Sine

We begin with a number x in the interval $[-1, 1]$. Although there are infinitely many numbers at which the sine function takes on the value x, only one of these numbers lies in the interval $[-\tfrac{1}{2}\pi, \tfrac{1}{2}\pi]$. This unique number is written *arc sin x* and is called the *arc sine of x*. More simply,

(7.3.1) $y = \text{arc sin } x \quad \text{iff} \quad \sin y = x \quad \text{and} \quad y \in [-\tfrac{1}{2}\pi, \tfrac{1}{2}\pi].$

The function

$$y = \text{arc sin } x$$

is called the *inverse sine function*. It is, of course, *not* the inverse of the entire sine function but rather the inverse of

$$y = \sin x, \qquad x \in [-\tfrac{1}{2}\pi, \tfrac{1}{2}\pi].$$

The graphs are given in Figures 7.3.1 and 7.3.2.

$y = \sin x, \quad x \in [-\tfrac{1}{2}\pi, \tfrac{1}{2}\pi]$

FIGURE 7.3.1

$y = \text{arc sin } x$

FIGURE 7.3.2

Since the derivative of the sine function,

$$\frac{d}{dx}(\sin x) = \cos x,$$

does not take on the value 0 in the *open* interval $(-\frac{1}{2}\pi, \frac{1}{2}\pi)$, the arc sine function is differentiable on the *open* interval $(-1, 1)$. We can find its derivative as follows:

$$y = \text{arc sin } x,$$
$$\sin y = x,$$
$$\cos y \frac{dy}{dx} = 1,$$
$$\frac{dy}{dx} = \frac{1}{\cos y} = \frac{1}{\sqrt{1 - x^2}}. \quad \square$$

In short

(7.3.2)
$$\frac{d}{dx}(\text{arc sin } x) = \frac{1}{\sqrt{1 - x^2}}.$$

Problem.   Differentiate

$$f(x) = \text{arc sin }(3x^2).$$

SOLUTION.   Once again the chain rule comes into play:

$$f'(x) = \frac{1}{\sqrt{1 - (3x^2)^2}} \frac{d}{dx}(3x^2) = \frac{6x}{\sqrt{1 - 9x^4}}. \quad \square$$

Problem.   Take $a > 0$ and show that

(7.3.3)
$$\int \frac{dx}{\sqrt{a^2 - x^2}} = \text{arc sin } \frac{x}{a} + C.$$

SOLUTION.   We change variables so that the a^2 in the denominator becomes 1. We set

$$au = x; \quad a\, du = dx.$$

Then

$$\int \frac{dx}{\sqrt{a^2 - x^2}} = \int \frac{a\, du}{\sqrt{a^2 - a^2 u^2}} \underset{\text{since } a > 0}{=} \int \frac{du}{\sqrt{1 - u^2}} = \text{arc sin } u + C = \text{arc sin } \frac{x}{a} + C. \quad \square$$

To find

$$\int \text{arc sin } x\, dx$$

we integrate by parts. We set

$$u = \arc\sin x, \qquad dv = dx,$$
$$du = \frac{dx}{\sqrt{1-x^2}}, \qquad v = x.$$

This gives

$$\int \arc\sin x \, dx = \int u \, dv = uv - \int v \, du = x \arc\sin x - \int \frac{x}{\sqrt{1-x^2}} \, dx.$$

As you can verify,

$$-\int \frac{x}{\sqrt{1-x^2}} \, dx = \sqrt{1-x^2} + C$$

so that

(7.3.4)
$$\int \arc\sin x \, dx = x \arc\sin x + \sqrt{1-x^2} + C.$$

The Arc Tangent

This time we begin with an arbitrary real number x. Although there are infinitely many numbers at which the tangent function takes on the value x, only one of these numbers lies in the interval $(-\tfrac{1}{2}\pi, \tfrac{1}{2}\pi)$. This unique number is written as *arc tan x* and is called the *arc tangent of x*. More simply,

(7.3.5)
$$y = \arc\tan x \qquad \text{iff} \qquad \tan y = x \quad \text{and} \quad y \in (-\tfrac{1}{2}\pi, \tfrac{1}{2}\pi).$$

$y = \tan x, \quad x \in (-\tfrac{1}{2}\pi, \tfrac{1}{2}\pi)$

FIGURE 7.3.3

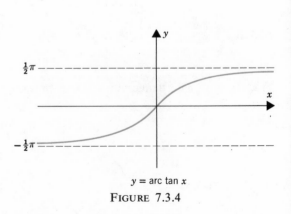

$y = \arc\tan x$

FIGURE 7.3.4

The function

$$y = \text{arc tan } x$$

is called the *inverse tangent function*. It is, of course, *not* the inverse of the entire tangent function but rather the inverse of

$$y = \tan x, \qquad x \in (-\tfrac{1}{2}\pi, \tfrac{1}{2}\pi).$$

The graphs of these two functions are given in Figures 7.3.3 and 7.3.4.
   Since the derivative of the tangent function,

$$\frac{d}{dx}(\tan x) = \sec^2 x = \frac{1}{\cos^2 x},$$

is never 0, the inverse tangent function is everywhere differentiable. Now let's find its derivative:

$$y = \text{arc tan } x,$$
$$\tan y = x,$$
$$\sec^2 y \, \frac{dy}{dx} = 1,$$
$$\frac{dy}{dx} = \frac{1}{\sec^2 y} = \cos^2 y = \frac{1}{1 + x^2}. \quad \square$$

We have found that

(7.3.6)
$$\boxed{\frac{d}{dx}(\text{arc tan } x) = \frac{1}{1 + x^2}.}$$

Problem.   Differentiate

$$f(x) = \text{arc tan } (ax^2 + bx + c).$$

SOLUTION.   Once again it's the chain rule:

$$f'(x) = \frac{1}{1 + (ax^2 + bx + c)^2} \frac{d}{dx}(ax^2 + bx + c) = \frac{2ax + b}{1 + (ax^2 + bx + c)^2}. \quad \square$$

Problem.   Show that

(7.3.7)
$$\boxed{\int \frac{dx}{a^2 + x^2} = \frac{1}{a} \text{ arc tan } \frac{x}{a} + C.}$$

SOLUTION.   We change variables so that the a^2 in the denominator becomes 1. We set

$$au = x; \qquad a \, du = dx.$$
$$\int \frac{dx}{a^2 + x^2} = \int \frac{a \, du}{a^2 + a^2 u^2} = \frac{1}{a} \int \frac{du}{1 + u^2}$$
$$= \frac{1}{a} \text{ arc tan } u + C = \frac{1}{a} \text{ arc tan } \frac{x}{a} + C. \quad \square$$

Problem.   Evaluate

$$\int_0^2 \frac{dx}{4 + x^2}.$$

SOLUTION.   By the last problem

$$\int \frac{dx}{4 + x^2} = \int \frac{dx}{2^2 + x^2} = \frac{1}{2} \text{ arc tan } \frac{x}{2} + C$$

so that

$$\int_0^2 \frac{dx}{4 + x^2} = \left[\frac{1}{2} \text{ arc tan } \frac{x}{2} \right]_0^2 = \frac{1}{2} \text{ arc tan } 1 = \frac{\pi}{8}. \quad \square$$

You can find

$$\int \text{ arc tan } x \, dx$$

by integrating by parts. Set

$$u = \text{arc tan } x, \qquad dv = dx,$$

$$du = \frac{dx}{1 + x^2}, \qquad v = x,$$

and you'll see that

(7.3.8)
$$\int \text{ arc tan } x \, dx = x \text{ arc tan } x - \tfrac{1}{2} \log (1 + x^2) + C.$$

A word of caution. While it is true that

$$\sin (\text{arc sin } x) = x$$

for all x in the domain of the arc sine, it is not true that

$$\text{arc sin } (\sin x) = x$$

for all x in the domain of the sine function. The relation

$$\text{arc sin } (\sin x) = x$$

holds only for $x \in [-\tfrac{1}{2}\pi, \tfrac{1}{2}\pi]$. For example,

$$\text{arc sin } (\sin \tfrac{1}{4}\pi) = \tfrac{1}{4}\pi \qquad \text{but} \qquad \text{arc sin } (\sin 2\pi) = \text{arc sin } 0 = 0.$$

The arc sine of the sine of 2π is that number in the interval $[-\tfrac{1}{2}\pi, \tfrac{1}{2}\pi]$ at which the sine takes on the same value as it does at 2π.

Similar remarks hold for the tangent and the arc tangent; namely, the relation

$$\tan (\text{arc tan } x) = x$$

is valid for all real x, but the relation

$$\text{arc tan } (\tan x) = x$$

holds only for $x \in (-\tfrac{1}{2}\pi, \tfrac{1}{2}\pi)$.

There are four other trigonometric inverses:

the *arc cosine,* which is the inverse of

$$y = \cos x, \qquad x \in [0, \pi];$$

the *arc cotangent,* which is the inverse of

$$y = \cot x, \qquad x \in (0, \pi);$$

the *arc secant,* which is the inverse of

$$y = \sec x, \qquad x \in [0, \tfrac{1}{2}\pi) \cup (\tfrac{1}{2}\pi, \pi];$$

the *arc cosecant,* which is the inverse of

$$y = \operatorname{cosec} x, \qquad x \in [-\tfrac{1}{2}\pi, 0) \cup (0, \tfrac{1}{2}\pi].$$

Of these four functions only the first two are much used. Figure 7.3.5 shows them all in terms of right triangles.†

$y = $ arc cos x

$y = $ arc cot x

$y = $ arc sec x

$y = $ arc cosec x

FIGURE 7.3.5

Exercises

Evaluate.

*1. arc sin 1.   2. arc tan 1.   *3. arc tan 0.

*4. arc cos 0.   5. arc tan (-1).   *6. arc sin $(\tfrac{1}{2}\sqrt{2})$.

*7. arc cos $(\tfrac{1}{2}\sqrt{2})$.   8. arc sin $\tfrac{1}{2}$.   *9. arc cos $\tfrac{1}{2}$.

*10. arc sin $(-\tfrac{1}{2})$.   11. arc sin $(\sin \tfrac{5}{4}\pi)$.   *12. arc tan $[\sin (-3\pi)]$.

---

† A word on notation: some American and British textbooks write $\sin^{-1} x$ for arc sin x, $\cos^{-1} x$ for arc cos x, $\tan^{-1} x$ for arc tan x, etc. We won't use this notation.

Differentiate.

*13. $y = \text{arc tan } (x + 1)$.   14. $y = \text{arc tan } \sqrt{x}$.
*15. $y = \text{arc sin } x^2$.   16. $f(x) = e^x \text{ arc sin } x$.
*17. $f(x) = x \text{ arc sin } 2x$.   18. $f(x) = e^{\text{arc tan } x}$.
*19. $u = (\text{arc sin } x)^2$.   20. $v = \text{arc tan } (e^x)$.

*21. $y = \dfrac{\text{arc tan } x}{x}$.   22. $y = \text{arc tan } \dfrac{2}{x}$.

*23. $f(x) = \sqrt{\text{arc tan } 2x}$.   24. $f(x) = \log (\text{arc tan } x)$.
*25. $y = \text{arc tan } (\log x)$.   26. $y = \text{arc tan } (\sin x)$.

*27. $\theta = \text{arc sin } (\sqrt{1 - r^2})$.   28. $\theta = \text{arc sin } \left(\dfrac{r}{r + 1}\right)$.

*29. $\theta = \text{arc tan } \left(\dfrac{c + r}{1 - cr}\right)$.   30. $\theta = \text{arc tan } \left(\dfrac{1}{1 + r^2}\right)$.

*31. $f(x) = \sqrt{c^2 - x^2} + c \text{ arc sin } \dfrac{x}{c} \cdot$†   32. $f(x) = \frac{1}{3} \text{ arc sin } (3x - 4x^2)$.

*33. $y = \dfrac{x}{\sqrt{c^2 - x^2}} - \text{arc sin } \dfrac{x}{c} \cdot$†   34. $y = x\sqrt{c^2 - x^2} + c^2 \text{ arc sin } \dfrac{x}{c} \cdot$†

Determine with the aid of the appropriate right triangle.

*35. $\tan (\text{arc sin } x)$.   36. $\sin (\text{arc tan } x)$.
*37. $\tan (\text{arc cos } x)$.   38. $\cot (\text{arc cosec } x)$.

39. Complete the proof that

$$\int \text{arc tan } x \, dx = x \text{ arc tan } x - \tfrac{1}{2} \log (1 + x^2) + C.$$

40. Show that for $a \neq 0$

(7.3.9) $$\int \frac{dx}{a^2 + (x + b)^2} = \frac{1}{a} \text{ arc tan } \left(\frac{x + b}{a}\right) + C.$$

Evaluate.

*41. $\displaystyle\int_0^1 \frac{dx}{1 + x^2} \cdot$ 42. $\displaystyle\int_{-1}^1 \frac{dx}{1 + x^2} \cdot$ *43. $\displaystyle\int_0^{1/\sqrt{2}} \frac{dx}{\sqrt{1 - x^2}} \cdot$

*44. $\displaystyle\int_0^1 \frac{dx}{\sqrt{4 - x^2}} \cdot$ 45. $\displaystyle\int_0^5 \frac{dx}{25 + x^2} \cdot$ *46. $\displaystyle\int_{-4}^4 \frac{dx}{16 + x^2} \cdot$

*47. $\displaystyle\int_0^{3/2} \frac{dx}{9 + 4x^2} \cdot$ 48. $\displaystyle\int_2^5 \frac{dx}{9 + (x - 2)^2} \cdot$ *49. $\displaystyle\int_{-3}^{-2} \frac{dx}{\sqrt{4 - (x + 3)^2}} \cdot$

*50. Show that

(7.3.10) $$\frac{d}{dx} (\text{arc sec } x) = \frac{1}{x\sqrt{x^2 - 1}}, \qquad \text{for } x > 1.$$

---

† Take $c > 0$.

Determine the following derivatives.

*51. $\dfrac{d}{dx} (\text{arc } \cos x),$ for $0 < x < 1.$

52. $\dfrac{d}{dx} (\text{arc } \cot x),$ for $x > 0.$

*53. $\dfrac{d}{dx} (\text{arc } \text{cosec } x),$ for $x > 1.$

7.4 Additional Exercises

Find the absolute extreme values with x restricted to $[0, \pi]$.

*1. $y = x + \sin 2x.$ 2. $y = x + \cos 2x.$ *3. $y = x - \sin 2x.$
*4. $y = x - \cos 2x.$ 5. $y = 10e^{-2x} \sin 2x.$ *6. $y = 10e^{-2x} \cos 2x.$

*7. Find the maximum value of $y = a \sin x + b \cos x.$ $(a > 0, b > 0)$
 8. Find the maximum value of the slope of the curve in Exercise 7.
 9. A particle moves along a coordinate line so that at time t it has coordinate
    $x(t)$, velocity $v(t)$, and acceleration $a(t)$. Find

   *(a) $x(1)$ given that $x(0) = 0$ and $v(t) = 1/(t^2 + 1).$
    (b) $a(1)$ given that $v(t) = 1/(t^2 + 1).$
   *(c) $v(0)$ given that $v(1) = 1$ and $a(t) = (\text{arc } \tan t)/(t^2 + 1).$
    (d) $x(1)$ given that $x(0) = 0,$ $v(0) = 1,$ and $a(t) = 1/(t^2 + 1).$
   *(e) $a(0)$ given that $x(t) = \text{arc } \tan (1 + t).$

Find the area below the graph.

*10. $y = \dfrac{x}{x^2 + 1},$ $x \in [0, 1].$ 11. $y = \dfrac{1}{x^2 + 1},$ $x \in [0, 1].$

*12. $y = \dfrac{1}{\sqrt{1 - x^2}},$ $x \in [0, \tfrac{1}{2}].$ 13. $y = \dfrac{x}{\sqrt{1 - x^2}},$ $x \in [0, \tfrac{1}{2}].$

*14. $y = \text{arc } \sin x,$ $x \in [0, \tfrac{1}{2}\sqrt{3}].$ 15. $y = \text{arc } \tan x,$ $x \in [0, 1].$
*16. $y = e^x \sin x,$ $x \in [0, \tfrac{1}{4}\pi].$ 17. $y = e^{2x} \cos 2x,$ $x \in [0, \tfrac{1}{8}\pi].$

 18. A ball is thrown from ground level. Take the origin as the point of release,
     take the x-axis as ground level, measure distances in feet, and take the path
     of the ball as given by the equation $y = x - \tfrac{1}{100}x^2.$ (a) At what angle is the ball
     released? (b) At what angle will it strike a wall 75 feet away? (c) At what
     angle will it fall on a horizontal roof 16 feet high?
 19. A balloon, which leaves the ground 500 feet from an observer, rises vertically
     at the rate of 140 feet per minute. At what rate is the inclination of the ob-
     server's line of sight increasing when the balloon is 500 feet above the
     ground?
*20. The base of an isosceles triangle is 6 feet. If the altitude is 4 feet and increas-
     ing at the rate of 2 inches per minute, at what rate is the vertex angle chang-
     ing?

21. As a boy winds up the cord, his kite is moving horizontally at a height of 60 feet with a speed of 10 feet per minute. How fast is the inclination of the cord changing when its length is 100 feet?

*22. A revolving searchlight which is $\frac{1}{2}$ mile from shore makes 1 revolution per minute. How fast is the light traveling along the straight beach when at a distance of 1 mile from the nearest point of the shore?

*23. A tapestry 7 feet in height is hung on a wall so that its lower edge is 9 feet above an observer's eye. At what distance from the wall should he stand in order to obtain the most favorable view?   HINT: The vertical angle subtended by the tapestry in the eye of the observer must be a maximum.

*24. Find the dimensions of the cylinder of maximum volume which can be inscribed in a sphere of radius 6 inches. (Use the central angle θ subtended by the radius of the base.)

25. Solve Exercise 24 if the curved surface of the cylinder is to be a maximum.

*26. A body of weight W is dragged along a horizontal plane by means of a force P whose line of action makes an angle θ with the plane. The magnitude of the force is given by the equation

$$P = \frac{mW}{m \sin \theta + \cos \theta},$$

where m denotes the coefficient of friction. For what value of θ is the pull a minimum?

*27. If a projectile is fired from O so as to strike an inclined plane which makes a constant angle α with the horizontal at O, the range is given by the formula

$$R = \frac{2v^2 \cos \theta \sin (\theta - \alpha)}{g \cos^2 \alpha},$$

where v and g are constants and θ is the angle of elevation. Calculate the value of θ which gives the maximum range.

Optional | 28. Set

$$F(x) = \begin{cases} \sin (1/x), & x \neq 0 \\ 0, & x = 0 \end{cases}, \qquad G(x) = \begin{cases} x \sin (1/x), & x \neq 0 \\ 0, & x = 0 \end{cases},$$
$$H(x) = \begin{cases} x^2 \sin (1/x), & x \neq 0 \\ 0, & x = 0 \end{cases}.$$

(a) Sketch a figure displaying the general nature of the graph of F.
(b) Sketch a figure displaying the general nature of the graph of G.
(c) Sketch a figure displaying the general nature of the graph of H.
(d) Which of these functions is continuous at 0?
(e) Which of these functions is differentiable at 0?

7.5   Simple Harmonic Motion

Sines and cosines play a central role in the study of all oscillations. Wherever there are waves to analyze (sound waves, radio waves, light waves, etc.) sines and cosines hold the center stage.

The simplest wave motion arises when an object moves back and forth along a coordinate line in such a way that its acceleration remains a constant negative multiple of its displacement:

$$a(t) = -kx(t).$$

Such motion is called *simple harmonic* motion.

Since

$$a(t) = x''(t),$$

we have

$$x''(t) = -kx(t)$$

and thus

$$x''(t) + kx(t) = 0.$$

To emphasize that k is positive, we set $k = \beta^2$. The equation then takes the form

(7.5.1) $\boxed{x''(t) + \beta^2 x(t) = 0.}$

It is easy to show that any function of the form

$$x(t) = C_2 \sin(\beta t + C_1)$$

satisfies the differential equation. Differentiation gives

$$x'(t) = C_2 \beta \cos(\beta t + C_1),$$
$$x''(t) = -C_2 \beta^2 \sin(\beta t + C_1) = -\beta^2 x(t)$$

so that

$$x''(t) + \beta^2 x(t) = 0.$$

What is not easy to see, but equally true, is that every solution of Equation (7.5.1) can be written in the form

(7.5.2) $\boxed{x(t) = C_2 \sin(\beta t + C_1).}$

This we ask you to take on faith. (Or consult a text on differential equations.)

Let's study the function of (7.5.2). Observe first that

$$x(t + 2\pi/\beta) = C_2 \sin[\beta(t + 2\pi/\beta) + C_1] = C_2 \sin(\beta t + C_1) = x(t).$$

This says that the motion is *periodic* and the *period* is $2\pi/\beta$:

$$p = \frac{2\pi}{\beta}.$$ (Figure 7.5.1)

If we measure t in seconds, then every $2\pi/\beta$ seconds the motion repeats itself. The reciprocal $\beta/2\pi$ gives the number of complete cycles per second. This number is called the *frequency*:

$$f = \frac{\beta}{2\pi}.$$

Since $\sin(\beta t + C_1)$ oscillates between -1 and 1, the product

$$x(t) = C_2 \sin(\beta t + C_1)$$

oscillates between $-|C_2|$ and $|C_2|$. The number $|C_2|$ is called the *amplitude* of the motion:

$$a = |C_2|. \qquad\qquad \text{(Figure 7.5.1)}$$

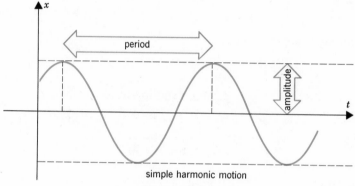

FIGURE 7.5.1

Problem.   Find an equation for the motion if the period is $2\pi/3$ and at time $t = 0$, $x = 1$ and $v = 3$.

SOLUTION.   We begin by setting

$$x(t) = C_2 \sin(\beta t + C_1).$$

Without loss of generality we take $C_2 \geq 0$ and $0 \leq C_1 < 2\pi$.† In general the period is $2\pi/\beta$, so that here

$$\frac{2\pi}{\beta} = \frac{2\pi}{3} \quad \text{and thus} \quad \beta = 3.$$

Consequently we have

$$x(t) = C_2 \sin(3t + C_1),$$

and by differentiation

$$v(t) = 3C_2 \cos(3t + C_1).$$

The conditions at $t = 0$ give

$$1 = x(0) = C_2 \sin C_1, \qquad 3 = v(0) = 3C_2 \cos C_1$$

and therefore

$$1 = C_2 \sin C_1, \qquad 1 = C_2 \cos C_1.$$

---

† In the exercises you are asked to justify this step.

Adding the squares, we have

$$2 = C_2^2 \sin^2 C_1 + C_2^2 \cos^2 C_1 = C_2^2$$

and therefore

$$C_2 = \sqrt{2}.$$

To find C_1 we note that

$$1 = \sqrt{2} \sin C_1, \qquad 1 = \sqrt{2} \cos C_1.$$

These equations are satisfied by setting

$$C_1 = \tfrac{1}{4}\pi.$$

The equation of motion can be written

$$x(t) = \sqrt{2} \sin (3t + \tfrac{1}{4}\pi). \quad \square$$

Problem.   An object in simple harmonic motion passes through the central point $x = 0$ at time 0 and every second thereafter. Find an equation for the motion if $v(0) = -4$.

SOLUTION

$$x(t) = C_2 \sin (\beta t + C_1).$$

On each complete cycle the object must pass through the central point twice —once going one way, once going the other way. The period is thus 2 seconds:

$$\frac{2\pi}{\beta} = 2 \quad \text{and therefore} \quad \beta = \pi.$$

We now know that

$$x(t) = C_2 \sin (\pi t + C_1)$$

and consequently

$$v(t) = \pi C_2 \cos (\pi t + C_1).$$

The initial conditions can be written

$$0 = x(0) = C_2 \sin C_1, \qquad -4 = v(0) = \pi C_2 \cos C_1$$

so that

$$0 = C_2 \sin C_1, \qquad -\frac{4}{\pi} = C_2 \cos C_1.$$

Taking $C_2 \geq 0$ and $0 \leq C_1 < 2\pi$, we have

$$C_2 = \frac{4}{\pi}, \qquad C_1 = \pi. \qquad \text{(check this out)}$$

The equation of motion can be written

$$x(t) = \frac{4}{\pi} \sin (\pi t + \pi). \quad \square$$

Problem.   A coil spring hangs naturally to a length l_0. When a mass m is attached to it, the spring stretches l_1 inches. The mass is later pulled down by hand an additional x_0 inches and released. What is the resulting motion?

SOLUTION.   Throughout we refer to Figure 7.5.2, taking the downward direction as positive.

natural length	equilibrium with mass m	release position	later
I	II	III	IV

FIGURE 7.5.2

We begin by analyzing the forces acting on the mass at general position x (Stage IV). First there is the weight of the object:

$$F_1 = mg.$$

This is a downward force, and by our choice of coordinate system, positive. Then there is the restoring force of the spring. This force, by Hooke's law, is proportional to the total displacement $l_1 + x$ and acts in the opposite direction:

$$F_2 = -k(l_1 + x) \qquad \text{with } k > 0.$$

If we neglect resistance, then these are the only forces acting on the object. Under these conditions the total force is

$$F = F_1 + F_2 = mg - k(l_1 + x),$$

which we rewrite as

(1) $$F = (mg - kl_1) - kx.$$

At stage II (Figure 7.5.2) there was equilibrium. The force of gravity, mg, plus the force of the spring, $-kl_1$, must have been 0:

$$mg - kl_1 = 0.$$

Equation (1) can therefore be simplified to

$$F = -kx.$$

Using Newton's

$$F = ma \qquad \text{(force} = \text{mass} \times \text{acceleration)}$$

we have

$$ma = -kx \quad \text{and thus} \quad a = -\frac{k}{m}x.$$

At any time t,

$$x''(t) = -\frac{k}{m}x(t).$$

Since $k/m > 0$, we can set $\beta = \sqrt{k/m}$ and write

$$x''(t) = -\beta^2 x(t).$$

The motion of the mass is simple harmonic motion. □

Exercises

*1. An object is in simple harmonic motion. Find an equation for the motion if the period is $\frac{1}{4}\pi$ and at time $t = 0$, $x = 1$ and $v = 0$. What is the amplitude? What is the frequency?

 2. An object is in simple harmonic motion. Find an equation for the motion if the frequency is $1/\pi$ and at time $t = 0$, $x = 0$ and $v = -2$. What is the amplitude? What is the period?

*3. An object is in simple harmonic motion with period p and amplitude a. What is its velocity at the central point $x = 0$?

 4. An object is in simple harmonic motion with period p. Find the amplitude if at $x = x_0$, $v = \pm v_0$.

*5. An object in simple harmonic motion passes through the central point $x = 0$ at time $t = 0$ and every 3 seconds thereafter. Find the equation of motion if $v(0) = 5$.

 6. Find the position at which an object in simple harmonic motion

$$x(t) = C_2 \sin (\beta t + C_1)$$

attains (a) maximum speed, (b) zero speed, (c) maximum acceleration, (d) zero acceleration.

*7. Where does the object of Exercise 6 take on half of its maximum speed?

 8. Show that in setting $x(t) = C_2 \sin (\beta t + C_1)$ there is no loss of generality in taking $C_1 \in [0, 2\pi)$ and $C_2 \geq 0$.†

---

† Namely, show that every expression of the form $C_2 \sin (\beta t + C_1)$ can be written in the form $B \sin (\beta t + A)$ with $A \in [0, 2\pi)$ and $B \geq 0$.

9. Show that harmonic motion

$$x(t) = C_2 \sin (\beta t + C_1)$$

can also be written in the form

$$x(t) = C_4 \cos (\beta t + C_3)$$

and also in the form

$$x(t) = C_5 \cos \beta t + C_6 \sin \beta t.$$

*10. What is $x(t)$ for the object of mass m discussed in the spring problem?

11. Given that $a(t) = 8 - 4x(t)$ with $x(0) = 0$ and $v(0) = 0$, show that the motion is simple harmonic motion with the center at $x = 2$. Find the amplitude and the period.

Optional | **7.6   The Hyperbolic Sine and Cosine**

The *hyperbolic sine* and *hyperbolic cosine* are the functions

(7.6.1) $\boxed{\sinh x = \tfrac{1}{2}(e^x - e^{-x}) \quad \text{and} \quad \cosh x = \tfrac{1}{2}(e^x + e^{-x}).}$

The reasons for these names will become apparent as we go on.
   Since

$$\frac{d}{dx} (\sinh x) = \frac{d}{dx} \left[\tfrac{1}{2}(e^x - e^{-x}) \right] = \tfrac{1}{2}(e^x + e^{-x})$$

and

$$\frac{d}{dx} (\cosh x) = \frac{d}{dx} \left[\tfrac{1}{2}(e^x + e^{-x}) \right] = \tfrac{1}{2}(e^x - e^{-x}),$$

we have

(7.6.2) $\boxed{\dfrac{d}{dx} (\sinh x) = \cosh x \quad \text{and} \quad \dfrac{d}{dx} (\cosh x) = \sinh x.}$

In short, each of these functions is the derivative of the other.

The Graphs

We begin with the hyperbolic sine. Since

$$\sinh (-x) = \tfrac{1}{2}(e^{-x} - e^x) = -\tfrac{1}{2}(e^x - e^{-x}) = -\sinh x,$$

the hyperbolic sine is an odd function. Its graph is therefore symmetric about the origin. Since

$$\frac{d}{dx} (\sinh x) = \cosh x = \tfrac{1}{2}(e^x + e^{-x}) > 0 \qquad \text{for all real } x,$$

the hyperbolic sine increases everywhere. From

$$\frac{d^2}{dx^2}(\sinh x) = \frac{d}{dx}(\cosh x) = \sinh x = \tfrac{1}{2}(e^x - e^{-x})$$

you can see that

$$\frac{d^2}{dx^2}(\sinh x) \quad \text{is} \quad \begin{cases} \text{negative,} & \text{for} \quad x < 0 \\ 0, & \text{at} \quad x = 0 \\ \text{positive,} & \text{for} \quad x > 0 \end{cases}.$$

The graph is therefore concave down on $(-\infty, 0]$ and concave up on $[0, \infty)$. The point $(0, \sinh 0) = (0, 0)$ is a point of inflection. The slope at the origin is $\cosh 0 = 1$. A sketch of the graph appears in Figure 7.6.1. $\quad\square$

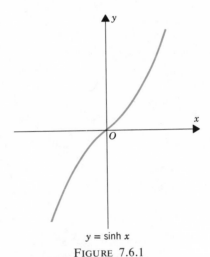

$y = \sinh x$

FIGURE 7.6.1

We turn now to the hyperbolic cosine. Since

$$\cosh(-x) = \tfrac{1}{2}(e^{-x} + e^x) = \tfrac{1}{2}(e^x + e^{-x}) = \cosh x,$$

the hyperbolic cosine is an even function. Its graph is therefore symmetric about the y-axis. Since

$$\frac{d}{dx}(\cosh x) = \sinh x,$$

you can see that

$$\frac{d}{dx}(\cosh x) \quad \text{is} \quad \begin{cases} \text{negative,} & \text{for} \quad x < 0 \\ 0, & \text{at} \quad x = 0 \\ \text{positive,} & \text{for} \quad x > 0 \end{cases}.$$

The function therefore decreases on $(-\infty, 0]$ and increases on $[0, \infty)$. The number

$$\cosh 0 = \tfrac{1}{2}(e^0 + e^{-0}) = \tfrac{1}{2}(1 + 1) = 1$$

Optional | is a minimum. There are no other extreme values. Since

$$\frac{d^2}{dx^2}(\cosh x) = \frac{d}{dx}(\sinh x) = \cosh x > 0 \qquad \text{for all real } x,$$

the graph is everywhere concave up. For a picture of the graph see Figure 7.6.2. □

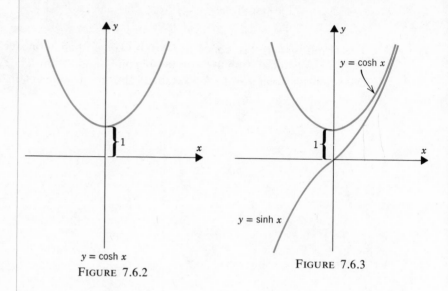

$y = \cosh x$
FIGURE 7.6.2

FIGURE 7.6.3

Figure 7.6.3 shows a comparison between the two graphs. The graph of $y = \cosh x$ remains above the graph of $y = \sinh x$,

$$\cosh x - \sinh x = \tfrac{1}{2}(e^x + e^{-x}) - \tfrac{1}{2}(e^x - e^{-x}) = e^{-x} > 0,$$

but the difference, e^{-x}, tends to 0 as x tends to $+\infty$.

Identities

These functions satisfy identities which are similar to those satisfied by the "circular" sine and cosine.

(7.6.3)

$$\cosh^2 t - \sinh^2 t = 1,$$
$$\sinh(t + s) = \sinh t \cosh s + \cosh t \sinh s,$$
$$\cosh(t + s) = \cosh t \cosh s + \sinh t \sinh s,$$
$$\sinh 2t = 2 \sinh t \cosh t,$$
$$\cosh 2t = \cosh^2 t + \sinh^2 t.$$

The verification of these identities is left to you as an exercise.

Optional | **Relation to the Hyperbola $x^2 - y^2 = 1$**

That
$$\cosh^2 t - \sinh^2 t = 1$$
means that points of the form $P(\cosh t, \sinh t)$ all lie on the hyperbola
$$x^2 - y^2 = 1.$$

This in itself might justify calling these functions hyperbolic, but, as you will see in a moment, there is actually a much deeper reason.

Let's return briefly to the unit circle $x^2 + y^2 = 1$ and the ordinary sine and cosine. If t lies between 0 and 2π, then $\tfrac{1}{2}t$ gives the area of the sector shaded in Figure 7.6.4.

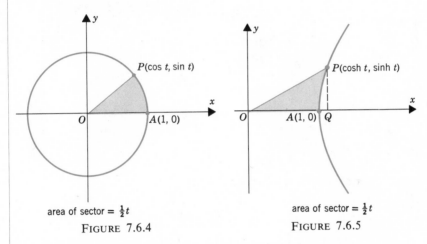

area of sector = $\tfrac{1}{2}t$ area of sector = $\tfrac{1}{2}t$
FIGURE 7.6.4   FIGURE 7.6.5

We have an entirely analogous situation in the case of the hyperbola $x^2 - y^2 = 1$ and the hyperbolic sine and cosine.

(7.6.4) | If $t > 0$, then $\tfrac{1}{2}t$ gives the area of the sector shaded in Figure 7.6.5.

PROOF. Let's call the area of the shaded sector $A(t)$. It's not hard to see that
$$A(t) = \tfrac{1}{2}\cosh t \sinh t - \int_1^{\cosh t} \sqrt{x^2 - 1}\, dx.$$

The first term, $\tfrac{1}{2}\cosh t \sinh t$, gives the area of the triangle OPQ, and the integral
$$\int_1^{\cosh t} \sqrt{x^2 - 1}\, dx$$
gives the area of the unshaded portion of the triangle. We wish to show that
$$A(t) = \tfrac{1}{2}t \qquad \text{for all } t \geq 0.$$

Optional | Differentiation of $A(t)$ gives

$$A'(t) = \frac{1}{2}\left[\cosh t \frac{d}{dt}(\sinh t) + \sinh t \frac{d}{dt}(\cosh t)\right]$$

$$-\frac{d}{dt}\left(\int_1^{\cosh t} \sqrt{x^2 - 1}\, dx\right)$$

and therefore

(1) $\qquad A'(t) = \frac{1}{2}(\cosh^2 t + \sinh^2 t) - \frac{d}{dt}\left(\int_1^{\cosh t} \sqrt{x^2 - 1}\, dx\right).$

In Section 5.9 we showed that, in general,

$$\frac{d}{dt}\left(\int_a^{g(t)} f(x)\, dx\right) = f(g(t))g'(t).$$

In this case, then,

$$\frac{d}{dt}\left(\int_1^{\cosh t} \sqrt{x^2 - 1}\, dx\right) = \sqrt{\cosh^2 t - 1} \frac{d}{dt}(\cosh t)$$

$$= \sinh t \cdot \sinh t = \sinh^2 t.$$

Substitution of this last expression into Equation (1) gives

$$A'(t) = \frac{1}{2}(\cosh^2 t + \sinh^2 t) - \sinh^2 t = \frac{1}{2}(\cosh^2 t - \sinh^2 t) = \frac{1}{2}.$$

It's not hard to see that $A(0) = 0$:

$$A(0) = \frac{1}{2}\cosh 0 \sinh 0 - \int_1^{\cosh 0} \sqrt{x^2 - 1}\, dx = 0.$$

The two conditions

$$A'(t) = \frac{1}{2} \quad \text{and} \quad A(0) = 0$$

together imply that

$$A(t) = \frac{1}{2}t. \quad \square$$

Exercises

Differentiate.

*1. $y = \sinh x^2$.

2. $y = \cosh(x + a)$.

*3. $y = \sqrt{\cosh ax}$.

4. $y = (\sinh ax)(\cosh ax)$.

*5. $y = \dfrac{\sinh x}{\cosh x - 1}$.

6. $y = \dfrac{\sinh x}{x}$.

*7. $y = a \sinh bx - b \cosh ax$.

8. $y = e^x(\cosh x + \sinh x)$.

*9. $y = \log|\sinh ax|$.

10. $y = \log|1 - \cosh ax|$.

Optional | Verify the following identities.

11. $\cosh^2 t - \sinh^2 t = 1$.
12. $\sinh(t + s) = \sinh t \cosh s + \cosh t \sinh s$.
13. $\cosh(t + s) = \cosh t \cosh s + \sinh t \sinh s$.
14. $\sinh 2t = 2 \sinh t \cosh t$.
15. $\cosh 2t = \cosh^2 t + \sinh^2 t$.

Find the absolute extreme values.

*16. $y = 5 \cosh x + 4 \sinh x$.   17. $y = -5 \cosh x + 4 \sinh x$.
*18. $y = 4 \cosh x + 5 \sinh x$.

19. Show that for each positive integer n

$$(\cosh x + \sinh x)^n = \cosh nx + \sinh nx.$$

20. Verify that $y = A \cosh cx + B \sinh cx$ satisfies the differential equation

$$y'' - c^2 y = 0.$$

*21. Find the solution of the differential equation

$$y'' - 9y = 0$$

which satisfies $y(0) = 2$ and $y'(0) = 1$.

22. Find the solution of the differential equation

$$4y'' - y = 0$$

which satisfies $y(0) = 1$ and $y'(0) = 2$.

7.7 Other Hyperbolic Functions

The hyperbolic tangent is defined by setting

$$\tanh x = \frac{\sinh x}{\cosh x} = \frac{e^x - e^{-x}}{e^x + e^{-x}}.$$

There is also a *hyperbolic cotangent*

$$\coth x = \frac{\cosh x}{\sinh x},$$

a *hyperbolic secant*

$$\operatorname{sech} x = \frac{1}{\cosh x},$$

and a *hyperbolic cosecant*

$$\operatorname{cosech} x = \frac{1}{\sinh x}.$$

Optional | The derivatives are as follows:

(7.7.1)

$$\frac{d}{dx}(\tanh x) = \operatorname{sech}^2 x,$$

$$\frac{d}{dx}(\coth x) = -\operatorname{cosech}^2 x,$$

$$\frac{d}{dx}(\operatorname{sech} x) = -\operatorname{sech} x \tanh x,$$

$$\frac{d}{dx}(\operatorname{cosech} x) = -\operatorname{cosech} x \coth x.$$

These formulas are easy to verify. For instance,

$$\frac{d}{dx}(\tanh x) = \frac{d}{dx}\left(\frac{\sinh x}{\cosh x}\right) = \frac{\cosh x \dfrac{d}{dx}(\sinh x) - \sinh x \dfrac{d}{dx}(\cosh x)}{\cosh^2 x}$$

$$= \frac{\cosh^2 x - \sinh^2 x}{\cosh^2 x} = \frac{1}{\cosh^2 x} = \operatorname{sech}^2 x.$$

We leave it to you to verify the other formulas. ☐

Let's analyze the hyperbolic tangent a little further. Since

$$\tanh(-x) = \frac{\sinh(-x)}{\cosh(-x)} = \frac{-\sinh x}{\cosh x} = -\tanh x,$$

the hyperbolic sine is an odd function and thus its graph is symmetric about the origin. Since

$$\frac{d}{dx}(\tanh x) = \operatorname{sech}^2 x > 0 \qquad \text{for all real } x,$$

the function is everywhere increasing. From the relation

$$\tanh x = \frac{e^x - e^{-x}}{e^x + e^{-x}} = 1 - \frac{2}{e^{2x} + 1}$$

you can see that $\tanh x$ always remains between -1 and 1. Moreover, it tends to 1 as x tends to $+\infty$ and it tends to -1 as x tends to $-\infty$. To check on the concavity of the graph, we take the second derivative:

$$\frac{d^2}{dx^2}(\tanh x) = \frac{d}{dx}(\operatorname{sech}^2 x) = 2\operatorname{sech} x \frac{d}{dx}(\operatorname{sech} x)$$

$$= 2\operatorname{sech} x \,(-\operatorname{sech} x \tanh x)$$

$$= -2\operatorname{sech}^2 x \tanh x.$$

Since

$$\tanh x = \frac{e^x - e^{-x}}{e^x + e^{-x}} \quad \text{is} \quad \begin{cases} \text{negative}, & \text{for} \quad x < 0 \\ \qquad\quad 0, & \text{at} \quad x = 0 \\ \text{positive}, & \text{for} \quad x > 0 \end{cases},$$

Optional | you can see that

$$\frac{d^2}{dx^2}(\tanh x) \quad \text{is} \quad \begin{cases} \text{positive,} & \text{for} \quad x < 0 \\ 0, & \text{at} \quad x = 0 \\ \text{negative,} & \text{for} \quad x > 0 \end{cases}.$$

The graph is therefore concave up on $(-\infty, 0]$ and concave down on $[0, \infty)$. The point $(0, \tanh 0) = (0, 0)$ is a point of inflection. At the origin the slope is

$$\text{sech}^2\, 0 = \frac{1}{\cosh^2 0} = 1.$$

For a picture of the graph see Figure 7.7.1.

$$y = \tanh x$$

FIGURE 7.7.1

The Hyperbolic Inverses

The hyperbolic inverses that are important to us are the *inverse hyperbolic sine*, the *inverse hyperbolic cosine*, and the *inverse hyperbolic tangent*. These functions

$$y = \sinh^{-1} x, \quad y = \cosh^{-1} x, \quad y = \tanh^{-1} x$$

are the inverses of

$$y = \sinh x, \quad y = \cosh x \, (x \geq 0), \quad y = \tanh x$$

respectively.

Theorem 7.7.2

(i) $\sinh^{-1} x = \log(x + \sqrt{x^2 + 1}).$ (x real)

(ii) $\cosh^{-1} x = \log(x + \sqrt{x^2 - 1}).$ ($x \geq 1$)

(iii) $\tanh^{-1} x = \frac{1}{2} \log\left(\frac{1 + x}{1 - x}\right).$ ($-1 < x < 1$)

PROOF.   To show (i), we set

$$y = \sinh^{-1} x$$

and note that

$$\sinh y = x.$$

Optional | This gives in sequence

$$\tfrac{1}{2}(e^y - e^{-y}) = x, \qquad e^y - e^{-y} = 2x, \qquad e^{2y} - 2xe^y - 1 = 0.$$

This last equation is a quadratic equation in e^y. From the general quadratic formula we get

$$e^y = \tfrac{1}{2}(2x \pm \sqrt{4x^2 + 4}) = x \pm \sqrt{x^2 + 1}.$$

Since $e^y > 0$, the minus sign on the right is impossible. Consequently we have

$$e^y = x + \sqrt{x^2 + 1}$$

and, taking logs,

$$y = \log(x + \sqrt{x^2 + 1}).$$

To show (ii), we set

$$y = \cosh^{-1} x, \qquad x \geq 1$$

and note that

$$\cosh y = x \quad \text{and} \quad y \geq 0.$$

This gives in sequence

$$\tfrac{1}{2}(e^y + e^{-y}) = x, \qquad e^y + e^{-y} = 2x, \qquad e^{2y} - 2xe^y + 1 = 0.$$

Again we have a quadratic in e^y. Here the general quadratic formula gives

$$e^y = \tfrac{1}{2}(2x \pm \sqrt{4x^2 - 4}) = x \pm \sqrt{x^2 - 1}.$$

Since y is nonnegative,

$$e^y = x \pm \sqrt{x^2 - 1}$$

cannot be less than 1. This renders the negative sign impossible (check this out) and leaves

$$e^y = x + \sqrt{x^2 - 1}$$

as the only possibility. Taking logs, we get

$$y = \log(x + \sqrt{x^2 - 1}).$$

The proof of (iii) is left as an exercise. $\square$

Exercises

Differentiate.

*1. $y = \tanh^2 x$.   2. $y = \tanh^2 3x$.

*3. $y = \log(\tanh x)$.   4. $y = \tanh(\log x)$.

*5. $y = \sinh(\arctan e^{2x})$.   6. $y = \operatorname{sech}(3x^2 + 1)$.

Optional

*7. $y = \coth(\sqrt{x^2 + 1})$.   8. $y = \log(\operatorname{sech} x)$.

*9. $y = \dfrac{\operatorname{sech} x}{1 + \cosh x}$.   10. $y = \dfrac{\cosh x}{1 + \operatorname{sech} x}$.

Verify the following differentiation formulas.

11. $\dfrac{d}{dx}(\coth x) = -\operatorname{cosech}^2 x$.   12. $\dfrac{d}{dx}(\operatorname{sech} x) = -\operatorname{sech} x \tanh x$.

13. $\dfrac{d}{dx}(\operatorname{cosech} x) = -\operatorname{cosech} x \coth x$.

14. Show that

$$\tanh(t + s) = \frac{\tanh t + \tanh s}{1 + \tanh t \tanh s}.$$

*15. Given that $\tanh x_0 = \frac{4}{5}$, find

(a) $\operatorname{sech} x_0$.   HINT: $1 - \tanh^2 x = \operatorname{sech}^2 x$.
Then find
(b) $\cosh x_0$.   (c) $\sinh x_0$.   (d) $\coth x_0$.   (e) $\operatorname{cosech} x_0$.

16. Given that $\tanh t_0 = -\frac{5}{12}$, evaluate the remaining hyperbolic functions at t_0.

17. Show that

$$\text{if} \quad x^2 \geq 1 \quad \text{then} \quad x - \sqrt{x^2 - 1} \leq 1.$$

18. Show that

$$\tanh^{-1} x = \frac{1}{2} \log\left(\frac{1 + x}{1 - x}\right).$$

19. Show that

(7.7.3) $\boxed{\dfrac{d}{dx}(\sinh^{-1} x) = \dfrac{1}{\sqrt{x^2 + 1}}.}$

20. Show that

(7.7.4) $\boxed{\dfrac{d}{dx}(\cosh^{-1} x) = \dfrac{1}{\sqrt{x^2 - 1}}.}$

21. Show that

(7.7.5) $\boxed{\dfrac{d}{dx}(\tanh^{-1} x) = \dfrac{1}{1 - x^2}.}$

Optional

22. Sketch the graphs of
    (a) $y = \sinh^{-1} x$.   (b) $y = \cosh^{-1} x$.   (c) $y = \tanh^{-1} x$.

23. Given that

$$\tan \phi = \sinh x,$$

    show that

    (a) $\dfrac{d\phi}{dx} = \operatorname{sech} x$.   (b) $x = \log(\sec \phi + \tan \phi)$.

    (c) $\dfrac{dx}{d\phi} = \sec \phi$.

The Technique of
Integration

8.1 A Short Table of Integrals; Review

For review and ready reference we begin with a short table of integrals. A more
extended table appears at the end of the book.

1. $\int \alpha \, dx = \alpha x + C.$

2. $\int x^n \, dx = \dfrac{1}{n+1} x^{n+1} + C, \qquad n \neq -1.$

3. $\int \dfrac{dx}{x} = \log |x| + C.$ $\;= \ln|x|$

4. $\int e^x \, dx = e^x + C.$

5. $\int p^x \, dx = \dfrac{p^x}{\log p} + C.$ $\;= \dfrac{e^x}{\ln p}$ $\qquad \int a^v = \dfrac{a^v}{\ln a}$

6. $\int \underset{\ln x}{\log x} \, dx = x \underset{\ln x}{\log x} - x + C.$

7. $\int \cos x \, dx = \sin x + C.$

8. $\int \sin x \, dx = -\cos x + C.$

9. $\int \sec^2 x \, dx = \tan x + C.$

10. $\int \operatorname{cosec}^2 x \, dx = -\cot x + C.$

11. $\int \sec x \tan x \, dx = \sec x + C.$

12. $\int \csc x \cot x \, dx = -\csc x + C.$

13. $\int \tan x \, dx = \log |\sec x| + C.$

14. $\int \cot x \, dx = \log |\sin x| + C.$

15. $\int \sec x \, dx = \log |\sec x + \tan x| + C.$

16. $\int \csc x \, dx = \log |\csc x - \cot x| + C.$

17. $\int \dfrac{dx}{\sqrt{1 - x^2}} = \arcsin x + C.$

18. $\int \dfrac{dx}{1 + x^2} = \arctan x + C.$

19. $\int \arcsin x \, dx = x \arcsin x + \sqrt{1 - x^2} + C.$

20. $\int \arctan x \, dx = x \arctan x - \tfrac{1}{2} \log (1 + x^2) + C.$

21. $\int \dfrac{dx}{x^2 - 1} = \dfrac{1}{2} \log \left| \dfrac{x - 1}{x + 1} \right| + C.$ $= \dfrac{1}{2a} \ln \left| \dfrac{x-a}{x+a} \right| + c$

$x^2 - a^2$

Only Formula 21 is new. To derive it observe that

$$\frac{1}{x^2 - 1} = \frac{1}{2} \left(\frac{1}{x - 1} - \frac{1}{x + 1} \right).$$

It follows then that

$$\int \frac{dx}{x^2 - 1} = \frac{1}{2} \int \frac{dx}{x - 1} - \frac{1}{2} \int \frac{dx}{x + 1}$$

$$= \frac{1}{2} \log |x - 1| - \frac{1}{2} \log |x + 1| + C = \frac{1}{2} \log \left| \frac{x - 1}{x + 1} \right| + C. \quad \square$$

In the spirit of review we work out a few integrals.

Problem. Find

$$\int x \tan x^2 \, dx.$$

SOLUTION.   Set

$$u = x^2; \quad du = 2x \, dx, \quad \tfrac{1}{2} du = x \, dx$$

$$\int x \tan x^2 \, dx = \tfrac{1}{2} \int \tan u \, du = \tfrac{1}{2} \log |\sec u| + C = \tfrac{1}{2} \log |\sec x^2| + C. \quad \square$$

Problem. Compute

$$\int_0^1 \frac{e^x}{e^x + 2}\, dx.$$

SOLUTION. First we calculate the indefinite integral

$$\int \frac{e^x}{e^x + 2}\, dx.$$

Set

$$u = e^x + 2; \qquad du = e^x\, dx.$$

$$\int \frac{e^x}{e^x + 2}\, dx = \int \frac{du}{u} = \log|u| + C = \log|e^x + 2| + C.$$

Consequently

$$\int_0^1 \frac{e^x}{e^x + 2}\, dx = \left[\log|e^x + 2|\right]_0^1 = \log(e + 2) - \log 3 = \log[\tfrac{1}{3}(e + 2)]. \quad \square$$

The next two examples should reinforce the idea of integration by parts.

Problem. Find

$$\int \log(x^2 + 1)\, dx.$$

SOLUTION. Set

$$u = \log(x^2 + 1), \qquad dv = dx,$$

$$du = \frac{2x}{x^2 + 1}\, dx, \qquad v = x.$$

$$\int \log(x^2 + 1)\, dx = \int u\, dv = uv - \int v\, du = x\log(x^2 + 1) - \int \frac{2x^2}{x^2 + 1}\, dx.$$

Since

$$\int \frac{2x^2}{x^2 + 1}\, dx = \int \frac{2(x^2 + 1)}{x^2 + 1}\, dx - \int \frac{2}{x^2 + 1}\, dx = 2x - 2\arctan x + C,$$

we have

$$\int \log(x^2 + 1)\, dx = x\log(x^2 + 1) - 2x + 2\arctan x + C. \quad \square$$

Problem. Find

$$\int x \arctan x\, dx.$$

SOLUTION.　We first set

$$u = \text{arc tan } x, \qquad dv = x\, dx,$$

$$du = \frac{1}{x^2 + 1}\, dx, \qquad v = \tfrac{1}{2}x^2. + C$$

We could go on this way, but since later we have to integrate $v\, du$, we may as well choose v so as to make $v\, du$ as simple as possible. By adding $\tfrac{1}{2}$ to our initial choice of v, we have

$$v = \tfrac{1}{2}x^2 + \tfrac{1}{2} = \tfrac{1}{2}(x^2 + 1)$$

and

$$v\, du = \tfrac{1}{2}.$$

Now the integration is simple:

$$\int x \text{ arc tan } x\, dx = \int u\, dv = uv - \int v\, du$$

$$= \tfrac{1}{2}(x^2 + 1)\text{ arc tan } x - \int \tfrac{1}{2}\, dx$$

$$= \tfrac{1}{2}(x^2 + 1)\text{ arc tan } x - \tfrac{1}{2}x + C. \quad \square$$

The final problem involves a little algebra.

Problem.　Find

$$\int \frac{dx}{x^2 + 2x + 5}.$$

SOLUTION.　First complete the square in the denominator:

$$\int \frac{dx}{x^2 + 2x + 5} = \int \frac{dx}{(x^2 + 2x + 1) + 4} = \int \frac{dx}{(x + 1)^2 + 2^2} = *.$$

Now set

$$2u = x + 1; \qquad 2du = dx.$$

Then

$$* = \int \frac{2du}{4u^2 + 4} = \tfrac{1}{2}\int \frac{du}{u^2 + 1} = \tfrac{1}{2}\text{ arc tan } u + C$$

$$= \tfrac{1}{2}\text{ arc tan } [\tfrac{1}{2}(x + 1)] + C. \quad \square$$

Exercises

Work out the following integrals.

*1. $\displaystyle\int e^{2-x}\, dx.$　　　　2. $\displaystyle\int \frac{x}{\sqrt{1 - x^2}}\, dx.$　　　*3. $\displaystyle\int \cos \tfrac{2}{3}x\, dx.$

*4. $\displaystyle\int_0^c \frac{dx}{x^2 + c^2}.$　　　5. $\displaystyle\int_{-\pi/4}^{\pi/4} \frac{\sin x}{\cos^2 x}\, dx.$　　*6. $\displaystyle\int_1^2 \frac{e^{1/x}}{x^2}\, dx.$

*7. $\int \dfrac{dx}{5^x}$.

8. $\int \log 2x \, dx$.

*9. $\int x \log (x + 1) \, dx$.

*10. $\int_{\pi/6}^{\pi/3} \operatorname{cosec} x \, dx$.

11. $\int_0^1 \sin \pi x \, dx$.

*12. $\int_{-\pi/4}^{\pi/4} \dfrac{dx}{\cos^2 x}$.

*13. $\int \sec^2 (1 - x) \, dx$.

14. $\int \dfrac{e^{\sqrt{x}}}{\sqrt{x}} \, dx$.

*15. $\int x \, 5^x \, dx$.

*16. $\int_0^1 \dfrac{x^3}{1 + x^4} \, dx$.

17. $\int_0^t \sec \pi x \tan \pi x \, dx$.

*18. $\int_{\pi/6}^{\pi/3} \cot x \, dx$.

*19. $\int e^{-x} \sin 2x \, dx$.

20. $\int \dfrac{x^3}{\sqrt{1 - x^4}} \, dx$.

*21. $\int \dfrac{x}{\sqrt{1 - x^4}} \, dx$.

*22. $\int x \arcsin 2x^2 \, dx$.

23. $\int \dfrac{x}{x^2 + 1} \, dx$.

*24. $\int \dfrac{x}{(x + \alpha)^2 + a^2} \, dx$.

*25. $\int \dfrac{\sec^2 \theta}{\sqrt{3 \tan \theta + 1}} \, d\theta$.

26. $\int \dfrac{\sin \phi}{3 - 2 \cos \phi} \, d\phi$.

*27. $\int \dfrac{e^x}{ae^x - b} \, dx$.

*28. $\int a^x e^x \, dx$.

*29. $\int \dfrac{1 + \cos 2x}{\sin^2 2x} \, dx$.

30. $\int \dfrac{dx}{x^2 + 6x + 10}$.

*31. $\int \dfrac{dx}{x^2 - 4x + 13}$.

*32. $\int_0^{\pi/4} \dfrac{\sec^2 x \tan x}{\sqrt{2 + \sec^2 x}} \, dx$.

Optional | Work out the following integrals.

*33. $\int \sinh^2 x \, dx$.

34. $\int x \sinh x \, dx$.

*35. $\int \cos x \sinh x \, dx$.

*36. $\int \tanh^3 dx$.

37. $\int \sinh^3 x \, dx$.

*38. $\int e^{ax} \cosh ax \, dx$.

8.2 Partial Fractions

A rational function is by definition the quotient of two polynomials. Thus, for example,

$$\frac{1}{x^2 - 4}, \qquad \frac{2x^2 + 3}{x(x - 1)^2}, \qquad \frac{-2x}{(x + 1)(x^2 + 1)}, \qquad \frac{1}{x(x^2 + x + 1)},$$

$$\frac{3x^4 + x^3 + 20x^2 + 3x + 31}{(x + 1)(x^2 + 4)^2}, \qquad \frac{x^5}{x^2 - 1}$$

are all rational functions, but

$$\frac{1}{\sqrt{x}}, \qquad \log x, \qquad \frac{|x - 2|}{x^2}$$

are not rational functions.

To integrate a rational function it's usually necessary to first rewrite it as a polynomial (which may be identically 0) plus fractions of the form

(8.2.1)
$$\boxed{\frac{A}{(x - \alpha)^k} \quad \text{and} \quad \frac{Bx + C}{(x^2 + \beta x + \gamma)^k}}$$

with the quadratic $x^2 + \beta x + \gamma$ irreducible (i.e., not factorable into linear terms with real coefficients)†. Such fractions are called *partial fractions*.

It is shown in algebra that every rational function can be written in such a way. Here are some examples.

Example 1. (*The denominator splits into distinct linear factors.*) For

$$\frac{1}{x^2 - 4} = \frac{1}{(x - 2)(x + 2)},$$

we write

$$\frac{1}{x^2 - 4} = \frac{A}{x - 2} + \frac{B}{x + 2}.$$

Clearing fractions, we have

$$1 = A(x + 2) + B(x - 2),$$
$$1 = (A + B)x + 2A - 2B.$$

Since this last equation is an identity, we can equate coefficients:

$$A + B = 0,$$
$$2A - 2B = 1.$$

Solving the equations simultaneously, we obtain

$$A = \tfrac{1}{4}, \quad B = -\tfrac{1}{4}.$$

The desired decomposition is

$$\frac{1}{x^2 - 4} = \frac{1}{4(x - 2)} - \frac{1}{4(x + 2)}.$$

You can check this by carrying out the subtraction on the right. □

> In general, each distinct linear factor $x - c$ in the denominator gives rise to a term of the form
>
> $$\frac{A}{x - c}$$
>
> in the decomposition.

Example 2. (*The denominator has a repeated linear factor.*) For

$$\frac{2x^2 + 3}{x(x - 1)^2},$$

---

† $\beta^2 - 4\gamma < 0.$

we write

$$\frac{2x^2 + 3}{x(x - 1)^2} = \frac{A}{x} + \frac{B}{x - 1} + \frac{C}{(x - 1)^2}.$$

This leads to

$$2x^2 + 3 = A(x - 1)^2 + Bx(x - 1) + Cx,$$
$$2x^2 + 3 = (A + B)x^2 - (2A + B - C)x + A,$$

and thus to the equations

$$A + B = 2,$$
$$2A + B - C = 0,$$
$$A = 3.$$

These equations give

$$A = 3, \quad B = -1, \quad C = 5.$$

The decomposition is therefore

$$\frac{2x^2 + 3}{x(x - 1)^2} = \frac{3}{x} - \frac{1}{x - 1} + \frac{5}{(x - 1)^2}. \quad \square$$

In general, each factor of the form $(x - c)^k$ in the denominator gives rise to an expression of the form

$$\frac{A_1}{x - c} + \frac{A_2}{(x - c)^2} + \cdots + \frac{A_k}{(x - c)^k}$$

in the decomposition.

Example 3. (*The denominator has an irreducible quadratic factor.*) For

$$\frac{-2x}{(x + 1)(x^2 + 1)},$$

we write

$$\frac{-2x}{(x + 1)(x^2 + 1)} = \frac{A}{x + 1} + \frac{Bx + C}{x^2 + 1}$$

and obtain

$$-2x = A(x^2 + 1) + (Bx + C)(x + 1),$$
$$-2x = (A + B)x^2 + (B + C)x + A + C.$$

This gives

$$A + B = 0,$$
$$B + C = -2,$$
$$A + C = 0.$$

Here the solutions are

$$A = 1, \quad B = -1, \quad C = -1$$

and we have the decomposition

$$\frac{-2x}{(x + 1)(x^2 + 1)} = \frac{1}{x + 1} - \frac{x + 1}{x^2 + 1}. \quad \square$$

Example 4. *(The denominator has an irreducible quadratic factor.)* For

$$\frac{1}{x(x^2 + x + 1)}$$

we write

$$\frac{1}{x(x^2 + x + 1)} = \frac{A}{x} + \frac{Bx + C}{x^2 + x + 1}$$

and obtain

$$1 = A(x^2 + x + 1) + (Bx + C)x,$$
$$1 = (A + B)x^2 + (A + C)x + A.$$

Here

$$A + B = 0,$$
$$A + C = 0,$$
$$A = 1.$$

We conclude that

$$A = 1, \quad B = -1, \quad C = -1,$$

and therefore

$$\frac{1}{x(x^2 + x + 1)} = \frac{1}{x} - \frac{x + 1}{x^2 + x + 1}. \quad \square$$

In general, each irreducible quadratic factor $x^2 + \beta x + \gamma$ in the denominator gives rise to a term of the form

$$\frac{Ax + B}{x^2 + \beta x + \gamma}$$

in the decomposition.

Example 5. *(The denominator has a repeated irreducible quadratic factor.)* For

$$\frac{3x^4 + x^3 + 20x^2 + 3x + 31}{(x + 1)(x^2 + 4)^2}$$

we write

$$\frac{3x^4 + x^3 + 20x^2 + 3x + 31}{(x + 1)(x^2 + 4)^2} = \frac{A}{x + 1} + \frac{Bx + C}{x^2 + 4} + \frac{Dx + E}{(x^2 + 4)^2}.$$

This gives

$3x^4 + x^3 + 20x^2 + 3x + 31$
$$= A(x^2 + 4)^2 + (Bx + C)(x + 1)(x^2 + 4) + (Dx + E)(x + 1).$$

We can rewrite the right-hand side as

$(A + B)x^4 + (B + C)x^3 + (8A + 4B + C + D)x^2$
$$+ (4B + 4C + D + E)x + (16A + 4C + E).$$

This leads to equations

$$A + B = 3,$$
$$B + C = 1,$$
$$8A + 4B + C + D = 20,$$
$$4B + 4C + D + E = 3,$$
$$16A + 4C + E = 31.$$

With a little patience you can see that

$$A = 2, \quad B = 1, \quad C = 0, \quad D = 0, \quad \text{and} \quad E = -1.$$

This gives the decomposition

$$\frac{3x^4 + x^3 + 20x^2 + 3x + 31}{(x + 1)(x^2 + 4)^2} = \frac{2}{x + 1} + \frac{x}{x^2 + 4} - \frac{1}{(x^2 + 4)^2}. \quad \square$$

In general, each multiple irreducible quadratic factor $(x^2 + \beta x + \gamma)^k$ in the denominator gives rise to an expression of the form

$$\frac{A_1 x + B_1}{x^2 + \beta x + \gamma} + \frac{A_2 x + B_2}{(x^2 + \beta x + \gamma)^2} + \cdots + \frac{A_k x + B_k}{(x^2 + \beta x + \gamma)^k}$$

in the decomposition.

A polynomial appears in the decomposition when the degree of the numerator is greater than or equal to that of the denominator. .

Example 6. (*The decomposition contains a polynomial.*) For

$$\frac{x^5 + 2}{x^2 - 1},$$

first carry out the suggested division:

$$
\begin{array}{r}
x^3 + x \\
x^2 - 1 \overline{) x^5 + 2} \\
\underline{x^5 - x^3 } \\
x^3 \\
\underline{x^3 - x } \\
x + 2.
\end{array}
$$

This gives

$$\frac{x^5 + 2}{x^2 - 1} = x^3 + x + \frac{x + 2}{x^2 - 1}.$$

Since the denominator of the fraction splits into linear factors, we write

$$\frac{x + 2}{x^2 - 1} = \frac{A}{x + 1} + \frac{B}{x - 1}.$$

This gives

$$x + 2 = A(x - 1) + B(x + 1),$$
$$x + 2 = (A + B)x - (A - B).$$

Here

$$A + B = 1,$$
$$A - B = -2,$$

so that

$$A = -\tfrac{1}{2}, \quad B = \tfrac{3}{2}.$$

The decomposition takes the form

$$\frac{x^5 + 2}{x^2 - 1} = x^3 + x - \frac{1}{2(x + 1)} + \frac{3}{2(x - 1)}. \quad \square$$

REMARK. The numerators for a partial fraction decomposition can sometimes be obtained more quickly by substituting the zeros of the denominator after clearing fractions. In Example 1 we set

$$\frac{1}{x^2 - 4} = \frac{A}{x - 2} + \frac{B}{x + 2}$$

and, clearing fractions, obtained

$$1 = A(x + 2) + B(x - 2).$$

Substitution of $x = 2$ gives $1 = 4A$, $A = \tfrac{1}{4}$; substitution of $x = -2$ gives $1 = -4B$, $B = -\tfrac{1}{4}$.

In Example 3 we began with

$$\frac{-2x}{(x + 1)(x^2 + 1)} = \frac{A}{x + 1} + \frac{Bx + C}{x^2 + 1}$$

and cleared fractions to obtain

$$-2x = A(x^2 + 1) + (Bx + C)(x + 1).$$

The original denominator is zero only at $x = -1$. Substitution of $x = -1$ gives $2 = 2A$, $A = 1$. The remaining coefficients can be found by the usual method: with $A = 1$

$$-2x = 1(x^2 + 1) + (Bx + C)(x + 1)$$

$$-x^2 - 2x - 1 = Bx^2 + (B + C)x + C.$$

Equating coefficients we have $B = -1$ and $C = -1$. $\square$

    We have been decomposing rational functions into partial fractions so as to be able to integrate them. Here we carry out the integrations leaving some of the details to you.

Example 1'

$$\int \frac{dx}{x^2 - 4} = \frac{1}{4} \int \left(\frac{1}{x - 2} - \frac{1}{x + 2} \right) dx$$

$$= \frac{1}{4} (\log |x - 2| - \log |x + 2|) + C = \frac{1}{4} \log \left| \frac{x - 2}{x + 2} \right| + C. \quad \square$$

Example 2'

$$\int \frac{2x^2 + 3}{x(x - 1)^2} \, dx = \int \left[\frac{3}{x} - \frac{1}{x - 1} + \frac{5}{(x - 1)^2} \right] dx$$

$$= 3 \log |x| - \log |x - 1| - \frac{5}{x - 1} + C. \quad \square$$

Example 3'

$$\int \frac{-2x}{(x + 1)(x^2 + 1)} \, dx = \int \left(\frac{1}{x + 1} - \frac{x + 1}{x^2 + 1} \right) dx.$$

Since

$$\int \frac{dx}{x + 1} = \log |x + 1| + C_1$$

and

$$\int \frac{x + 1}{x^2 + 1} \, dx = \frac{1}{2} \int \frac{2x}{x^2 + 1} \, dx + \int \frac{dx}{x^2 + 1} = \frac{1}{2} \log (x^2 + 1) + \arctan x + C_2,$$

we have

$$\int \frac{-2x}{(x + 1)(x^2 + 1)} \, dx = \log |x + 1| - \frac{1}{2} \log (x^2 + 1) - \arctan x + C. \quad \square$$

Example 4'

$$\int \frac{dx}{x(x^2 + x + 1)} = \int \left(\frac{1}{x} - \frac{x + 1}{x^2 + x + 1} \right) dx = \log |x| - \int \frac{x + 1}{x^2 + x + 1} \, dx = *.$$

To compute the remaining integral, note that

$$\int \frac{x + 1}{x^2 + x + 1} \, dx = \frac{1}{2} \int \frac{2x + 1}{x^2 + x + 1} \, dx + \frac{1}{2} \int \frac{dx}{x^2 + x + 1}.$$

The first integral is a logarithm:

$$\frac{1}{2} \int \frac{2x + 1}{x^2 + x + 1} \, dx = \frac{1}{2} \log (x^2 + x + 1) + C_1.$$

To compute the second integral, we complete the square in the denominator:

$$\frac{1}{2} \int \frac{dx}{x^2 + x + 1} = \frac{1}{2} \int \frac{dx}{(x + \frac{1}{2})^2 + (\sqrt{3}/2)^2} = \frac{1}{\sqrt{3}} \arc \tan \left[\frac{2}{\sqrt{3}} \left(x + \frac{1}{2} \right) \right] + C_2.$$

Combining results, we have

$$* = \log |x| - \frac{1}{2} \log (x^2 + x + 1) - \frac{1}{\sqrt{3}} \arc \tan \left[\frac{2}{\sqrt{3}} \left(x + \frac{1}{2} \right) \right] + C. \quad \square$$

Example 5 '

$$\int \frac{3x^4 + x^3 + 20x^2 + 3x + 31}{(x + 1)(x^2 + 4)^2} \, dx = \int \left[\frac{2}{x + 1} + \frac{x}{x^2 + 4} - \frac{1}{(x^2 + 4)^2} \right] dx.$$

The first two fractions are easy to integrate:

$$\int \frac{2}{x + 1} \, dx = 2 \log |x + 1| + C_1,$$

$$\int \frac{x}{x^2 + 4} \, dx = \frac{1}{2} \int \frac{2x}{x^2 + 4} \, dx = \frac{1}{2} \log (x^2 + 4) + C_2.$$

The integral of the last fraction is of the form

$$\int \frac{dx}{(x^2 + c^2)^n}.$$

All such integrals can be calculated by setting $c \tan u = x$:

(8.2.2)

$$\boxed{\int \frac{dx}{(x^2 + c^2)^n} = \frac{1}{c^{2n-1}} \int \cos^{2(n-1)} u \, du.}^{\dagger}$$

$c \tan u = x$

The integrations of such trigonometric powers is discussed in detail in the next section. Here we have

$$\int \frac{dx}{(x^2 + 4)^2} = \frac{1}{8} \int \cos^2 u \, du$$

$2 \tan u = x$

$$= \frac{1}{16} \int (1 + \cos 2u) \, du$$

half-angle formula

$$= \frac{1}{16} u + \frac{1}{32} \sin 2u + C_3$$

---

† The proof of this reduction formula is left to you in the exercises.

$$= \frac{1}{16} u + \frac{1}{16} \sin u \cos u + C_3$$

$\sin 2u = 2 \sin u \cos u$
$$= \frac{1}{16} \arctan \tfrac{1}{2}x + \frac{1}{16} \left(\frac{x}{\sqrt{x^2 + 4}} \right) \left(\frac{2}{\sqrt{x^2 + 4}} \right) + C_3$$

$$= \frac{1}{16} \arctan \tfrac{1}{2}x + \frac{1}{8} \left(\frac{x}{x^2 + 4} \right) + C_3.$$

The integral we want is therefore

$$2 \log |x + 1| + \frac{1}{2} \log (x^2 + 4) - \frac{1}{8} \left(\frac{x}{x^2 + 4} \right) - \frac{1}{16} \arctan \tfrac{1}{2}x + C. \quad \square$$

Example 6 '

$$\int \frac{x^5 + 2}{x^2 - 1} \, dx = \int \left[x^3 + x - \frac{1}{2(x + 1)} + \frac{3}{2(x - 1)} \right] dx$$

$$= \tfrac{1}{4}x^4 + \tfrac{1}{2}x^2 - \tfrac{1}{2} \log |x + 1| + \tfrac{3}{2} \log |x - 1| + C. \quad \square$$

Exercises

Work out the following integrals.

*1. $\displaystyle\int \frac{7}{(x - 2)(x + 5)} \, dx.$

2. $\displaystyle\int \frac{x}{(x + 1)(x + 2)(x + 3)} \, dx.$

*3. $\displaystyle\int \frac{x^2 + 1}{x(x^2 - 1)} \, dx.$

*4. $\displaystyle\int \frac{2x^2 + 3}{x^2(x - 1)} \, dx.$

5. $\displaystyle\int \frac{x^5}{x - 2} \, dx.$

*6. $\displaystyle\int \frac{x^5}{(x - 2)^2} \, dx.$

*7. $\displaystyle\int \frac{x + 3}{x^2 - 3x + 2} \, dx.$

8. $\displaystyle\int \frac{x^2 + 3}{x^2 - 3x + 2} \, dx.$

*9. $\displaystyle\int \frac{dx}{(x - 1)^3}.$

*10. $\displaystyle\int \frac{x^2}{(x - 1)^2(x + 1)} \, dx.$

11. $\displaystyle\int \frac{x^3 + 4x^2 - 4x - 1}{(x^2 + 1)^2} \, dx.$

*12. $\displaystyle\int \frac{2x - 1}{(x + 1)^2(x - 2)^2} \, dx.$

*13. $\displaystyle\int \frac{dx}{x^4 - 16}.$

14. $\displaystyle\int \frac{x}{x^3 - 1} \, dx.$

*15. $\displaystyle\int \frac{dx}{x^4 + 4}.$

*16. $\displaystyle\int_0^4 \frac{dx}{x^2 + 16}.$

17. $\displaystyle\int_0^4 \frac{dx}{(x^2 + 16)^2}.$

18. $\displaystyle\int_1^2 \frac{dx}{x^2 + 2x + 2}.$

19. Show that

$$\text{if} \quad y = \frac{1}{x^2 - 1} \quad \text{then} \quad \frac{d^n y}{dx^n} = \frac{1}{2} (-1)^n n! \left[\frac{1}{(x - 1)^{n+1}} - \frac{1}{(x + 1)^{n+1}} \right].$$

20. Verify reduction formula (8.2.2).

Optional   *21. It is known that m parts of chemical A combine with n parts of chemical B to produce a compound C. Suppose that the rate at which C is produced varies directly with the product of the amounts of A and B present at that instant. Find the amount of C produced in t minutes from an initial mixing of A_0 pounds of A with B_0 pounds of B, given that

(a) $n = m$, $A_0 = B_0$, and A_0 pounds of C are produced in the first minute.

(b) $n = m$, $A_0 = \quad B_0$, and A_0 pounds of C are produced in the first minute.

(c) $n \neq m$, $A_0 = B_0$, and A_0 pounds of C are produced in the first minute.

HINT: Denote by $A(t)$, $B(t)$, and $C(t)$ the amounts of A, B, and C present at time t. Observe that

$$C'(t) = kA(t)B(t).$$

Note that

$$A_0 - A(t) = \frac{m}{m+n} C(t) \quad \text{and} \quad B_0 - B(t) = \frac{n}{m+n} C(t)$$

and thus

$$C'(t) = k \left[A_0 - \frac{m}{m+n} C(t) \right] \left[B_0 - \frac{n}{m+n} C(t) \right].$$

8.3   Powers and Products of Sines and Cosines

I. We begin by explaining how to calculate integrals of the form

$$\int \sin^m x \cos^n x \, dx \quad \text{with } m \text{ or } n \text{ odd.}$$

Suppose that n is odd. If $n = 1$, the integral is of the form

(1) $$\int \sin^m x \cos x \, dx = \frac{1}{m+1} \sin^{m+1} x + C.$$

If $n > 1$, write

$$\cos^n x = \cos^{n-1} x \cos x.$$

Since $n - 1$ is even, $\cos^{n-1} x$ can be expressed in powers of $\sin^2 x$ by noting that

$$\cos^2 x = 1 - \sin^2 x.$$

The integral then takes the form

$$\int (\text{sum of powers of } \sin x) \cdot \cos x \, dx,$$

which can be broken up into integrals of form (1).

Similarly if m is odd, write

$$\sin^m x = \sin^{m-1} x \sin x$$

and use the substitution

$$\sin^2 x = 1 - \cos^2 x.$$

Example

$$\int \sin^2 x \cos^5 x \, dx = \int \sin^2 x \cos^4 x \cos x \, dx$$

$$= \int \sin^2 x \, (1 - \sin^2 x)^2 \cos x \, dx$$

$$= \int (\sin^2 x - 2 \sin^4 x + \sin^6 x) \cos x \, dx$$

$$= \int \sin^2 x \cos x \, dx - 2 \int \sin^4 x \cos x \, dx + \int \sin^6 x \cos x \, dx$$

$$= \tfrac{1}{3} \sin^3 x - \tfrac{2}{5} \sin^5 x + \tfrac{1}{7} \sin^7 x + C. \quad \square$$

Example

$$\int \sin^5 x \, dx = \int \sin^4 x \sin x \, dx$$

$$= \int (1 - \cos^2 x)^2 \sin x \, dx$$

$$= \int (1 - 2 \cos^2 x + \cos^4 x) \sin x \, dx$$

$$= \int \sin x \, dx - 2 \int \cos^2 x \sin x \, dx + \int \cos^4 x \sin x \, dx$$

$$= -\cos x + \tfrac{2}{3} \cos^3 x - \tfrac{1}{5} \cos^5 x + C. \quad \square$$

II. To calculate integrals of the form

$$\boxed{\int \sin^m x \cos^n x \, dx \quad \text{with } m \text{ and } n \text{ both even}}$$

use the following identities:

$$\sin x \cos x = \tfrac{1}{2} \sin 2x, \quad \sin^2 x = \tfrac{1}{2} - \tfrac{1}{2} \cos 2x, \quad \cos^2 x = \tfrac{1}{2} + \tfrac{1}{2} \cos 2x.$$

Example

$$\int \cos^2 x \, dx = \int (\tfrac{1}{2} + \tfrac{1}{2} \cos 2x) \, dx$$

$$= \tfrac{1}{2} \int dx + \tfrac{1}{2} \int \cos 2x \, dx = \tfrac{1}{2}x + \tfrac{1}{4} \sin 2x + C. \quad \square$$

Example

$$\int \sin^2 x \cos^2 x \, dx = \tfrac{1}{4} \int \sin^2 2x \, dx$$

$$= \tfrac{1}{4} \int (\tfrac{1}{2} - \tfrac{1}{2} \cos 4x) \, dx$$

$$= \tfrac{1}{8} \int dx - \tfrac{1}{8} \int \cos 4x \, dx = \tfrac{1}{8}x - \tfrac{1}{32} \sin 4x + C. \quad \square$$

Example

$$\int \sin^4 x \cos^2 x \, dx = \int (\sin x \cos x)^2 \sin^2 x \, dx$$

$$= \int \tfrac{1}{4} \sin^2 2x \, (\tfrac{1}{2} - \tfrac{1}{2} \cos 2x) \, dx$$

$$= \tfrac{1}{8} \int \sin^2 2x \, dx - \tfrac{1}{8} \int \sin^2 2x \cos 2x \, dx$$

$$= \tfrac{1}{8} \int (\tfrac{1}{2} - \tfrac{1}{2} \cos 4x) \, dx - \tfrac{1}{8} \int \sin^2 2x \cos 2x \, dx$$

$$= \tfrac{1}{16}x - \tfrac{1}{64} \sin 4x - \tfrac{1}{48} \sin^3 2x + C. \quad \square$$

III. Finally we come to integrals of the form

$$\int \sin mx \cos nx \, dx, \quad \int \sin mx \sin nx \, dx,$$

$$\int \cos mx \cos nx \, dx.$$

If $m = n$, there is no difficulty. Suppose that $m \neq n$. For the first integral use the identity

(1) $$\sin mx \cos nx = \tfrac{1}{2} \sin [(m + n)x] + \tfrac{1}{2} \sin [(m - n)x].$$

You can confirm this identity by using the usual expansions for $\sin (\theta_1 + \theta_2)$ and $\sin (\theta_1 - \theta_2)$. Set

$$\theta_1 = mx \quad \text{and} \quad \theta_2 = nx,$$

and you'll find that

$$\sin [(m + n)x] = \sin mx \cos nx + \cos mx \sin nx,$$
$$\sin [(m - n)x] = \sin mx \cos nx - \cos mx \sin nx.$$

Add these two equations, divide by 2, and you'll get identity (1).
   Integration of (1) gives

$$\int \sin mx \cos nx \, dx = \frac{1}{2} \int \sin [(m + n)x] \, dx + \frac{1}{2} \int \sin [(m - n)x] \, dx$$

$$= -\frac{\cos [(m + n)x]}{2(m + n)} - \frac{\cos [(m - n)x]}{2(m - n)} + C.$$

Similarly,

$$\int \sin mx \sin nx \, dx = -\frac{\sin [(m+n)x]}{2(m+n)} + \frac{\sin [(m-n)x]}{2(m-n)} + C,$$

$$\int \cos mx \cos nx \, dx = \frac{\sin [(m+n)x]}{2(m+n)} + \frac{\sin [(m-n)x]}{2(m-n)} + C.$$

Exercises

Work out the following integrals.

*1. $\int \sin^3 x \, dx$.   2. $\int \sin^2 x \, dx$.   *3. $\int \cos^4 x \sin^3 x \, dx$.

*4. $\int_0^{\pi/2} \cos^3 x \, dx$.   5. $\int_0^{\pi} \sin^4 x \, dx$.   *6. $\int \sin 2x \cos 3x \, dx$.

*7. $\int_0^{\pi/2} \cos^4 x \, dx$.   8. $\int \cos 2x \sin 3x \, dx$.   *9. $\int \sin^3 x \cos^2 x \, dx$.

*10. $\int \sin^3 x \cos^3 x \, dx$.   11. $\int \cos 2x \cos 3x \, dx$.   *12. $\int_0^{\pi/2} \cos^5 x \, dx$.

*13. $\int \sin^5 x \cos^2 x \, dx$.   14. $\int \sin^3 3x \, dx$.   *15. $\int \sin^2 x \cos^4 x \, dx$.

Calculate the following integrals by a trigonometric substitution.

*16. $\int \dfrac{dx}{(x^2+1)^3}$.   17. $\int \dfrac{dx}{(x^2+4)^3}$.

*18. $\int \dfrac{dx}{[(x+1)^2+1]^2}$.   19. $\int \dfrac{dx}{[(2x+1)^2+9]^2}$.

Evaluate the following integrals. They are important in applied mathematics.

*20. $\dfrac{1}{\pi} \int_0^{2\pi} \cos^2 nx \, dx$.   21. $\dfrac{1}{\pi} \int_0^{2\pi} \sin^2 nx \, dx$.

*22. $\int_0^{2\pi} \sin mx \cos nx \, dx, \quad m \neq n$.   23. $\int_0^{2\pi} \sin mx \sin nx \, dx, \quad m \neq n$.

*24. $\int_0^{2\pi} \cos mx \cos nx \, dx, \quad m \neq n$.

8.4   Other Trigonometric Powers

I. First we take up integrals of the form

$$\boxed{\int \tan^n x \, dx, \quad \int \cot^n x \, dx.}$$

To integrate $\tan^n x$, set

$$\tan^n x = \tan^{n-2} x \tan^2 x = (\tan^{n-2} x)(\sec^2 x - 1) = \tan^{n-2} x \sec^2 x - \tan^{n-2} x.$$

To integrate $\cot^n x$, set

$$\cot^n x = \cot^{n-2} x \cot^2 x = (\cot^{n-2} x)(\operatorname{cosec}^2 x - 1) =$$
$$= \cot^{n-2} x \operatorname{cosec}^2 x - \cot^{n-2} x.$$

Example

$$\int \tan^6 x \, dx = \int (\tan^4 x \sec^2 x - \tan^4 x) \, dx$$

$$= \int (\tan^4 x \sec^2 x - \tan^2 x \sec^2 x + \tan^2 x) \, dx$$

$$= \int (\tan^4 x \sec^2 x - \tan^2 x \sec^2 x + \sec^2 x - 1) \, dx$$

$$= \int \tan^4 x \sec^2 x \, dx - \int \tan^2 x \sec^2 x \, dx + \int \sec^2 x \, dx - \int dx$$

$$= \tfrac{1}{5} \tan^5 x - \tfrac{1}{3} \tan^3 x + \tan x - x + C. \quad \square$$

II. Now consider the integrals

$$\boxed{\int \sec^n x \, dx, \quad \int \operatorname{cosec}^n x \, dx.}$$

For even powers write

$$\sec^n x = \sec^{n-2} x \sec^2 x = (\tan^2 x + 1)^{(n-2)/2} \sec^2 x$$

and

$$\operatorname{cosec}^n x = \operatorname{cosec}^{n-2} x \operatorname{cosec}^2 x = (\cot^2 x + 1)^{(n-2)/2} \operatorname{cosec}^2 x.$$

Example

$$\int \sec^4 x \, dx = \int \sec^2 x \sec^2 x \, dx = \int (\tan^2 x + 1) \sec^2 x \, dx$$

$$= \int \tan^2 x \sec^2 x \, dx + \int \sec^2 x \, dx = \tfrac{1}{3} \tan^3 x + \tan x + C. \quad \square$$

The odd powers you can integrate by parts. For $\sec^n x$ set

$$u = \sec^{n-2} x, \quad dv = \sec^2 x \, dx$$

and, when $\tan^2 x$ appears, use the identity

$$\tan^2 x = \sec^2 x - 1.$$

You can handle $\operatorname{cosec}^n x$ in a similar manner.

Example.   For

$$\int \sec^3 x \, dx$$

set

$$u = \sec x, \qquad\qquad dv = \sec^2 x \, dx,$$
$$du = \sec x \tan x \, dx, \qquad v = \tan x.$$

$$\int \sec^3 x \, dx = \int \sec x \sec^2 x \, dx$$

$$= \int u \, dv = uv - \int v \, du$$

$$= \sec x \tan x - \int \tan^2 x \sec x \, dx$$

$$= \sec x \tan x - \int (\sec^2 x - 1) \sec x \, dx$$

$$= \sec x \tan x - \int \sec^3 x \, dx + \int \sec x \, dx$$

$$= \sec x \tan x - \int \sec^3 x \, dx + \log |\sec x + \tan x|.$$

$$2 \int \sec^3 x \, dx = \sec x \tan x + \log |\sec x + \tan x|,$$

$$\int \sec^3 x \, dx = \tfrac{1}{2} \sec x \tan x + \tfrac{1}{2} \log |\sec x + \tan x|.$$

Now add the arbitrary constant:

$$\int \sec^3 x \, dx = \tfrac{1}{2} \sec x \tan x + \tfrac{1}{2} \log |\sec x + \tan x| + C. \quad \square$$

III. Finally we come to integrals of the form

$$\int \tan^m x \sec^n x \, dx, \quad \int \cot^m x \operatorname{cosec}^n x \, dx.$$

When n is even, write

$$\tan^m x \sec^n x = \tan^m x \sec^{n-2} x \sec^2 x$$

and express $\sec^{n-2} x$ entirely in terms of $\tan^2 x$ using

$$\sec^2 x = \tan^2 x + 1.$$

Example

$$\int \tan^5 x \sec^4 x \, dx = \int \tan^5 x \sec^2 x \sec^2 x \, dx$$

$$= \int \tan^5 x (\tan^2 x + 1) \sec^2 x \, dx$$

$$= \int \tan^7 x \sec^2 x \, dx + \int \tan^5 x \sec^2 x \, dx$$

$$= \tfrac{1}{8} \tan^8 x + \tfrac{1}{6} \tan^6 x + C. \quad \square$$

When n and m are both odd, write

$$\tan^m x \sec^n x = \tan^{m-1} x \sec^{n-1} x \sec x \tan x$$

and express $\tan^{m-1} x$ entirely in terms of $\sec^2 x$ using

$$\tan^2 x = \sec^2 x - 1.$$

Example

$$
\begin{aligned}
\int \tan^5 x \sec^3 x \, dx &= \int \tan^4 x \sec^2 x \sec x \tan x \, dx \\
&= \int (\sec^2 x - 1)^2 \sec^2 x \sec x \tan x \, dx \\
&= \int (\sec^6 x - 2 \sec^4 x + \sec^2 x) \sec x \tan x \, dx \\
&= \tfrac{1}{7} \sec^7 x - \tfrac{2}{5} \sec^5 x + \tfrac{1}{3} \sec^3 x + C. \quad \square
\end{aligned}
$$

Finally, if n is odd and m is even, write the product entirely in terms of $\sec x$ and integrate by parts.

Example

$$
\begin{aligned}
\int \tan^2 x \sec x \, dx &= \int (\sec^2 x - 1) \sec x \, dx \\
&= \int (\sec^3 x - \sec x) \, dx = \int \sec^3 x \, dx - \int \sec x \, dx.
\end{aligned}
$$

We have computed each of these integrals by parts before:

$$\int \sec^3 x \, dx = \tfrac{1}{2} \sec x \tan x + \tfrac{1}{2} \log |\sec x + \tan x| + C$$

and

$$\int \sec x \, dx = \log |\sec x + \tan x| + C.$$

It follows that

$$\int \tan^2 x \sec x \, dx = \tfrac{1}{2} \sec x \tan x - \tfrac{1}{2} \log |\sec x + \tan x| + C. \quad \square$$

You can handle integrals of $\cot^m x \operatorname{cosec}^n x$ in a similar manner.

Exercises

*1. $\displaystyle\int \tan^3 x \, dx.$ 2. $\displaystyle\int \tan^2 (x + 1) \, dx.$ *3. $\displaystyle\int \sec^2 (x + 1) \, dx.$

*4. $\displaystyle\int \cot^3 x \, dx.$ 5. $\displaystyle\int \tan^2 x \sec^2 x \, dx.$ *6. $\displaystyle\int \cot^5 x \, dx.$

*7. $\displaystyle\int \tan^4 x \sec^4 x \, dx.$ 8. $\displaystyle\int \sec^4 4x \, dx.$ *9. $\displaystyle\int \tan^5 3x \, dx.$

*10. $\displaystyle\int \operatorname{cosec}^4 \tfrac{1}{4}x \, dx.$ 11. $\displaystyle\int \cot^3 2x \, dx.$ *12. $\displaystyle\int \cot^2 x \sec x \, dx.$

*13. $\displaystyle\int \cot^2 x \operatorname{cosec}^2 x \, dx.$ 14. $\displaystyle\int \operatorname{cosec}^3 x \, dx.$ *15. $\displaystyle\int \sec^5 x \, dx.$

8.5   Integrals Involving $\sqrt{a^2 \pm x^2}$ and $\sqrt{x^2 \pm a^2}$

Such integrals can usually be calculated by a trigonometric substitution:

$$\text{for} \quad \sqrt{a^2 - x^2} \quad \text{set} \quad a \sin u = x,$$
$$\text{for} \quad \sqrt{a^2 + x^2} \quad \text{set} \quad a \tan u = x,$$
$$\text{for} \quad \sqrt{x^2 - a^2} \quad \text{set} \quad a \sec u = x.$$

In each case take $a > 0$.

Problem.   Find

$$\int \frac{dx}{(a^2 - x^2)^{3/2}}.$$

SOLUTION.   Set

$$a \sin u = x; \quad\quad a \cos u \, du = dx.$$

$$\int \frac{dx}{(a^2 - x^2)^{3/2}} = \int \frac{a \cos u}{(a^2 - a^2 \sin^2 u)^{3/2}} \, du$$

$$= \frac{1}{a^2} \int \frac{\cos u}{\cos^3 u} \, du$$

$$= \frac{1}{a^2} \int \sec^2 u \, du$$

$$= \frac{1}{a^2} \tan u + C = \frac{x}{a^2 \sqrt{a^2 - x^2}} + C. \quad \square$$

Problem.   Find

$$\int \sqrt{a^2 + x^2} \, dx.$$

SOLUTION.   Set

$$a \tan u = x; \quad\quad a \sec^2 u \, du = dx.$$

$$\int \sqrt{a^2 + x^2} \, dx = \int \sqrt{a^2 + a^2 \tan^2 u} \, a \sec^2 u \, du$$

$$= a^2 \int \sqrt{1 + \tan^2 u} \, \sec^2 u \, du$$

$$= a^2 \int \sec u \cdot \sec^2 u \, du$$

$$= a^2 \int \sec^3 u \, du$$

by Section 8.4 $\xrightarrow{}$ $\overset{\downarrow}{=} \dfrac{a^2}{2} (\sec u \tan u + \log |\sec u + \tan u|) + C$

$$= \dfrac{a^2}{2} \left[\dfrac{\sqrt{a^2 + x^2}}{a} \left(\dfrac{x}{a} \right) + \log \left| \dfrac{\sqrt{a^2 + x^2}}{a} + \dfrac{x}{a} \right| \right] + C$$

$$= \tfrac{1}{2} x \sqrt{a^2 + x^2} + \tfrac{1}{2} a^2 \log (x + \sqrt{a^2 + x^2}) - \tfrac{1}{2} a^2 \log a + C.$$

We can absorb the constant $-\tfrac{1}{2} a^2 \log a$ in C and write

(8.5.1) $\qquad \boxed{\displaystyle\int \sqrt{a^2 + x^2} \, dx = \tfrac{1}{2} x \sqrt{a^2 + x^2} + \tfrac{1}{2} a^2 \log (x + \sqrt{a^2 + x^2}) + C.}$

This is one of the standard formulas. □

Now a slight variation.

Problem. Find

$$\int \dfrac{dx}{x \sqrt{4x^2 + 9}}.$$

SOLUTION. Set

$$3 \tan u = 2x; \qquad 3 \sec^2 u \, du = 2 \, dx, \quad \tfrac{3}{2} \sec^2 u \, du = dx.$$

$$\int \dfrac{dx}{x \sqrt{4x^2 + 9}} = \int \dfrac{\tfrac{3}{2} \sec^2 u}{\tfrac{3}{2} \tan u \cdot 3 \sec u} \, du$$

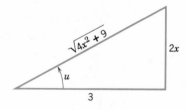

$$= \dfrac{1}{3} \int \dfrac{\sec u}{\tan u} \, du$$

$$= \dfrac{1}{3} \int \operatorname{cosec} u \, du$$

$$= \dfrac{1}{3} \log |\operatorname{cosec} u - \cot u| + C.$$

$$= \dfrac{1}{3} \log \left| \dfrac{\sqrt{4x^2 + 9} - 3}{2x} \right| + C. □$$

Exercises

Work out the following integrals.

*1. $\displaystyle\int \dfrac{dx}{\sqrt{a^2 - x^2}}.$

2. $\displaystyle\int \dfrac{dx}{(x^2 + 2)^{3/2}}.$

*3. $\displaystyle\int \dfrac{dx}{(5 - x^2)^{3/2}}.$

*4. $\displaystyle\int \dfrac{x}{\sqrt{x^2 - 4}} \, dx.$

5. $\displaystyle\int_1^2 \sqrt{x^2 - 1} \, dx.$

*6. $\displaystyle\int \dfrac{x}{\sqrt{4 - x^2}} \, dx.$

*7. $\displaystyle\int \dfrac{x^2}{\sqrt{x^2 - 4}} \, dx.$

8. $\displaystyle\int \dfrac{x^2}{\sqrt{4 - x^2}} \, dx.$

*9. $\displaystyle\int \dfrac{x^2}{\sqrt{4 + x^2}} \, dx.$

*10. $\displaystyle\int \frac{x}{(1 - x^2)^{3/2}}\, dx.$ 11. $\displaystyle\int \frac{x^2}{(1 - x^2)^{3/2}}\, dx.$ *12. $\displaystyle\int \frac{x}{a^2 + x^2}\, dx.$

*13. $\displaystyle\int x\sqrt{4 - x^2}\, dx.$ 14. $\displaystyle\int \frac{e^x}{\sqrt{9 - e^{2x}}}\, dx.$ *15. $\displaystyle\int \frac{x^2}{(x^2 + 8)^{3/2}}\, dx.$

*16. $\displaystyle\int \frac{\sqrt{1 - x^2}}{x^4}\, dx.$ 17. $\displaystyle\int \frac{\sqrt{x^2 - 1}}{x}\, dx.$ *18. $\displaystyle\int \frac{dx}{\sqrt{x^2 + a^2}}.$

*19. $\displaystyle\int \frac{dx}{\sqrt{x^2 - a^2}}.$ 20. $\displaystyle\int \sqrt{a^2 - x^2}\, dx.$ *21. $\displaystyle\int \sqrt{x^2 - a^2}\, dx.$

*22. $\displaystyle\int \frac{dx}{x\sqrt{a^2 - x^2}}.$ 23. $\displaystyle\int \frac{dx}{x^2\sqrt{a^2 + x^2}}.$ *24. $\displaystyle\int \frac{dx}{x^2\sqrt{a^2 - x^2}}.$

*25. $\displaystyle\int \frac{dx}{x^2\sqrt{x^2 - a^2}}.$ 26. $\displaystyle\int \frac{dx}{x\sqrt{a^2 + x^2}}.$ *27. $\displaystyle\int \frac{dx}{x\sqrt{x^2 - a^2}}.$

8.6   Rational Expressions in sin *x* and cos *x*

There is a substitution which makes it possible to integrate all rational expressions of sin *x* and cos *x*. The result is significant enough to be cast as a theorem.

Theorem 8.6.1

If $f(x)$ is a rational expression in sin *x* and cos *x*, then the substitution

$$2 \text{ arc tan } u = x$$

transforms the integral

$$\int f(x)\, dx$$

into the integral of a rational function of *u*.

PROOF.   Set

$$2 \text{ arc tan } u = x; \qquad \frac{2}{1 + u^2}\, du = dx$$

and observe that $u = \tan \frac{1}{2}x$. The substitution gives

$$\int f(x)\, dx = 2 \int \frac{f(2 \text{ arc tan } u)}{1 + u^2}\, du.$$

Let's assume now that $f(x)$ is rational in cos *x* and sin *x*. To prove the theorem, we need to show that $f(2 \text{ arc tan } u)$ is rational in *u*. This we can do by showing that sin (2 arc tan *u*) and cos (2 arc tan *u*) are both rational in *u*. This follows directly:

$$\sin{(2\arctan u)} = \sin x = 2\sin{\tfrac{1}{2}x}\cos{\tfrac{1}{2}x}$$

$$= 2\left(\frac{u}{\sqrt{1+u^2}}\right)\left(\frac{1}{\sqrt{1+u^2}}\right)$$

$$= 2\left(\frac{u}{1+u^2}\right).$$

$$\cos{(2\arctan u)} = \cos x = \cos^2{\tfrac{1}{2}x} - \sin^2{\tfrac{1}{2}x}$$

$$= \frac{1}{1+u^2} - \frac{u^2}{1+u^2}$$

$$= \frac{1-u^2}{1+u^2}. \quad \square$$

In proving the theorem we showed that the substitution

$$\boxed{2\arctan u = x}$$

had the following consequences:

$$(8.6.2) \qquad \boxed{\sin{(2\arctan u)} = \frac{2u}{1+u^2}, \quad \cos{(2\arctan u)} = \frac{1-u^2}{1+u^2}.}$$

Keep this in mind as you work the problems.

Problem.   Find

$$\int \frac{dx}{1+2\cos x}.$$

SOLUTION.   Set

$$2\arctan u = x; \qquad \frac{2}{1+u^2}\,du = dx.$$

With $u = \tan{\tfrac{1}{2}x}$,

$$\int \frac{dx}{1+2\cos x} = \int \frac{2}{1+u^2}\left[\frac{1}{1+2\cos{(2\arctan u)}}\right]du$$

$$= \int \frac{2}{1+u^2}\left[\frac{1}{1+2\left(\dfrac{1-u^2}{1+u^2}\right)}\right]du = -2\int \frac{du}{u^2-3}$$

$$= \frac{1}{\sqrt{3}}\int\left(\frac{1}{u+\sqrt{3}} - \frac{1}{u-\sqrt{3}}\right)du$$

$$= \frac{1}{\sqrt{3}}\left(\log{|u+\sqrt{3}|} - \log{|u-\sqrt{3}|}\right) + C$$

$$= \frac{1}{\sqrt{3}}\log{\left|\frac{u+\sqrt{3}}{u-\sqrt{3}}\right|} + C = \frac{1}{\sqrt{3}}\log{\left|\frac{\tan{\tfrac{1}{2}x}+\sqrt{3}}{\tan{\tfrac{1}{2}x}-\sqrt{3}}\right|} + C. \quad \square$$

Since the other trigonometric functions—tan x, cot x, sec x, cosec x—are all rational combinations of sin x and cos x, the substitution 2 arc tan $u = x$ also works for rational expressions involving these other functions.

Problem. Find

$$\int \frac{\sec x}{2 \tan x + \sec x - 1} \, dx.$$

SOLUTION. Set

$$2 \text{ arc tan } u = x; \qquad \frac{2}{1 + u^2} \, du = dx.$$

With $u = \tan \frac{1}{2}x$,

$$\int \frac{\sec x}{2 \tan x + \sec x - 1} \, dx = \int \frac{\dfrac{1}{\cos x}}{2 \dfrac{\sin x}{\cos x} + \dfrac{1}{\cos x} - 1} \, dx$$

$$= \int \frac{dx}{2 \sin x - \cos x + 1}$$

$$= \int \frac{\dfrac{2}{1 + u^2}}{2 \sin (2 \text{ arc tan } u) - \cos (2 \text{ arc tan } u) + 1} \, du$$

$$= \int \frac{\dfrac{2}{1 + u^2}}{2 \cdot \dfrac{2u}{1 + u^2} - \dfrac{1 - u^2}{1 + u^2} + 1} \, du = \int \frac{du}{u(u + 2)}$$

$$= \frac{1}{2} \int \left(\frac{1}{u} - \frac{1}{u + 2} \right) du = \frac{1}{2} \log \left| \frac{u}{u + 2} \right| + C$$

$$= \frac{1}{2} \log \left| \frac{\tan \frac{1}{2}x}{\tan \frac{1}{2}x + 2} \right| + C. \quad \square$$

Exercises

Work out the following integrals.

*1. $\displaystyle\int \frac{dx}{1 + \cos x}$.

2. $\displaystyle\int \frac{dx}{1 - \sin x}$.

*3. $\displaystyle\int \frac{dx}{1 - \cos x}$.

*4. $\displaystyle\int_0^{\pi/2} \frac{dx}{3 + 2 \cos x}$.

5. $\displaystyle\int_0^{\pi/2} \frac{dx}{3 + \cos x}$.

*6. $\displaystyle\int \frac{\sin x}{2 - \sin x} \, dx$.

*7. $\displaystyle\int \frac{dx}{5 + 4 \cos x}$.

8. $\displaystyle\int \frac{dx}{1 + \tan x}$.

*9. $\displaystyle\int \frac{dx}{5 \sec x - 3}$.

*10. $\displaystyle\int \frac{\cos x}{1 - \cos x} \, dx$.

11. $\displaystyle\int \frac{1 - \cos x}{1 + \sin x} \, dx$.

*12. $\displaystyle\int \frac{1 + \sin x}{1 + \cos x} \, dx$.

8.7 Some Rationalizing Substitutions

Problem. Find

$$\int \frac{dx}{1 + \sqrt{x}}.$$

SOLUTION. To rationalize the integrand we set

$$u^2 = x; \qquad 2u \, du = dx.$$

With $u = \sqrt{x}$,

$$\int \frac{dx}{1 + \sqrt{x}} = \int \frac{2u}{1 + u} \, du = \int \left(2 - \frac{2}{1 + u}\right) du$$

$$= 2u - 2 \log |1 + u| + C = 2\sqrt{x} - 2 \log |1 + \sqrt{x}| + C. \quad \square$$

Problem. Find

$$\int \frac{x^{1/2}}{4(1 + x^{3/4})} \, dx.$$

SOLUTION. Here we set

$$u^4 = x; \qquad 4u^3 \, du = dx.$$

With $u = x^{1/4}$,

$$\int \frac{x^{1/2}}{4(1 + x^{3/4})} \, dx = \int \frac{(u^2)(4u^3)}{4(1 + u^3)} \, du = \int \frac{u^5}{1 + u^3} \, du$$

$$= \int \left(u^2 - \frac{u^2}{1 + u^3}\right) du = \tfrac{1}{3}u^3 - \tfrac{1}{3} \log |1 + u^3| + C$$

$$= \tfrac{1}{3}x^{3/4} - \tfrac{1}{3} \log |1 + x^{3/4}| + C. \quad \square$$

Problem. Find

$$\int \sqrt{1 - e^x} \, dx.$$

SOLUTION. To rationalize the integrand we set

$$u^2 = 1 - e^x.$$

To find dx in terms of u and du we solve the equation for x:

$$1 - u^2 = e^x, \qquad \log (1 - u^2) = x, \qquad -\frac{2u}{1 - u^2} \, du = dx.$$

The rest is straightforward:

$$\int \sqrt{1 - e^x} \, dx = \int u \left(-\frac{2u}{1 - u^2}\right) du = \int \frac{2u^2}{u^2 - 1} \, du$$

$$= \int \left(2 + \frac{1}{u - 1} - \frac{1}{u + 1}\right) du$$

$$= 2u + \log |u - 1| - \log |u + 1| + C$$

$$= 2u + \log \left| \frac{u - 1}{u + 1} \right| + C$$

$$= 2\sqrt{1 - e^x} + \log \left| \frac{\sqrt{1 - e^x} - 1}{\sqrt{1 - e^x} + 1} \right| + C. \quad \square$$

Exercises

Work out the following integrals.

*1. $\int \dfrac{dx}{1 - \sqrt{x}}$.

2. $\int \dfrac{\sqrt{x}}{1 + x} \, dx$.

*3. $\int \sqrt{1 + e^x} \, dx$.

4. $\int \dfrac{dx}{x(x^{1/3} - 1)}$.

*5. $\int x\sqrt{1 + x} \, dx$.   [(a) set $u^2 = 1 + x$; (b) set $u = 1 + x$].

6. $\int x^2\sqrt{1 + x} \, dx$.   ["   "   "   "   ].

*7. $\int (x + 2)\sqrt{x - 1} \, dx$.

8. $\int (x - 1)\sqrt{x + 2} \, dx$.

*9. $\int \dfrac{x^3}{(1 + x^2)^3} \, dx$.

10. $\int x(1 + x)^{1/3} \, dx$.

*11. $\int \dfrac{\sqrt{x}}{\sqrt{x} - 1} \, dx$.

12. $\int \dfrac{x}{\sqrt{x} + 1} \, dx$.

*13. $\int \dfrac{\sqrt{x - 1} + 1}{\sqrt{x - 1} - 1} \, dx$.

14. $\int \dfrac{1 - e^x}{1 + e^x} \, dx$.

*15. $\int \dfrac{dx}{\sqrt{1 + e^x}}$.

16. $\int \dfrac{dx}{1 + e^{-x}}$.

*17. $\int \dfrac{x}{\sqrt{x + 4}} \, dx$.

18. $\int \dfrac{x + 1}{x\sqrt{x - 2}} \, dx$.

*19. $\int 2x^2(4x + 1)^{-5/2} \, dx$.

20. $\int x^2\sqrt{x - 1} \, dx$.

*21. $\int \dfrac{x}{(ax + b)^{3/2}} \, dx$.

22. $\int \dfrac{x}{\sqrt{ax + b}} \, dx$.

8.8   Approximate Integration.

To evaluate a definite integral by the formula

$$\int_a^b f(x) \, dx = F(b) - F(a)$$

we must be able to find an antiderivative F and we must be able to evaluate this antiderivative both at a and at b. When this is not possible, the method fails.

The method fails even for such simple-looking integrals as

$$\int_0^1 \sqrt{x}\,\sin x\,dx \quad \text{and} \quad \int_0^1 e^{-x^2}\,dx.$$

There are no *elementary functions* with derivatives $\sqrt{x}\,\sin x$ and e^{-x^2}.

Here we take up some simple numerical methods for estimating definite integrals—methods that you can use whether or not you can find an antiderivative. All the methods we describe involve only simple arithmetic and are ideally suited to the electronic computer.

We focus now on

$$\int_a^b f(x)\,dx.$$

As usual, we suppose that f is continuous on $[a, b]$ and, for pictorial convenience, assume that f is positive. We begin by subdividing $[a, b]$ into n nonoverlapping subintervals each of length $(b - a)/n$:

$$[a, b] = [x_0, x_1] \cup \cdots \cup [x_{i-1}, x_i] \cup \cdots \cup [x_{n-1}, x_n],$$

with

$$\Delta x_i = \frac{b - a}{n}.$$

FIGURE 8.8.1

The region Ω_i pictured in Figure 8.8.1 can be approximated in several ways.

(1) By the left endpoint rectangle (Figure 8.8.2):

FIGURE 8.8.2

$$\text{area} = f(x_{i-1})\,\Delta x_i = f(x_{i-1})\left(\frac{b - a}{n}\right).$$

$-1 - \sqrt{3}$

$-(1 + \sqrt{3})$

(2) By the right endpoint rectangle (Figure 8.8.3):

FIGURE 8.8.3

$$\text{area} = f(x_i)\,\Delta x_i = f(x_i)\left(\frac{b-a}{n}\right).$$

(3) By the midpoint rectangle (Figure 8.8.4):

FIGURE 8.8.4

$$\text{area} = f\!\left(\frac{x_i + x_{i-1}}{2}\right)\Delta x_i = f\!\left(\frac{x_i + x_{i-1}}{2}\right)\!\left(\frac{b-a}{n}\right).$$

(4) By a trapezoid (Figure 8.8.5):

FIGURE 8.8.5

$$\text{area} = \frac{1}{2}\,[f(x_i) + f(x_{i-1})]\Delta x_i = \frac{1}{2}\,[f(x_i) + f(x_{i-1})]\!\left(\frac{b-a}{n}\right).$$

(5) By a parabolic region (Figure 8.8.6): take the parabola $y = Ax^2 + Bx + C$ which passes through the three points indicated.

x_{i-1} $\dfrac{x_i + x_{i-1}}{2}$ x_i

FIGURE 8.8.6

$$\text{area} = \frac{1}{6}\left[f(x_{i-1}) + 4f\left(\frac{x_i + x_{i-1}}{2}\right) + f(x_i)\right]\Delta x_i$$
$$= \left[f(x_{i-1}) + 4f\left(\frac{x_i + x_{i-1}}{2}\right) + f(x_i)\right]\left(\frac{b - a}{6n}\right).$$

You can verify this formula by doing Exercises 7 and 8. (If the three points are collinear, the parabola degenerates to a straight line and the parabolic region becomes a trapezoid. The formula then gives the area of the trapezoid.)

The approximations to Ω_i just considered yield the following estimates for

$$\int_a^b f(x)\,dx.$$

(1) The left endpoint estimate:

$$L_n = \frac{b - a}{n}\left[f(x_0) + f(x_1) + \cdots + f(x_{n-1})\right].$$

(2) The right endpoint estimate:

$$R_n = \frac{b - a}{n}\left[f(x_1) + f(x_2) + \cdots + f(x_n)\right].$$

(3) The midpoint estimate:

$$M_n = \frac{b - a}{n}\left[f\left(\frac{x_1 + x_0}{2}\right) + \cdots + f\left(\frac{x_n + x_{n-1}}{2}\right)\right].$$

(4) The trapezoidal estimate (*trapezoidal rule*):

$$T_n = \frac{b - a}{2n}\left[f(x_0) + 2f(x_1) + \cdots + 2f(x_{n-1}) + f(x_n)\right].$$

(5) The parabolic estimate (*Simpson's rule*):

$$S_n = \frac{b - a}{6n}\left\{f(x_0) + f(x_n) + 2[f(x_1) + \cdots + f(x_{n-1})]\right.$$
$$\left. + 4\left[f\left(\frac{x_1 + x_0}{2}\right) + \cdots + f\left(\frac{x_n + x_{n-1}}{2}\right)\right]\right\}.$$

As an example, we estimate

$$\log 2 = \int_1^2 \frac{dx}{x}$$

by each of these methods. Here we have

$$f(x) = \frac{1}{x}, \quad [a, b] = [1, 2].$$

If we take $n = 5$, then each subinterval has length

$$\frac{b - a}{n} = \frac{2 - 1}{5} = \frac{1}{5}$$

and the partition points are

$$x_0 = \tfrac{5}{5}, \quad x_1 = \tfrac{6}{5}, \quad x_2 = \tfrac{7}{5}, \quad x_3 = \tfrac{8}{5}, \quad x_4 = \tfrac{9}{5}, \quad x_5 = \tfrac{10}{5}.$$

This gives

$$L_5 = \tfrac{1}{5}(\tfrac{5}{5} + \tfrac{5}{6} + \tfrac{5}{7} + \tfrac{5}{8} + \tfrac{5}{9}) = (\tfrac{1}{5} + \tfrac{1}{6} + \tfrac{1}{7} + \tfrac{1}{8} + \tfrac{1}{9}),$$
$$R_5 = \tfrac{1}{5}(\tfrac{5}{6} + \tfrac{5}{7} + \tfrac{5}{8} + \tfrac{5}{9} + \tfrac{5}{10}) = (\tfrac{1}{6} + \tfrac{1}{7} + \tfrac{1}{8} + \tfrac{1}{9} + \tfrac{1}{10}),$$
$$M_5 = \tfrac{1}{5}(\tfrac{10}{11} + \tfrac{10}{13} + \tfrac{10}{15} + \tfrac{10}{17} + \tfrac{10}{19}) = 2(\tfrac{1}{11} + \tfrac{1}{13} + \tfrac{1}{15} + \tfrac{1}{17} + \tfrac{1}{19}),$$
$$T_5 = \tfrac{1}{10}(\tfrac{5}{5} + \tfrac{10}{6} + \tfrac{10}{7} + \tfrac{10}{8} + \tfrac{10}{9} + \tfrac{5}{10}) = (\tfrac{1}{10} + \tfrac{1}{6} + \tfrac{1}{7} + \tfrac{1}{8} + \tfrac{1}{9} + \tfrac{1}{20}),$$
$$S_5 = \tfrac{1}{30}[\tfrac{5}{5} + \tfrac{5}{10} + 2(\tfrac{5}{6} + \tfrac{5}{7} + \tfrac{5}{8} + \tfrac{5}{9}) + 4(\tfrac{10}{11} + \tfrac{10}{13} + \tfrac{10}{15} + \tfrac{10}{17} + \tfrac{10}{19})].$$

If you carry out the computations and round off to the nearest hundredth, you will have the following estimates:

$$L_5 \cong 0.75, \quad R_5 \cong 0.65, \quad M_5 \cong 0.69, \quad T_5 \cong 0.70, \quad S_5 \cong 0.69.$$

Since the integrand $1/x$ decreases throughout the interval $[1, 2]$, you can expect the left endpoint estimate, 0.75, to be too large and you can expect the right endpoint estimate, 0.65, to be too small. The other estimates should be better.

Table 1 at the end of the book gives $\log 2 \cong 0.693$. The estimate 0.69 is correct to the nearest hundredth.

Problem.　Estimate

$$\int_0^2 \sqrt{4 + x^3} \, dx$$

by the trapezoidal rule. Take $n = 4$.

SOLUTION.　Each subinterval has length

$$\frac{b - a}{n} = \frac{2 - 0}{4} = \frac{1}{2}.$$

The partition points are

$$x_0 = 0, \quad x_1 = \tfrac{1}{2}, \quad x_2 = 1, \quad x_3 = \tfrac{3}{2}, \quad x_4 = 2.$$

Consequently

$$T_4 = \tfrac{1}{4}[f(0) + 2f(\tfrac{1}{2}) + 2f(1) + 2f(\tfrac{3}{2}) + f(2)].$$

With

$$f(0) = 2.000,$$
$$f(\tfrac{1}{2}) = \sqrt{4 + \tfrac{1}{8}} = \sqrt{4.125} \cong 2.031,$$
$$f(1) = \sqrt{5} \cong 2.236,$$
$$f(\tfrac{3}{2}) = \sqrt{4 + \tfrac{27}{8}} = \sqrt{7.375} \cong 2.716,$$
$$f(2) = \sqrt{12} \cong 3.464,$$

we have

$$T_4 \cong \tfrac{1}{4}(2.000 + 4.062 + 4.472 + 5.432 + 3.464) \cong 4.858. \quad \square$$

Problem.   Estimate

$$\int_0^2 \sqrt{4 + x^3}\, dx$$

by Simpson's rule. Take $n = 2$.

SOLUTION.   There are two intervals each of length

$$\frac{b - a}{n} = \frac{2 - 0}{2} = 1.$$

We have

$$x_0 = 0, \quad x_1 = 1, \quad x_2 = 2.$$
$$\frac{x_0 + x_1}{2} = \frac{1}{2}, \qquad \frac{x_1 + x_2}{2} = \frac{3}{2}.$$

Simpson's rule yields

$$S_2 = \tfrac{1}{6}[f(0) + f(2) + 2f(1) + 4f(\tfrac{1}{2}) + 4f(\tfrac{3}{2})],$$

which, in the light of the square root estimates given in the last problem, gives

$$S_2 \cong \tfrac{1}{6}(2.000 + 3.464 + 4.472 + 8.124 + 10.864) \cong 4.821. \quad \square$$

Exercises

(In each of the numerical computations below round off your final answer to the nearest hundredth.)

*1. Estimate

$$\int_0^{12} x^2\, dx.$$

(a) Use the left endpoint estimate, $n = 12$.
(b) Use the right endpoint estimate, $n = 12$.
(c) Use the midpoint estimate, $n = 6$.
(d) Use the trapezoidal rule, $n = 12$.
(e) Use Simpson's rule, $n = 6$.
Check your results by performing the integration.

2. Estimate

$$\int_0^1 \sin^2 \pi x \, dx.$$

  (a) Use the midpoint estimate, $n = 3$.
  (b) Use the trapezoidal rule, $n = 6$.
  (c) Use Simpson's rule, $n = 3$.
Check your results by performing the integration.

*3. Estimate

$$\int_0^3 \frac{dx}{1 + x^3}.$$

  (a) Use the left endpoint estimate, $n = 6$.
  (b) Use the right endpoint estimate, $n = 6$.
  (c) Use the midpoint estimate, $n = 3$.
  (d) Use the trapezoidal rule, $n = 6$.
  (e) Use Simpson's rule, $n = 3$.

4. Estimate

$$\int_0^\pi \frac{\sin x}{\pi + x} \, dx.$$

  (a) Use the trapezoidal rule, $n = 6$.   (b) Use Simpson's rule, $n = 3$.

*5. Find the approximate value of π by estimating the integral

$$\frac{\pi}{4} = \arctan 1 = \int_0^1 \frac{dx}{1 + x^2}.$$

  (a) Use the trapezoidal rule, $n = 4$.   (b) Use Simpson's rule, $n = 4$.

6. Estimate

$$\int_0^2 \frac{dx}{\sqrt{4 + x^3}}.$$

  (a) Use the trapezoidal rule, $n = 4$.   (b) Use Simpson's rule, $n = 2$.

7. Show that there is one and only one curve of the form $y = Ax^2 + Bx + C$ through three distinct points with different x-coordinates.

8. Show that the function $g(x) = Ax^2 + Bx + C$ satisfies the condition

$$\int_a^b g(x) \, dx = \frac{b - a}{6} \left[g(a) + 4g\left(\frac{a + b}{2}\right) + g(b) \right]$$

for every interval $[a, b]$.

8.9   Additional Exercises

Work out the following integrals and check your answers by differentiation.

*1. $\displaystyle\int 10^{nx} \, dx.$ 2. $\displaystyle\int e^x \tan e^x \, dx.$ *3. $\displaystyle\int \sqrt{2x + 1} \, dx.$

*4. $\int x\sqrt{2x+1}\,dx.$

5. $\int \tan\left(\dfrac{\pi}{n}x\right)dx.$

*6. $\int \dfrac{dx}{\sqrt{x+1}-\sqrt{x}}.$

*7. $\int \dfrac{dx}{a^2x^2+b^2}.$

8. $\int \sin 2x \cos x\,dx.$

*9. $\int \dfrac{e^{-\sqrt{x}}}{\sqrt{x}}\,dx.$

*10. $\int \sec 2x\,dx.$

11. $\int \dfrac{dx}{\sqrt{1-e^{2x}}}.$

*12. $\int x2^x\,dx.$

*13. $\int \sin^2\left(\dfrac{\pi}{n}x\right)dx.$

14. $\int \dfrac{dx}{x^3-1}.$

*15. $\int \dfrac{dx}{a\sqrt{x}+b}.$

*16. $\int (1-\sec x)^2\,dx.$

17. $\int \dfrac{\sin 3x}{2+\cos 3x}\,dx.$

*18. $\int \dfrac{\sqrt{a-x}}{\sqrt{a+x}}\,dx.$ $(a>0)$

*19. $\int \dfrac{\sin x}{\cos^3 x}\,dx.$

20. $\int \dfrac{\cos x}{\sin^3 x}\,dx.$

*21. $\int \dfrac{\sin^5 x}{\cos^7 x}\,dx.$

*22. $\int \dfrac{dx}{\sin x \cos x}.$

23. $\int \dfrac{\sqrt{a+x}}{\sqrt{a-x}}\,dx.$

*24. $\int \dfrac{1-\sin 2x}{1+\sin 2x}\,dx.$

*25. $\int \dfrac{\sqrt{a^2-x^2}}{x^2}\,dx.$

26. $\int \log\sqrt{x+1}\,dx.$

*27. $\int \dfrac{x}{(x+1)^2}\,dx.$

*28. $\int \dfrac{dx}{\cos x-\sin x}.$

29. $\int a^{2x}\,dx.$

*30. $\int \log(ax+b)\,dx.$

*31. $\int x\log(ax+b)\,dx.$

32. $\int (\tan x+\cot x)^2\,dx.$

*33. $\int \dfrac{\sin\sqrt{x}}{\sqrt{x}}\,dx.$

34. $\int \log(x\sqrt{x})\,dx.$

*35. $\int \dfrac{x^2}{1+x^2}\,dx.$

*36. $\int \sqrt{\dfrac{x^2}{9}-1}\,dx.$

37. $\int \dfrac{-x^2}{\sqrt{1-x^2}}\,dx.$

*38. $\int e^x\sin\pi x\,dx.$

*39. $\int \dfrac{dx}{1+3\sin x}.$

40. $\int x\tan^2\pi x\,dx.$

*41. $\int \dfrac{x^3}{\sqrt{1+x^2}}\,dx.$

*42. $\int x\log\sqrt{x^2+1}\,dx.$

43. $\int x\arctan(x-3)\,dx.$

*44. $\int \dfrac{dx}{2-\sqrt{x}}.$

45. $\int \dfrac{2}{x(1+x^2)}\,dx.$

*46. $\int \dfrac{dx}{\sqrt{2x-x^2}}.$

*47. $\int \dfrac{\cos^4 x}{\sin^2 x}\,dx.$

48. $\int \dfrac{x-3}{x^2(x+1)}\,dx.$

*49. $\int \dfrac{\sqrt{x^2+4}}{x}\,dx.$

*50. $\int \dfrac{e^x}{\sqrt{e^x+1}}\,dx.$

51. $\int \dfrac{dx}{x\sqrt{9-x^2}}.$

*52. $\int \dfrac{dx}{e^x-2e^{-x}}.$

*53. $\int \sin^5\left(\dfrac{x}{2}\right)dx.$

54. $\int \dfrac{dx}{2x^2-2x+1}.$

*55. $\int \sin 2x\cos 3x\,dx.$

56. $\int (\sin^2 x-\cos x)^2\,dx.$

*57. $\int \log(1-\sqrt{x})\,dx.$

58. $\int \dfrac{3}{\sqrt{2-3x-4x^2}}\,dx.$

*59. $\int \dfrac{\sin x}{\cos^2 x - 2 \cos x + 3} \, dx.$

60. $\int \left[\dfrac{x}{\sqrt{a^2 - x^2}} - \arcsin\left(\dfrac{x}{a}\right) \right] dx.$

Optional   *61. $\int \sinh^2 x \, dx.$

62. $\int 2x \sinh x \, dx.$

*63. $\int e^{-x} \cosh x \, dx.$

64. $\int \tanh^2 2x \, dx.$

The Conic Sections

9

If a "double right circular cone" is cut by a plane, the resulting intersection is called a *conic section* or, more briefly, a *conic*. In Figure 9.1.1 we depict three important cases.

ellipse   parabola   hyperbola

FIGURE 9.1.1

By choosing a plane perpendicular to the axis of the cone, we can obtain a circle. The other possibilities are: a point, a line, or a pair of lines. Try to visualize this.

The study of conic sections from this point of view goes back to Apollonius of Perga, a Greek of the third century B.C. He wrote eight books on the subject.

For our purposes, it is useful to dispense with cones and three-dimensional geometry and, instead, define ellipse, parabola, and hyperbola entirely in terms of plane geometry. We begin by deriving a formula for the distance between a point and a line.

9.2   The Distance Between a Point and a Line; Translations

First we compute the distance between a line and the origin.

Theorem 9.2.1

The distance between the line $l: Ax + By + C = 0$ and the origin is given by the formula

$$d(0, l) = \frac{|C|}{\sqrt{A^2 + B^2}}.$$

PROOF.   If the line passes through the origin, then $C = 0$ and the formula holds. Suppose that the line l does not pass through the origin and thus that $C \neq 0$. To find the desired distance, we will get the equation of the line n which passes through the origin and is normal to l. (See Figure 9.2.1.) From this equation we will obtain the point Q at which l and n intersect. The distance we want is the distance between O and Q.

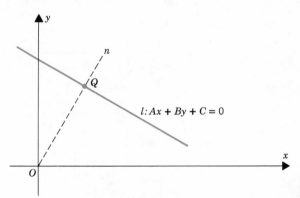

FIGURE 9.2.1

Since l has equation

$$Ax + By + C = 0,$$

n has equation

$$Bx - Ay = 0. \qquad\qquad \text{(check this out)}$$

Solve the two equations simultaneously and you'll see that

$$Q = \left(-\frac{AC}{A^2 + B^2}, \ -\frac{BC}{A^2 + B^2} \right).$$

The distance between O and Q is therefore

$$\sqrt{A^2C^2/(A^2 + B^2)^2 + B^2C^2/(A^2 + B^2)^2}.$$

This expression can be simplified to

$$\frac{|C|}{\sqrt{A^2 + B^2}}. \quad \square$$

To find the distance between the line l: $Ax + By + C = 0$ and an arbitrary point $P_0(x_0, y_0)$, we could proceed as we just did, but the computations would be cumbersome. The computations are much easier if we first "translate" the coordinate system.

Translations

In Figure 9.2.2, you can see a rectangular coordinate system and a point $O'(x_0, y_0)$. Think of the Oxy system as a rigid frame and in your mind slide it along the plane, without tilting it, until the origin O falls on the point O'. (Figure 9.2.3) Such a motion, called a *translation,* produces a new coordinate system $O'XY$.

FIGURE 9.2.2

FIGURE 9.2.3

A point P now has two pairs of coordinates: a pair (x, y) with respect to the Oxy system and a pair (X, Y) with respect to the $O'XY$ system. To see the relation between these coordinates, note that, starting at O, we can reach P by first going to O' and then going on to P:

$$x = x_0 + X, \qquad y = y_0 + Y.$$

FIGURE 9.2.4

Translations are often used to simplify geometric arguments. See, for example, the proof of the next theorem.

Theorem 9.2.2

The distance between the line l: $Ax + By + C = 0$ and the point $P_0(x_0, y_0)$ is given by the formula

$$d(P_0, l) = \frac{|Ax_0 + By_0 + C|}{\sqrt{A^2 + B^2}}.$$

PROOF. We translate the Oxy coordinate system to obtain a new system $O'XY$ with O' falling on the point P_0. The new coordinates are now related to the old coordinates by the equations

$$x = x_0 + X, \qquad y = y_0 + Y.$$

In the xy-coordinates, l has equation

$$Ax + By + C = 0.$$

In the XY-coordinates, l has equation

$$A(x_0 + X) + B(y_0 + Y) + C = 0.$$

This we can write as

$$AX + BY + K = 0 \qquad \text{with} \quad K = Ax_0 + By_0 + C.$$

The distance we want is the distance between the line with equation $AX + BY + K = 0$ and the new origin O'. By Theorem 9.2.1, this distance is

$$\frac{|K|}{\sqrt{A^2 + B^2}}.$$

Since $K = Ax_0 + By_0 + C$, we have

$$d(P_0, l) = \frac{|Ax_0 + By_0 + C|}{\sqrt{A^2 + B^2}}. \quad \square$$

Problem. Find the distance between the line

$$l: 3x + 4y - 5 = 0$$

and (a) the origin, (b) the point $P(-6, 2)$.

SOLUTION

(a) $d(0, l) = \dfrac{|C|}{\sqrt{A^2 + B^2}} = \dfrac{|-5|}{\sqrt{3^2 + 4^2}} = \dfrac{5}{\sqrt{25}} = 1.$

(b) $d(P, l) = \dfrac{|Ax_0 + By_0 + C|}{\sqrt{A^2 + B^2}} = \dfrac{|3(-6) + 4(2) - 5|}{\sqrt{3^2 + 4^2}} = \dfrac{15}{\sqrt{25}} = 3. \quad \square$

Exercises

1. Find the distance between
   *(a) the line $5x + 12y + 2 = 0$ and the origin.
   (b) the line $5x + 12y + 2 = 0$ and the point $P(1, -3)$.
   *(c) the line $2x - 3y + 1 = 0$ and the point $P(2, 1)$.
   (d) the line $y = 2x + 3$ and the origin.
   *(e) the line $y = 2x$ and the point $P(3, 5)$.
2. Which of the points $(0, 1)$, $(1, 0)$, and $(-1, 1)$ is closest to $l: 8x + 7y - 6 = 0$? Which is farthest from l?
3. Compute the area of the triangle
   *(a) with vertices $(1, -2)$, $(-1, 3)$, $(2, 4)$.
   (b) with vertices $(-1, 1)$, $(3, \sqrt{2})$, $(\sqrt{2}, -1)$.
4. Show that the equation of a line

$$Ax + By + C = 0$$

can always be written in the *normal form*:

$$x \cos \alpha + y \sin \alpha = p \qquad \text{with } p \geq 0.$$

What is the geometric significance of p? of α?
*5. A ray l is rotating clockwise about the point $Q(-b^2, 0)$ at the rate of one revolution per minute. How fast is the distance between l and the origin changing at the moment that l has slope $\frac{3}{4}$?
6. A ray l is rotating counterclockwise about the point $Q(-b^2, 0)$ at the rate of $\frac{1}{2}$ revolution per minute. What is the slope of l at the moment that l is receding most quickly from the point $P(b^2, -1)$?

9.3 The Parabola

We begin with a line l and a point F not on l.

(9.3.1) | The set of points P equidistant from F and l is called a *parabola*.

See Figure 9.3.1.

a parabola

FIGURE 9.3.1

The line l is called the *directrix* of the parabola, and the point F is called the *focus*. (You will see why later on.) The line through F which is perpendicular to l is called the *axis* of the parabola. The point at which the axis intersects the parabola is called the *vertex*. (See Figure 9.3.2.)

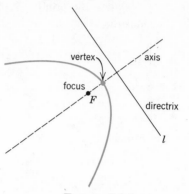

FIGURE 9.3.2

The equation of a parabola is particularly simple if we place the vertex at the origin and the focus along one of the coordinate axes. Suppose for a moment that the focus F is on the y-axis. Then F has coordinates of the form $(0, c)$, and, with the vertex at the origin, the directrix must have equation $y = -c$. (See Figure 9.3.3.)

FIGURE 9.3.3

By definition every point $P(x, y)$ which lies on this parabola has the property that

$$d(P, F) = d(P, l).$$

Since

$$d(P, F) = \sqrt{x^2 + (y - c)^2} \quad \text{and} \quad d(P, l) = |y + c|,$$

we must have

$$\sqrt{x^2 + (y - c)^2} = |y + c|,$$

$$x^2 + (y - c)^2 = |y + c|^2 = (y + c)^2,$$
$$x^2 + y^2 - 2cy + c^2 = y^2 + 2cy + c^2,$$
$$x^2 = 4cy.$$

You have just seen that the equation

(**9.3.2**) $\qquad\qquad\boxed{x^2 = 4cy}\qquad\qquad$ (Figure 9.3.4)

represents a parabola with vertex at the origin and focus at $(0, c)$. If the vertex remains at the origin and the focus is placed at $(c, 0)$, the equation takes the form

(**9.3.3**) $\qquad\qquad\boxed{y^2 = 4cx.}\qquad\qquad$ (Figure 9.3.5)

$x^2 = 4cy$

FIGURE 9.3.4

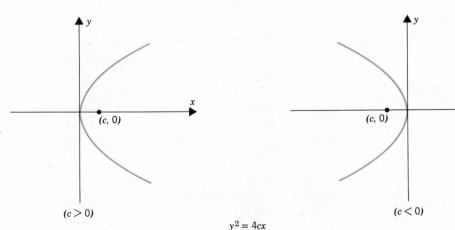

$y^2 = 4cx$

FIGURE 9.3.5

Every parabola with a vertical axis is a translation of a parabola of the form $x^2 = 4cy$, and every parabola with a horizontal axis is a translation of a parabola of the form $y^2 = 4cx$. The equation

$$(x - x_0)^2 = 4c(y - y_0)$$

represents a parabola with vertex (x_0, y_0) and focus $(x_0, y_0 + c)$. The axis is vertical. The equation

$$(y - y_0)^2 = 4c(x - x_0)$$

represents a parabola with vertex (x_0, y_0) and focus $(x_0 + c, y_0)$. The axis is horizontal.

Problem.   Identify the curve

$$(x - 1)^2 = 8(y + 3).$$

SOLUTION.   We rewrite the equation as

$$(x - 1)^2 = 8[y - (-3)].$$

This is the parabola

$$x^2 = 8y \qquad\qquad (c = 2)$$

translated so that the vertex falls at $(1, -3)$. The axis is vertical and the focus is at $(1, -1)$.   □

Problem.   Identify the curve

$$(y - 1)^2 = 8(x + 3).$$

SOLUTION.   We rewrite the equation as

$$(y - 1)^2 = 8[x - (-3)].$$

This is the parabola

$$y^2 = 8x \qquad\qquad (c = 2)$$

translated so that the vertex falls at $(-3, 1)$. The axis is horizontal and the focus is at $(-1, 1)$.   □

Problem.   Identify the curve

$$y = x^2 + 2x - 2.$$

SOLUTION.   We first complete the square on the right by adding 3 to both sides of the equation:

$$y + 3 = x^2 + 2x + 1 = (x + 1)^2.$$

This gives

$$(x + 1)^2 = y + 3,$$
$$[x - (-1)]^2 = y - (-3).$$

This is the parabola

$$x^2 = y \qquad\qquad (c = \tfrac{1}{4})$$

translated so that the vertex falls at $(-1, -3)$. The axis is vertical and the focus is at $(-1, -\frac{11}{4})$. $\square$

By the method of the last problem one can show that every quadratic

$$y = Ax^2 + Bx + C$$

represents a parabola with vertical axis. It looks like $\vee$ if A is positive and like $\wedge$ if A is negative.

Parabolic Reflectors

Parabolic curves play a prominent role in the design of searchlights and telescopes. To show how this comes about, we take a parabola and choose the coordinate system so that the equation of the parabola takes the form $x^2 = 4cy$. We can express y in terms of x by writing

$$y = \frac{x^2}{4c}.$$

Since

$$\frac{dy}{dx} = \frac{2x}{4c} = \frac{x}{2c},$$

the tangent at the point $P(x_0, y_0)$ has slope $m = x_0/2c$ and the tangent line has equation

$$(y - y_0) = \frac{x_0}{2c}(x - x_0).$$

For the rest we refer to Figure 9.3.6.

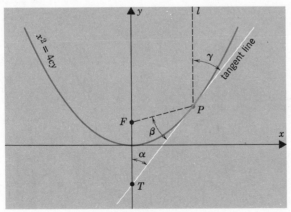

FIGURE 9.3.6

To find the coordinates of the point marked T in the figure, we set $x = 0$ and solve for y. Doing this, we find that T has coordinates $(0, y_0 - x_0^2/2c)$. Since the point

(x_0, y_0) is on the parabola, we know that $x_0^2 = 4cy_0$, and therefore we can rewrite the coordinates of T as $(0, -y_0)$. Since the focus F has coordinates $(0, c)$,

$$d(F, T) = c + y_0.$$

Now

$$d(F, P) = \sqrt{x_0^2 + (y_0 - c)^2} = \sqrt{4cy_0 + (y_0 - c)^2} = \sqrt{(y_0 + c)^2} = c + y_0.$$

$$x_0^2 = 4cy_0$$

We have shown that

$$d(F, T) = d(F, P).$$

It follows that the triangle *TFP* is isosceles and that the base angles marked α and β in the figure are equal. With l parallel to the y-axis,

$$\alpha = \gamma \quad \text{and thus} \quad \gamma = \beta.$$

From physics we know that when light is reflected, the angle of incidence equals the angle of reflection. We can take either γ or β as the angle of incidence; the other will then be the angle of reflection. The fact that $\gamma = \beta$ has important optical consequences. It means that *light from a source at the focus is reflected in a beam parallel to the axis*. It also means that *a beam of light parallel to the axis is reflected by the parabola through the focus*. The parabolic mirror of a searchlight uses the first principle, that of a reflecting telescope the second. (See Figure 9.3.7.)

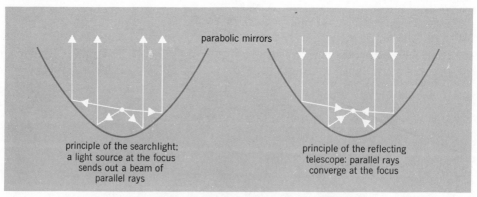

parabolic mirrors

principle of the searchlight:
a light source at the focus
sends out a beam of
parallel rays

principle of the reflecting
telescope: parallel rays
converge at the focus

FIGURE 9.3.7

Parabolic Trajectories

In the early part of the seventeenth century Galileo Galilei observed the motion of stones thrown from the tower of Pisa and noted that their trajectory was parabolic (in the shape of a parabola). By very simple calculus, together with some simplifying physical assumptions, we obtain results which agree with Galileo's observations.

Consider a projectile fired at angle θ from a point (x_0, y_0) with initial velocity of v_0 feet per second. (Figure 9.3.8)

FIGURE 9.3.8

The horizontal component of v_0 is $v_0 \cos \theta$, and the vertical component of v_0 is $v_0 \sin \theta$. (Figure 9.3.9)

FIGURE 9.3.9

Let's neglect air resistance and the curvature of the earth. Under these circumstances there is no acceleration in the x-direction:

$$x''(t) = 0.$$

The only acceleration in the y-direction is due to gravity:

$$y''(t) = -32.$$

From the first equation,

$$x'(t) = C$$

and, since $x'(0) = v_0 \cos \theta$,

$$x'(t) = v_0 \cos \theta.$$

Integrating again,

$$x(t) = (v_0 \cos \theta)t + C.$$

Since $x(0) = x_0$,

(1) $$x(t) = (v_0 \cos \theta)t + x_0.$$

From $y''(t) = -32$, it follows that

$$y'(t) = -32t + C.$$

Since $y'(0) = v_0 \sin \theta$,

$$y'(t) = -32t + v_0 \sin \theta.$$

Integrating again,

$$y(t) = -16t^2 + (v_0 \sin \theta)t + C$$

and, since $y(0) = y_0$,

(2) $$y(t) = -16t^2 + (v_0 \sin \theta)t + y_0.$$

From (1)

$$t = \frac{1}{v_0 \cos \theta}[x(t) - x_0].$$

If you substitute this value of t in (2), you will find that

$$y(t) = -\frac{16 \sec^2 \theta}{v_0^2}[x(t) - x_0]^2 + \tan \theta\,[x(t) - x_0] + y_0.$$

The trajectory (the path followed by the projectile) is the curve

(9.3.4) $$\boxed{y = -\frac{16}{v_0^2}\sec^2 \theta\,[x - x_0]^2 + \tan \theta\,[x - x_0] + y_0.}$$

This is a quadratic in x and therefore a parabola.   □

Exercises

1. Sketch the parabola and give an equation for it:
   *(a) vertex $(0, 0)$, focus $(2, 0)$.   *(b) vertex $(0, 0)$, focus $(-2, 0)$.
   (c) vertex $(0, 0)$, focus $(0, 2)$.   *(d) vertex $(1, 2)$, focus $(3, 2)$.
   (e) vertex $(1, 2)$, focus $(1, 3)$.   *(f) focus $(1, 1)$, directrix $y = -1$.
   (g) focus $(1, 1)$, directrix $x = 2$.   *(h) focus $(2, -2)$, directrix $x = 5$.
2. Find the vertex, focus, axis, and directrix; then sketch the parabola.
   *(a) $y^2 = 2x$.   (b) $x^2 = -5y$.
   *(c) $y^2 = 2(x - 1)$.   (d) $2y = 4x^2 - 1$.
   *(e) $(x + 2)^2 = 8y - 12$.   (f) $y - 3 = 2(x - 1)^2$.
   *(g) $y = x^2 + x + 1$.   (h) $x = y^2 + y + 1$.
3. Find an equation for the indicated parabola:
   *(a) focus $(1, 2)$, directrix $x + y + 1 = 0$.
   (b) vertex $(2, 0)$, directrix $2x - y = 0$.
   *(c) vertex $(2, 0)$, focus $(0, 2)$.
4. Show that every parabola has an equation of the form

$$(\alpha x + \beta y)^2 = \gamma x + \delta y + \epsilon \qquad \text{with} \quad \alpha^2 + \beta^2 \neq 0.$$

   HINT: Take $l: Ax + By + C = 0$ as the directrix and $F(a, b)$ as the focus.
5. Show that not every equation of the form

$$(\alpha x + \beta y)^2 = \gamma x + \delta y + \epsilon \quad \text{with} \quad \alpha^2 + \beta^2 \neq 0$$

represents a parabola.

6. Identify the graph of $(x - y)^2 = b$ for different values of b.

*7. Find an equation for the parabola which has directrix $y = 1$, axis $x = 2$, and passes through the point $(5, 6)$.

8. Find an equation for the parabola which has horizontal axis, vertex $(-1, 1)$, and passes through the point $(-6, 13)$.

In Exercises 9 through 14 we measure distance in feet and time in seconds. We neglect air resistance and the curvature of the earth. We take O as the origin, the x-axis as ground level, and consider a projectile fired from O at an angle θ with initial velocity v_0.

*9. Find an equation for the trajectory.

10. What is the maximum height attained by the projectile?

*11. Find the range of the projectile.

12. How many seconds after firing does the impact take place?

*13. How should θ be chosen so as to maximize the range?

14. How should θ be chosen so that the range becomes r?

15. (a) Show that every parabola with axis parallel to the y-axis has an equation of the form $y = Ax^2 + Bx + C$.

   *(b) Find the vertex, the focus, and the directrix of such a parabola.

9.4 The Ellipse

Start with two points F_1, F_2 and a number k greater than the distance between them.

(9.4.1)

> The set of all points P such that
> $$d(P, F_1) + d(P, F_2) = k$$
> is called an *ellipse*. F_1 and F_2 are called the *foci*.

The idea is illustrated in Figure 9.4.1. A string is looped over tacks placed at the foci. The pencil placed in the loop traces out an ellipse.

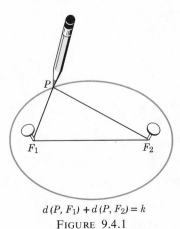

$d(P, F_1) + d(P, F_2) = k$

FIGURE 9.4.1

To derive the usual equation of an ellipse we set $k = 2a$ and place the foci along the x-axis at equal distances from the origin. (Figure 9.4.2) A point $P(x, y)$ will lie on the ellipse iff

$$d(P, F_1) + d(P, F_2) = 2a.$$

With F_1 at $(-c, 0)$ and F_2 at $(c, 0)$ we must have

$$\sqrt{(x + c)^2 + y^2} + \sqrt{(x - c)^2 + y^2} = 2a.$$

By transferring the second term to the right-hand side and squaring both sides, we obtain

$$(x + c)^2 + y^2 = 4a^2 + (x - c)^2 + y^2 - 4a\sqrt{(x - c)^2 + y^2},$$

which, as you can check, reduces to

$$\frac{x^2}{a^2} + \frac{y^2}{a^2 - c^2} = 1. \quad \square$$

It is customary to set $b = \sqrt{a^2 - c^2}$ and then write the equation as

(9.4.2)
$$\boxed{\frac{x^2}{a^2} + \frac{y^2}{b^2} = 1.}$$

The relationship between a, b, and c is illustrated in Figure 9.4.3.

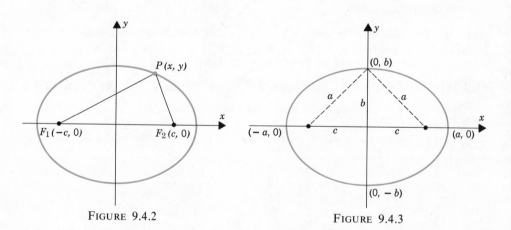

FIGURE 9.4.2   FIGURE 9.4.3

The line segment that joins the *vertices* $(-a, 0)$ and $(a, 0)$ is called the *major axis*; the line segment that joins the *vertices* $(0, -b)$ and $(0, b)$ is called the *minor axis*. The number $2a$ gives the length of the major axis, and the number $2b$ gives the length of the minor axis.

Example. The equation

$$16x^2 + 25y^2 = 400$$

can be written

$$\frac{x^2}{25} + \frac{y^2}{16} = 1. \qquad \text{(divide by 400)}$$

Here $a = 5$, $b = 4$, and $c = \sqrt{a^2 - b^2} = \sqrt{9} = 3$. We are dealing with an ellipse with foci at $(-3, 0)$ and $(3, 0)$. The major axis has length $2a = 10$, and the minor axis has length $2b = 8$. (Figure 9.4.4) $\square$

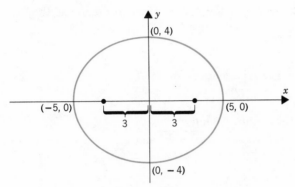

FIGURE 9.4.4

If the foci of the ellipse are placed on the y-axis, F_1 at $(0, -c)$ and F_2 at $(0, c)$ the equation takes the form

(9.4.3)
$$\frac{x^2}{b^2} + \frac{y^2}{a^2} = 1.$$

Once again $b = \sqrt{a^2 - c^2}$. The major axis is now vertical, and the minor axis is horizontal. $\square$

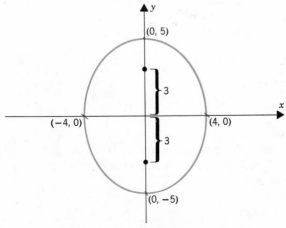

FIGURE 9.4.5

Example.   The ellipse

$$\frac{x^2}{16} + \frac{y^2}{25} = 1 \qquad \text{(Figure 9.4.5)}$$

has the same size and the same shape as

$$\frac{x^2}{25} + \frac{y^2}{16} = 1,$$

but this time the foci are on the *y*-axis. The coordinates of the foci are $(0, -3)$ and $(0, 3)$.   □

The ellipses

$$\frac{x^2}{a^2} + \frac{y^2}{b^2} = 1 \quad \text{and} \quad \frac{x^2}{b^2} + \frac{y^2}{a^2} = 1$$

are centered at the origin. Equations

$$\frac{(x - x_0)^2}{a^2} + \frac{(y - y_0)^2}{b^2} = 1 \quad \text{and} \quad \frac{(x - x_0)^2}{b^2} + \frac{(y - y_0)^2}{a^2} = 1$$

represent ellipses of the same shape and size centered at (x_0, y_0).   □

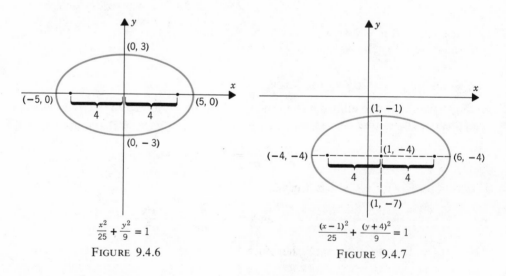

$$\frac{x^2}{25} + \frac{y^2}{9} = 1$$

FIGURE 9.4.6

$$\frac{(x-1)^2}{25} + \frac{(y+4)^2}{9} = 1$$

FIGURE 9.4.7

Example.   The ellipse

$$\frac{(x - 1)^2}{25} + \frac{(y + 4)^2}{9} = 1$$

is the ellipse

$$\frac{x^2}{25} + \frac{y^2}{9} = 1$$

translated so that the center falls at $(1, -4)$. The major axis has length $2a = 10$, and the minor axis has length $2b = 6$. Since $c = \sqrt{a^2 - b^2} = \sqrt{25 - 9} = 4$, one of the foci lies 4 units to the left of the center [at $(-3, -4)$], and the other lies 4 units to the right of center [at $(5, -4)$]. See Figures 9.4.6 and 9.4.7. $\square$

Example.   To identify the curve

$$4x^2 - 8x + y^2 + 4y - 8 = 0,$$

we write

$$4(x^2 - 2x + \quad) + (y^2 + 4y + \quad) = 8.$$

By completing the squares within the parentheses, we get

$$4(x^2 - 2x + 1) + (y^2 + 4y + 4) = 16,$$
$$4(x - 1)^2 + (y + 2)^2 = 16,$$
$$\frac{(x - 1)^2}{4} + \frac{(y + 2)^2}{16} = 1.$$

This is an ellipse with a vertical major axis. The center is at $(1, -2)$. The major axis has length $2a = 8$, and the minor axis has length $2b = 4$. Since $c = \sqrt{a^2 - b^2} = \sqrt{12} = 2\sqrt{3}$, one of the foci lies $2\sqrt{3}$ units below the center [at $(1, -2 - 2\sqrt{3})$] and the other lies $2\sqrt{3}$ units above the center [at $(1, -2 + 2\sqrt{3})$]. $\square$

The Reflecting Property of the Ellipse

Like the parabola, the ellipse has an interesting reflecting property. To derive it, we consider the ellipse

$$\frac{x^2}{a^2} + \frac{y^2}{b^2} = 1.$$

Differentiation with respect to x gives

$$\frac{2x}{a^2} + \frac{2y}{b^2}\frac{dy}{dx} = 0 \quad \text{and thus} \quad \frac{dy}{dx} = -\frac{b^2 x}{a^2 y}.$$

The slope at the point (x_0, y_0) is therefore

$$-\frac{b^2 x_0}{a^2 y_0}$$

and the tangent line has equation

$$y - y_0 = -\frac{b^2 x_0}{a^2 y_0}(x - x_0).$$

We can rewrite this last equation as

$$(b^2 x_0)x + (a^2 y_0)y - a^2 b^2 = 0.$$

We can now show the following:

(9.4.4) At each point P of the ellipse, the focal radii $\overline{F_1P}$ and $\overline{F_2P}$ make equal angles with the tangent.

PROOF.   If P lies on the x-axis, the focal radii are coincident and there is nothing to show. To visualize the argument for P not on the x-axis, see Figure 9.4.8. To show that $\overline{F_1P}$ and $\overline{F_2P}$ make equal angles with the tangent we need only show that the triangles PT_1F_1 and PT_2F_2 are similar. We can do this by showing that

$$\frac{d(T_1, F_1)}{d(F_1, P)} = \frac{d(T_2, F_2)}{d(F_2, P)}$$

or equivalently by showing that

$$\frac{|-b^2x_0c - a^2b^2|}{\sqrt{(x_0 + c)^2 + y_0^2}} = \frac{|b^2x_0c - a^2b^2|}{\sqrt{(x_0 - c)^2 + y_0^2}}.$$

The validity of this last equation can be seen by canceling the factor b^2 and then squaring. This gives

$$\frac{(x_0c + a^2)^2}{(x_0 + c)^2 + y_0^2} = \frac{(x_0c - a^2)^2}{(x_0 - c)^2 + y_0^2},$$

which can be simplified to

$$(a^2 - c^2)x_0^2 + a^2y_0^2 = a^2(a^2 - c^2)$$

and thus to

$$\frac{x_0^2}{a^2} + \frac{y_0^2}{b^2} = 1.$$

This last equation holds since the point $P(x_0, y_0)$ is on the ellipse. $\square$

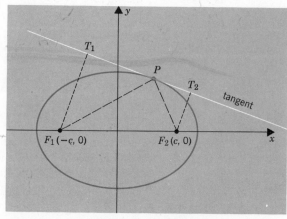

FIGURE 9.4.8

The result we just proved has the following physical consequence:

(9.4.5) | An elliptical mirror takes light or sound originating at one focus and converges it at the other focus.

Exercises

For each of the following ellipses (a) find the foci, (b) find the length of the major axis, (c) find the length of the minor axis, and then (d) sketch the figure.

*1. $\dfrac{x^2}{9} + \dfrac{y^2}{4} = 1.$ 2. $\dfrac{x^2}{4} + \dfrac{y^2}{9} = 1.$

*3. $3x^2 + 2y^2 = 12.$ 4. $(x - 1)^2 + 4y^2 = 64.$

*5. $3x^2 + 4y^2 - 12 = 0.$ 6. $4x^2 + y^2 - 6y + 5 = 0.$

*7. $4(x - 1)^2 + y^2 = 64.$ 8. $16(x - 2)^2 + 25(y - 3)^2 = 400.$

Find an equation for the ellipse which satisfies the given conditions.

*9. Foci at $(-1, 0)$, $(1, 0)$; major axis 6.

10. Foci at $(0, -1)$, $(0, 1)$; major axis 6.

*11. Foci at $(3, 1)$, $(9, 1)$; major axis 10.

12. Foci at $(1, 3)$, $(1, 9)$; minor axis 8.

*13. Focus at $(1, 1)$; center at $(1, 3)$; major axis 10.

14. Center at $(2, 1)$; vertices at $(2, 6)$ and $(1, 1)$.

*15. Major axis 10; vertices at $(3, 2)$ and $(3, -4)$.

*16. Find the area enclosed by the ellipse

$$\frac{x^2}{a^2} + \frac{y^2}{b^2} = 1.$$

17. Show that in an ellipse the product of the distances between the foci and a tangent tc the ellipse is always b^2, the square of one half the length of the minor axis.

The *eccentricity e* of an ellipse is the ratio of the distance between the foci to the length of the major axis:

(9.4.6) | $e = \dfrac{c}{a}.$

*18. What happens to the ellipse if a remains constant but e tends to 0?

19. What happens to the ellipse if a remains constant but e tends to 1?

*20. Find an equation for the ellipse in standard position which has major axis 10 and eccentricity $\frac{1}{2}$.

21. Find an equation for the ellipse in standard position which has minor axis $2\sqrt{3}$ and eccentricity $\frac{1}{2}$.

22. Like the parabola the ellipse can be defined in terms of one focus F and a directrix line l. Show that the ellipse

$$\frac{x^2}{a^2} + \frac{y^2}{a^2 - c^2} = 1$$

is the set of points $P(x, y)$ such that

$$d(P, F) = e\, d(P, l),$$

where F is $(c, 0)$, $e = c/a$, and $l: x = a^2/c$. Note that $0 < e < 1$.

9.5 The Hyperbola

Start with two points F_1, F_2 and take a positive number k less than the distance between them.

(9.5.1)

> The set of all points P such that
> $$|d(P, F_1) - d(P, F_2)| = k$$
> is called a *hyperbola*. F_1 and F_2 are called the *foci*.

To derive the usual equation for the hyperbola, we proceed as we did with the ellipse. We set $k = 2a$ and place the foci along the x-axis at equal distances from the origin. (Figure 9.5.1)

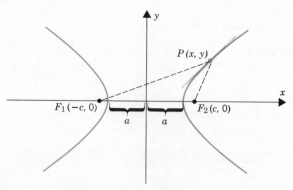

FIGURE 9.5.1

A point $P(x, y)$ will lie on the hyperbola iff

$$|d(P, F_1) - d(P, F_2)| = 2a.$$

With F_1 at $(-c, 0)$ and F_2 at $(c, 0)$ we must have

$$\sqrt{(x + c)^2 + y^2} - \sqrt{(x - c)^2 + y^2} = \pm 2a.$$

Transferring the second term to the right and squaring both sides, we obtain

$$(x + c)^2 + y^2 = 4a^2 \pm 4a\sqrt{(x - c)^2 + y^2} + (x - c)^2 + y^2,$$

which, as you can check, reduces to

$$\frac{x^2}{a^2} - \frac{y^2}{c^2 - a^2} = 1. \quad \square$$

In this setting it is customary to set $b = \sqrt{c^2 - a^2}$ and write

(9.5.2)
$$\boxed{\frac{x^2}{a^2} - \frac{y^2}{b^2} = 1.}$$

The hyperbola together with the relationship between $a, b,$ and c is illustrated in Figure 9.5.2. As the figure indicates, the hyperbola remains between the lines

$$y = \frac{b}{a}x \quad \text{and} \quad y = -\frac{b}{a}x.$$

These lines are called the *asymptotes* of the hyperbola. As $|x|$ increases without bound, the separation between the hyperbola and the asymptotes tends to zero. To see this, solve the equation

$$\frac{x^2}{a^2} - \frac{y^2}{b^2} = 1$$

for y. This gives

$$y = \pm\sqrt{\frac{b^2}{a^2}x^2 - b^2}.$$

For large $|x|$, b^2 is negligible compared to $(b^2/a^2)x^2$, and y is therefore approximately

$$\pm\sqrt{\frac{b^2}{a^2}x^2} = \pm\frac{b}{a}x. \quad \square$$

The line determined by the foci of a hyperbola intersects the hyperbola at two points called the *vertices*. The line segment that joins the vertices is called the *transverse* axis. The midpoint of the transverse axis is called the *center* of the hyperbola.

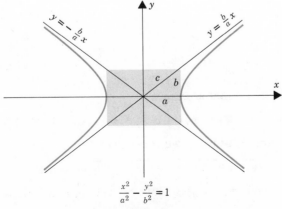

$$\frac{x^2}{a^2} - \frac{y^2}{b^2} = 1$$

FIGURE 9.5.2

In Figure 9.5.2, the vertices are $(\pm a, 0)$, the transverse axis has length $2a$, and the center is at the origin.

Example 1.   In the case of

$$\frac{x^2}{1} - \frac{y^2}{3} = 1, \qquad\qquad \text{(Figure 9.5.3)}$$

$a = 1, b = \sqrt{3}, c = \sqrt{1 + 3} = 2$. The equation represents the hyperbola with foci at $(\pm 2, 0)$ and transverse axis of length 2. The center is at the origin, the vertices are at $(\pm 1, 0)$, and the asymptotes are

$$y = \pm\sqrt{3}x. \quad \square$$

FIGURE 9.5.3

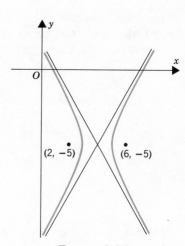

FIGURE 9.5.4

Example 2.   The equation

$$\frac{(x - 4)^2}{1} - \frac{(y + 5)^2}{3} = 1 \qquad\qquad \text{(Figure 9.5.4)}$$

represents the hyperbola of Example 1 translated so that the center falls at the point $(4, -5)$. The foci are now at $(2, -5)$, $(6, -5)$, and the vertices are at $(3, -5)$, $(5, -5)$. The asymptotes are

$$y + 5 = \pm\sqrt{3}(x - 4). \quad \square$$

Example 3.   The equation

$$\frac{y^2}{1} - \frac{x^2}{3} = 1 \qquad\qquad \text{(Figure 9.5.5)}$$

is the equation of Example 1 with x and y interchanged. We still have a hyperbola with center at the origin. The foci are now at $(0, \pm 2)$, and the vertices are at $(0, \pm 1)$. The asymptotes are

$$x = \pm \sqrt{3} y. \quad \square$$

Example 4. The equation

$$\frac{(y - 5)^2}{1} - \frac{(x - 2)^2}{3} = 1 \qquad \text{(Figure 9.5.6)}$$

represents the hyperbola of Example 3 translated so that the center falls at $(2, 5)$. $\quad \square$

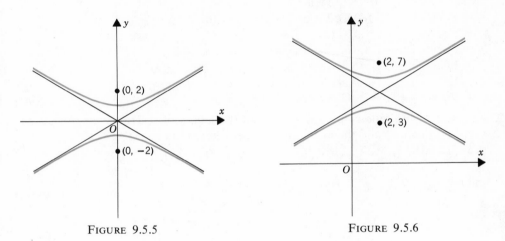

FIGURE 9.5.5   FIGURE 9.5.6

The Tangent to a Hyperbola

Proceeding as in the case of the ellipse, one can show that, if (x_0, y_0) with $y_0 \neq 0$ is a point of the hyperbola

$$\frac{x^2}{a^2} - \frac{y^2}{b^2} = 1,$$

then the slope at (x_0, y_0) is given by the number

$$m = \frac{b^2 x_0}{a^2 y_0} \qquad \text{(Exercise 20)}$$

and the tangent line has equation

$$(b^2 x_0) x - (a^2 y_0) y - a^2 b^2 = 0. \qquad \text{(Exercise 21)}$$

Using this information, one can show that

(9.5.3) | *at each point P of the hyperbola the tangent bisects the angle between the focal radii* $\overline{F_1 P}$ *and* $\overline{F_2 P}$. | (Exercise 22)

An Application to Range Finding

There is a simple application of the hyperbola to range-finding. If observers, located at two listening posts at a known distance apart, time the firing of a gun, the time difference multiplied by the velocity of sound gives the value of $2a$ and hence determines a hyperbola on which the gun must be located. A third listening post gives two more hyperbolas. The gun is found where the hyperbolas intersect.

Exercises

Find an equation for the indicated hyperbola.

*1. Foci at $(-5, 0)$, $(5, 0)$; transverse axis 6.

 2. Foci at $(-13, 0)$, $(13, 0)$; transverse axis 10.

*3. Foci at $(0, -13)$, $(0, 13)$; transverse axis 10.

 4. Foci at $(0, -13)$, $(0, 13)$; transverse axis 24.

*5. Foci at $(-5, 1)$, $(5, 1)$; transverse axis 6.

 6. Foci at $(-3, 1)$, $(7, 1)$; transverse axis 6.

*7. Foci at $(-1, -1)$, $(1, 1)$; transverse axis 2.

For each of the following hyperbolas find the length of the transverse axis, the vertices, the foci, and the asymptotes. Then sketch the figure.

*8. $x^2 - y^2 = 1$.

 9. $y^2 - x^2 = 1$.

*10. $\dfrac{x^2}{9} - \dfrac{y^2}{16} = 1$.

 11. $\dfrac{x^2}{16} - \dfrac{y^2}{9} = 1$.

 12. $\dfrac{y^2}{16} - \dfrac{x^2}{9} = 1$.

 13. $\dfrac{y^2}{9} - \dfrac{x^2}{16} = 1$.

*14. $\dfrac{(x-1)^2}{9} - \dfrac{(y-3)^2}{16} = 1$.

 15. $\dfrac{(x-1)^2}{16} - \dfrac{(y-3)^2}{9} = 1$.

*16. $-3x^2 + y^2 - 6x = 0$.

 17. $4x^2 - 8x - y^2 + 6y - 1 = 0$.

*18. Find the length of the transverse axis, the vertices, the foci, and the asymptotes of the hyperbola with equation $xy = 1$. HINT: Define new XY-coordinates by setting $x = X + Y$ and $y = X - Y$.

 19. Find the area of the region that lies between the right-hand branch of the hyperbola

$$\frac{x^2}{a^2} - \frac{y^2}{b^2} = 1$$

and the vertical line $x = 2a$.

 20. Show that, if $y_0 \neq 0$, then the slope of the hyperbola

$$\frac{x^2}{a^2} - \frac{y^2}{b^2} = 1$$

at the point (x_0, y_0) is given by

$$m = \frac{b^2 x_0}{a^2 y_0}.$$

21. Show that the tangent line at (x_0, y_0) has equation

$$(b^2 x_0)x - (a^2 y_0)y - a^2 b^2 = 0.$$

22. Show that the tangent to the hyperbola at the point P bisects the angle between the focal radii $\overline{F_1 P}$ and $\overline{F_2 P}$.

23. Like the parabola and the ellipse, the hyperbola can be defined in terms of one focus F and a directrix line l. Show that the hyperbola

$$\frac{x^2}{a^2} - \frac{y^2}{c^2 - a^2} = 1$$

is the set of all points $P(x, y)$ such that

$$d(P, F) = e\, d(P, l),$$

where F is $(c, 0)$, $e = c/a$, and $l{:}x = a^2/c$. Note that $e > 1$.

9.6   Additional Exercises

Describe the following curves in detail.

*1. $x^2 - 4y - 4 = 0$.   2. $3x^2 + 2y^2 - 6 = 0$.

*3. $x^2 - 4y^2 - 10x + 41 = 0$.   4. $9x^2 - 4y^2 - 18x - 8y - 31 = 0$.

*5. $x^2 + 3y^2 + 6x + 8 = 0$.   6. $x^2 - 10x - 8y + 41 = 0$.

*7. $y^2 + 4y + 2x + 1 = 0$.   8. $9x^2 + 4y^2 - 18x - 8y - 23 = 0$.

*9. $9x^2 + 25y^2 + 100y + 99 = 0$.   10. $7x^2 - y^2 + 42x + 14y + 21 = 0$.

*11. $7x^2 - 5y^2 + 14x - 40y = 118$.   12. $2x^2 - 3y^2 + 4\sqrt{3}x - 6\sqrt{3}y = 9$.

*13. $(x^2 - 4y)(4x^2 + 9y^2 - 36) = 0$.   14. $(x^2 - 4y)(x^2 - 4y^2) = 0$.

9.7   Rotations; Eliminating the *xy*-Term

We begin by referring to Figure 9.7.1:

FIGURE 9.7.1

From the figure

$$\cos \theta = \frac{x}{r} \quad \text{and} \quad \sin \theta = \frac{y}{r}.$$

Thus

(9.7.1) $$\boxed{x = r \cos \theta \quad \text{and} \quad y = r \sin \theta.}$$

These equations come up repeatedly in calculus. We will consider them in more detail in Section 11.1 when we take up polar coordinates.

Consider now a rectangular coordinate system Oxy. If we rotate this system α radians about the origin, we obtain a new coordinate system OXY. See Figure 9.7.2.

FIGURE 9.7.2   FIGURE 9.7.3

A point P will now have two pairs of rectangular coordinates:

(x, y) in the Oxy system   and (X, Y) in the OXY system.

Here we investigate the relation between (x, y) and (X, Y). With P as in Figure 9.7.3,

$$x = r \cos (\alpha + \beta), \quad y = r \sin (\alpha + \beta)$$

and

$$X = r \cos \beta, \quad Y = r \sin \beta.$$

Since

$$\cos (\alpha + \beta) = \cos \alpha \cos \beta - \sin \alpha \sin \beta,$$
$$\sin (\alpha + \beta) = \sin \alpha \cos \beta + \cos \alpha \sin \beta,$$

we have

$$x = r \cos (\alpha + \beta) = (\cos \alpha) r \cos \beta - (\sin \alpha) r \sin \beta,$$
$$y = r \sin (\alpha + \beta) = (\sin \alpha) r \cos \beta + (\cos \alpha) r \sin \beta,$$

and therefore

(9.7.2) $$\boxed{x = (\cos \alpha)X - (\sin \alpha)Y \quad \text{and} \quad y = (\sin \alpha)X + (\cos \alpha)Y.}$$

These formulas give the algebraic consequences of a rotation of α radians.

Eliminating the *xy*-Term

Rotations of the coordinate system enable us to simplify equations of the second degree by eliminating the *xy*-term; that is, if in the Oxy system, S has an equation of the form

(1) $ax^2 + bxy + cy^2 + dx + ey + f = 0$ with $b \neq 0$,

then there exists a coordinate system $OXY,$ differing from Oxy by a rotation, such that in the OXY system S has an equation of the form

$$AX^2 + CY^2 + DX + EY + F = 0.$$

To see this, substitute

$$x = (\cos \alpha)X - (\sin \alpha)Y, \qquad y = (\sin \alpha)X + (\cos \alpha)Y$$

in Equation (1). This will give you a second-degree equation in X and Y in which the coefficient of XY is

$$-2a \cos \alpha \sin \alpha + b(\cos^2 \alpha - \sin^2 \alpha) + 2c \cos \alpha \sin \alpha.$$

This can be simplified to

$$(c - a) \sin 2\alpha = b \cos 2\alpha.$$

To eliminate the XY term we must have this coefficient equal to zero, that is, we must have

$$(a - c) \sin 2\alpha = b \cos 2\alpha.$$

There are two possibilities here: $a = c, a \neq c$.

I. If $a = c$, then

$$\cos 2\alpha = 0,$$
$$2\alpha = \tfrac{1}{2}\pi + n\pi, \qquad (n \text{ an arbitrary integer})$$
$$\alpha = \tfrac{1}{4}\pi + \tfrac{1}{2}n\pi.$$

Thus, if $a = c,$ we can eliminate the XY term by setting

(9.7.3) $\boxed{\alpha = \tfrac{1}{4}\pi.}$ (choose $n = 0$)

II. If $a \neq c$, then

$$\tan 2\alpha = \frac{b}{a - c},$$

$$2\alpha = \arctan\left(\frac{b}{a - c}\right) + n\pi,$$

$$\alpha = \tfrac{1}{2}\arctan\left(\frac{b}{a - c}\right) + \tfrac{1}{2}n\pi.$$

Thus, if $a \neq c,$ we can eliminate the XY term by setting

(9.7.4) $\boxed{\alpha = \tfrac{1}{2}\arctan\left(\dfrac{b}{a - c}\right).}$ (choose $n = 0$)

Example.   In the case of

$$xy - 2 = 0,$$

we have $a = c$ and thus can choose $\alpha = \frac{1}{4}\pi$. Setting

$$x = (\cos\tfrac{1}{4}\pi)X - (\sin\tfrac{1}{4}\pi)Y = \tfrac{1}{2}\sqrt{2}(X - Y),$$
$$y = (\sin\tfrac{1}{4}\pi)X + (\cos\tfrac{1}{4}\pi)Y = \tfrac{1}{2}\sqrt{2}(X + Y),$$

we find that $xy - 2 = 0$ becomes

$$\tfrac{1}{2}(X^2 - Y^2) - 2 = 0,$$

which can be written

$$\frac{X^2}{4} - \frac{Y^2}{4} = 1.$$

This is the equation of a hyperbola in standard position in the OXY system. The graph is shown in Figure 9.7.4.   □

FIGURE 9.7.4

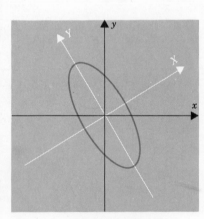

FIGURE 9.7.5

Example.   In the case of

$$11x^2 + 4\sqrt{3}xy + 7y^2 - 1 = 0,$$

we have $a = 11$, $b = 4\sqrt{3}$, and $c = 7$. Thus we can choose

$$\alpha = \tfrac{1}{2}\arctan\left(\frac{b}{a - c}\right) = \tfrac{1}{2}\arctan\sqrt{3} = \tfrac{1}{6}\pi.$$

Setting

$$x = (\cos\tfrac{1}{6}\pi)X - (\sin\tfrac{1}{6}\pi)Y = \tfrac{1}{2}(\sqrt{3}X - Y),$$
$$y = (\sin\tfrac{1}{6}\pi)X + (\cos\tfrac{1}{6}\pi)Y = \tfrac{1}{2}(X + \sqrt{3}Y),$$

we find after simplification that our initial equation becomes

$$13X^2 + 5Y^2 - 1 = 0,$$

which we can write as

$$\frac{X^2}{(1/\sqrt{13})^2} + \frac{Y^2}{(1/\sqrt{5})^2} = 1.$$

This is the equation of an ellipse. Its graph is pictured in Figure 9.7.5.　□

Exercises

For each of the equations below, (a) find a rotation $\alpha \in (-\frac{1}{4}\pi, \frac{1}{4}\pi]$ that eliminates the xy-term; (b) rewrite the equation in terms of the new coordinate system; (c) sketch the graph displaying both coordinate systems.

*1. $xy = 1$.
 2. $xy - y + x = 1$.
*3. $11x^2 + 10\sqrt{3}xy + y^2 - 4 = 0$.
 4. $52x^2 - 72xy + 73y^2 - 100 = 0$.
*5. $x^2 - 2xy + y^2 + x + y = 0$.
 6. $3x^2 + 2\sqrt{3}xy + y^2 - 2x + 2\sqrt{3}y = 0$.
*7. $x^2 + 2\sqrt{3}xy + 3y^2 + 2\sqrt{3}x - 2y = 0$.
 8. $2x^2 + 4\sqrt{3}xy + 6y^2 + (8 - \sqrt{3})x + (8\sqrt{3} + 1)y + 8 = 0$.

For each of the equations below, find a rotation $\alpha \in (-\frac{1}{4}\pi, \frac{1}{4}\pi]$ that eliminates the xy-term. Then find $\cos \alpha$ and $\sin \alpha$.

*9. $x^2 + xy + Kx + Ly + M = 0$.
10. $5x^2 + 24xy + 12y^2 + Kx + Ly + M = 0$.

Optional | **Supplement to Section 9.7**

The Second-Degree Equation

It is possible to draw general conclusions about the graph of a second-degree equation

$$ax^2 + bxy + cy^2 + dx + ey + f = 0, \qquad a, b, c \text{ not all } 0,$$

just from its *discriminant*

$$\Delta = b^2 - 4ac.$$

There are three cases.

CASE 1.　If $\Delta < 0$, the graph is an ellipse, a circle, a point, or empty.

CASE 2.　If $\Delta > 0$, the graph is a hyperbola or a pair of intersecting lines.

Optional | CASE 3.   If $\Delta = 0$, the graph is a parabola, a line, a pair of lines, or empty.

Below we outline how these assertions may be verified. A useful first step is to rotate the coordinate system so that the equation takes the form

(1) $$AX^2 + CY^2 + DX + EY + F = 0.$$

An elementary but time-consuming computation shows that the discriminant is unchanged by a rotation, so that in this instance we have

$$\Delta = b^2 - 4ac = -4AC.$$

Moreover, A and C cannot both be zero. If $\Delta < 0$, then $AC > 0$ and we can rewrite (1) as

$$\frac{X^2}{C} + \frac{D}{AC} X + \frac{Y^2}{A} + \frac{EY}{AC} + \frac{F}{AC} = 0.$$

By completing the squares, we obtain an equation of the form

$$\frac{(X - \alpha)^2}{(\sqrt{|C|})^2} + \frac{(Y - \beta)^2}{(\sqrt{|A|})^2} = K.$$

If $K > 0$, we have an ellipse or a circle. If $K = 0$, we have the point (α, β). If $K < 0$, the set is empty.

If $\Delta > 0$, then $AC < 0$. Proceeding as before, we obtain an equation of the form

$$\frac{(X - \alpha)^2}{(\sqrt{|C|})^2} - \frac{(Y - \beta)^2}{(\sqrt{|A|})^2} = K.$$

If $K \neq 0$, we have a hyperbola. If $K = 0$, the equation becomes

$$\left(\frac{X - \alpha}{\sqrt{|C|}} - \frac{Y - \beta}{\sqrt{|A|}}\right)\left(\frac{X - \alpha}{\sqrt{|C|}} + \frac{Y - \beta}{\sqrt{|A|}}\right) = 0,$$

so that we have a pair of lines intersecting at the point (α, β).

If $\Delta = 0$, then $AC = 0$, so that either $A = 0$ or $C = 0$. Since A and C are not both zero, there is no loss in generality in assuming that $A \neq 0$ and $C = 0$. In this case Equation (1) reduces to

$$AX^2 + DX + EY + F = 0.$$

Dividing by A, we obtain an equation of the form

$$X^2 + HX + IY + J = 0.$$

Completion of the square gives us

$$(X - \alpha)^2 = \beta Y + K.$$

If $\beta \neq 0$, we have a parabola. If $\beta = 0$ and $K = 0$, we have a line. If $\beta = 0$ and $K > 0$, we have a pair of parallel lines. If $\beta = 0$ and $K < 0$, the set is empty. $\square$

Volume, Work, and Other Applications of the Integral

10.1 The Average Value of a Continuous Function

Take a function f continuous on an interval $[a, b]$. What number should we call the *average* of f on $[a, b]$?

We could define the average as $\frac{1}{2}[f(a) + f(b)]$, but that would take into consideration only what happens at the endpoints. We want a notion of average that takes into consideration what happens throughout the interval $[a, b]$.

To see how to proceed, let's look at averages in a more familiar setting. Think of a student taking tests. During the first marking period he takes N_1 tests, during the next marking period N_2 tests, and so on, until the last marking period when he takes N_n tests. Now let N be the total number of tests taken by the student:

$$N = N_1 + \cdots + N_n.$$

If we give equal weight to each of the tests, then the student's average has the following two properties:

(i) lowest of N grades $\leq$ average on N tests $\leq$ highest of N grades;

(ii) $\begin{pmatrix} \text{average on} \\ N \text{ tests} \end{pmatrix} \cdot N = \begin{pmatrix} \text{average on} \\ \text{first } N_1 \text{ tests} \end{pmatrix} \cdot N_1 + \cdots + \begin{pmatrix} \text{average on} \\ \text{final } N_n \text{ tests} \end{pmatrix} \cdot N_n.$

Property (i) is obvious. Property (ii) holds since each side of the equation represents the total number of points scored.

Now let's return to the function f. In view of (i) and (ii), we make two requirements of the number which we call the average of f:

Requirement 1

Minimum value of $f \le$ average of $f \le$ maximum value of f.

Requirement 2

If $P = \{x_0, x_1, \ldots, x_n\}$ is a partition of $[a, b]$, then

$$\left(\begin{array}{c}\text{average of } f \\ \text{on } [a, b]\end{array}\right)(b - a) = \left(\begin{array}{c}\text{average of } f \\ \text{on } [x_0, x_1]\end{array}\right)\Delta x_1 + \cdots + \left(\begin{array}{c}\text{average of } f \\ \text{on } [x_{n-1}, x_n]\end{array}\right)\Delta x_n.$$

To meet Requirements 1 and 2, there is only one way to define average, and that is to set

(10.1.1)
$$\text{average of } f \text{ on } [a, b] = \frac{1}{b - a}\int_a^b f(x)\, dx.$$

PROOF.   Take an arbitrary partition $P = \{x_0, x_1, \ldots, x_n\}$ of $[a, b]$. For each subinterval $[x_{i-1}, x_i]$, Requirement 1 gives

$$m_i \le \text{average of } f \text{ on } [x_{i-1}, x_i] \le M_i,$$

where, as usual,

$$m_i = \text{minimum value of } f \text{ on } [x_{i-1}, x_i]$$

and

$$M_i = \text{maximum value of } f \text{ on } [x_{i-1}, x_i].$$

Multiplying by Δx_i, we have

$$m_i \Delta x_i \le \left(\begin{array}{c}\text{average of } f \\ \text{on } [x_{i-1}, x_i]\end{array}\right)\Delta x_i \le M_i \Delta x_i.$$

Summing from $i = 1$ to $i = n$, we get

$$L_f(P) \le \left(\begin{array}{c}\text{average of } f \\ \text{on } [x_0, x_1]\end{array}\right)\Delta x_1 + \cdots + \left(\begin{array}{c}\text{average of } f \\ \text{on } [x_{n-1}, x_n]\end{array}\right)\Delta x_n \le U_f(P).\dagger$$

Requirement 2 now gives

$$L_f(P) \le \left(\begin{array}{c}\text{average of } f \\ \text{on } [a, b]\end{array}\right)(b - a) \le U_f(P).$$

$\dagger$ $L_f(P)$ and $U_f(P)$ are the lower and upper sums discussed in Section 5.2.

Since P was chosen arbitrarily, this inequality must hold for all partitions P. There is only one number with this property: the integral of f on $[a, b]$.† Thus we must have

$$\left(\begin{array}{c}\text{average of } f \\ \text{on } [a, b]\end{array}\right)(b - a) = \int_a^b f(x)\ dx$$

and

$$\text{average of } f \text{ on } [a, b] = \frac{1}{b - a}\int_a^b f(x)\ dx. \quad \square$$

This sense of average provides a powerful and intuitive way of viewing the definite integral. Think for a moment in terms of area. If f is constant and positive on $[a, b]$, then Ω, the region below the graph, is a rectangle. Its area is given by the formula

$$\text{area of } \Omega = (\text{constant value of } f \text{ on } [a, b]) \cdot (b - a).$$

If f is now allowed to vary continuously on $[a, b]$, then we have

$$\text{area of } \Omega = \int_a^b f(x)\ dx.$$

In terms of average values the formula reads

(10.1.2) $\boxed{\textit{area of } \Omega = (\textit{average value of } f \textit{ on } [a, b]) \cdot (b - a).}$

area = (constant value of f)•$(b - a)$

FIGURE 10.1.1

area = (average value of f)•$(b - a)$

FIGURE 10.1.2

Think now in terms of motion. If an object moves along a line with constant speed $|v|$ during the time interval $[a, b]$, then

$$\text{the distance traveled} = (\text{the constant value of } |v| \text{ on } [a, b]) \cdot (b - a).$$

If the speed $|v|$ varies, then we have

$$\text{distance traveled} = \int_a^b |v(t)|\ dt$$

and the formula reads

(10.1.3) $\boxed{\textit{distance traveled} = (\textit{average value of } |v| \textit{ on } [a, b]) \cdot (b - a).}$

---

† See Section 5.2.

In general we have the identity

(10.1.4) $$\int_a^b f(x)\,dx = (average\ value\ of\ f\ on\ [a,\ b]) \cdot (b - a).$$

Theorem 10.1.5.   The First Mean-Value Theorem for Integrals

If f is continuous on $[a,\ b]$, then there is a number c in $(a,\ b)$ such that

$$\int_a^b f(x)\,dx = f(c)(b - a).$$

In other words, there is a number c in $(a,\ b)$ at which f takes on its average value.

PROOF.   Apply the Mean-Value Theorem to the function

$$F(x) = \int_a^x f(t)\,dt \qquad \text{on } [a,\ b]. \quad \square$$

To see this theorem geometrically, look at Figure 10.1.3. Here we are assuming that f is positive. The theorem says that somewhere in $(a,\ b)$ there exists a number c such that the rectangle displayed has the same area as the region under the curve.

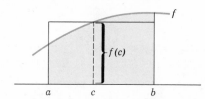

$f(c)$ is the average value of f:
the area of the rectangle = the area under the curve
FIGURE 10.1.3

Exercises

Determine the average value on the indicated interval and find a point in this interval at which the function takes on this average value.

*1. $f(x) = mx + b, \quad x \in [0,\ c]$.   2. $f(x) = x^2, \quad x \in [-1,\ 1]$.
*3. $f(x) = x^3, \quad x \in [-1,\ 1]$.   4. $f(x) = x^{-1}, \quad x \in [1,\ e]$.

Find the average value.

*5. $f(x) = e^{kx}, \quad x \in [0,\ 1]$.   6. $f(x) = \dfrac{a}{a^2 + x^2}, \quad x \in [0,\ a]$.

*7. $f(x) = \sqrt{a^2 - x^2}, \quad x \in [-a,\ a]$.   8. $f(x) = \tan \tfrac{1}{4}\pi x, \quad x \in [0,\ 1]$.

*9. What are the averages of the sine and cosine on an interval of length 2π?
10. Find the average value of $\sin^2 x$ between $x = 0$ and $x = \pi$. (This average is frequently used in the theory of alternating currents.)

11. Given that f is continuous on $[a, b]$, compare

$$f(b)(b - a) \quad \text{and} \quad \int_a^b f(x)\, dx$$

if f is (a) constant on $[a, b]$; (b) increasing on $[a, b]$; (c) decreasing on $[a, b]$.

12. In Chapter 4, we viewed $[f(b) - f(a)]/(b - a)$ as the average rate of change of f on $[a, b]$ and $f'(t)$ as the instantaneous rate of change at time t. If our new sense of average is to be consistent with the old one, we must have

$$\frac{f(b) - f(a)}{b - a} = \text{average of } f' \text{ on } [a, b].$$

Prove that this is the case.

13. Show that the average slope of the logarithm curve from $x = a$ to $x = b$ is

$$\frac{1}{b - a} \log\left(\frac{b}{a}\right).$$

14. Determine whether the assertion is true or false.

   (a) $\left(\begin{array}{c}\text{the average of } f + g \\ \text{on } [a, b]\end{array}\right) = \left(\begin{array}{c}\text{the average of } f \\ \text{on } [a, b]\end{array}\right) + \left(\begin{array}{c}\text{the average of } g \\ \text{on } [a, b]\end{array}\right).$

   (b) $\left(\begin{array}{c}\text{the average of } \alpha f \\ \text{on } [a, b]\end{array}\right) = \alpha \left(\begin{array}{c}\text{the average of } f \\ \text{on } [a, b]\end{array}\right).$

   (c) $\left(\begin{array}{c}\text{the average of } fg \\ \text{on } [a, b]\end{array}\right) = \left(\begin{array}{c}\text{the average of } f \\ \text{on } [a, b]\end{array}\right)\left(\begin{array}{c}\text{the average of } g \\ \text{on } [a, b]\end{array}\right).$

*15. A stone falls from rest in a vacuum for x seconds. (a) Compare its terminal velocity to its average velocity; (b) compare its average velocity during the first $\frac{1}{2}x$ seconds to its average velocity during the next $\frac{1}{2}x$ seconds.

16. Express the meaning of area formulas 5.10.1 and 5.10.2 in terms of averages.

17. Let f be continuous. Show that, if f is an odd function, then its average value on every interval of the form $[-a, a]$ is zero.

*18. Find the average vertical and horizontal dimensions of the ellipse

$$\frac{x^2}{a^2} + \frac{y^2}{b^2} = 1.$$

Optional 19. Prove that two distinct continuous functions cannot have the same average on every interval.

20. Let f be continuous on $[a, b]$. Let $a < c < b$. Prove that

$$f(c) = \lim_{h \downarrow 0} (\text{average of } f \text{ on } [c - h, c + h]).$$

10.2 Volume: Parallel Cross Sections

We begin with a solid as pictured in Figure 10.2.1. As in the figure we suppose that the solid lies entirely between $x = a$ and $x = b$. By $A(x)$ we mean the area of the cross section that has coordinate x.

FIGURE 10.2.1

FIGURE 10.2.2

If the cross-sectional area, $A(x)$, varies continuously with x, then we can find the volume of the solid by integrating $A(x)$ from a to b:

(10.2.1)
$$V = \int_a^b A(x)\, dx.$$

PROOF.   Let T be the solid in question and let $P = \{x_0, x_1, \ldots, x_n\}$ be a partition of $[a, b]$. For each subinterval $[x_{i-1}, x_i]$ let m_i and M_i be respectively the minimum and maximum values of A on $[x_{i-1}, x_i]$. Let T_i be that portion of the solid (Figure 10.2.2) that corresponds to the interval $[x_{i-1}, x_i]$, and let V_i be its volume. T_i contains a solid of cross-sectional area m_i and thickness Δx_i and is contained in a solid of cross-sectional area M_i and thickness Δx_i. This suggests that

$$m_i \Delta x_i \leq V_i \leq M_i \Delta x_i.$$

Summing these inequalities from $i = 1$ to $i = n$, we get

(1) $$L_A(P) \leq V_1 + \cdots + V_n \leq U_A(P).$$

Since

$$T = T_1 \cup T_2 \cup \cdots \cup T_n$$

and the T_i do not overlap, V, the total volume of T, must satisfy

$$V = V_1 + \cdots + V_n.$$

Inequality (1) now implies that

$$L_A(P) \leq V \leq U_A(P).$$

Since this last inequality must hold for every partition P, we must have

$$V = \int_a^b A(x)\, dx. \quad \square$$

REMARK.   For solids of constant cross-sectional area (such as those of Figure 10.2.3), the volume is simply the cross-sectional area times the thickness:

$$V = (\text{constant cross-sectional area}) \cdot (b - a).$$

FIGURE 10.2.3

When the cross-sectional area varies, we have the formula

$$V = \int_a^b A(x)\, dx,$$

which says that

(10.2.2) | $V = $ (average cross-sectional area) $\cdot (b - a)$.

Problem.   Find the volume of a cone that has base radius r and height h.

SOLUTION.   As a basis for the coordinate system we choose the line that passes through the vertex of the cone and the center of the base. (See Figure 10.2.4.)

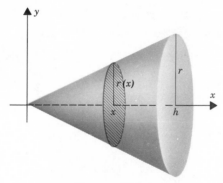

FIGURE 10.2.4

The cross section with coordinate x is a circular disc. Its radius $r(x)$ satisfies the equation

$$\frac{r(x)}{x} = \frac{r}{h}.$$

This you can see by similar triangles. Since

$$A(x) = \pi[r(x)]^2 \qquad \text{and} \qquad r(x) = \frac{r}{h}x,$$

we have

$$A(x) = \frac{\pi r^2}{h^2} x^2.$$

The volume of the cone is thus

$$\int_0^h \frac{\pi r^2}{h^2} x^2 \, dx = \frac{\pi r^2}{h^2} \left[\frac{x^3}{3}\right]_0^h = \frac{1}{3} \pi r^2 h. \quad \square$$

The next result is more general and contains the cone as a special instance.

Solids of Revolution: Disc Method

Suppose that f is nonnegative and continuous on $[a, b]$. (See Figure 10.2.5.) If we revolve the region below the graph of f about the x-axis, we obtain a solid. The volume of this solid is given by the formula

(10.2.3)
$$V = \int_a^b \pi [f(x)]^2 \, dx.$$

PROOF.   The cross section with coordinate x is a circular disc of radius $f(x)$. The cross-sectional area is thus $\pi[f(x)]^2$. The result follows from Formula 10.2.1. $\square$

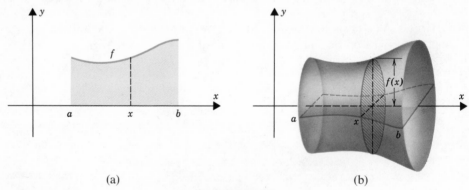

(a)   (b)

FIGURE 10.2.5

As special instances of such solids of revolution we have the cone and the sphere.

Formula 10.2.3 Applied to the Cone

To generate a cone of base radius r and height h, we can take

$$f(x) = \frac{r}{h} x, \qquad 0 \le x \le h$$

and revolve the region below the graph about the x-axis. Formula 10.2.3 then gives

$$V = \int_0^h \pi \frac{r^2}{h^2} x^2 \, dx = \tfrac{1}{3} \pi r^2 h. \quad \square$$

Formula 10.2.3 Applied to the Sphere

A sphere of radius r can be obtained by revolving the region below the graph of

$$f(x) = \sqrt{r^2 - x^2}, \qquad -r \le x \le r$$

about the x-axis. Consequently, we have

$$\text{volume of sphere} = \int_{-r}^{r} \pi(r^2 - x^2)\, dx = \pi \left[r^2 x - \tfrac{1}{3}x^3 \right]_{-r}^{r} = \tfrac{4}{3}\pi r^3.$$

Incidentally, this result was obtained by Archimedes in the third century B.C. □

FIGURE 10.2.6

Problem. Find the volume of a triangular pyramid with base area B and height h.

SOLUTION. In Figure 10.2.6 you can see a coordinate line that passes through the apex and is perpendicular to the base. To the apex we assign coordinate 0, and to the base point coordinate h. The cross section at a distance of x units from the apex ($0 \le x \le h$) is a triangle similar to the base triangle. The corresponding sides of these triangles are proportional, the factor of proportionality being x/h. This means that the areas are also proportional, but now the factor of proportionality is $(x/h)^2$†. Thus we have

$$A(x) = B\,\frac{x^2}{h^2} \qquad \text{and} \qquad V = \int_0^h A(x)\, dx = \frac{B}{h^2} \int_0^h x^2\, dx = \tfrac{1}{3}Bh.$$

This formula too was discovered in ancient times. It was known to Eudoxos of Cnidos (408–355 B.C.). □

Problem. The base of a solid is the region bounded by the ellipse

$$\frac{x^2}{a^2} + \frac{y^2}{b^2} = 1.$$

---

† If the base triangle has dimensions a, b, c, then its area is $ab \sin \theta$ where θ is the included angle. For the x-level triangle the dimensions are $a(x/h)$, $b(x/h)$, $c(x/h)$, and the area is $a(x/h)b(x/h) \sin \theta = ab \sin \theta \, (x/h)^2$.

Find the volume of the solid given that the cross sections perpendicular to the x-axis are equilateral triangles.

FIGURE 10.2.7

SOLUTION.   Take x as in Figure 10.2.7. The cross section with coordinate x is an equilateral triangle with side $\overline{PQ}$. The equation of the ellipse can be rewritten as

$$y^2 = \frac{b^2}{a^2}(a^2 - x^2).$$

Since

$$\text{length of } \overline{PQ} = 2y = \frac{2b}{a}\sqrt{a^2 - x^2},$$

the equilateral triangle has area

$$A(x) = \frac{\sqrt{3}b^2}{a^2}(a^2 - x^2).†$$

To find the volume we integrate $A(x)$ from $-a$ to a:

$$V = \int_{-a}^{a} A(x)\,dx = \frac{\sqrt{3}b^2}{a^2}\int_{-a}^{a}(a^2 - x^2)\,dx$$

$$= \frac{\sqrt{3}b^2}{a^2}\left[a^2x - \frac{x^3}{3}\right]_{-a}^{a} = \frac{\sqrt{3}b^2}{a^2}\left(\frac{4}{3}a^3\right) = \frac{4}{3}\sqrt{3}\,ab^2. \quad \square$$

Problem.   The base of a solid is the region between the parabolas

$$x = y^2 \quad \text{and} \quad x = 3 - 2y^2.$$

Find the volume of the solid if the cross sections perpendicular to the x-axis are squares.

SOLUTION.   The two parabolas intersect at $x = 1$. (See Figure 10.2.8.) From $x = 0$ to $x = 1$, the x cross section has area

$$A(x) = (2y)^2 = 4y^2 = 4x.$$

---

† In general, the area of an equilateral triangle is $\frac{1}{4}\sqrt{3}s^2$, where s is the length of a side.

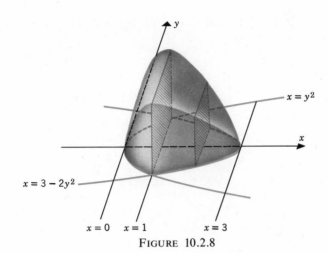

FIGURE 10.2.8

The volume from $x = 0$ to $x = 1$ is

$$V_1 = \int_0^1 4x \, dx = 4 \left[\tfrac{1}{2} x^2 \right]_0^1 = 2.$$

From $x = 1$ to $x = 3$, the x cross section has area

$$A(x) = (2y)^2 = 4y^2 = 2(3 - x).$$

(Here we are measuring the span across the second parabola $x = 3 - 2y^2$.) The volume from $x = 1$ to $x = 3$ is

$$V_2 = \int_1^3 2(3 - x) \, dx = 2 \left[-\tfrac{1}{2}(3 - x)^2 \right]_1^3 = 4.$$

The total volume is

$$V = V_1 + V_2 = 6. \quad \square$$

Problem.   Let Ω be the region between

$$\sqrt{x} + \sqrt{y} = 1 \quad \text{and} \quad x + y = 1.$$

Find the volume of the solid generated by revolving Ω about the x-axis.

SOLUTION.   We refer to Figure 10.2.9. In terms of x, the upper boundary becomes

$$y = 1 - x, \qquad 0 \le x \le 1$$

and the lower boundary

$$y = (1 - \sqrt{x})^2, \qquad 0 \le x \le 1.$$

The solid enclosed by revolving the upper boundary has volume

$$V_1 = \int_0^1 \pi y^2 \, dx = \pi \int_0^1 (1 - x)^2 \, dx = \tfrac{1}{3}\pi.$$

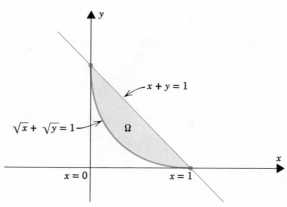

FIGURE 10.2.9

The solid enclosed by revolving the lower boundary has volume

$$V_2 = \int_0^1 \pi y^2 \, dx$$

$$= \pi \int_0^1 (1 - \sqrt{x})^4 \, dx$$

$$= \pi \int_0^1 (1 - 4x^{1/2} + 6x - 4x^{3/2} + x^2) \, dx = \tfrac{1}{15}\pi.$$

The volume of the solid generated by Ω is the difference

$$V = V_1 - V_2 = \tfrac{1}{3}\pi - \tfrac{1}{15}\pi = \tfrac{4}{15}\pi. \quad \square$$

Problem. Find the volume if the region OAB of Figure 10.2.10 is revolved about AB.

SOLUTION. The curve OB given by

$$y = x^2, \qquad 0 \le x \le 2$$

is also given by

$$x = \sqrt{y}, \qquad 0 \le y \le 4. \qquad \text{(Figure 10.2.11)}$$

The horizontal cross section at height y is a circular disc of radius $2 - \sqrt{y}$. The area of this disc is

$$A(y) = \pi(2 - \sqrt{y})^2 = \pi(4 - 4\sqrt{y} + y).$$

We can find the volume we want by integrating $A(y)$ from $y = 0$ to $y = 4$:

$$V = \int_0^4 A(y) \, dy = \int_0^4 \pi(4 - 4\sqrt{y} + y) \, dy = \tfrac{8}{3}\pi. \quad \square$$

FIGURE 10.2.10

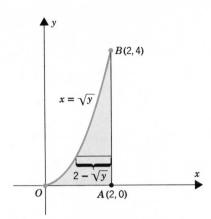

FIGURE 10.2.11

Exercises

Sketch the graph and find the volume generated by revolving the region below it about the x-axis.

*1. $f(x) = x^2$, $x \in [0, 1]$.   2. $f(x) = x^3$, $x \in [1, 2]$.
*3. $f(x) = e^x$, $x \in [0, 1]$.   4. $f(x) = \sin x$, $x \in [0, \pi]$.
*5. $f(x) = x^{2/3}$, $x \in [0, 2]$.   6. $f(x) = \sec \frac{1}{2}\pi x$, $x \in [-\frac{1}{2}, \frac{1}{2}]$.
*7. $f(x) = 1/x$, $x \in [1, 2]$.   8. $f(x) = \log x$, $x \in [1, e]$.

Sketch the graphs and find the volume generated by revolving the region between them about the x-axis.

*9. $f(x) = \sqrt{x}$, $x \in [0, 1]$; $g(x) = x^2$, $x \in [0, 1]$.
10. $f(x) = \sqrt{1 - x^2}$, $g(x) = \frac{1}{2}$.
*11. $f(x) = \sin x$, $x \in [0, \pi]$; $g(x) = x$, $x \in [0, \pi]$.

12. Find the volume of the solid generated by revolving the ellipse

$$\frac{x^2}{a^2} + \frac{y^2}{b^2} = 1$$

about the x-axis.

*13. Derive a formula for the volume of the frustum of a cone in terms of its height h, the lower base radius R, and the upper base radius r. (See Figure 10.2.12.)

FIGURE 10.2.12

*14. Find the volume of the solid generated by revolving the equilateral triangle with vertices $(0, 0)$, $(a, 0)$, $(\frac{1}{2}a, \frac{1}{2}\sqrt{3}a)$ about the x-axis.

15. The base of a solid is the circle $x^2 + y^2 = r^2$. Find the volume of the solid given that the cross sections perpendicular to the x-axis are (a) squares. (b) equilateral triangles.

*16. The base of a solid is the region bounded by the ellipse

$$\frac{x^2}{a^2} + \frac{y^2}{b^2} = 1.$$

Find the volume of the solid, given that the cross sections perpendicular to the x-axis are (a) isosceles right triangles each with hypotenuse on the xy-plane, (b) squares, (c) triangles of height 2.

*17. A hemispheric basin of radius r feet is being used to store water. To what percent of capacity is it filled when the water is (a) $\frac{1}{2}r$ feet deep, (b) $\frac{1}{3}r$ feet deep?

18. A sphere of radius r is cut by two parallel planes: in the one case, a units above the equator; in the other case, b units above the equator. Assume that $a < b$ and find the volume of that portion of the sphere that lies between the two planes.

*19. The base of a solid is the region between the parabolas $x = y^2$ and $x = 3 - 2y^2$. Find the volume if the cross sections perpendicular to the x-axis are, (a) rectangles of height h, (b) equilateral triangles, (c) isosceles right triangles each with hypotenuse on the xy-plane.

FIGURE 10.2.13

*20. Find the volume when the region OAB of Figure 10.2.13 is revolved about (a) the x-axis.   (b) AB.   (c) CA.   (d) the y-axis.

21. Find the volume when the region OAC of Figure 10.2.13 is revolved about (a) the y-axis.   (b) CA.   (c) AB.   (d) the x-axis.

*22. Find the volume of the intersection of two cylinders of radius r if the axes meet at right angles.

10.3   Volume: The Shell Method

First we have to go a little further into the theory of integration. If f and g are both continuous on $[a, b]$, then so is their product. The definite integral

$$\int_a^b f(x)g(x)\,dx$$

is then the only number I which satisfies the inequality

$$L_{fg}(P) \le I \le U_{fg}(P) \qquad \text{for all partitions } P \text{ of } [a, b].$$

This you already know. For some applications (including the one we're after here) we need a slightly different way of getting to

$$\int_a^b f(x)g(x)\,dx.$$

Take two functions f and g which are continuous and nonnegative on $[a, b]$. Let $P = \{x_0, x_1, \ldots, x_n\}$ be a partition of $[a, b]$. For each subinterval $[x_{i-1}, x_i]$ let

$$M_i(f) = \text{max value of } f \text{ on } [x_{i-1}, x_i], \quad M_i(g) = \text{max value of } g \text{ on } [x_{i-1}, x_i]$$
$$m_i(f) = \text{min value of } f \text{ on } [x_{i-1}, x_i], \quad m_i(g) = \text{min value of } g \text{ on } [x_{i-1}, x_i].$$

Now form the sums

$$U_{f,g}^*(P) = M_1(f)M_1(g)\,\Delta x_1 + M_2(f)M_2(g)\,\Delta x_2 + \cdots + M_n(f)M_n(g)\,\Delta x_n,$$
$$L_{f,g}^*(P) = m_1(f)m_1(g)\,\Delta x_1 + m_2(f)m_2(g)\,\Delta x_2 + \cdots + m_n(f)m_n(g)\,\Delta x_n.$$

Although $U_{f,g}^*(P)$ and $L_{f,g}^*(P)$ are not the usual upper and lower sums,† in this setting they do play a similar role.

Theorem 10.3.1   (Bliss)††

If f and g are continuous and nonnegative on $[a, b]$, then the definite integral

$$\int_a^b f(x)g(x)\,dx$$

is the only number I which satisfies the inequality

$$L_{f,g}^*(P) \le I \le U_{f,g}^*(P) \qquad \text{for all partitions } P \text{ of } [a, b].$$

---

† For example, in the case of $U_{f,g}^*(P)$ the coefficient of Δx_i is the product

$$M_i(f)M_i(g) = (\text{max value of } f \text{ on } [x_{i-1}, x_i])(\text{max value of } g \text{ on } [x_{i-1}, x_i]).$$

In the case of the usual upper sum $U_{fg}(P)$ the coefficient of Δx_i is

$$M_i(fg) = \text{max value of } fg \text{ on } [x_{i-1}, x_i].$$

These are in general not the same. (See Exercise 17.)
†† This is a simplified version of a theorem first proved by the American mathematician G. A. Bliss (1876–1951).

This result is a little too technical to be proved here. You can, however, find a proof in Appendix B.6.   □

To describe the shell method of computing volumes, we begin with a solid cylinder of radius R and height h, and from it we cut out a cylindrical core of radius r. (Figure 10.3.1)

FIGURE 10.3.1

Since the original cylinder had volume $\pi R^2 h$ and the piece removed had volume $\pi r^2 h$, the cylindrical shell that remains has volume

(10.3.2) $$\pi R^2 h - \pi r^2 h = \pi h (R + r)(R - r).$$

We'll use this shortly.

Consider now a function f which is nonnegative on an interval $[a, b]$. For convenience assume that $a \geq 0$. If the region below the graph is revolved about the y-axis, then a solid T is generated. See Figure 10.3.2.

FIGURE 10.3.2

The volume of this solid is given by the *shell method formula*:

(10.3.3) $$V = \int_a^b 2\pi x f(x) \; dx.$$

PROOF.   We take a partition $P = \{x_0, x_1, \ldots, x_n\}$ of $[a, b]$ and concentrate on what's happening on the ith interval $[x_{i-1}, x_i]$.

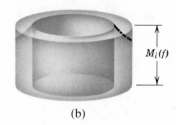

(a)   (b)

FIGURE 10.3.3

Let T_i be the solid generated by the ith interval, and let V_i be its volume. T_i contains a cylindrical shell of height $m_i(f)$, outer radius x_i, and inner radius x_{i-1}. [Part (a) of Figure 10.3.3] On the other hand, T_i is itself contained in a cylindrical shell of height $M_i(f)$, outer radius x_i, and inner radius x_{i-1}. [Part (b) of Figure 10.3.3] It follows now (from 10.3.2) that

$$\pi m_i(f)[x_i + x_{i-1}][x_i - x_{i-1}] \leq V_i \leq \pi M_i(f)[x_i + x_{i-1}][x_i - x_{i-1}].$$

Note that

$$2x_{i-1} \leq x_i + x_{i-1} \leq 2x_i$$

and as usual set $x_i - x_{i-1} = \Delta x_i$. This gives

(1) $$m_i(f)[2\pi x_{i-1}]\Delta x_i \leq V_i \leq M_i(f)[2\pi x_i]\Delta x_i.$$

Since for the function $g(x) = 2\pi x$

$$2\pi x_{i-1} = m_i(g) \quad \text{and} \quad 2\pi x_i = M_i(g),$$

you can rewrite (1) to read

$$m_i(f)m_i(g)\Delta x_i \leq V_i \leq M_i(f)M_i(g)\Delta x_i.$$

If you sum these inequalities from $i = 1$ to $i = n$, you'll find that the total volume

$$V = V_1 + V_2 + \cdots + V_n$$

must satisfy the inequality

$$L_{f,g}(P) \leq V \leq U_{f,g}(P).$$

Since the partition P was chosen arbitrarily, the inequality must hold for all partitions P. It follows from Theorem 10.3.1 that

$$V = \int_a^b f(x)g(x) \, dx.$$

Since $g(x) = 2\pi x$, we have

$$V = \int_a^b f(x)[2\pi x] \, dx = \int_a^b 2\pi x f(x) \, dx. \quad \square$$

REMARK.   The product $2\pi x f(x)$ gives the lateral area of a cylindrical surface of radius x and height $f(x)$. The shell method formula

$$V = \int_a^b 2\pi x f(x)\, dx$$

says that:

(10.3.4) | volume = (average lateral area) $\cdot (b - a)$.

Problem. Use the shell method to find a formula for the volume of a cone.

SOLUTION.   A cone of base radius r and height h can be obtained by revolving the region below the graph of

$$f(x) = -\frac{h}{r} x + h, \qquad 0 \le x \le r$$

about the y-axis. (You can convince yourself of this by drawing your own diagram.) The shell method then gives

$$V = \int_0^r 2\pi x \left(-\frac{h}{r} x + h \right) dx.$$

A short computation yields the familiar formula

$$V = \tfrac{1}{3}\pi r^2 h. \quad \square$$

Problem. Use the shell method to find the volume of the ring generated by revolving the circular disc

$$(x - c)^2 + y^2 \le r^2$$

of Figure 10.3.4 about the y-axis.

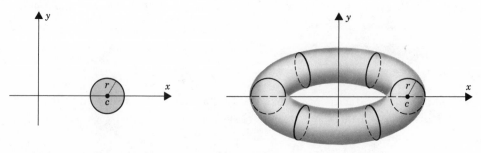

FIGURE 10.3.4

SOLUTION.   The upper half of this ring is generated by the region below the graph of

$$f(x) = \sqrt{r^2 - (x - c)^2}, \qquad c - r \le x \le c + r.$$

For the volume of the entire ring we have

$$V = 2 \int_{c-r}^{c+r} 2\pi x \sqrt{r^2 - (x - c)^2} \, dx = 2c\pi^2 r^2. \quad \square$$

set $r \sin u = x - c$.

Problem.   Let Ω be the region between

$$\sqrt{x} + \sqrt{y} = 1 \quad \text{and} \quad x + y = 1.$$

Find the volume of the solid generated by revolving Ω about the y-axis.

SOLUTION.   We refer to Figure 10.2.9. In terms of x, the upper boundary becomes

$$y = 1 - x, \qquad 0 \le x \le 1$$

and the lower boundary

$$y = (1 - \sqrt{x})^2, \qquad 0 \le x \le 1.$$

The solid enclosed by revolving the upper boundary has volume

$$V_1 = \int_0^1 2\pi xy \, dx = \int_0^1 2\pi x(1 - x) \, dx = \tfrac{1}{3}\pi.$$

The solid enclosed by revolving the lower boundary has volume

$$V_2 = \int_0^1 2\pi xy \, dx = \int_0^1 2\pi x(1 - \sqrt{x})^2 \, dx = \tfrac{1}{15}\pi.$$

The volume of the solid generated by Ω is the difference

$$V = V_1 - V_2 = \tfrac{1}{3}\pi - \tfrac{1}{15}\pi = \tfrac{4}{15}\pi. \quad \square$$

FIGURE 10.3.5

Problem.   Use the shell method to find the volume when the region OAB of Figure 10.2.10 is revolved about AB.

SOLUTION.   Introduce a new variable t as in Figure 10.3.5. The height of the curve t units from AB is $(2 - t)^2$. Since t ranges from $t = 0$ to $t = 2$,

$$V = \int_0^2 2\pi t(2 - t)^2 \, dt = 2\pi \int_0^2 (4t - 4t^2 + t^3) \, dt = \tfrac{8}{3}\pi. \quad \square$$

Exercises

Use the shell method to find the volume when the region below the graph is revolved about the y-axis.

*1. $f(x) = x^2, \quad x \in [0, 1]$.

*3. $f(x) = e^x, \quad x \in [0, 1]$.

*5. $f(x) = x^{2/3}, \quad x \in [0, 2]$.

*7. $f(x) = 1/x, \quad x \in [1, 2]$.

*9. $f(x) = \sqrt{x}, \quad x \in [0, 1]$.

2. $f(x) = x^3, \quad x \in [1, 2]$.

4. $f(x) = \sin x, \quad x \in [0, \pi]$.

6. $f(x) = \sin x^2, \quad x \in [0, \tfrac{1}{2}\sqrt{\pi}]$.

8. $f(x) = e^{x^2}, \quad x \in [0, 1]$.

10. $f(x) = \log x, \quad x \in [1, e]$.

Use the shell method to find the volume when the region between the graphs is revolved about the y-axis.

*11. $f(x) = \sqrt{x}, \quad x \in [0, 1]; \quad g(x) = x^2, \quad x \in [0, 1]$.

12. $f(x) = \sqrt{1 - x^2}, \quad g(x) = \tfrac{1}{2}$.

*13. $f(x) = \sin x, \quad x \in [0, \pi]; \quad g(x) = x, \quad x \in [0, \pi]$.

14. Find the volume enclosed when the ellipse

$$\frac{x^2}{a^2} + \frac{y^2}{b^2} = 1$$

is revolved about the y-axis (a) by the shell method; (b) by the disc method.

*15. Find the volume enclosed when the equilateral triangle with vertices $(0, 0)$, $(a, 0)$, $(\tfrac{1}{2}a, \tfrac{1}{2}\sqrt{3}a)$ is revolved about the y-axis.

16. The region between the x-axis and the graph of

$$y = \sin x, \qquad x \in [2n\pi, 2(n + 1)\pi]$$

is revolved about the y-axis. Find the volume of the solid generated.

17. Show that, if f and g are continuous and nonnegative on $[x_{i-1}, x_i]$, then

$$M_i(fg) \leq M_i(f)M_i(g).$$

Show that this inequality does not necessarily hold if f and g are allowed to take on negative values.

10.4   The Notion of Work

If an object moves a *distance* s in a fixed direction subject to a *constant force* f applied *in the direction of the motion,* then the *work* done by f is defined in physics by the equation

(10.4.1) $\boxed{W = f \cdot s.}$ (work = force · distance)

If force is measured in pounds and distance in feet, the units of work are called foot-pounds. For example, if a constant force of 500 pounds is applied to pushing a car 60 feet, the work done by the pushing force is said to be 30,000 foot-pounds.

If we coordinatize the line along which the motion takes place, assigning coordinate a to the initial point and coordinate b to the terminal point $(b > a)$, then we have

$$\text{work} = (\text{constant force}) \cdot (b - a).$$

In other words,

$$\text{work} = (\text{constant value of } f \text{ on } [a, b]) \cdot (b - a).$$

If f, the force applied, varies continuously, then we replace the constant value of f by its average value and define

$$\text{work} = (\text{average value of } f \text{ on } [a, b]) \cdot (b - a).$$

Thus, in the case of a variable force, we have

(10.4.2)
$$\boxed{\text{work} = \int_a^b f(x)\, dx.}$$
(Figure 10.4.1)

constant force
$W = f \cdot (b - a)$

variable force
$W = \int_a^b f(x)\, dx$

FIGURE 10.4.1

Stretching a Spring

A simple instance of a variable force is offered by the behavior of a steel spring. According to Hooke's law (Robert Hooke, 1635–1703), a spring stretched x units beyond its natural length exerts a force which is proportional to x:

$$f(x) = kx.$$

The constant k depends on the spring and on the units used.

Problem. A certain spring exerts a force of 10 pounds when stretched $\frac{1}{3}$ foot beyond its natural length. What is the work done in stretching the spring $\frac{1}{10}$ foot beyond its natural length? What is the work done in stretching it an additional $\frac{1}{10}$ foot?

SOLUTION. We know that the force function is of the form

$$f(x) = kx.$$

Since $f(\tfrac{1}{3}) = 10$, we have

$$k \cdot \tfrac{1}{3} = 10,$$

so that $k = 30$. Thus, for this particular spring, the force function is

$$f(x) = 30x.$$

The work done in the initial stretching is

$$W_1 = \int_0^{1/10} 30x \, dx = 30 \left[\tfrac{1}{2} x^2 \right]_0^{1/10} = \tfrac{3}{20} \text{ foot-pounds.}$$

The work done in the next stretching is

$$W_2 = \int_{1/10}^{2/10} 30x \, dx = 30 \left[\tfrac{1}{2} x^2 \right]_{1/10}^{2/10} = \tfrac{9}{20} \text{ foot-pounds.} \quad \square$$

Pumping out a Tank

To lift an object we must counteract the force of gravity. Consequently, the work done in lifting an object is given by the formula

work = (weight of the object) · (distance lifted).

If we lift a leaking sand bag or pump out a water tank from above, the calculation of work is more complicated. In the first instance the weight varies during the motion. (There's less sand in the bag as we keep lifting.) In the second instance the distance varies. (Water at the top of the tank doesn't have to be pumped as far as water at the bottom.) The sand bag problem is left to the exercises. Here we take up the second problem.

FIGURE 10.4.2

Figure 10.4.2 depicts a storage tank filled to within a feet of the top with some liquid, not necessarily water. Assume that the liquid is homogeneous and weighs σ pounds per cubic foot. Suppose now that this storage tank is pumped out from above until the level of the liquid drops to b feet below the top of the tank. How much work has been done?

We can answer this question by the methods of integral calculus. For each $x \in [a, b]$, we let

$A(x)$ = cross-sectional area x feet below the top of the tank,

$s(x)$ = distance that the x-level must be lifted.

We let $P = \{x_0, x_1, \ldots, x_n\}$ be an arbitrary partition of $[a, b]$ and focus our attention on the ith subinterval $[x_{i-1}, x_i]$. (Figure 10.4.3)

FIGURE 10.4.3

Now set

$$m_i(A) = \text{min value of } A \text{ on } [x_{i-1}, x_i],$$
$$M_i(A) = \text{max value of } A \text{ on } [x_{i-1}, x_i],$$
$$m_i(s) = \text{min value of } s \text{ on } [x_{i-1}, x_i],$$
$$M_i(s) = \text{max value of } s \text{ on } [x_{i-1}, x_i].$$

V_i, the volume of the liquid between x_{i-1} and x_i, satisfies the inequality

$$m_i(A)\Delta x_i \leq V_i \leq M_i(A)\Delta x_i. \qquad \text{(explain)}$$

Its weight, w_i, therefore satisfies the inequality

$$\sigma m_i(A)\Delta x_i \leq w_i \leq \sigma M_i(A)\Delta x_i.$$

W_i, the work necessary to lift this weight to the top of the tank, must in turn satisfy the inequality

$$\sigma m_i(s)m_i(A)\Delta x_i \leq W_i \leq \sigma M_i(s)M_i(A)\Delta x_i. \qquad \text{(explain)}$$

Setting

$$f(x) = \sigma s(x),$$

we have with obvious notation

$$m_i(f)m_i(A)\Delta x_i \leq W_i \leq M_i(f)M_i(A)\Delta x_i.$$

If you now sum this last inequality from $i = 1$ to $i = n$, then you'll see that the total work done

$$W = W_1 + W_2 + \cdots + W_n$$

must satisfy the inequality

$$L_{f,A}(P) \le W \le U_{f,A}(P). \qquad \text{(Section 10.3)}$$

Since this must hold for all partitions P, you can conclude from Theorem 10.3.1 that

$$W = \int_a^b f(x)A(x)\, dx.$$

Since $f(x) = \sigma s(x)$,

(10.4.3)
$$\boxed{W = \int_a^b \sigma s(x)A(x)\, dx.} \quad \square$$

We use this formula in the next problem.

Problem. A hemispherical water tank of radius 10 feet is being pumped out. Find the work done in lowering the water level from 2 feet below the top of the tank to 4 feet below the top of the tank given that the pump is placed (a) at the top of the tank, (b) 3 feet above the top of the tank.

SOLUTION. As the weight of water take 62.5 pounds per cubic foot. It's not hard to see that the cross section x feet from the top of the tank is a disc of radius $\sqrt{100 - x^2}$. Its area is therefore

$$A(x) = \pi(100 - x^2).$$

For part (a) we have $s(x) = x$, so that

$$W = \int_2^4 62.5\pi x(100 - x^2)\, dx = 33{,}750\pi \text{ foot-pounds.}$$

For part (b) we have $s(x) = x + 3$, so that

$$W = \int_2^4 62.5\pi(x + 3)(100 - x^2)\, dx = 67{,}750\pi \text{ foot-pounds.} \quad \square$$

Exercises

*1. A weight of 200 pounds stretches a certain spring 4 feet beyond its natural length. Find the work done in stretching the spring (a) 1 foot beyond its natural length, (b) $1\frac{1}{2}$ feet beyond its natural length.

2. A certain spring has natural length l. Given that W is the work done in stretching the spring from l feet to $l + a$ feet, find the work done in stretching the spring (a) from l feet to $l + 2a$ feet, (b) from l feet to $l + na$ feet, (c) from $l + a$ feet to $l + 2a$ feet, (d) from $l + a$ feet to $l + na$ feet.

*3. Find the natural length of a heavy metal spring, given that the work done in stretching it from a length of 2 feet to a length of 3 feet is one-half the work done in stretching it from a length of 3 feet to a length of 4 feet.

4. A vertical cylindrical tank of radius 2 feet and height 6 feet is full of water. Find the work done in pumping out the water (a) to an outlet at the top of the

tank, (b) to a level 5 feet above the top of the tank. (Assume that the water weighs 62.5 pounds per cubic foot.)

*5. A horizontal cylindrical tank of radius 3 feet and length 8 feet is half full of oil weighing 60 pounds per cubic foot. What is the work done in pumping out the oil (a) to the top of the tank, (b) to a level 4 feet above the top of the tank?

*6. What is the work done by gravity if the tank of exercise 5 is completely drained through an opening at the bottom?

7. A conical container (vertex down) of radius r feet and height h feet is full of a liquid weighing σ pounds per cubic foot. Find the work done in pumping out the top $\frac{1}{2}h$ feet of liquid (a) to the top of the tank, (b) to a level k feet above the top of the tank.

8. What is the work done by gravity if the tank of exercise 7 is completely drained through an opening at the bottom?

*9. The force of gravity exerted by the earth on a mass m at a distance r from the center of the earth is given by Newton's formula

$$F = -G\,\frac{mM}{r^2}$$

where M is the mass of the earth and G is a gravitational constant. Find the work done by gravity in pulling a mass m from $r = r_1$ to $r = r_2$.

10. A box that weighs w pounds is dropped to the floor from a height of d feet. (a) What is the work done by gravity? (b) Show that the work is the same if the box slides to the floor along a smooth inclined plane.

*11. A 100 pound bag of sand is lifted for 2 seconds at the rate of 4 feet per second. Find the work done in lifting the bag if the sand leaks out at the rate of $1\frac{1}{2}$ pounds per second.

12. A water container initially weighing w pounds is hoisted by a crane at the rate of n feet per second. What is the work done if the tank is raised m feet and the water leaks out constantly at the rate of p gallons per second? (Assume that the water weighs 8.3 pounds per gallon.)

*13. A rope of length l feet that weighs σ pounds per foot is lying on the ground. What is the work done in lifting the rope so that it hangs from a beam (a) l feet high, (b) $2l$ feet high?

14. An 800-pound steel beam hangs from a 50-foot cable which weighs 6 pounds per foot. Find the work done in winding 20 feet of the cable about a steel drum.

10.5 Fluid Pressure

Here we take up the study of fluid pressure and discuss how to calculate the force exerted by a fluid on a vertical wall. By pressure we mean *force per unit area*.

When a horizontal surface is submerged in a liquid, the force against it is the total weight of the fluid above it. If the horizontal surface has area A, if the depth is h, and if the density of the fluid is σ, then the force is given by the formula

(10.5.1) $\boxed{f = \sigma h A.}$

In this discussion we'll measure area in square feet, depth in feet, and density in pounds per cubic foot. The force is then expressed in pounds.

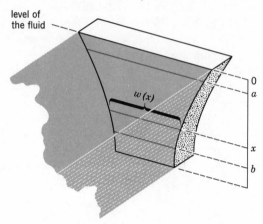

FIGURE 10.5.1

In Figure 10.5.1 we have represented a vertical wall. By $w(x)$ we mean the width of the wall at depth x. To determine the force exerted on the wall by the fluid from depth a to depth b, we let $P = \{x_0, x_1, \ldots, x_n\}$ be a partition of $[a, b]$ and focus our attention on that portion of the wall which corresponds to the ith subinterval $[x_{i-1}, x_i]$. (See Figure 10.5.2.) With pressure the same in all directions (the physicists tell us that this is correct, and the skin diver confirms it), we can estimate the force against this ith strip. If we take A_i as the area of this strip, then f_i, the force against this strip, satisfies the inequality

(1) $$\sigma x_{i-1} A_i \le f_i \le \sigma x_i A_i.$$ (explain)

Now set

$$m_i(w) = \text{min value of } w \text{ on } [x_{i-1}, x_i],$$
$$M_i(w) = \text{max value of } w \text{ on } [x_{i-1}, x_i],$$

and note that

$$m_i(w)\,\Delta x_i \le A_i \le M_i(w)\,\Delta x_i.$$

FIGURE 10.5.2

Combining this with (1), we have

$$\sigma x_{i-1} m_i(w)\,\Delta x_i \le f_i \le \sigma x_i M_i(w)\,\Delta x_i.$$

By setting $k(x) = \sigma x$, we have, with obvious notation,

$$m_i(k)m_i(w)\,\Delta x_i \le f_i \le M_i(k)M_i(w)\,\Delta x_i.$$

Summing from $i = 1$ to $i = n$, you can see that the total force

$$f = f_1 + f_2 + \cdots + f_n$$

must satisfy the inequality

$$L_{k,w}^*(P) \le f \le U_{k,w}^*(P) \qquad\qquad \text{(Section 10.3)}$$

and from the arbitrariness of P conclude (from Theorem 10.3.1) that

$$f = \int_a^b k(x)w(x)\,dx.$$

Since we chose $k(x) = \sigma x$, we have

(10.5.2)
$$\boxed{f = \int_a^b \sigma x w(x)\,dx.}\qquad \square$$

FIGURE 10.5.3

Problem.   A circular water main (Figure 10.5.3) 6 feet in diameter is half full of water. Find the total force on the gate that closes the main.

SOLUTION.   Here

$$w(x) = \sqrt{9 - x^2} \quad \text{and} \quad \sigma = 62.5 \text{ pounds per cubic foot.}$$

The force exerted against the main is

$$f = \int_0^3 (62.5)x(2\sqrt{9 - x^2})\,dx = 62.5 \int_0^3 2x\sqrt{9 - x^2}\,dx = 1125 \text{ pounds.} \quad \square$$

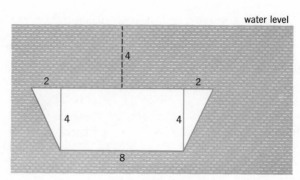

FIGURE 10.5.4

Problem.   The trapezoidal gate of a dam is shown in Figure 10.5.4. Find the force on the gate when the surface of the water is 4 feet above the top of the gate.

SOLUTION.   First we find the width of the gate x feet below the water level. By similar triangles (see Figure 10.5.5)

$$t = \tfrac{1}{2}(8 - x) \quad \text{so that} \quad w(x) = 8 + 2t = 16 - x.$$

The force against the gate is

$$f = \int_4^8 62.5x(16 - x)\, dx = 14{,}666\tfrac{2}{3} \text{ pounds.} \quad \square$$

FIGURE 10.5.5

Exercises

*1. Each end of a horizontal oil tank is an ellipse with horizontal axis 12 feet long and vertical axis 6 feet long. Calculate the force on one end when the tank is half full of oil weighing 60 pounds per cubic foot.

2. Each vertical end of a vat is a segment of a parabola (vertex down) 8 feet across the top and 16 feet deep. Calculate the force on an end when the vat is full of liquid weighing 70 pounds per cubic foot.

*3. The vertical ends of a water trough are isosceles right triangles with the 90° angle at the bottom. Calculate the force on each triangle when the trough is full of water given that the legs of the triangle are 8 feet long.

4. The vertical ends of a water trough are isosceles triangles 5 feet across the top and 5 feet deep. Calculate the force on an end when the trough is full of water.

*5. A horizontal cylindrical tank of diameter 8 feet is half full of oil weighing 60 pounds per cubic foot. Calculate the force on one end.

6. Calculate the force on one end if the tank of Exercise 5 is full.

*7. A rectangular gate in a vertical dam is 10 feet wide and 6 feet high. Find (a) the force on the gate when the level of the water is 8 feet above the top of the gate; (b) how much higher must the water rise to double the force found in (a)?

8. A vertical cylindrical tank of diameter 30 feet and height 50 feet is full of oil weighing 60 pounds per cubic foot. Find the force on the curved surface.

10.6 Revenue Streams

In Section 6.6 we talked about continuous compounding and you saw that the present value of A dollars t years from now is given by the formula

$$P.V. = Ae^{-rt},$$

where r is the rate of continuous compounding.

Suppose now that revenue flows continually at a constant rate of R dollars per year for n years. What is the present value of such a revenue stream?

To answer this, we let $P = \{t_0, t_1, \ldots, t_n\}$ be a partition of the time interval $[0, n]$ and focus on what happens during a typical time subinterval $[t_{i-1}, t_i]$.

The revenue that flows during that time interval is the product $R\,\Delta t_i$. To get the present value of this revenue $(P.V.)_i$, we must discount it, by more than $e^{-rt_{i-1}}$ but by less than e^{-rt_i}. (Explain.) This says that

$$R\,\Delta t_i e^{-rt_i} \le (P.V.)_i \le R\,\Delta t_i e^{-rt_{i-1}},$$

which we can rewrite as

(1) $$Re^{-rt_i}\,\Delta t_i \le (P.V.)_i \le Re^{-rt_{i-1}}\,\Delta t_i.$$

Setting $f(t) = Re^{-rt}$ we have in our usual notation

(2) $$m_i(f)\,\Delta t_i \le (P.V.)_i \le M_i(f)\,\Delta t_i.$$

The present value of the entire stream of revenues is the sum

$$P.V. = (P.V.)_1 + (P.V.)_2 + \cdots + (P.V.)_n.$$

By summing (2) from $i = 1$ to $i = n$, you can see that

$$L_f(P) \le P.V. \le U_f(P),$$

and, since P was chosen arbitrarily, that

$$P.V. = \int_0^n f(t)\,dt.$$

Since $f(t) = Re^{-rt}$,

(10.6.1)
$$P.V. = \int_0^n Re^{-rt} \, dt. \quad \square$$

Problem 1. What is the present value of an 8-year stream of revenues expected to flow continually at the rate of $100 per year? Assume continuous compounding at 5%.

SOLUTION

$$P.V. = \int_0^8 100e^{-0.05t} \, dt = 2000(1 - e^{-0.40}) \cong 2000(1 - 0.670) = 666.$$

— by Table 2 at the end of the book

The present value is about $666. $\square$

Revenues usually do not flow at a constant rate R but rather at a time-dependent rate $R(t)$. In general, $R(t)$ tends to increase when business is good and tends to decrease when business is poor. "Growth companies" are so dubbed because of their continually increasing $R(t)$. The "cyclical companies" owe their name to a fluctuating $R(t)$; $R(t)$ increases for a while, then decreases for a while, and then the cycle is supposedly repeated.

If we postulate a varying $R(t)$, then the present value of an n-year stream is given by the formula

(10.6.2)
$$P.V. = \int_0^n R(t)e^{-rt} \, dt.$$

Here, as before, r is the rate of continuous compounding used to discount future dollars.

You should be able to convince yourself of the validity of this formula on your own. A proof is outlined in the exercises.

Problem 2. What is the present value of an 8-year stream of revenues which is expected to flow at the rate of

$$R(t) = (1 + 0.04t)100$$

dollars per year? As in Problem 1, assume continuous compounding at 5%.

SOLUTION

$$P.V. = \int_0^8 (1 + 0.04t)100e^{-0.05t} \, dt = \int_0^8 100e^{-0.05t} \, dt + \int_0^8 4te^{-0.05t} \, dt.$$

From Problem 1,

$$\int_0^8 100e^{-0.05t} \, dt \cong 666.$$

To evaluate the second integral, we first find

$$\int 4te^{-0.05t}\, dt$$

using integration by parts. We set

$$u = 4t, \qquad dv = e^{-0.05t}\, dt,$$
$$du = 4dt, \qquad v = -20e^{-0.05t}.$$

$$\int 4te^{-0.05t}\, dt = \int u\, dv = uv - \int v\, du$$
$$= -80te^{-0.05t} + \int 80e^{-0.05t}\, dt$$
$$= -80te^{-0.05t} - 1600e^{-0.05t} + C$$
$$= -80(t + 20)e^{-0.05t} + C.$$

$$\int_0^8 4te^{-0.05t}\, dt = \left[-80(t + 20)e^{-0.05t} \right]_0^8$$
$$= -80[28e^{-0.4} - 20] \cong -80[28(0.670) - 20] \cong 99.20.$$

The present value is about $765.20.   □

Exercises

Let S be a constant revenue stream of 1000 dollars per year.

*1. What is the present value of the first 4 years of revenue? Assume continuous compounding (a) at 4%; (b) at 8%.
2. What is the present value of the fifth year of revenue with continuous compounding (a) at 4%? (b) at 8%?

Let S be a revenue stream with $R(t) = 1000 + 60t$ dollars per year.

3. What is the present value of the first 2 years of revenue? Assume continuous compounding (a) at 5%; (b) at 10%.
*4. What is the present value of the third year of revenue with continuous compounding (a) at 5%? (b) at 10%?

Let S be a revenue stream with $R(t) = 1000 + 80t$ dollars per year.

*5. What is the present value of the first 4 years of revenue? Assume continuous compounding (a) at 6%; (b) at 8%.
6. What is the present value of the fourth year of revenue with continuous compounding at (a) 6%? (b) 8%?

7. Prove the validity of (10.6.2). As we did for (10.6.1), partition the interval $[0, n]$. Take the ith subinterval and argue that

$$m_i(R)m_i(f)\, \Delta t_i \le (P.V.)_i \le M_i(R)M_i(f)\, \Delta t_i,$$

where $f(t) = e^{-rt}$.

10.7   Additional Exercises

*1. The height at time t of a bullet fired at time $t = 0$ at an angle θ is given by the formula $h(t) = (v_0 \sin \theta)t - \frac{1}{2}gt^2$. What is the average of h with respect to t during the flight of the bullet?

2. Let f and g be two continuous functions with the same average value on $[a, b]$. Show that $f(x)$ need not equal $g(x)$ for all $x \in [a, b]$ but that it must equal $g(x)$ for some $x \in [a, b]$.

The region bounded by the following curves is revolved about the x-axis. Find the volume of the solid generated.

*3. $ay^2 = x^3$, $y = 0$, $x = a$.
4. $y^2 = (2 - x)^3$, $y = 0$, $x = 0$, $x = 1$.
*5. $\sqrt{x} + \sqrt{y} = \sqrt{a}$, $x = 0$, $y = 0$.
6. $x^{2/3} + y^{2/3} = a^{2/3}$.
*7. $(x/a)^2 + (y/b)^{2/3} = 1$.
8. $a^2y^2 = x^2(a^2 - x^2)$.

The region bounded by the following curves is revolved about the y-axis. Find the volume of the solid generated.

*9. $y = x^3$, $y = 0$, $x = 2$.   10. $y^2 = 9 - x$, $x = 0$.
*11. $(x/a)^2 + (y/b)^{2/3} = 1$.   12. $x^2 = 16 - y$, $y = 0$.

13. The base of a solid is the region between the curves $y = \sin x$ and $y = \cos x$ with $x \in [0, \frac{1}{2}\pi]$. Find the volume of the solid if the cross sections perpendicular to the x-axis are *(a) rectangles of height h, (b) squares, (c) isosceles triangles of height h, *(d) equilateral triangles.

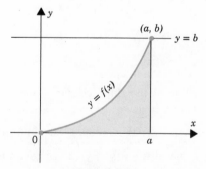

FIGURE 10.7.1

14. Represent by an integral the volume of the solid obtained by revolving the region shaded in Figure 10.7.1
   *(a) about the x-axis, first by the disc method and then by the shell method.
    (b) about the y-axis, first by the disc method and then by the shell method.
   *(c) about the line $x = a$, first by the disc method and then by the shell method.

(d) about the line $y = b$, first by the disc method and then by the shell method.

*15. A load of weight w is lifted from the bottom of a shaft h feet deep. Find the work done if the rope used to hoist the load weighs σ pounds per foot.

16. A wedge is cut from a cylinder of radius r by two planes, one perpendicular to the axis of the cylinder and the other passing through a diameter of the section made by the first plane and inclined to this plane at an angle θ. Find the volume of the wedge.

Polar Coordinates; Parametric Equations

11

11.1 Polar Coordinates

The purpose of coordinates is to fix position with respect to a given frame of reference. When we use rectangular coordinates, our frame of reference is a pair of lines which intersect at right angles. For a *polar coordinate system,* the frame of reference is a point O that we call the *pole* and a ray that emanates from it that we call the *polar axis.* (Figure 11.1.1)

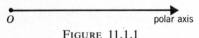

O polar axis

FIGURE 11.1.1

In Figure 11.1.2 we have drawn two more rays from the pole. One lies at an angle of θ radians from the polar axis. We call it *ray θ.* The opposite ray lies at an angle of $\theta + \pi$ radians. We call it *ray $\theta + \pi$.*

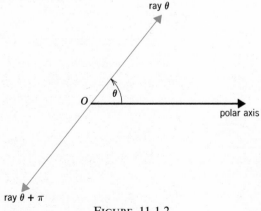

FIGURE 11.1.2

429

Figure 11.1.3 shows some points along these same rays, labeled with *polar coordinates*.

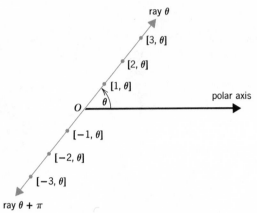

FIGURE 11.1.3

(11.1.1) | In general, a point is given *polar coordinates* $[r,\ \theta]$ iff it lies at a distance $|r|$ from the pole
along the ray θ, if $r \geq 0$,   and   along the ray $\theta + \pi$, if $r < 0$.

Figure 11.1.4 shows the point $[2,\frac{2}{3}\pi]$ at a distance of 2 units from the pole along the ray $\frac{2}{3}\pi$. The point $[-2,\frac{2}{3}\pi]$ also lies 2 units from the pole, not along the ray $\frac{2}{3}\pi$, but along the opposite ray.

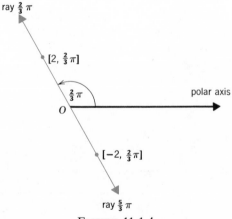

FIGURE 11.1.4

Unfortunately, polar coordinates are not unique. Many pairs $[r,\ \theta]$ can represent the same point.

1. If $r = 0$, it does not matter how we choose θ. The resulting point is still the pole:

(11.1.2) $O = [0, \theta]$ for all θ.

2. Geometrically there is no distinction between angles that differ by an integral multiple of 2π. Consequently, as suggested in Figure 11.1.5,

(11.1.3) $[r, \theta] = [r, \theta + 2n\pi]$ for all integers n.

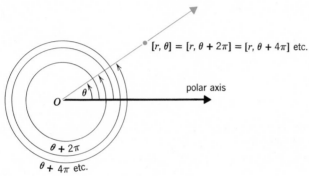

FIGURE 11.1.5

3. Adding π to the second coordinate is equivalent to changing the sign of the first coordinate:

(11.1.4) $[r, \theta + \pi] = [-r, \theta].$ (Figure 11.1.6)

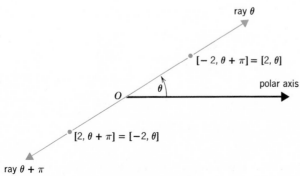

FIGURE 11.1.6

Relation to Rectangular Coordinates

In Figure 11.1.7 we have superimposed a polar coordinate system on a rectangular coordinate system. We have placed the pole at the origin and the polar axis along the positive x-axis.

FIGURE 11.1.7   FIGURE 11.1.8

The relation between polar coordinates $[r, \theta]$ and rectangular coordinates (x, y) can now be written

(11.1.5)
$$x = r \cos \theta, \qquad y = r \sin \theta.$$

PROOF. If $r = 0$, the formulas hold, since the point $[r, \theta]$ is then the origin and both x and y are 0:

$$0 = 0 \cos \theta, \qquad 0 = 0 \sin \theta.$$

For $r > 0$, we refer to Figure 11.1.8. From the figure,

$$\cos \theta = \frac{x}{r}, \qquad \sin \theta = \frac{y}{r}$$

and therefore

$$x = r \cos \theta, \qquad y = r \sin \theta.$$

Suppose now that $r < 0$. Since $[r, \theta] = [-r, \theta + \pi]$ and $-r > 0$, we know from the previous case that

$$x = -r \cos (\theta + \pi) \quad \text{and} \quad y = -r \sin (\theta + \pi).$$

Since

$$\cos (\theta + \pi) = \cos \theta \cos \pi - \sin \theta \sin \pi = -\cos \theta$$

and

$$\sin (\theta + \pi) = \sin \theta \cos \pi + \cos \theta \sin \pi = -\sin \theta,$$

you can see that

$$x = -r(-\cos \theta) = r \cos \theta, \qquad y = -r(-\sin \theta) = r \sin \theta.$$

Once again the formulas hold. $\square$

From the relations we just proved you can see that, subject to the obvious exclusions,

(11.1.6)
$$\boxed{\tan \theta = \frac{y}{x}}$$

and, under all circumstances,

(11.1.7)
$$\boxed{x^2 + y^2 = r^2.}$$ (check this out)

Problem.   Find the rectangular coordinates of the point $[-2, \frac{1}{3}\pi]$.

SOLUTION.   The relations

$$x = r \cos \theta, \qquad y = r \sin \theta$$

give

$$x = -2 \cos \tfrac{1}{3}\pi = -2(\tfrac{1}{2}) = -1$$

and

$$y = -2 \sin \tfrac{1}{3}\pi = -2(\tfrac{1}{2}\sqrt{3}) = -\sqrt{3}. \quad \square$$

Problem.   Find all possible polar coordinates for the point with rectangular coordinates $(-2, 2\sqrt{3})$.

SOLUTION.   Here

$$-2 = r \cos \theta, \qquad 2\sqrt{3} = r \sin \theta$$

so that

$$r^2 = r^2 \cos^2 \theta + r^2 \sin^2 \theta = (-2)^2 + (2\sqrt{3})^2 = 4 + 12 = 16.$$

If we set $r = 4$, then we have

$$-2 = 4 \cos \theta, \qquad 2\sqrt{3} = 4 \sin \theta$$

and thus

$$-\tfrac{1}{2} = \cos \theta, \qquad \tfrac{1}{2}\sqrt{3} = \sin \theta.$$

This means that we can take

$$\theta = \tfrac{2}{3}\pi \quad \text{or more generally} \quad \theta = \tfrac{2}{3}\pi + 2n\pi.$$

If we set $r = -4$, then we can take

$$\theta = \tfrac{2}{3}\pi + \pi = \tfrac{5}{3}\pi \quad \text{or more generally} \quad \theta = \tfrac{5}{3}\pi + 2n\pi. \quad \square$$

Let's specify some simple sets in polar coordinates.

(1) The circle of radius a, $x^2 + y^2 = a^2$, becomes $r = a$. The interior of the circle is given by $0 \le r < a$ and the exterior by $r > a$.
(2) The line through the origin with inclination α is given by the equation $\theta = \alpha$.
(3) The vertical line $x = a$ becomes $r \cos \theta = a$.
(4) The horizontal line $y = b$ becomes $r \sin \theta = b$.
(5) The line $Ax + By + C = 0$ becomes $r(A \cos \theta + B \sin \theta) + C = 0$. $\quad \square$

Problem. Find an equation for the equilateral hyperbola

$$x^2 - y^2 = a^2$$

in terms of polar coordinates.

SOLUTION

$$r^2 \cos^2 \theta - r^2 \sin^2 \theta = a^2,$$
$$r^2(\cos^2 \theta - \sin^2 \theta) = a^2,$$
$$r^2 \cos 2\theta = a^2. \quad \square$$

Problem. Show that the equation

$$r = 2a \cos \theta$$

represents a circle.

SOLUTION.   Multiplication by r gives

$$r^2 = 2ar \cos \theta,$$
$$x^2 + y^2 = 2ax,$$
$$x^2 - 2ax + y^2 = 0,$$
$$x^2 - 2ax + a^2 + y^2 = a^2,$$
$$(x - a)^2 + y^2 = a^2.$$

This is a circle of radius a centered at the point with rectangular coordinates $(a, 0)$. $\quad \square$

Symmetry

Symmetry with respect to each of the coordinate axes and with respect to the origin is illustrated in Figures 11.1.9, 11.1.10, and 11.1.11. The coordinates marked are, of course, not the only ones possible.

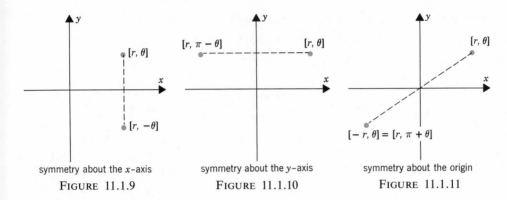

symmetry about the x-axis

FIGURE 11.1.9

symmetry about the y-axis

FIGURE 11.1.10

symmetry about the origin

FIGURE 11.1.11

Problem. Test the lemniscate

$$r^2 = \cos 2\theta$$

for symmetry.

SOLUTION. Since

$$\cos [2(-\theta)] = \cos (-2\theta) = \cos 2\theta,$$

you can see that, if $[r, \theta]$ is on the curve, then so is $[r, -\theta]$. This says that the curve is symmetric about the x-axis.

Since

$$\cos [2(\pi - \theta)] = \cos (2\pi - 2\theta) = \cos (-2\theta) = \cos 2\theta,$$

you can see that, if $[r, \theta]$ is on the curve, then so is $[r, \pi - \theta]$. The curve is therefore symmetric about the y-axis.

Being symmetric about both axes, the curve must also be symmetric about the origin. You can also verify this directly by noting that

$$\cos [2(\pi + \theta)] = \cos (2\pi + 2\theta) = \cos 2\theta,$$

so that, if $[r, \theta]$ lies on the curve, then so does $[r, \pi + \theta]$. A sketch of the lemniscate appears in Figure 11.1.12.   □

$r^2 = \cos 2\theta$
lemniscate

FIGURE 11.1.12

Exercises

Plot the following points:

1. $[1, \frac{1}{3}\pi]$.   2. $[1, \frac{1}{2}\pi]$.   3. $[-1, \frac{1}{3}\pi]$.   4. $[-1, -\frac{1}{3}\pi]$.
5. $[4, \frac{5}{4}\pi]$.   6. $[-2, 0]$.   7. $[-\frac{1}{2}, \pi]$.   8. $[\frac{1}{3}, \frac{2}{3}\pi]$.

Find the rectangular coordinates of each of the following points:

*9. $[2, 3\pi]$.   10. $[4, \frac{1}{6}\pi]$.   *11. $[-3, -\frac{1}{3}\pi]$.   12. $[-1, \frac{1}{4}\pi]$.
*13. $[-1, -\pi]$.   14. $[2, 0]$.   *15. $[3, \frac{1}{2}\pi]$.   16. $[3, -\frac{1}{2}\pi]$.

The following points are given in rectangular coordinates. Find all possible polar coordinates for each point.

*17. $(0, 1)$.   18. $(1, 0)$.   *19. $(-3, 0)$.
20. $(4, 4)$.   *21. $(2, -2)$.   22. $(3, -3\sqrt{3})$.
*23. $(4\sqrt{3}, 4)$.   24. $(\sqrt{3}, -1)$.   *25. $(-1, \sqrt{3})$.

26. Show that the distance between $[r_1, \theta_1]$ and $[r_2, \theta_2]$ is

$$\sqrt{r_1{}^2 + r_2{}^2 - 2r_1 r_2 \cos(\theta_1 - \theta_2)}.$$

[HINT: Change to rectangular coordinates.]

Test these curves for symmetry about the coordinate axes and the origin.

*27. $r = 2 + \cos\theta$.   28. $r = \cos 2\theta$.   *29. $r(\sin\theta + \cos\theta) = 1$.
*30. $r^2 = \sin\theta$.   31. $r^2 \sin 2\theta = 1$.   *32. $r^2 \cos 2\theta = 1$.

Express the following equations in terms of polar coordinates.

*33. $2xy = 1$.   34. $y = mx$.
*35. $x^2 + y^2 = 4$.   36. $x^2 + (y - b)^2 = b^2$.
*37. $x^2 + y^2 + ax = a\sqrt{x^2 + y^2}$.   38. $(x^2 + y^2)^2 = a^2(x^2 - y^2)$.

Identify the following curves. Change to rectangular coordinates.

*39. $r \sin\theta = 4$.   40. $r \cos\theta = 4$.   *41. $\theta = \frac{1}{3}\pi$.
*42. $\theta^2 = \frac{1}{9}\pi^2$.   43. $r = 2(1 - \cos\theta)^{-1}$.   *44. $r = 4\sin(\theta + \pi)$.

11.2 Graphing in Polar Coordinates

In rectangular coordinates the simplest equations to graph take the form

$$y = y_0 \quad \text{and} \quad x = x_0.$$

The first of these equations represents a horizontal line and the second a vertical line. The polar counterparts of these equations take the form

$$r = r_0 \quad \text{and} \quad \theta = \theta_0.$$

The first of these equations represents a circle. (Figure 11.2.1) The center is at the pole, and the radius is $|r_0|$. The second equation represents a straight line. (Figure 11.2.2) The line passes through the pole at an angle θ_0 from the polar axis.

FIGURE 11.2.1　　　　　　　　　　　　　FIGURE 11.2.2

In earlier chapters we graphed equations of the form $y = f(x)$ in rectangular coordinates. Here we graph some equations of the form $r = f(\theta)$ in polar coordinates.

We begin with the equation

$$r = \theta, \qquad \theta \geq 0.$$

Its graph is a nonending spiral, the famous *spiral of Archimedes*. It is shown in detail from $\theta = 0$ to $\theta = 2\pi$ in Figure 11.2.3.

$r = \theta, \quad \theta \geq 0$
spiral of Archimedes
FIGURE 11.2.3

Problem.　Sketch the graph of the equation

$$r = a(1 + \cos \theta). \qquad\qquad (a > 0)$$

SOLUTION.　Since the cosine has period 2π, we will concern ourselves only with θ between $-\pi$ and π. Outside that interval the curve repeats itself.

Since the cosine function is even, that is, $\cos(-\theta) = \cos\theta$, the value of r corresponding to $-\theta$ is the same as the value corresponding to θ. This tells us that the curve is symmetric about the x-axis. All we have to do, then, is sketch the curve from 0 to π (that is, in the upper half plane). The rest we can fill in by symmetry.

TABLE 11.2.1

θ	r
0	$2a$
$\frac{1}{6}\pi$	$(1 + \frac{1}{2}\sqrt{3})a$
$\frac{1}{3}\pi$	$\frac{3}{2}a$
$\frac{1}{2}\pi$	a
$\frac{2}{3}\pi$	$\frac{1}{2}a$
π	0

Table 11.2.1 lists some representative values of θ from 0 to π and the corresponding values of r obtained from the equation $r = a(1 + \cos \theta)$. It is not hard to see that as θ increases from 0 to π, $\cos \theta$ decreases from 1 to 0, and r itself decreases from $2a$ to 0. From 0 to π the graph must look as in Figure 11.2.4.

FIGURE 11.2.4

Filling in the lower part by symmetry (remember, the curve is symmetric about the x-axis), we have the complete curve as shown in Figure 11.2.5. This heart-shaped curve is called a *cardioid*. □

$$r = a(1 + \cos \theta)$$

FIGURE 11.2.5

Problem.   Sketch the three-leaved rose

$$r = a \sin 3\theta. \qquad\qquad (a > 0)$$

SOLUTION.   As θ increases from 0 to $\frac{1}{6}\pi$, r increases from 0 to a:

$$a \sin 3(\tfrac{1}{6}\pi) = a \sin \tfrac{1}{2}\pi = a.$$

As θ increases from $\frac{1}{6}\pi$ to $\frac{1}{3}\pi$, r decreases back to 0:

$$a \sin 3(\tfrac{1}{3}\pi) = a \sin \pi = 0. \qquad\qquad \text{(Figure 11.2.6)}$$

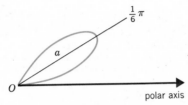

FIGURE 11.2.6

As θ increases from $\frac{1}{3}\pi$ to $\frac{1}{3}\pi + \frac{1}{6}\pi = \frac{1}{2}\pi$, r decreases from 0 to $-a$:

$$a \sin 3(\tfrac{1}{2}\pi) = -a.$$

As θ increases from $\frac{1}{2}\pi$ to $\frac{1}{2}\pi + \frac{1}{6}\pi = \frac{2}{3}\pi$, r increases back to 0:

$$a \sin 3(\tfrac{2}{3}\pi) = a \sin 2\pi = 0.\dagger \qquad\qquad \text{(Figure 11.2.7)}$$

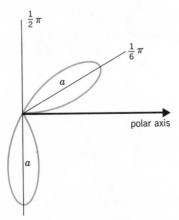

FIGURE 11.2.7

---

† For θ between $\frac{1}{3}\pi$ and $\frac{2}{3}\pi$, $r = a \sin 3\theta$ is negative and the points appear on the opposite ray.

As θ increases from $\frac{2}{3}\pi$ to $\frac{2}{3}\pi + \frac{1}{6}\pi = \frac{5}{6}\pi$, r increases from 0 to a:

$$a \sin 3(\tfrac{5}{6}\pi) = a \sin \tfrac{5}{2}\pi = a \sin \tfrac{1}{2}\pi = a.$$

As θ increases from $\frac{5}{6}\pi$ to π, r decreases back to 0:

$$a \sin 3\pi = a \sin \pi = 0. \qquad\qquad \text{(Figure 11.2.8)}$$

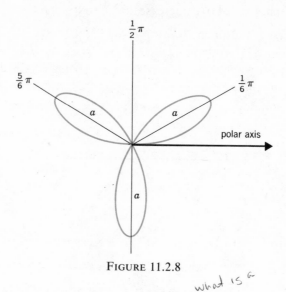

FIGURE 11.2.8

what is a

From here on, the curve repeats itself. The value $\theta + \pi$ gives rise to the same point on the curve as did θ:

$$[a \sin 3(\theta + \pi), \ \theta + \pi] = [-a \sin 3\theta, \ \theta + \pi] = [a \sin 3\theta, \ \theta].$$

The last equality follows from the fact that in general $[r, \ \theta + \pi] = [-r, \ \theta]$.   □

Exercises

Graph the following polar equations.

*1. $r = 2$.   2. $r = -3$.   *3. $\theta = \frac{1}{6}\pi$.   4. $\theta = -\frac{1}{6}\pi$.
*5. $r = 2\theta$, $\ 0 \le \theta \le 2\pi$.   6. $r = \frac{1}{2}\theta$, $\ 0 \le \theta \le 2\pi$.
*7. $r = \cos \theta$.   8. $r = \sin \theta$.   *9. $r = 2 \cos \theta$.   10. $r = 2 \sin \theta$.

Graph the following, taking $a > 0$.

*11. The three-leaved rose $r = a \cos 3\theta$.
12. The parabola $r = a \sec^2 \frac{1}{2}\theta$.
*13. The four-leaved rose $r = a \sin 2\theta$.
14. The four-leaved rose $r = a \cos 2\theta$.
*15. The two-leaved lemniscate $r^2 = a^2 \sin 2\theta$.
16. The eight-leaved rose $r = a \sin 4\theta$.

Optional | 17. Let $e > 0$. Show that the polar equation

$$r = \frac{ed}{1 - e \cos \theta} \quad \text{represents} \quad \begin{cases} \text{an ellipse} & \text{if } e < 1 \\ \text{a parabola} & \text{if } e = 1 \\ \text{a hyperbola} & \text{if } e > 1 \end{cases}.$$

11.3　The Intersection of Polar Curves

The fact that a single point has many pairs of polar coordinates can cause complications. In particular, it means that a point $[r_1, \theta_1]$ can lie on a curve given by a polar equation although its coordinates r_1 and θ_1 do not satisfy the equation. For example, the coordinates of $[2, \pi]$ do not satisfy the equation

$$r^2 = 4 \cos \theta.$$
$$(r^2 = 2^2 = 4 \quad \text{but} \quad 4 \cos \theta = 4 \cos \pi = -4.)$$

Nevertheless the point $[2, \pi]$ does lie on the curve $r^2 = 4 \cos \theta$. It lies on the curve because

$$[2, \pi] = [-2, 0]$$

and the coordinates of $[-2, 0]$ satisfy the equation:

$$r^2 = (-2)^2 = 4, \quad 4 \cos \theta = 4 \cos 0 = 4. \quad \square$$

The difficulties are compounded when we deal with two or more curves. Here is an example.

Problem.　Find the points where the cardioids

$$r = a(1 - \cos \theta) \quad \text{and} \quad r = a(1 + \cos \theta)$$

intersect.

SOLUTION.　We begin by solving the two equations simultaneously. Adding these equations, we get $2r = 2a$ and thus $r = a$. This tells us that $\cos \theta = 0$ and therefore $\theta = \frac{1}{2}\pi + n\pi$. The points $[a, \frac{1}{2}\pi + n\pi]$ all lie on both curves. Not all of these points are distinct, however:

for n even, $[a, \frac{1}{2}\pi + n\pi] = [a, \frac{1}{2}\pi]$　and　for n odd, $[a, \frac{1}{2}\pi + n\pi] = [a, \frac{3}{2}\pi]$.

In short, by solving the two equations simultaneously we have arrived at two common points:

$$[a, \tfrac{1}{2}\pi] = (0, a) \quad \text{and} \quad [a, \tfrac{3}{2}\pi] = (0, -a).$$

There is, however, a third point at which the curves intersect, and that is the origin O. (Figure 11.3.1)

The origin clearly lies on both curves:

for $r = a(1 - \cos \theta)$　take $\theta = 0, 2\pi$, etc.

for $r = a(1 + \cos \theta)$　take $\theta = \pi, 3\pi$, etc.

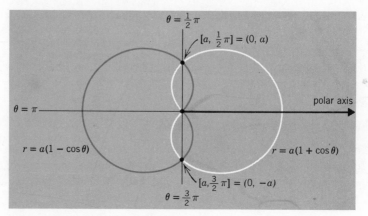

FIGURE 11.3.1

The reason that the origin does not appear when we solve the two equations simultaneously is that the curves do not pass through the origin "simultaneously"; that is, they do not pass through the origin for the same values of θ. Think of each of the equations

$$r = a(1 - \cos \theta) \quad \text{and} \quad r = a(1 + \cos \theta)$$

as giving the position of an object at time θ. At the points we found by solving the two equations simultaneously, the objects collide. (They both arrive there at the same time.) At the origin the situation is different. Both objects pass through the origin, but no collision takes place because the objects pass through the origin at *different* times. □

Exercises

Determine whether the point lies on the curve.

*1. $r^2 \cos \theta = 1$; $[1, \pi]$.
 2. $r^2 = \cos 2\theta$; $[1, \frac{1}{4}\pi]$.
*3. $r = \sin \frac{1}{3}\theta$; $[\frac{1}{2}, \frac{1}{2}\pi]$.
 4. $r^2 = \sin 3\theta$; $[1, -\frac{5}{6}\pi]$.

 5. Show that the point $[2, \pi]$ lies on both $r^2 = 4 \cos \theta$ and $r = 3 + \cos \theta$.

Find the points at which the curves intersect. Express your answers in rectangular coordinates.

*6. $r = \cos^2 \theta$, $r = -1$.
 7. $r = 2 \sin \theta$, $r = 2 \cos \theta$.

*8. $r = \dfrac{1}{1 - \cos \theta}$, $r \sin \theta = b$.
 9. $r^2 = \sin \theta$, $r = 2 - \sin \theta$.

*10. $r = 1 - \cos \theta$, $r = \cos \theta$.
 11. $r = 1 - \cos \theta$, $r = \sin \theta$.

11.4 Area in Polar Coordinates

Here we develop a technique for calculating the area of a region the boundary of which is given in polar coordinates.

As a start, we suppose that α and β are two real numbers with

$$\alpha < \beta \leq \alpha + 2\pi.$$

We take ρ as a function continuous and nonnegative on $[\alpha, \beta]$ and ask for the area of the region Γ which is bounded by

the rays $\quad \theta = \alpha, \quad \theta = \beta \quad$ and $\quad$ the curve $\quad r = \rho(\theta)$.

Such a region is portrayed in Figure 11.4.1.

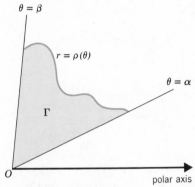

FIGURE 11.4.1

It's not hard to show that the area of Γ is given by the formula

(11.4.1)
$$A = \int_{\alpha}^{\beta} \tfrac{1}{2}[\rho(\theta)]^2 \, d\theta.$$

PROOF. As usual we begin by partitioning the interval in question. We take $P = \{\theta_0, \theta_1, \ldots, \theta_n\}$ as a partition of $[\alpha, \beta]$ and direct our attention to what happens between θ_{i-1} and θ_i.

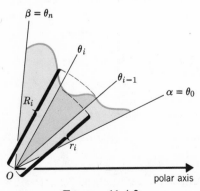

FIGURE 11.4.2

We set

$$r_i = \text{min value of } \rho \text{ on } [\theta_{i-1}, \theta_i] \quad \text{and} \quad R_i = \text{max value of } \rho \text{ on } [\theta_{i-1}, \theta_i].$$

That portion of Γ which lies between θ_{i-1} and θ_i contains a circular sector of radius r_i and central angle $\Delta\theta_i = \theta_i - \theta_{i-1}$ and is itself contained in a circular sector of radius R_i and central angle $\Delta\theta_i = \theta_i - \theta_{i-1}$. (See Figure 11.4.2.) Its area A_i must therefore satisfy the inequality

$$\tfrac{1}{2}r_i^2\,\Delta\theta_i \le A_i \le \tfrac{1}{2}R_i^2\,\Delta\theta_i.\dagger$$

By summing these inequalities from $i = 1$ to $i = n$, you can see that the total area A must satisfy the inequality

(1) $$\qquad\qquad L_f(P) \le A \le U_f(P)$$

with

$$f(\theta) = \tfrac{1}{2}[\rho(\theta)]^2.$$

Since (1) must hold for every partition P of $[a, b]$, we must have

$$A = \int_\alpha^\beta f(\theta)\,d\theta = \int_\alpha^\beta \tfrac{1}{2}[\rho(\theta)]^2\,d\theta. \quad \square$$

REMARK.   When ρ is constant, Γ is a circular sector and

the area of Γ = (constant value of $\tfrac{1}{2}\rho^2$ on $[\alpha, \beta]$) $\cdot (\beta - \alpha)$.

When ρ varies continuously,

(11.4.2)   the area of Γ = (average value of $\tfrac{1}{2}\rho^2$ on $[\alpha, \beta]$) $\cdot (\beta - \alpha)$.

Problem.   Calculate the area of the region Γ pictured in Figure 11.4.3.

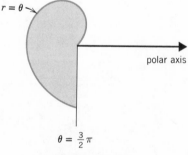

$r = \theta$

polar axis

$\theta = \dfrac{3}{2}\pi$

FIGURE 11.4.3

---

† The area of a circular sector $= \tfrac{1}{2}r^2\theta$ where r is the radius and θ is the central angle.

SOLUTION

$$\text{area of } \Gamma = \int_0^{3\pi/2} \tfrac{1}{2}\theta^2 \, d\theta = \left[\tfrac{1}{6}\theta^3 \right]_0^{3\pi/2} = \tfrac{9}{16}\pi^3. \quad \square$$

Problem. Calculate the area enclosed by the cardioid

$$r = 1 - \cos \theta. \qquad\qquad \text{(Figure 11.3.1)}$$

SOLUTION

$$\text{area} = \int_0^{2\pi} \tfrac{1}{2}(1 - \cos \theta)^2 \, d\theta$$

$$= \tfrac{1}{2} \int_0^{2\pi} (1 - 2 \cos \theta + \cos^2 \theta) \, d\theta$$

$$= \tfrac{1}{2} \left[\int_0^{2\pi} (1 - 2 \cos \theta) \, d\theta + \int_0^{2\pi} \cos^2 \theta \, d\theta \right].$$

Since

$$\int_0^{2\pi} (1 - 2 \cos \theta) \, d\theta = \left[\theta - 2 \sin \theta \right]_0^{2\pi} = 2\pi$$

and

$$\int_0^{2\pi} \cos^2 \theta \, d\theta = \int_0^{2\pi} \tfrac{1}{2}(1 + \cos 2\theta) \, d\theta = \left[\tfrac{1}{2}\theta + \tfrac{1}{4} \sin 2\theta \right]_0^{2\pi} = \pi,$$

we have

$$\text{area} = \tfrac{3}{2}\pi. \quad \square$$

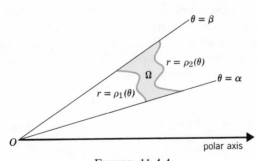

FIGURE 11.4.4

A slightly more complicated type of region is pictured in Figure 11.4.4. We can calculate the area of such a region Ω by taking the area up to $r = \rho_2(\theta)$ and subtracting from it the area up to $r = \rho_1(\theta)$. This gives the formula

(11.4.3)
$$\text{area of } \Omega = \int_\alpha^\beta \tfrac{1}{2}([\rho_2(\theta)]^2 - [\rho_1(\theta)]^2) \, d\theta.$$

Exercises

Calculate the area enclosed by the following curves. Take $a > 0$.

*1. $r = a \cos \theta, \quad -\frac{1}{2}\pi \le \theta \le \frac{1}{2}\pi$.
 2. $r = a \cos 3\theta, \quad -\frac{1}{6}\pi \le \theta \le \frac{1}{6}\pi$.
*3. $r = a\sqrt{\cos 2\theta}, \quad -\frac{1}{4}\pi \le \theta \le \frac{1}{4}\pi$.
 4. $r = a(1 + \cos 3\theta), \quad -\frac{1}{3}\pi \le \theta \le \frac{1}{3}\pi$.
*5. $r^2 = a^2 \sin^2 \theta$.
 6. $r^2 = a^2 \sin^2 2\theta$.

Calculate the area of the region bounded by the following.

*7. $r = \tan 2\theta$ and the rays $\theta = 0, \quad \theta = \frac{1}{8}\pi$.
 8. $r = \cos \theta, \quad r = \sin \theta$, and the rays $\theta = 0, \quad \theta = \frac{1}{4}\pi$.
*9. $r = 2 \cos \theta, \quad r = \cos \theta$, and the rays $\theta = 0, \quad \theta = \frac{1}{4}\pi$.
 10. $r = 1 + \cos \theta, \quad r = \cos \theta$, and the rays $\theta = 0, \quad \theta = \frac{1}{2}\pi$.
*11. $r = a(4 \cos \theta - \sec \theta)$ and the rays $\theta = 0, \quad \theta = \frac{1}{4}\pi$.
 12. The parabola $r = \frac{1}{2} \sec^2 \frac{1}{2}\theta$ and the vertical line through the origin.

Find the area of the region bounded by the graphs of the following.

*13. $r = e^\theta, \quad 0 \le \theta \le \pi; \qquad r = \theta, 0 \le \theta \le \pi; \qquad$ the rays $\theta = 0, \quad \theta = \pi$.
 14. $r = e^\theta, \quad 2\pi \le \theta \le 3\pi; \quad r = \theta, 0 \le \theta \le \pi; \qquad$ the rays $\theta = 0, \quad \theta = \pi$.
*15. $r = e^\theta, \quad 0 \le \theta \le \pi; \qquad r = e^{\theta/2}, \quad 0 \le \theta \le \pi; \qquad$ the rays $\theta = 2\pi, \quad \theta = 3\pi$.
 16. $r = e^\theta, \quad 0 \le \theta \le \pi; \qquad r = e^\theta, 2\pi \le \theta \le 3\pi; \qquad$ the rays $\theta = 0, \quad \theta = \pi$.

11.5 Curves Given Parametrically

We begin with a pair of functions x and y differentiable on some interval I. At the endpoints we require only continuity.

For each t in I we can interpret $(x(t), y(t))$ as the point with x-coordinate $x(t)$ and y-coordinate $y(t)$. As t then ranges over I, the point $(x(t), y(t))$ traces out a curve in the plane. The curve so traced out is said to be given *parametrically* by the functions x and y.

Problem. An object moves so that at time t it has coordinates

$$x(t) = t + 1, \quad y(t) = 2t - 5.$$

Find the path generated by the motion.

SOLUTION. Here we can express $y(t)$ in terms of $x(t)$:

$$y(t) = 2[x(t) - 1] - 5 = 2x(t) - 7.$$

The path generated is the line

$$y = 2x - 7. \quad \square$$

Problem.   An object moves so that at time t it has coordinates

$$x(t) = \sin^2 t, \quad y(t) = \cos t.$$

Find the path generated by the motion.

SOLUTION.   Here

$$x(t) = \sin^2 t = 1 - \cos^2 t = 1 - [y(t)]^2$$

so that the path generated would seem to be the parabola $x = 1 - y^2$. However, this is not quite right. The parametrization restricts $x(t)$ to $[0, 1]$ and $y(t)$ to $[-1, 1]$. The actual path (Figure 11.5.1) is only a small piece of the parabola:

$$x = 1 - y^2, \quad -1 \le y \le 1. \quad \square$$

$$x = 1 - y^2$$

(0, 1)

(1, 0)

(0, −1)

$$x(t) = \sin^2 t, \quad y(t) = \cos t$$

FIGURE 11.5.1

Problem.   Find the curve generated by

$$x(t) = a \cos t, \quad y(t) = a \sin t.$$

SOLUTION.   Here

$$[x(t)]^2 + [y(t)]^2 = a^2 \cos^2 t + a^2 \sin^2 t = a^2$$

so that the curve is the circle

$$x^2 + y^2 = a^2. \quad \square$$

Problem.   Find the curve generated by

$$x(t) = a \cos t, \quad y(t) = b \sin t.$$

SOLUTION.   Here

$$\frac{[x(t)]^2}{a^2} + \frac{[y(t)]^2}{b^2} = \cos^2 t + \sin^2 t = 1$$

so that the curve is the ellipse

$$\frac{x^2}{a^2} + \frac{y^2}{b^2} = 1. \quad \square$$

A single curve can be parametrized in many different ways. As an example take the ellipse

$$\frac{x^2}{a^2} + \frac{y^2}{b^2} = 1$$

and think of t as time measured in seconds. A particle whose position is given by the equations

$$x(t) = a \cos t, \quad y(t) = b \sin t \qquad \text{with } t \in [0, 2\pi]$$

traverses the ellipse in a counterclockwise manner. It begins at the point $(a, 0)$ and makes a full circuit in 2π seconds. If the equations of motion are

$$x(t) = a \cos 2\pi t, \quad y(t) = -b \sin 2\pi t \qquad \text{with } t \in [0, 1],$$

the particle still travels the same ellipse, but in a different manner. Once again it starts at $(a, 0)$, but this time it moves clockwise and makes the full circuit in only one second. If the equations of motion are

$$x(t) = a \sin 4\pi t, \quad y(t) = b \cos 4\pi t \qquad \text{with } t \in [0, \infty),$$

the motion begins at $(0, b)$ and goes on in perpetuity. The motion is clockwise, a complete circuit taking place every half second.

Problem.   Two particles start at the same instant, the first along the linear path

$$x_1(t) = \tfrac{16}{3} - \tfrac{8}{3}t, \quad y_1(t) = 4t - 5 \qquad \text{with } t \geq 0$$

and the second along the elliptical path

$$x_2(t) = 2 \sin \tfrac{1}{2}\pi t, \quad y_2(t) = -3 \cos \tfrac{1}{2}\pi t \qquad \text{with } t \geq 0.$$

(i) At what points, if any, do the paths intersect?
(ii) At what points, if any, do the particles collide?

SOLUTION.   To see where the paths intersect, we first express them both as equations in x and y. The linear path can be written as

$$3x + 2y - 6 = 0, \qquad x \leq \tfrac{16}{3}$$

and the elliptical path as

$$\frac{x^2}{4} + \frac{y^2}{9} = 1.$$

Solving the two equations simultaneously, we get

$$x = 2, \quad y = 0 \qquad \text{and} \qquad x = 0, \quad y = 3.$$

This means that the paths intersect at the points $(2, 0)$ and $(0, 3)$. This answers part (i).

Now for part (ii). The first particle passes through (2, 0) only when

$$x_1(t) = \tfrac{16}{3} - \tfrac{8}{3}t = 2 \quad \text{and} \quad y_1(t) = 4t - 5 = 0.$$

As you can check, this happens only when $t = \tfrac{5}{4}$. When $t = \tfrac{5}{4}$, the second particle is elsewhere. Hence no collision takes place at (2, 0). There is however a collision at (0, 3) because both particles get there at exactly the same time, $t = 2$:

$$x_1(2) = 0 = x_2(2), \quad y_1(2) = 3 = y_2(2). \quad \square$$

Tangents

The problem of finding tangent lines to a parametrized curve

$$C: x(t), y(t) \qquad \text{with } t \in I$$

is complicated by the fact that such a curve can intersect itself. This means that at a given point a curve can have

(i) one tangent,   (ii) two or more tangents,   or   (iii) no tangent at all.

We illustrate these possibilities in Figure 11.5.2.

one tangent

two tangents

no tangent

FIGURE 11.5.2

To make sure that at least one tangent line exists at each point we will make the additional *assumption* that

(11.5.1) | x' and y' are continuous   and $[x'(t)]^2 + [y'(t)]^2 \neq 0.$ |

Now choose a point (x_0, y_0) on the curve C and a time t_0 at which

$$x(t_0) = x_0 \quad \text{and} \quad y(t_0) = y_0.$$

What we want is the slope of the curve as it passes through the point (x_0, y_0) at time t_0.† To find this slope, we assume that $x'(t_0) \neq 0$. With $x'(t_0) \neq 0$, we can be

---

† It could pass through the point (x_0, y_0) at other times also.

sure that for h sufficiently small

$$x(t_0 + h) - x(t_0) \neq 0. \qquad \text{(why?)}$$

We can therefore form the quotient

$$\frac{y(t_0 + h) - y(t_0)}{x(t_0 + h) - x(t_0)}.$$

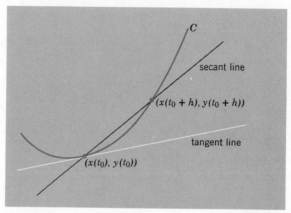

FIGURE 11.5.3

This quotient is the slope of the secant line pictured in Figure 11.5.3. The limit of this quotient as h tends to zero is the slope we want. Since

$$\frac{y(t_0 + h) - y(t_0)}{x(t_0 + h) - x(t_0)} = \frac{(1/h)[\,y(t_0 + h) - y(t_0)]}{(1/h)[x(t_0 + h) - x(t_0)]} \rightarrow \frac{y'(t_0)}{x'(t_0)},$$

we have

(11.5.2) $$\boxed{m = \frac{y'(t_0)}{x'(t_0)}.}$$

As an equation for the tangent line, we can write

$$y - y(t_0) = \frac{y'(t_0)}{x'(t_0)}\,[x - x(t_0)],$$

which simplifies to

$$y'(t_0)[x - x(t_0)] - x'(t_0)[\,y - y(t_0)] = 0,$$

and thus to

(11.5.3) $$\boxed{y'(t_0)[x - x_0] - x'(t_0)[\,y - y_0] = 0.}$$

We derived this equation under the assumption that $x'(t_0) \neq 0$. If $x'(t_0) = 0$,

Equation (11.5.3) still makes sense. It is simply $y'(t_0)[x - x_0] = 0$, which, since $y'(t_0) \neq 0$,† can be simplified to

(11.5.4) $$\boxed{x = x_0.}$$

In this instance the line is vertical, and we say that the curve has a *vertical tangent*.

Problem. Find the tangent(s) to the curve

$$x(t) = t^3, \quad y(t) = 1 - t$$

at the point $(8, -1)$.

SOLUTION. Since the curve passes through the point $(8, -1)$ only when $t = 2$, there can be only one tangent line at that point. With

$$x(t) = t^3, \quad y(t) = 1 - t$$

we have

$$x'(t) = 3t^2, \quad y'(t) = -1$$

and therefore

$$x'(2) = 12, \quad y'(2) = -1.$$

The tangent line has equation

$$(-1)[x - 8] - (12)[y - (-1)] = 0.$$

This reduces to

$$x + 12y + 4 = 0. \quad \square$$

Problem. Find the points of the curve

$$x(t) = 3 - 4 \sin t, \quad y(t) = 4 + 3 \cos t$$

at which there is (i) a horizontal tangent, (ii) a vertical tangent.

SOLUTION. Since

$$x'(t) = -4 \cos t \quad \text{and} \quad y'(t) = -3 \sin t,$$

the derivatives are never 0 simultaneously. To find the points at which there is a horizontal tangent, we set $y'(t) = 0$. This gives $t = n\pi$. Horizontal tangents occur therefore at points of the form $(x(n\pi), y(n\pi))$. Since

$$x(n\pi) = 3 - 4 \sin n\pi = 3 \quad \text{and} \quad y(n\pi) = 4 + 3 \cos n\pi = \begin{cases} 7, & n \text{ even} \\ 1, & n \text{ odd} \end{cases},$$

there are horizontal tangents at $(3, 7)$ and $(3, 1)$.

---

† We are assuming that $[x'(t)]^2 + [y'(t)]^2$ is never 0. Since $x'(t_0) = 0$, $y'(t_0) \neq 0$.

To find the vertical tangents, we set $x'(t) = 0$. This gives $t = \frac{1}{2}\pi + n\pi$. Vertical tangents occur therefore at points of the form $(x(\frac{1}{2}\pi + n\pi), y(\frac{1}{2}\pi + n\pi))$. Since

$$x(\tfrac{1}{2}\pi + n\pi) = 3 - 4 \sin\,(\tfrac{1}{2}\pi + n\pi) = \begin{cases} -1, & n \text{ even} \\ 7, & n \text{ odd} \end{cases}$$

and

$$y(\tfrac{1}{2}\pi + n\pi) = 4 + 3 \cos\,(\tfrac{1}{2}\pi + n\pi) = 4,$$

there are vertical tangents at $(-1, 4)$ and $(7, 4)$.   □

Problem.   Find the tangent(s) to the curve

$$x(t) = t^2 - 2t + 1, \quad y(t) = t^4 - 4t^2 + 4$$

at the point $(1, 4)$.

SOLUTION.   The curve passes through the point $(1, 4)$ when $t = 0$ and when $t = 2$. (Check this out.) Differentiation gives

$$x'(t) = 2t - 2, \quad y'(t) = 4t^3 - 8t.$$

At $t = 0$,

$$x'(t) \neq 0 \quad \text{and} \quad y'(t) = 0,$$

so that the tangent line is horizontal.

At $t = 2$,

$$x'(t) = 2 \quad \text{and} \quad y'(t) = 16,$$

and the tangent line takes the form

$$y - 4 = 8(x - 1). \quad \square$$

We can apply these ideas to a curve given in polar coordinates by first using the relations

$$x = r \cos\,\theta, \quad y = r \sin\,\theta.$$

The polar curve $r = f(\theta)$ can then be written

$$x(\theta) = f(\theta) \cos\,\theta, \quad y(\theta) = f(\theta) \sin\,\theta.$$

Problem.   Find the slope of the curve

$$r = a\theta \qquad \text{at } \theta = \tfrac{1}{2}\pi.$$

SOLUTION

$$x(\theta) = r \cos\,\theta = a\theta \cos\,\theta, \quad y(\theta) = r \sin\,\theta = a\theta \sin\,\theta.$$

Now we differentiate:

$$x'(\theta) = -a\theta \sin\,\theta + a \cos\,\theta, \quad y'(\theta) = a\theta \cos\,\theta + a \sin\,\theta.$$

Thus

$$x'(\tfrac{1}{2}\pi) = -\tfrac{1}{2}\pi a, \quad y'(\tfrac{1}{2}\pi) = a \qquad \text{and} \qquad \text{slope} = \frac{y'(\tfrac{1}{2}\pi)}{x'(\tfrac{1}{2}\pi)} = -\frac{2}{\pi}. \quad \square$$

Problem. Find the points of the cardioid

$$r = 1 - \cos \theta$$

at which there is a vertical tangent.

SOLUTION. Since the cosine function has period 2π, we need only concern ourselves with θ in $[0, 2\pi)$. Parametrically we have

$$x(\theta) = (1 - \cos \theta) \cos \theta, \quad y(\theta) = (1 - \cos \theta) \sin \theta.$$

Differentiating and simplifying, we obtain

$$x'(\theta) = (2 \cos \theta - 1) \sin \theta, \quad y'(\theta) = (1 - \cos \theta)(1 + 2 \cos \theta).$$

The only numbers in the interval $[0, 2\pi)$ at which x' is zero and y' is not are $\frac{1}{3}\pi$, π, and $\frac{5}{3}\pi$. We have a vertical tangent at each of the following points:

$$[\tfrac{1}{2}, \tfrac{1}{3}\pi], \quad [2, \pi], \quad [\tfrac{1}{2}, \tfrac{5}{3}\pi].$$

In rectangular coordinates, these points are:

$$(\tfrac{1}{4}, \tfrac{1}{4}\sqrt{3}), \quad (-2, 0), \quad (\tfrac{1}{4}, -\tfrac{1}{4}\sqrt{3}). \quad \square$$

This section has been introductory. We return to curves and their parametrizations in Chapter 15.

Exercises

Express the curve in terms of x and y.

*1. $x(t) = t^2$, $y(t) = 2t + 1$.   2. $x(t) = t^2$, $y(t) = t^4 + 1$.
*3. $x(t) = at + b$, $y(t) = t^3$.   4. $x(t) = 2 \cos t$, $y(t) = 3 \sin t$.
*5. $x(t) = 2t + 1$, $y(t) = e^{t+1}$.   6. $x(t) = \cos^2 t$, $y(t) = \sin t$.

Find an equation in x and y for the tangent line.

*7. $x(t) = t$, $y(t) = t^3 - 1$; $t = 1$.   8. $x(t) = t^2$, $y(t) = t + 5$; $t = 2$.
*9. $x(t) = 2t$, $y(t) = \cos \pi t$; $t = 0$.   10. $x(t) = 2t - 1$, $y(t) = t^4$; $t = 1$.
*11. $x(t) = t^2$, $y(t) = (2 - t)^2$; $t = \frac{1}{2}$.   12. $x(t) = 1/t$, $y(t) = t^2 + 1$; $t = 1$.
*13. $x(t) = \cos^3 t$, $y(t) = \sin^3 t$; $t = \frac{1}{4}\pi$.   14. $x(t) = e^t$, $y(t) = 3e^{-t}$; $t = 0$.

Find an equation in x and y for the tangent line.

*15. $r = 4 - 2 \sin \theta$, $\theta = 0$.   16. $r = 4 \cos 2\theta$, $\theta = \frac{1}{2}\pi$.

*17. $r = \dfrac{4}{5 - \cos \theta}$, $\theta = \frac{1}{2}\pi$.   18. $r = \dfrac{5}{4 - \cos \theta}$, $\theta = \frac{1}{6}\pi$.

*19. $r = \dfrac{\sin \theta - \cos \theta}{\sin \theta + \cos \theta}$, $\theta = 0$.   20. $r = \dfrac{\sin \theta + \cos \theta}{\sin \theta - \cos \theta}$, $\theta = \frac{1}{2}\pi$.

21. In Section 3.9 we said that the graph of $f(x) = x^{1/3}$ has a vertical tangent at the origin. Verify this by the methods of this section.

FIGURE 11.5.4

22. Let $P = [r_1, \theta_1]$ be a point on a polar curve $r = f(\theta)$ as in Figure 11.5.4. Show that, if $f'(\theta_1) = 0$, then the tangent at P is perpendicular to the line segment $\overline{OP}$.

Find the points (x, y) at which the curve has (a) a horizontal tangent, (b) a vertical tangent. Then sketch the curve.

*23. $x(t) = 3t - t^3, \quad y(t) = t + 1.$　　　　24. $x(t) = t^2 - 2t, \quad y(t) = t^3 - 12t.$
*25. $x(t) = 3 - 4 \sin t, \quad y(t) = 4 + 3 \cos t.$　　26. $x(t) = \sin 2t, \quad y(t) = \sin t.$

*27. Two particles start at the same instant, the first along the elliptical path

$$x_1(t) = 2 - 3 \cos \pi t, \quad y_1(t) = 3 + 7 \sin \pi t \qquad \text{with } t \geq 0$$

and the second along the parabolic path

$$x_2(t) = 3t + 2, \quad y_2(t) = -\tfrac{7}{15}(3t + 1)^2 + \tfrac{157}{15} \qquad \text{with } t \geq 0.$$

(a) At what points, if any, do these paths intersect?
(b) At what points, if any, will the particles collide?

~~Optional~~ 11.6　The Cycloid; A Note on Area

This section is set as a collection of exercises.

The Cycloid

Take a circle of radius r and choose a point P on the circumference. Set the circle on the x-axis so that P is at the origin. Now roll the circle along the x-axis. The path described by the point P is called a *cycloid*. Figure 11.6.1 shows the position of P after a turn of t radians.

1. Show that the cycloid is given parametrically by the equations

$$x(t) = r(t - \sin t), \quad y(t) = r(1 - \cos t).$$

2. At the end of each arch the cycloid comes to a cusp and there is no tangent line. Verify that x' and y' are both 0 at the end of each arch.
3. Show that the line tangent to the cycloid at P (a) passes through the

Optional | top of the circle and (b) intersects the *x*-axis at an angle of $\frac{1}{2}(\pi - t)$ radians.

*4. Express the cycloid from $t = 0$ to $t = \pi$ by an equation in x and y.

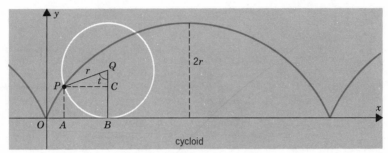

FIGURE 11.6.1

A Note on Area

Begin with a curve C which is the graph of a nonnegative function f defined on an interval $[a, b]$. Suppose now that C is also given parametrically:

$$C: x(t), y(t) \qquad \text{with } t \in [c, d].$$

Assume that x' and y are continuous and that $x(c) = a$ and $x(d) = b$.

5. Show that

(**11.6.1**)
$$\boxed{\text{the area below } C = \int_c^d y(t)\,x'(t)\,dt.}$$

6. Show that the area below one arch of the cycloid is three times the area of the rolling circle.

*7. Sketch the curve

$$x(t) = rt, \quad y(t) = r(1 - \cos t) \qquad \text{with } t \in [0, 2\pi]$$

and find the area below it.

8. Calculate the area of an ellipse

$$\frac{x^2}{a^2} + \frac{y^2}{b^2} = 1$$

using

$$x(t) = -a \cos t, \quad y(t) = b \sin t \qquad \text{with } t \in [0, \pi]$$

as a parametrization for the upper half

$$y = \frac{b}{a} \sqrt{a^2 - x^2}, \qquad x \in [-a, a].$$

Optional | What difficulty arises if we use the parametrization

$$x(t) = a \cos t, \quad y(t) = b \sin t \quad \text{with } t \in [0, \pi]?$$

*9. Find the area enclosed by the astroid

$$x(t) = a \cos^3 t, \quad y(t) = a \sin^3 t. \qquad \text{(Figure 11.7.3)}$$

11.7 The Least Upper Bound Axiom; Arc Length

Before we can define what we mean by the *length* of a curve, we have to look a little deeper into the real number system.

The Least Upper Bound Axiom

We begin with a set S of real numbers. As we indicated in Chapter 1, a number M is called an *upper bound* for S iff

$$x \leq M \qquad \text{for all } x \in S.$$

Of course, not all sets of real numbers have upper bounds. Those that do are said to be *bounded above*.

It is obvious that every set that has a largest element has an upper bound; if b is the largest element of S, then

$$x \leq b \qquad \text{for all } x \in S,$$

and therefore b is an upper bound for S. The converse is false: for example, the sets

$$(-\infty, 0) \quad \text{and} \quad \left\{ \frac{1}{2}, \frac{2}{3}, \frac{3}{4}, \cdots, \frac{n}{n+1}, \cdots \right\}$$

both have upper bounds (2 for instance), but neither has a largest element.

Let's return to the set $(-\infty, 0)$. While $(-\infty, 0)$ does not have a largest element, the set of its upper bounds, $[0, \infty)$, does have a least element, namely 0. A similar remark can be made about the second set. While the set of quotients

$$\frac{n}{n+1} = 1 - \frac{1}{n+1}$$

does not have a greatest element, the set of its upper bounds, $[1, \infty)$, does have a least element, namely 1. Along these lines there is a key *assumption* that we make about the real number system. It is called the *least upper bound axiom*.

Axiom 11.7.1 The Least Upper Bound Axiom

Every nonempty set of real numbers that has an upper bound has a *least* upper bound.

The least upper bound of a set has a property that deserves particular attention. The idea is this: the fact that M is the least upper bound of the set S does not tell us that M is in S (it need not be), but it does tell us that we can approximate M as closely as we wish by members of S.

Theorem 11.7.2

If M is the least upper bound of the set S and ϵ is positive, then there is a number s in S such that

$$M - \epsilon < s \leq M.$$

PROOF. Since all numbers in S are less than or equal to M, the right-hand side of the inequality causes no difficulty. All we have to show therefore is that there exists some number s in S such that

$$M - \epsilon < s.$$

Suppose on the contrary that no such number exists. In that case

$$x \leq M - \epsilon \quad \text{for all } x \in S,$$

and $M - \epsilon$ becomes an upper bound for S. But this cannot happen, for it makes $M - \epsilon$ an upper bound that is less than M, and M is by assumption the *least* upper bound. $\square$

Similar observations can be made about lower bounds. In the first place, a number m is called a *lower bound* for S iff

$$m \leq x \quad \text{for all } x \in S.$$

Sets that have lower bounds are said to be *bounded below*. Not all sets have lower bounds, but those that do, have *greatest lower bounds*. This need not be taken as an axiom. Using the least upper bound axiom we can prove it as a theorem.

Theorem 11.7.3

Every nonempty set of real numbers that has a lower bound has a *greatest* lower bound.

PROOF. Suppose that S is nonempty and that it has a lower bound x. Then

$$x \leq s \quad \text{for all } s \in S.$$

It follows that

$$-s \leq -x \quad \text{for all } s \in S;$$

that is,

$$\{-s : s \in S\} \quad \text{has an upper bound } -x.$$

From the least upper bound axiom we conclude that $\{-s: s \in S\}$ has a least upper bound. Call it x_0. Since

$$-s \le x_0 \qquad \text{for all } s \in S,$$

we see that

$$-x_0 \le s \qquad \text{for all } s \in S,$$

and thus $-x_0$ is a lower bound for S. We now assert that $-x_0$ is the greatest lower bound of the set S. To see this, note that, if there existed x_1 satisfying

$$-x_0 < x_1 \le s \qquad \text{for all } s \in S,$$

then we would have

$$-s \le -x_1 < x_0 \qquad \text{for all } s \in S,$$

and thus x_0 would not be the *least* upper bound of $\{-s: s \in S\}$. $\square$

As in the case of the least upper bound, the greatest lower bound of a set need not be in the set, but it can be approximated as closely as we wish by members of the set. In short, we have the following theorem, the proof of which is left as an exercise.

Theorem 11.7.4

If m is the greatest lower bound of the set S and ϵ is positive, then there is a number s in S such that

$$m \le s < m + \epsilon.$$

Arc Length Defined

We come now to the notion of arc length. In Figure 11.7.1 we have sketched a curve C which we assume is parametrized by a pair of continuously differentiable functions†:

$$C: x(t), y(t) \qquad \text{with } t \in [a, b].$$

What we want to do is assign a length to the curve C. Here our experience in Chapter 5 can be used as a model.

To decide what should be meant by the area of a region Ω, we approximated Ω by the union of a finite number of rectangles. To decide what should be meant by the length of the curve C, we will approximate C by the union of a finite number of line segments.

---

† By this we mean functions that have continuous first derivatives.

What is the length of this curve?

FIGURE 11.7.1

Each point t in $[a, b]$ gives rise to a point $P = P(x(t), y(t))$ that lies on the curve C. By choosing a finite number of points in $[a, b]$,

$$a = t_0 < t_1 < \cdots < t_{i-1} < t_i < \cdots < t_{n-1} < t_n = b,$$

we obtain a finite number of points of C,

$$P_0, P_1, \ldots, P_{i-1}, P_i, \ldots, P_{n-1}, P_n.$$

We now join these points consecutively by line segments and call the resulting path

$$L = \overline{P_0 P_1} \cup \cdots \cup \overline{P_{i-1} P_i} \cup \cdots \cup \overline{P_{n-1} P_n},$$

a *polygonal path* inscribed in the curve C. (See Figure 11.7.2.)

a polygonal path inscribed in the curve C

FIGURE 11.7.2

The length of such a polygonal path is the sum of the distances between consecutive vertices:

$$\text{length of } L = d(P_0, P_1) + \cdots + d(P_{i-1}, P_i) + \cdots + d(P_{n-1}, P_n).$$

The path L serves as an approximation to the curve C, but obviously a better approximation can be obtained by adding more vertices to L. At this point let's ask ourselves exactly what it is we require of the number that we shall call the length of C. Certainly we require that

$$l(L) \leq \text{the length of } C \qquad \text{for each } L \text{ inscribed in } C.$$

But that is not enough. There is another requirement that seems reasonable. If we can choose L to approximate C as closely as we wish, then we should be able to choose L so that $l(L)$ approximates the length of C as closely as we wish. Namely, for each positive ϵ there should exist a polygonal path L such that

$$(\text{length of } C) - \epsilon < l(L) \leq \text{length of } C.$$

Theorem 11.7.2 tells us that we can achieve this result by defining the length of C as the least upper bound of all the $l(L)$. This is in fact what we do:

Definition 11.7.5

$$\text{length of } C = \left\{ \begin{array}{l} \text{the least upper bound of the set of all} \\ \text{lengths of polygonal paths inscribed in } C \end{array} \right].$$

Arc-Length Formulas

Now that we have explained what we mean by the length of a curve C, it is time to describe a practical way to compute it. The basic result is easy to state. If C is parametrized by a pair of continuously differentiable functions

$$C: x(t),\ y(t) \qquad \text{with } t \in [a, b],$$

then

(11.7.6)
$$\text{length of } C = \int_a^b \sqrt{[x'(t)]^2 + [y'(t)]^2}\, dt.$$

Obviously this is not something to be taken on faith. It has to be proven. We will do so, but not until Chapter 15. Right now we assume that the result is true and carry out some computations.

Example. If C is a circle of radius r, say

$$C: \quad x(t) = r\cos t, \quad y(t) = r\sin t \qquad \text{with } t \in [0, 2\pi],$$

the formula yields

$$l(C) = \int_0^{2\pi} \sqrt{r^2 \sin^2 t + r^2 \cos^2 t}\, dt = \int_0^{2\pi} r\, dt = 2\pi r,$$

which is reassuring. $\square$

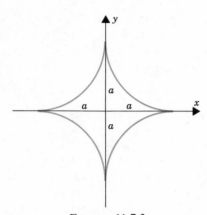

FIGURE 11.7.3

Example. In the case of the astroid

$$C: \quad x(t) = a \cos^3 t, \quad y(t) = a \sin^3 t$$

shown in Figure 11.7.3, we have

$$[x'(t)]^2 + [y'(t)]^2 = 9a^2 \sin^2 t \cos^2 t,$$

so that

$$l(C) = 4 \int_0^{\pi/2} 3a \sin t \cos t \, dt = 12a \left[\tfrac{1}{2} \sin^2 t \right]_0^{\pi/2} = 6a. \quad \Box$$

Suppose now that the curve C is actually the graph of a continuously differentiable function

$$y = f(x), \quad x \in [a, b].$$

As a parametrization for C we can take

$$x(t) = t, \quad y(t) = f(t) \quad \text{with } t \in [a, b].$$

Since

$$x'(t) = 1 \quad \text{and} \quad y'(t) = f'(t),$$

the arc length formula reduces to

$$l(C) = \int_a^b \sqrt{1 + [f'(t)]^2} \, dt.$$

Replacing t by x we can write

(11.7.7) $\boxed{\text{length of the graph of } f = \int_a^b \sqrt{1 + [f'(x)]^2} \, dx.}$

A direct derivation of this formula is outlined in Exercise 36.

Example. For

$$f(x) = x^2, \qquad x \in [0, 1]$$

the graph is a parabolic arc from $(0, 0)$ to $(1, 1)$. The length of this arc is

$$\int_0^1 \sqrt{1 + [f'(x)]^2} \, dx = \int_0^1 \sqrt{1 + 4x^2} \, dx = 2 \int_0^1 \sqrt{(\tfrac{1}{2})^2 + x^2} \, dx$$

$$\underset{\text{by } 8.5.1}{=} \left[x\sqrt{(\tfrac{1}{2})^2 + x^2} + (\tfrac{1}{2})^2 \log (x + \sqrt{(\tfrac{1}{2})^2 + x^2}) \right]_0^1$$

$$= \tfrac{1}{2}\sqrt{5} + \tfrac{1}{4} \log (2 + \sqrt{5}). \quad \square$$

Curves given in polar coordinates can be measured in a similar manner. Suppose that C is given by a polar equation

$$r = \rho(\theta), \qquad \alpha \le \theta \le \beta.$$

Then we have the parametric representation

$$x(\theta) = \rho(\theta) \cos \theta, \quad y(\theta) = \rho(\theta) \sin \theta \qquad \text{with } \theta \in [\alpha, \beta].$$

With

$$x'(\theta) = -\rho(\theta) \sin \theta + \rho'(\theta) \cos \theta \quad \text{and} \quad y'(\theta) = \rho(\theta) \cos \theta + \rho'(\theta) \sin \theta,$$

you'll find, after simplification, that

$$[x'(\theta)]^2 + [y'(\theta)]^2 = [\rho(\theta)]^2 + [\rho'(\theta)]^2.$$

Consequently, we have the formula

(11.7.8)
$$l(C) = \int_\alpha^\beta \sqrt{[\rho(\theta)]^2 + [\rho'(\theta)]^2} \, d\theta.$$

Example. As a polar equation for the circle of radius ρ we have simply

$$r = \rho.$$

Since in this case ρ is a constant, its derivative is zero. The circumference of the circle is therefore

$$\int_0^{2\pi} \sqrt{\rho^2} \, d\theta = \int_0^{2\pi} \rho \, d\theta = 2\pi\rho. \quad \square$$

Example. In the case of the cardioid

$$r = 1 - \cos \theta,$$

we have

$$\rho(\theta) = 1 - \cos \theta, \qquad \rho'(\theta) = \sin \theta.$$

Here

$$[\rho(\theta)]^2 + [\rho'(\theta)]^2 = 1 - 2\cos\theta + \cos^2\theta + \sin^2\theta = 2(1 - \cos\theta).$$

The identity

$$\tfrac{1}{2}(1 - \cos\theta) = \sin^2\tfrac{1}{2}\theta$$

gives

$$[\rho(\theta)]^2 + [\rho'(\theta)]^2 = 4\sin^2\tfrac{1}{2}\theta.$$

The length of the cardioid is therefore

$$\int_0^{2\pi} \sqrt{[\rho(\theta)]^2 + [\rho'(\theta)]^2}\,d\theta = \int_0^{2\pi} 2\sin\tfrac{1}{2}\theta\,d\theta = 4\int_0^{2\pi} \tfrac{1}{2}\sin\tfrac{1}{2}\theta\,d\theta$$

$$= 4\left[-\cos\tfrac{1}{2}\theta\right]_0^{2\pi} = 8. \quad \square$$

Exercises

Find the least upper bound for each of the following sets.

*1. (0, 2). 2. [0, 2].
*3. $\{x: x^2 < 4\}$. (the set of all x such that $x^2 < 4$)
 4. $\{x: x^2 < 2\}$. *5. $\{x: x^2 \geq 2\}$.
 6. $\{0.9, 0.99, 0.999, \ldots\}$. *7. $\{x: \log x < 1\}$.

Find the greatest lower bound for each of the following sets.

*8. (0, 2). 9. [0, 2].
10. $\{x: \log x > 0\}$. *11. $\{2\tfrac{1}{2}, 2\tfrac{1}{3}, 2\tfrac{1}{4}, \ldots\}$.

Find the length of the graph.

*12. $f(x) = mx + b$, $x \in [c, d]$.
 13. $f(x) = x^{3/2}$, $x \in [0, 44]$.
*14. $f(x) = \tfrac{1}{3}\sqrt{x}\,(x - 3)$, $x \in [0, 3]$.
 15. $f(x) = 2e^{x/2}$, $x \in [0, \log 8]$.
*16. $f(x) = \log(\sec x)$, $x \in [-\tfrac{1}{4}\pi, \tfrac{1}{4}\pi]$.
 17. $f(x) = \tfrac{1}{2}(e^x + e^{-x})$, $x \in [0, \log 2]$.
*18. $f(x) = e^x$, $x \in [0, \tfrac{1}{2}\log 3]$.
 19. $f(x) = \tfrac{1}{3}(x^2 + 2)^{3/2}$, $x \in [0, 1]$.
*20. $f(x) = \tfrac{2}{3}(x - 1)^{3/2}$, $x \in [1, 2]$.
 21. $f(x) = \tfrac{1}{2}x^2$, $x \in [0, 1]$.
*22. $f(x) = \tfrac{1}{2}x\sqrt{3 - x^2} + \tfrac{3}{2}\arcsin(\tfrac{1}{3}\sqrt{3}x)$, $x \in [0, 1]$.

Find the length of the curve.

 23. $x(t) = t^2$, $y(t) = 2t$ with $t \in [0, \sqrt{3}]$.
*24. $x(t) = t - 1$, $y(t) = \tfrac{1}{2}t^2$ with $t \in [0, 1]$.
 25. $x(t) = t^2$, $y(t) = t^3$ with $t \in [0, 1]$.

*26. $x(t) = e^t \sin t, \quad y(t) = e^t \cos t \qquad$ with $t \in [0, \pi]$.

27. $x(t) = \cos t + t \sin t, \quad y(t) = \sin t - t \cos t \qquad$ with $t \in [0, \pi]$.

*28. $r = e^\theta, \quad 0 \leq \theta \leq 4\pi$. (logarithmic spiral)

29. $r = ae^\theta, \quad -2\pi \leq \theta \leq 2\pi$.

*30. $r = e^{2\theta}, \quad 0 \leq \theta \leq 2\pi$.

31. $r = a\theta, \quad 0 \leq \theta \leq 2\pi$. (spiral of Archimedes)

*32. $r = 2a \sec \theta, \quad 0 \geq \theta \geq \frac{1}{4}\pi$.

33. Prove Theorem 11.7.4.

34. At time t a particle has position

$$x(t) = 1 - \cos t, \quad y(t) = t - \sin t.$$

Find the total distance traveled between $t = 0$ and $t = 2\pi$.

*35. At time t a particle has position

$$x(t) = 1 + \arctan t, \quad y(t) = 1 - \log\sqrt{1 + t^2}.$$

Find the total distance traveled between $t = 0$ and $t = 1$.

FIGURE 11.7.4

36. Figure 11.7.4 shows the graph of a continuously differentiable function f from $x = a$ to $x = b$ together with a polygonal approximation. Show that the length of this polygonal approximation can be written as the following Riemann sum:

$$\sqrt{1 + [f'(x_1^*)]^2} \, \Delta x_1 + \sqrt{1 + [f'(x_2^*)]^2} \, \Delta x_2 + \cdots + \sqrt{1 + [f'(x_n^*)]^2} \, \Delta x_n.$$

As $\|P\| = \max \Delta x_i$ tends to 0, such Riemann sums tend to

$$\int_a^b \sqrt{1 + [f'(x)]^2} \, dx.$$

Optional | *37. Let f be a positive nonconstant function with the property that for every interval $[a, b]$ the length of the graph equals the area of the region under the graph. Find f, given that $f(0) = 1$.

Optional

38. Show that a homogeneous, flexible, inelastic rope hanging from two fixed points assumes the shape of a catenary:

$$f(x) = a \cosh\left(\frac{x}{a}\right) = \frac{a}{2}(e^{x/a} + e^{-x/a}).$$

HINT: Refer to Figure 11.7.5. That part of the rope which corresponds to the interval $[0, x]$ is subject to the following forces:

(1) its weight, which is proportional to its length;
(2) a horizontal pull at 0, $p(0)$;
(3) a tangential pull at x, $p(x)$.

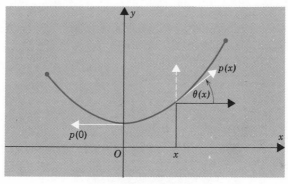

FIGURE 11.7.5

Balancing the vertical forces, we have

$$k \int_0^x \sqrt{1 + [f'(t)]^2}\, dt = p(x) \sin \theta. \quad \text{(weight = vertical pull at } x)$$

Balancing the horizontal forces, we have

$$p(0) = p(x) \cos \theta. \quad \text{(pull at 0 = horizontal pull at } x)$$

11.8   Area of a Surface of Revolution

Let f be a nonnegative function with derivative f' continuous on $[a, b]$. By revolving the graph of f from $x = a$ to $x = b$ about the x-axis, we obtain a surface of revolution. (See Figure 11.8.1.) The area of this surface is given by the formula

(11.8.1)
$$S = \int_a^b 2\pi f(x) \sqrt{1 + [f'(x)]^2}\, dx.$$

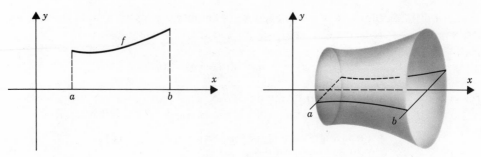

FIGURE 11.8.1

PROOF.   We begin with a partition $P = \{x_0, x_1, \ldots, x_n\}$ of $[a, b]$ and focus our attention on that portion of the surface which is generated by the ith subinterval $[x_{i-1}, x_i]$. Call the area of this portion S_i. (Figure 11.8.2) The arc which generates this portion has length

$$s_i = \int_{x_{i-1}}^{x_i} \sqrt{1 + [f'(x)]^2}\, dx.$$

Its minimum distance from the x-axis is

$$m_i(f) = \text{min value of } f \text{ on } [x_{i-1}, x_i],$$

and its maximum distance from the x-axis is

$$M_i(f) = \text{max value of } f \text{ on } [x_{i-1}, x_i].$$

FIGURE 11.8.2

If we replace this arc by an arc of the same length but closer to the x-axis, then we can expect the surface area to be diminished. If, on the other hand, we replace this arc by an arc of the same length but further away from the x-axis, then we can expect the surface area to be increased. This reasoning leads to the inequality

(1) $$2\pi m_i(f)s_i \leq S_i \leq 2\pi M_i(f)s_i.$$

The lower bound for S_i is the surface area generated by a horizontal line segment of length s_i at a distance $m_i(f)$ from the x-axis. The upper bound is the surface area generated by a horizontal line segment of length s_i at a distance $M_i(f)$ from the x-axis. (Figure 11.8.3)

cylindrical surface   surface area S_i cylindrical surface
surface area $= 2\pi m_i(f)s_i$ surface area $= 2\pi M_i(f)s_i$

FIGURE 11.8.3

If we set

$$g(x) = \sqrt{1 + [f'(x)]^2},$$

then we have

$$s_i = \int_{x_{i-1}}^{x_i} g(x)\,dx$$

and consequently the inequality

$$m_i(g)\,\Delta x_i \le s_i \le M_i(g)\,\Delta x_i. \quad \text{(Property IV, Section 5.9)}$$

Combining this with (1), we have

$$2\pi m_i(f)m_i(g)\,\Delta x_i \ge S_i \ge 2\pi M_i(f)M_i(g)\,\Delta x_i.$$

Summing these inequalities from $i = 1$ to $i = n$, you can see that S, the total surface area generated by the graph, must satisfy the inequality

$$L^*_{2\pi f,g}(P) \ge S \ge U^*_{2\pi f,g}(P). \quad\quad\quad \text{(Section 10.3)}$$

Since P was chosen arbitrarily, this must hold for all partitions P. It follows from Theorem 10.3.1 that

$$S = \int_a^b 2\pi f(x)g(x)\,dx = \int_a^b 2\pi f(x)\sqrt{1 + [f'(x)]^2}\,dx. \quad \square$$

Problem. Find a formula for the lateral area of a cone.

SOLUTION. A cone of base radius r and height h can be generated by revolving the graph of

$$f(x) = \frac{r}{h}x, \quad\quad x \in [0, h]$$

about the x-axis.

FIGURE 11.8.4

For such a cone we have

$$\text{lateral area} = \int_0^h 2\pi f(x)\sqrt{1 + [f'(x)]^2}\,dx$$

$$= 2\pi \int_0^h \frac{r}{h} x \sqrt{1 + \frac{r^2}{h^2}}\,dx$$

$$= \frac{2\pi r \sqrt{h^2 + r^2}}{h^2} \int_0^h x\,dx = \pi r \sqrt{h^2 + r^2}. \quad \Box$$

Problem.   Find the area of the surface generated by revolving the graph of

$$f(x) = \sin x, \qquad x \in [0, \tfrac{1}{2}\pi]$$

about the x-axis.

SOLUTION.   In this case we have

$$f(x) = \sin x, \qquad f'(x) = \cos x$$

and thus

$$\text{surface area} = \int_0^{\pi/2} 2\pi \sin x \sqrt{1 + \cos^2 x}\,dx.$$

To calculate this integral, set

$$u = \cos x; \qquad du = -\sin x\,dx, \quad -2\pi\,du = 2\pi \sin x\,dx.$$

At $x = 0$, $u = 1$; at $x = \tfrac{1}{2}\pi$, $u = 0$. Therefore

$$\int_0^{\pi/2} 2\pi \sin x \sqrt{1 + \cos^2 x}\,dx = -2\pi \int_1^0 \sqrt{1 + u^2}\,du$$

$$= 2\pi \int_0^1 \sqrt{1 + u^2}\,du$$

$$\underset{\underset{\text{by 8.5.1}}{\uparrow}}{=} 2\pi \left[\tfrac{1}{2} u \sqrt{1 + u^2} + \tfrac{1}{2} \log (u + \sqrt{1 + u^2}) \right]_0^1$$

$$= \pi[\sqrt{2} + \log (1 + \sqrt{2})]. \quad \Box$$

Suppose now that the graph of a nonnegative function f is given parametrically:

$$\text{graph of } f: x(t), y(t) \qquad \text{with } t \in [c, d].$$

If x' and y' are continuous and x' remains positive, then the surface area generated by revolving the graph about the x-axis is given by the formula

(11.8.2)

$$S = \int_c^d 2\pi y(t)\sqrt{[x'(t)]^2 + [y'(t)]^2}\, dt.$$

PROOF. Set

$$a = x(c), \quad b = x(d).$$

Note that

$$f(x(t)) = y(t) \quad \text{and} \quad f'(x(t))x'(t) = y'(t).$$

By 11.8.1

$$S = \int_a^b 2\pi f(u)\sqrt{1 + [f'(u)]^2}\, du.$$

The substitution

$$u = x(t); \qquad du = x'(t)\, dt$$

gives

$$\begin{aligned}
S &= \int_c^d 2\pi f(x(t))\sqrt{1 + [f'(x)]^2}\, x'(t)\, dt \\
&= \int_c^d 2\pi y(t)\sqrt{1 + \left[\frac{y'(t)}{x'(t)}\right]^2}\, x'(t)\, dt \\
&= \int_c^d 2\pi y(t)\sqrt{[x'(t)]^2 + [y'(t)]^2}\, dt. \quad \square
\end{aligned}$$

Problem. Derive a formula for the surface area of a sphere.

SOLUTION. We can generate a sphere of radius r by revolving the arc

$$x(t) = r\cos t, \quad y(t) = r\sin t \qquad \text{with } t \in [0, \pi]$$

about the x-axis. Formula 11.8.2 then gives

$$\begin{aligned}
S &= 2\pi \int_0^\pi r\sin t\sqrt{r^2(\sin^2 t + \cos^2 t)}\, dt \\
&= 2\pi r^2 \int_0^\pi \sin t\, dt = 2\pi r^2\left[-\cos t\right]_0^\pi = 4\pi r^2. \quad \square
\end{aligned}$$

Problem. Find the surface area when the curve

$$y^2 - 2\log y = 4x \qquad \text{from} \quad y = 1 \text{ to } y = 2$$

is revolved about the x-axis.

SOLUTION. We can represent the curve parametrically by setting

$$x(t) = \tfrac{1}{4}(t^2 - 2 \log t), \quad y(t) = t \qquad \text{with } t \in [1, 2].$$

Here

$$x'(t) = \tfrac{1}{2}(t - t^{-1}), \quad y'(t) = 1$$

and

$$\sqrt{[x'(t)]^2 + [y'(t)]^2} = \tfrac{1}{2}(t + t^{-1}).$$

It follows that

$$S = \int_{1}^{2} 2\pi t [\tfrac{1}{2}(t + t^{-1})] \, dt = \int_{1}^{2} \pi(t^2 + 1) \, dt = \pi \left[\tfrac{1}{3}t^3 + t \right]_{1}^{2} = \tfrac{10}{3} \pi. \quad \Box$$

Exercises

Find the surface area when the curve is revolved about the x-axis.

*1. $f(x) = r, \quad x \in [0, 1]$.   2. $f(x) = \sqrt{x}, \quad x \in [1, 2]$.

*3. $y = \cos x, \quad x \in [-\tfrac{1}{2}\pi, \tfrac{1}{2}\pi]$.   4. $f(x) = 2\sqrt{1-x}, \quad x \in [-1, 0]$.

*5. $f(x) = \tfrac{1}{3}x^3, \quad x \in [0, 2]$.   6. $f(x) = |x - 3|, \quad x \in [0, 6]$.

*7. $4y = x^3, \quad x \in [0, 1]$.   8. $y^2 = 9x, \quad x \in [0, 4]$.

*9. $x(t) = 2t, \quad y(t) = 3t \qquad \text{with } t \in [0, 1]$.

10. $x(t) = t^2, \quad y(t) = 2t \qquad \text{with } t \in [0, \sqrt{3}]$.

*11. $x^2 + y^2 = r^2, \quad x \in [-\tfrac{1}{2}r, \tfrac{1}{2}r]$.   12. $y = \tfrac{1}{2}(e^x + e^{-x}), \quad x \in [0, 1]$.

*13. $x(t) = a \cos^3 t, \quad y(t) = a \sin^3 t \qquad \text{with } t \in [0, \tfrac{1}{2}\pi]$.

14. $x(t) = e^t \sin t, \quad y(t) = e^t \cos t \qquad \text{with } t \in [0, \tfrac{1}{2}\pi]$.

*15. $r = \rho(\theta), \quad \theta \in [\theta_1, \theta_2]$.   16. $r = e^\theta, \quad \theta \in [0, \tfrac{1}{2}\pi]$.

*17. Find a formula for the lateral surface area of the frustum of a cone. (Figure 10.2.12)

18. Find the surface area when the curve

$$2x = y\sqrt{y^2 - 1} + \log |y - \sqrt{y^2 - 1}| \qquad \text{from } y = 2 \text{ to } y = 5$$

is revolved about the x-axis.

*19. Find the surface area when the curve

$$6a^2 xy = y^4 + 3a^4 \qquad \text{from } y = a \text{ to } y = 3a$$

is revolved about the x-axis.

20. Where in the proof of 11.8.2 did we use the hypothesis that x' remained positive?

*21. Find the surface area of the ellipsoid obtained by revolving the ellipse

$$\frac{x^2}{a^2} + \frac{y^2}{b^2} = 1 \qquad\qquad (0 < b < a)$$

(a) about its major axis; (b) about its minor axis.

11.9   Some Additional Exercises

*1. Sketch the curve $r^2 = 4 \sin 2\theta$ and find the area it encloses.

2. Sketch the curve $r = 2 + \sin 3\theta$ and find the area it encloses.

*3. Sketch the region bounded by the curve $r = \sec\theta + \tan\theta$, the polar axis, and the ray $\theta = \frac{1}{4}\pi$. Then find its area.

Find the length of the curve.

4. $y = \log(\operatorname{cosec} x)$ from $x = \frac{1}{6}\pi$ to $x = \frac{1}{2}\pi$.

*5. $y^2 = x^3$ from $x = 0$ to $x = \frac{5}{9}$.   (upper branch)

6. $9y^2 = 4(1 + x^2)^3$ from $x = 0$ to $x = 3$.   (upper branch)

*7. $ay^2 = x^3$ from $x = 0$ to $x = 5a$.   (upper branch)

8. $y = \log(1 - x^2)$ from $x = 0$ to $x = \frac{1}{2}$.

*9. $x^{2/3} + y^{2/3} = a^{2/3}$.

10. Find the length of the parabolic arc $r = 2(1 + \cos\theta)^{-1}$, $\theta \in [0, \frac{1}{2}\pi]$.

*11. Find the length of the first arch of the cycloid:

$$x(t) = r(t - \sin t), \quad y(t) = r(1 - \cos t) \qquad \text{with } t \in [0, 2\pi].$$

12. Find the length of the polar curve $r = a \sin^3 \frac{1}{3}\theta$, $\theta \in [0, 3\pi]$.

Find the surface area when the curve is revolved about the x-axis.

$SA = \int 2\pi \sqrt{2pr}\sqrt{1?}$

$= p$

*13. $y^2 = 2px$ from $x = 0$ to $x = 4p$.

14. $y^2 = 4x$ from $x = 0$ to $x = 24$.

*15. $6a^2 xy = x^4 + 3a^4$ from $x = a$ to $x = 2a$.

16. $3x^2 + 4y^2 = 3a^2$.

*17. Find the surface area when the curve

$$x(t) = r(t - \sin t), \quad y(t) = r(1 - \cos t) \qquad \text{with } t \in [0, 2\pi]$$

is revolved about the x-axis.

Optional | 18. Find the length of that portion of the curve

$$(x/a)^{2/3} + (y/b)^{2/3} = 1$$

which lies in the first quadrant (a) if $a = b$; (b) if $a \neq b$.

Sequences; Indeterminate Forms; Improper Integrals

12.1 Sequences of Real Numbers

So far our attention has been fixed on functions defined on an interval or on the union of intervals. Here we study functions defined on the set of positive integers.

Definition 12.1.1

A function that takes on real values and is defined on the set of positive integers is called a *sequence of real numbers*.

The functions defined on the set of positive integers by setting

$$a(n) = n - 1, \qquad b(n) = \frac{n^2 - 1}{n}, \qquad c(n) = \sqrt{\log n}, \qquad d(n) = \frac{e^n}{n}$$

are all sequences of real numbers.

The notions developed for functions in general carry over to sequences. For example, a sequence is said to be bounded (bounded above, bounded below) iff its range is bounded (bounded above, bounded below). If a and b are sequences, then their linear combination $\alpha a + \beta b$ and their product ab are also sequences:

$$(\alpha a + \beta b)(n) = \alpha a(n) + \beta b(n) \qquad \text{and} \qquad (ab)(n) = a(n) \cdot b(n).$$

If the sequence b does not take on the value 0, then the reciprocal $1/b$ is a sequence and so is the quotient a/b:

$$\frac{1}{b}(n) = \frac{1}{b(n)} \qquad \text{and} \qquad \frac{a}{b}(n) = \frac{a(n)}{b(n)}.$$

The number $a(n)$, called *the nth term of the sequence a*, is usually written a_n.

The sequence itself can then be denoted by writing either

$$\{a_1, a_2, a_3, \ldots\} \quad \text{or} \quad \{a_n\}.$$

For instance, the sequence of reciprocals

$$a(n) = 1/n \qquad \text{for each positive integer } n$$

can be defined by setting

$$a_n = 1/n$$

or by writing

$$\{1, \tfrac{1}{2}, \tfrac{1}{3}, \ldots\} \quad \text{or} \quad \{1/n\}.$$

The sequence of roots

$$a(n) = 10^{1/n} \qquad \text{for each positive integer } n$$

can be indicated by setting

$$a_n = 10^{1/n}$$

or by writing

$$\{10, 10^{1/2}, 10^{1/3}, \ldots\} \quad \text{or} \quad \{10^{1/n}\}.$$

In our abbreviated notation, we have

$$\alpha\{a_n\} + \beta\{b_n\} = \{\alpha a_n + \beta b_n\}, \qquad \{a_n\}\{b_n\} = \{a_n b_n\}$$

and, provided that none of the b_n are zero,

$$\frac{1}{\{b_n\}} = \{1/b_n\} \quad \text{and} \quad \frac{\{a_n\}}{\{b_n\}} = \{a_n/b_n\}.$$

We illustrate the notation by some trivial assertions.

(1) $$5\{\tfrac{1}{2}, \tfrac{1}{4}, \tfrac{1}{8}, \tfrac{1}{16}, \ldots\} = \{\tfrac{5}{2}, \tfrac{5}{4}, \tfrac{5}{8}, \tfrac{5}{16}, \ldots\}$$

or, equivalently,

$$5\{1/2^n\} = \{5/2^n\}.$$

(2) $$\{1, 2, 3, \ldots\} + \{1, \tfrac{1}{2}, \tfrac{1}{3}, \ldots\} = \{2, 2\tfrac{1}{2}, 3\tfrac{1}{3}, \ldots\}$$

or, equivalently,

$$\{n\} + \{1/n\} = \{n + 1/n\}.$$

(3) $$\{n\}\{n + 1\} = \{n^2 + n\}, \qquad \frac{\{2^n\}}{\{n\}} = \{2^n/n\}.$$

(4) The sequence $\{1/n\}$ is bounded below by 0 and bounded above by 1:

$$0 \leq 1/n \leq 1 \qquad \text{for all } n.$$

(5) The sequence $\{2^n\}$ is bounded below by 2:

$$2 \le 2^n \quad \text{for all } n.$$

It is not bounded above: there is no fixed number M which satisfies

$$2^n \le M \quad \text{for all } n.$$

(6) The sequence

$$a_n = (-1)^n 2^n$$

can be written

$$\{-2, 4, -8, 16, -32, 64, \ldots\}.$$

It is unbounded below and unbounded above.

Definition 12.1.2

A sequence $\{a_n\}$ is said to be

  (i) *increasing*   iff $a_n < a_{n+1}$ for each positive integer n,
  (ii) *nondecreasing*   iff $a_n \le a_{n+1}$ for each positive integer n,
  (iii) *decreasing*   iff $a_n > a_{n+1}$ for each positive integer n,
  (iv) *nonincreasing*   iff $a_n \ge a_{n+1}$ for each positive integer n.

If any of these four properties holds, the sequence is said to be *monotonic*.

The sequences

$$\{1, \tfrac{1}{2}, \tfrac{1}{3}, \ldots\}, \quad \{2, 4, 8, 16, \ldots\}, \quad \{2, 2, 4, 4, 8, 8, 16, 16, \ldots\}$$

are all monotonic, but

$$\{1, \tfrac{1}{2}, 1, \tfrac{1}{3}, 1, \tfrac{1}{4}, \ldots\}$$

is not.

The next examples are less trivial.

Example. The sequence

$$a_n = \frac{n}{n+1}$$

is increasing. It is bounded below by $\tfrac{1}{2}$ and above by 1.

PROOF. Since

$$\frac{a_{n+1}}{a_n} = \frac{n+1}{n+2} \cdot \frac{n+1}{n} = \frac{n^2 + 2n + 1}{n^2 + 2n} > 1,$$

we have

$$a_n < a_{n+1}.$$

This tells us that the sequence is increasing. The bounds are obvious.   □

Example.   The sequence

$$a_n = \frac{2^n}{n!}$$

is nonincreasing. It decreases for $n \geq 2$.

PROOF.   The first two terms are equal:

$$a_1 = \frac{2^1}{1!} = 2 = \frac{2^2}{2!} = a_2.$$

For $n \geq 2$ the sequence decreases:

$$\frac{a_{n+1}}{a_n} = \frac{2^{n+1}}{(n + 1)!} \cdot \frac{n!}{2^n} = \frac{2}{n + 1} < 1 \qquad \text{if } n \geq 2. \quad □$$

Example.   If $|c| > 1$, the sequence

$$a_n = |c|^n$$

increases without bound.

PROOF.   Assume that $|c| > 1$. Then

$$\frac{a_{n+1}}{a_n} = \frac{|c|^{n+1}}{|c|^n} = |c| > 1.$$

This shows that the sequence increases. To show the unboundedness, we take an arbitrary positive number M and show that there exists a positive integer k for which

$$|c|^k \geq M.$$

A suitable k is one such that

$$k \geq \frac{\log M}{\log |c|},$$

for then

$$k \log |c| \geq \log M, \quad \log |c|^k \geq \log M, \quad \text{and thus} \quad |c|^k \geq M. \quad □$$

Since sequences are defined on the set of positive integers and not on an interval, they are not directly susceptible to the methods of calculus. Fortunately, we can often get around this difficulty by dealing initially, not with the sequence itself, but with a function which is defined on all of $[1, \infty)$ and agrees with the given sequence on the integers.

We illustrate this idea in the next two examples. In the first example, we study the sequence

$$a_n = \frac{n}{e^n}$$

by directing our attention first to the function

$$f(x) = \frac{x}{e^x}.$$

Example. The sequence

$$a_n = \frac{n}{e^n}$$

decreases. It is bounded above by $1/e$ and below by 0.

PROOF. We begin by setting

$$f(x) = \frac{x}{e^x}.$$

Differentiation gives

$$f'(x) = \frac{e^x - xe^x}{e^{2x}} = \frac{1-x}{e^x}.$$

Since $f'(x)$ is 0 at $x = 1$ and negative for $x > 1$, f decreases on $[1, \infty)$. Obviously then the sequence $a_n = n/e^n$ decreases.

Since the sequence decreases, its first term $a_1 = 1/e$ is the largest. Thus $1/e$ is certainly an upper bound. Since all the terms of the sequence are positive, the number 0 is a lower bound. $\square$

Example. The sequence

$$a_n = n^{1/n}$$

decreases for $n \geq 3$.

PROOF. We could compare a_n with a_{n+1} directly, but it is easier to consider the function

$$f(x) = x^{1/x}$$

instead. Since

$$f(x) = e^{(1/x)\log x},$$

we have

$$f'(x) = e^{(1/x)\log x}\frac{d}{dx}\left(\frac{1}{x}\log x\right) = x^{1/x}\left(\frac{1-\log x}{x^2}\right).$$

For $x > e$, $f'(x) < 0$. This tells us that f decreases on $[e, \infty)$. Since $3 > e$, f decreases on $[3, \infty)$. It follows that $\{a_n\}$ decreases for $n \geq 3$. □

Exercises

Discuss the boundedness and monotonicity of the sequence whose nth term is given by the following.

*1. $\dfrac{2}{n}$.

2. $\dfrac{(-1)^n}{n}$.

*3. $\sqrt{n}$.

*4. $\dfrac{n + (-1)^n}{n}$.

5. $\dfrac{n - 1}{n}$.

*6. $(0.9)^n$.

*7. $\dfrac{n^2}{n + 1}$.

8. $\dfrac{n + 1}{n^2}$.

*9. $\dfrac{4n}{\sqrt{4n^2 + 1}}$.

*10. $\dfrac{4^n}{2^n + 100}$.

11. $\dfrac{2^n}{4^n + 1}$.

*12. $\dfrac{10^{10}\sqrt{n}}{n + 1}$.

*13. $(-1)^n \sqrt{n}$.

14. $\dfrac{n^2 + 1}{3n + 2}$.

*15. $\log \left(\dfrac{2n}{n + 1} \right)$.

*16. $(-\tfrac{1}{2})^n$.

17. $\dfrac{(n + 1)^2}{n^2}$.

*18. $\sqrt{4 - \dfrac{1}{n}}$.

*19. $\dfrac{2^n - 1}{2^n}$.

20. $\cos n\pi$.

*21. $\dfrac{1}{n} - \dfrac{1}{n + 1}$.

*22. $\dfrac{\log (n + 2)}{n + 2}$.

23. $\log \left(\dfrac{n + 1}{n} \right)$.

*24. $\dfrac{\sqrt{n + 1}}{\sqrt{n}}$.

*25. $\dfrac{3^n}{(n + 1)^2}$.

26. $\dfrac{1 - (\tfrac{1}{2})^n}{(\tfrac{1}{2})^n}$.

*27. $\dfrac{(-2)^n}{n^{10}}$.

28. Show that linear combinations and products of bounded sequences are bounded.
29. Show that the sequence $a_n = (c^n + d^n)^{1/n}$ with $0 < c < d$ is bounded and monotonic.

12.2 The Limit of a Sequence

The gist of

$$\lim_{x \to c} f(x) = l$$

is that we can make $f(x)$ as close as we wish to the number l simply by requiring that x be sufficiently close to c. The gist of

$$\lim_{n \to \infty} a_n = l$$

(read "the limit of a_n as n tends to infinity is l") is that we can make a_n as close as we wish to the number l simply by requiring that n be sufficiently large.

Definition 12.2.1 Limit

$$\lim_{n\to\infty} a_n = l \quad \text{iff} \quad \begin{cases} \text{for each } \epsilon > 0 \text{ there exists an integer } k > 0 \text{ such that} \\ \quad \text{if} \quad n \geq k, \quad \text{then} \quad |a_n - l| < \epsilon. \end{cases}$$

Example

$$\lim_{n\to\infty} \frac{1}{n} = 0.$$

PROOF. We take $\epsilon > 0$ and note that, if $k > 1/\epsilon$, then $1/k < \epsilon$. Consequently, for $n \geq k$, we have

$$\frac{1}{n} \leq \frac{1}{k} < \epsilon \quad \text{and thus} \quad \left| \frac{1}{n} - 0 \right| < \epsilon. \quad \square$$

Example

$$\lim_{n\to\infty} \frac{2n - 1}{n} = 2.$$

PROOF. Let $\epsilon > 0$. We must show that there exists an integer k such that

$$\left| \frac{2n - 1}{n} - 2 \right| < \epsilon \quad \text{for all } n \geq k.$$

Since

$$\left| \frac{2n - 1}{n} - 2 \right| = \left| \frac{2n - 1 - 2n}{n} \right| = \left| -\frac{1}{n} \right| = \frac{1}{n},$$

again we need only choose $k > 1/\epsilon$. $\square$

The next example justifies the familiar statement

$$\tfrac{1}{3} = 0.333 \cdots .$$

Example. The decimal fractions

$$a_n = 0.\overset{n}{\overbrace{33 \cdots 3}}$$

tend to $\tfrac{1}{3}$ as a limit:

$$\lim_{n\to\infty} a_n = \tfrac{1}{3}.$$

PROOF. Let $\epsilon > 0$. In the first place

(1) $$\left| a_n - \frac{1}{3} \right| = \left| 0.\overset{n}{\overbrace{33 \cdots 3}} - \frac{1}{3} \right| = \left| \frac{0.\overset{n}{\overbrace{99 \cdots 9}} - 1}{3} \right| = \frac{1}{3} \cdot \frac{1}{10^n} < \frac{1}{10^n}.$$

Now choose k so that $1/10^k < \epsilon$. If $n \geq k$, then by (1)

$$\left| a_n - \frac{1}{3} \right| < \frac{1}{10^n} \leq \frac{1}{10^k} < \epsilon. \quad \square$$

Limit Theorems

The limit process for sequences is so similar to the limit process you've already studied that you may find you can prove many of the limit theorems yourself. In any case, try to come up with your own proofs and refer to these only if necessary.

Theorem 12.2.2   Uniqueness of Limit

If

$$\lim_{n \to \infty} a_n = l \quad \text{and} \quad \lim_{n \to \infty} a_n = m,$$

then

$$l = m.$$

PROOF.   If $l \neq m$, then

$$\tfrac{1}{2}|l - m| > 0.$$

From

$$\lim_{n \to \infty} a_n = l \quad \text{and} \quad \lim_{n \to \infty} a_n = m,$$

we conclude that there exists k_1 such that

$$\text{if} \quad n \geq k_1, \quad \text{then} \quad |a_n - l| < \tfrac{1}{2}|l - m|$$

and there exists k_2 such that

$$\text{if} \quad n \geq k_2, \quad \text{then} \quad |a_n - m| < \tfrac{1}{2}|l - m|.$$

For $n \geq \max\{k_1, k_2\}$ we have

$$|a_n - l| + |a_n - m| < |l - m|.$$

But by the triangle inequality we have

$$|l - m| = |(l - a_n) + (a_n - m)| \leq |l - a_n| + |a_n - m| = |a_n - l| + |a_n - m|.$$

Combining the last two statements, we have

$$|l - m| < |l - m|.$$

The hypothesis $l \neq m$ has led to an absurdity. We conclude that $l = m$. $\quad \square$

Definition 12.2.3

A sequence that has a limit is said to be *convergent*. A sequence that has no limit is said to be *divergent*.

Instead of writing

$$\lim_{n\to\infty} a_n = l,$$

we'll often write

$$a_n \to l \qquad \text{(read ``} a_n \text{ converges to } l\text{'')}$$

or more fully

$$a_n \to l \quad \text{as} \quad n \to \infty.$$

Theorem 12.2.4

Every convergent sequence is bounded.

PROOF.   Assume that

$$a_n \to l$$

and choose any positive number: 1, for instance. Using 1 as ϵ, you can see that there must exist k such that

$$|a_n - l| < 1 \qquad \text{for all } n \geq k.$$

This means that

$$|a_n| < 1 + |l| \qquad \text{for all } n \geq k,$$

and consequently

$$|a_n| \leq \max \{|a_1|, |a_2|, \ldots, |a_{k-1}|, 1 + |l|\} \qquad \text{for all } n.$$

This proves that $\{a_n\}$ is bounded. $\square$

Since every convergent sequence is bounded, a sequence which is not bounded cannot be convergent; namely,

(12.2.5)   every unbounded sequence is divergent.

Each of the sequences

$$a_n = \tfrac{1}{2}n, \quad b_n = \frac{n^2}{n + 1}, \quad c_n = n \log n$$

is unbounded. Each of these sequences is therefore divergent.

Boundedness does not imply convergence. As a counter-example, consider the oscillating sequence

$$\{1, 0, 1, 0, \ldots\}.$$

This sequence is certainly bounded (above by 1 and below by 0), but obviously it does not converge.

Boundedness together with monotonicity does imply convergence.

Theorem 12.2.6.

A bounded nondecreasing sequence converges to the least upper bound of its range; a bounded nonincreasing sequence converges to the greatest lower bound of its range.

PROOF. Suppose that $\{a_n\}$ is bounded and nondecreasing. Let ϵ be an arbitrary positive number. If l is the least upper bound of the range of this sequence, then there must exist a_k such that

$$l - \epsilon < a_k.$$

(Otherwise $l - \epsilon$ would be an upper bound less than l.) Since the sequence is nondecreasing,

$$a_k \leq a_n \qquad \text{for all } n \geq k.$$

It follows that

$$l - \epsilon < a_n \leq l \qquad \text{for all } n \geq k.$$

This shows that

$$|a_n - l| < \epsilon \qquad \text{for all } n \geq k$$

and proves that

$$a_n \to l.$$

The nonincreasing case can be handled in a similar manner.   □

Example.   Take the sequence

$$\{(3^n + 4^n)^{1/n}\}.$$

Since

$$3 = (3^n)^{1/n} < (3^n + 4^n)^{1/n} < (2 \cdot 4^n)^{1/n} = 2^{1/n} \cdot 4 \leq 8,$$

the sequence is bounded. Note that

$$\begin{aligned}
(3^n + 4^n)^{(n+1)/n} &= (3^n + 4^n)^{1/n}(3^n + 4^n) \\
&= (3^n + 4^n)^{1/n}3^n + (3^n + 4^n)^{1/n}4^n \\
&> (3^n)^{1/n}3^n + (4^n)^{1/n}4^n > 3 \cdot 3^n + 4 \cdot 4^n = 3^{n+1} + 4^{n+1}.
\end{aligned}$$

Taking the $(n + 1)$st root of both extremes, we get

$$(3^n + 4^n)^{1/n} > (3^{n+1} + 4^{n+1})^{1/(n+1)}.$$

The sequence is decreasing. Being bounded, it must be convergent. (Later you will be asked to show that the limit is 4.)   □

Theorem 12.2.7

Let α be a real number. If $a_n \to l$ and $b_n \to m$, then

(i) $a_n + b_n \to l + m$,   (ii) $\alpha a_n \to \alpha l$,   (iii) $a_n b_n \to lm$.

If, in addition, $m \neq 0$ and b_n is never 0, then

$$(iv)\ \frac{1}{b_n} \to \frac{1}{m} \quad \text{and} \quad (v)\ \frac{a_n}{b_n} \to \frac{l}{m}.$$

PROOF.   Parts (i) and (ii) are left as exercises. To prove (iii), we set $\epsilon > 0$. For each n,

$$|a_n b_n - lm| = |(a_n b_n - a_n m) + (a_n m - lm)|$$
$$\leq |a_n||b_n - m| + |m||a_n - l|.$$

Since $\{a_n\}$ is convergent, $\{a_n\}$ is bounded; that is, there exists $M > 0$ such that

$$|a_n| \leq M \quad \text{for all } n.$$

Since $|m| < |m| + 1$, we have

(1) $$|a_n b_n - lm| \leq M|b_n - m| + (|m| + 1)|a_n - l|.^\dagger$$

Since $b_n \to m$, we know that there exists k_1 such that

$$\text{if } n \geq k_1, \quad \text{then} \quad |b_n - m| < \frac{\epsilon}{2M}.$$

Since $a_n \to l$, we know that there exists k_2 such that

$$\text{if } n \geq k_2, \quad \text{then} \quad |a_n - l| < \frac{\epsilon}{2(|m| + 1)}.$$

For $n \geq \max\{k_1, k_2\}$ both conditions hold, and consequently

$$M|b_n - m| + (|m| + 1)|a_n - l| < \frac{\epsilon}{2} + \frac{\epsilon}{2} = \epsilon.$$

In view of (1), we can conclude that

$$\text{if } n \geq \max\{k_1, k_2\}, \quad \text{then} \quad |a_n b_n - lm| < \epsilon.$$

---

† Soon we'll want to divide by the coefficient of $|a_n - l|$. We have replaced $|m|$ by $|m| + 1$ because $|m|$ can be zero.

This proves that

$$a_n b_n \to lm.$$

To prove (iv), once again we set $\epsilon > 0$. In the first place

$$\left| \frac{1}{b_n} - \frac{1}{m} \right| = \left| \frac{m - b_n}{b_n m} \right| = \frac{|b_n - m|}{|b_n|\,|m|}.$$

Since $b_n \to m$ and $|m|/2 > 0$, there exists k_1 such that

$$\text{if} \quad n \geq k_1, \qquad \text{then} \quad |b_n - m| < \frac{|m|}{2}.$$

This tells us that for $n \geq k_1$ we have

$$|b_n| > \frac{|m|}{2} \qquad \text{and thus} \qquad \frac{1}{|b_n|} < \frac{2}{|m|}.$$

Thus for $n \geq k_1$ we have

(2) $$\left| \frac{1}{b_n} - \frac{1}{m} \right| \leq \frac{2}{|m|^2} |b_n - m|.$$

Since $b_n \to m$, there exists k_2 such that

$$\text{if} \quad n \geq k_2, \qquad \text{then} \quad |b_n - m| < \frac{\epsilon|m|^2}{2}.$$

Thus for $n \geq k_2$ we have

$$\frac{2}{|m|^2} |b_n - m| < \epsilon.$$

In view of (2), we can be sure that

$$\text{if} \quad n \geq \max\{k_1, k_2\}, \qquad \text{then} \quad \left| \frac{1}{b_n} - \frac{1}{m} \right| < \epsilon.$$

This proves that

$$\frac{1}{b_n} \to \frac{1}{m}.$$

The proof of (v) is now easy:

$$\frac{a_n}{b_n} = a_n \cdot \frac{1}{b_n} \to l \cdot \frac{1}{m} = \frac{l}{m}. \qquad \square$$

We are now in a position to handle any rational sequence

$$a_n = \frac{\alpha_k n^k + \alpha_{k-1} n^{k-1} + \cdots + \alpha_0}{\beta_j n^j + \beta_{j-1} n^{j-1} + \cdots + \beta_0}.$$

To determine the behavior of such a sequence we need only divide both numerator and denominator by the highest power of n that occurs.

Example

$$\frac{3n^4 - 2n^2 + 1}{n^5 - 3n^3} = \frac{3/n - 2/n^3 + 1/n^5}{1 - 3/n^2} \to \frac{0}{1} = 0. \quad \square$$

Example

$$\frac{1 - 4n^7}{n^7 + 12n} = \frac{1/n^7 - 4}{1 + 12/n^6} \to \frac{-4}{1} = -4. \quad \square$$

Example

$$\frac{n^4 - 3n^2 + n + 2}{n^3 + 7n} = \frac{1 - 3/n^2 + 1/n^3 + 2/n^4}{1/n + 7/n^3}.$$

Since the numerator tends to 1 and the denominator tends to 0, the sequence is unbounded. Therefore it cannot converge. $\square$

Theorem 12.2.8

$$a_n \to l \quad \text{iff} \quad a_n - l \to 0 \quad \text{iff} \quad |a_n - l| \to 0.$$

The proof is left as an exercise. $\square$

Theorem 12.2.9   The Pinching Theorem for Sequences

Suppose that for all n sufficiently large

$$a_n \le b_n \le c_n.$$

If $a_n \to l$ and $c_n \to l$, then $b_n \to l$.

Once again the proof is left as an exercise. $\square$

As an immediate and obvious consequence of the pinching theorem we have the following corollary.

(12.2.10)

Suppose that for all n sufficiently large
$$|b_n| \le |c_n|.$$
If $|c_n| \to 0$, then $|b_n| \to 0$.

Example

$$\frac{\cos \pi n}{n} \to 0 \quad \text{since} \quad \left|\frac{\cos \pi n}{n}\right| \le \frac{1}{n} \quad \text{and} \quad \frac{1}{n} \to 0. \quad \square$$

Example

$$\sqrt{4 + \left(\frac{1}{n}\right)^2} \to 2$$

since

$$2 \leq \sqrt{4 + \left(\frac{1}{n}\right)^2} \leq \sqrt{4 + 4\left(\frac{1}{n}\right) + \left(\frac{1}{n}\right)^2} = 2 + \frac{1}{n} \quad \text{and} \quad 2 + \frac{1}{n} \to 2. \quad \square$$

Example

(12.2.11)

$$\lim_{n \to \infty} \left(1 + \frac{1}{n}\right)^n = e.$$

PROOF.   You have already seen that

$$\left(1 + \frac{1}{n}\right)^n \leq e \leq \left(1 + \frac{1}{n}\right)^{n+1}. \qquad \text{(Theorem 6.5.12)}$$

From this it follows that

$$\left|\left(1 + \frac{1}{n}\right)^n - e\right| \leq \left|\left(1 + \frac{1}{n}\right)^n - \left(1 + \frac{1}{n}\right)^{n+1}\right|$$

$$= \left(1 + \frac{1}{n}\right)^n \left|1 - \left(1 + \frac{1}{n}\right)\right| = \left(1 + \frac{1}{n}\right)^n \left(\frac{1}{n}\right) \leq \frac{e}{n}.$$

In short, we have the estimate

$$\left|\left(1 + \frac{1}{n}\right)^n - e\right| \leq \frac{e}{n}.$$

Since $e/n \to 0$, you can see that

$$\left[\left(1 + \frac{1}{n}\right)^n - e\right] \to 0 \quad \text{and thus by Theorem 12.2.8} \quad \left(1 + \frac{1}{n}\right)^n \to e. \quad \square$$

The sequences

$$\left\{\cos \frac{\pi}{n}\right\}, \quad \left\{\log\left(\frac{n}{n+1}\right)\right\}, \quad \{e^{1/n}\}, \quad \left\{\tan\left(\sqrt{\frac{\pi^2 n^2 - 8}{16 n^2}}\right)\right\}$$

are all of the form $\{f(c_n)\}$ with f a continuous function. Such sequences are easy to deal with. The basic idea is this: when a continuous function is applied to a convergent sequence, the result is itself a convergent sequence. More precisely, we have the following theorem.

Theorem 12.2.12

Suppose that

$$c_n \to c$$

and that, for each n, c_n is in the domain of f. If f is continuous at c, then

$$f(c_n) \to f(c).$$

PROOF.   We assume that f is continuous at c and take $\epsilon > 0$. From the continuity of f at c we know that there exists $\delta > 0$ such that

$$\text{if} \quad |x - c| < \delta, \qquad \text{then} \quad |f(x) - f(c)| < \epsilon.$$

Since $c_n \to c$, we know that there exists a positive integer k such that

$$\text{if} \quad n \geq k, \qquad \text{then} \quad |c_n - c| < \delta.$$

It follows therefore that

$$\text{if} \quad n \geq k, \qquad \text{then} \quad |f(c_n) - f(c)| < \epsilon. \quad \square$$

Examples

(1) Since

$$\frac{\pi}{n} \to 0$$

and the cosine function is continuous at 0,

$$\cos \frac{\pi}{n} \to \cos 0 = 1. \quad \square$$

(2) Since

$$\frac{n}{n + 1} = \frac{1}{1 + 1/n} \to 1$$

and the logarithm function is continuous at 1,

$$\log \left(\frac{n}{n + 1} \right) \to \log 1 = 0. \quad \square$$

(3) Since

$$\frac{1}{n} \to 0$$

and the exponential function is continuous at 0,

$$e^{1/n} \to e^0 = 1. \quad \square$$

(4) Since

$$\frac{\pi^2 n^2 - 8}{16 n^2} = \frac{\pi^2 - 8/n^2}{16} \to \frac{\pi^2}{16}$$

and the function $f(x) = \tan \sqrt{x}$ is continuous at $\pi^2/16$,

$$\tan \left(\sqrt{\frac{\pi^2 n^2 - 8}{16 n^2}} \right) \to \tan \left(\sqrt{\frac{\pi^2}{16}} \right) = \tan \frac{\pi}{4} = 1. \quad \square$$

(5) Since

$$\frac{2n + 1}{n} + \left(5 - \frac{1}{n^2} \right) \to 7$$

and the square-root function is continuous at 7, we have

$$\sqrt{\frac{2n + 1}{n} + \left(5 - \frac{1}{n^2} \right)} \to \sqrt{7}. \quad \square$$

(6) Since the absolute-value function is everywhere continuous,

$$a_n \to l \quad \text{implies} \quad |a_n| \to |l|. \quad \square$$

Exercises

State whether or not the sequence converges and, if it does, find the limit.

*1. $\dfrac{2}{n}$.

2. $\dfrac{(-1)^n}{n}$.

*3. $\sqrt{n}$.

*4. $\dfrac{n-1}{n}$.

5. $\dfrac{n+(-1)^n}{n}$.

*6. $\sin \dfrac{\pi}{2n}$.

*7. $\dfrac{n^2}{n+1}$.

8. $\dfrac{n+1}{n^2}$.

*9. $\dfrac{4n}{\sqrt{n^2+1}}$.

*10. $\dfrac{4^n}{2^n+10^6}$.

11. $\dfrac{2^n}{4^n+1}$.

*12. $\dfrac{10^{10}\sqrt{n}}{n+1}$.

*13. $(-1)^n \sqrt{n}$.

14. 2^n.

*15. $\log\left(\dfrac{2n}{n+1}\right)$.

*16. $(-\tfrac{1}{2})^n$.

17. $\dfrac{n^4-1}{n^4+n-6}$.

*18. $\dfrac{n^5}{17n^4+12}$.

*19. $\dfrac{(n+1)^2}{n^2}$.

20. $\sqrt{4-\dfrac{1}{n}}$.

*21. $\dfrac{2^n-1}{2^n}$.

*22. $\dfrac{n^2}{\sqrt{2n^4+1}}$.

23. $\dfrac{1}{n}-\dfrac{1}{n+1}$.

*24. $\cos n\pi$.

*25. $(0.9)^n$.

26. $(0.9)^{-n}$.

*27. $\log n - \log(n+1)$.

*28. $\dfrac{\sqrt{n+1}}{2\sqrt{n}}$.

29. $\dfrac{\sqrt{n}\sin(e^n\pi)}{n+1}$.

*30. $\left(1+\dfrac{2}{n}\right)^n$.

31. Prove that, if $a_n \to l$ and $b_n \to m$, then $a_n + b_n \to l + m$.

32. Let α be a real number. Prove that, if $a_n \to l$, then $\alpha a_n \to \alpha l$.

33. Prove that

$$\left(1+\frac{1}{n}\right)^{n+1} \to e \quad \text{given that} \quad \left(1+\frac{1}{n}\right)^n \to e.$$

34. Prove that a bounded nonincreasing sequence converges to the greatest lower bound of its range.

35. Prove the pinching theorem for sequences.

12.3 Some Important Limits

Our purpose here is to familiarize you with some limits that are particularly important in calculus and to give you more experience with limit arguments.

(12.3.1)

> If $x > 0$, then
> $$x^{1/n} \to 1 \quad \text{as} \quad n \to \infty.$$

$\log_{10}$

PROOF. Note that

$$\log (x^{1/n}) = \frac{1}{n} \log x \to 0.$$

Since the exponential function is continuous at 0, it follows from Theorem 12.2.12 that

$$x^{1/n} = e^{(1/n)\log x} \to 1. \quad \Box$$

(12.3.2)

> If $|x| < 1$, then
>
> $$x^n \to 0 \qquad \text{as} \quad n \to \infty.$$

PROOF. Take $|x| < 1$ and observe that $\{|x|^n\}$ is a decreasing sequence:

$$|x|^{n+1} = |x| \, |x|^n < |x|^n.$$

Now let $\epsilon > 0$. By 12.3.1

$$\epsilon^{1/n} \to 1 \qquad \text{as} \quad n \to \infty.$$

Thus there exists an integer $k > 0$ such that

$$|x| < \epsilon^{1/k}. \qquad \text{(explain)}$$

Obviously then

$$|x|^k < \epsilon.$$

Since $\{|x|^n\}$ is a decreasing sequence,

$$|x^n| = |x|^n < \epsilon \qquad \text{for all } n \geq k. \quad \Box$$

(12.3.3)

> For each real x
>
> $$\frac{x^n}{n!} \to 0 \qquad \text{as} \quad n \to \infty.$$

PROOF. Choose an integer k such that

$$k > |x|.$$

For $n > k$,

$$n! = \underbrace{[n(n - 1) \cdots (k + 1)]}_{n - k \text{ terms}} k! > (k^{n-k})(k!) = (k^n)(k^{-k})(k!).$$

Consequently

$$\frac{|x^n|}{n!} = \frac{|x|^n}{n!} < \frac{|x|^n}{(k^n)(k^{-k})(k!)} = \left(\frac{|x|}{k}\right)^n \frac{k^k}{k!}.$$

Since $|x|/k < 1$,

$$\left(\frac{|x|}{k}\right)^n \to 0 \qquad \text{and therefore} \qquad \left(\frac{|x|}{k}\right)^n \frac{k^k}{k!} \to 0.$$

It follows that

$$\frac{|x^n|}{n!} \to 0. \quad \square$$

(12.3.4)

> For each $\alpha > 0$
>
> $$\frac{1}{n^\alpha} \to 0 \quad \text{as} \quad n \to \infty.$$

PROOF.   Since $\alpha > 0$, there exists an odd positive integer p such that

$$\frac{1}{p} < \alpha.$$

Obviously

$$0 < \frac{1}{n^\alpha} = \left(\frac{1}{n}\right)^\alpha \le \left(\frac{1}{n}\right)^{1/p}.$$

Since $1/n \to 0$ and $f(x) = x^{1/p}$ is continuous at 0, we have

$$\left(\frac{1}{n}\right)^{1/p} \to 0 \quad \text{and thus by the pinching theorem} \quad \frac{1}{n^\alpha} \to 0. \quad \square$$

(12.3.5)

> $$\frac{\log n}{n} \to 0 \quad \text{as} \quad n \to \infty.$$

PROOF.   A routine proof can be based on L'Hospital's rule (12.6.3), but that is not available to us yet. We will base our argument on the integral representation of the logarithm:

$$\frac{\log n}{n} = \frac{1}{n}\int_1^n \frac{dt}{t} \le \frac{1}{n}\int_1^n \frac{dt}{\sqrt{t}} = \frac{2}{n}(\sqrt{n} - 1) = 2\left(\frac{1}{\sqrt{n}} - \frac{1}{n}\right) \to 0. \quad \square$$

(12.3.6)

> $$n^{1/n} \to 1 \quad \text{as} \quad n \to \infty.$$

PROOF.   We have

$$n^{1/n} = e^{(1/n)\log n}.$$

Since

$$(1/n)\log n \to 0 \qquad\qquad\qquad\qquad (12.3.5)$$

and the exponential function is continuous at 0, it follows from (12.2.12) that

$$n^{1/n} \to e^0 = 1. \quad \square$$

(12.3.7)

> For each real x
>
> $$\left(1 + \frac{x}{n}\right)^n \to e^x \quad \text{as} \quad n \to \infty.$$

PROOF.   For $x = 0$, the result is obvious. For $x \neq 0$,

$$\log \left(1 + \frac{x}{n}\right)^n = n \log \left(1 + \frac{x}{n}\right) = x \left[\frac{\log (1 + x/n) - \log 1}{x/n}\right].$$

The crux here is to recognize the bracketed expression as a difference quotient for the logarithm function. Once we see this, we can easily obtain

$$\lim_{n \to \infty} \left[\frac{\log (1 + x/n) - \log 1}{x/n}\right] = \lim_{h \to 0} \left[\frac{\log (1 + h) - \log 1}{h}\right] = 1.^\dagger$$

This tells us that

$$\log \left(1 + \frac{x}{n}\right)^n \to x,$$

and consequently

$$\left(1 + \frac{x}{n}\right)^n = e^{\log (1 + x/n)^n} \to e^x. \quad \square$$

Exercises

State whether or not the sequence converges as $n \to \infty$; if it does, find the limit.

*1. $2^{2/n}$.

2. $e^{-\alpha/n}$.

*3. $\left(\frac{2}{n}\right)^n$.

*4. $\dfrac{\log_{10} n}{n}$.

5. $\dfrac{\log (n + 1)}{n}$.

*6. $\log \left(\dfrac{n + 1}{n}\right)$.

*7. $\dfrac{x^{100n}}{n!}$.

8. $n^{\alpha/n}$, $\alpha > 0$.

*9. $\dfrac{3^n}{4^n}$.

*10. $\dfrac{3^{n+1}}{4^{n-1}}$.

11. $n^{1/(n+2)}$.

*12. $(n + 2)^{1/n}$.

*13. $\left(1 - \dfrac{1}{n}\right)^n$:

14. $(n + 2)^{1/(n+2)}$.

*15. $\displaystyle\int_0^n e^{-x} \, dx$.

*16. $\displaystyle\int_{-n}^0 e^{2x} \, dx$.

17. $\displaystyle\int_0^n e^{-nx} \, dx$.

*18. $\displaystyle\int_{-1/n}^{1/n} \sin x^2 \, dx$.

*19. $\displaystyle\int_{-1+1/n}^{1-1/n} \dfrac{dx}{\sqrt{1 - x^2}}$.

20. $\displaystyle\int_n^{n+1} e^{-x^2} \, dx$.

*21. $\displaystyle\int_{1/n}^1 \dfrac{dx}{\sqrt{x}}$.

*22. $\dfrac{n!}{2^n}$.

23. $n^2 \sin n\pi$.

*24. $n^2 \sin \dfrac{\pi}{n}$.

25. Show that

$$\lim_{n \to \infty} (\sqrt{n + 1} - \sqrt{n}) = 0.$$

$\dagger$ For each $t > 0$

$$\lim_{h \to 0} \frac{\log (t + h) - \log t}{h} = \frac{d}{dt} (\log t) = \frac{1}{t}.$$

26. Show that

$$\text{if} \quad 0 < c < d, \quad \text{then} \quad (c^n + d^n)^{1/n} \to d.$$

*27. Find $\lim\limits_{n \to \infty} 2n \sin(\pi/n)$. What is the geometric significance of this limit?

HINT: think of a regular polygon of n sides inscribed in the unit circle.

Find the following limits.

*28. $\lim\limits_{n \to \infty} \dfrac{1 + 2 + \cdots + n}{n^2}$.   HINT: $1 + 2 + \cdots + n = \dfrac{n(n + 1)}{2}$.

29. $\lim\limits_{n \to \infty} \dfrac{1^2 + 2^2 + \cdots + n^2}{(1 + n)(2 + n)}$.   HINT: $1^2 + 2^2 + \cdots + n^2 = \dfrac{n(n + 1)(2n + 1)}{6}$.

*30. $\lim\limits_{n \to \infty} \dfrac{1^3 + 2^3 + \cdots + n^3}{2n^4 + n - 1}$.   HINT: $1^3 + 2^3 + \cdots + n^3 = \dfrac{n^2(n + 1)^2}{4}$.

31. A sequence $\{a_n\}$ is said to be a *Cauchy sequence* [after the French mathematician Augustin-Louis (baron de) Cauchy, 1789–1857] iff

(12.3.8)
$$\boxed{\begin{array}{l} \text{for each } \epsilon > 0 \text{ there exists a positive integer } k \text{ such that} \\[4pt] |a_n - a_m| < \epsilon \quad \text{for all} \quad m, n \geq k. \end{array}}$$

Show that

(12.3.9) $\boxed{\text{every convergent sequence is a Cauchy sequence.}}^{\dagger}$

Optional | 32. (*Arithmetic means.*) Let

$$m_n = \frac{1}{n}(a_1 + a_2 + \cdots + a_n).$$

(a) Prove that, if $\{a_n\}$ is increasing, then $\{m_n\}$ is increasing.
(b) Prove that, if $a_n \to 0$, then $m_n \to 0$.
    HINT: Choose an integer $j > 0$ such that, if $n \geq j$, then $a_n < \epsilon/2$. Then for $n \geq j$,

$$|m_n| < \frac{|a_1 + a_2 + \cdots + a_j|}{n} + \frac{\epsilon}{2}\left(\frac{n - j}{n}\right).$$

State whether or not the indicated sequence converges and, if it does, find the limit.

*33. $\left(\dfrac{n + 1}{n + 2}\right)^n$.   34. $\left(1 + \dfrac{1}{n}\right)^{n^2}$.

*35. $\left(1 + \dfrac{1}{n^2}\right)^n$.   36. $\dfrac{n^n}{2^{n^2}}$.

*37. $\left(t + \dfrac{x}{n}\right)^n$, $t > 0, x > 0$.   38. $\displaystyle\int_{-n}^{n} \dfrac{dx}{1 + x^2}$.

$\dagger$ It is also true that every Cauchy sequence is convergent, but this is more difficult to prove.

12.4 Limits as $x \to \pm\infty$

Now that you are familiar with convergence for sequences, limits as $x \to \infty$ should present no difficulties. To say that

$$\lim_{x \to \infty} f(x) = l$$

[read "the limit of $f(x)$ as x tends to infinity is l"] is to say that we can make the difference between $f(x)$ and l as small as we wish simply by requiring that x be sufficiently large. More precisely,

(12.4.1) $\quad \lim\limits_{x \to \infty} f(x) = l \quad$ iff $\quad \begin{cases} \text{for each } \epsilon > 0 \text{ there exists } K > 0 \text{ such that} \\ \quad \text{if } x \geq K, \qquad \text{then } |f(x) - l| < \epsilon. \end{cases}$

Here are some simple examples:

$$\lim_{x \to \infty} \frac{1}{x} = 0, \qquad \lim_{x \to \infty} \frac{x - 1}{2x + 5} = \lim_{x \to \infty} \frac{1 - 1/x}{2 + 5/x} = \frac{1}{2}, \qquad \lim_{x \to \infty} e^{1/x} = 1.$$

It is easy to interpret the idea geometrically. To say that $\lim_{x \to \infty} f(x) = A$ (see Figure 12.4.1) is to say that, as x becomes arbitrarily large, the graph of f approaches the horizontal line $y = A$. The line is called a *horizontal asymptote*.

FIGURE 12.4.1

Limits as $x \to \infty$ are unique (if they exist), and the usual rules for computation hold. For example, if

$$\lim_{x \to \infty} f(x) = l \quad \text{and} \quad \lim_{x \to \infty} g(x) = m,$$

then

$$\lim_{x \to \infty} [\alpha f(x) + \beta g(x)] = \alpha l + \beta m, \quad \lim_{x \to \infty} [f(x)g(x)] = lm,$$

and if $m \neq 0$

$$\lim_{x \to \infty} \frac{f(x)}{g(x)} = \frac{l}{m}.$$

Moreover, the pinching theorem is valid. Suppose that for all x sufficiently large $h(x) \leq f(x) \leq g(x)$. If $\lim_{x \to \infty} h(x) = l = \lim_{x \to \infty} g(x)$, then $\lim_{x \to \infty} f(x) = l$.

We illustrate these ideas by carrying out some computations.

Problem.   Find

$$\lim_{x \to \infty} \frac{\sin^2 x}{x^2}.$$

SOLUTION.   For $x \neq 0$,

$$0 \leq \frac{\sin^2 x}{x^2} \leq \frac{1}{x^2}. \qquad \text{(since } |\sin x| \leq 1\text{)}$$

Since obviously

$$\lim_{x \to \infty} \frac{1}{x^2} = 0,$$

we have

$$\lim_{x \to \infty} \frac{\sin^2 x}{x^2} = 0. \quad \square$$

Problem.   Find

$$\lim_{x \to \infty} \frac{x(e^{-x} + 1)}{2x - 1}.$$

SOLUTION.   For x different from 0 and $\frac{1}{2}$,

$$\frac{x(e^{-x} + 1)}{2x - 1} = \frac{e^{-x} + 1}{2 - 1/x}.$$

Since

$$\lim_{x \to \infty} (e^{-x} + 1) = 1 \quad \text{and} \quad \lim_{x \to \infty} (2 - 1/x) = 2,$$

we have

$$\lim_{x \to \infty} \frac{x(e^{-x} + 1)}{2x - 1} = \frac{1}{2}. \quad \square$$

Problem.   Find

$$\lim_{x \to \infty} \frac{x + \log x}{x \log x}.$$

SOLUTION.   For x different from 0 and 1,

$$\frac{x + \log x}{x \log x} = \frac{x}{x \log x} + \frac{\log x}{x \log x} = \frac{1}{\log x} + \frac{1}{x}.$$

Since obviously

$$\lim_{x \to \infty} \frac{1}{\log x} = 0 \quad \text{and} \quad \lim_{x \to \infty} \frac{1}{x} = 0,$$

we have

$$\lim_{x \to \infty} \frac{x + \log x}{x \log x} = 0. \quad \square$$

There is a simple relation between limits at ∞ and right-hand limits at 0. Set $x = 1/t$, and you'll see that

(12.4.2)
$$\boxed{\lim_{x \to \infty} f(x) = l \quad \text{iff} \quad \lim_{t \downarrow 0} f(1/t) = l.}$$

We'll use this idea in the next section.

Limits as $x \to -\infty$ are defined by setting

(12.4.3)
$$\boxed{\lim_{x \to -\infty} f(x) = l \quad \text{iff} \quad \begin{cases} \text{for each } \epsilon > 0 \text{ there exist } K < 0 \text{ such that} \\ \quad \text{if} \quad x \leq K, \quad \text{then} \quad |f(x) - l| < \epsilon. \end{cases}}$$

The behavior of such limits is entirely analogous to the case $x \to \infty$. As simple examples, we have

$$\lim_{x \to -\infty} e^x = 0 \quad \text{and} \quad \lim_{x \to -\infty} \frac{3 - x}{2 + x} = \lim_{x \to -\infty} \frac{3/x - 1}{2/x + 1} = -1.$$

In the case of

$$h(x) = \begin{cases} -1, & x < 0 \\ 1, & x > 0 \end{cases},$$

we have

$$\lim_{x \to \infty} h(x) = 1 \quad \text{but} \quad \lim_{x \to -\infty} h(x) = -1. \quad \square$$

Exercises

Find the following limits.

*1. $\lim\limits_{x \to \infty} \dfrac{3}{x}$.

2. $\lim\limits_{x \to \infty} \dfrac{x}{x^2 + 1}$.

*3. $\lim\limits_{x \to \infty} \dfrac{x - 1}{x + 1}$.

*4. $\lim\limits_{x \to \infty} \dfrac{2 \sin x}{x}$.

5. $\lim\limits_{x \to \infty} \log \left(\dfrac{2x}{x + 1} \right)$.

*6. $\lim\limits_{x \to \infty} 10^{1/x}$.

*7. $\lim\limits_{x \to \infty} \dfrac{(ax - b)^2}{(a - x)(b - x)}$.

8. $\lim\limits_{x \to \infty} \dfrac{(x - a)(x - b)}{(ax + b)^2}$.

*9. $\lim\limits_{x \to \infty} \sin \dfrac{1}{x}$.

*10. $\lim\limits_{x \to \infty} \left(1 - \sin \dfrac{1}{x} \right)^2$.

11. $\lim\limits_{x \to \infty} \left(x \sin \dfrac{1}{x} \right)$.

*12. $\lim\limits_{x \to \infty} \dfrac{x^2 e^{-x}}{2 - x^2}$.

*13. $\lim\limits_{x \to \infty} \displaystyle\int_0^x \frac{dt}{1 + t^2}$.

14. $\lim\limits_{x \to \infty} \displaystyle\int_e^x \frac{dt}{t(\log t)^2}$.

*15. $\lim\limits_{x \to \infty} \left(x - x \cos \dfrac{1}{x} \right)$.

16. $\lim\limits_{x \to \infty} (\sqrt{x + 1} - \sqrt{x})$.

*17. $\lim\limits_{x \to \infty} (\sqrt{x^2 + x} - \sqrt{x^2 + 1})$.

18. $\lim\limits_{x \to -\infty} \dfrac{\sqrt{x^2 + 1}}{2x}$.

Sketch the following curves specifying the horizontal asymptotes.

*19. $y = \dfrac{x}{x^2 + 1}$.

20. $y = \dfrac{x^2}{x^2 + 1}$.

*21. $y = \dfrac{1 - 2x^2}{x^2 + 4}$.

*22. $y = \dfrac{x}{\sqrt{4x^2 + 1}}$.

23. $y = \dfrac{2x}{\sqrt{x^2 + 1}}$.

*24. $y = \sqrt{x + 4} - \sqrt{x}$.

25. (a) Prove that, if f is increasing and bounded on $[a, \infty)$, then f has a limit as x tends to ∞.

(b) Prove that, if f is decreasing and bounded on $[a, \infty)$, then f has a limit as x tends to ∞.

26. Prove that, if f is continuous on $[a, \infty)$ and has a finite limit as x tends to ∞, then f is bounded on $[a, \infty)$.

12.5 The Indeterminate Form (0/0)

Here we are concerned with limits of quotients

$$\frac{f(x)}{g(x)}$$

where the numerator and denominator both tend to 0. For convenience we will dispense with the formality of writing

$$\lim_{x \to c} f(x) = 0, \qquad \lim_{x \to \infty} f(x) = 0, \qquad \text{etc.}$$

and write instead

$$\text{as} \quad x \to c, \quad f(x) \to 0,$$
$$\text{as} \quad x \to \infty, \quad f(x) \to 0, \quad \text{etc.}$$

Theorem 12.5.1 L'Hospital's Rule (0/0)†

Suppose that

$$f(x) \to 0 \quad \text{and} \quad g(x) \to 0$$

as $x \downarrow c, x \uparrow c, x \to c, x \to \infty,$ or $x \to -\infty.$

$$\text{If} \quad \frac{f'(x)}{g'(x)} \to l, \quad \text{then} \quad \frac{f(x)}{g(x)} \to l.$$

---

† Named after a Frenchman G. F. A. L'Hospital (1661–1704). The result was actually discovered by his teacher, Johann Bernoulli (1667–1748).

We will prove the validity of L'Hospital's rule a little later. First we demonstrate its usefulness.

Problem.   Find

$$\lim_{x \to \frac{1}{2}\pi} \frac{\cos x}{\pi - 2x}.$$

SOLUTION.   As $x \to \frac{1}{2}\pi$, the numerator $f(x) = \cos x$ and the denominator $g(x) = \pi - 2x$ both tend to zero, but it's not at all obvious what happens to the quotient

$$\frac{f(x)}{g(x)} = \frac{\cos x}{\pi - 2x}.$$

Therefore we test the quotient of derivatives:

$$\frac{f'(x)}{g'(x)} = \frac{-\sin x}{-2} = \frac{\sin x}{2} \to \frac{1}{2} \qquad \text{as} \quad x \to \frac{1}{2}\pi.$$

It follows from L'Hospital's rule that

$$\frac{\cos x}{\pi - 2x} \to \frac{1}{2} \qquad \text{as} \quad x \to \frac{1}{2}\pi.$$

We can express all this on just one line using $*$ to indicate the differentiation of numerator and denominator:

$$\lim_{x \to \frac{1}{2}\pi} \frac{\cos x}{\pi - 2x} \overset{*}{=} \lim_{x \to \frac{1}{2}\pi} \frac{-\sin x}{-2} = \lim_{x \to \frac{1}{2}\pi} \frac{\sin x}{2} = \frac{1}{2}. \quad \square$$

Problem.   Find

$$\lim_{x \downarrow 0} \frac{x}{\sin \sqrt{x}}.$$

SOLUTION.   As $x \downarrow 0$, both numerator and denominator tend to 0. Since

$$\frac{f'(x)}{g'(x)} = \frac{1}{(\cos \sqrt{x})(1/2\sqrt{x})} = \frac{2\sqrt{x}}{\cos \sqrt{x}} \to 0 \quad \text{as} \quad x \downarrow 0,$$

it follows from L'Hospital's rule that

$$\frac{x}{\sin \sqrt{x}} \to 0 \qquad \text{as} \quad x \downarrow 0.$$

For short we can write

$$\lim_{x \downarrow 0} \frac{x}{\sin \sqrt{x}} \overset{*}{=} \lim_{x \downarrow 0} \frac{2\sqrt{x}}{\cos \sqrt{x}} = 0. \quad \square$$

Sometimes it's necessary to differentiate numerator and denominator more than once. The next problem gives such an instance.

Problem.   Find

$$\lim_{x \to 0} \frac{e^x - x - 1}{x^2}.$$

SOLUTION.   As $x \to 0$, both numerator and denominator tend to 0. Here

$$\frac{f'(x)}{g'(x)} = \frac{e^x - 1}{2x}.$$

Since both numerator and denominator still tend to 0, we differentiate again:

$$\frac{f''(x)}{g''(x)} = \frac{e^x}{2}.$$

Since this last quotient tends to $\frac{1}{2}$, we can conclude that

$$\frac{e^x - 1}{2x} \to \frac{1}{2} \quad \text{and therefore} \quad \frac{e^x - x - 1}{x^2} \to \frac{1}{2}.$$

For short we can write

$$\lim_{x \to 0} \frac{e^x - x - 1}{x^2} \stackrel{*}{=} \lim_{x \to 0} \frac{e^x - 1}{2x} \stackrel{*}{=} \lim_{x \to 0} \frac{e^x}{2} = \frac{1}{2}. \quad \square$$

Problem.   Find

$$\lim_{x \downarrow 0} (1 + x)^{1/x}.$$

SOLUTION.   As the expression stands, it is not amenable to L'Hospital's rule. To make it so, we take its logarithm. This gives

$$\lim_{x \downarrow 0} \log (1 + x)^{1/x} = \lim_{x \downarrow 0} \frac{\log (1 + x)}{x} \stackrel{*}{=} \lim_{x \downarrow 0} \frac{1}{1 + x} = 1.$$

Since $\log (1 + x)^{1/x}$ tends to 1, $(1 + x)^{1/x}$ must itself tend to e. $\square$

Problem.   Find

$$\lim_{n \to \infty} (\cos \pi/n)^n.$$

SOLUTION.   To apply the methods of this section we replace the integer variable n by the real variable x and examine the behavior of

$$(\cos \pi/x)^x \quad \text{as} \quad x \to \infty.$$

Taking the logarithm of this expression, we have

$$\log (\cos \pi/x)^x = x \log (\cos \pi/x) = \frac{\log (\cos \pi/x)}{1/x},$$

and therefore

$$\lim_{x \to \infty} \log (\cos \pi/x)^x = \lim_{x \to \infty} \frac{\log (\cos \pi/x)}{1/x}$$

$$\stackrel{*}{=} \lim_{x \to \infty} \frac{(-\sin \pi/x)(-\pi/x^2)}{(\cos \pi/x)(-1/x^2)} = \lim_{x \to \infty} (-\pi \tan \pi/x) = 0.$$

It follows then that $(\cos \pi/x)^x$ tends to 1 and therefore $(\cos \pi/n)^n$ also tends to 1. $\square$

To prove the validity of L'Hospital's rule, we need first a generalization of the mean-value theorem.

Theorem 12.5.2 The Cauchy Mean-Value Theorem†

Suppose that f and g are differentiable on (a, b) and continuous on $[a, b]$. If g' is never 0 in (a, b), then there is a number c in (a, b) such that

$$\frac{f'(c)}{g'(c)} = \frac{f(b) - f(a)}{g(b) - g(a)}.$$

PROOF. We can prove this by applying the mean-value theorem to the function

$$G(x) = [g(b) - g(a)][f(x) - f(a)] - [g(x) - g(a)][f(b) - f(a)].$$

Since

$$G(a) = 0 \quad \text{and} \quad G(b) = 0,$$

there exists (by the mean-value theorem) a number c in (a, b) such that

$$G'(c) = 0.$$

Since in general

$$G'(x) = [g(b) - g(a)]f'(x) - g'(x)[f(b) - f(a)],$$

we must have

$$[g(b) - g(a)]f'(c) - g'(c)[f(b) - f(a)] = 0,$$

and therefore

$$[g(b) - g(a)]f'(c) = g'(c)[f(b) - f(a)].$$

Since g' is never 0 in (a, b),

$$g'(c) \neq 0 \quad \text{and} \quad g(b) - g(a) \neq 0.$$
$$\underset{\uparrow}{\qquad} \text{— explain}$$

We can therefore divide by these numbers and obtain

$$\frac{f'(c)}{g'(c)} = \frac{f(b) - f(a)}{g(b) - g(a)}. \quad \square$$

Now we prove L'Hospital's rule for the case $x \downarrow c$. We assume that, as $x \downarrow c$,

$$f(x) \to 0, \quad g(x) \to 0, \quad \text{and} \quad \frac{f'(x)}{g'(x)} \to l$$

and show that

$$\frac{f(x)}{g(x)} \to l.$$

---

† Another contribution of A. L. Cauchy, after whom Cauchy sequences were named.

PROOF.   The fact that

$$\frac{f'(x)}{g'(x)} \to l \quad \text{as} \quad x \downarrow c$$

assures us that both f' and g' exist on a set of the form $(c, c + h]$ and that g' is not zero there. By setting $f(c) = 0$ and $g(c) = 0$, we ensure that f and g are both continuous on $[c, c + h]$. We can now apply the Cauchy mean-value theorem and conclude that there exists a number c_h between $c + h$ such that

$$\frac{f'(c_h)}{g'(c_h)} = \frac{f(c + h) - f(c)}{g(c + h) - g(c)} = \frac{f(c + h)}{g(c + h)} \, .$$

The result is now obvious.   □

Here is a proof of L'Hospital's rule for the case $x \to \infty$.

PROOF.   The key here is to set $x = 1/t$:

$$\lim_{x \to \infty} \frac{f'(x)}{g'(x)} = \lim_{t \downarrow 0} \frac{f'(1/t)}{g'(1/t)} = \lim_{t \downarrow 0} \frac{-t^{-2}f'(1/t)}{-t^{-2}g'(1/t)} = \lim_{t \downarrow 0} \frac{f(1/t)}{g(1/t)} = \lim_{x \to \infty} \frac{f(x)}{g(x)} \, . \quad \square$$

$$\text{by L'Hospital's rule for the case } t \downarrow 0$$

CAUTION.   L'Hospital's rule does not apply when the numerator or the denominator has a finite nonzero limit. For example,

$$\lim_{x \to 0} \frac{x}{x + \cos x} = \frac{0}{1} = 0.$$

A blind application of L'Hospital's rule would lead to

$$\lim_{x \to 0} \frac{x}{x + \cos x} \overset{*}{=} \lim_{x \to 0} \frac{1}{1 - \sin x} = 1.$$

This is wrong.   □

Exercises

Find the indicated limit.

*1. $\displaystyle\lim_{x \downarrow 0} \frac{\sin x}{\sqrt{x}}$.

2. $\displaystyle\lim_{x \to 1} \frac{\log x}{1 - x}$.

*3. $\displaystyle\lim_{x \to 0} \frac{e^x - 1}{\log (1 + x)}$.

*4. $\displaystyle\lim_{x \to a} \frac{x - a}{x^n - a^n}$.

5. $\displaystyle\lim_{x \to \frac{1}{2}\pi} \frac{\cos x}{\sin 2x}$.

*6. $\displaystyle\lim_{x \to 0} \frac{e^x - 1}{x(1 + x)}$.

*7. $\displaystyle\lim_{x \to 0} \frac{1 - \cos x}{3x}$.

8. $\displaystyle\lim_{x \to 1} \frac{x^{1/2} - x^{1/4}}{x - 1}$.

*9. $\displaystyle\lim_{x \to 0} \frac{e^x - e^{-x}}{\sin x}$.

*10. $\displaystyle\lim_{x \to 0} \frac{x + \sin \pi x}{x - \sin \pi x}$.

11. $\displaystyle\lim_{x \to 0} \frac{\tan \pi x}{e^x - 1}$.

*12. $\displaystyle\lim_{x \to 0} \frac{a^x - (a + 1)^x}{x}$.

*13. $\displaystyle\lim_{x \to 0} \frac{1 + x - e^x}{x(e^x - 1)}$.

14. $\displaystyle\lim_{x \to 0} \frac{\sqrt{a + x} - \sqrt{a - x}}{x}$.

*15. $\lim\limits_{x\to 0}\dfrac{x-\tan x}{x-\sin x}.$

16. $\lim\limits_{x\to 0}\dfrac{\log(\sec x)}{x^2}.$

*17. $\lim\limits_{x\uparrow 1}\dfrac{\sqrt{1-x^2}}{\sqrt{1-x^3}}.$

*18. $\lim\limits_{x\to 0}\left(\dfrac{1}{\sin x}-\dfrac{1}{x}\right).$

19. $\lim\limits_{x\to\infty}\left(1+\dfrac{1}{x}\right)^x.$

*20. $\lim\limits_{x\to 0}(e^x+x)^{1/x}.$

*21. $\lim\limits_{x\to 1}\left(\dfrac{1}{\log x}-\dfrac{x}{x-1}\right).$

22. $\lim\limits_{x\to\frac{1}{4}\pi}\dfrac{\sec^2 x-2\tan x}{1+\cos 4x}.$

*23. $\lim\limits_{x\downarrow 0}\dfrac{\sqrt{x}}{\sqrt{x}+\sin\sqrt{x}}.$

24. $\lim\limits_{x\to 0}\dfrac{xe^{nx}-x}{1-\cos nx}.$

*25. $\lim\limits_{x\to 0}\dfrac{x-\arcsin x}{\sin^3 x}.$

26. $\lim\limits_{x\to\frac{1}{2}\pi}\dfrac{\log(\sin x)}{(\pi-2x)^2}.$

Find the limit of the sequence.

*27. $\lim\limits_{n\to\infty}n(\pi/2-\arctan n).$

28. $\lim\limits_{n\to\infty}[\log(1-1/n)\operatorname{cosec}1/n].$

29. Find the fallacy:

$$\lim\limits_{x\to 0}\frac{2+x+\sin x}{x^3+x-\cos x}\overset{*}{=}\lim\limits_{x\to 0}\frac{1+\cos x}{3x^2+1+\sin x}\overset{*}{=}\lim\limits_{x\to 0}\frac{-\sin x}{6x+\cos x}=\frac{0}{1}=0.$$

30. Given that f is continuous, use L'Hospital's rule to determine

$$\lim\limits_{x\to 0}\left(\frac{1}{x}\int_0^x f(t)\,dt\right).$$

Optional | *31. Calculate

$$\lim\limits_{x\to 0}\frac{\sinh x-\sin x}{\sin^3 x}.$$

12.6 Infinite Limits; The Indeterminate Form (∞/∞)

It's convenient at times to allow infinite limits. Here is the basic idea. We write

$$\lim\limits_{x\to c}f(x)=\infty$$

iff we can make $f(x)$ as large as we wish simply by requiring that x be sufficiently close to c. More formally, we have

(12.6.1) | $\lim\limits_{x\to c}f(x)=\infty$ iff $\begin{cases}\text{for each }M>0\text{ there exists }\delta>0\text{ such that,}\\ \quad\text{if }\;0<|x-c|<\delta,\quad\text{ then }\;f(x)\geq M.\end{cases}$

Examples

$$\lim\limits_{x\to c}\frac{1}{|x-c|}=\infty,\qquad \lim\limits_{x\to c}\frac{1}{(\sqrt{x}-\sqrt{c})^2}=\infty,\qquad \lim\limits_{x\to c}\frac{1}{\sin^2(x-c)}=\infty. \quad\square$$

One-sided infinite limits

$$\lim_{x \downarrow c} f(x) = \infty \quad \text{and} \quad \lim_{x \uparrow c} f(x) = \infty$$

are defined in a similar manner. We won't take time to spell out the details. The idea must be obvious. It must also be obvious that all of these notions can be extended to allow for a limit of $-\infty$.

Geometrically, limits of the form

$$\lim_{x \to c} f(x) = \pm\infty, \qquad \lim_{x \downarrow c} f(x) = \pm\infty, \qquad \lim_{x \uparrow c} f(x) = \pm\infty$$

signal the presence of a *vertical asymptote*. We give some pictorial examples in Figures 12.6.1–12.6.4.

$$\lim_{x \to c} f(x) = \infty$$

FIGURE 12.6.1

$$\lim_{x \to c} g(x) = -\infty.$$

FIGURE 12.6.2

$$\lim_{x \downarrow c} f(x) = \infty, \quad \lim_{x \downarrow c} g(x) = -\infty$$

FIGURE 12.6.3

$$\lim_{x \uparrow c} f(x) = \infty, \quad \lim_{x \uparrow c} g(x) = -\infty$$

FIGURE 12.6.4

We have yet to discuss limits of the form

$$\lim_{x \to \pm\infty} f(x) = \pm\infty.$$

There are four possibilities here. We will spell out details only for one of them. By writing

$$\lim_{x \to \infty} f(x) = \infty$$

we mean that we can make $f(x)$ as large as we wish simply by requiring that x be sufficiently large:

$$(12.6.2) \qquad \lim_{x \to \infty} f(x) = \infty \quad \text{iff} \quad \begin{cases} \text{for each } M > 0 \text{ there exists } K > 0 \text{ such that,} \\ \qquad \text{if } \ x \geq K, \qquad \text{then } \ f(x) \geq M. \end{cases}$$

As simple examples you can take

$$\lim_{x \to \infty} x = \infty, \qquad \lim_{x \to \infty} e^x = \infty, \qquad \lim_{x \to \infty} \log x = \infty, \qquad \lim_{x \to \infty} (x + 1/x) = \infty.$$

To illustrate these ideas further we calculate some limits which are not quite so obvious.

Problem. Find

$$\lim_{x \to \infty} x(2 + \sin x).$$

SOLUTION. Since $-1 \leq \sin x$, we have

$$1 \leq 2 + \sin x$$

and thus for $x > 0$

$$x \leq x(2 + \sin x).$$

Since $\lim_{x \to \infty} x = \infty$, it's clear that

$$\lim_{x \to \infty} x(2 + \sin x) = \infty. \quad \square$$

Problem. Find

$$\lim_{x \to \infty} \frac{1 - x^2}{2x + 1}.$$

SOLUTION

$$\lim_{x \to \infty} \frac{1 - x^2}{2x + 1} = \lim_{x \to \infty} \frac{1/x^2 - 1}{2/x + 1/x^2} = -\infty. \quad \square$$

Problem. Find

$$\lim_{x \to \infty} \log \left(\frac{x}{x^2 + 1} \right).$$

SOLUTION.   Note first that

$$\lim_{x \to \infty} \frac{x}{x^2 + 1} = \lim_{x \to \infty} \frac{1}{x + 1/x} = 0.$$

It follows that

$$\lim_{x \to \infty} \log \left(\frac{x}{x^2 + 1} \right) = -\infty. \quad \square$$

Problem.   Find

$$\lim_{x \to \infty} \left(x \sin \frac{1}{x} \right).$$

SOLUTION

$$\lim_{x \to \infty} \left(x \sin \frac{1}{x} \right) = \lim_{x \to \infty} \frac{\sin 1/x}{1/x} = \lim_{t \downarrow 0} \frac{\sin t}{t} = 1. \quad \square$$

Problem.   Find

$$\lim_{x \to 0} \left(\frac{1}{x} - \frac{1}{x^2} \right).$$

SOLUTION

$$\lim_{x \to 0} \left(\frac{1}{x} - \frac{1}{x^2} \right) = \lim_{x \to 0} \frac{x - 1}{x^2} = -\infty. \quad \square$$

We come now to limits of quotients

$$\frac{f(x)}{g(x)}$$

where numerator and denominator both tend to ∞. We take l as a fixed real number.

Theorem 12.6.3   L'Hospital's Rule (∞/∞)

Suppose that

$$f(x) \to \infty \quad \text{and} \quad g(x) \to \infty$$

as $x \uparrow c, x \downarrow c, x \to c, x \to \infty,$ or $x \to -\infty.$

$$\text{If} \quad \frac{f'(x)}{g'(x)} \to l, \quad \text{then} \quad \frac{f(x)}{g(x)} \to l.$$

While the proof of L'Hospital's rule in this setting is a little more complicated than it was in the $(0/0)$ case,† the application of the rule is much the same.

Problem. Show that

(12.6.4)
$$\lim_{x \to \infty} \frac{\log x}{x} = 0.$$

SOLUTION.   Both numerator and denominator tend to ∞ with x. L'Hospital's rule gives

$$\lim_{x \to \infty} \frac{\log x}{x} \overset{*}{=} \lim_{x \to \infty} \frac{1}{x} = 0. \quad \square$$

Problem. Show that

(12.6.5)
$$\lim_{x \to \infty} \frac{x^k}{e^x} = 0.$$

SOLUTION.   Here we differentiate numerator and denominator k times:

$$\lim_{x \to \infty} \frac{x^k}{e^x} \overset{*}{=} \lim_{x \to \infty} \frac{kx^{k-1}}{e^x} \overset{*}{=} \lim_{x \to \infty} \frac{k(k-1)x^{k-2}}{e^x} \overset{*}{=} \cdots \overset{*}{=} \lim_{x \to \infty} \frac{k!}{e^x} = 0. \quad \square$$

Problem. Show that

(12.6.6)
$$\lim_{x \downarrow 0} x^x = 1.$$

SOLUTION.   We first take the logarithm of x^x and then apply L'Hospital's rule:

$$\lim_{x \downarrow 0} \log x^x = \lim_{x \downarrow 0} (x \log x) = \lim_{x \downarrow 0} \frac{\log x}{1/x} \overset{*}{=} \lim_{x \downarrow 0} \frac{1/x}{-1/x^2} = \lim_{x \downarrow 0} (-x) = 0.$$

Since $\log x^x$ tends to 0, x^x must tend to 1. $\quad \square$

Problem. Find the limit of the sequence

$$a_n = \frac{\log n}{\sqrt{n}} \qquad \text{as} \quad n \to \infty.$$

SOLUTION.   To bring into play the methods of calculus we investigate

$$\lim_{x \to \infty} \frac{\log x}{\sqrt{x}}$$

† We omit the proof.

instead. Since both numerator and denominator tend to ∞ with x, we try L'Hospital's rule:

$$\lim_{x\to\infty} \frac{\log x}{\sqrt{x}} \overset{*}{=} \lim_{x\to\infty} \frac{1/x}{1/2\sqrt{x}} = \lim_{x\to\infty} \frac{2}{\sqrt{x}} = 0.$$

The limit of the sequence must also be 0. $\square$

CAUTION. L'Hospital's rule does not apply when the numerator or denominator has a finite nonzero limit. While

$$\lim_{x\downarrow 0} \frac{1+x}{\sin x} = \infty,$$

a misapplication of L'Hospital's rule can lead to

$$\lim_{x\downarrow 0} \frac{1+x}{\sin x} \overset{*}{=} \lim_{x\downarrow 0} \frac{1}{-\cos x} = -1. \square$$

Exercises

Find the indicated limit.

*1. $\lim\limits_{x\uparrow\frac{1}{2}\pi} \tan x.$

2. $\lim\limits_{x\downarrow 0} \log \dfrac{1}{x}.$

*3. $\lim\limits_{x\to\infty} \dfrac{x^3-1}{2-x}.$

*4. $\lim\limits_{x\to\infty} \dfrac{x^3}{1-x^3}.$

5. $\lim\limits_{x\to\infty} \left(x^2 \sin \dfrac{1}{x}\right).$

*6. $\lim\limits_{x\to\infty} \dfrac{\log x^k}{x}.$

*7. $\lim\limits_{x\uparrow\frac{1}{2}\pi} \dfrac{\tan 5x}{\tan x}.$

8. $\lim\limits_{x\to 0} (x \log |\sin x|).$

*9. $\lim\limits_{x\downarrow 0} x^{2x}.$

*10. $\lim\limits_{x\to 0} [x (\log |x|)^2].$

11. $\lim\limits_{x\to\infty} \left(x \sin \dfrac{\pi}{x}\right).$

*12. $\lim\limits_{x\downarrow 0} \dfrac{\log x}{\cot x}.$

*13. $\lim\limits_{x\to\infty} \dfrac{\sqrt{1+x^2}}{x}.$

14. $\lim\limits_{x\to\infty} \left(\dfrac{1}{x} \displaystyle\int_0^x e^{t^2}dt\right).$

*15. $\lim\limits_{x\downarrow 0} x^{\sin x}.$

*16. $\lim\limits_{x\to 0} |\sin x|^x.$

17. $\lim\limits_{x\to 1} x^{1/(x-1)}.$

*18. $\lim\limits_{x\to 0} \left[\dfrac{1}{\sin^2 x} - \dfrac{1}{x^2}\right].$

*19. $\lim\limits_{x\to\infty} \left(\cos \dfrac{1}{x}\right)^x.$

20. $\lim\limits_{x\to\frac{1}{2}\pi} |\sec x|^{\cos x}.$

*21. $\lim\limits_{x\to\infty} (x^2+a^2)^{(1/x)^2}.$

*22. $\lim\limits_{x\to 0} \left[\dfrac{1}{\log(1+x)} - \dfrac{1}{x}\right].$

23. $\lim\limits_{x\to 1} \left[\dfrac{x}{x-1} - \dfrac{1}{\log x}\right].$

*24. $\lim\limits_{x\to\infty} \log \left(\dfrac{x^2-1}{x^2+1}\right)^3.$

25. $\lim\limits_{x\to\infty} (\sqrt{x^2+2x} - x).$

Find the limit of the sequence.

*26. $\lim\limits_{n\to\infty} \left(\dfrac{1}{n} \log \dfrac{1}{n}\right).$

27. $\lim\limits_{n\to\infty} \dfrac{n^2 \log n}{e^n}.$

*28. $\lim\limits_{n\to\infty} (\log n)^{1/n}.$

*29. $\lim\limits_{n\to\infty} \dfrac{n^k}{2^n}.$

30. $\lim\limits_{n\to\infty} (n^2+n)^{1/n}.$

*31. $\lim\limits_{n\to\infty} n^{\sin(\pi/n)}.$

Sketch the following curves specifying all asymptotes.

*32. $y = \dfrac{1}{x-4}$.

33. $y = \dfrac{x}{x-4}$.

*34. $y = \dfrac{x}{x^2-4}$.

*35. $y = \dfrac{x^2}{x^2-4}$.

36. $y = \sqrt{\dfrac{x}{x-2}}$.

*37. $y = \dfrac{\sqrt{x^2+1}}{3x}$.

*38. $y = xe^{-x}$.

39. $y = x^2e^{-x}$.

*40. $y = \dfrac{\log x}{x}$.

41. Show that for each positive integer k

$$\lim_{n\to\infty} \frac{(\log n)^k}{n} = 0.$$

42. Find the fallacy:

$$\lim_{x\downarrow 0} \frac{x^2}{\sin x} \overset{*}{=} \lim_{x\downarrow 0} \frac{2x}{\cos x} \overset{*}{=} \lim_{x\downarrow 0} \frac{2}{-\sin x} = -\infty.$$

12.7   Improper Integrals

We begin with a function f continuous on an unbounded interval $[a, \infty)$. For each number $b > a$ we can form the integral

$$\int_a^b f(x)\, dx.$$

If, as b tends to ∞, this integral tends to a finite limit l,

$$\lim_{b\to\infty} \int_a^b f(x)\, dx = l,$$

then we write

$$\int_a^\infty f(x)\, dx = l$$

and call

$$\int_a^\infty f(x)\, dx$$

a *convergent improper integral*. Otherwise we say that

$$\int_a^\infty f(x)\, dx$$

is a *divergent improper integral*. Improper integrals of the form

$$\int_{-\infty}^b f(x)\, dx$$

are introduced in a similar manner.

Examples

(1) $\displaystyle\int_1^\infty e^{-x}\,dx = \frac{1}{e}.$ (2) $\displaystyle\int_1^\infty \frac{dx}{x}$ diverges.

(3) $\displaystyle\int_1^\infty \frac{dx}{x^2} = 1.$ (4) $\displaystyle\int_1^\infty \cos \pi x\,dx$ diverges.

VERIFICATION

(1) $\displaystyle\int_1^\infty e^{-x}\,dx = \lim_{b\to\infty}\int_1^b e^{-x}\,dx = \lim_{b\to\infty}\left[-e^{-x}\right]_1^b = \lim_{b\to\infty}\left(\frac{1}{e}-\frac{1}{e^b}\right) = \frac{1}{e}.$

(2) $\displaystyle\int_1^\infty \frac{dx}{x} = \lim_{b\to\infty}\int_1^b \frac{dx}{x} = \lim_{b\to\infty}\log b = \infty.$

(3) $\displaystyle\int_1^\infty \frac{dx}{x^2} = \lim_{b\to\infty}\int_1^b \frac{dx}{x^2} = \lim_{b\to\infty}\left[-\frac{1}{x}\right]_1^b = \lim_{b\to\infty}\left(1-\frac{1}{b}\right) = 1.$

(4) Note first that

$$\int_1^b \cos \pi x\,dx = \left[\frac{1}{\pi}\sin \pi x\right]_1^b = \frac{1}{\pi}\sin \pi b.$$

As b tends to ∞, $\sin \pi b$ oscillates between -1 and 1. The integral

$$\int_1^b \cos \pi x\,dx$$

oscillates between $-1/\pi$ and $1/\pi$. It does not converge.   □

The usual formulas for area and volume are extended to the unbounded case by means of improper integrals.

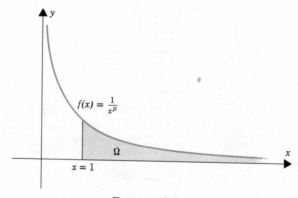

FIGURE 12.7.1

Example. If Ω is the region below the graph of

$$f(x) = \frac{1}{x^p}, \quad x \geq 1$$ (Figure 12.7.1)

then we have

$$\text{area of } \Omega = \left\{ \begin{array}{ll} \dfrac{1}{p-1}, & \text{if } p > 1 \\ \infty, & \text{if } p \le 1 \end{array} \right].$$

This comes about from setting

$$\text{area of } \Omega = \lim_{b \to \infty} \int_1^b \frac{dx}{x^p} = \int_1^\infty \frac{dx}{x^p}.$$

For $p \ne 1$, we have

$$\int_1^\infty \frac{dx}{x^p} = \lim_{b \to \infty} \int_1^b \frac{dx}{x^p} = \lim_{b \to \infty} \frac{1}{1-p}(b^{1-p} - 1) = \left\{ \begin{array}{ll} \dfrac{1}{p-1}, & \text{if } p > 1 \\ \infty, & \text{if } p < 1 \end{array} \right].$$

For $p = 1$, we have

$$\int_1^\infty \frac{dx}{x^p} = \int_1^\infty \frac{dx}{x} = \infty,$$

as you have seen already. ☐

Example. From the last example you know that the region below the graph of

$$f(x) = \frac{1}{x}, \qquad x \ge 1$$

has infinite area. Suppose that this region with infinite area is revolved about the x-axis. (See Figure 12.7.2.) What is the volume of the resulting solid? It may surprise you somewhat, but the volume is not infinite. In fact, it is π:

$$V = \pi \int_1^\infty \frac{dx}{x^2} = \pi \cdot 1 = \pi. \quad \Box$$

$$f(x) = \frac{1}{x}$$

FIGURE 12.7.2

For future reference we record the following:

(12.7.1)　　　$\displaystyle \int_1^\infty \frac{dx}{x^p}$ converges for $p > 1$ and diverges for $p \le 1$.

It's often difficult to determine the convergence or divergence of a given integral by direct methods, but we can usually gain some information by comparison with integrals of known behavior.

(12.7.2)

> (*A comparison test*) Suppose that f and g are continuous and that
> $$0 \le f(x) \le g(x) \qquad \text{for all } x \in [a, \infty).$$
> (i) If $\displaystyle\int_a^\infty g(x)\,dx$ converges, then $\displaystyle\int_a^\infty f(x)\,dx$ converges.
>
> (ii) If $\displaystyle\int_a^\infty f(x)\,dx$ diverges, then $\displaystyle\int_a^\infty g(x)\,dx$ diverges.

The proof is left as an exercise. □

Example. The improper integral

$$\int_1^\infty \frac{dx}{\sqrt{1 + x^3}}$$

converges since

$$\frac{1}{\sqrt{1 + x^3}} < \frac{1}{x^{3/2}} \quad \text{for } x \in [1, \infty) \qquad \text{and} \qquad \int_1^\infty \frac{dx}{x^{3/2}} \text{ converges.} \quad \square$$

Example. The improper integral

$$\int_1^\infty \frac{dx}{\sqrt{1 + x^2}}$$

diverges since

$$\frac{1}{1 + x} \le \frac{1}{\sqrt{1 + x^2}} \quad \text{for } x \in [1, \infty) \quad \text{and} \quad \int_1^\infty \frac{dx}{1 + x} \text{ diverges.} \quad \square$$

Suppose now that f is continuous on $(-\infty, \infty)$. The *improper integral*

$$\int_{-\infty}^\infty f(x)\,dx$$

is said to *converge* iff both

$$\int_{-\infty}^0 f(x)\,dx \quad \text{and} \quad \int_0^\infty f(x)\,dx$$

converge. If

$$\int_{-\infty}^0 f(x)\,dx = l \quad \text{and} \quad \int_0^\infty f(x)\,dx = m,$$

then we set

$$\int_{-\infty}^\infty f(x)\,dx = l + m.$$

Improper integrals can also arise on bounded intervals. Suppose that f is continuous on the half-open interval $[a, b)$ but unbounded there. See Figure 12.7.3. If

$$\lim_{\epsilon \downarrow 0} \int_a^{b-\epsilon} f(x)\, dx = l \qquad (l \text{ finite}),$$

then we write

$$\int_a^{b-} f(x)\, dx = l$$

and call

$$\int_a^{b-} f(x)\, dx$$

a *convergent improper integral.* Otherwise we say that

$$\int_a^{b-} f(x)\, dx$$

is a *divergent improper integral.*

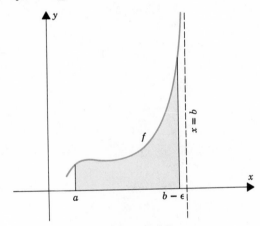

FIGURE 12.7.3

Functions continuous but unbounded on intervals of the form $(a, b]$ lead to limits of the form

$$\lim_{\epsilon \downarrow 0} \int_{a+\epsilon}^{b} f(x)\, dx$$

and thus to improper integrals

$$\int_{a+}^{b} f(x)\, dx.$$

Examples

(1) $\displaystyle \int_0^{1-} (1 - x)^{-2/3}\, dx = 3.$ \qquad (2) $\displaystyle \int_{0+}^{1} \frac{dx}{x}$ diverges.

Verification

(1)
$$\int_0^{1-} (1-x)^{-2/3}\, dx = \lim_{\epsilon \downarrow 0} \int_0^{1-\epsilon} (1-x)^{-2/3}\, dx$$

$$= \lim_{\epsilon \downarrow 0} \left[-3(1-x)^{1/3} \right]_0^{1-\epsilon} = \lim_{\epsilon \downarrow 0} 3(1 - \epsilon^{1/3}) = 3.$$

(2)
$$\int_{0+}^1 \frac{dx}{x} = \lim_{\epsilon \downarrow 0} \int_\epsilon^1 \frac{dx}{x} = \lim_{\epsilon \downarrow 0} (-\log \epsilon) = \infty. \quad \square$$

Exercises

Evaluate the improper integrals that converge.

*1. $\displaystyle\int_1^\infty \frac{dx}{x^2}$.

2. $\displaystyle\int_0^\infty \frac{dx}{1+x^2}$.

*3. $\displaystyle\int_0^\infty \frac{dx}{4+x^2}$.

*4. $\displaystyle\int_0^\infty e^{px}\, dx$, $p > 0$.

5. $\displaystyle\int_0^\infty e^{-px}\, dx$, $p > 0$.

*6. $\displaystyle\int_{0+}^1 \frac{dx}{\sqrt{x}}$.

7. $\displaystyle\int_{0+}^8 \frac{dx}{x^{2/3}}$.

*8. $\displaystyle\int_{0+}^1 \frac{dx}{x^2}$.

*9. $\displaystyle\int_0^{1-} \frac{dx}{\sqrt{1-x^2}}$.

10. $\displaystyle\int_0^{1-} \frac{dx}{\sqrt{1-x}}$.

*11. $\displaystyle\int_0^{2-} \frac{x}{\sqrt{4-x^2}}\, dx$.

*12. $\displaystyle\int_0^{a-} \frac{dx}{\sqrt{a^2-x^2}}$.

13. $\displaystyle\int_c^\infty \frac{\log x}{x}\, dx$.

*14. $\displaystyle\int_e^\infty \frac{dx}{x \log x}$.

*15. $\displaystyle\int_e^\infty \frac{dx}{x(\log x)^2}$.

16. $\displaystyle\int_{0+}^1 x \log x\, dx$.

*17. $\displaystyle\int_{-\infty}^\infty \frac{dx}{1+x^2}$.

*18. $\displaystyle\int_2^\infty \frac{dx}{x^2-1}$.

19. $\displaystyle\int_{-\infty}^0 xe^x\, dx$.

*20. $\displaystyle\int_0^3 \frac{dx}{\sqrt[3]{3x-1}}$.

*21. $\displaystyle\int_1^\infty \frac{dx}{x(x+1)}$.

22. $\displaystyle\int_{3+}^5 \frac{x}{\sqrt{x^2-9}}\, dx$.

*23. $\displaystyle\int_1^\infty \frac{x}{(1+x^2)^2}\, dx$.

*24. Let Ω be the region under the curve $y = e^{-x}$, $x \geq 0$. (a) Sketch Ω. (b) Find the area of Ω. (c) Find the volume of the solid obtained by revolving Ω about the x-axis. (d) Find the volume obtained by revolving Ω about the y-axis. (e) Find the surface area of the solid in part (c).

25. Find the area of the region bounded below by $y(x^2 + 1) = x$, above by $xy = 1$, and to the left by $x = 1$.

*26. Let Ω be the region below the curve $y = x^{-1/4}$, $0 < x \leq 1$. (a) Sketch Ω. (b) Find the area of Ω. (c) Find the volume of the solid obtained by revolving Ω about the x-axis. (d) Find the volume of the solid obtained by revolving Ω about the y-axis.

27. Prove the validity of the comparison test (12.7.2).

Use the comparison test (12.7.2) to determine which of the following integrals converge.

*28. $\displaystyle\int_1^\infty \frac{x}{\sqrt{1+x^5}}\,dx.$ 29. $\displaystyle\int_1^\infty e^{-x^2}\,dx.$ *30. $\displaystyle\int_0^\infty (1+x^5)^{-1/6}\,dx.$

*31. $\displaystyle\int_\pi^\infty \frac{\sin^2 2x}{x^2}\,dx.$ 32. $\displaystyle\int_1^\infty \frac{\log x}{x^2}\,dx.$ *33. $\displaystyle\int_e^\infty \frac{dx}{\sqrt{x+1}\,\log x}.$

34. (*Useful later*) Let f be a function, continuous, decreasing, and positive on $[1, \infty)$. Show that

$$\int_1^\infty f(x)\,dx \quad \text{converges} \quad \text{iff} \quad \text{the sequence} \quad \left\{ \int_1^n f(x)\,dx \right\} \quad \text{converges}.$$

The present value of a perpetual stream of revenues which flows continually at the rate of $R(t)$ dollars per year is given by the formula

$$P.V. = \int_0^\infty R(t)\,e^{-rt}\,dt,$$

where r is the rate of continuous compounding.

*35. What is the present value of such a stream if $R(t)$ is constantly \$100 per year and r is 5%?

36. What is the present value if $R(t) = 100 + 60t$ dollars per year and r is 10%?

Infinite Series

13

13.1 Sigma Notation

To indicate the sequence

$$\{1, \tfrac{1}{2}, \tfrac{1}{4}, \ldots\}$$

we can set $a_n = (\tfrac{1}{2})^{n-1}$ and write

$$\{a_1, a_2, a_3, \ldots\},$$

but we can also set $b_n = (\tfrac{1}{2})^n$ and write

$$\{b_0, b_1, b_2, \ldots\}.$$

More generally, we can set $c_n = (\tfrac{1}{2})^{n-p}$ and write

$$\{c_p, c_{p+1}, c_{p+2}, \ldots\}.$$

In this chapter we will often begin with an index other than 1.

The symbol Σ is the capital Greek letter "sigma." We write

(1)
$$\sum_{k=0}^{n} a_k$$

(read "the sum of the a sub k from k equals 0 to n") to indicate the sum

$$a_0 + a_1 + \cdots + a_n.$$

More generally, if $n \geq m$, we write

(2)
$$\sum_{k=m}^{n} a_k$$

to indicate the sum

$$a_m + a_{m+1} + \cdots + a_n.$$

In (1) and (2) the ketter "k" is being used as a "dummy" variable. That is, it can be replaced by any letter not already engaged. For instance,

$$\sum_{i=3}^{7} a_i, \quad \sum_{j=3}^{7} a_j, \quad \sum_{k=3}^{7} a_k$$

can all be used to indicate the sum

$$a_3 + a_4 + a_5 + a_6 + a_7.$$

Translating

$$(a_0 + \cdots + a_n) + (b_0 + \cdots + b_n) = (a_0 + b_0) + \cdots + (a_n + b_n),$$
$$\alpha(a_0 + \cdots + a_n) = \alpha a_0 + \cdots + \alpha a_n,$$
$$(a_0 + \cdots + a_m) + (a_{m+1} + \cdots + a_n) = a_0 + \cdots + a_n,$$

into the Σ-notation, we have

$$\sum_{k=0}^{n} a_k + \sum_{k=0}^{n} b_k = \sum_{k=0}^{n} (a_k + b_k), \quad \alpha \sum_{k=0}^{n} a_k = \sum_{k=0}^{n} \alpha a_k,$$
$$\sum_{k=0}^{m} a_k + \sum_{k=m+1}^{n} a_k = \sum_{k=0}^{n} a_k.$$

At times it's convenient to change indices. In this connection note that

$$\sum_{k=j}^{n} a_k = \sum_{k=0}^{n-j} a_{k+j}.$$

Both expressions are abbreviations for $a_j + a_{j+1} + \cdots + a_n$.

You can familiarize yourself further with this notation by doing the exercises below, but first one more remark. If all the a_k are equal to some fixed number x, then

$$\sum_{k=0}^{n} a_k \quad \text{can be written} \quad \sum_{k=0}^{n} x.$$

Obviously then

$$\sum_{k=0}^{n} x = \overbrace{x + x + \cdots + x}^{n+1} = (n+1)x.$$

In particular

$$\sum_{k=0}^{n} 1 = n + 1.$$

Exercises

1. Let p be a positive integer. Show that

$$a_n \to l \quad \text{iff} \quad a_{n-p} \to l.$$

2. Evaluate the following expressions:

*(a) $\displaystyle\sum_{k=0}^{2} (3k + 1)$.   (b) $\displaystyle\sum_{k=0}^{3} 2^k$.   *(c) $\displaystyle\sum_{k=0}^{3} (-1)^k 2^k$.

*(d) $\displaystyle\sum_{k=0}^{3} (-1)^k 2^{k+1}$.   (e) $\displaystyle\sum_{k=2}^{5} (-1)^{k+1} 2^{k-1}$.   *(f) $\displaystyle\sum_{k=1}^{4} \frac{1}{2^k}$.

*(g) $\displaystyle\sum_{k=2}^{5} \frac{1}{k!}$.   (h) $\displaystyle\sum_{k=0}^{3} (\tfrac{1}{2})^{2k}$.   *(i) $\displaystyle\sum_{k=1}^{3} (-1)^{k+1} (\tfrac{1}{2})^{2k-1}$.

3. Show that

$$\sum_{k=1}^{n} \frac{1}{\sqrt{k}} \ge \sqrt{n}.$$

4. Express

$$\sum_{k=1}^{n} \frac{a_k}{10^k}$$

as a decimal fraction, given that each a_k is an integer from 0 to 9.

5. Express in sigma notation:
   *(a) the lower sum $m_1 \Delta x_1 + m_2 \Delta x_2 + \cdots + m_n \Delta x_n$.
   (b) the upper sum $M_1 \Delta x_1 + M_2 \Delta x_2 + \cdots + M_n \Delta x_n$.
   *(c) the Riemann sum $f(x_1^*)\Delta x_1 + f(x_2^*)\Delta x_2 + \cdots + f(x_n^*)\Delta x_n$.

6. Express in sigma notation:
   *(a) $a^5 + a^4 b + a^3 b^2 + a^2 b^3 + ab^4 + b^5$.
   (b) $a^5 - a^4 b + a^3 b^2 - a^2 b^3 + ab^4 - b^5$.
   *(c) $a^n + a^{n-1}b + \cdots + ab^{n-1} + b^n$.
   (d) $a_0 x^4 + a_1 x^3 + a_2 x^2 + a_3 x + a_4$.
   *(e) $a_0 x^n + a_1 x^{n-1} + \cdots + a_{n-1}x + a_n$.
   (f) $1 - 2x + 3x^2 - 4x^3 + 5x^4$.

7. Show that for $x \ne 1$

$$\sum_{k=0}^{n} x^k = \frac{1 - x^{n+1}}{1 - x}.$$

8. Verify by induction:

(a) $\displaystyle\sum_{k=1}^{n} k = \tfrac{1}{2}(n)(n + 1)$.   (b) $\displaystyle\sum_{k=1}^{n} (2k - 1) = n^2$.

(c) $\displaystyle\sum_{k=1}^{n} k^2 = \tfrac{1}{6}(n)(n + 1)(2n + 1)$.   (d) $\displaystyle\sum_{k=1}^{n} k^3 = \left(\sum_{k=1}^{n} k\right)^2$.

13.2 Infinite Series

While it is possible to form the sum of two numbers, or three numbers, or a hundred numbers, or a thousand numbers, it is impossible to form the sum of an infinite number of numbers. Mathematicians have responded to this impossibility by devising the theory of *infinite series*.

To define such a series we begin with an infinite sequence $\{a_0, a_1, a_2, \ldots\}$ of real numbers. Although we can't form the sum of all the a_k (there is an infinite number of them), we can form the sequence

$$s_0 = a_0, \quad s_1 = a_0 + a_1, \quad s_2 = a_0 + a_1 + a_2, \quad s_3 = a_0 + a_1 + a_2 + a_3, \quad \text{etc.}$$

The nth term of this sequence

$$s_n = a_0 + a_1 + \cdots + a_n = \sum_{k=0}^{n} a_k$$

is called the *n*th *partial sum*.

Definition 13.2.1

The *infinite series*

$$\sum_{k=0}^{\infty} a_k \qquad (\text{sometimes written } a_0 + a_1 + a_2 + a_3 + \cdots)$$

is the sequence $\{s_0, s_1, s_2, s_3, \ldots\}$ of partial sums. We say that

$$\textit{the infinite series} \quad \sum_{k=0}^{\infty} a_k \quad \textit{converges to the number } l,$$

write

$$\sum_{k=0}^{\infty} a_k = l,$$

and call l the *sum* of the series iff

$$s_n = \sum_{k=0}^{n} a_k \to l \qquad \text{as} \quad n \to \infty.$$

We say that

$$\textit{the infinite series} \quad \sum_{k=0}^{\infty} a_k \quad \textit{diverges}$$

iff the sequence of partial sums diverges.

Let's look at some examples.

Example. We begin by examining the series

$$\sum_{k=0}^{\infty} 2^k.$$

Here the partial sums take the form

$$s_n = \sum_{k=0}^{n} 2^k = 1 + 2 + \cdots + 2^n.$$

The sequence of partial sums is unbounded and therefore cannot converge. (12.2.5) The series is thus divergent. □

Example. Now let's prove that

$$\sum_{k=0}^{\infty} \frac{1}{(k+1)(k+2)} = 1.$$

To do this, we must show that

$$s_n = \sum_{k=0}^{n} \frac{1}{(k+1)(k+2)} \to 1.$$

Since

$$\frac{1}{(k+1)(k+2)} = \frac{1}{k+1} - \frac{1}{k+2},$$

you can see that

$$s_n = \frac{1}{1 \cdot 2} + \frac{1}{2 \cdot 3} + \cdots + \frac{1}{(n+1)(n+2)}$$

$$= \left(\frac{1}{1} - \frac{1}{2}\right) + \left(\frac{1}{2} - \frac{1}{3}\right) + \cdots + \left(\frac{1}{n+1} - \frac{1}{n+2}\right).$$

The sum on the right telescopes to give

$$s_n = 1 - \frac{1}{n+2}.$$

It is now obvious that $s_n \to 1$. □

Example. The series

$$\sum_{k=1}^{\infty} (-1)^k$$

diverges. Here

$$s_n = -1 \quad \text{for } n \text{ odd} \quad \text{and} \quad s_n = 0 \quad \text{for } n \text{ even}.$$

Consequently $\{s_n\}$ does not converge. □

We come now to a series which has many applications.

(13.2.2)

> *The Geometric Series*
>
> (i) If $|x| < 1$, then $\sum_{k=0}^{\infty} x^k = \dfrac{1}{1-x}$.
>
> (ii) If $|x| \geq 1$, then $\sum_{k=0}^{\infty} x^k$ diverges.

PROOF. The nth partial sum of the geometric series

$$\sum_{k=0}^{\infty} x^k$$

is the number

(1) $$s_n = 1 + x + \cdots + x^n.$$

Multiplication by x gives

$$x s_n = x + x^2 + \cdots + x^{n+1}.$$

We now subtract the second equation from the first and obtain

$$(1-x)s_n = 1 - x^{n+1}.$$

For $x \neq 1$ this gives

(2) $$s_n = \frac{1 - x^{n+1}}{1 - x}.$$

If $|x| < 1$, then $x^{n+1} \to 0$ and thus

$$s_n \to \frac{1}{1-x}.$$

This proves (i). Now let's prove (ii). For $x = 1$, we use equation (1) and deduce that $s_n = n + 1$. Obviously $\{s_n\}$ diverges. For $|x| \geq 1$, $x \neq 1$, we use equation (2). Since in this instance $\{x^{n+1}\}$ diverges, $\{s_n\}$ diverges. $\square$

As a special case of the geometric series, we have

$$\sum_{k=0}^{\infty} \frac{1}{2^k} = \frac{1}{1 - \frac{1}{2}} = 2.$$

By beginning the summation at $k = 1$, instead of at $k = 0$, we find that

(13.2.3)

$$\sum_{k=1}^{\infty} \frac{1}{2^k} = 1.$$

The partial sums of this series

$$s_1 = \tfrac{1}{2},$$
$$s_2 = \tfrac{1}{2} + \tfrac{1}{4} = \tfrac{3}{4},$$
$$s_3 = \tfrac{1}{2} + \tfrac{1}{4} + \tfrac{1}{8} = \tfrac{7}{8},$$
$$s_4 = \tfrac{1}{2} + \tfrac{1}{4} + \tfrac{1}{8} + \tfrac{1}{16} = \tfrac{15}{16},$$
$$s_5 = \tfrac{1}{2} + \tfrac{1}{4} + \tfrac{1}{8} + \tfrac{1}{16} + \tfrac{1}{32} = \tfrac{31}{32},$$

etc.

are illustrated in Figure 13.2.1. Each new partial sum lies halfway between the previous one and the number 1.

FIGURE 13.2.1

The geometric series plays an important role in our decimal system. To see how this comes about, take the geometric series and consider the case $x = \tfrac{1}{10}$. Here

$$\sum_{k=0}^{\infty} \frac{1}{10^k} = \frac{1}{1 - \tfrac{1}{10}} = \frac{10}{9} \quad \text{and} \quad \sum_{k=1}^{\infty} \frac{1}{10^k} = \frac{1}{9}.$$

The nth partial sum in this last instance is

$$s_n = \frac{1}{10} + \frac{1}{10^2} + \cdots + \frac{1}{10^n}.$$

Now take the series

$$\sum_{k=1}^{\infty} \frac{a_k}{10^k} \quad \text{with} \quad a_k \in \{1, 2, \ldots, 9\}.$$

The partial sums here are

$$t_n = \frac{a_1}{10} + \frac{a_2}{10^2} + \cdots + \frac{a_n}{10^n} \le 9\left(\frac{1}{10} + \frac{1}{10^2} + \cdots + \frac{1}{10^n}\right) = 9s_n.$$

Since $\{s_n\}$ converges, $\{s_n\}$ is bounded. This tells us that $\{t_n\}$ is bounded. Since $\{t_n\}$ is nondecreasing, $\{t_n\}$ is itself convergent; that is, the series

$$\sum_{k=1}^{\infty} \frac{a_k}{10^k}$$

converges. The sum of this series is what we mean by the decimal

$$0.a_1a_2a_3 \cdots.$$

The geometric series arises naturally in many settings. Here is a somewhat off-beat example.

FIGURE 13.2.2

Example. The hour hand of a clock travels $\frac{1}{12}$ as fast as the minute hand. At 2 o'clock (Figure 13.2.2) the minute hand points to 12 and the hour hand points to 2. By the time the minute hand reaches 2, the hour hand points to $2 + \frac{1}{6}$. By the time the minute hand reaches $2 + \frac{1}{6}$, the hour hand points to

$$2 + \frac{1}{6} + \frac{1}{6 \cdot 12}.$$

By the time the minute hand reaches

$$2 + \frac{1}{6} + \frac{1}{6 \cdot 12},$$

the hour hand points to

$$2 + \frac{1}{6} + \frac{1}{6 \cdot 12} + \frac{1}{6 \cdot 12^2}$$

and so on. In general, by the time the minute hand reaches

$$2 + \frac{1}{6} + \frac{1}{6 \cdot 12} + \cdots + \frac{1}{6 \cdot 12^{n-1}} = 2 + \frac{1}{6} \sum_{k=0}^{n-1} \frac{1}{12^k},$$

the hour hand points to

$$2 + \frac{1}{6} + \frac{1}{6 \cdot 12} + \cdots + \frac{1}{6 \cdot 12^{n-1}} + \frac{1}{6 \cdot 12^n} = 2 + \frac{1}{6} \sum_{k=0}^{n} \frac{1}{12^k}.$$

The two hands coincide when both of them point to the limiting value

$$2 + \frac{1}{6} \sum_{k=0}^{\infty} \frac{1}{12^k} = 2 + \frac{1}{6} \left(\frac{1}{1 - \frac{1}{12}} \right) = 2 + \frac{2}{11}.$$

This happens at $2 + \frac{2}{11}$ o'clock. $\square$

We will return to the geometric series later. Right now we turn our attention to some results about series in general.

(13.2.4)

> If
> $$\sum_{k=0}^{\infty} a_k = l \quad \text{and} \quad \sum_{k=0}^{\infty} b_k = m,$$
> then
> $$\sum_{k=0}^{\infty} (a_k + b_k) = l + m \quad \text{and} \quad \sum_{k=0}^{\infty} \alpha a_k = \alpha l \quad \text{for each real } \alpha.$$

Proof.   Let

$$s_n = \sum_{k=0}^{n} a_k, \qquad t_n = \sum_{k=0}^{n} b_k,$$

$$u_n = \sum_{k=0}^{n} (a_k + b_k), \qquad v_n = \sum_{k=0}^{n} \alpha a_k.$$

Note that

$$u_n = s_n + t_n \quad \text{and} \quad v_n = \alpha s_n.$$

Since $s_n \to l$ and $t_n \to m$, we have

$$u_n \to l + m \quad \text{and} \quad v_n \to \alpha l. \quad \square$$

Theorem 13.2.5

The kth *term* of a convergent series tends to 0; namely,

$$\text{if} \quad \sum_{k=0}^{\infty} a_k \quad \text{converges,} \qquad \text{then} \quad a_k \to 0.$$

Proof.   For each n, set

$$s_n = \sum_{k=0}^{n} a_k.$$

To say that the series converges to some number l is to say that $s_n \to l$. Obviously then $s_{n-1} \to l$. Since $a_n = s_n - s_{n-1}$, we have $a_n \to l - l = 0$. $\quad \square$

It follows, of course, from Theorem 13.2.5 that

(13.2.6)

> if $\quad a_k \not\to 0, \qquad$ then $\displaystyle\sum_{k=0}^{\infty} a_k \quad$ diverges.

Series such as

$$\sum_{k=0}^{\infty} \sin k \quad \text{and} \quad \sum_{k=0}^{\infty} \frac{k}{k+1}$$

are therefore divergent.

CAUTION.   Theorem 13.2.5 does *not* say that, if $a_k \to 0$, then $\sum\limits_{k=0}^{\infty} a_k$ converges. There are divergent series for which $a_k \to 0$.

Example.   In the case of

$$\sum_{k=1}^{\infty} \frac{1}{\sqrt{k}}$$

we have $a_k = \dfrac{1}{\sqrt{k}} \to 0$ but $s_n = \dfrac{1}{\sqrt{2}} + \dfrac{1}{\sqrt{3}} + \cdots + \dfrac{1}{\sqrt{n}} > \dfrac{n}{\sqrt{n}} = \sqrt{n}.$

The sequence of partial sums is unbounded, and thus the series diverges.   □

Exercises

Find the sum of the series.

*1. $\sum\limits_{k=3}^{\infty} \dfrac{1}{(k+1)(k+2)}.$ 2. $\sum\limits_{k=0}^{\infty} \dfrac{1}{(k+3)(k+4)}.$ *3. $\sum\limits_{k=1}^{\infty} \dfrac{1}{2k(k+1)}.$

*4. $\sum\limits_{k=0}^{\infty} \dfrac{3}{10^k}.$ 5. $\sum\limits_{k=0}^{\infty} \dfrac{12}{100^k}.$ *6. $\sum\limits_{k=0}^{\infty} \dfrac{67}{1000^k}.$

*7. $\sum\limits_{k=0}^{\infty} \dfrac{(-1)^k}{5^k}.$ 8. $\sum\limits_{k=0}^{\infty} \left(\dfrac{3}{4}\right)^k.$ *9. $\sum\limits_{k=0}^{\infty} \dfrac{3^k+4^k}{5^k}.$

*10. $\sum\limits_{k=0}^{\infty} \dfrac{1-2^k}{3^k}.$ 11. $\sum\limits_{k=0}^{\infty} \left(\dfrac{25}{10^k} - \dfrac{6}{100^k}\right).$ *12. $\sum\limits_{k=3}^{\infty} \dfrac{1}{2^{k-1}}.$

*13. $\sum\limits_{k=0}^{\infty} \dfrac{1}{2^{k+3}}.$ 14. $\sum\limits_{k=0}^{\infty} \dfrac{2^{k+3}}{3^k}.$

Write the decimal fraction as an infinite series and express the sum as the quotient of two integers.

*15. 0.777 · · · · .   16. 0.999 · · · · .   *17. 0.2424 · · · · .
*18. 0.8989 · · · · .   19. 0.112112112 · · · · .   *20. 0.315315315 · · · · .
*21. 0.624545 · · · · .   22. 0.112019019 · · · · .

23. Using series, show that every repeating decimal represents a rational number (the quotient of two integers).
24. Show that

$$\sum_{k=0}^{\infty} a_k = l \quad \text{iff} \quad \sum_{k=j+1}^{\infty} a_k = l - (a_0 + \cdots + a_j).$$

Derive these results from the geometric series.

25. $\sum\limits_{k=0}^{\infty} (-1)^k x^k = \dfrac{1}{1+x}, \quad |x| < 1.$ 26. $\sum\limits_{k=0}^{\infty} (-1)^k x^{2k} = \dfrac{1}{1+x^2}, \quad |x| < 1.$

27. Find a series expansion for

     *(a) $\dfrac{x}{1 + x}$ valid for $|x| < 1$.   (b) $\dfrac{x}{1 + x^2}$ valid for $|x| < 1$.

     *(c) $\dfrac{1}{1 - x^2}$ valid for $|x| < 1$.   (d) $\dfrac{1}{1 + 4x^2}$ valid for $|x| < \dfrac{1}{2}$.

28. At some time between 4 and 5 o'clock the minute hand is directly above the hour hand. Express this time as a geometric series. What is the sum of this series?

*29. A ball dropped from a height of h feet is known to rebound to a height of σh feet where σ is a positive constant less than 1. Find the total distance traveled by the ball if dropped initially from a height of h_0 feet.

*30. How much money do you have to deposit at $r\%$ interest compounded annually to enable your descendants to withdraw n_1 dollars at the end of the first year, n_2 dollars at the end of the second year, n_3 dollars at the end of the third year, and so on in perpetuity? Express your answer as an infinite series.

31. Sum the series you obtained in exercise 30 setting
     (a) $r = 5$, $n_k = 5000(\tfrac{1}{2})^{k-1}$.   (b) $r = 6$, $n_k = 1000(0.8)^{k-1}$.

Optional | 32. Show that

$$\sum_{k=1}^{\infty} kx^{k-1} = \frac{1}{(1 - x)^2} \quad \text{for} \quad |x| < 1.$$

     HINT: verify that s_n, the nth partial sum, satisfies the identity

$$(1 - x)^2 s_n = 1 - (n + 1)x^n + nx^{n+1}.$$

13.3 The Integral Test; Comparison Theorems

Here and in the next section we direct our attention to series with nonnegative terms: $a_k \geq 0$ for all k. For such series the following theorem is fundamental.

Theorem 13.3.1

A series with nonnegative terms converges iff the sequence of partial sums is bounded.

PROOF. If the sequence of partial sums is not bounded, it cannot converge. (12.2.5) Suppose now that the sequence of partial sums is bounded. Since the terms are nonnegative, the sequence of partial sums is nondecreasing. By Theorem 12.2.6 it must converge. □

Under certain conditions the convergence or divergence of a series can be deduced from the convergence or divergence of a closely related improper integral.

Theorem 13.3.2   The Integral Test

If f is continuous, decreasing, and positive on $[1, \infty)$, then

$$\sum_{k=1}^{\infty} f(k) \quad \text{converges} \quad \text{iff} \quad \int_{1}^{\infty} f(x)\, dx \quad \text{converges.}$$

PROOF.   In Exercise 34, Section 12.7, you were asked to show that with f continuous, decreasing, and positive on $[1, \infty)$

$$\int_{1}^{\infty} f(x)\, dx \quad \text{converges} \quad \text{iff} \quad \text{the sequence} \quad \left\{ \int_{1}^{n} f(x)\, dx \right\} \quad \text{converges.}$$

We assume this result and base our proof on the behavior of the sequence. To visualize our argument see Figure 13.3.1.

FIGURE 13.3.1

Since f decreases on the interval $[1, n]$,

$$f(2) + \cdots + f(n) \quad \text{is a lower sum for } f \text{ on } [1, n]$$

and

$$f(1) + \cdots + f(n - 1) \quad \text{is an upper sum for } f \text{ on } [1, n].$$

Consequently

$$f(2) + \cdots + f(n) \leq \int_{1}^{n} f(x)\, dx \quad \text{and} \quad \int_{1}^{n} f(x)\, dx \leq f(1) + \cdots + f(n - 1).$$

If the sequence

$$\left\{ \int_{1}^{n} f(x)\, dx \right\}$$

converges, it is bounded. By the first inequality the series itself must be bounded and therefore convergent. Suppose now that the sequence

$$\left\{ \int_{1}^{n} f(x)\, dx \right\}$$

diverges. Since f is positive,

$$\int_{1}^{n} f(x)\, dx < \int_{1}^{n+1} f(x)\, dx.$$

This tells us that the sequence of integrals is monotonic. Being divergent and monotonic, it must be unbounded. By the second inequality, the series itself must be unbounded and therefore divergent. □

Applying the Integral Test

Example. (*The Harmonic Series*)

(13.3.3)
$$\sum_{k=1}^{\infty} \frac{1}{k} = 1 + \frac{1}{2} + \frac{1}{3} + \frac{1}{4} + \cdots \quad \text{diverges.}$$

PROOF. The function $f(x) = 1/x$ is continuous, decreasing, and positive on $[1, \infty)$. The series

$$\sum_{k=1}^{\infty} \frac{1}{k} \quad \text{diverges}$$

since the improper integral

$$\int_{1}^{\infty} \frac{dx}{x} \quad \text{diverges.†} \quad □$$

The next example gives a more general result.

Example. (*The p-series*)

(13.3.4)
$$\sum_{k=1}^{\infty} \frac{1}{k^p} = 1 + \frac{1}{2^p} + \frac{1}{3^p} + \frac{1}{4^p} + \cdots \quad \text{converges} \quad \text{iff} \quad p > 1.$$

PROOF. The function $f(x) = 1/x^p$ is continuous, decreasing, and positive on $[1, \infty)$. The series

$$\sum_{k=1}^{\infty} \frac{1}{k^p} \quad \text{converges} \quad \text{iff} \quad p > 1$$

since the improper integral

$$\int_{1}^{\infty} \frac{dx}{x^p} \quad \text{converges} \quad \text{iff} \quad p > 1.†† \quad □$$

Example. Here we show that the series

$$\sum_{k=1}^{\infty} \frac{1}{k \log (k + 1)} = \frac{1}{\log 2} + \frac{1}{2 \log 3} + \frac{1}{3 \log 4} + \cdots \quad \text{diverges.}$$

We begin by setting

$$f(x) = \frac{1}{x \log (x + 1)}.$$

---

† (12.7.1).
†† (12.7.1).

Since f is continuous, decreasing, and positive on $[1, \infty)$, we can use the integral test. Note that

$$\int_1^b \frac{dx}{x \log (x + 1)} > \int_1^b \frac{dx}{(x + 1) \log (x + 1)}$$

$$= \left[\log (\log (x + 1)) \right]_1^b = \log (\log (b + 1)) - \log (\log 2).$$

As $b \to \infty$, $\log (\log (b + 1)) \to \infty$. This shows that

$$\int_1^\infty \frac{dx}{x \log (x + 1)} \quad \text{diverges.}$$

It follows that the series diverges. $\square$

REMARK ON NOTATION. You have seen that for each $j \geq 0$

$$\sum_{k=0}^\infty a_k \quad \text{converges} \quad \text{iff} \quad \sum_{k=j+1}^\infty a_k \quad \text{converges.}$$

(Exercise 24, Section 13.2) This tells you that, in determining whether or not a series converges, it does not matter where we begin the summation.† Where detailed indexing would contribute nothing, we will omit it and write

$$\Sigma \, a_k$$

without specifying where the summation begins. For instance, it makes sense to say that

$$\Sigma \frac{1}{k^2} \quad \text{converges} \quad \text{and} \quad \Sigma \frac{1}{k} \quad \text{diverges}$$

without specifying where we begin the summation.

Much of the work on series with nonnegative terms is based on comparison with series of known behavior. The basic comparison theorem is extremely simple.

Theorem 13.3.5   The Basic Comparison Theorem

Let $\Sigma \, a_k$ be a series with nonnegative terms.

(i) $\Sigma \, a_k$ converges if there exists a convergent series $\Sigma \, c_k$ with nonnegative terms such that

$$a_k \leq c_k \quad \text{for all } k \text{ sufficiently large;}$$

(ii) $\Sigma \, a_k$ diverges if there exists a divergent series $\Sigma \, d_k$ with nonnegative terms such that

$$d_k \leq a_k \quad \text{for all } k \text{ sufficiently large.}$$

---

† In the convergent case it does, however, affect the sum. Thus, for example,

$$\sum_{k=0}^\infty \frac{1}{2^k} = 2, \quad \sum_{k=1}^\infty \frac{1}{2^k} = 1, \quad \sum_{k=2}^\infty \frac{1}{2^k} = \frac{1}{2} \quad \text{and so forth.}$$

PROOF. The proof is just a matter of noting that in the first instance

$$\Sigma\ a_k \quad \text{and} \quad \Sigma\ c_k \quad \text{are both bounded,}$$

and in the second instance

$$\Sigma\ a_k \quad \text{and} \quad \Sigma\ d_k \quad \text{are both unbounded.}$$

The details are left to you.   □

Examples

1. $\Sigma\ \dfrac{1}{2k^3 + 1}$ converges by comparison with $\Sigma\ \dfrac{1}{k^3}$:

$$\frac{1}{2k^3 + 1} < \frac{1}{k^3} \quad \text{and} \quad \Sigma\ \frac{1}{k^3} \quad \text{converges.}$$

2. $\Sigma\ \dfrac{1}{3k + 1}$ diverges by comparison with $\Sigma\ \dfrac{1}{3(k + 1)}$:

$$\frac{1}{3k + 1} > \frac{1}{3(k + 1)} \quad \text{and} \quad \Sigma\ \frac{1}{3(k + 1)} = \tfrac{1}{3}\Sigma\ \frac{1}{k + 1} \quad \text{diverges.}$$

3. $\Sigma\ \dfrac{k^3}{k^5 + 5k^4 + 7}$ converges by comparison with $\Sigma\ \dfrac{1}{k^2}$:

$$\frac{k^3}{k^5 + 5k^4 + 7} < \frac{k^3}{k^5} = \frac{1}{k^2} \quad \text{and} \quad \Sigma\ \frac{1}{k^2} \quad \text{converges.}$$

4. $\Sigma\ \dfrac{1}{\log (k + 6)}$ diverges by comparison with $\Sigma\ \dfrac{1}{k}$:

$$\frac{1}{\log (k + 6)} > \frac{1}{k} \quad \text{for } k \geq 3 \quad \text{and} \quad \Sigma\ \frac{1}{k} \quad \text{diverges.} \quad □$$

We come now to a somewhat more sophisticated comparison theorem.

Theorem 13.3.6   The Limit Comparison Theorem

Let $\Sigma\ a_k$ and $\Sigma\ b_k$ be series with positive terms. If $a_k/b_k \rightarrow l$ where l is some positive number, then

$$\Sigma\ a_k \quad \text{converges} \quad \text{iff} \quad \Sigma\ b_k \quad \text{converges.}$$

PROOF. Suppose that

$$\frac{a_k}{b_k} \rightarrow l \quad \text{and} \quad l > 0.$$

Now choose $\epsilon > 0$ so that $l - \epsilon > 0$. Since the quotients a_k/b_k converge to l, we know that for all k sufficiently large (for all k greater than some k_0)

$$\left| \frac{a_k}{b_k} - l \right| < \epsilon.$$

For such k we have

$$l - \epsilon < \frac{a_k}{b_k} < l + \epsilon$$

and thus

$$(l - \epsilon)b_k < a_k < (l + \epsilon)b_k.$$

This last inequality is what we needed:

if $\Sigma\, a_k$ converges, then $\Sigma\,(l - \epsilon)b_k$ converges, and thus $\Sigma\, b_k$ converges;

if $\Sigma\, b_k$ converges, then $\Sigma\,(l + \epsilon)b_k$ converges, and thus $\Sigma\, a_k$ converges.   □

The key to applying the limit comparison test to a series $\Sigma\, a_k$ is to gauge the order of magnitude of the a_k for large k.

Examples

1. $\Sigma \dfrac{3k - 2}{k^2 + 1}$ diverges by limit comparison with $\Sigma \dfrac{1}{k}$:

$$\frac{3k - 2}{k^2 + 1} \div \frac{1}{k} = \frac{3k^2 - 2k}{k^2 + 1} = \frac{3 - 2/k}{1 + 1/k^2} \to 3.$$

2. $\Sigma \dfrac{\sqrt{k} + 100}{3k^2\sqrt{k} + 2\sqrt{k}}$ converges by limit comparison with $\Sigma \dfrac{1}{k^2}$:

$$\frac{\sqrt{k} + 100}{3k^2\sqrt{k} + 2\sqrt{k}} \div \frac{1}{k^2} = \frac{k^2\sqrt{k} + 100k^2}{3k^2\sqrt{k} + 2\sqrt{k}} = \frac{1 + 100/\sqrt{k}}{3 + 2/k^2} \to \frac{1}{3}.$$

3. $\Sigma \sin \dfrac{\pi}{k}$ diverges by limit comparison with $\Sigma \dfrac{1}{k}$:

$$\sin \frac{\pi}{k} \div \frac{1}{k} = \frac{\sin (\pi/k)}{1/k} = \pi \left[\frac{\sin (\pi/k)}{\pi/k} \right] \to \pi$$

since $(\sin x)/x \to 1$ as $x \to 0$.†   □

Exercises

Determine whether the series converges or diverges.

*1. $\Sigma \dfrac{k}{k^3 + 1}$.   2. $\Sigma \dfrac{1}{3k + 2}$.   *3. $\Sigma \dfrac{1}{(2k + 1)^2}$.

*4. $\Sigma \dfrac{\log k}{k}$.   5. $\Sigma \dfrac{1}{\sqrt{k} + 1}$.   *6. $\Sigma \dfrac{1}{k^2 + 1}$.

*7. $\Sigma \dfrac{1}{1 + 2\log k}$.   8. $\Sigma\, (\tfrac{3}{4})^{-k}$.   *9. $\Sigma \dfrac{\log \sqrt{k}}{k}$.

*10. $\Sigma \dfrac{2}{k\,(\log k)^2}$.   11. $\Sigma \dfrac{1}{2 + 3^{-k}}$.   *12. $\Sigma \dfrac{1}{k\log k}$.

---

† (7.1.1).

*13. $\Sigma \dfrac{2k+5}{5k^3+3k^2}$.   14. $\Sigma \dfrac{k^4-1}{3k^2+5}$.   *15. $\Sigma \dfrac{7k+2}{2k^5+7}$.

*16. $\Sigma \dfrac{2k+1}{\sqrt{k^4+1}}$.   17. $\Sigma \dfrac{2k+1}{\sqrt{k^3+1}}$.   *18. $\Sigma \dfrac{2k+1}{\sqrt{k^5+1}}$.

*19. $\Sigma\, ke^{-k^2}$.   20. $\Sigma \dfrac{1}{\sqrt{2k(k+1)}}$.   *21. $\Sigma\, k^2\, 2^{-k^3}$.

13.4 The Root Test; The Ratio Test

Comparison with the geometric series

$$\Sigma\, x^k$$

and with the p-series

$$\Sigma \frac{1}{k^p}$$

leads to two important tests for convergence: the root test and the ratio test.

Theorem 13.4.1 The Root Test

Let $\Sigma\, a_k$ be a series with nonnegative terms and suppose that

$$(a_k)^{1/k} \to \rho.$$

If $\rho < 1$, $\Sigma\, a_k$ converges. If $\rho > 1$, $\Sigma\, a_k$ diverges. If $\rho = 1$, the test is inconclusive.

PROOF.   We suppose first that $\rho < 1$ and choose μ so that

$$\rho < \mu < 1.$$

Since $(a_k)^{1/k} \to \rho$, we have

$$(a_k)^{1/k} < \mu \qquad \text{for all } k \text{ sufficiently large.} \qquad \text{(explain)}$$

Thus

$$a_k < \mu^k \qquad \text{for all } k \text{ sufficiently large.}$$

Since $\Sigma\, \mu^k$ converges (a geometric series with $0 < \mu < 1$), we know by the comparison theorem that $\Sigma\, a_k$ converges.

We suppose now that $\rho > 1$ and choose μ so that

$$\rho > \mu > 1.$$

Since $(a_k)^{1/k} \to \rho$, we have

$$(a_k)^{1/k} > \mu \qquad \text{for all } k \text{ sufficiently large.} \qquad \text{(explain)}$$

Thus

$$a_k > \mu^k \qquad \text{for all } k \text{ sufficiently large.}$$

Since $\Sigma\ \mu^k$ diverges (a geometric series with $\mu > 1$), the comparison theorem tells us that $\Sigma\ a_k$ diverges.

To see the inconclusiveness of the root test when $\rho = 1$, note that $(a_k)^{1/k} \to 1$ for both $\Sigma\ (1/k)^2$ and $\Sigma\ 1/k$.† The first series converges, but the second diverges.   □

Applying the Root Test

Example.   In the case of

$$\Sigma\ \frac{1}{(\log k)^k},$$

we have

$$(a_k)^{1/k} = \frac{1}{\log k} \to 0.$$

The series converges.   □

Example.   In the case of

$$\Sigma\ \frac{2^k}{k^3},$$

we have

$$(a_k)^{1/k} = 2\left(\frac{1}{k}\right)^{3/k} = 2\left[\left(\frac{1}{k}\right)^{1/k}\right]^3 \to 2 \cdot 1^3 = 2.$$

The series diverges.   □

Example.   In the case of

$$\Sigma\ \left(1 - \frac{1}{k}\right)^k,$$

we have

$$(a_k)^{1/k} = 1 - \frac{1}{k} \to 1.$$

Here the root test is inconclusive. Since $a_k = (1 - 1/k)^k$ tends to $1/e$ and not to 0, we know from (13.2.6) that the series diverges.   □

---

† In the first instance

$$(a_k)^{1/k} = \left(\frac{1}{k^2}\right)^{1/k} = \left(\frac{1}{k^{1/k}}\right)^2 \to 1^2 = 1;$$

in the second instance

$$(a_k)^{1/k} = \left(\frac{1}{k}\right)^{1/k} = \frac{1}{k^{1/k}} \to 1.$$

Theorem 13.4.2   The Ratio Test

Let $\Sigma\, a_k$ be a series with positive terms and suppose that

$$\frac{a_{k+1}}{a_k} \to \lambda.$$

If $\lambda < 1$, $\Sigma\, a_k$ converges. If $\lambda > 1$, $\Sigma\, a_k$ diverges. If $\lambda = 1$, the test is inconclusive.

PROOF.   We suppose first that $\lambda < 1$ and choose μ so that

$$\lambda < \mu < 1.$$

Since

$$\frac{a_{k+1}}{a_k} \to \lambda,$$

we know that there exists $k_0 > 0$ such that

$$\text{if } k \geq k_0, \quad \text{then} \quad \frac{a_{k+1}}{a_k} < \mu.$$

This gives

$$a_{k_0+1} < \mu a_{k_0}, \qquad a_{k_0+2} < \mu a_{k_0+1} < \mu^2 a_{k_0},$$

and more generally,

$$a_{k_0+j} < \mu^j a_k.$$

In other words, for $k > k_0$ we have

$$(1) \qquad\qquad a_k < \mu^{k-k_0}\, a_{k_0} = \frac{a_{k_0}}{\mu^{k_0}}\, \mu^k.$$

Since $\mu < 1$,

$$\Sigma\, \frac{a_{k_0}}{\mu^{k_0}}\, \mu^k = \frac{a_{k_0}}{\mu^{k_0}}\, \Sigma\, \mu^k \quad \text{converges.}$$

Recalling (1), you can see by the comparison theorem that $\Sigma\, a_k$ converges. The proof of the rest of the theorem is left to the exercises.   □

Applying the Ratio Test

Example.   The convergence of the series

$$\Sigma\, \frac{1}{k!}$$

is easy to verify by the ratio test:

$$\frac{a_{k+1}}{a_k} = \frac{1}{(k+1)!} \cdot \frac{k!}{1} = \frac{1}{k+1} \to 0.   □$$

Example. In the case of

$$\Sigma \frac{k}{10^k},$$

we have

$$\frac{a_{k+1}}{a_k} = \frac{k + 1}{10^{k+1}} \cdot \frac{10^k}{k} = \frac{1}{10} \frac{k + 1}{k} \rightarrow \frac{1}{10}.$$

The series therefore converges.†   □

Example. In the case of

$$\Sigma \frac{k^k}{k!},$$

we have

$$\frac{a_{k+1}}{a_k} = \frac{(k + 1)^{k+1}}{(k + 1)!} \cdot \frac{k!}{k^k} = \left(\frac{k + 1}{k}\right)^k = \left(1 + \frac{1}{k}\right)^k \rightarrow e.$$

Since $e > 1$, the series diverges.   □

Example. For the series

$$\Sigma \frac{1}{2k + 1}$$

the ratio test is inconclusive:

$$\frac{a_{k+1}}{a_k} = \frac{1}{2(k + 1) + 1} \cdot \frac{2k + 1}{1} = \frac{2k + 1}{2k + 3} = \frac{2 + 1/k}{2 + 3/k} \rightarrow 1.$$

Therefore we have to look further. Comparison with the harmonic series shows that the series diverges:

$$\frac{1}{2k + 1} > \frac{1}{2(k + 1)} \quad \text{and} \quad \Sigma \frac{1}{2(k + 1)} \quad \text{diverges.}   □$$

In general the root test is used only when powers are involved. The ratio test is particularly effective with factorials and with combinations of powers and factorials. If the terms are rational functions of k, the ratio test is inconclusive and the root test is difficult to apply. Rational terms are most easily handled by comparison or limit comparison with a p-series, $\Sigma \, 1/k^p$. If the terms have the configuration of a derivative, you may be able to apply the integral test.

Exercises

Determine whether the series converges or diverges.

*1. $\Sigma \dfrac{10^k}{k!}$.   2. $\Sigma \dfrac{1}{k \, 2^k}$.   *3. $\Sigma \dfrac{1}{k^k}$.

† This series can be summed explicitly. See Exercise 28.

*4. $\Sigma \left(\dfrac{k}{2k+1}\right)^k$.

5. $\Sigma \dfrac{k!}{100^k}$.

*6. $\Sigma \dfrac{(\log k)^2}{k}$.

*7. $\Sigma \dfrac{k^2+2}{k^3+6k}$.

8. $\Sigma k(\tfrac{2}{3})^k$.

*9. $\Sigma \dfrac{1}{(\log k)^k}$.

*10. $\Sigma \dfrac{1}{(\log k)^{10}}$.

11. $\Sigma \dfrac{1}{1+\sqrt{k}}$.

*12. $\Sigma \dfrac{2k+\sqrt{k}}{k^3+\sqrt{k}}$.

*13. $\Sigma \dfrac{k!}{10^{4k}}$.

14. $\Sigma \dfrac{k^2}{e^k}$.

*15. $\Sigma \dfrac{\sqrt{k}}{k^2+1}$.

*16. $\Sigma \dfrac{2^k k!}{k^k}$.

17. $\Sigma \dfrac{k!}{(k+2)!}$.

*18. $\Sigma \dfrac{1}{k}\left(\dfrac{1}{\log k}\right)^{3/2}$.

*19. $\Sigma \dfrac{1}{k}\left(\dfrac{1}{\log k}\right)^{1/2}$.

20. $\Sigma \dfrac{1}{\sqrt{k^3-1}}$.

*21. $\Sigma \left(\dfrac{k}{k+100}\right)^k$.

*22. $\Sigma \dfrac{(k!)^2}{(2k)!}$.

23. $\Sigma k^{-(1+1/k)}$.

*24. $\Sigma \dfrac{11}{1+100^{-k}}$.

*25. $\Sigma \dfrac{\log k}{e^k}$.

26. $\Sigma \dfrac{k!}{k^k}$.

*27. $\Sigma \dfrac{\log k}{k^2}$.

28. Determine the sum of the series

$$\sum_{k=1}^{\infty} \frac{k}{10^k}.$$

HINT: Use exercise 32 of Section 13.2.

29. Complete the proof of the ratio test.
   (a) Prove that, if $\lambda > 1$, then $\Sigma\, a_k$ diverges.
   (b) Prove that, if $\lambda = 1$, the ratio test is inconclusive.
       HINT: Consider $\Sigma\,(1/k)$ and $\Sigma\,(1/k^2)$.

13.5 Absolute and Conditional Convergence; Alternating Series

In this section we consider series which have both positive and negative terms.

Absolute and Conditional Convergence

Let $\Sigma\, a_k$ be a series with both positive and negative terms. One way to show that $\Sigma\, a_k$ converges is to show that $\Sigma\, |a_k|$ converges.

Theorem 13.5.1

If $\Sigma\, |a_k|$ converges,   then $\Sigma\, a_k$ converges.

PROOF. For each k,

$$-|a_k| \le a_k \le |a_k| \quad \text{and therefore} \quad 0 \le a_k + |a_k| \le 2|a_k|.$$

If $\Sigma\,|a_k|$ converges, then $\Sigma\, 2|a_k| = 2\,\Sigma\,|a_k|$ converges, and therefore, by the comparison test, $\Sigma \, (a_k + |a_k|)$ converges. Since

$$a_k = (a_k + |a_k|) - |a_k|,$$

we can conclude that $\Sigma \, a_k$ converges. $\square$

A series $\Sigma \, a_k$ for which $\Sigma \, |a_k|$ converges is called *absolutely convergent*. The theorem we just proved says that

(13.5.2) | if $\Sigma \, a_k$ converges absolutely,   then $\Sigma \, a_k$ converges.

As you will see a little later, the converse is false. There are series that converge but do not converge absolutely. Such series are called *conditionally convergent*.

Here are some examples that will help you get some feeling for these ideas.

Example

$$1 - \frac{1}{2^2} + \frac{1}{3^2} - \frac{1}{4^2} + \frac{1}{5^2} - \frac{1}{6^2} + \ \cdots.$$

If we replace each of the terms by its absolute value, we obtain the series

$$1 + \frac{1}{2^2} + \frac{1}{3^2} + \frac{1}{4^2} + \frac{1}{5^2} + \frac{1}{6^2} + \ \cdots.$$

This is a p-series with $p = 2$. It is therefore convergent. This means that the initial series is absolutely convergent. $\square$

Example

$$1 - \frac{1}{2} - \frac{1}{2^2} + \frac{1}{2^3} - \frac{1}{2^4} - \frac{1}{2^5} + \frac{1}{2^6} - \frac{1}{2^7} - \frac{1}{2^8} + \ \cdots.$$

If we replace each of the terms by its absolute value, we obtain the series

$$1 + \frac{1}{2} + \frac{1}{2^2} + \frac{1}{2^3} + \frac{1}{2^4} + \frac{1}{2^5} + \frac{1}{2^6} + \frac{1}{2^7} + \frac{1}{2^8} + \ \cdots.$$

This is a convergent geometric series. The initial series is therefore absolutely convergent. $\square$

Example

$$1 - \frac{1}{2} + \frac{1}{3} - \frac{1}{4} + \frac{1}{5} - \frac{1}{6} + \ \cdots.$$

This series is only conditionally convergent. It is convergent (see the next theorem), but it is not absolutely convergent: if we replace each term by its absolute value, we obtain the divergent harmonic series

$$1 + \frac{1}{2} + \frac{1}{3} + \frac{1}{4} + \frac{1}{5} + \frac{1}{6} + \ \cdots. \dagger \quad \square$$

---

† In section 13.8 you'll be in a position to evaluate

$$1 - \tfrac{1}{2} + \tfrac{1}{3} - \tfrac{1}{4} + \tfrac{1}{5} - \tfrac{1}{6} + \ \cdots$$

directly. You'll see that the sum is log 2.

Example

$$\frac{1}{2} - 1 + \frac{1}{3} - 1 + \frac{1}{4} - 1 + \frac{1}{5} - 1 + \cdots.$$

Here the terms do not tend to zero. The series is divergent. □

Alternating Series

A series in which successive terms have opposite signs is called an *alternating series*. As examples of alternating series we have

$$1 - \frac{1}{2} + \frac{1}{3} - \frac{1}{4} + \frac{1}{5} - \frac{1}{6} + \cdots,$$

$$1 - \frac{1}{\sqrt{2}} + \frac{1}{\sqrt{3}} - \frac{1}{\sqrt{4}} + \frac{1}{\sqrt{5}} - \frac{1}{\sqrt{6}} + \cdots,$$

$$1 - \frac{2}{\log 2} + \frac{3}{\log 3} - \frac{4}{\log 4} + \frac{5}{\log 5} - \frac{6}{\log 6} + \cdots.$$

The series

$$1 - \frac{1}{2} - \frac{1}{3} + \frac{1}{4} - \frac{1}{5} - \frac{1}{6} + \cdots$$

is not an alternating series because there are consecutive terms with the same sign.

Theorem 13.5.3   Alternating Series†

Let $\{a_k\}$ be a decreasing sequence of positive numbers.

$$\text{If} \quad a_k \to 0, \qquad \text{then} \quad \sum_{k=0}^{\infty} (-1)^k a_k \quad \text{converges.}$$

PROOF.   First we look at the even partial sums, s_{2m}. Since

$$s_{2m} = (a_0 - a_1) + (a_2 - a_3) + \cdots + (a_{2m-2} - a_{2m-1}) + a_{2m}$$

is the sum of positive numbers, the even partial sums are all positive. Since

$$s_{2m+2} = s_{2m} - a_{2m+1} + a_{2m+2} \qquad \text{and} \qquad a_{2m+1} > a_{2m+2},$$

we have

$$s_{2m+2} < s_{2m}.$$

This means that the sequence of even partial sums is decreasing. Being bounded below by 0, it is convergent; say,

$$s_{2m} \to l.$$

---

† This theorem dates back to Leibniz. He observed the result in 1705.

Now

$$s_{2m+1} = s_{2m} - a_{2m+1}.$$

Since $a_{2m+1} \to 0$, we also have

$$s_{2m+1} \to l.$$

With both the even and the odd partial sums tending to l, it is obvious that all the partial sums must tend to l. $\square$

The theorem we just proved makes it clear that each of the following series converges:

$$1 - \frac{1}{2} + \frac{1}{3} - \frac{1}{4} + \frac{1}{5} - \frac{1}{6} + \cdots,$$

$$1 - \frac{1}{\sqrt{2}} + \frac{1}{\sqrt{3}} - \frac{1}{\sqrt{4}} + \frac{1}{\sqrt{5}} - \frac{1}{\sqrt{6}} + \cdots,$$

$$1 - \frac{1}{2!} + \frac{1}{3!} - \frac{1}{4!} + \frac{1}{5!} - \frac{1}{6!} + \cdots.$$

The first two series are only conditionally convergent. The third series converges absolutely.

Rearrangements

A *rearrangement* of a series Σa_k is a series which has exactly the same terms but in a different order. Thus, for example,

$$1 + \frac{1}{3^3} - \frac{1}{2^2} + \frac{1}{5^5} - \frac{1}{4^4} + \frac{1}{7^7} - \frac{1}{6^6} + \cdots$$

and

$$1 + \frac{1}{3^3} + \frac{1}{5^5} - \frac{1}{2^2} - \frac{1}{4^4} + \frac{1}{7^7} + \frac{1}{9^9} - \cdots$$

are both rearrangements of

$$1 - \frac{1}{2^2} + \frac{1}{3^3} - \frac{1}{4^4} + \frac{1}{5^5} - \frac{1}{6^6} + \frac{1}{7^7} - \cdots.$$

In 1867 Riemann published a theorem on rearrangements of series which underscores the importance of distinguishing between absolute convergence and conditional convergence. According to this thereom all rearrangements of an absolutely convergent series converge absolutely to the same sum. In sharp contrast, a series which is only conditionally convergent can be rearranged to converge to any number we please. It can also be arranged to diverge to $+\infty$, or to diverge to $-\infty$, or even to oscillate between any two bounds we choose.†

---

† For a complete proof see pp. 138–139, 318–320 in Konrad Knopp's *Theory and Applications of Infinite Series* (Second English Edition), Blackie & Son Limited, London, 1951.

Exercises

Test these series for (a) absolute convergence, (b) conditional convergence.

*1. $1 + (-1) + 1 + \cdots + (-1)^k + \cdots$.

2. $\dfrac{1}{4} - \dfrac{1}{6} + \dfrac{1}{8} - \dfrac{1}{10} + \cdots + \dfrac{(-1)^k}{2k} + \cdots$.

*3. $\dfrac{1}{2} - \dfrac{2}{3} + \dfrac{3}{4} - \dfrac{4}{5} + \cdots + (-1)^k \dfrac{k}{k+1} + \cdots$.

4. $\dfrac{1}{2 \log 2} - \dfrac{1}{3 \log 3} + \dfrac{1}{4 \log 4} - \dfrac{1}{5 \log 5} + \cdots + (-1)^k \dfrac{1}{k \log k} + \cdots$.

*5. $\Sigma (-1)^k \dfrac{\log k}{k}$.   6. $\Sigma (-1)^k \dfrac{k}{\log k}$.   *7. $\Sigma \left(\dfrac{1}{k} - \dfrac{1}{k!} \right)$.

*8. $\Sigma \dfrac{k^3}{2^k}$.   9. $\Sigma (-1)^k \dfrac{1}{2k+1}$.   *10. $\Sigma (-1)^k \dfrac{(k!)^2}{(2k)!}$.

*11. $\Sigma \dfrac{k!}{(-2)^k}$.   12. $\Sigma \sin \left(\dfrac{k\pi}{4} \right)$.   *13. $\Sigma \dfrac{(-1)^k}{\sqrt{k} + \sqrt{k+1}}$.

*14. $\Sigma \sin \left(\dfrac{\pi}{4k^2} \right)$.   15. $\Sigma (-1)^k \dfrac{k}{k^2+1}$.   *16. $\Sigma \dfrac{(-1)^k}{\sqrt{k(k+1)}}$.

*17. $\Sigma (-1)^k \dfrac{k}{2^k}$.   18. $\Sigma \left(\dfrac{1}{\sqrt{k}} - \dfrac{1}{\sqrt{k+1}} \right)$.

*19. $\dfrac{1}{2} - \dfrac{1}{3} - \dfrac{1}{4} + \dfrac{1}{5} - \dfrac{1}{6} - \dfrac{1}{7} + \cdots + \dfrac{1}{3k+2} - \dfrac{1}{3k+3} - \dfrac{1}{3k+4} + \cdots$.

20. $\dfrac{2 \cdot 3}{4 \cdot 5} - \dfrac{5 \cdot 6}{7 \cdot 8} + \cdots + (-1)^k \dfrac{(3k+2)(3k+3)}{(3k+4)(3k+5)} + \cdots$.

21. Verify that the series

$$1 - \frac{1}{2} + \frac{1}{2} - \frac{1}{3} + \frac{1}{2} - \frac{1}{3} + \frac{1}{3} - \frac{1}{4} + \frac{1}{3} - \frac{1}{4} + \frac{1}{3} - \frac{1}{4} + \cdots$$

diverges and explain how this does not violate the theorem on alternating series.

22. (*Important*) *An Estimate for Alternating Series.* If $\{a_k\}$ is a decreasing sequence of positive numbers, then the alternating series

$$\sum_{k=0}^{\infty} (-1)^k a_k = a_0 - a_1 + a_2 - a_3 + a_4 - a_5 + a_6 - \cdots$$

converges to some number l. Show that this number l lies between consecutive partial sums s_n, s_{n+1} and that $|s_n - l| < a_{n+1}$.

Optional | 23. In Section 9 of this chapter we will prove that, if $\Sigma a_k x_1{}^k$ converges, then $\Sigma a_k x^k$ converges absolutely for $|x| < |x_1|$. Try to prove this now.

Optional | 24. Indicate how a conditionally convergent series may be rearranged
(a) to converge to an arbitrary real number l; (b) to diverge to $+\infty$;
(c) to diverge to $-\infty$.
HINT: collect the positive terms

$$p_1, p_2, p_3, \ldots, p_k, \ldots$$

and also the negative terms

$$n_1, n_2, n_3, \ldots, n_k, \ldots$$

in the order in which they appear in the original series.

13.6   Taylor Polynomials in x; Taylor Series in Powers of x

Taylor Polynomials in x

We begin with a function f that is differentiable at 0. The linear function that best
approximates f at 0 is the function

$$P_1(x) = f(0) + f'(0)x.$$

At 0 it has the same value as f and also the same derivative:

$$P_1(0) = f(0), \quad P_1'(0) = f'(0).$$

If f has two derivatives at 0, then we can get a better approximation by using the
quadratic polynomial

$$P_2(x) = f(0) + f'(0)x + f''(0)\, \frac{x^2}{2!}.$$

Here, as you can check,

$$P_2(0) = f(0), \quad P_2'(0) = f'(0), \quad \text{and} \quad P_2''(0) = f''(0).$$

If f has three derivatives at 0, then we can use the cubic polynomial

$$P_3(x) = f(0) + f'(0)x + \frac{f''(0)}{2!}\, x^2 + \frac{f'''(0)}{3!}\, x^3$$

and thereby take the third derivative into account. Here

$$P_3(0) = f(0), \quad P_3'(0) = f'(0), \quad P_3''(0) = f''(0), \quad \text{and} \quad P_3'''(0) = f'''(0).$$

More generally, for a function f with n derivatives at 0, we can use the polynomial

$$P_n(x) = f(0) + f'(0)x + \frac{f''(0)}{2!}\, x^2 + \cdots + \frac{f^{(n)}(0)}{n!}\, x^n.$$

This is the polynomial of degree n whose first n derivatives agree with those of
f at 0:

$$P_n(0) = f(0), \quad P_n'(0) = f'(0), \quad P_n''(0) = f''(0), \quad \ldots, P_n^{(n)}(0) = f^{(n)}(0).$$

The polynomials $P_1(x)$, $P_2(x)$, $\ldots$, $P_n(x)$ that we have just described are
called *Taylor polynomials*, after the English mathematician Brook Taylor
(1685–1731). The groundwork for all that we discuss here was laid by Taylor in
1712.

The fact that the Taylor polynomials

$$P_n(x) = f(0) + f'(0)x + \frac{f''(0)}{2!} x^2 + \cdots + \frac{f^{(n)}(0)}{n!} x^n$$

approximate $f(x)$ is of little practical use unless we know something about the accuracy of the approximations. The key to this question of accuracy is Taylor's theorem.

Theorem 13.6.1 Taylor's Theorem

Let I be an interval containing the number 0. If f has n continuous derivatives on I, then for each x in I

$$f(x) = f(0) + f'(0)x + \frac{f''(0)}{2!} x^2 + \cdots + \frac{f^{(n-1)}(0)}{(n-1)!} x^{n-1} + R_n(x)$$

where the remainder $R_n(x)$ is given by the formula

$$R_n(x) = \frac{1}{(n-1)!} \int_0^x f^{(n)}(t)(x-t)^{n-1} \, dt.$$

PROOF.   Integration by parts (see Exercise 27 at the end of the section) gives

$$f'(0)x = \int_0^x f'(t) \, dt - \int_0^x f''(t)(x-t) \, dt,$$

$$\frac{f''(0)}{2!} x^2 = \int_0^x f''(t)(x-t) \, dt - \frac{1}{2!} \int_0^x f'''(t)(x-t)^2 \, dt,$$

$$\frac{f'''(0)}{3!} x^3 = \frac{1}{2!} \int_0^x f'''(t)(x-t)^2 \, dt - \frac{1}{3!} \int_0^x f^{(4)}(t)(x-t)^3 \, dt,$$

$$\vdots$$

$$\frac{f^{(n-1)}(0)}{(n-1)!} x^{n-1} = \frac{1}{(n-2)!} \int_0^x f^{(n-1)}(t)(x-t)^{n-2} \, dt - \frac{1}{(n-1)!} \int_0^x f^{(n)}(t)(x-t)^{n-1} \, dt.$$

We now add these equations. The sum on the left is simply

$$f'(0)x + \frac{f''(0)}{2!} x^2 + \frac{f'''(0)}{3!} x^3 + \cdots + \frac{f^{(n-1)}(0)}{(n-1)!} x^{n-1}.$$

The sum on the right telescopes to

$$\int_0^x f(t) \, dt - \frac{1}{(n-1)!} \int_0^x f^{(n)}(t)(x-t)^{n-1} \, dt = f(x) - f(0) - R_n(x).$$

The result is now obvious. $\square$

Taylor's formula can be written

$$f(x) = P_{n-1}(x) + R_n(x)$$

where

$$R_n(x) = \frac{1}{(n-1)!} \int_0^x f^{(n)}(t)(x-t)^{n-1}\, dt.$$

For most purposes it is convenient to write the remainder $R_n(x)$ in the manner of Lagrange†:

(13.6.2) $\boxed{R_n(x) = \frac{f^{(n)}(c_n)}{n!}\, x^n \quad \text{with } c_n \text{ between 0 and } x.}$

So as not to obscure the main ideas by too many technical arguments, we have placed the proof of Lagrange's form of the remainder in the supplement to this section.

Taylor Series in Powers of x

If we adopt the conventions that $0! = 1$ and that the zeroth derivative of f is f itself, then we can write Taylor's formula in the form

$$f(x) = \sum_{k=0}^{n-1} \frac{f^{(k)}(0)}{k!}\, x^k + R_n(x).$$

Of special interest is the case in which the function has derivatives of all orders. In this case Taylor's formula is valid for all positive integers n. If it happens that $R_n(x) \to 0$, then

$$f(x) = \lim_{n \to \infty} \sum_{k=0}^{n-1} \frac{f^{(k)}(0)}{k!}\, x^k$$

and we have the *Taylor series expansion*

(13.6.3) $\boxed{f(x) = \sum_{k=0}^{\infty} \frac{f^{(k)}(0)}{k!}\, x^k.}$

Such Taylor series are often called Maclaurin series after Colin Maclaurin, a Scottish mathematician (1698–1746). The name Maclaurin remains attached to these series, although Taylor considered them some twenty years before Maclaurin.

It is time for some examples.

Example

(13.6.4) $\boxed{e^x = \sum_{k=0}^{\infty} \frac{x^k}{k!} = 1 + x + \frac{x^2}{2!} + \frac{x^3}{3!} + \cdots \qquad \text{for all real } x.}$

PROOF.- Let x be a fixed real number. Setting

$$f(t) = e^t,$$

---

† Joseph Louis Lagrange, French mathematician (1736–1813).

we find that

$$f^{(k)}(t) = e^t \qquad \text{for all } k.$$

This tells us that for all k

$$f^{(k)}(0) = e^0 = 1.$$

The coefficients in (13.6.4) are thus the Taylor coefficients.

To verify that the Taylor series does converge to e^x, we must verify that

$$R_n(x) \to 0.$$

To this end we let M be the maximum value of the exponential function on the interval joining 0 to x. For all t in that interval

$$|f^{(n)}(t)| = e^t \leq M.$$

Using Lagrange's form of the remainder, we have

$$|R_n(x)| = \frac{|e^{c_n}|}{n!} |x|^n \leq M \frac{|x|^n}{n!}.$$

By (12.3.3) we know that $|x|^n/n! \to 0$. It follows that $R_n(x) \to 0$. $\square$

To say that

$$e^x = \sum_{k=0}^{\infty} \frac{x^k}{k!} = 1 + x + \frac{x^2}{2!} + \frac{x^3}{3!} + \cdots$$

is to say that the Taylor polynomials

$$P_0(x) = 1,$$
$$P_1(x) = 1 + x,$$
$$P_2(x) = 1 + x + \frac{x^2}{2!},$$
$$P_3(x) = 1 + x + \frac{x^2}{2!} + \frac{x^3}{3!},$$
$$\text{etc.}$$

converge to e^x. The convergence is illustrated in Figure 13.6.1.

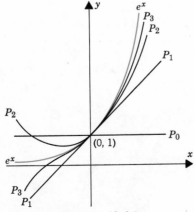

FIGURE 13.6.1

Example

(13.6.5) $$\sin x = \sum_{k=0}^{\infty} \frac{(-1)^k}{(2k+1)!} x^{2k+1} = x - \frac{x^3}{3!} + \frac{x^5}{5!} - \frac{x^7}{7!} + \cdots \qquad \text{for all real } x$$

and

(13.6.6) $$\cos x = \sum_{k=0}^{\infty} \frac{(-1)^k}{(2k)!} x^{2k} = 1 - \frac{x^2}{2!} + \frac{x^4}{4!} - \frac{x^6}{6!} + \cdots \qquad \text{for all real } x.$$

PROOF.   We focus on the sine function and leave the cosine as an exercise. We take x as a fixed real number and let

$$f(t) = \sin t.$$

Differentiation gives

$$f'(t) = \cos t, \qquad f''(t) = -\sin t, \qquad f'''(t) = -\cos t, \qquad f^{(4)}(t) = \sin t.$$

More generally,

$$f^{(2k)}(t) = (-1)^k \sin t, \qquad f^{(2k+1)}(t) = (-1)^k \cos t.$$

Since

$$f^{(2k)}(0) = 0 \qquad \text{and} \qquad f^{(2k+1)}(0) = (-1)^k,$$

you can see that the coefficients in (13.6.5) are actually the Taylor coefficients. Since for all t, $|f^{(n)}(t)| \le 1$, Lagrange's form of the remainder gives

$$|R_n(x)| = \frac{|f^{(n)}(c_n)|}{n!} |x|^n \le \frac{|x|^n}{n!}.$$

Since $|x|^n/n! \to 0$, we have $R_n(x) \to 0$. This tells us that the series does converge to $\sin x$.   □

It's important to keep in mind the role played by the remainder term $R_n(x)$. We can form a Taylor series

$$\sum_{k=0}^{\infty} \frac{f^{(k)}(0)}{k!} x^k$$

for any function f with derivatives of all orders at $x = 0$, but such a series need not converge for any $x \ne 0$. Even if it does converge, the sum need not be $f(x)$. (See Exercise 28.) The Taylor series converges to $f(x)$ only if the remainder term $R_n(x)$ tends to 0.

Some Numerical Calculations

The fact that the Taylor polynomial

$$P_{n-1}(x) = f(0) + f'(0)x + \frac{f''(0)}{2!} x^2 + \cdots + \frac{f^{(n-1)}(x)}{(n-1)!} x^{n-1}$$

approximates $f(x)$ to within $R_n(x)$ enables us to calculate certain functional values as accurately as we wish. Below we give some sample calculations. For ready reference we list some values of $n!$ and $1/n!$ in table form.

TABLE 13.6.1

$n!$
$2! = 2$
$3! = 6$
$4! = 24$
$5! = 120$
$6! = 720$
$7! = 5,040$
$8! = 40,320$

TABLE 13.6.2

$\dfrac{1}{n!}$	
$0.50000 = \dfrac{1}{2!} = 0.50000$	$0.00833 < \dfrac{1}{5!} < 0.00834$
$0.16666 < \dfrac{1}{3!} < 0.16667$	$0.00138 < \dfrac{1}{6!} < 0.00139$
$0.04166 < \dfrac{1}{4!} < 0.04167$	$0.00019 < \dfrac{1}{7!} < 0.00020$

Problem.   Estimate e within 0.001.

SOLUTION.   You are already familiar with the series expansion

$$e^x = 1 + x + \frac{x^2}{2!} + \frac{x^3}{3!} + \cdots + \frac{x^{n-1}}{(n-1)!} + \cdots .$$

Taking $x = 1$, we have

$$e = 1 + 1 + \frac{1}{2!} + \frac{1}{3!} + \cdots + \frac{1}{(n-1)!} + \cdots .$$

To estimate e within 0.001, we need only consider the expansion to the point where $|R_n(1)| < 0.001$. Since

$$|R_7(1)| = \frac{e^{c_7}}{7!} < \frac{e}{7!} < 3(0.0002) = 0.0006,$$
$$\underset{\text{\small $0 < c_7 < 1$ and e^x increases}}{\uparrow}$$

we know that

$$P_6(1) = 1 + 1 + \frac{1}{2!} + \frac{1}{3!} + \frac{1}{4!} + \frac{1}{5!} + \frac{1}{6!} = \frac{1957}{720}$$

differs from e by less than 0.0006. To turn this into a decimal estimate for e, note that

$$2.7180 < \frac{1957}{720} < 2.7181.$$

Within the prescribed limits of accuracy we can take $e \cong 2.718$. $\square$

Problem.   Estimate $e^{0.2}$ within 0.001.

SOLUTION.   Here we take

$$1 + 0.2 + \frac{(0.2)^2}{2!} + \frac{(0.2)^3}{3!} + \cdots + \frac{(0.2)^{n-1}}{(n-1)!} + \cdots$$

to the point where $|R_n(0.2)| < 0.001$. Since

$$|R_4(0.2)| = \frac{|e^{c_4}|}{4!}\,|0.2|^4 \le \frac{e^{0.2}16}{240{,}000} < \frac{(3)(16)}{240{,}000} < 0.001,$$

we can conclude that

$$1 + 0.2 + \frac{(0.2)^2}{2!} + \frac{(0.2)^3}{3!} = \frac{7326}{6000} = 1.221$$

differs from $e^{0.2}$ by less than 0.001. $\square$

Problem.   Estimate sin 0.5 within 0.001.

SOLUTION.   Recall the series expansion

$$\sin x = x - \frac{x^3}{3!} + \frac{x^5}{5!} - \frac{x^7}{7!} + \cdots$$

and the remainder estimate

$$|R_n(x)| \le \frac{|x|^n}{n!}.$$

Note that

$$|R_5(0.5)| \le \frac{(0.5)^5}{5!} < \frac{1}{32}\,(0.00834) < 0.0003.$$

It follows that

$$\overset{\text{the coefficient of } x^4 \text{ is } 0}{P_4(0.5) = P_3(0.5)} = 0.5 - \frac{(0.5)^3}{3!} = \frac{23}{48}$$

differs from sin 0.5 by less than 0.0003. To obtain a decimal estimate note that

$$0.4791 < \tfrac{23}{48} < 0.4792.$$

Within the prescribed limits of accuracy we can take sin $0.5 \cong 0.479$. $\square$

Exercises

Find the Taylor polynomial $P_5(x)$ for each of the following functions.

*1. e^{-x}.   2. e^{2x}.   *3. arc tan x.   4. arc sin x.
*5. $(1 + x)^{-1}$.   6. tan x.   *7. x^2 tan x.   8. e^x sin x.

Find the Taylor polynomial $P_4(x)$ for each of the following functions.

*9. sec x.   10. $x - \cos x$.   *11. log $(1 + x)$.   12. $\sqrt{1 + x}$.
*13. log cos x.   14. $\sqrt{\cos x}$.   *15. e^x tan x.   16. $\sqrt{\sec x}$.

Use Taylor polynomials to estimate the following within 0.01.

*17. $\sqrt{e}$.   18. $\sin 0.3$.   *19. $\cos 1$.   20. $\cos 0.5$.

*21. arc $\sin 0.9$ 22. $\sin 1$.   *23. $e^{2.1}$.   24. $e^{-1.4}$.

25. Show that a polynomial $P(x) = a_0 + a_1 x + \cdots + a_n x^n$ is its own Taylor series.

26. Show that

$$\cos x = \sum_{k=0}^{\infty} \frac{(-1)^k}{(2k)!} x^{2k} \qquad \text{for all real } x.$$

27. Verify the identity

$$\frac{f^{(k)}(0)}{k!} x^k = \frac{1}{(k-1)!} \int_0^x f^{(k)}(t)(x-t)^{k-1} \, dt - \frac{1}{k!} \int_0^x f^{(k+1)}(t)(x-t)^k \, dt$$

by computing the second integral by parts.

28. Show that

the Taylor series $\displaystyle\sum_{n=0}^{\infty} \frac{f^{(n)}(0)}{n!} x^n$ of the function $f(x) = \left\{ \begin{array}{cc} e^{-1/x^2}, & x \neq 0 \\ 0, & x = 0 \end{array} \right]$

is identically zero and hence does not represent the function except at $x = 0$.

Optional | 29. Find the Taylor polynomial

     *(a) $P_5(x)$ for $\sinh x$.   (b) $P_4(x)$ for $\cosh x$.

30. Show that e is irrational by following these steps.
(1) Take the expansion

$$e = \sum_{k=0}^{\infty} \frac{1}{k!}$$

and show that the qth partial sum

$$s_q = \sum_{k=0}^{q} \frac{1}{k!}$$

satisfies the inequality

$$0 < q! \, (e - s_q) < \frac{1}{q}.$$

(2) Show that $q! \, s_q$ is an integer and argue that, if e were of the form p/q, then we would have $q! \, (e - s_q)$ as a positive integer less than 1.

Supplement to Section 13.6

Before proving the validity of Lagrange's form of the remainder, we prove a simple extension of the first mean-value theorem for integrals.

Theorem 13.6.7

If f and g are continuous on $[a, b]$ and g keeps a constant sign on $[a, b]$, then

$$\int_a^b f(t)g(t)\,dt = f(c)\int_a^b g(t)\,dt$$

for some number c in the interval $[a, b]$.

PROOF.   Set

m = minimum value of f on $[a, b]$, M = maximum value of f on $[a, b]$.

For each number t in $[a, b]$

$$m \leq f(t) \leq M$$

and, if g remains nonnegative,

$$mg(t) \leq f(t)g(t) \leq Mg(t).$$

(If g remains nonpositive, the inequality is reversed and the rest of the argument must be adjusted accordingly.)

By integrating the last inequality from a to b, we get

$$\int_a^b mg(t)\,dt \leq \int_a^b f(t)g(t)\,dt \leq \int_a^b Mg(t)\,dt$$

and therefore

$$m\int_a^b g(t)\,dt \leq \int_a^b f(t)g(t)\,dt \leq M\int_a^b g(t)\,dt.$$

If the integral of g is 0, then

$$\int_a^b f(t)g(t)\,dt = 0$$

and the result holds. If the integral of g is not 0, then

$$m \leq \frac{\displaystyle\int_a^b f(t)g(t)\,dt}{\displaystyle\int_a^b g(t)\,dt} \leq M,$$

and since f, being continuous, skips no values, there must exist c in $[a, b]$ such that

$$f(c) = \frac{\displaystyle\int_a^b f(t)g(t)\,dt}{\displaystyle\int_a^b g(t)\,dt}.$$

Obviously then

$$\int_a^b f(t)g(t)\,dt = f(c)\int_a^b g(t)\,dt. \quad \square$$

Lagrange's form of the remainder is now easy to obtain. For $x > 0$,

$$R_n(x) = \frac{1}{(n-1)!}\int_0^x f^{(n)}(t)(x-t)^{n-1}\,dt$$

$$\underset{\uparrow}{=} \frac{f^{(n)}(c_n)}{(n-1)!}\int_0^x (x-t)^{n-1}\,dt = \frac{f^{(n)}(c_n)}{(n-1)!}\left[-\frac{(x-t)^n}{n}\right]_0^x = \frac{f^{(n)}(c_n)}{n!}x^n.$$

here is where we use the
last theorem

For $x < 0$,

$$R_n(x) = \frac{1}{(n-1)!}\int_0^x f^{(n)}(t)(x-t)^{n-1}\,dt$$

$$= \frac{1}{(n-1)!}\int_x^0 f^{(n)}(t)[-(x-t)^{n-1}]\,dt$$

$$= \frac{f^{(n)}(c_n)}{(n-1)!}\int_x^0 [-(x-t)^{n-1}]\,dt = \frac{f^{(n)}(c_n)}{(n-1)!}\left[\frac{(x-t)^n}{n}\right]_x^0 = \frac{f^{(n)}(c_n)}{n!}x^n. \quad \square$$

13.7 Taylor Polynomials in $x - a$; Taylor Series in Powers of $x - a$

So far we've considered series expansions only in powers of x. Here we generalize to expansions in powers of $x - a$ where a is an arbitrary real number. We begin with a more general version of Taylor's theorem.

Theorem 13.7.1 Taylor's Theorem

Let I be an interval containing the number a. If g has n continuous derivatives on I, then

$$g(x) = g(a) + g'(a)(x-a) + \cdots + \frac{g^{(n-1)}(a)}{(n-1)!}(x-a)^{n-1} + R_n(x)$$

where

$$R_n(x) = \frac{1}{(n-1)!}\int_a^x g^{(n)}(s)(x-s)^{n-1}\,ds.$$

In this more general setting the Lagrange form of the remainder can be written

$$(13.7.2) \qquad \boxed{R_n(x) = \frac{g^{(n)}(c_n)}{n!}(x-a)^n \qquad \text{with } c_n \text{ between } a \text{ and } x.}$$

If $R_n(x) \to 0$, then

$$g(x) = g(a) + g'(a)(x - a) + \frac{g''(a)}{2!}(x - a)^2 + \frac{g'''(a)}{3!}(x - a)^3 + \cdots.$$

In sigma notation

(13.7.3)
$$g(x) = \sum_{k=0}^{\infty} \frac{g^{(k)}(a)}{k!}(x - a)^k.$$

This is called the Taylor series expansion of $g(x)$ *in powers of* $(x - a)$.

All this is entirely analogous to what you saw in the last section. These results follow from those by setting

$$f(x - a) = g(x).$$

Note that then $f^{(k)}(x - a) = g^{(k)}(x)$ and therefore $f^{(k)}(0) = g^{(k)}(a)$. □

Problem. Expand

$$g(x) = e^{\frac{1}{2}x} \qquad \text{in powers of } x - 3$$

and show that the series represents the function for all real x.

SOLUTION. Differentiation gives

$$g'(x) = \tfrac{1}{2}e^{\frac{1}{2}x}, \qquad g''(x) = \tfrac{1}{4}e^{\frac{1}{2}x}, \qquad g'''(x) = \tfrac{1}{8}e^{\frac{1}{2}x}$$

and, in general,

$$g^{(k)}(x) = \frac{1}{2^k}\, e^{\frac{1}{2}x}.$$

The Taylor series for $g(x) = e^{\frac{1}{2}x}$ in powers of $x - 3$ takes the form

$$\sum_{k=0}^{\infty} \frac{g^{(k)}(3)}{k!}(x - 3)^k = \sum_{k=0}^{\infty} \frac{e^{\frac{3}{2}}}{2^k k!}(x - 3)^k = e^{\frac{3}{2}} \sum_{k=0}^{\infty} \frac{1}{2^k k!}(x - 3)^k.$$

To show that this series represents the function for all real x, we must show that the remainder $R_n(x)$ tends to zero for all real x. The remainder may be written

$$R_n(x) = \frac{g^{(n)}(c_n)}{n!}(x - 3)^n = \frac{e^{\frac{1}{2}c_n}}{2^n n!}(x - 3)^n \quad \text{with } c_n \text{ between } x \text{ and } 3.$$

For $x \le 3$ we have

$$|R_n(x)| = \frac{e^{\frac{1}{2}c_n}}{2^n n!}|x - 3|^n \le \frac{e^{\frac{3}{2}x}}{2^n n!}|x - 3|^n \to 0,$$

and for $x \ge 3$ we have

$$|R_n(x)| = \frac{e^{\frac{1}{2}c_n}}{2^n n!}|x - 3|^n \le \frac{e^{\frac{1}{2}x}}{2^n n!}|x - 3|^n \to 0. \quad \square$$

REMARK.   An easier way to expand $e^{\frac{1}{2}x}$ in powers of $x - 3$ is to note that

$$\tfrac{1}{2}x = \tfrac{3}{2} + \tfrac{1}{2}(x - 3).$$

We can therefore write

$$e^{\frac{1}{2}x} = e^{\frac{3}{2}+\frac{1}{2}(x-3)} = e^{\frac{3}{2}}e^{\frac{1}{2}(x-3)}.$$

Since

$$e^t = \sum_{k=0}^{\infty} \frac{1}{k!} t^k \qquad \text{for all real } t, \quad \text{(the exponential series)}$$

we have here

$$e^{\frac{1}{2}x} = e^{\frac{3}{2}}e^{\frac{1}{2}(x-3)} = e^{\frac{3}{2}} \sum_{k=0}^{\infty} \frac{1}{k!} \left(\frac{x-3}{2}\right)^k = e^{\frac{3}{2}} \sum_{k=0}^{\infty} \frac{1}{2^k k!} (x-3)^k \qquad \text{for all real } x. \quad \square$$

Problem.   Take $a \neq b$ and expand

$$g(x) = \frac{1}{b-x} \quad \text{in powers of } x - a.$$

For what values of x does the series represent the function?

SOLUTION.   Differentiation gives

$$g'(x) = \frac{1}{(b-x)^2}, \qquad g''(x) = \frac{2}{(b-x)^3}, \qquad g'''(x) = \frac{3 \cdot 2}{(b-x)^4}$$

and, in general,

$$g^{(k)}(x) = \frac{k!}{(b-x)^{k+1}}.$$

The Taylor series in powers of $x - a$ can be written

$$\sum_{k=0}^{\infty} \frac{g^{(k)}(a)}{k!} (x-a)^k = \sum_{k=0}^{\infty} \frac{1}{(b-a)^{k+1}} (x-a)^k = \frac{1}{b-a} \sum_{k=0}^{\infty} \frac{1}{(b-a)^k} (x-a)^k.$$

The problem of applicability remains. To determine the values of x for which this series represents the function, we would have to determine the values of x for which the remainder $R_n(x)$ tends to zero. We can avoid such calculations by noting that

$$g(x) = \frac{1}{b-x} = \frac{1}{(b-a) - (x-a)} = \frac{1}{b-a} \left[\frac{1}{1 - \left(\dfrac{x-a}{b-a}\right)} \right].$$

Since

$$\frac{1}{1-t} = \sum_{k=0}^{\infty} t^k \qquad \text{for } |t| < 1, \qquad \text{(the geometric series)}$$

you can see that

$$g(x) = \frac{1}{b-a} \sum_{k=0}^{\infty} \left(\frac{x-a}{b-a}\right)^k = \frac{1}{b-a} \sum_{k=0}^{\infty} \frac{1}{(b-a)^k} (x-a)^k \qquad \text{for } \left|\frac{x-a}{b-a}\right| < 1.$$

Since

$$\left| \frac{x-a}{b-a} \right| < 1 \quad \text{iff} \quad |x-a| < |b-a| \quad \text{iff} \quad a - |b-a| < x < a + |b-a|,$$

the series represents the function on the interval $(a - |b-a|, a + |b-a|)$. $\square$

Exercises

*1. Expand $3x^3 - 2x^2 + 4x + 1$ in powers of $x - 1$.
  2. Expand $x^4 - x^3 + x^2 - x + 1$ in powers of $x - 2$.
*3. Expand $2x^5 + x^2 - 3x - 5$ in powers of $x + 1$.
  4. Expand x^{-1} in powers of $x - 1$.
*5. Expand $(1 + x)^{-1}$ in powers of $x - 1$.
  6. Expand $(b + x)^{-1}$ in powers of $x - a$.
*7. (a) Verify the Taylor expansion

$$e^x = e^a \sum_{k=0}^{\infty} \frac{(x-a)^k}{k!}.$$

  (b) Write the remainder $R_n(x)$ in the form of Lagrange.
  (c) Show that $R_n(x) \to 0$ for all real x.
  (d) Use the Taylor expansion to show that $e^{x_1 + x_2} = e^{x_1} e^{x_2}$.
8. (a) Derive the Taylor expansions

$$\sin x = \sin a + (x-a) \cos a - \frac{(x-a)^2}{2!} \sin a - \frac{(x-a)^3}{3!} \cos a + \cdots,$$

$$\cos x = \cos a - (x-a) \sin a - \frac{(x-a)^2}{2!} \cos a + \frac{(x-a)^3}{3!} \sin a + \cdots.$$

  (b) Show that both series are absolutely convergent for all real x.
  (c) As noted earlier (Section 13.5) Riemann proved that the order of the terms
      of an absolutely convergent series may be changed without altering the
      sum of the series. Use Riemann's discovery and the Taylor expansions of
      part (a) to derive the addition formulas

$$\sin (x_1 + x_2) = \sin x_1 \cos x_2 + \cos x_1 \sin x_2,$$
$$\cos (x_1 + x_2) = \cos x_1 \cos x_2 - \sin x_1 \sin x_2.$$

9. Set $f(x) = \log x$ and choose $a > 0$.
  *(a) Find the nth derivative $f^{(n)}(a)$.
  *(b) Expand $\log x$ in powers of $x - a$.
  (c) Write the remainder $R_n(x)$ in the form of Lagrange.
  (d) Write the remainder $R_n(x)$ in its integral form.
  (e) Show that the expansion in powers of $x - a$ is valid for all $x \in (0, 2a]$.
      HINT: for $x \in [a, 2a]$ use Lagrange's form of $R_n(x)$; for $x \in (0, a)$ use the
      integral representation.

13.8   The Logarithm and the Arc Tangent; Computing π

Series expansions in powers of x for $\log(1 + x)$ and arc tan x can be derived by the methods of Section 13.6. It's easier, however, to follow the method described below.

A Series for log (1 + x)

For $x \neq 1$

$$1 + x + x^2 + \cdots + x^{n-1} = \frac{1 - x^n}{1 - x} = \frac{1}{1 - x} - \frac{x^n}{1 - x},$$

see proof of 13.2.2

and thus

$$\frac{1}{1 - x} = 1 + x + x^2 + \cdots + x^{n-1} + \frac{x^n}{1 - x}.$$

Setting $x = -t$, we have

$$\frac{1}{1 + t} = 1 + (-t) + (-t)^2 + \cdots + (-t)^{n-1} + \frac{(-t)^n}{1 + t}.$$

It follows that

$$\log(1 + x) = \int_0^x \frac{dt}{1 + t} = x - \frac{x^2}{2} + \frac{x^3}{3} - \cdots + (-1)^{n-1}\frac{x^n}{n} + T_n(x),$$

where

$$T_n(x) = (-1)^n \int_0^x \frac{t^n}{1 + t}\, dt.$$

Below we show that

$$T_n(x) \to 0 \qquad \text{for} \quad -1 < x \leq 1.$$

For $0 \leq x \leq 1$, we have

$$|T_n(x)| = \int_0^x \frac{t^n}{1 + t}\, dt \leq \int_0^x t^n\, dt = \frac{x^{n+1}}{n + 1} \to 0.$$

For $-1 < x < 0$, we have

$$|T_n(x)| = \left| \int_0^x \frac{t^n}{1 + t}\, dt \right| \leq \int_x^0 \frac{|t|^n}{|1 + t|}\, dt = \int_x^0 \frac{(-t)^n}{1 + t}\, dt$$

$$\leq \int_x^0 \frac{(-t)^n}{1 + x}\, dt = \frac{1}{1 + x}\left[\frac{(-x)^{n+1}}{n + 1} \right] \to 0.$$

You have now seen that $T_n(x) \to 0$ for $-1 < x \leq 1$. It follows that

(13.8.1) $\log(1 + x) = x - \dfrac{x^2}{2} + \dfrac{x^3}{3} - \dfrac{x^4}{4} + \cdots$ for $-1 < x \leq 1$.

At $x = 1$ the series gives an intriguing result:

$$\log 2 = 1 - \tfrac{1}{2} + \tfrac{1}{3} - \tfrac{1}{4} + \cdots .$$

There is no hope of extending the relation to $x = -1$ since $\log 0$ is not defined.

Problem. Estimate $\log 1.3$ within 0.01 by using (13.8.1).

SOLUTION

$$\log 1.3 = \log (1 + 0.3) = 0.3 - \tfrac{1}{2}(0.3)^2 + \tfrac{1}{3}(0.3)^3 - \tfrac{1}{4}(0.3)^4 + \cdots .$$

Since this is an alternating series and the terms form a decreasing sequence, we know that $\log 1.3$ lies between consecutive odd and even partial sums.

The first term less than 0.01 is

$$\tfrac{1}{3}(0.3)^3 = \tfrac{1}{3}(0.027) = 0.009.$$

The relation

$$0.3 - \tfrac{1}{2}(0.3)^2 < \log 1.3 < 0.3 - \tfrac{1}{2}(0.3)^2 + \tfrac{1}{3}(0.3)^3$$

gives

$$0.255 < \log 1.3 < 0.264.$$

Within the prescribed limits of accuracy we can take

$$\log 1.3 \cong 0.26. \quad \square\dagger$$

A Series for arc tan x

The development here is so similar to the one we just used for $\log (1 + x)$ that we can leave out some of the details. We begin with

$$\frac{1}{1 + t^2} = \frac{1}{1 - (-t^2)} = 1 + (-t^2) + (-t^2)^2 + \cdots + (-t^2)^{n-1} + \frac{(-t^2)^n}{1 + t^2}.$$

Since

$$\arctan x = \int_0^x \frac{dt}{1 + t^2},$$

we have

$$\arctan x = x + (-1)\frac{x^3}{3} + (-1)^2\frac{x^5}{5} + \cdots + (-1)^{n-1}\frac{x^{2n-1}}{2n-1} + S_n(x),$$

where

$$S_n(x) = (-1)^n \int_0^x \frac{t^{2n}}{1 + t^2}\, dt.$$

---

† A much more effective tool for computing logarithms is discussed in the exercises.

If $x \in [-1, 1]$, then $S_n(x) \to 0$. (Exercise 12) It follows that

(13.8.2)
$$\text{arc tan } x = x - \frac{x^3}{3} + \frac{x^5}{5} - \frac{x^7}{7} + \cdots \qquad \text{for} \quad -1 \leq x \leq 1.$$

This series was known to the Scottish mathematician James Gregory in 1671.
Since arc tan $1 = \frac{1}{4}\pi$, we have

$$\tfrac{1}{4}\pi = 1 - \tfrac{1}{3} + \tfrac{1}{5} - \tfrac{1}{7} + \tfrac{1}{9} - \cdots .$$

Leibniz seems to have found this formula in 1673. It is an elegant formula for π, but it converges too slowly to be efficient for computational purposes.

Computing π

A more convenient way to compute π is to use the relation

(13.8.3)
$$\tfrac{1}{4}\pi = 4 \text{ arc tan } \tfrac{1}{5} - \text{arc tan } \tfrac{1}{239}.$$

This relation was discovered in 1706 by John Machin, a Scotsman. It can be verified by repeated application of the addition formula

$$\tan (A + B) = \frac{\tan A + \tan B}{1 - \tan A \tan B}. \qquad \text{(See Exercise 14.)}$$

The arc tangent series (13.8.2) gives

$$\text{arc tan } \tfrac{1}{5} = \tfrac{1}{5} - \tfrac{1}{3}(\tfrac{1}{5})^3 + \tfrac{1}{5}(\tfrac{1}{5})^5 - \tfrac{1}{7}(\tfrac{1}{5})^7 + \cdots$$

and

$$\text{arc tan } \tfrac{1}{239} = \tfrac{1}{239} - \tfrac{1}{3}(\tfrac{1}{239})^3 + \tfrac{1}{5}(\tfrac{1}{239})^5 - \tfrac{1}{7}(\tfrac{1}{239})^7 + \cdots .$$

Since these are alternating series and the terms in each case form a decreasing sequence, we know that

$$\tfrac{1}{5} - \tfrac{1}{3}(\tfrac{1}{5})^3 \leq \text{arc tan } \tfrac{1}{5} \leq \tfrac{1}{5} - \tfrac{1}{3}(\tfrac{1}{5})^3 + \tfrac{1}{5}(\tfrac{1}{5})^5$$

and

$$\tfrac{1}{239} - \tfrac{1}{3}(\tfrac{1}{239})^3 \leq \text{arc tan } \tfrac{1}{239} \leq \tfrac{1}{239}.$$

With these inequalities together with (13.8.3) and a little arithmetic, you can verify that

$$3.14 < \pi < 3.147.$$

By using six terms of the series for arc tan $\tfrac{1}{5}$ and still only two of the series for arc tan $\tfrac{1}{239}$, one can show that

$$3.14159262 < \pi < 3.14159267.$$

Greater accuracy can be obtained by taking more terms into account. Here for instance is π to to twenty decimal places:

$$\pi = 3.14159\ 26535\ 89793\ 23846 \cdot\ \cdot\ \cdot\ .$$

Exercises

Estimate the following within 0.01 by using (13.8.1) or (13.8.2).

*1. log 1.1.   2. log 1.2.   *3. log 0.8.
4. log 1.4.   *5. arc tan 0.4.   6. arc tan 0.5.

*7. The series we derived for log $(1 + x)$ converges too slowly to be of much practical use. The following logarithm series converges much more quickly:

$$(13.8.4) \quad \log\left(\frac{1 + x}{1 - x}\right) = 2\left(x + \frac{x^3}{3} + \frac{x^5}{5} + \cdot\ \cdot\ \cdot\ \right) \quad \text{for } -1 < x < 1.$$

Derive this series expansion and verify the interval of convergence.

8. Set $x = \frac{1}{3}$ and use the first three nonzero terms of (13.8.4) to estimate log 2. Check this estimate against the table in the back of the book.

*9. Use the first two nonzero terms of (13.8.4) to estimate log 1.3. Check this estimate against the table in the back of the book and compare it to the estimate we obtained earlier by using the first three terms of (13.8.1).

10. Substitute $x = 1/(2n + 1)$ in (13.8.4) to verify that

$$\log (n + 1) = \log n + 2\left[\frac{1}{2n + 1} + \frac{1}{3(2n + 1)^3} + \frac{1}{5(2n + 1)^5} + \cdot\ \cdot\ \cdot\ \right].$$

11. You estimated log 2 in exercise 8. Now use exercise 10 and the properties of the logarithm to estimate *(a) log 3. (b) log 5. *(c) log 7.

12. Prove that

$$\text{if } \ x \in [-1, 1], \quad \text{then} \quad s_n(x) = (-1)^n \int_0^x \frac{t^{2n}}{1 + t^2}\, dt \to 0.$$

*13. Take $a > 0$ and expand log $(a + x)$ in powers of x. For what numbers x is the expansion valid?

*14. Verify that

$$\tfrac{1}{4}\pi = 4 \text{ arc tan } \tfrac{1}{5} - \text{ arc tan } \tfrac{1}{239}$$

by the following three-step computation: (a) compute tan (2 arc tan $\frac{1}{5}$); (b) then tan (4 arc tan $\frac{1}{5}$); (c) finally tan (4 arc tan $\frac{1}{5}$ − arc tan $\frac{1}{239}$).

13.9　Power Series, Part I

You've become familiar with Taylor series

$$\sum_{k=0}^{\infty} \frac{f^{(k)}(a)}{k!}\, (x - a)^k \quad \text{and} \quad \sum_{k=0}^{\infty} \frac{f^{(k)}(0)}{k!}\, x^k.$$

Here we study series of the form

$$\sum_{k=0}^{\infty} a_k(x - a)^k \quad \text{and} \quad \sum_{k=0}^{\infty} a_k x^k$$

without regard to how the coefficients have been generated. Such series are called *power series*.

Since a simple translation converts

$$\sum_{k=0}^{\infty} a_k(x - a)^k \quad \text{into} \quad \sum_{k=0}^{\infty} a_k x^k,$$

we can focus our attention on

$$\sum_{k=0}^{\infty} a_k x^k.$$

When detailed indexing is unnecessary, we will omit it and write

$$\sum a_k x^k.$$

We launch our discussion with a definition.

Definition 13.9.1

A power series $\sum a_k x^k$ is said to converge

(i) at x_1 iff $\sum a_k x_1^k$ converges.
(ii) on the set S iff $\sum a_k x^k$ converges for each $x \in S$.

For determining at what numbers a power series converges, the following theorem is fundamental.

Theorem 13.9.2

If $\sum a_k x^k$ converges at $x_1 \neq 0$, then it converges absolutely for $|x| < |x_1|$.
If $\sum a_k x^k$ diverges at x_1, then it diverges for $|x| > |x_1|$.

PROOF.   If $\sum a_k x_1^k$ converges, then $a_k x_1^k \to 0$. In particular, for k sufficiently large,

$$|a_k x_1^k| \leq 1$$

and thus

$$|a_k x^k| = |a_k x_1^k| \left| \frac{x}{x_1} \right|^k \leq \left| \frac{x}{x_1} \right|^k.$$

For $|x| < |x_1|$, we have

$$\left| \frac{x}{x_1} \right| < 1.$$

The convergence of $\Sigma \, |a_k x^k|$ follows by comparison with the geometric series. This proves the first statement.

Suppose now that $\Sigma \, a_k x_1^k$ diverges. By the previous argument, there cannot exist x with $|x| > |x_1|$ such that $\Sigma \, a_k x^k$ converges. The existence of such an x would imply the absolute convergence of $\Sigma \, a_k x_1^k$. This proves the second statement.   □

From the theorem we just proved you can see that for a power series there are exactly three possibilities:

Possibility I. *The series converges only at 0.* This is what happens with

$$\Sigma \, k^k x^k.$$

The kth term, $k^k x^k$, tends to 0 only if $x = 0$.

Possibility II. *The series converges everywhere absolutely.* This is what happens with the exponential series

$$\Sigma \, \frac{x^k}{k!}.$$

Possibility III. *There exists a positive number r such that the series converges absolutely for $|x| < r$ and diverges for $|x| > r$.* This is what happens with the geometric series

$$\Sigma \, x^k.$$

In this instance, there is absolute convergence for $|x| < 1$ and divergence for $|x| > 1$.

Associated with each case is a *radius of convergence*:

In case I, we say that the radius of convergence is 0.
In case II, we say that the radius of convergence is ∞.
In case III, we say that the radius of convergence is r.

The three cases are pictured in Figure 13.9.1.

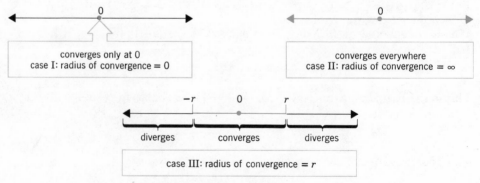

FIGURE 13.9.1

In general the behavior of a power series at $-r$ and at r is not predictable. The series

$$\Sigma \left(\frac{1}{k}\right)^{2} x^{k}, \qquad \Sigma \frac{(-1)^{k}}{k} x^{k}, \qquad \Sigma \frac{1}{k} x^{k}, \qquad \Sigma x^{k}$$

all have radius of convergence 1, but, while the first series converges on all of $[-1, 1]$, the second series converges only on $(-1, 1]$, the third only on $[-1, 1)$, and the fourth only on $(-1, 1)$.

The maximal interval on which a power series converges is called the *interval of convergence*. For a series with infinite radius of convergence, the interval of convergence is $(-\infty, \infty)$. For a series with radius of convergence r, the interval of convergence can be $[-r, r]$, $(-r, r]$, $[-r, r)$, or $(-r, r)$ as you saw. For a series with radius of convergence 0, the interval of convergence reduces to a point, $\{0\}$.

Problem. Verify that

$$(1) \qquad\qquad \Sigma \frac{(-1)^{k}}{k} x^{k}$$

has interval of convergence $(-1, 1]$.

SOLUTION.   First we show that the radius of convergence is 1. We do this by examining the series

$$(2) \qquad\qquad \Sigma \left| \frac{(-1)^{k}}{k} x^{k} \right| = \Sigma \frac{1}{k} |x|^{k}.$$

We will use the ratio test. We set

$$b_{k} = \frac{1}{k} |x|^{k}$$

and note that

$$\frac{b_{k+1}}{b_{k}} = \frac{k}{k+1} \frac{|x|^{k+1}}{|x|^{k}} = \frac{k}{k+1} |x| \rightarrow |x|.$$

By the ratio test, series (2) converges for $|x| < 1$ and diverges for $|x| > 1$.† This shows that series (1) converges absolutely for $|x| < 1$ and diverges for $|x| > 1$. The radius of convergence is therefore 1.

Now we test the endpoints $x = -1$ and $x = 1$. At $x = -1$

$$\Sigma \frac{(-1)^{k}}{k} x^{k} \qquad \text{becomes} \qquad \Sigma \frac{(-1)^{k}}{k} (-1)^{k} = \Sigma \frac{1}{k}.$$

This is the harmonic series which, as you know, diverges. At $x = 1$,

$$\Sigma \frac{(-1)^{k}}{k} x^{k} \qquad \text{becomes} \qquad \Sigma \frac{(-1)^{k}}{k}.$$

This is a convergent alternating series.

---

† We could also have used the root test:

$$(b_{k})^{1/k} = \left| \frac{1}{k} \right|^{1/k} |x| = \frac{1}{k^{1/k}} |x| \rightarrow |x|.$$

We have shown that series (1) converges absolutely for $|x| < 1$, diverges at -1 and converges at 1. The interval of convergence is $(-1, 1]$. □

Problem. Verify that

$$(1) \qquad \Sigma \frac{1}{k^2} x^k$$

has interval of convergence $[-1, 1]$.

SOLUTION.   First we examine the series

$$(2) \qquad \Sigma \left| \frac{1}{k^2} x^k \right| = \Sigma \frac{1}{k^2} |x|^k.$$

Here again we use the ratio test. We set

$$b_k = \frac{1}{k^2} |x|^k$$

and note that

$$\frac{b_{k+1}}{b_k} = \frac{k^2}{(k+1)^2} \frac{|x|^{k+1}}{|x|^k} = \left(\frac{k}{k+1} \right)^2 |x| \to |x|.$$

By the ratio test, (2) converges for $|x| < 1$ and diverges for $|x| > 1$.† This shows that (1) converges absolutely for $|x| < 1$ and diverges for $|x| > 1$. The radius of convergence is therefore 1.

Now for the endpoints. At $x = -1$

$$\Sigma \frac{1}{k^2} x^k \qquad \text{becomes} \qquad \Sigma \frac{1}{k^2} (-1)^k,$$

which expanded looks like this:

$$-1 + \tfrac{1}{4} - \tfrac{1}{9} + \tfrac{1}{16} - \cdots .$$

This is a convergent alternating series. At $x = 1$,

$$\Sigma \frac{1}{k^2} x^k \qquad \text{becomes} \qquad \Sigma \frac{1}{k^2}.$$

This is a convergent p-series. The interval of convergence is therefore the entire closed interval $[-1, 1]$. □

Problem. Find the interval of convergence of

$$(1) \qquad \Sigma \frac{k}{6^k} x^k.$$

---

† Once again we could have used the root test:

$$(b_k)^{1/k} = \frac{1}{k^{2/k}} |x| \to |x|.$$

SOLUTION.   We begin by examining the series

(2) $$\Sigma \left| \frac{k}{6^k} x^k \right| = \Sigma \frac{k}{6^k} |x|^k.$$

We set

$$b_k = \frac{k}{6^k} |x|^k$$

and apply the root test. (The ratio test will also work.) Since

$$b^{1/k} = \tfrac{1}{6} k^{1/k} |x| \to \tfrac{1}{6} |x|,$$

you can see that (2) converges

$$\text{for} \quad \tfrac{1}{6} |x| < 1 \qquad\qquad\qquad (\text{for } |x| < 6)$$

and diverges

$$\text{for} \quad \tfrac{1}{6} |x| > 1. \qquad\qquad\qquad (\text{for } |x| > 6)$$

This shows that (1) converges absolutely for $|x| < 6$ and diverges for $|x| > 6$. The radius of convergence is 6.

It is easy to see that (1) diverges both at -6 and at 6. The interval of convergence is therefore $(-6, 6)$.   □

Problem.   Find the interval of convergence of

(1) $$\Sigma \frac{(2k)!}{(3k)!} x^k.$$

SOLUTION.   We begin by examining

(2) $$\Sigma \left| \frac{(2k)!}{(3k)!} x^k \right| = \Sigma \frac{(2k)!}{(3k)!} |x|^k.$$

Set

$$b_k = \frac{(2k)!}{(3k)!} |x|^k.$$

When factorials are involved, it is generally easier to use the ratio test. Note that

$$\frac{b_{k+1}}{b_k} = \frac{[2(k+1)]!\,(3k)!}{[3(k+1)]!\,(2k)!} \frac{|x|^{k+1}}{|x|^k} = \frac{(2k+2)(2k+1)}{(3k+3)(3k+2)(3k+1)} |x|.$$

As $k \to \infty$, this ratio tends to 0 no matter what x is. By the ratio test, (2) converges for all x and therefore (1) converges absolutely for all x. The radius of convergence is ∞ and the interval of convergence is $(-\infty, \infty)$.   □

Problem.   Find the interval of convergence of

$$\Sigma \, (\tfrac{1}{2}k)^k x^k.$$

SOLUTION.   Since $(\frac{1}{2}k)^k x^k \to 0$ only if $x = 0$, there is no need to invoke the ratio test or the root test. By Theorem 13.2.5 the series can converge only at $x = 0$. There it converges trivially.   □

Exercises

Find the interval of convergence.

*1. $\Sigma\, kx^k$.

2. $\Sigma\, \dfrac{1}{k}\, x^k$.

*3. $\Sigma\, \dfrac{1}{(2k)!}\, x^k$.

*4. $\Sigma\, \dfrac{2^k}{k^2}\, x^k$.

5. $\Sigma\, (-k)^{2k} x^{2k}$.

*6. $\Sigma\, \dfrac{(-1)^k}{\sqrt{k}}\, x^k$.

*7. $\Sigma\, \dfrac{1}{k2^k}\, x^k$.

8. $\Sigma\, \dfrac{1}{k^2 2^k}\, x^k$.

*9. $\Sigma\, \dfrac{1}{k(k+1)}\, x^k$.

*10. $\Sigma\, \dfrac{k^2}{1+k^2}\, x^k$.

11. $\Sigma\, \dfrac{2^k}{\sqrt{k}}\, x^k$.

*12. $\Sigma\, \dfrac{1}{\log k}\, x^k$.

*13. $\Sigma\, \dfrac{k-1}{k}\, x^k$.

14. $\Sigma\, ka^k x^k$.

*15. $\Sigma\, \dfrac{k}{10^k}\, x^k$.

*16. $\Sigma\, \dfrac{(-1)^k}{k^k}\, (x-2)^k$.

17. $\Sigma\, \dfrac{(-1)^k a^k}{k^2}\, (x-a)^k$.

*18. $\Sigma\, \dfrac{1}{(\log k)^k}\, (x-1)^k$.

19. $\Sigma\, \dfrac{2^{1/k} \pi^k}{k(k+1)(k+2)}\, (x-2)^k$.

*20. $\Sigma\, (-1)^k (\frac{2}{3})^k\, (x+1)^k$.

21. $\Sigma\, \dfrac{\log k}{2^k}\, (x-2)^k$.

22. Let $\Sigma\, a_k x^k$ be a power series, and let r be its radius of convergence.
    (a) Given that $|a_k|^{1/k} \to \rho$, show that, if $\rho \neq 0$, then $r = 1/\rho$ and, if $\rho = 0$, then $r = \infty$.
    (b) Given that $|a_{k+1}/a_k| \to \lambda$, show that, if $\lambda \neq 0$, then $r = 1/\lambda$ and, if $\lambda = 0$, then $r = \infty$.

13.10 Power Series, Part II

In the interior of its interval of convergence a power series represents an infinitely differentiable function. As a first step toward proving this, we have the following theorem.

Theorem 13.10.1

If

$$\Sigma\, a_k x^k \quad \text{converges on } (-c, c),$$

then

$$\Sigma\, \frac{d}{dx}\, (a_k x^k) = \Sigma\, ka_k x^{k-1} \quad \text{also converges on } (-c, c).$$

PROOF.   Let's assume that $\Sigma\, a_k x^k$ converges on $(-c, c)$. By Theorem 13.9.2 we know that it converges there absolutely.

Now take $t \in (-c, c)$ and choose $\epsilon > 0$ such that

$$|t| < |t| + \epsilon < |c|.$$

Since $|t| + \epsilon$ is within the interval of convergence,

$$\Sigma\, |a_k(|t| + \epsilon)^k| \quad \text{converges.}$$

In Exercise 13 you are asked to show that for all k sufficiently large

$$|kt^{k-1}| \le (|t| + \epsilon)^k.$$

From this inequality you can see that

$$|ka_k t^{k-1}| \le |a_k(|t| + \epsilon)^k|.$$

Since $\Sigma\, |a_k(|t| + \epsilon)^k|$ converges, it follows that

$$\Sigma\, |ka_k t^{k-1}| \quad \text{converges.} \quad \square$$

Repeated application of this theorem shows that the series

$$\Sigma\, \frac{d^2}{dx^2}\,(a_k x^k) = \Sigma\, k(k-1)a_k x^{k-2},$$

$$\Sigma\, \frac{d^3}{dx^3}\,(a_k x^k) = \Sigma\, k(k-1)(k-2)a_k x^{k-3},$$

$$\text{etc.,}$$

also converge on $(-c, c)$. For example, since the geometric series

$$\Sigma\, x^k = 1 + x + x^2 + x^3 + x^4 + \cdots + x^n + \cdots$$

converges on $(-1, 1)$, you know now that

$$\Sigma\, \frac{d}{dx}\,(x^k) = \Sigma\, kx^{k-1} = 1 + 2x + 3x^2 + 4x^3 + \cdots + nx^{n-1} + \cdots$$

$$\Sigma\, \frac{d^2}{dx^2}\,(x^k) = \Sigma\, k(k-1)x^{k-2} = 2 + 6x + 12x^2 + \cdots + n(n-1)x^{n-2} + \cdots,$$

$$\text{etc.,}$$

also converge on $(-1, 1)$. $\square$

Suppose now that

$$\Sigma\, a_k x^k \quad \text{converges on } (-c, c).$$

Then, as you just saw,

$$\Sigma\, ka_k x^{k-1} \quad \text{also converges on } (-c, c).$$

Using the first series, we can define a function f on $(-c, c)$ by setting

$$f(x) = \Sigma\, a_k x^k \qquad \text{for all } x \text{ in } (-c, c).$$

Using the second series, we can also define a function g by setting

$$g(x) = \Sigma\, ka_k x^{k-1} \qquad \text{for all } x \text{ in } (-c, c).$$

The crucial point is that

$$f'(x) = g(x).$$

For emphasis we cast this as a theorem.

Theorem 13.10.2   The Differentiability of Power-Series

If

$$f(x) = \sum_{k=0}^{\infty} a_k x^k \qquad \text{for all } x \text{ in } (-c, c),$$

then f is differentiable on $(-c, c)$ and

$$f'(x) = \sum_{k=1}^{\infty} ka_k x^{k-1} \qquad \text{for all } x \text{ in } (-c, c).$$

By applying this theorem to f', you will see that f' is itself differentiable. This in turn implies that f'' is differentiable, and so on. In short, f has derivatives of all orders.

We can summarize this way:

> In the interior of its interval of convergence a power series defines an infinitely differentiable function, the derivatives of which can be obtained by differentiating term by term.

The term-by-term differentiation refers to the fact that

$$\frac{d}{dx}\left(\Sigma\, a_k x^k\right) = \Sigma\, \frac{d}{dx}\left(a_k x^k\right).$$

For a proof of the differentiability theorem see the supplement at the end of the section. Right now we give some examples.

Example.   You already know that

$$\frac{d}{dx}\,(e^x) = e^x.$$

You can see this directly by differentiating the exponential series:

$$\frac{d}{dx}\,(e^x) = \frac{d}{dx}\left(\sum_{k=0}^{\infty} \frac{x^k}{k!}\right) = \sum_{k=0}^{\infty} \frac{d}{dx}\left(\frac{x^k}{k!}\right) = \sum_{k=1}^{\infty} \frac{x^{k-1}}{(k-1)!} = \sum_{k=0}^{\infty} \frac{x^k}{k!} = e^x. \quad \square$$

Example.   You know that

$$\sin x = x - \frac{x^3}{3!} + \frac{x^5}{5!} - \frac{x^7}{7!} + \frac{x^9}{9!} - \cdots$$

and

$$\cos x = 1 - \frac{x^2}{2!} + \frac{x^4}{4!} - \frac{x^6}{6!} + \frac{x^8}{8!} - \cdots .$$

The relations

$$\frac{d}{dx}(\sin x) = \cos x, \qquad \frac{d}{dx}(\cos x) = -\sin x$$

can be easily verified by differentiating the series term by term:

$$\frac{d}{dx}(\sin x) = 1 - \frac{3x^2}{3!} + \frac{5x^4}{5!} - \frac{7x^6}{7!} + \frac{9x^8}{9!} - \cdots$$

$$= 1 - \frac{x^2}{2!} + \frac{x^4}{4!} - \frac{x^6}{6!} + \frac{x^8}{8!} - \cdots = \cos x,$$

$$\frac{d}{dx}(\cos x) = -\frac{2x}{2!} + \frac{4x^3}{4!} - \frac{6x^5}{6!} + \frac{8x^7}{8!} - \cdots = -x + \frac{x^3}{3!} - \frac{x^5}{5!} + \frac{x^7}{7!} - \cdots$$

$$= -\left(x - \frac{x^3}{3!} + \frac{x^5}{5!} - \frac{x^7}{7!} + \cdots\right) = -\sin x. \quad \square$$

Example. We can sum the series

$$\sum_{n=1}^{\infty} \frac{x^n}{n}, \qquad x \in (-1, 1)$$

by setting

$$g(x) = \sum_{n=1}^{\infty} \frac{x^n}{n}, \qquad x \in (-1, 1)$$

and noting that

$$g'(x) = \sum_{n=1}^{\infty} \frac{nx^{n-1}}{n} = \sum_{n=1}^{\infty} x^{n-1} = \sum_{n=0}^{\infty} x^n = \frac{1}{1-x}.$$

Since

$$g(0) = 0 \quad \text{and} \quad g'(x) = \frac{1}{1-x},$$

we must have

$$g(x) = -\log|1-x|. \quad \square$$

Power series can also be integrated term by term.

Theorem 13.10.3   Term-by-Term Integration

If

$$f(x) = \Sigma\, a_k x^k \quad \text{converges on } (-c, c),$$

then

$$g(x) = \Sigma\, \frac{a_k}{k+1} x^{k+1} \quad \text{converges on } (-c, c) \quad \text{and} \quad \int f(x)\, dx = g(x) + C.$$

PROOF.   If $\Sigma\, a_k x^k$ converges on $(-c, c)$, then $\Sigma\, |a_k x^k|$ converges on $(-c, c)$. Since

$$\left|\frac{a_k}{k+1}\, x^k\right| \le |a_k x^k|,$$

we know by comparison that

$$\Sigma\, \left|\frac{a_k}{k+1}\, x^k\right| \quad \text{also converges on } (-c, c).$$

It follows that

$$x\, \Sigma\, \frac{a_k}{k+1}\, x^k = \Sigma\, \frac{a_k}{k+1}\, x^{k+1} \quad \text{converges on } (-c, c).$$

With

$$f(x) = \Sigma\, a_k x^k \quad \text{and} \quad g(x) = \Sigma\, \frac{a_k}{k+1}\, x^{k+1},$$

we know from the differentiability theorem that

$$g'(x) = f(x) \quad \text{and therefore} \quad \int f(x)\, dx = g(x) + C. \quad \square$$

The idea of term-by-term integration can also be expressed by writing

(13.10.4)
$$\int (\Sigma\, a_k x^k)\, dx = \left(\Sigma\, \frac{a_k}{k+1}\, x^{k+1}\right) + C.$$

For definite integrals, we have

(13.10.5)
$$\int_a^b (\Sigma\, a_k x^k)\, dx = \Sigma\, \left(\int_a^b a_k x^k\, dx\right) = \Sigma\, \frac{a_k}{k+1}\, (b^{k+1} - a^{k+1})$$

so long, of course, as $[a, b]$ is contained in the interval of convergence.

Here are some examples of term-by-term integration.

Example.   By integrating

$$\frac{1}{1+x} = \frac{1}{1-(-x)} = \sum_{k=0}^{\infty} (-1)^k x^k, \qquad |x| < 1,$$

we get

$$\log(1 + x) = \int \left(\sum_{k=0}^{\infty} (-1)^k x^k\right) dx = \left(\sum_{k=0}^{\infty} \frac{(-1)^k}{k+1}\, x^{k+1}\right) + C, \qquad |x| < 1.$$

The constant C is 0 as you can tell from the fact that the series on the right and $\log(1 + x)$ are both 0 at $x = 0$. $\square$

Example.   The arc tangent series can also be obtained by term-by-term integration:

$$\frac{1}{1 + x^2} = \frac{1}{1 - (-x^2)} = \sum_{k=0}^{\infty} (-1)^k x^{2k}$$

and therefore

$$\text{arc tan } x = \int \left(\sum_{k=0}^{\infty} (-1)^k x^{2k} \right) dx = \left(\sum_{k=0}^{\infty} \frac{(-1)^k}{2k + 1} x^{2k+1} \right) + C.$$

The constant C is again 0, this time because the series on the right and arc tan x are both 0 at $x = 0$. □

Term-by-term integration can be a useful technique even if at the beginning there are no series in sight. Suppose that you are trying to evaluate an integral

$$\int_a^b f(x) \, dx$$

but cannot find an antiderivative. If you know that $f(x)$ has a convergent power series expansion, then you can evaluate the integral in question by first forming the series and then integrating term by term.

Example.   We will apply these ideas to the evaluation of

$$\int_0^1 e^{-x^2} \, dx.$$

Since

$$e^y = 1 + y + \frac{y^2}{2!} + \frac{y^3}{3!} + \frac{y^4}{4!} + \cdots,$$

we have

$$e^{-x^2} = 1 - x^2 + \frac{x^4}{2!} - \frac{x^6}{3!} + \frac{x^8}{4!} - \cdots.$$

Term-by-term integration gives

$$\int_0^1 e^{-x^2} \, dx = \left[x - \frac{x^3}{3} + \frac{x^5}{5 \cdot 2!} - \frac{x^7}{7 \cdot 3!} + \frac{x^9}{9 \cdot 4!} - \cdots \right]_0^1$$

$$= 1 - \frac{1}{3} + \frac{1}{5 \cdot 2!} - \frac{1}{7 \cdot 3!} + \frac{1}{9 \cdot 4!} - \cdots.$$

This last series is an alternating series of the type we have discussed. It is not hard to see that

$$1 - \frac{1}{3} + \frac{1}{5 \cdot 2!} - \frac{1}{7 \cdot 3!} \leq \int_0^1 e^{-x^2} \, dx \leq 1 - \frac{1}{3} + \frac{1}{5 \cdot 2!} - \frac{1}{7 \cdot 3!} + \frac{1}{9 \cdot 4!}$$

and therefore

$$0.74 < \int_0^1 e^{-x^2}\, dx < 0.75.$$

You can obtain greater accuracy by using more terms.   □

It is time to relate Taylor series

$$\sum_{k=0}^{\infty} \frac{f^{(k)}(0)}{k!}\, x^k$$

to power series in general. The relation is very simple:

| On its interval of convergence a power series is the Taylor series of its sum. |

To see this, all you have to do is differentiate

$$f(x) = a_0 + a_1 x + a_2 x^2 + \cdots + a_k x^k + \cdots$$

term by term. By doing this you'll find that $f^{(k)}(0) = k!\, a_k$ and therefore

$$a_k = \frac{f^{(k)}(0)}{k!}.\quad □$$

Exercises

1. Expand the following in powers of x basing your calculations on the geometric series

$$\frac{1}{1-x} = 1 + x + x^2 + \cdots + x^n + \cdots .$$

*(a) $\dfrac{1}{(1-x)^2}$.   (b) $\dfrac{1}{(1-x)^3}$.   *(c) $\dfrac{1}{(1-x)^k}$.

*(d) $\log(1-x)$.   (e) $\log(1-x^2)$.   *(f) $\log(2-3x)$.

2. Given that

$$\tan x = x + \tfrac{1}{3}x^3 + \tfrac{2}{15}x^5 + \tfrac{17}{315}x^7 + \cdots ,$$

find expansions for *(a) $\sec^2 x$ and (b) $\log \cos x$.

3. Find $f^{(9)}(0)$ for *(a) $f(x) = x^2 \sin x$ and (b) $f(x) = x \cos x^2$.

Estimate within 0.01.

*4. $\displaystyle\int_0^1 e^{-x^3}\, dx.$ 5. $\displaystyle\int_0^1 \sin x^2\, dx.$ *6. $\displaystyle\int_0^1 \sin \sqrt{x}\, dx.$

*7. $\displaystyle\int_0^1 \arctan x^2\, dx.$ 8. $\displaystyle\int_0^1 x^4 e^{-x^2}\, dx.$ *9. $\displaystyle\int_1^2 \frac{1-\cos x}{x}\, dx.$

*10. $\displaystyle\int_0^{1/3} e^x \log(1+x)\, dx.$ 11. $\displaystyle\int_0^1 x \sin \sqrt{x}\, dx.$

12. Show that, if $\Sigma\ a_k x^k$ and $\Sigma\ b_k x^k$ both converge to the same sum on some interval, then $a_k = b_k$ for each k.
13. Show that, if $\epsilon > 0$, then

$$|kt^{k-1}| < (|t| + \epsilon)^k \qquad \text{for all } k \text{ sufficiently large.}$$

HINT: $k^{1/k}|t|^{(k-1)/k} = k^{1/k}|t|^{1-(1/k)} \to |t|$.

Optional | **Supplement to Section 13.10**

Proof of the Differentiability of Power Series

We begin by setting

$$f(x) = \sum_{k=0}^{\infty} a_k x^k \quad \text{and} \quad g(x) = \sum_{k=1}^{\infty} k a_k x^{k-1}.$$

Taking x from $(-c, c)$, we want to show that

$$\lim_{h \to 0} \frac{f(x + h) - f(x)}{h} = g(x).$$

With $x + h$ in $(-c, c)$, $h \neq 0$, we have

$$\left| g(x) - \frac{f(x + h) - f(x)}{h} \right| = \left| \sum_{k=1}^{\infty} k a_k x^{k-1} - \sum_{k=0}^{\infty} \frac{a_k (x + h)^k - a_k x^k}{h} \right|$$

$$= \left| \sum_{k=1}^{\infty} k a_k x^{k-1} - \sum_{k=1}^{\infty} a_k \left[\frac{(x + h)^k - x^k}{h} \right] \right| .$$

By the mean-value theorem we know that there exists t_k between x and $x + h$ such that

$$\frac{(x + h)^k - x^k}{h} = k t_k^{k-1}.$$

Using this result, we have

$$\left| g(x) - \frac{f(x + h) - f(x)}{h} \right| = \left| \sum_{k=1}^{\infty} k a_k x^{k-1} - \sum_{k=1}^{\infty} k a_k t_k^{k-1} \right|$$

$$= \left| \sum_{k=1}^{\infty} k a_k (x^{k-1} - t_k^{k-1}) \right|$$

$$= \left| \sum_{k=2}^{\infty} k a_k (x^{k-1} - t_k^{k-1}) \right| .$$

Again by the mean-value theorem we know that there exists p_{k-1} between x and t_k such that

$$\frac{x^{k-1} - t_k^{k-1}}{x - t_k} = (k - 1) p_{k-1}^{k-2}$$

Now

$$|x^{k-1} - t_k^{k-1}| = |x - t_k| |(k - 1) p_{k-1}^{k-2}|$$

Optional | Since $|x - t_k| < |h|$ and $|p_{k-1}| \leq |\alpha|$, where $|\alpha| = \max\{|x|, |x + h|\}$, we have

$$|x^{k-1} - t_k^{k-1}| \leq |h| \, |(k - 1)\alpha^{k-2}|.$$

Thus

$$\left| g(x) - \frac{f(x + h) - f(x)}{h} \right| \leq |h| \sum_{k=2}^{\infty} |k(k - 1)a_k\alpha^{k-2}|.$$

Since the series on the right converges, we have

$$\lim_{h \to 0} \left(|h| \sum_{k=2}^{\infty} |k(k - 1)a_k\alpha^{k-2}| \right) = 0,$$

so that

$$\lim_{h \to 0} \left| g(x) - \frac{f(x + h) - f(x)}{h} \right| = 0$$

and

$$\lim_{h \to 0} \frac{f(x + h) - f(x)}{h} = g(x). \quad \square$$

13.11 Problems on the Binomial Series

Through a collection of problems we invite you to derive on your own the basic properties of one of the most celebrated series of all—the *binomial series*.

Start with the binomial $1 + x$ (2 terms). Choose a real number α and form the function

$$f(x) = (1 + x)^\alpha.$$

Problem 1. Show that

$$\frac{f^{(k)}(0)}{k!} = \frac{\alpha[\alpha - 1][\alpha - 2] \cdots [\alpha - (k - 1)]}{k!}.$$

The number you just obtained is the coefficient of x^k in the expansion of $(1 + x)^\alpha$. It is called the kth *binomial coefficient* and is usually denoted by $\binom{\alpha}{k}$:

$$\binom{\alpha}{k} = \frac{\alpha[\alpha - 1][\alpha - 2] \cdots [\alpha - (k - 1)]}{k!}.\,^{\dagger}$$

Problem 2. Show that the binomial series

$$\Sigma \binom{\alpha}{k} x^k$$

---

† By special definition $\binom{\alpha}{0} = 1$.

has radius of convergence 1. HINT: Use the ratio test.

From Problem 2 you know that the binomial series converges on the open interval $(-1, 1)$ and defines there an infinitely differentiable function. The next thing to show is that this function (the one defined by the series) is actually $(1 + x)^\alpha$. To do this, you first need some other results.

Problem 3.   Verify the identity

$$(k + 1) \binom{\alpha}{k + 1} + k \binom{\alpha}{k} = \alpha \binom{\alpha}{k}.$$

Problem 4.   Use the identity of Problem 3 to show that the sum of the binomial series

$$\phi(x) = \sum_{k=0}^{\infty} \binom{\alpha}{k} x^k, \qquad x \in (-1, 1)$$

satisfies the differential equation

$$(1 + x)\phi'(x) = \alpha \phi(x).$$

You are now in a position to prove the main result.

Problem 5.   Show that

$$(1 + x)^\alpha = \sum_{k=0}^{\infty} \binom{\alpha}{k} x^k \qquad \text{for} \quad -1 < x < 1.$$

You can probably get a better feeling for the series by writing out the first few binomial coefficients:

(13.11.1) $$(1 + x)^\alpha = 1 + \alpha x + \frac{\alpha(\alpha - 1)}{2!} x^2 + \frac{\alpha(\alpha - 1)(\alpha - 2)}{3!} x^3 + \cdots.$$

Problem 6.   Using (13.11.1) as a model, expand each of the following in powers of x up to x^4.

*(a) $\sqrt{1 + x}$.   (b) $\dfrac{1}{\sqrt{1 + x}}$.   *(c) $\sqrt{1 + x^2}$.   (d) $\dfrac{1}{\sqrt{1 - x^2}}$.

Problem 7.   Use series to estimate the following within 0.01.
*(a) $\sqrt{98}$.   HINT: $\sqrt{98} = (100 - 2)^{1/2} = 10(1 - \frac{1}{50})^{1/2}$.
  (b) $\sqrt[3]{9}$.   *(c) $\sqrt[4]{620}$.   (d) $\sqrt[5]{36}$.   *(e) $17^{-1/4}$.

13.12 Additional Exercises

State whether the indicated sequence converges, and if it does, find the limit.

*1. $3^{1/n}$.

2. $n2^{1/n}$.

*3. $\cos n\pi \sin n\pi$.

*4. $\dfrac{(n+1)(n+2)}{(n+3)(n+4)}$.

5. $\left(\dfrac{n}{1+n}\right)^{1/n}$.

*6. $\dfrac{n^2+5n+1}{n^3+1}$.

*7. $\cos \dfrac{\pi}{n} \sin \dfrac{\pi}{n}$.

8. $\dfrac{n^\pi}{1-n^\pi}$.

*9. $\left(2+\dfrac{1}{n}\right)^n$.

*10. $\dfrac{\pi}{n} \cos \dfrac{\pi}{n}$.

11. $\dfrac{\log[n(n+1)]}{n}$.

*12. $\left[\log\left(1+\dfrac{1}{n}\right)\right]^n$.

*13. $\dfrac{\pi}{n} \log \dfrac{n}{\pi}$.

14. $\dfrac{\pi}{n} e^{\pi/n}$.

*15. $\dfrac{n}{\pi} \sin n\pi$.

Sum the following series.

*16. $\displaystyle\sum_{k=0}^{\infty} \left(\dfrac{1}{4}\right)^k$.

17. $\displaystyle\sum_{k=0}^{\infty} \left(\dfrac{3}{4}\right)^{k+1}$.

*18. $\displaystyle\sum_{k=0}^{\infty} (-1)^k \left(\dfrac{1}{2}\right)^k$.

*19. $\displaystyle\sum_{k=0}^{\infty} \dfrac{(\log 2)^k}{k!}$.

20. $\displaystyle\sum_{k=1}^{\infty} \left(\dfrac{1}{k} - \dfrac{1}{k+1}\right)$.

*21. $\displaystyle\sum_{k=2}^{\infty} \left(\dfrac{1}{k^2} - \dfrac{1}{(k+1)^2}\right)$.

*22. $\displaystyle\sum_{k=0}^{\infty} x^{5k+1}$.

23. $\displaystyle\sum_{k=0}^{\infty} 2x^{3k+2}$.

*24. $\displaystyle\sum_{k=1}^{\infty} \tfrac{3}{2} x^{2k-1}$.

*25. $\displaystyle\sum_{k=1}^{\infty} \dfrac{1}{(k-1)!} x^k$.

26. $\displaystyle\sum_{k=1}^{\infty} \dfrac{k^2}{k!}$.

*27. $\displaystyle\sum_{k=1}^{\infty} \dfrac{1}{k(k+1)(k+2)}$.

Test for (a) absolute convergence, (b) conditional convergence.

*28. $1 + \dfrac{1}{3} + \dfrac{1}{5} + \cdots + \dfrac{1}{2n+1} + \cdots$.

29. $\dfrac{1}{1\cdot 3} + \dfrac{1}{3\cdot 5} + \dfrac{1}{5\cdot 7} + \cdots + \dfrac{1}{(2n+1)(2n+3)} + \cdots$.

*30. $\dfrac{1}{2\cdot 3} - \dfrac{1}{3\cdot 4} + \dfrac{1}{4\cdot 5} - \cdots + \dfrac{(-1)^{n+1}}{(n+1)(n+2)} + \cdots$.

31. $\dfrac{1}{2\log 2} + \dfrac{1}{3\log 3} + \dfrac{1}{4\log 4} + \cdots + \dfrac{1}{n\log n} + \cdots$.

*32. $1 - \dfrac{1}{3} + \dfrac{1}{5} - \cdots + (-1)^n \dfrac{1}{2n+1} + \cdots$.

33. $100 - \dfrac{100^2}{2!} + \dfrac{100^3}{3!} - \cdots + (-1)^{n+1} \dfrac{100^n}{n!} + \cdots$.

*34. $1 - \dfrac{2}{3} + \dfrac{3}{3^2} - \cdots + (-1)^{n-1} \dfrac{n}{3^{n-1}} + \cdots$.

35. $\dfrac{3}{4} + 2\left(\dfrac{3}{4}\right)^2 + 3\left(\dfrac{3}{4}\right)^3 + \cdots + n\left(\dfrac{3}{4}\right)^n + \cdots$.

*36. $\dfrac{1}{\sqrt{2 \cdot 3}} - \dfrac{1}{\sqrt{3 \cdot 4}} + \dfrac{1}{\sqrt{4 \cdot 5}} - \cdots + \dfrac{(-1)^{n-1}}{\sqrt{(n+1)(n+2)}} + \cdots$.

37. $\dfrac{1}{5} - \dfrac{1}{\sqrt{5}} + \dfrac{1}{\sqrt[3]{5}} - \cdots + \dfrac{(-1)^{n-1}}{\sqrt[n]{5}} + \cdots$.

Find the interval of convergence. *(ratio test)*

*38. $\Sigma \dfrac{5^k}{k}(x-2)^k$.

39. $\Sigma \dfrac{(-1)^k}{3^k} x^{k+1}$.

*40. $\Sigma (k+1)k(x-1)^{2k}$.

*41. $\Sigma (-1)^k 4^k x^{2k}$.

42. $\Sigma \dfrac{k}{2k+1} x^{2k+1}$.

*43. $\Sigma \dfrac{(-1)^k k}{3^{2k}} x^k$.

*44. $\Sigma \dfrac{k!}{2}(x+1)^k$.

45. $\Sigma \dfrac{(-1)^k}{5^{k+1}}(x-2)^k$.

*46. $\Sigma \dfrac{1}{2^{k!}}(x-2)^k$.

Expand in powers of x. *(binomial)*

*47. a^x.

48. $e^{\sin x}$.

*49. $\log(1+x^2)$.

Estimate within 0.01.

*50. $\displaystyle\int_0^{1/2} \dfrac{dx}{1+x^4}$.

51. $e^{2/3}$.

*52. $\sqrt[3]{60}$.

53. $\displaystyle\int_0^1 x \sin x^4 \, dx$.

know binomial series

$1 - (-x^4)$

$\int 1 + (-x^4) + (-x^4)^2 + \cdots$

Vectors

14

14.1 Cartesian Space Coordinates

To introduce a Cartesian coordinate system in space we begin with a plane Cartesian coordinate system $O\text{-}xy$. Through the point O, which we continue to call the origin, we pass a third line, perpendicular to the other two. This third line we call the z-axis. We coordinatize it using the same scale as before, assigning z-coordinate 0 to the origin O.

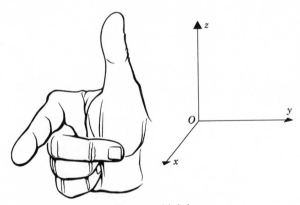

FIGURE 14.1.1

For later convenience we orient the z-axis so that $O\text{-}xyz$ forms a "right-handed" system. That is, if the index finger of the right hand points along the positive x-axis and the middle finger along the positive y-axis, then the thumb will point along the positive z-axis. See Figure 14.1.1.

A point $P(x_0, y_0, z_0)$ is shown in Figure 14.1.2. A plane through P perpendicular to the x-axis will intersect the x-axis at $x = x_0$; a plane through P perpendicular to the y-axis will intersect the y-axis at $y = y_0$; a plane through P perpendicular to the z-axis will intersect the z-axis at $z = z_0$.

FIGURE 14.1.2

The Distance Formula

The distance $d(P_1, P_2)$ between two points $P_1(x_1, y_1, z_1)$ and $P_2(x_2, y_2, z_2)$ can be found by applying the Pythagorean theorem twice.

FIGURE 14.1.3

With Q and R as in Figure 14.1.3, P_1P_2R and P_1RQ are both right triangles. From the first triangle

$$[d(P_1, P_2)]^2 = [d(P_1, R)]^2 + [d(R, P_2)]^2,$$

and from the second triangle

$$[d(P_1, R)]^2 = [d(Q, R)]^2 + [d(P_1, Q)]^2.$$

Combining equations,

$$[d(P_1, P_2)]^2 = [d(Q, R)]^2 + [d(P_1, Q)]^2 + [d(R, P_2)]^2$$
$$= (x_2 - x_1)^2 + (y_2 - y_1)^2 + (z_2 - z_1)^2.$$

Taking square roots, we have the distance formula:

(14.1.1) $$d(P_1, P_2) = \sqrt{(x_2 - x_1)^2 + (y_2 - y_1)^2 + (z_2 - z_1)^2}.$$

Symmetry

You are already familiar with two kinds of symmetry: symmetry about a line and symmetry about a point. In space we can also speak of symmetry about a plane. The idea is obvious.

Exercises

1. Sketch a right-handed coordinate system and plot the following points.
   (a) $A(1, 0, 0)$.   (b) $B(0, 1, 0)$.   (c) $C(0, 0, 1)$.
   (d) $D(1, 1, 1)$.   (e) $E(-1, -1, -1)$.   (f) $F(1, -1, 1)$.
   (g) $G(-1, 1, 2)$.   (h) $H(1, -2, 1)$.   (i) $I(-2, 1, -1)$.
2. Find the distance between the given points.
   *(a) $P(1, 0, 0)$ and $Q(0, 1, 0)$.   (b) $P(-1, 1, 2)$ and $Q(1, -2, 1)$.
   *(c) $P(4, 2, 5)$ and $Q(1, 4, -1)$.   (d) $P(x_1, x_2, 0)$ and $Q(y_1, y_2, 0)$.
*3. Find an equation for the sphere of radius r centered at
   (a) the origin.   (b) $Q(a_1, a_2, a_3)$.
4. Find an equation for the plane through $Q(a_1, a_2, a_3)$ that is parallel to
   *(a) the xy-plane.   (b) the xz-plane.   *(c) the yz-plane.
*5. Find the distance between an arbitrary point $P(a, b, c)$ and each of the coordinate axes. (Draw figures.)
6. What are the coordinates of the point halfway between $P(a_1, a_2, a_3)$ and $Q(b_1, b_2, b_3)$?
*7. What are the coordinates of the point two-thirds of the way from $P(a_1, a_2, a_3)$ to $Q(b_1, b_2, b_3)$?
8. Let $P(x_0, y_0, z_0)$ be as in Figure 14.1.2. Find the coordinates of Q given that the line segment $\overline{PQ}$ is symmetric
   *(a) about the xy-plane.   (b) about the xz-plane.
   (c) about the yz-plane.   *(d) about the origin.
   Draw the figure in each case.

14.2 Displacements and Forces

Our purpose here is to motivate the notion of vector. A quick reading of this section will do.

A displacement along a coordinate line can be specified by a real number and depicted by an arrow. For a displacement of a_1 units we can use the number a_1 and an arrow that begins at some number x and ends at $x + a_1$. If $a_1 > 0$ (and we follow convention), the arrow will point to the right, and, if $a_1 < 0$, the arrow will point to the left.

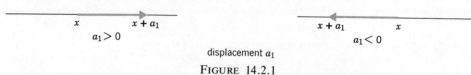

displacement a_1

FIGURE 14.2.1

Displacements in the xy-plane are more interesting because instead of having two possible directions there are now an infinite number of directions.

a displacement (a_1, a_2) in the xy-plane

FIGURE 14.2.2

The most general displacements take place in space. Here ordered triples of real numbers come into play. A displacement of a_1 units in x-coordinate, a_2 units in y-coordinate, and a_3 units in z-coordinate can be indicated by an arrow that begins at some point (x, y, z) and ends at $(x + a_1, y + a_2, z + a_3)$.

a displacement (a_1, a_2, a_3) in space

FIGURE 14.2.3

We can follow one displacement by another. A displacement (a_1, a_2, a_3) followed by a displacement (b_1, b_2, b_3) results in a total displacement $(a_1 + b_1, a_2 + b_2, a_3 + b_3)$. We can express this by writing

$$(a_1, a_2, a_3) + (b_1, b_2, b_3) = (a_1 + b_1, a_2 + b_2, a_3 + b_3).$$

We can picture the first displacement by an arrow that begins at some point $P(x, y, z)$ and ends at $Q(x + a_1, y + a_2, z + a_3)$, the second displacement by an arrow that begins at $Q(x + a_1, y + a_2, z + a_3)$ and ends at $R(x + a_1 + b_1, y + a_2 + b_2, z + a_3 + b_3)$, and the resultant displacement by an arrow that begins at $P(x, y, z)$ and ends at $R(x + a_1 + b_1, y + a_2 + b_2, z + a_3 + b_3)$. The three arrows then form a triangular pattern that is easy to remember. See Figure 14.2.4.

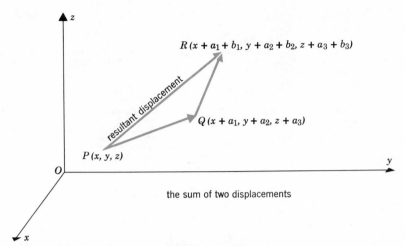

the sum of two displacements

FIGURE 14.2.4

From a displacement (a_1, a_2, a_3) and a real number α, we can form a new displacement $(\alpha a_1, \alpha a_2, \alpha a_3)$. We call this new displacement α times the initial one and write

$$\alpha(a_1, a_2, a_3) = (\alpha a_1, \alpha a_2, \alpha a_3).$$

The effect of multiplying a displacement by a number α is to change its magnitude by a factor of $|\alpha|$, keeping the same direction if $\alpha > 0$ and reversing it if $\alpha < 0$. We illustrate this idea in Figure 14.2.5.

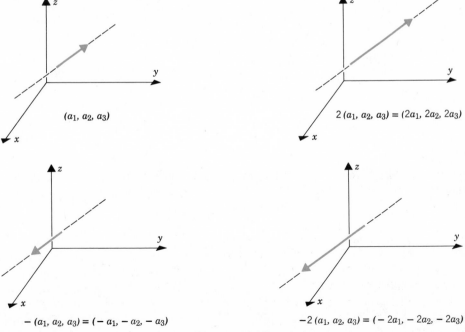

FIGURE 14.2.5

The displacement

$$2(a_1, a_2, a_3) = (2a_1, 2a_2, 2a_3)$$

is twice as long as (a_1, a_2, a_3) and has the same direction; the displacement

$$-(a_1, a_2, a_3) = (-a_1, -a_2, -a_3)$$

has the same length as (a_1, a_2, a_3) but the opposite direction; the displacement

$$-2(a_1, a_2, a_3) = (-2a_1, -2a_2, -2a_3)$$

is twice as long as (a_1, a_2, a_3) and has the opposite direction.

This same pattern

$$(a_1, a_2, a_3) + (b_1, b_2, b_3) = (a_1 + b_1, a_2 + b_2, a_3 + b_3)$$

$$\alpha(a_1, a_2, a_3) = (\alpha a_1, \alpha a_2, \alpha a_3)$$

arises naturally in other settings; for example, in the analysis of forces. To specify a force completely, we need to know its componenets along the x, y, and z axes. If these are respectively a_1, a_2, a_3, then the force is totally specified by the ordered triple (a_1, a_2, a_3). See Figure 14.2.6.

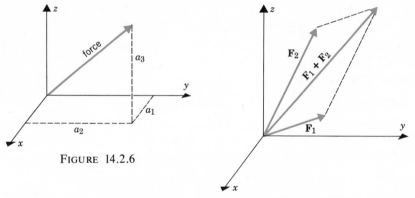

FIGURE 14.2.6

FIGURE 14.2.7

If two forces

$$F_1 = (a_1, a_2, a_3) \quad \text{and} \quad F_2 = (b_1, b_2, b_3)$$

are applied simultaneously at the same point, the effect is the same as that produced by the single force

$$F_3 = (a_1 + b_1, a_2 + b_2, a_3 + b_3).$$

We call F_3 the *resultant* or *total force* and write

$$F_1 + F_2 = F_3.$$

For the ordered triples,

$$(a_1, a_2, a_3) + (b_1, b_2, b_3) = (a_1 + b_1, a_2 + b_2, a_3 + b_3).$$

Pictorially we have the usual force diagram, Figure 14.2.7. It is a parallelogram with the sides representing

$$F_1 = (a_1, a_2, a_3) \quad \text{and} \quad F_2 = (b_1, b_2, b_3)$$

and the diagonal representing

$$F_3 = F_1 + F_2 = (a_1 + b_1, a_2 + b_2, a_3 + b_3).$$

For any force $F = (a_1, a_2, a_3)$ and a real number α, the force αF is defined by the equation

$$\alpha F = (\alpha a_1, \alpha a_2, \alpha a_3).$$

Thus once again we have

$$\alpha(a_1, a_2, a_3) = (\alpha a_1, \alpha a_2, \alpha a_3).$$

14.3   Vectors

The algebra of number triples which we introduced in the last section to discuss displacements and forces is so prodigiously rich in applications that it has found a firm place in the world of science and engineering and has generated much mathematics. It is to this mathematics that we now turn.

Definition 14.3.1

Ordered triples of real numbers subject to addition

$$(a_1, a_2, a_3) + (b_1, b_2, b_3) = (a_1 + b_1, a_2 + b_2, a_3 + b_3)$$

and multiplication by scalars (real numbers)

$$\alpha(a_1, a_2, a_3) = (\alpha a_1, \alpha a_2, \alpha a_3)$$

are called *vectors*.†

Two vectors are called *equal* iff they are identical; namely

$$(a_1, a_2, a_3) = (b_1, b_2, b_3) \quad \text{iff} \quad a_1 = b_1, a_2 = b_2, a_3 = b_3.$$

Geometrically we can depict vectors by arrows. To depict the vector (a_1, a_2, a_3) we can choose any initial point $Q = Q(x, y, z)$ and use the arrow

$$\overrightarrow{QR} \quad \text{with} \quad R = R(x + a_1, y + a_2, z + a_3). \quad \text{(Figure 14.3.1)}$$

Usually we choose the origin as the initial point and use the arrow

$$\overrightarrow{OP} \quad \text{with} \quad P = P(a_1, a_2, a_3).††$$

---

† This is not the most general notion of *vector*, but it suffices for our purpose.
†† The ordered triple $(0, 0, 0)$ will not be represented by an arrow but simply by the origin.

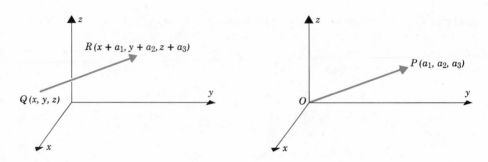

representations of (a_1, a_2, a_3)

FIGURE 14.3.1

We will use boldface letters to denote vectors. Thus for

$$\mathbf{a} = (a_1, a_2, a_3) \quad \text{and} \quad \mathbf{b} = (b_1, b_2, b_3)$$

we have

$$\mathbf{a} + \mathbf{b} = (a_1 + b_1, a_2 + b_2, a_3 + b_3),$$

and, if α is a scalar (a real number),

$$\alpha\mathbf{a} = (\alpha a_1, \alpha a_2, \alpha a_3).$$

For the zero vector $(0, 0, 0)$ we will use the symbol $\mathbf{0}$. Obviously

$$0\mathbf{a} = \mathbf{0} \qquad \text{for all vectors } \mathbf{a}.$$

By the vector $-\mathbf{b}$ we mean $(-1)\mathbf{b}$; that is,

$$-(b_1, b_2, b_3) = (-b_1, -b_2, -b_3).$$

By $\mathbf{a} - \mathbf{b}$ we mean $\mathbf{a} + (-\mathbf{b})$; that is,

$$(a_1, a_2, a_3) - (b_1, b_2, b_3) = (a_1 - b_1, a_2 - b_2, a_3 - b_3).$$

Problem.   Given that

$$\mathbf{a} = (1, -1, -2), \quad \mathbf{b} = (2, 3, -1), \quad \mathbf{c} = (8, 7, 1),$$

find

(a) $\mathbf{a} - \mathbf{b}$.   (b) $2\mathbf{a} + \mathbf{b}$.   (c) $3\mathbf{a} - 7\mathbf{b}$.   (d)$2\mathbf{a} + 3\mathbf{b} - \mathbf{c}$.

SOLUTION

(a) $\mathbf{a} - \mathbf{b} = (1, -1, 2) - (2, 3, -1) = (1 - 2, -1 - 3, 2 + 1) = (-1, -4, 3)$.
(b) $2\mathbf{a} + \mathbf{b} = 2(1, -1, 2) + (2, 3, -1) = (2, -2, 4) + (2, 3, -1) = (4, 1, 3)$.
(c) $3\mathbf{a} - 7\mathbf{b} = 3(1, -1, 2) - 7(2, 3, -1)$
$\qquad = (3, -3, 6) - (14, 21, -7) = (-11, -24, 13)$.
(d) $2\mathbf{a} + 3\mathbf{b} - \mathbf{c} = 2(1, -1, 2) + 3(2, 3, -1) - (8, 7, 1)$
$\qquad\qquad = (2, -2, 4) + (6, 9, -3) - (8, 7, 1) = (0, 0, 0) = \mathbf{0}$. $\square$

The addition of vectors can be visualized as the "addition" of arrows by the parallelogram law (as in the case of forces). If you picture **a**, **b**, and **a** + **b** all as arrows emanating from the same point, say the origin, then **a** + **b** acts as the diagonal of the parallelogram generated by **a** and **b**. (See Figure 14.3.2.)

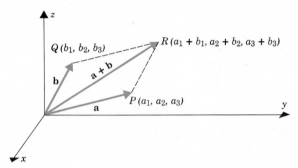

FIGURE 14.3.2

The addition of vectors can also be visualized as the "tail-to-head" addition of arrows (as in the case of displacements). If, instead of starting **b** at the origin, you start **b** at the tip of **a**, then **a** + **b** goes from the tail of **a** to the tip of **b**. (See Figure 14.3.3.) These two pictorial representations of vector addition lead to the same result.

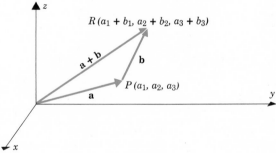

FIGURE 14.3.3

Definition 14.3.2

Two nonzero vectors **a** and **b** are said to be *parallel* provided that

$$\mathbf{a} = \alpha \mathbf{b} \qquad \text{for some real number } \alpha.$$

If $\alpha > 0$, **a** and **b** are said to have the *same direction*; if $\alpha < 0$, they are said to have *opposite directions*.

In the case of

$$\mathbf{a} = (2, -2, 6), \quad \mathbf{b} = (1, -1, 3), \quad \mathbf{c} = (-1, 1, -3)$$

we have

$$\mathbf{a} = 2\mathbf{b} \quad \text{and} \quad \mathbf{a} = -2\mathbf{c}.$$

This tells us that $\mathbf{a}$ and $\mathbf{b}$ are parallel and have the same direction whereas $\mathbf{a}$ and $\mathbf{c}$, though parallel, have opposite directions. (See Figure 14.3.4.)

$$\mathbf{a} = (2, -2, 6) \qquad \mathbf{b} = (1, -1, 3) \qquad \mathbf{c} = (-1, 1, -3)$$

some parallel vectors

FIGURE 14.3.4

Definition 14.3.2 did not include the zero vector $\mathbf{0}$. By special convention, $\mathbf{0}$ is called *parallel* to *every vector*. We will come back to this idea later.

Definition 14.3.3

The norm of a vector $\mathbf{a} = (a_1, a_2, a_3)$ is the number

$$\|\mathbf{a}\| = \sqrt{a_1^2 + a_2^2 + a_3^2}.$$

The norm of $\mathbf{a}$ is also called the *length* or *magnitude* of $\mathbf{a}$. The reason for this is simple: if we represent the vector $\mathbf{a}$ by an arrow, say

$$\overrightarrow{QR} \quad \text{with} \quad Q = Q(x, y, z) \quad \text{and} \quad R = R(x + a_1, y + a_2, z + a_3),$$

then $\|\mathbf{a}\|$ gives the length of $\overrightarrow{QR}$.

The norm properties of vectors are very similar to the absolute value properties of real numbers. In particular, we have

(14.3.4)

> (1) $\|\mathbf{a}\| \geq 0$ and $\|\mathbf{a}\| = 0$ iff $\mathbf{a} = \mathbf{0}$.
> (2) $\|\alpha\mathbf{a}\| = |\alpha|\,\|\mathbf{a}\|$.
> (3) $\|\mathbf{a} + \mathbf{b}\| \leq \|\mathbf{a}\| + \|\mathbf{b}\|$.   (the triangle inequality)

Property (1) is obvious. Property (2) is easy to verify:

$$\|\alpha\mathbf{a}\| = \sqrt{(\alpha a_1)^2 + (\alpha a_2)^2 + (\alpha a_3)^2} = |\alpha|\sqrt{a_1^2 + a_2^2 + a_3^2} = |\alpha|\,\|\mathbf{a}\|.$$

We prove Property (3), the triangle inequality, in the next section, where we have "dot products" at our disposal. A proof at this time would be laborious.

You can interprete the triangle inequality as saying that the length of a side of a triangle cannot exceed the sum of the lengths of the other two sides. (See Figure 14.3.5.)

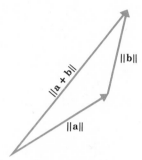

FIGURE 14.3.5

Problem.   Given that

$$\mathbf{a} = (1, -2, 3) \quad \text{and} \quad \mathbf{b} = (-4, 1, 0),$$

find

(a) $\|\mathbf{a}\|$.

(b) $\|\mathbf{b}\|$.

(c) $\|\mathbf{a} + \mathbf{b}\|$.

(d) $\|\mathbf{a} - \mathbf{b}\|$.

(e) $\|-7\mathbf{a}\|$.

(f) $\|2\mathbf{a} - 3\mathbf{b}\|$.

SOLUTION

(a) $\|\mathbf{a}\| = \sqrt{1^2 + (-2)^2 + (3)^2} = \sqrt{1 + 4 + 9} = \sqrt{14}.$

(b) $\|\mathbf{b}\| = \sqrt{(-4)^2 + 1^2 + 0^2} = \sqrt{16 + 1 + 0} = \sqrt{17}.$

(c) $\|\mathbf{a} + \mathbf{b}\| = \|(-3, -1, 3)\| = \sqrt{(-3)^2 + (-1)^2 + (3)^2} = \sqrt{9 + 1 + 9} = \sqrt{19}.$

(d) $\|\mathbf{a} - \mathbf{b}\| = \|(5, -3, 3)\| = \sqrt{(5)^2 + (-3)^2 + (3)^2} = \sqrt{25 + 9 + 9} = \sqrt{43}.$

(e) $\|-7\mathbf{a}\| = |-7|\,\|\mathbf{a}\| = 7\sqrt{14}.$

(f) $\|2\mathbf{a} - 3\mathbf{b}\| = \|2(1, -2, 3) - 3(-4, 1, 0)\| = \|(14, -7, 6)\|$
$$= \sqrt{(14)^2 + (-7)^2 + (6)^2} = \sqrt{196 + 49 + 36} = \sqrt{281}. \quad \square$$

To multiply a nonzero vector by a nonzero scalar α is to change its length by a factor of $|\alpha|$,

$$\|\alpha\mathbf{a}\| = |\alpha|\,\|\mathbf{a}\|,$$

keeping the same direction if $\alpha > 0$ and reversing the direction if $\alpha < 0$. The vector obtained from **a** simply by reversing its direction is the vector $(-1)\mathbf{a} = -\mathbf{a}$. (See Figure 14.3.6.)

FIGURE 14.3.6

Problem. Let **a** be a nonzero vector. Find the *unit vector* (the vector of norm 1) which has the same direction as **a**.

SOLUTION.   If **b** is the required vector, then

$$\|\mathbf{b}\| = 1 \qquad \text{and} \qquad \mathbf{b} = \alpha\mathbf{a} \quad \text{for some } \alpha > 0.$$

It follows that

$$1 = \|\mathbf{b}\| = \|\alpha\mathbf{a}\| = |\alpha|\,\|\mathbf{a}\| = \alpha\|\mathbf{a}\|.$$

This says that

$$\alpha = \frac{1}{\|\mathbf{a}\|} \quad \text{and consequently} \quad \mathbf{b} = \frac{\mathbf{a}}{\|\mathbf{a}\|}. \quad \square$$

While

$$\frac{\mathbf{a}}{\|\mathbf{a}\|}$$

is the unit vector in the direction of **a**,

$$-\frac{\mathbf{a}}{\|\mathbf{a}\|}$$

is the unit vector in the opposite direction.
Since

$$\mathbf{a} - \mathbf{b} = \mathbf{a} + (-\mathbf{b})$$

we can draw the vector $\mathbf{a} - \mathbf{b}$ by drawing $-\mathbf{b}$ and adding it to the vector **a**. (Figure 14.3.7.)

FIGURE 14.3.7

We can obtain the same result more easily by noting that $\mathbf{a} - \mathbf{b}$ is the vector that we must add to $\mathbf{b}$ to obtain $\mathbf{a}$. (Figure 14.3.8)

FIGURE 14.3.8

We single out for special attention the vectors

$$\mathbf{i} = (1, 0, 0), \quad \mathbf{j} = (0, 1, 0), \quad \mathbf{k} = (0, 0, 1).$$

These vectors all have norm 1 and, if pictured as emanating from the origin, lie along the positive coordinate axes. They are called *unit coordinate vectors*. (See Figure 14.3.9.)

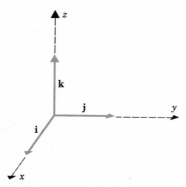

FIGURE 14.3.9

Every vector can be expressed as a linear combination of the unit coordinate vectors:

(14.3.5) | for $\mathbf{a} = (a_1, a_2, a_3)$ we have $\mathbf{a} = a_1\mathbf{i} + a_2\mathbf{j} + a_3\mathbf{k}.$

PROOF

$$(a_1, a_2, a_3) = (a_1, 0, 0) + (0, a_2, 0) + (0, 0, a_3)$$
$$= a_1(1, 0, 0) + a_2(0, 1, 0) + a_3(0, 0, 1)$$
$$= a_1\mathbf{i} + a_2\mathbf{j} + a_3\mathbf{k}. \quad \square$$

If $\mathbf{a} = a_1\mathbf{i} + a_2\mathbf{j} + a_3\mathbf{k}$, the numbers a_1, a_2, a_3 are called the *scalar components* of $\mathbf{a}$ or more simply the *components* of $\mathbf{a}$. The vectors $a_1\mathbf{i}, a_2\mathbf{j}, a_3\mathbf{k}$ are called the *vector components* of $\mathbf{a}$.

Problem.   Given that

$$\mathbf{a} = 3\mathbf{i} - \mathbf{j} + \mathbf{k} \quad \text{and} \quad \mathbf{b} = 2\mathbf{i} + 3\mathbf{j} - \mathbf{k},$$

(1) express $2\mathbf{a} - \mathbf{b}$ as a linear combination of $\mathbf{i}, \mathbf{j}, \mathbf{k}$.

(2) compute $\|2\mathbf{a} - \mathbf{b}\|$.

(3) find the unit vector $\mathbf{c}$ in the direction of $2\mathbf{a} - \mathbf{b}$.

SOLUTION

$$
\begin{aligned}
(1) \qquad 2\mathbf{a} - \mathbf{b} &= 2(3\mathbf{i} - \mathbf{j} + \mathbf{k}) - (2\mathbf{i} + 3\mathbf{j} - \mathbf{k}) \\
&= 6\mathbf{i} - 2\mathbf{j} + 2\mathbf{k} - 2\mathbf{i} - 3\mathbf{j} + \mathbf{k} \\
&= 4\mathbf{i} - 5\mathbf{j} + 3\mathbf{k}.
\end{aligned}
$$

(2) $\quad \|2\mathbf{a} - \mathbf{b}\| = \|4\mathbf{i} - 5\mathbf{j} + 3\mathbf{k}\| = \sqrt{16 + 25 + 9} = \sqrt{50} = 5\sqrt{2}.$

(3) $\qquad \mathbf{c} = \dfrac{2\mathbf{a} - \mathbf{b}}{\|2\mathbf{a} - \mathbf{b}\|} = \dfrac{1}{5\sqrt{2}}(4\mathbf{i} - 5\mathbf{j} + 3\mathbf{k}). \quad \square$

Exercises

Simplify the following linear combinations.

*1. $(2\mathbf{i} - \mathbf{j} + \mathbf{k}) + (\mathbf{i} - 3\mathbf{j} + 5\mathbf{k})$.   2. $(6\mathbf{j} - \mathbf{k}) + (3\mathbf{i} - \mathbf{j} + 2\mathbf{k})$.

*3. $2(\mathbf{j} + \mathbf{k}) - 3(\mathbf{i} + \mathbf{j} - 2\mathbf{k})$.   4. $2(\mathbf{i} - \mathbf{j}) + 6(2\mathbf{i} + \mathbf{j} - 2\mathbf{k})$.

Compute the norm of the following vectors.

*5. $3\mathbf{i} + 4\mathbf{j}$.   6. $\mathbf{i} - \mathbf{j}$.

*7. $2\mathbf{i} + \mathbf{j} - 2\mathbf{k}$.   8. $6\mathbf{i} + 2\mathbf{j} - \mathbf{k}$.

*9. $\frac{1}{2}(\mathbf{i} + 4\mathbf{j}) - (\frac{3}{2}\mathbf{i} + \mathbf{k})$.   10. $(\mathbf{i} - \mathbf{j}) + 2(\mathbf{j} - \mathbf{i}) + (\mathbf{k} - \mathbf{j})$.

*11. Let

$$
\begin{aligned}
\mathbf{a} &= \mathbf{i} - \mathbf{j} + 2\mathbf{k}, \qquad \mathbf{b} = 2\mathbf{i} - \mathbf{j} + 2\mathbf{k}, \\
\mathbf{c} &= 3\mathbf{i} - 3\mathbf{j} + 6\mathbf{k}, \quad \mathbf{d} = -2\mathbf{i} + 2\mathbf{j} - 4\mathbf{k}.
\end{aligned}
$$

(a) Which vectors are parallel?

(b) Which vectors have the same direction?

(c) Which vectors have opposite directions?

*12. Label the vectors.

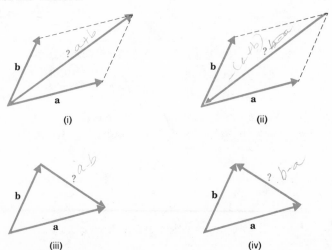

(i)   (ii)

(iii)   (iv)

13. (*Important*) Prove the following version of the triangle inequality:

$$\big| \|\mathbf{a}\| - \|\mathbf{b}\| \big| \le \|\mathbf{a} - \mathbf{b}\|.$$

    HINT: Apply the previously stated triangle inequality to $\mathbf{a} = (\mathbf{a} - \mathbf{b}) + \mathbf{b}$.

*14. Given $\mathbf{a} = (1, 1, 1)$, $\mathbf{b} = (-1, 3, 2)$, $\mathbf{c} = (-3, 0, 1)$, $\mathbf{d} = (4, -1, 1)$,

    (a) express $\mathbf{a} + 2\mathbf{b} + 3\mathbf{c} + 4\mathbf{d}$ as a linear combination of $\mathbf{i}, \mathbf{j}, \mathbf{k}$.

    (b) find scalars A, B, C such that $\mathbf{d} = A\mathbf{a} + B\mathbf{b} + C\mathbf{c}$.

15. Find α given that $3\mathbf{i} + \mathbf{j} - \mathbf{k}$ and $\alpha\mathbf{i} - 4\mathbf{j} + 4\mathbf{k}$ are parallel.

*16. Find α given that $3\mathbf{i} + \mathbf{j}$ and $\alpha\mathbf{j} - \mathbf{k}$ have the same length.

17. Find the unit vector in the direction of $\mathbf{i} - 2\mathbf{j} + 2\mathbf{k}$.

18. Find α given that $\|\alpha\mathbf{i} + (\alpha - 1)\mathbf{j} + (\alpha + 1)\mathbf{k}\| = 2$.

*19. Find the vector of norm 2 in the direction of $\mathbf{i} + 2\mathbf{j} - \mathbf{k}$.

20. Find the vectors of norm 2 parallel to $3\mathbf{j} + 2\mathbf{k}$.

14.4 The Dot Product

We come now to a notion widely used in geometry and in physics.

Definition 14.4.1

For arbitrary vectors

$$\mathbf{a} = a_1\mathbf{i} + a_2\mathbf{j} + a_3\mathbf{k} \qquad \text{and} \qquad \mathbf{b} = b_1\mathbf{i} + b_2\mathbf{j} + b_3\mathbf{k}$$

we define

$$\mathbf{a} \cdot \mathbf{b} = a_1 b_1 + a_2 b_2 + a_3 b_3.$$

The number $\mathbf{a} \cdot \mathbf{b}$, read "a dot b", is called the *dot product* of $\mathbf{a}$ and $\mathbf{b}$.

Examples.   For

$$\mathbf{a} = 2\mathbf{i} - \mathbf{j} + 3\mathbf{k}, \qquad \mathbf{b} = -3\mathbf{i} + \mathbf{j} + 4\mathbf{k}, \qquad \mathbf{c} = \mathbf{i} - \mathbf{j}$$

$\mathbf{a} \cdot \mathbf{b} = (2)(-3) + (-1)(1) + (3)(4) = -6 - 1 + 12 = 5,$

$\mathbf{a} \cdot \mathbf{c} = (2)(1) + (-1)(-1) + (3)(0) = 2 + 1 = 3,$

$\mathbf{b} \cdot \mathbf{c} = (-3)(1) + (1)(-1) + (4)(0) = -3 - 1 = -4. \quad \square$

    The dot product is also called the *scalar product* or *inner product*. We'll generally call it the dot product and speak of "dotting" $\mathbf{a}$ with $\mathbf{b}$. If we dot a vector with itself, we obtain the square of the norm of the vector:

(14.4.2) $$\boxed{\mathbf{a} \cdot \mathbf{a} = \|\mathbf{a}\|^2.}$$

$$[\mathbf{a} \cdot \mathbf{a} = a_1 a_1 + a_2 a_2 + a_3 a_3 = a_1^2 + a_2^2 + a_3^2 = \|\mathbf{a}\|^2.]$$

Consequently

(14.4.3) $$\boxed{\mathbf{a} \cdot \mathbf{a} \ge 0 \qquad \text{and} \qquad \mathbf{a} \cdot \mathbf{a} = 0 \quad \text{iff} \quad \mathbf{a} = \mathbf{0}.}$$

FIGURE 14.4.1

For a geometric interpretation of $\mathbf{a} \cdot \mathbf{b}$ see the triangle of Figure 14.4.1. Since the sides of the triangle have lengths $\|\mathbf{a}\|$, $\|\mathbf{b}\|$, and $\|\mathbf{a} - \mathbf{b}\|$, the law of cosines gives

$$\|\mathbf{a} - \mathbf{b}\|^2 = \|\mathbf{a}\|^2 + \|\mathbf{b}\|^2 - 2\|\mathbf{a}\|\,\|\mathbf{b}\| \cos \theta.$$

Consequently

$$
\begin{aligned}
2\|\mathbf{a}\|\,\|\mathbf{b}\| \cos \theta &= \|\mathbf{a}\|^2 + \|\mathbf{b}\|^2 - \|\mathbf{a} - \mathbf{b}\|^2 \\
&= a_1^2 + a_2^2 + a_3^2 + b_1^2 + b_2^2 + b_3^2 - (a_1 - b_1)^2 - (a_2 - b_2)^2 - (a_3 - b_3)^2 \\
&= 2(a_1 b_1 + a_2 b_2 + a_3 b_3) = 2(\mathbf{a} \cdot \mathbf{b}),
\end{aligned}
$$

so that

(14.4.4)
$$\boxed{\mathbf{a} \cdot \mathbf{b} = \|\mathbf{a}\|\,\|\mathbf{b}\| \cos \theta.}$$

If neither $\mathbf{a}$ nor $\mathbf{b}$ is the zero vector, we can divide this equation by $\|\mathbf{a}\|\,\|\mathbf{b}\|$ and write

(14.4.5)
$$\boxed{\cos \theta = \frac{\mathbf{a} \cdot \mathbf{b}}{\|\mathbf{a}\|\,\|\mathbf{b}\|}.}$$

The number θ in $[0, \pi]$ which satisfies this equation is called *the angle between* $\mathbf{a}$ *and* $\mathbf{b}$. If either $\mathbf{a}$ or $\mathbf{b}$ is the zero vector, the angle between $\mathbf{a}$ and $\mathbf{b}$ is not defined.

If $\mathbf{a} \cdot \mathbf{b} = 0$, which happens iff $\theta = \frac{1}{2}\pi$ or one of the vectors is the zero vector, then we say that $\mathbf{a}$ and $\mathbf{b}$ are perpendicular.† In symbols

(14.4.6)
$$\boxed{\mathbf{a} \cdot \mathbf{b} = 0 \qquad \text{iff} \qquad \mathbf{a} \perp \mathbf{b}.}$$
(Figure 14.4.2)

$\mathbf{a} \cdot \mathbf{b} = 0$

FIGURE 14.4.2

---

† This makes the zero vector both parallel and perpendicular to every vector. There is, however, no contradiction since we never apply the notion of angle to the zero vector.

Obviously the coordinate vectors $\mathbf{i}$, $\mathbf{j}$, $\mathbf{k}$ are mutually perpendicular.

Problem. Find the angle between

$$\mathbf{a} = 2\mathbf{i} + \mathbf{j} + \mathbf{k} \quad \text{and} \quad \mathbf{b} = \mathbf{i} + \mathbf{j} - 3\mathbf{k}.$$

SOLUTION

$$\mathbf{a} \cdot \mathbf{b} = (2)(1) + (1)(1) + (1)(-3) = 2 + 1 - 3 = 0.$$

Since the dot product is 0, the two vectors must be perpendicular. The angle between them is $\frac{1}{2}\pi$ radians. $\square$

Problem. Find the angle between

$$\mathbf{a} = 2\mathbf{i} + 3\mathbf{j} + 2\mathbf{k} \quad \text{and} \quad \mathbf{b} = \mathbf{i} + 2\mathbf{j} - \mathbf{k}.$$

SOLUTION

$$\mathbf{a} \cdot \mathbf{b} = (2)(1) + (3)(2) + (2)(-1) = 6,$$
$$\|\mathbf{a}\| = \sqrt{2^2 + 3^2 + 2^2} = \sqrt{17},$$
$$\|\mathbf{b}\| = \sqrt{1^2 + 2^2 + (-1)^2} = \sqrt{6}.$$

In general

$$\cos\theta = \frac{\mathbf{a} \cdot \mathbf{b}}{\|\mathbf{a}\|\|\mathbf{b}\|},$$

so that here

$$\cos\theta = \frac{6}{\sqrt{17}\sqrt{6}} = \frac{1}{17}\sqrt{102} \cong \frac{10.1}{17} \cong 0.594.$$

From Table 4 at the end of the book you can see that θ is about 0.93 radians. $\square$

Problem. Find α given that

$$\mathbf{i} + 2\mathbf{j} \quad \text{and} \quad 3\mathbf{i} - \alpha\mathbf{j} + \mathbf{k}$$

are perpendicular.

SOLUTION. Since the two vectors are perpendicular, their dot product

$$(\mathbf{i} + 2\mathbf{j}) \cdot (3\mathbf{i} - \alpha\mathbf{j} + \mathbf{k}) = (1)(3) + (2)(-\alpha) + (0)(1) = 3 - 2\alpha$$

is zero and $\alpha = \frac{3}{2}$. $\square$

Note that the dot product has the following properties:

(14.4.7)

$\mathbf{a} \cdot \mathbf{b} = \mathbf{b} \cdot \mathbf{a}$	(the dot product is commutative)
$\alpha\mathbf{a} \cdot \beta\mathbf{b} = \alpha\beta(\mathbf{a} \cdot \mathbf{b})$	(scalars can be factored out)
$\mathbf{a} \cdot (\mathbf{b} + \mathbf{c}) = \mathbf{a} \cdot \mathbf{b} + \mathbf{a} \cdot \mathbf{c}$	(the dot distributes over addition)
$|\mathbf{a} \cdot \mathbf{b}| \leq \|\mathbf{a}\|\|\mathbf{b}\|.$	(Schwarz's inequality)

The first three properties follow easily from the definition. Schwarz's inequality is obvious from the relation

$$\mathbf{a} \cdot \mathbf{b} = \|\mathbf{a}\| \|\mathbf{b}\| \cos \theta$$

together with the observation that $|\cos \theta| \leq 1$. $\square$

Problem.   Show that

$$(\mathbf{a} + \mathbf{b}) \cdot \mathbf{c} = \mathbf{a} \cdot \mathbf{c} + \mathbf{b} \cdot \mathbf{c}.$$

SOLUTION

$$
\begin{aligned}
(\mathbf{a} + \mathbf{b}) \cdot \mathbf{c} &= \mathbf{c} \cdot (\mathbf{a} + \mathbf{b}) &&\text{(by commutativity)}\\
&= \mathbf{c} \cdot \mathbf{a} + \mathbf{c} \cdot \mathbf{b} &&\text{(by distributivity)}\\
&= \mathbf{a} \cdot \mathbf{c} + \mathbf{b} \cdot \mathbf{c}. \quad \square &&\text{(by commutativity)}
\end{aligned}
$$

We are now in a position to prove the triangle inequality

$$\|\mathbf{a} + \mathbf{b}\| \leq \|\mathbf{a}\| + \|\mathbf{b}\|.$$

PROOF

$$
\begin{aligned}
\|\mathbf{a} + \mathbf{b}\|^2 &= (\mathbf{a} + \mathbf{b}) \cdot (\mathbf{a} + \mathbf{b})\\
&= \mathbf{a} \cdot \mathbf{a} + \mathbf{b} \cdot \mathbf{a} + \mathbf{a} \cdot \mathbf{b} + \mathbf{b} \cdot \mathbf{b}\\
&= \|\mathbf{a}\|^2 + 2(\mathbf{a} \cdot \mathbf{b}) + \|\mathbf{b}\|^2\\
&\leq \|\mathbf{a}\|^2 + 2\|\mathbf{a}\| \|\mathbf{b}\| + \|\mathbf{b}\|^2\\
&= (\|\mathbf{a}\| + \|\mathbf{b}\|)^2
\end{aligned}
$$

so that

$$\|\mathbf{a} + \mathbf{b}\| \leq \|\mathbf{a}\| + \|\mathbf{b}\|.   \square$$

We go back now to the formula

$$\mathbf{a} \cdot \mathbf{b} = \|\mathbf{a}\| \|\mathbf{b}\| \cos \theta.$$

If $\mathbf{b}$ is a nonzero vector, the number

$$\frac{\mathbf{a} \cdot \mathbf{b}}{\|\mathbf{b}\|} = \|\mathbf{a}\| \cos \theta \qquad \text{(see Figure 14.4.3)}$$

measures the "advance" of $\mathbf{a}$ in the direction of $\mathbf{b}$. It is called the (*scalar*) *component* of $\mathbf{a}$ in the direction of $\mathbf{b}$.

FIGURE 14.4.3

In abbreviated form we write

(14.4.8)

$$\text{comp}_{\mathbf{b}}\, \mathbf{a} = \frac{\mathbf{a} \cdot \mathbf{b}}{\|\mathbf{b}\|} = \|\mathbf{a}\| \cos \theta.$$

Note that

$$\text{comp}_{\mathbf{b}}\, \mathbf{a} \quad \text{is} \quad \left\{ \begin{array}{lll} \text{positive,} & \text{if} & 0 \le \theta < \tfrac{1}{2}\pi \\ \text{zero,} & \text{if} & \theta = \tfrac{1}{2}\pi \\ \text{negative,} & \text{if} & \tfrac{1}{2}\pi < \theta \le \pi \end{array} \right]. \qquad \text{(see Figure 14.4.4)}$$

$\text{comp}_{\mathbf{b}}\, \mathbf{a} > 0$

$\text{comp}_{\mathbf{b}}\, \mathbf{a} = 0$

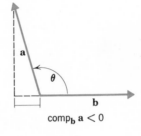
$\text{comp}_{\mathbf{b}}\, \mathbf{a} < 0$

FIGURE 14.4.4

Problem.   Given that

$$\mathbf{a} = \mathbf{j} - 2\mathbf{i} + \mathbf{k}, \qquad \mathbf{b} = 4\mathbf{i} - 3\mathbf{j} + \mathbf{k}, \qquad \mathbf{c} = 3\mathbf{i} - 4\mathbf{j} + 12\mathbf{k},$$

compute $\text{comp}_{\mathbf{b}}\, \mathbf{a}$ and $\text{comp}_{\mathbf{c}}\, \mathbf{a}$.

SOLUTION.   We begin by rewriting $\mathbf{a}$ as $-2\mathbf{i} + \mathbf{j} + \mathbf{k}$.

$$\text{comp}_{\mathbf{b}}\, \mathbf{a} = \frac{\mathbf{a} \cdot \mathbf{b}}{\|\mathbf{b}\|} = \frac{(-2)(4) + (1)(-3) + (1)(1)}{\sqrt{(4)^2 + (-3)^2 + (1)^2}} = -\frac{10}{\sqrt{26}} = -\frac{5}{13}\sqrt{26}.$$

$$\text{comp}_{\mathbf{c}}\, \mathbf{a} = \frac{\mathbf{a} \cdot \mathbf{c}}{\|\mathbf{c}\|} = \frac{(-2)(3) + (1)(-4) + (1)(12)}{\sqrt{(3)^2 + (-4)^2 + (12)^2}} = \frac{2}{13}. \quad \square$$

For each vector

$$\mathbf{a} = a_1 \mathbf{i} + a_2 \mathbf{j} + a_3 \mathbf{k}$$

we have

$$\text{comp}_{\mathbf{i}}\, \mathbf{a} = \frac{\mathbf{a} \cdot \mathbf{i}}{\|\mathbf{i}\|} = a_1, \qquad \text{comp}_{\mathbf{j}}\, \mathbf{a} = \frac{\mathbf{a} \cdot \mathbf{j}}{\|\mathbf{j}\|} = a_2, \qquad \text{comp}_{\mathbf{k}}\, \mathbf{a} = \frac{\mathbf{a} \cdot \mathbf{k}}{\|\mathbf{k}\|} = a_3.$$

These results agree with our previous use of the word "component."†

---

† At the end of Section 14.3.

FIGURE 14.4.5

In Figure 14.4.5 we have portrayed a nonzero vector **a**. The angles α, β, γ that the vector makes with the coordinate vectors are called the *direction angles* of **a** and their cosines are called the direction cosines. Since

$$\cos \alpha = \frac{\mathbf{a} \cdot \mathbf{i}}{\|\mathbf{a}\| \|\mathbf{i}\|} = \frac{a_1}{\|\mathbf{a}\|}, \quad \cos \beta = \frac{\mathbf{a} \cdot \mathbf{j}}{\|\mathbf{a}\| \|\mathbf{j}\|} = \frac{a_2}{\|\mathbf{a}\|}, \quad \cos \gamma = \frac{\mathbf{a} \cdot \mathbf{k}}{\|\mathbf{a}\| \|\mathbf{k}\|} = \frac{a_3}{\|\mathbf{a}\|},$$

you can see that

(14.4.9)
$$\mathbf{a} = \|\mathbf{a}\|(\cos \alpha\, \mathbf{i} + \cos \beta\, \mathbf{j} + \cos \gamma\, \mathbf{k}).$$

Taking the norm of both sides, we have

$$\|\mathbf{a}\| = \|\mathbf{a}\| \|\cos \alpha\, \mathbf{i} + \cos \beta\, \mathbf{j} + \cos \gamma\, \mathbf{k}\|,$$
$$1 = \|\cos \alpha\, \mathbf{i} + \cos \beta\, \mathbf{j} + \cos \gamma\, \mathbf{k}\| = \sqrt{\cos^2 \alpha + \cos^2 \beta + \cos^2 \gamma},$$

and therefore

(14.4.10)
$$\cos^2 \alpha + \cos^2 \beta + \cos^2 \gamma = 1.$$

The sum of the squares of the direction cosines is always 1.
If **u** is a unit vector, the relation

$$\mathbf{u} = \|\mathbf{u}\|(\cos \alpha\, \mathbf{i} + \cos \beta\, \mathbf{j} + \cos \gamma\, \mathbf{k})$$

gives

(14.4.11)
$$\mathbf{u} = \cos \alpha\, \mathbf{i} + \cos \beta\, \mathbf{j} + \cos \gamma\, \mathbf{k}.$$

For a unit vector, the scalar components are the direction cosines.

Problem.   Find the direction cosines of

$$\mathbf{a} = \mathbf{i} + 2\mathbf{j} - 2\mathbf{k}.$$

What are the direction angles?

SOLUTION.   Here

$$\|\mathbf{a}\| = \sqrt{1^2 + 2^2 + (-2)^2} = 3$$

so that

$$\cos \alpha = \frac{a_1}{\|\mathbf{a}\|} = \tfrac{1}{3}, \qquad \cos \beta = \frac{a_2}{\|\mathbf{a}\|} = \tfrac{2}{3}, \qquad \cos \gamma = \frac{a_3}{\|\mathbf{a}\|} = -\tfrac{2}{3}.$$

$\alpha = \operatorname{arc\,cos} \tfrac{1}{3} \cong 1.23$ radians,   (about 71 degrees)

$\beta = \operatorname{arc\,cos} \tfrac{2}{3} \cong 0.84$ radians,   (about 48 degrees)

$\gamma = \operatorname{arc\,cos}\left(-\tfrac{2}{3}\right) \cong 1.57 + \operatorname{arc\,cos} \tfrac{2}{3} \cong 2.41$ radians.   (about 138 degrees)

Problem.   Find the vector of norm 3 with direction angles

$$\alpha = \tfrac{1}{4}\pi, \quad \beta = \tfrac{2}{3}\pi, \quad \gamma = \tfrac{1}{3}\pi.$$

SOLUTION

$$\mathbf{a} = \|\mathbf{a}\|(\cos \alpha \, \mathbf{i} + \cos \beta \, \mathbf{j} + \cos \gamma \, \mathbf{k})$$
$$= 3(\tfrac{1}{2}\sqrt{2}\,\mathbf{i} - \tfrac{1}{2}\mathbf{j} + \tfrac{1}{2}\mathbf{k}) = \tfrac{3}{2}\sqrt{2}\,\mathbf{i} - \tfrac{3}{2}\mathbf{j} + \tfrac{3}{2}\mathbf{k}. \quad \square$$

Exercises

1. Simplify
   *(a) $(3\mathbf{a} \cdot \mathbf{b}) - (\mathbf{a} \cdot 2\mathbf{b})$.
   (b) $(\mathbf{a} - \mathbf{b}) \cdot \mathbf{c} + \mathbf{b} \cdot (\mathbf{c} + \mathbf{a})$.
   *(c) $\mathbf{a} \cdot (\mathbf{a} - \mathbf{b}) + \mathbf{b} \cdot (\mathbf{b} + \mathbf{a})$.
   (d) $\mathbf{a} \cdot (\mathbf{a} + 2\mathbf{c}) + (2\mathbf{b} - \mathbf{a}) \cdot (\mathbf{a} + 2\mathbf{c}) - 2\mathbf{b} \cdot (\mathbf{a} + 2\mathbf{c})$.

2. Taking

$$\mathbf{a} = 2\mathbf{i} + \mathbf{j}, \qquad \mathbf{b} = 3\mathbf{i} - \mathbf{j} + 2\mathbf{k}, \qquad \mathbf{c} = 4\mathbf{i} + 3\mathbf{k}$$

   *(a) compute the three dot products, $\mathbf{a} \cdot \mathbf{b}$, $\mathbf{a} \cdot \mathbf{c}$, $\mathbf{b} \cdot \mathbf{c}$;
   (b) find the cosines of the angles between these vectors;
   *(c) find the component of $\mathbf{a}$ in the direction of $\mathbf{b}$ and in the direction of $\mathbf{c}$;
   (d) find the component of $\mathbf{b}$ in the direction of $\mathbf{a}$ and in the direction of $\mathbf{c}$.

*3. Find the unit vector with direction angles $\tfrac{1}{3}\pi, \tfrac{1}{4}\pi, \tfrac{2}{3}\pi$.

4. Find the vector of norm 2 with direction angles $\tfrac{1}{4}\pi, \tfrac{1}{4}\pi, \tfrac{1}{2}\pi$.

5. Find the angle between the vectors:
   *(a) $3\mathbf{i} - \mathbf{j} - 2\mathbf{k}, \mathbf{i} + 2\mathbf{j} - 3\mathbf{k}$.   (b) $2\mathbf{i} - 3\mathbf{j} + \mathbf{k}, -3\mathbf{i} + \mathbf{j} + 9\mathbf{k}$.

6. Find the direction angles of the vector $\mathbf{i} - \mathbf{j} + \sqrt{2}\mathbf{k}$.

7. Find a relation between $\operatorname{comp}_{\mathbf{c}} \mathbf{a}$, $\operatorname{comp}_{\mathbf{c}} \mathbf{b}$, and $\operatorname{comp}_{\mathbf{c}} (\mathbf{a} + \mathbf{b})$.

*8. What can you conclude about $\mathbf{a}$ and $\mathbf{b}$ given that
   (a) $\|\mathbf{a}\|^2 + \|\mathbf{b}\|^2 = \|\mathbf{a} + \mathbf{b}\|^2$?   (b) $\|\mathbf{a}\|^2 + \|\mathbf{b}\|^2 = \|\mathbf{a} - \mathbf{b}\|^2$?
   HINT: Draw figures.

*9. (*Important*) Show that $\mathbf{a} \cdot \mathbf{b} = \mathbf{a} \cdot \mathbf{c}$ does not necessarily imply that $\mathbf{b} = \mathbf{c}$.
   Draw a figure illustrating this for nonzero vectors.

10. Verify that

$$\mathbf{a} \perp \mathbf{b} \quad \text{iff} \quad \|\mathbf{a} + \mathbf{b}\| = \|\mathbf{a} - \mathbf{b}\|$$

by showing that

$$4(\mathbf{a} \cdot \mathbf{b}) = \|\mathbf{a} + \mathbf{b}\|^2 - \|\mathbf{a} - \mathbf{b}\|^2.$$

11. Show that each vector $\mathbf{a}$ can be written

(14.4.12)

$$\mathbf{a} = (\mathbf{a} \cdot \mathbf{i})\mathbf{i} + (\mathbf{a} \cdot \mathbf{j})\mathbf{j} + (\mathbf{a} \cdot \mathbf{k})\mathbf{k}.$$

12. What are the direction angles of $-\mathbf{a}$ if the direction angles of $\mathbf{a}$ are α, β, γ?
13. Under what conditions does $|\mathbf{a} \cdot \mathbf{b}| = \|\mathbf{a}\|\,\|\mathbf{b}\|$?
14. If a force $\mathbf{F}$ is applied throughout a displacement $\mathbf{r}$, then the work done is given by the formula

$$W = (\text{comp}_{\mathbf{r}}\, \mathbf{F})\|\mathbf{r}\|.$$

  *(a) Express the work W as a dot product.
  (b) What is the work done if $\mathbf{F} \perp \mathbf{r}$?
  *(c) Two forces of the same magnitude $\mathbf{F}_1$ and $\mathbf{F}_2$ are applied throughout a displacement $\mathbf{r}$ at angles θ_1 and θ_2 respectively. Compare the work done by $\mathbf{F}_1$ to that done by $\mathbf{F}_2$ if (i) $\theta_1 = -\theta_2$. (ii) $\theta_1 = \frac{1}{3}\pi$, $\theta_2 = \frac{1}{6}\pi$.
  (d) What is the work done by $\mathbf{F}$ if the object moves completely around a triangle? Justify your answer.
*15. Find all the numbers x for which $x\mathbf{i} + 11\mathbf{j} - 3\mathbf{k} \perp 2x\mathbf{i} - x\mathbf{j} - 5\mathbf{k}$.
16. Find all the numbers x for which the angle between $\mathbf{c} = x\mathbf{i} + \mathbf{j} + \mathbf{k}$ and $\mathbf{d} = \mathbf{i} + x\mathbf{j} + \mathbf{k}$ is $\frac{1}{3}\pi$.
*17. Find the angle between the diagonal of a cube and one of its edges.
18. Find the angle between the diagonal of a cube and the diagonal of one of its faces.
19. Show that $\frac{1}{4}\pi$, $\frac{1}{6}\pi$, $\frac{2}{3}\pi$ cannot be the direction angles of a vector.
20. Let $\mathbf{b}$ be a nonzero vector. Express the vector $\mathbf{a}$ as the sum of a vector parallel to $\mathbf{b}$ and a vector perpendicular to $\mathbf{b}$. HINT: Set $\mathbf{a} = s\mathbf{b} + (\mathbf{a} - s\mathbf{b})$ and choose s so that $\mathbf{a} - s\mathbf{b} \perp \mathbf{b}$.

14.5 Lines

Vectors to us are simply ordered triples of real numbers, and, as you've seen, they can be given many interpretations. When vectors are used to indicate position.in three-dimensional space, they are called *position vectors*. In general, we picture position vectors as emanating from the origin. In this section we will use position vectors to characterize lines.

We begin with the idea that two distinct points determine a line. In Figure 14.5.1 we have marked two points and the line l that they determine. To obtain a vector characterization of l, we choose the vectors $\mathbf{r}_0$ and $\mathbf{d}$ as in Figure 14.5.2. Since we began with two distinct points, the vector $\mathbf{d}$ is nonzero.

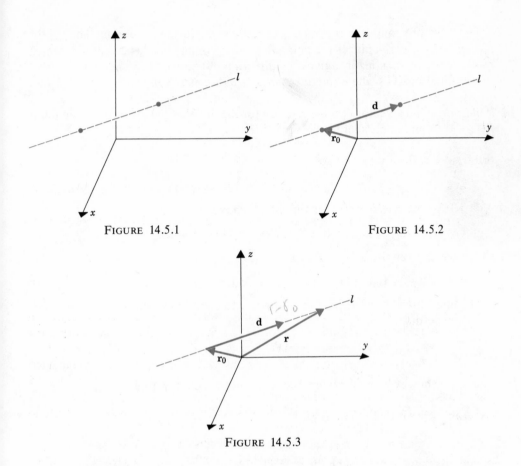

FIGURE 14.5.1   FIGURE 14.5.2

FIGURE 14.5.3

In Figure 14.5.3 we have drawn an additional vector $\mathbf{r}$. What we want is a necessary and sufficient condition on the vector $\mathbf{r}$ for the tip of $\mathbf{r}$ to fall on the line l. This is not hard to find: the tip of $\mathbf{r}$ will fall on l iff

$$\mathbf{r} - \mathbf{r}_0 \quad \text{and} \quad \mathbf{d} \quad \text{are parallel};$$

this in turn will happen iff

$$\mathbf{r} - \mathbf{r}_0 = t\mathbf{d} \quad \text{for some real } t \quad \textit{like } \alpha$$

or, equivalently, iff

$$\mathbf{r} = \mathbf{r}_0 + t\mathbf{d} \quad \text{for some real } t.$$

In view of this we call the equation

(14.5.1) $$\boxed{\mathbf{r}(t) = \mathbf{r}_0 + t\mathbf{d}, \qquad t \text{ real}}$$

a *vector parametrization* for the line l. By varying t, we vary the vector $\mathbf{r}(t)$, but its tip remains on the line l. As t ranges over the set of real numbers, the tip of $\mathbf{r}(t)$ traces out the line l.

The roles of $\mathbf{r}_0$ and $\mathbf{d}$ in this matter deserve some further attention. The vector $\mathbf{r}_0 = x_0\mathbf{i} + y_0\mathbf{j} + z_0\mathbf{k}$ places the point $P(x_0, y_0, z_0)$ on the line. The vector $\mathbf{d}$, which as we remarked is nonzero, gives the direction of the line. It is called a *direction vector* for l, and its scalar components d_1, d_2, d_3 are *direction numbers* for the line.

Problem. Find a vector parametrization for the line that passes through the point $P(1, -1, 2)$ and is parallel to the vector $2\mathbf{i} - 3\mathbf{j} + \mathbf{k}$.

SOLUTION. Here we can take

$$\mathbf{r}_0 = \mathbf{i} - \mathbf{j} + 2\mathbf{k} \quad \text{and} \quad \mathbf{d} = 2\mathbf{i} - 3\mathbf{j} + \mathbf{k}.$$

As a vector parametrization for the line we have

$$\mathbf{r}(t) = (\mathbf{i} - \mathbf{j} + 2\mathbf{k}) + t(2\mathbf{i} - 3\mathbf{j} + \mathbf{k}),$$

which we can rewrite as

$$\mathbf{r}(t) = (1 + 2t)\mathbf{i} - (1 + 3t)\mathbf{j} + (2 + t)\mathbf{k}. \quad \square$$

The line which passes through the origin with direction vector $\mathbf{d}$ can be parametrized by setting

(14.5.2)
$$\boxed{\mathbf{r}(t) = t\mathbf{d}.}$$

There are, however, many other ways to parametrize this line.

Problem. Given that

$$\mathbf{r}(t) = \mathbf{r}_0 + t\mathbf{d}$$

parametrizes the line through the origin with direction vector $\mathbf{d}$, what can we conclude about $\mathbf{r}_0$?

SOLUTION. At first glance it may appear that $\mathbf{r}_0$ has to be $\mathbf{0}$, but this is not true. From the fact that the line passes through the origin we cannot conclude that $\mathbf{r}_0$ is $\mathbf{0}$ but only that there exists a real number t_0 such that

$$\mathbf{r}_0 + t_0\mathbf{d} = \mathbf{0}.$$

This tells us that $\mathbf{r}_0 = -t_0\mathbf{d}$ and thus that $\mathbf{r}_0$ is a scalar multiple of $\mathbf{d}$. $\quad \square$

There are other ways to represent lines. The line

$$l: \mathbf{r}(t) = \mathbf{r}_0 + t\mathbf{d}$$

which passes through $P(x_0, y_0, z_0)$ with direction numbers d_1, d_2, d_3 can also be represented parametrically by the three scalar functions

(14.5.3)
$$\boxed{x(t) = x_0 + d_1t, \quad y(t) = y_0 + d_2t, \quad z(t) = z_0 + d_3t.}$$

These are just the scalar components of $\mathbf{r}(t)$.

Problem. Find a scalar parametrization for the line which passes through $P(-1, 4, 2)$ with direction numbers $(1, 2, 3)$.

SOLUTION

$$x(t) = -1 + t, \quad y(t) = 4 + 2t, \quad z(t) = 2 + 3t. \quad \square$$

Problem. Find the direction numbers of the line

$$x(t) = 3 - t, \quad y(t) = 2 + 4t, \quad z(t) = 1 - 5t.$$

What other direction numbers can be used for the same line?

SOLUTION. The direction numbers are $-1, 4, -5$. All multiples

$$-k, 4k, -5k \quad \text{with} \quad k \neq 0$$

can be used as direction numbers for the same line. $\quad \square$

   If none of the direction numbers is zero, then you can solve each of the equations in (14.5.3) for t:

$$t = \frac{x(t) - x_0}{d_1} = \frac{y(t) - y_0}{d_2} = \frac{z(t) - z_0}{d_3}.$$

Elimination of the parameter t gives

(14.5.4)

$$\frac{x - x_0}{d_1} = \frac{y - y_0}{d_2} = \frac{z - z_0}{d_3}.$$

These equations are called the *symmetric equations* for the line. They are valid only if none of the direction numbers is zero.

Problem. Find a set of symmetric equations for the line which passes through the points $P(x_0, y_0, z_0)$ and $Q(x_1, y_1, z_1)$. Under what conditions are the symmetric equations valid?

SOLUTION. As direction numbers we can take the triple

$$x_1 - x_0, \quad y_1 - y_0, \quad z_1 - z_0.$$

We can then write the symmetric equations as

$$\frac{x - x_0}{x_1 - x_0} = \frac{y - y_0}{y_1 - y_0} = \frac{z - z_0}{z_1 - z_0}.$$

or equivalently as

$$\frac{x - x_1}{x_1 - x_0} = \frac{y - y_1}{y_1 - y_0} = \frac{z - z_1}{z_1 - z_0}.$$

These equations are valid provided that $x_1 \neq x_0, y_1 \neq y_0$, and $z_1 \neq z_0$. $\quad \square$

Two distinct lines

$$l_1: \mathbf{r}(t) = \mathbf{r}_0 + t\mathbf{d}, \quad l_2: \mathbf{R}(u) = \mathbf{R}_0 + u\mathbf{D}$$

intersect iff there are numbers t and u at which

$$\mathbf{r}(t) = \mathbf{R}(u).$$

Problem.   Find the point at which the lines

$$l_1: \mathbf{r}(t) = (\mathbf{i} - 6\mathbf{j} + 2\mathbf{k}) + t(\mathbf{i} + 2\mathbf{j} + \mathbf{k}), \quad l_2: \mathbf{R}(u) = (4\mathbf{j} + \mathbf{k}) + u(2\mathbf{i} + \mathbf{j} + 2\mathbf{k})$$

intersect.

SOLUTION.   We set

$$\mathbf{r}(t) = \mathbf{R}(u)$$

and solve for t and u:

$$(\mathbf{i} - 6\mathbf{j} + 2\mathbf{k}) + t(\mathbf{i} + 2\mathbf{j} + \mathbf{k}) = (4\mathbf{j} + \mathbf{k}) + u(2\mathbf{i} + \mathbf{j} + 2\mathbf{k}),$$
$$\mathbf{i} - 6\mathbf{j} + 2\mathbf{k} + t\mathbf{i} + 2t\mathbf{j} + t\mathbf{k} = 4\mathbf{j} + \mathbf{k} + 2u\mathbf{i} + u\mathbf{j} + 2u\mathbf{k},$$
$$(1 + t - 2u)\mathbf{i} + (-10 + 2t - u)\mathbf{j} + (1 + t - 2u)\mathbf{k} = 0.$$

This tells us that

$$1 + t - 2u = 0,$$
$$-10 + 2t - u = 0,$$
$$1 + t - 2u = 0.$$

Note that the first and third equations are the same. Solving the first two equations simultaneously, we obtain $t = 7$, $u = 4$. As you can verify,

$$\mathbf{r}(7) = 8\mathbf{i} + 8\mathbf{j} + 9\mathbf{k} = \mathbf{R}(4).$$

The two lines intersect at the tip of this vector, which is the point $P(8, 8, 9)$ □

In the setting of plane geometry we can think of two lines as parallel iff they do not intersect. This point of view is not satisfactory in three-dimensional space. See Figure 12.5.4.

FIGURE 14.5.4

The lines l_1 and l_2 marked in the figure do not intersect, and yet we would hesitate to call them parallel. We can get around this difficulty by using direction vectors; we will call two lines *parallel* iff their direction vectors are parallel.

If two lines intersect, we can compute the angle between them by finding the angle between their direction vectors. Depending upon our choice of direction vectors ($\mathbf{d}$ or $-\mathbf{d}$, for example) there are two such angles, θ and $\pi - \theta$. We choose the smaller of the two, the one with nonnegative cosine.

Problem.   Find the angle between the lines

$$l_1: \mathbf{r}(t) = (\mathbf{i} - 6\mathbf{j} + 2\mathbf{k}) + t(\mathbf{i} + 2\mathbf{j} + \mathbf{k}), \quad l_2: \mathbf{R}(u) = (4\mathbf{j} + \mathbf{k}) + u(2\mathbf{i} + \mathbf{j} + 2\mathbf{k}).$$

SOLUTION.   As direction vectors we have

$$\mathbf{d} = \mathbf{i} + 2\mathbf{j} + \mathbf{k} \quad \text{and} \quad \mathbf{D} = 2\mathbf{i} + \mathbf{j} + 2\mathbf{k}.$$

As you can check,

$$\cos \theta = \frac{|\mathbf{d} \cdot \mathbf{D}|}{\|\mathbf{d}\| \|\mathbf{D}\|} = \frac{\sqrt{6}}{3} \cong 0.816 \quad \text{so that} \quad \theta \cong 0.62 \text{ radians.} \quad \square$$

If two parallel vectors are both depicted by arrows emanating from the origin, then the two arrows will both fall on the same line. Because of this, parallel vectors are often called *collinear vectors*.

Exercises

*1. Which of the points $P(1, 2, 0)$, $Q(-5, 1, 5)$, and $R(-4, 2, 5)$ lie on the line

$$l: \mathbf{r}(t) = (\mathbf{i} + 2\mathbf{j}) + t(6\mathbf{i} + \mathbf{j} - 5\mathbf{k})?$$

*2. Determine which of the following lines are parallel:

$$l_1: \mathbf{r}_1(t) = (\mathbf{i} + 2\mathbf{k}) + t(\mathbf{i} - 2\mathbf{j} + 3\mathbf{k}),$$
$$l_2: \mathbf{r}_2(u) = (\mathbf{i} + 2\mathbf{k}) + u(\mathbf{i} + 2\mathbf{j} - 3\mathbf{k}),$$
$$l_3: \mathbf{r}_3(v) = (6\mathbf{i} - \mathbf{j}) - v(2\mathbf{i} + 4\mathbf{j} + 6\mathbf{k}),$$
$$l_4: \mathbf{r}_4(w) = (\tfrac{1}{2} + \tfrac{1}{2}w)\mathbf{i} - w\mathbf{j} + (1 + \tfrac{3}{2}w)\mathbf{k}.$$

*3. The vector $\mathbf{d}$ serves as a direction vector for the line $l: \mathbf{r}(t) = \mathbf{r}_0 + t\mathbf{d}$. What other vectors could be used as direction vectors for this same line?

4. Find a vector parametrization for the line that
   *(a) contains $P(3, 1, 0)$ and is parallel to the line $\mathbf{r}(t) = (\mathbf{i} - \mathbf{j}) + t\mathbf{k}$.
   (b) contains $P(1, -1, 2)$ and is parallel to the line $\mathbf{r}(t) = t(3\mathbf{i} - \mathbf{j} + \mathbf{k})$.
   (c) contains the origin and $Q(x_1, y_1, z_1)$.
   *(d) contains $P(x_0, y_0, z_0)$ and $Q(x_1, y_1, z_1)$.

5. Find a set of scalar parametric equations for the line that
   *(a) contains $P(1, 0, 3)$ and $Q(2, -1, 4)$.
   (b) contains $P(x_0, y_0, z_0)$ and $Q(x_1, y_1, z_1)$.
   *(c) contains $P(2, -2, 3)$ and is perpendicular to the xz-plane.

6. Give a vector parametrization for the line that contains $P(-1, 2, -3)$ and is parallel to

$$\frac{x+1}{2} = \frac{y-3}{1} = \frac{z}{4}.$$

*7. Write a set of symmetric equations for the line that contains the origin and the point $P(x_0, y_0, z_0)$.

8. Where does the line

$$\frac{x-x_0}{d_1} = \frac{y-y_0}{d_2} = \frac{z-z_0}{d_3}$$

intersect the xy-plane?

9. Find the point where l_1 and l_2 intersect and give the angle of intersection.
   (a) l_1: $\mathbf{r}_1(t) = \mathbf{i} + t\mathbf{j}$,
       l_2: $\mathbf{r}_2(u) = \mathbf{j} + u(\mathbf{i} + \mathbf{j})$.
   *(b) l_1: $\mathbf{r}_1(t) = (\mathbf{i} - 4\sqrt{3}\mathbf{j}) + t(\mathbf{i} + \sqrt{3}\mathbf{j})$,
       l_2: $\mathbf{r}_2(u) = (4\mathbf{i} + 3\sqrt{3}\mathbf{j}) + u(\mathbf{i} - \sqrt{3}\mathbf{j})$.
   (c) l_1: $\mathbf{r}_1(t) = -6\mathbf{k} + t(\mathbf{i} - \mathbf{j} + 2\mathbf{k})$,
       l_2: $\mathbf{r}_2(u) = \mathbf{i} + \mathbf{j} - u(\mathbf{i} - 3\mathbf{j} - 2\mathbf{k})$.
   *(d) l_1: $x_1(t) = 3 + t$, $y_1(t) = 1 - t$, $z_1(t) = 5 + 2t$.
       l_2: $x_2(u) = 1$, $y_2(u) = 4 + u$, $z_2(u) = 2 + u$..

*10. What can you conclude about the lines

$$\frac{x-x_0}{d_1} = \frac{y-y_0}{d_2} = \frac{z-z_0}{d_3}, \qquad \frac{x-x_0}{D_1} = \frac{y-y_0}{D_2} = \frac{z-z_0}{D_3}$$

from knowing that

$$d_1 D_1 + d_2 D_2 + d_3 D_3 = 0?$$

*11. What can you conclude about the lines

$$\frac{x-x_0}{d_1} = \frac{y-y_0}{d_2} = \frac{z-z_0}{d_3}, \qquad \frac{x-x_1}{D_1} = \frac{y-y_1}{D_2} = \frac{z-z_1}{D_3}$$

from knowing that

$$\frac{d_1}{D_1} = \frac{d_2}{D_2} = \frac{d_3}{D_3}?$$

12. Let P_0, P_1 be two distinct points and let $\mathbf{r}_0, \mathbf{r}_1$ be the position vectors that they determine:

$$\mathbf{r}_0 = \overrightarrow{OP_0}, \quad \mathbf{r}_1 = \overrightarrow{OP_1}.$$

As t ranges over the set of real numbers,

$$\mathbf{r}(t) = \mathbf{r}_0 + t(\mathbf{r}_1 - \mathbf{r}_0)$$

traces out the line determined by P_0 and P_1. Restrict t so that $\mathbf{r}(t)$ traces out only the line segment $P_0 P_1$.

*13. Given the line

$$l: \mathbf{r}(t) = \mathbf{r}_0 + t\mathbf{d},$$

(a) find the scalar t_0 for which the vector $\mathbf{r}(t_0)$ is perpendicular to l.

(b) find the parametrizations

$$\mathbf{R}(t) = \mathbf{R}_0 + t\mathbf{D}$$

of l in which $\mathbf{R}_0 \perp l$ and $\|\mathbf{D}\| = 1$.

*14. Use Exercise 13 to find the distance between the line

$$l: \mathbf{r}(t) = \mathbf{r}_0 + t\mathbf{d}$$

and the origin given that

(a) $\mathbf{r}_0 = \mathbf{j}$ and $\mathbf{d} = \mathbf{i} - \mathbf{j}.$ (b) $\mathbf{r}_0 = \sqrt{3}\mathbf{i}$ and $\mathbf{d} = \frac{1}{3}\sqrt{3}(\mathbf{i} + \mathbf{j} + \mathbf{k}).$

14.6 Planes

We derived the equation of a line in space by using a point from it and a nonzero vector parallel to it, called a direction vector. We can derive the equation of a plane in space by using a point from it and a nonzero vector *perpendicular* to it, called a *normal vector*.

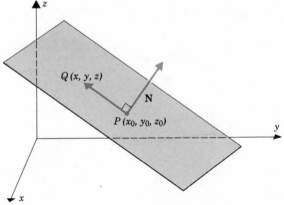

FIGURE 14.6.1

Let the point be $P(x_0, y_0, z_0)$ and let the normal vector be

$$\mathbf{N} = A\mathbf{i} + B\mathbf{j} + C\mathbf{k}.$$

For convenience we depict $\mathbf{N}$ as an arrow starting at $P(x_0, y_0, z_0)$. See Figure 14.6.1. The point $Q(x, y, z)$ of the figure will lie on the given plane iff the normal $\mathbf{N}$ is perpendicular to the vector

$$\overrightarrow{PQ} = (x - x_0)\mathbf{i} + (y - y_0)\mathbf{j} + (z - z_0)\mathbf{k}.$$

This happens iff

$$\mathbf{N} \cdot \overrightarrow{PQ} = 0$$

and thus iff

(14.6.1) $$A(x - x_0) + B(y - y_0) + C(z - z_0) = 0.$$

This is the usual *xyz*-equation for the plane that contains $P(x_0, y_0, z_0)$ and has normal $\mathbf{N} = A\mathbf{i} + B\mathbf{j} + C\mathbf{k}$. Note that the A, B, C of (14.6.1) are the components of the normal vector and hence direction numbers for the normal line.

Problem.   Write an equation for the plane that contains $P(1, 0, 2)$ and has normal $\mathbf{N} = 3\mathbf{i} - 2\mathbf{j} + \mathbf{k}$.

SOLUTION.   The general equation

$$A(x - x_0) + B(y - y_0) + C(z - z_0) = 0$$

becomes

$$3(x - 1) + (-2)(y - 0) + (1)(z - 2) = 0,$$

which simplifies to

$$3x - 2y + z - 5 = 0. \quad \square$$

Problem.   Find an equation for the plane that contains $P(1, 3, -2)$ and is perpendicular to the vector $\mathbf{i} - \mathbf{j}$.

SOLUTION.   We can take $\mathbf{i} - \mathbf{j}$ as the normal:

$$\mathbf{N} = \mathbf{i} - \mathbf{j}.$$

The equation becomes

$$1(x - 1) + (-1)(y - 3) + 0(z + 2) = 0.$$

This simplifies to

$$x - y + 2 = 0.$$

This last equation looks very much like the equation of a line. If our domain of discourse were the *xy*-plane, then the equation $x - y + 2 = 0$ would represent a line. In this case, however, our domain of discourse is three-space. Here the equation $x - y + 2 = 0$ represents the set of all points $Q(x, y, z)$, where $x - y + 2 = 0$ and z is unrestricted. This set forms a vertical plane which intersects the *xy*-plane in the line just mentioned. $\square$

We can write the equation of a plane entirely in vector notation. With

$$\mathbf{N} = A\mathbf{i} + B\mathbf{j} + C\mathbf{k}$$

and

$$\mathbf{r}_0 = x_0\mathbf{i} + y_0\mathbf{j} + z_0\mathbf{k}, \qquad \mathbf{r} = x\mathbf{i} + y\mathbf{j} + z\mathbf{k},$$

equation (14.6.1) becomes

(14.6.2) $$\boxed{\mathbf{N} \cdot (\mathbf{r} - \mathbf{r}_0) = 0.}$$

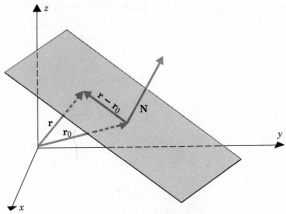

FIGURE 14.6.2

This vector equation represents the plane with normal $\mathbf{N}$ determined by the position vector $\mathbf{r}_0$. (See Figure 14.6.2.) If the plane passes through the origin, then we can take $\mathbf{r}_0 = \mathbf{0}$. Equation (14.6.2) then takes the form

(14.6.3) $$\boxed{\mathbf{N} \cdot \mathbf{r} = 0.}$$

Three vectors $\mathbf{a}, \mathbf{b}, \mathbf{c}$ are said to be *coplanar* iff there are scalars s, t, u not all zero such that

$$s\mathbf{a} + t\mathbf{b} + u\mathbf{c} = \mathbf{0}.$$

Problem. Show that, if $\mathbf{a}, \mathbf{b}, \mathbf{c}$ are coplanar, then the points

$$P(a_1, a_2, a_3), \quad Q(b_1, b_2, b_3), \quad R(c_1, c_2, c_3)$$

all lie on the same plane through the origin.

SOLUTION. If $\mathbf{a}, \mathbf{b}, \mathbf{c}$ are coplanar, then we can write

$$s\mathbf{a} + t\mathbf{b} + u\mathbf{c} = \mathbf{0} \quad \text{with } s, t, u \text{ not all } 0.$$

Let's assume that $s \neq 0$. The plane through the origin that contains Q and R has a vector equation of the form

$$\mathbf{N} \cdot \mathbf{r} = 0$$

Here $\mathbf{N}$ is a normal vector and $\mathbf{r}$ is the position vector of any point of the plane. Since Q and R lie on the plane,

$$\mathbf{N} \cdot \mathbf{b} = 0 \quad \text{and} \quad \mathbf{N} \cdot \mathbf{c} = 0.$$

Consequently,

$$0 = \mathbf{N} \cdot \mathbf{0} = \mathbf{N} \cdot (s\mathbf{a} + t\mathbf{b} + u\mathbf{c}) = s(\mathbf{N} \cdot \mathbf{a}) + t(\mathbf{N} \cdot \mathbf{b}) + u(\mathbf{N} \cdot \mathbf{c}) = s(\mathbf{N} \cdot \mathbf{a}).$$

With $s \neq 0$, we must have $\mathbf{N} \cdot \mathbf{a} = 0$. This means that $P(a_1, a_2, a_3)$ also lies on this same plane. The cases $t \neq 0$ and $u \neq 0$ can be handled in a similar manner. $\square$

Problem.   Show that every equation of the form

$$ax + by + cz + d = 0 \quad \text{with} \quad \sqrt{a^2 + b^2 + c^2} \neq 0$$

represents a plane in space.

SOLUTION.   Since $\sqrt{a^2 + b^2 + c^2} \neq 0$, the numbers a, b, c are not all zero, and therefore there exists a triple x_0, y_0, z_0 such that

$$ax_0 + by_0 + cz_0 + d = 0.\dagger$$

Equation

$$ax + by + cz + d = 0$$

can now be written

$$(ax + by + cz + d) - (ax_0 + by_0 + cz_0 + d) = 0,$$

and thus

$$a(x - x_0) + b(y - y_0) + c(z - z_0) = 0.$$

This equation (and hence the initial equation) represents the plane through the point $P(x_0, y_0, z_0)$ with normal $\mathbf{N} = a\mathbf{i} + b\mathbf{j} + c\mathbf{k}$. $\square$

If $\mathbf{N}$ is normal to a given plane, then all other normals to that plane are parallel to $\mathbf{N}$ and hence scalar multiples of $\mathbf{N}$. In particular there are only two normals of length 1:

$$\frac{\mathbf{N}}{\|\mathbf{N}\|} \quad \text{and} \quad -\frac{\mathbf{N}}{\|\mathbf{N}\|}.$$

These are called the *unit normals*.

Problem.   Find the unit normals for the plane

$$3x - 4y + 12z + 8 = 0.$$

SOLUTION.   We can take

$$\mathbf{N} = 3\mathbf{i} - 4\mathbf{j} + 12\mathbf{k}.$$

Since

$$\|\mathbf{N}\| = \sqrt{(3)^2 + (-4)^2 + (12)^2} = \sqrt{169} = 13,$$

$\dagger$ Would such a triple necessarily exist if $\sqrt{a^2 + b^2 + c^2}$ were zero?

the unit normals are

$$\frac{\mathbf{N}}{\|\mathbf{N}\|} = \frac{1}{13}\,(3\mathbf{i} - 4\mathbf{j} + 12\mathbf{k}) \quad \text{and} \quad -\frac{\mathbf{N}}{\|\mathbf{N}\|} = -\frac{1}{13}\,(3\mathbf{i} - 4\mathbf{j} + 12\mathbf{k}). \quad \square$$

FIGURE 14.6.3

Two planes are called *parallel* iff their normals are parallel. If two planes p_1, p_2 intersect, we can find the angle between them by finding the angle between their normals, $\mathbf{N}_1$, $\mathbf{N}_2$. (See Figure 14.6.3.) Depending on our choice of normals, there are two such angles, each the supplement of the other. We will choose the acute angle, the one with the nonnegative cosine:

(14.6.4)
$$\cos\theta = \frac{|\mathbf{N}_1 \cdot \mathbf{N}_2|}{\|\mathbf{N}_1\|\,\|\mathbf{N}_2\|}.$$

Problem. Let

$$p_1: 2(x - 1) - 3y + 5(z - 2) = 0, \quad p_2: -4x + 6y + 10z + 24 = 0,$$
$$p_3: 4x - 6y - 10z + 1 = 0, \quad\quad\quad p_4: 2x - 3y + 5z - 12 = 0.$$

(a) Indicate which planes are identical.
(b) Indicate which planes are distinct but parallel.
(c) Find the angle between p_1 and p_2.

SOLUTION

(a) p_1 and p_4 are identical, as you can verify by simplifying the equation of p_1.
(b) p_2 and p_3 are distinct but parallel. The planes are distinct since $P(0, 0, \frac{1}{10})$ lies on p_3 but not on p_2; they are parallel since the normals

$$-4\mathbf{i} + 6\mathbf{j} + 10\mathbf{k} \quad \text{and} \quad 4\mathbf{i} - 6\mathbf{j} - 10\mathbf{k}$$

are parallel.
(c) As normals we can take

$$\mathbf{N}_1 = 2\mathbf{i} - 3\mathbf{j} + 5\mathbf{k} \quad \text{and} \quad \mathbf{N}_2 = -4\mathbf{i} + 6\mathbf{j} + 10\mathbf{k}.$$

$$\cos\theta = \frac{|\mathbf{N}_1 \cdot \mathbf{N}_2|}{\|\mathbf{N}_1\|\,\|\mathbf{N}_2\|} = \frac{|-8 - 18 + 50|}{\sqrt{4 + 9 + 25}\sqrt{16 + 36 + 100}} = \frac{6}{19} \cong 0.316.$$

The angle between the planes is about 1.25 radians. $\square$

Problem. Show that the distance from the point $P_1(x_1, y_1, z_1)$ to the plane $Ax + By + Cz + D = 0$ is

$$\frac{|Ax_1 + By_1 + Cz_1 + D|}{\sqrt{A^2 + B^2 + C^2}}.$$

SOLUTION. Pick any point $P_0(x_0, y_0, z_0)$ in the plane. Obvious geometry shows that the desired distance is

$$|\text{comp}_N \overrightarrow{P_0P_1}| = \frac{|N \cdot \overrightarrow{P_0P_1}|}{\|N\|}$$

$$= \frac{|A(x_1 - x_0) + B(y_1 - y_0) + C(z_1 - z_0)|}{\|N\|}$$

$$= \frac{|Ax_1 + By_1 + Cz_1 - (Ax_0 + By_0 + Cz_0)|}{\|N\|}$$

$$= \frac{|Ax_1 + By_1 + Cz_1 + D|}{\sqrt{A^2 + B^2 + C^2}}. \quad \square$$

Exercises

*1. Which of the points $P(3, 2, 1)$, $Q(2, 3, 1)$, and $R(1, 4, 1)$ lie on the plane

$$3(x - 1) + 4y - 5(z + 2) = 0?$$

*2. Which of the points $P(2, 1, -2)$, $Q(2, 0, 0)$, $R(4, 1, -1)$, and $S(0, -1, -3)$ lie on the plane

$$N \cdot (r - r_0) = 0$$

if $N = i - 3j + k$ and $r_0 = 4i + j - k$?

*3. Write an equation for the plane that contains the point $P(2, 3, 4)$ and is perpendicular to $i - 4j + 3k$.

4. Write an equation for the plane that contains the point $P(1, -2, 3)$ and is perpendicular to $j + 2k$.

*5. Write an equation for the plane that passes through the point $(2, 1, 1)$ and is parallel to the plane $3x - 2y + 5z - 2 = 0$.

6. Write an equation for the plane that contains $P_0(x_0, y_0, z_0)$ and is perpendicular to $\overrightarrow{OP_0}$.

7. Show that the plane

$$\frac{x}{a} + \frac{y}{b} + \frac{z}{c} = 1$$

intersects the coordinate axes at $x = a, y = b, z = c$. This is called the *intercept form* of the equation of a plane.

8. Write the equation of the plane $4x + 5y - 6z = 60$ in intercept form.

*9. Where does the plane $3x - y + 4z + 2 = 0$ intersect the coordinate axes?

*10. Find the unit normals for the plane $2x - 3y + 7z - 3 = 0$.

Find the angle between the planes:

*11. $5(x - 1) - 3(y + 2) + 2z = 0$, $x + 3(y - 1) + 2(z + 4) = 0$.
  12. $2x - y + 3z = 5$, $5x + 5y - z = 1$.
*13. $x - y + z - 1 = 0$, $2x + y + 3z + 5 = 0$.

Determine whether or not the vectors are coplanar.

  14. $\mathbf{i}$, $\mathbf{i} - 2\mathbf{j}$, $3\mathbf{j} + \mathbf{k}$.   *15. $4\mathbf{j} - \mathbf{k}$, $3\mathbf{i} + \mathbf{j} + 2\mathbf{k}$, $\mathbf{0}$.
  16. $\mathbf{j} - \mathbf{k}$, $3\mathbf{i} - \mathbf{j} + 2\mathbf{k}$, $3\mathbf{i} - 2\mathbf{j} + 3\mathbf{k}$.   *17. $\mathbf{i} + \mathbf{j} + \mathbf{k}$, $2\mathbf{i} - \mathbf{j}$, $3\mathbf{i} - \mathbf{j} - \mathbf{k}$.

  18. Find the distance from $P(2, -1, 3)$ to the plane $2x + 4y - z + 1 = 0$.
*19. Find the distance between the parallel planes

$$Ax + By + Cz + D_1 = 0 \quad \text{and} \quad Ax + By + Cz + D_2 = 0.$$

  20. Find symmetric equations for the line that passes through $P_0(x_0, y_0, z_0)$ and is perpendicular to the plane $Ax + By + Cz + D = 0$.
  21. Show that two nonparallel lines

$$\mathbf{r}(t) = \mathbf{r}_0 + t\mathbf{d} \quad \text{and} \quad \mathbf{R}(t) = \mathbf{R}_0 + t\mathbf{D}$$

intersect iff the vectors $\mathbf{r}_0 - \mathbf{R}_0$, $\mathbf{d}$, and $\mathbf{D}$ are coplanar.

14.7 The Cross Product

Here we explore a notion that should be of special value to those of you who have an interest in physics, particularly in mechanics and in electromagnetism. In mechanics cross products come up in connection with angular momentum, torque, and other phenomena of turning. In electromagnetism cross products play an important role. In particular they give us a convenient formulation of the law of forces between moving charges.

For vectors

$$\mathbf{a} = a_1\mathbf{i} + a_2\mathbf{j} + a_3\mathbf{k} \quad \text{and} \quad \mathbf{b} = b_1\mathbf{i} + b_2\mathbf{j} + b_3\mathbf{k},$$

we define the *cross product* $\mathbf{a} \times \mathbf{b}$ as follows:

$$(14.7.1) \quad \boxed{\mathbf{a} \times \mathbf{b} = (a_2b_3 - a_3b_2)\mathbf{i} + (a_3b_1 - a_1b_3)\mathbf{j} + (a_1b_2 - a_2b_1)\mathbf{k}.}$$

While the dot product $\mathbf{a} \cdot \mathbf{b}$ is a scalar and is often called the scalar product of $\mathbf{a}$ and $\mathbf{b}$, the cross product $\mathbf{a} \times \mathbf{b}$ is is a vector and is often called the *vector product* of $\mathbf{a}$ and $\mathbf{b}$.

Example. For

$$\mathbf{a} = \mathbf{i} - 2\mathbf{j} + 3\mathbf{k} \quad \text{and} \quad \mathbf{b} = 2\mathbf{i} + \mathbf{j} - \mathbf{k}$$

we have

$$\mathbf{a} \times \mathbf{b} = (a_2b_3 - a_3b_2)\mathbf{i} + (a_3b_1 - a_1b_3)\mathbf{j} + (a_1b_2 - a_2b_1)\mathbf{k}$$
$$= [(-2)(-1) - (3)(1)]\mathbf{i} + [(3)(2) - (1)(-1)]\mathbf{j} + [(1)(1) - (-2)(2)]\mathbf{k}$$
$$= -\mathbf{i} + 7\mathbf{j} + 5\mathbf{k}. \quad \square$$

Example. For

$$\mathbf{a} = \mathbf{i} - \mathbf{j} \quad \text{and} \quad \mathbf{b} = \mathbf{i} + \mathbf{k}$$

we have

$$\mathbf{a} \times \mathbf{b} = (a_2b_3 - a_3b_2)\mathbf{i} + (a_3b_1 - a_1b_3)\mathbf{j} + (a_1b_2 - a_2b_1)\mathbf{k}$$
$$= [(-1)(1) - (0)(0)]\mathbf{i} + [(0)(1) - (1)(1)]\mathbf{j} + [(1)(0) - (-1)(1)]\mathbf{k}$$
$$= -\mathbf{i} - \mathbf{j} + \mathbf{k}. \quad \square$$

If you are familiar with determinants, you will find it easy to remember the defining formula by noting that

(14.7.2)
$$\mathbf{a} \times \mathbf{b} = \begin{vmatrix} \mathbf{i} & \mathbf{j} & \mathbf{k} \\ a_1 & a_2 & a_3 \\ b_1 & b_2 & b_3 \end{vmatrix} = \begin{vmatrix} a_2 & a_3 \\ b_2 & b_3 \end{vmatrix}\mathbf{i} - \begin{vmatrix} a_1 & a_3 \\ b_1 & b_3 \end{vmatrix}\mathbf{j} + \begin{vmatrix} a_1 & a_2 \\ b_1 & b_2 \end{vmatrix}\mathbf{k}.$$

Straightforward calculations that you can carry out yourself show that the cross product has the following properties:

(14.7.3)

$\mathbf{b} \times \mathbf{a} = -(\mathbf{a} \times \mathbf{b}).$	(anticommutative)
$\mathbf{a} \times \mathbf{a} = \mathbf{0}.$	(self-annihilating)
$\alpha\mathbf{a} \times \beta\mathbf{b} = \alpha\beta(\mathbf{a} \times \mathbf{b}).$	(scalars can be factored)
$\mathbf{a} \times (\mathbf{b} + \mathbf{c}) = (\mathbf{a} \times \mathbf{b}) + (\mathbf{a} \times \mathbf{c}).$	($\times$ distributes over $+$)
$\mathbf{a} \times \mathbf{b} \perp \mathbf{a}, \quad \mathbf{a} \times \mathbf{b} \perp \mathbf{b}.$	
$\|\mathbf{a} \times \mathbf{b}\|^2 = \|\mathbf{a}\|^2\|\mathbf{b}\|^2 - (\mathbf{a} \cdot \mathbf{b})^2.$	

We focus on the last two properties. The first of these tells us that $\mathbf{a} \times \mathbf{b}$ is perpendicular both to $\mathbf{a}$ and to $\mathbf{b}$. The other gives the length of $\mathbf{a} \times \mathbf{b}$. From that equation you can see that, if either $\mathbf{a}$ or $\mathbf{b}$ is $\mathbf{0}$, then $\mathbf{a} \times \mathbf{b} = \mathbf{0}$. If, on the other hand, neither $\mathbf{a}$ nor $\mathbf{b}$ is $\mathbf{0}$, then we can write

$$\mathbf{a} \cdot \mathbf{b} = \|\mathbf{a}\| \|\mathbf{b}\| \cos \theta.$$

This gives us

$$\|\mathbf{a} \times \mathbf{b}\|^2 = \|\mathbf{a}\|^2\|\mathbf{b}\|^2 - (\mathbf{a} \cdot \mathbf{b})^2$$
$$= \|\mathbf{a}\|^2\|\mathbf{b}\|^2 - \|\mathbf{a}\|^2\|\mathbf{b}\|^2 \cos^2 \theta$$
$$= \|\mathbf{a}\|^2\|\mathbf{b}\|^2(1 - \cos^2 \theta)$$
$$= \|\mathbf{a}\|^2\|\mathbf{b}\|^2 \sin^2 \theta.$$

Taking square roots, we have

(14.7.4) $$\|\mathbf{a} \times \mathbf{b}\| = \|\mathbf{a}\|\,\|\mathbf{b}\|\,\sin\theta.$$

This means (see Figure 14.7.1) that the length of $\mathbf{a} \times \mathbf{b}$ equals the area of the parallelogram generated by $\mathbf{a}$ and $\mathbf{b}$.

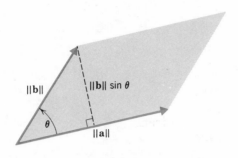

area = $\|\mathbf{a}\|\,\|\mathbf{b}\|\sin\theta = \|\mathbf{a}\times\mathbf{b}\|$

FIGURE 14.7.1

As must be evident,

(14.7.5) $\mathbf{a} \times \mathbf{b} = \mathbf{0}$ iff $\mathbf{a}$ and $\mathbf{b}$ are parallel.

We know that the cross product of $\mathbf{a}$ and $\mathbf{b}$ is perpendicular to $\mathbf{a}$, perpendicular to $\mathbf{b}$, and has length $\|\mathbf{a}\|\,\|\mathbf{b}\|\sin\theta$. Does this determine the vector completely? The answer is no. In general there are two such vectors, each the negative of the other. It can be argued that the vector we want, $\mathbf{a} \times \mathbf{b}$, is the one which makes

$$\mathbf{a}, \mathbf{b}, \mathbf{a} \times \mathbf{b}$$

a right-handed system. (See Figure 14.7.2.) If the index finger of the right hand points along $\mathbf{a}$ and the middle finger along $\mathbf{b}$, then the thumb will point along $\mathbf{a} \times \mathbf{b}$.

FIGURE 14.7.2

As you can verify,

$$\mathbf{i} \times \mathbf{j} = \mathbf{k}, \qquad \mathbf{j} \times \mathbf{k} = \mathbf{i}, \qquad \mathbf{k} \times \mathbf{i} = \mathbf{j}. \qquad \text{(Figure 14.7.3)}$$

FIGURE 14.7.3

An easy way to remember these products is to arrange $\mathbf{i}$, $\mathbf{j}$, $\mathbf{k}$ in cyclic order,

and note that

[each unit vector] $\times$ [the next one] = [the third one].

From this observation and properties

$$\mathbf{b} \times \mathbf{a} = -(\mathbf{a} \times \mathbf{b}),$$
$$\mathbf{a} \times \mathbf{a} = \mathbf{0},$$
$$\alpha \mathbf{a} \times \beta \mathbf{b} = \alpha \beta (\mathbf{a} \times \mathbf{b}),$$
$$\mathbf{a} \times (\mathbf{b} + \mathbf{c}) = (\mathbf{a} \times \mathbf{b}) + (\mathbf{a} \times \mathbf{c}),$$

we can compute cross products without having to appeal to the somewhat complicated defining formula.

Examples

$\mathbf{j} \times \mathbf{i} = -(\mathbf{i} \times \mathbf{j}) = -\mathbf{k}, \quad \mathbf{k} \times \mathbf{j} = -(\mathbf{j} \times \mathbf{k}) = -\mathbf{i}, \quad \mathbf{i} \times \mathbf{k} = -(\mathbf{k} \times \mathbf{i}) = -\mathbf{j}.$

$\mathbf{i} \times \mathbf{i} = \mathbf{j} \times \mathbf{j} = \mathbf{k} \times \mathbf{k} = \mathbf{0}.$

$2\mathbf{i} \times 3\mathbf{k} = 6(\mathbf{i} \times \mathbf{k}) = -6(\mathbf{k} \times \mathbf{i}) = -6\mathbf{j}.$

$\mathbf{k} \times (-3\mathbf{j} + 2\mathbf{k}) = -3(\mathbf{k} \times \mathbf{j}) + 2(\mathbf{k} \times \mathbf{k}) = 3(\mathbf{j} \times \mathbf{k}) + \mathbf{0} = 3\mathbf{i}.$

$$
\begin{aligned}
(\mathbf{i} + \mathbf{j}) \times (-2\mathbf{j} + \mathbf{k}) &= [(\mathbf{i} + \mathbf{j}) \times (-2\mathbf{j})] + [(\mathbf{i} + \mathbf{j}) \times \mathbf{k}] \\
&= [2\mathbf{j} \times (\mathbf{i} + \mathbf{j})] - [\mathbf{k} \times (\mathbf{i} + \mathbf{j})] \\
&= 2(\mathbf{j} \times \mathbf{i}) + 2(\mathbf{j} \times \mathbf{j}) - (\mathbf{k} \times \mathbf{i}) - (\mathbf{k} \times \mathbf{j}) \\
&= -2(\mathbf{i} \times \mathbf{j}) + \mathbf{0} - \mathbf{j} + (\mathbf{j} \times \mathbf{k}) \\
&= -2\mathbf{k} - \mathbf{j} + \mathbf{i} = \mathbf{i} - \mathbf{j} - 2\mathbf{k}. \quad \square
\end{aligned}
$$

Problem.   Prove that

(14.7.6)

$$\boxed{(\mathbf{a} + \mathbf{b}) \times \mathbf{c} = (\mathbf{a} \times \mathbf{c}) + (\mathbf{b} \times \mathbf{c}).}$$

PROOF

$$(\mathbf{a} + \mathbf{b}) \times \mathbf{c} = -[\mathbf{c} \times (\mathbf{a} + \mathbf{b})]$$
$$= -[(\mathbf{c} \times \mathbf{a}) + (\mathbf{c} \times \mathbf{b})]$$
$$= -(\mathbf{c} \times \mathbf{a}) - (\mathbf{c} \times \mathbf{b})$$
$$= (\mathbf{a} \times \mathbf{c}) + (\mathbf{b} \times \mathbf{c}). \quad \square$$

We use (14.7.6) in the next problem.

Problem. Compute $\mathbf{a} \times \mathbf{b}$ given that

$$\mathbf{a} = 2\mathbf{i} - \mathbf{k}, \qquad \mathbf{b} = \mathbf{i} - 3\mathbf{j} + 2\mathbf{k}.$$

SOLUTION

$$\mathbf{a} \times \mathbf{b} = (2\mathbf{i} - \mathbf{k}) \times (\mathbf{i} - 3\mathbf{j} + 2\mathbf{k})$$
$$= [2\mathbf{i} \times (\mathbf{i} - 3\mathbf{j} + 2\mathbf{k})] - [\mathbf{k} \times (\mathbf{i} - 3\mathbf{j} + 2\mathbf{k})]$$
$$= 2(\mathbf{i} \times \mathbf{i}) - 6(\mathbf{i} \times \mathbf{j}) + 4(\mathbf{i} \times \mathbf{k}) - (\mathbf{k} \times \mathbf{i}) + 3(\mathbf{k} \times \mathbf{j}) - 2(\mathbf{k} \times \mathbf{k})$$
$$= 0 - 6\mathbf{k} - 4\mathbf{j} - \mathbf{j} - 3\mathbf{i} - 0 = -3\mathbf{i} - 5\mathbf{j} - 6\mathbf{k}. \quad \square$$

Problem. Find the volume of the parallelepiped with edges $\mathbf{a}, \mathbf{b}, \mathbf{c}$ pictured in Figure 14.7.4.

FIGURE 14.7.4

SOLUTION. As a base we take the parallelogram with edges $\mathbf{b}$ and $\mathbf{c}$. The area of the base is $\|\mathbf{b} \times \mathbf{c}\|$. Since $\mathbf{b} \times \mathbf{c}$ is perpendicular to the base, the height is

$$|\text{comp}_{\mathbf{b} \times \mathbf{c}}\mathbf{a}| = \frac{|\mathbf{a} \cdot (\mathbf{b} \times \mathbf{c})|}{\|\mathbf{b} \times \mathbf{c}\|}.$$

Consequently

$$V = (\text{area of base}) (\text{height}) = |\mathbf{a} \cdot (\mathbf{b} \times \mathbf{c})|. \quad \square$$

The number $\mathbf{a} \cdot (\mathbf{b} \times \mathbf{c})$ that arose in the last problem is called the *triple scalar product* of the vectors $\mathbf{a}, \mathbf{b}, \mathbf{c}$. In terms of coordinates we have

$$\mathbf{a} \cdot (\mathbf{b} \times \mathbf{c}) = \mathbf{a} \cdot [(b_2c_3 - b_3c_2)\mathbf{i} + (b_3c_1 - b_1c_3)\mathbf{j} + (b_1c_2 - b_2c_1)\mathbf{k}]$$

and thus

(14.7.7) $\boxed{\mathbf{a} \cdot (\mathbf{b} \times \mathbf{c}) = a_1(b_2c_3 - b_3c_2) + a_2(b_3c_1 - b_1c_3) + a_3(b_1c_2 - b_2c_1).}$

Those of you familiar with determinants can note that

$$(14.7.8) \qquad \mathbf{a} \cdot (\mathbf{b} \times \mathbf{c}) = \begin{vmatrix} a_1 & a_2 & a_3 \\ b_1 & b_2 & b_3 \\ c_1 & c_2 & c_3 \end{vmatrix}.$$

Problem. Given that

$$\mathbf{a} \cdot (\mathbf{b} \times \mathbf{c}) = 0$$

what can you conclude about **a**, **b**, and **c**?

SOLUTION. If $\mathbf{a} \cdot (\mathbf{b} \times \mathbf{c}) = 0$, then the parallelepiped with edges **a**, **b**, **c** has volume 0. This can only happen if **a**, **b**, and **c** are coplanar. □

From the last problem you know that

$$(14.7.9) \qquad \mathbf{a} \cdot (\mathbf{b} \times \mathbf{c}) = 0 \qquad \text{iff} \qquad \text{\textbf{a}, \textbf{b}, \textbf{c} are coplanar.}$$

We can use this to find a convenient equation for the plane that passes through three noncollinear points P_1, P_2, P_3. A point P will lie on this plane iff the arrows

$$\overrightarrow{P_1P}, \ \overrightarrow{P_1P_2}, \ \overrightarrow{P_1P_3}$$

are coplanar. By (14.7.9) this will happen iff

$$(14.7.10) \qquad \overrightarrow{P_1P} \cdot (\overrightarrow{P_1P_2} \times \overrightarrow{P_1P_3}) = 0.$$

Problem. Find an equation in x, y, z for the plane that passes through the points $P_1(0, 1, 1)$, $P_2(1, 0, 1)$, $P_3(1, 1, 0)$.

SOLUTION. The point $P = P(x, y, z)$ will lie on this plane iff

$$\overrightarrow{P_1P} \cdot (\overrightarrow{P_1P_2} \times \overrightarrow{P_1P_3}) = 0.$$

Here

$$\overrightarrow{P_1P} = x\mathbf{i} + (y - 1)\mathbf{j} + (z - 1)\mathbf{k}, \qquad \overrightarrow{P_1P_2} = \mathbf{i} - \mathbf{j}, \qquad \overrightarrow{P_1P_3} = \mathbf{i} - \mathbf{k}.$$

A calculation that you can carry out using (14.7.7) gives the equation

$$x + y + z - 2 = 0. \quad \square$$

Exercises

Compute.

*1. $(\mathbf{i} + \mathbf{j}) \times (\mathbf{i} - \mathbf{j})$.　2. $(\mathbf{i} - \mathbf{j}) \times (\mathbf{j} - \mathbf{i})$.　*3. $(\mathbf{i} - \mathbf{j}) \times (\mathbf{j} - \mathbf{k})$.
*4. $\mathbf{j} \times (2\mathbf{i} - \mathbf{k})$.　5. $(2\mathbf{j} - \mathbf{k}) \times (\mathbf{i} - 3\mathbf{j})$.　*6. $\mathbf{i} \cdot (\mathbf{j} \times \mathbf{k})$.
*7. $\mathbf{j} \cdot (\mathbf{i} \times \mathbf{k})$.　8. $(\mathbf{j} \times \mathbf{i}) \cdot (\mathbf{i} \times \mathbf{k})$.　*9. $(\mathbf{i} \times \mathbf{j}) \times \mathbf{k}$.

Compute.

*10. $(\mathbf{i} + \mathbf{j} + \mathbf{k}) \times (2\mathbf{i} + \mathbf{k})$.

11. $(2\mathbf{i} - \mathbf{k}) \times (\mathbf{i} - 2\mathbf{j} + 2\mathbf{k})$.

*12. $[2\mathbf{i} + \mathbf{j}] \cdot [(\mathbf{i} - 3\mathbf{j} + \mathbf{k}) \times (4\mathbf{i} + \mathbf{k})]$.

13. $[(-2\mathbf{i} + \mathbf{j} - 3\mathbf{k}) \times \mathbf{i}] \times [\mathbf{i} + \mathbf{j}]$.

*14. $[(\mathbf{i} - \mathbf{j}) \times (\mathbf{j} - \mathbf{k})] \times [\mathbf{i} + 5\mathbf{k}]$.

15. $[\mathbf{i} - \mathbf{j}] \times [(\mathbf{j} - \mathbf{k}) \times (\mathbf{j} + 5\mathbf{k})]$.   *brackets to show you what 1st because x not associative*

16. Show that

$$\mathbf{b} \times \mathbf{a} = -(\mathbf{a} \times \mathbf{b})$$

and deduce that

$$\mathbf{a} \times \mathbf{a} = \mathbf{0}.$$

17. Show that

$$\mathbf{a} \times (\mathbf{b} + \mathbf{c}) = (\mathbf{a} \times \mathbf{b}) + (\mathbf{a} \times \mathbf{c}).$$

18. Show that

$$\mathbf{a} \times \mathbf{b} \perp \mathbf{a} \quad \text{and} \quad \mathbf{a} \times \mathbf{b} \perp \mathbf{b}.$$

19. Show that

$$\|\mathbf{a} \times \mathbf{b}\|^2 = \|\mathbf{a}\|^2 \|\mathbf{b}\|^2 - (\mathbf{a} \cdot \mathbf{b})^2.$$

20. Is $\times$ associative? Namely, is it always true that

$$\mathbf{a} \times (\mathbf{b} \times \mathbf{c}) = (\mathbf{a} \times \mathbf{b}) \times \mathbf{c}? \quad no$$

21. Use a cross product to find the area of triangle PQR.
    *(a) $P(0, 1, 0)$, $Q(-1, 1, 2)$, $R(2, 1, -1)$.
    (b) $P(1, 2, 3)$, $Q(-1, 3, 2)$, $R(3, -1, 2)$.

22. Find the volume of the parallelepiped with edges
    *(a) $\mathbf{i} + \mathbf{j}$, $2\mathbf{i} - \mathbf{k}$, $3\mathbf{j} + \mathbf{k}$.
    (b) $\mathbf{i} - 3\mathbf{j} + \mathbf{k}$, $2\mathbf{j} - \mathbf{k}$, $\mathbf{i} + \mathbf{j} - 2\mathbf{k}$.

23. Express $(\mathbf{a} + \mathbf{b}) \times (\mathbf{a} - \mathbf{b})$ as a scalar multiple of $\mathbf{a} \times \mathbf{b}$.

24. Find an equation in x, y, z for the plane that passes through the points $\overline{P_1 P_2} = 1\mathbf{i} + \mathbf{j} - \mathbf{k}$ $\overline{P_1 P_3} = 0\mathbf{i} + \mathbf{j}$
    *(a) $P_1(1, 0, 1)$, $P_2(2, 1, 0)$, $P_3(1, 1, 1)$.
    (b) $P_1(1, 1, 1)$, $P_2(2, -2, -1)$, $P_3(0, 2, 1)$.

25. Let l be the line determined by P_1, P_2 and let p be the plane determined by Q_1, Q_2, Q_3. Where, if anywhere, does l intersect p?
    *(a) $P_1(1, -1, 2)$, $P_2(-2, 3, 1)$; $Q_1(2, 0, -4)$, $Q_2(1, 2, 3)$, $Q_3(-1, 2, 1)$.
    (b) $P_1(4, -3, 1)$, $P_2(2, -2, 3)$; $Q_1(2, 0, -4)$, $Q_2(1, 2, 3)$, $Q_3(-1, 2, 1)$.

26. Let $\mathbf{a}, \mathbf{b}, \mathbf{c}$ be distinct nonzero vectors. Show that

$$\mathbf{a} \times \mathbf{b} = \mathbf{a} \times \mathbf{c} \quad \text{iff} \quad \mathbf{a} \text{ and } \mathbf{b} - \mathbf{c} \text{ are parallel.}$$

27. Sketch a figure depicting all those vectors $\mathbf{c}$ that satisfy $\mathbf{a} \times \mathbf{b} = \mathbf{a} \times \mathbf{c}$.

28. Some authors write

$$\mathbf{a} \times \mathbf{b} = \|\mathbf{a}\| \|\mathbf{b}\| \sin \theta \, \mathbf{u},$$

where θ is the angle between $\mathbf{a}$ and $\mathbf{b}$ and $\mathbf{u}$ is a unit vector. Find the vector $\mathbf{u}$ which makes the equation correct.

14.8 Some Geometry by Vector Methods

Try your hand at proving the following theorems by vector methods. Follow the hints if you like, but you may find it more interesting to disregard them and come up with proofs that are entirely your own.

Proposition 1   The diagonals of a parallelogram are perpendicular iff the parallelogram is a rhombus.

HINT FOR PROOF.   With $\mathbf{a}$ and $\mathbf{b}$ as in Figure 14.8.1 the diagonals are $\mathbf{a} + \mathbf{b}$ and $\mathbf{a} - \mathbf{b}$. Show that $(\mathbf{a} + \mathbf{b}) \cdot (\mathbf{a} - \mathbf{b}) = 0$ iff $\|\mathbf{a}\| = \|\mathbf{b}\|$. $\square$

FIGURE 14.8.1

FIGURE 14.8.2

Proposition 2   Every angle inscribed in a semicircle is a right angle.

HINT FOR PROOF.   Take $\mathbf{a}$ and $\mathbf{b}$ as in Figure 14.8.2 and show that $\mathbf{a} \cdot \mathbf{b} = 0$. $\square$

Proposition 3   In a parallelogram the sum of the squares of the lengths of the diagonals equals the sum of the squares of the lengths of the sides.

HINT FOR PROOF.   With $\mathbf{a}$ and $\mathbf{b}$ as in Figure 14.8.1 the diagonals are $\mathbf{a} + \mathbf{b}$ and $\mathbf{a} - \mathbf{b}$. Show that $\|\mathbf{a} + \mathbf{b}\|^2 + \|\mathbf{a} - \mathbf{b}\|^2 = 2\|\mathbf{a}\|^2 + 2\|\mathbf{b}\|^2$. $\square$

Proposition 4   The three altitudes of a triangle meet at one point.

HINT FOR PROOF.   As in Figure 14.8.3 assume that the altitudes drawn from P_1 and P_2 intersect at Q. Use the fact that $\overrightarrow{P_1Q} \perp \overrightarrow{OP_2}$ and $\overrightarrow{P_2Q} \perp \overrightarrow{OP_1}$ to show that $\overrightarrow{OQ} \perp \overrightarrow{P_1P_2}$. $\square$

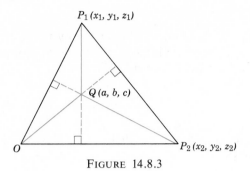

FIGURE 14.8.3

Proposition 5　The medians of a triangle meet at one point.

HINT FOR PROOF.　With l_1, l_2, l_3 as in Figure 14.8.4, l_1: $\mathbf{r}_1(t) = t(\mathbf{a} + \mathbf{b})$, l_2: $\mathbf{r}_2(u) = \frac{1}{2}\mathbf{b} + u(\mathbf{a} - \frac{1}{2}\mathbf{b})$, l_3: $\mathbf{r}_3(v) = \frac{1}{2}\mathbf{a} + v(\mathbf{b} - \frac{1}{2}\mathbf{a})$. Show that l_1 intersects both l_2 and l_3 at the same point.　□

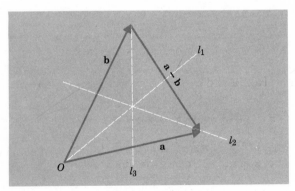

FIGURE 14.8.4

Proposition 6　(The law of sines)　If a triangle has sides $\mathbf{a}$, $\mathbf{b}$, $\mathbf{c}$ and opposite angles A, B, C, then

$$\frac{\sin A}{\|\mathbf{a}\|} = \frac{\sin B}{\|\mathbf{b}\|} = \frac{\sin C}{\|\mathbf{c}\|}.$$

HINT FOR PROOF.   With $\mathbf{a}, \mathbf{b}, \mathbf{c}$ as in Figure 14.8.5, $\mathbf{a} + \mathbf{b} + \mathbf{c} = \mathbf{0}$. Observe then that $\mathbf{a} \times [\mathbf{a} + \mathbf{b} + \mathbf{c}] = \mathbf{0}$ and $\mathbf{b} \times [\mathbf{a} + \mathbf{b} + \mathbf{c}] = \mathbf{0}$. $\square$

FIGURE 14.8.5

Proposition 7   If two distinct planes have a point in common, then they have a line in common.

HINT FOR PROOF.   As equations for the planes take $\mathbf{n} \cdot (\mathbf{r} - \mathbf{r}_0) = 0$ and $\mathbf{N} \cdot (\mathbf{R} - \mathbf{R}_0) = 0$. If the point $P(a_1, a_2, a_3)$ lies on both planes, then the vector $\mathbf{a} = a_1\mathbf{i} + a_2\mathbf{j} + a_3\mathbf{k}$ satisfies both equations. In that case we have $\mathbf{n} \cdot (\mathbf{a} - \mathbf{r}_0) = 0$ and $\mathbf{N} \cdot (\mathbf{a} - \mathbf{R}_0) = 0$. Consider the line $\mathbf{r}(t) = \mathbf{a} + t(\mathbf{n} \times \mathbf{N})$. $\square$

14.9   Additional Exercises

1. Throughout this exercise let

$$P_1 = P_1(3, 2, -1), P_2 = P_2(7, -5, 4), P_3 = P_3(6, 6, -4).$$

*(a) Find the length of the line segment that joins P_1 to P_2.
(b) What are the coordinates of the midpoint of this line segment?
*(c) If one of the points in (a) is the endpoint of a line segment and the other the midpoint, what are the coordinates of the other endpoint?
(d) Find an equation in x, y, z for the set of all points equidistant from P_1 and P_2.
*(e) Find scalar parametric equations for the line which passes through P_3 and is parallel to the line through P_1 and P_2.
(f) Find scalar parametric equations for the line which passes through P_3 and intersects the line through P_1 and P_2 at a right angle.
*(g) The lines in (f) determine a plane. Find an equation for this plane.
2. *(a) Are the points $(3, 2, -1)$, $(7, -5, 4)$, $(-2, 5, 0)$ collinear? That is, do they all lie on the same line?
(b) Are the points $(3, 2, -1)$, $(7, -5, 4)$, $(-2, 5, 0)$, $(2, 1, 6)$ coplanar? That is, do they all lie on the same plane?
3. Let l be the line $x + 1 = y + 2 = z + 1$ and let P be the point $(3, 1, -2)$. Find the point Q on l such that $\overrightarrow{PQ} \perp l$.

4. Write an equation for the plane which contains
   *(a) the points $(1, -2, 1)$, $(2, 0, 3)$, $(0, 1, -1)$.
   (b) the point $(2, 1, -3)$ and is perpendicular to the line

   $$\frac{x+1}{2} = \frac{y-1}{3} = \frac{z}{-4}.$$

   *(c) the point $(1, -2, -1)$ and is parallel to the plane $3x + 2y - z = -4$.
   (d) the lines $x = -1 + 2t$, $y = 2 + 3t$, $z = 1 + t$ and $x = -1 + u$, $y = 2 - u$, $z = 1 + 2u$.
   *(e) the point $(3, -1, 2)$ and the line $x = 2 + 2t$, $y = -1 + 3t$, $z = -2t$.
5. Verify that the plane $5x - 3y - z - 6 = 0$ contains the line $x = 1 + 2t$, $y = -1 + 3t$, $z = 2 + t$.
*6. The planes $2x + y - 3z + 6 = 0$ and $x + 4y - 5z - 7 = 0$ intersect in a line. Find scalar parametric equations for this line.
7. Where does the line

   $$\frac{x+2}{3} = \frac{y-1}{2} = \frac{z+6}{1}$$

   intersect the plane $2x + y - 3z + 6 = 0$?
*8. Find the distance from $(2, 1, -1)$ to the plane $x - 2y + 2z + 5 = 0$.
9. Write an equation for the plane which contains the line $x = -1 + 3t$, $y = 1 + 2t$, $z = 2 + 4t$ and is perpendicular to the plane with equation $2x + y - 3z + 4 = 0$.
*10. Find an equation for the plane which passes through the point $(2, 1, -3)$ and the line determined by the planes $3x + y - z = 2$ and $2x + y + 4z = 1$.
11. Given that $\mathbf{a} = 2\mathbf{i} - \mathbf{j} + 3\mathbf{k}$ and $\mathbf{b} = 3\mathbf{i} + 2\mathbf{j} - \mathbf{k}$, find all vectors of length 4 which are perpendicular to both $\mathbf{a}$ and $\mathbf{b}$.
12. The setting is the xy-plane.
    (a) Verify that the line l: $Ax + By + C = 0$ can be parametrized by setting $\mathbf{r}(t) = (-C/A)\mathbf{i} + (B\mathbf{i} - A\mathbf{j})t$.
    (b) Show that $A\mathbf{i} + B\mathbf{j}$ is normal to l.
    (c) Prove Theorem 9.2.1 by vector methods.
*13. Find the distance from the point $(4, 6, -4)$ to the line joining $(2, 2, 1)$ to $(4, 3, -1)$.
14. Show that the midpoints of the sides of a quadrilateral are always arrayed so as to be the four vertices of a parallelogram.
15. Test your capacity for detailed computation by trying to show that
    (a) $\mathbf{a} \times (\mathbf{b} \times \mathbf{c}) = (\mathbf{a} \cdot \mathbf{c})\mathbf{b} - (\mathbf{a} \cdot \mathbf{b})\mathbf{c}$.　　(b) $(\mathbf{a} \times \mathbf{b}) \times \mathbf{c} = (\mathbf{a} \cdot \mathbf{c})\mathbf{b} - (\mathbf{b} \cdot \mathbf{c})\mathbf{a}$.
16. Show that

    $$(\mathbf{a} \times \mathbf{b}) \times (\mathbf{c} \times \mathbf{d}) = (\mathbf{a} \cdot \mathbf{b} \times \mathbf{d})\mathbf{c} - (\mathbf{a} \cdot \mathbf{b} \times \mathbf{c})\mathbf{d}.$$

Vector Calculus

15.1 Vector Functions

If f_1, f_2, f_3 are real-valued functions defined on some interval I, then for each $t \in I$ we can form the vector

(15.1.1)
$$\boxed{\mathbf{f}(t) = f_1(t)\mathbf{i} + f_2(t)\mathbf{j} + f_3(t)\mathbf{k}}$$

and thereby create a *vector-valued function* $\mathbf{f}$. For short we will call such a function a *vector function*.

Example. Taking

$$f_1(t) = x_0 + d_1 t, \quad f_2(t) = y_0 + d_2 t, \quad f_3(t) = z_0 + d_3 t$$

we can form the vector function

$$\mathbf{f}(t) = (x_0 + d_1 t)\mathbf{i} + (y_0 + d_2 t)\mathbf{j} + (z_0 + d_3 t)\mathbf{k}.$$

If d_1, d_2, d_3 are not all 0, then $\mathbf{f}(t)$ traces out the line that passes through the point $P(x_0, y_0, z_0)$ and has direction numbers d_1, d_2, d_3. If d_1, d_2, d_3 are all 0, then we have the constant function

$$\mathbf{f}(t) = x_0\mathbf{i} + y_0\mathbf{j} + z_0\mathbf{k}. \quad \square$$

Example. From

$$f_1(t) = \cos t, \quad f_2(t) = \sin t, \quad f_3(t) = 0$$

we can form the function

$$\mathbf{f}(t) = \cos t\, \mathbf{i} + \sin t\, \mathbf{j}.$$

For each t

$$\|\mathbf{f}(t)\| = \sqrt{\cos^2 t + \sin^2 t} = 1.$$

As t ranges over the set of real numbers, the position vector $\mathbf{f}(t)$ remains in the xy-plane and its tip traces out the unit circle in a counterclockwise manner, effecting a complete revolution each time that t increases by 2π. (Figure 15.1.1) □

$$\mathbf{f}(t) = \cos t\,\mathbf{i} + \sin t\,\mathbf{j}$$
Figure 15.1.1

The ideas of limit, continuity, derivative, and integral are easily extended to vector functions. Suppose that

$$\mathbf{f}(t) = f_1(t)\mathbf{i} + f_2(t)\mathbf{j} + f_3(t)\mathbf{k}, \qquad t \in I.$$

We say that $\mathbf{f}$ has a *limit* at t_0 provided that each of its components f_1, f_2, f_3 has a limit at t_0; namely, for $\mathbf{L} = l_1\mathbf{i} + l_2\mathbf{j} + l_3\mathbf{k}$, we set

(15.1.2) $\qquad \lim_{t \to t_0} \mathbf{f}(t) = \mathbf{L} \qquad \text{iff} \qquad \lim_{t \to t_0} f_1(t) = l_1, \quad \lim_{t \to t_0} f_2(t) = l_2, \quad \lim_{t \to t_0} f_3(t) = l_3.$

We say that $\mathbf{f}$ is *continuous* at t_0 provided that each of its components is continuous at t_0; that is, provided that

(15.1.3) $\qquad \lim_{t \to t_0} f_1(t) = f_1(t_0), \quad \lim_{t \to t_0} f_2(t) = f_2(t_0), \quad \lim_{t \to t_0} f_3(t) = f_3(t_0).$

More briefly

(15.1.4) $\qquad \mathbf{f}$ is *continuous* at $t_0 \qquad$ iff $\qquad \lim_{t \to t_0} \mathbf{f}(t) = \mathbf{f}(t_0).$

Problem. Find $\lim_{t \to 0} \mathbf{f}(t)$ for

$$\mathbf{f}(t) = \cos (t + \pi)\,\mathbf{i} + 2\,\frac{\sin t}{t}\,\mathbf{j} + e^{-t^2}\,\mathbf{k}.$$

SOLUTION

$$\lim_{t \to 0} \mathbf{f}(t) = \lim_{t \to 0} \left[\cos{(t + \pi)} \, \mathbf{i} + 2 \frac{\sin t}{t} \, \mathbf{j} + e^{-t^2} \, \mathbf{k} \right]$$

$$= \left[\lim_{t \to 0} \cos{(t + \pi)} \right] \mathbf{i} + \left[\lim_{t \to 0} 2 \frac{\sin t}{t} \right] \mathbf{j} + \left[\lim_{t \to 0} e^{-t^2} \right] \mathbf{k}$$

$$= -\mathbf{i} + 2\mathbf{j} + \mathbf{k}. \quad \square$$

We say that **f** is *differentiable* at t provided that each of its components is differentiable at t. If this is the case, we write

$$\mathbf{f}'(t) = f_1'(t)\mathbf{i} + f_2'(t)\mathbf{j} + f_3'(t)\mathbf{k}$$

and call $\mathbf{f}'(t)$ the *derivative* of **f** at t. The derivative $\mathbf{f}'(t)$ can also be written as the limit of a *vector difference quotient*:

(15.1.5)
$$\mathbf{f}'(t) = \lim_{h \to 0} \frac{1}{h} \left[\mathbf{f}(t + h) - \mathbf{f}(t) \right].$$

PROOF

$$\mathbf{f}'(t) = f_1'(t)\mathbf{i} + f_2'(t)\mathbf{j} + f_3'(t)\mathbf{k}$$

$$= \left[\lim_{h \to 0} \frac{f_1(t + h) - f_1(t)}{h} \right] \mathbf{i} + \left[\lim_{h \to 0} \frac{f_2(t + h) - f_2(t)}{h} \right] \mathbf{j} + \left[\lim_{h \to 0} \frac{f_3(t + h) - f_3(t)}{h} \right] \mathbf{k}$$

$$= \lim_{h \to 0} \left[\frac{f_1(t + h) - f_1(t)}{h} \mathbf{i} + \frac{f_2(t + h) - f_2(t)}{h} \mathbf{j} + \frac{f_3(t + h) - f_3(t)}{h} \mathbf{k} \right]$$

$$= \lim_{h \to 0} \frac{1}{h} \left[\mathbf{f}(t + h) - \mathbf{f}(t) \right]. \quad \square$$

Interpretations of the vector derivative and applications of vector differentiation are postponed until Sections 15.3 and 15.4. Here we limit ourselves to computations.

Problem. Given that

$$\mathbf{f}(t) = t\mathbf{i} + \sqrt{t + 1}\,\mathbf{j} - e^t\mathbf{k},$$

find the following:

(1) the domain of **f**.   (2) $\mathbf{f}(0)$.   (3) $\mathbf{f}'(t)$.

(4) $\mathbf{f}'(0)$.   (5) $\|\mathbf{f}(t)\|$.   (6) $\mathbf{f}(t) \cdot \mathbf{f}'(t)$.

SOLUTION

(1) For a number to be in the domain of **f** it is necessary only that it be in the domain of each of its components. The domain of **f** is thus $[-1, \infty)$.

(2) $\mathbf{f}(0) = 0\,\mathbf{i} + \sqrt{0 + 1}\,\mathbf{j} - e^0\,\mathbf{k} = \mathbf{j} - \mathbf{k}$.

(3) $\mathbf{f}'(t) = \mathbf{i} + \dfrac{1}{2\sqrt{t + 1}}\mathbf{j} - e^t\,\mathbf{k}$.

(4) $\mathbf{f}'(0) = \mathbf{i} + \dfrac{1}{2\sqrt{0 + 1}}\mathbf{j} - e^0\,\mathbf{k} = \mathbf{i} + \tfrac{1}{2}\mathbf{j} - \mathbf{k}$.

(5) $\|\mathbf{f}(t)\| = \sqrt{t^2 + (\sqrt{t+1})^2 + (e^t)^2} = \sqrt{t^2 + t + 1 + e^{2t}}.$

(6) $\mathbf{f}(t) \cdot \mathbf{f}'(t) = (t\mathbf{i} + \sqrt{t+1}\,\mathbf{j} - e^t\,\mathbf{k}) \cdot \left(\mathbf{i} + \dfrac{1}{2\sqrt{t+1}}\,\mathbf{j} - e^t\,\mathbf{k}\right)$

$$= (t)(1) + (\sqrt{t+1})\left(\frac{1}{2\sqrt{t+1}}\right) + (-e^t)(-e^t) = t + \tfrac{1}{2} + e^{2t}. \quad \square$$

In differentiating vector functions we need not stop at the first derivative. If $\mathbf{f}'$ is itself differentiable, we can form the second derivative $\mathbf{f}''$ and so on.

Problem. Find $\mathbf{f}''(t)$ for $\mathbf{f}(t) = t \sin t \,\mathbf{i} + e^{-t}\,\mathbf{j} + t\mathbf{k}.$

SOLUTION

$$\mathbf{f}'(t) = (t \cos t + \sin t)\mathbf{i} - e^{-t}\mathbf{j} + \mathbf{k},$$
$$\mathbf{f}''(t) = (-t \sin t + \cos t + \cos t)\mathbf{i} + e^{-t}\mathbf{j} = (2 \cos t - t \sin t)\mathbf{i} + e^{-t}\mathbf{j}. \quad \square$$

Integration can also be handled component by component: set

(15.1.6) $\qquad \boxed{\displaystyle \int_a^b \mathbf{f}(t)\,dt = \left(\int_a^b f_1(t)\,dt\right)\mathbf{i} + \left(\int_a^b f_2(t)\,dt\right)\mathbf{j} + \left(\int_a^b f_3(t)\,dt\right)\mathbf{k}.}$

Problem. Find

$$\int_0^1 \mathbf{f}(t)\,dt \qquad \text{for} \qquad \mathbf{f}(t) = t\mathbf{i} + \sqrt{t+1}\,\mathbf{j} - e^t\,\mathbf{k}.$$

SOLUTION

$$\int_0^1 \mathbf{f}(t)\,dt = \left(\int_0^1 t\,dt\right)\mathbf{i} + \left(\int_0^1 \sqrt{t+1}\,dt\right)\mathbf{j} + \left(\int_0^1 (-e^t)\,dt\right)\mathbf{k}$$
$$= \left[\tfrac{1}{2}t^2\right]_0^1 \mathbf{i} + \left[\tfrac{2}{3}(t+1)^{3/2}\right]_0^1 \mathbf{j} + \left[-e^t\right]_0^1 \mathbf{k}$$
$$= \tfrac{1}{2}\mathbf{i} + \tfrac{2}{3}(2\sqrt{2} - 1)\mathbf{j} + (1 - e)\mathbf{k}. \quad \square$$

Problem. Find $\mathbf{f}(t)$ given that

$$\mathbf{f}'(t) = 2 \cos t \,\mathbf{i} - t \sin t^2 \,\mathbf{j} + 2t\mathbf{k} \quad \text{and} \quad \mathbf{f}(0) = \mathbf{i} + 3\mathbf{k}.$$

SOLUTION. By integrating $\mathbf{f}'(t)$ we get

$$\mathbf{f}(t) = (2 \sin t + c_1)\mathbf{i} + (\tfrac{1}{2}\cos t^2 + c_2)\mathbf{j} + (t^2 + c_3)\mathbf{k},$$

where c_1, c_2, c_3 are constants to be determined. Since

$$\mathbf{i} + 3\mathbf{k} = \mathbf{f}(0) = c_1\mathbf{i} + (\tfrac{1}{2} + c_2)\mathbf{j} + c_3\mathbf{k},$$

you can see that

$$c_1 = 1, \quad c_2 = -\tfrac{1}{2}, \quad c_3 = 3.$$

Thus

$$\mathbf{f}(t) = (2 \sin t + 1)\mathbf{i} + (\tfrac{1}{2} \cos t^2 - \tfrac{1}{2})\mathbf{j} + (t^2 + 3)\mathbf{k}. \quad \square$$

In passing, we point out that

(15.1.7)
$$\int_a^b [\mathbf{f}(t) + \mathbf{g}(t)] \, dt = \int_a^b \mathbf{f}(t) \, dt + \int_a^b \mathbf{g}(t) \, dt$$

and

(15.1.8)
$$\int_a^b [\alpha \mathbf{f}(t)] \, dt = \alpha \int_a^b \mathbf{f}(t) \, dt \qquad \text{for every scalar } \alpha.$$

Not so obvious is that

(15.1.9)
$$\int_a^b [\mathbf{c} \cdot \mathbf{f}(t)] \, dt = \mathbf{c} \cdot \int_a^b \mathbf{f}(t) \, dt \qquad \text{for every vector } \mathbf{c}$$

and

(15.1.10)
$$\left\| \int_a^b \mathbf{f}(t) \, dt \right\| \le \int_a^b \|\mathbf{f}(t)\| \, dt.$$

The proof of (15.1.9) is left as an exercise. Here we prove (15.1.10).

PROOF. Set

$$\mathbf{c} = \int_a^b \mathbf{f}(t) \, dt$$

and note that

$$\|\mathbf{c}\|^2 = \mathbf{c} \cdot \mathbf{c} = \mathbf{c} \cdot \int_a^b \mathbf{f}(t) \, dt = \int_a^b [\mathbf{c} \cdot \mathbf{f}(t)] \, dt \le \int_a^b \|\mathbf{c}\| \|\mathbf{f}(t)\| \, dt = \|\mathbf{c}\| \int_a^b \|\mathbf{f}(t)\| \, dt.$$

by (15.1.9)   by Schwarz's inequality (14.4.7)

$|\mathbf{a} \cdot \mathbf{b}| \le \|\mathbf{a}\| \|\mathbf{b}\|$

If $\mathbf{c} \ne \mathbf{0}$, we can divide by $\|\mathbf{c}\|$ and conclude that

$$\|\mathbf{c}\| \le \int_a^b \|\mathbf{f}(t)\| \, dt.$$

If $\mathbf{c} = \mathbf{0}$, the result is obvious in the first place. $\square$

Exercises

Differentiate.

*1. $\mathbf{f}(t) = (1 + 2t)\mathbf{i} + (3 - t)\mathbf{j} + (2 + 3t)\mathbf{k}.$
  2. $\mathbf{f}(t) = 2\mathbf{i} - \cos t \, \mathbf{k}.$

*3. $\mathbf{f}(t) = \sqrt{1-t}\,\mathbf{i} + \sqrt{1+t}\,\mathbf{j} + \dfrac{1-t}{1+t}\,\mathbf{k}$.

4. $\mathbf{f}(t) = \dfrac{1-t^2}{1+t^2}\,\mathbf{i} + \dfrac{2t}{1+t^2}\,\mathbf{j}$.

*5. $\mathbf{f}(t) = \sin t\,\mathbf{i} + \cos t\,\mathbf{j} + \tan t\,\mathbf{k}$.

6. $\mathbf{f}(t) = e^t\,\mathbf{i} + \log t\,\mathbf{j} + \arctan t\,\mathbf{k}$.

*7. $\mathbf{f}(t) = \arcsin t\,\mathbf{i} + \log(1+2t)\,\mathbf{j} + t^2\,\mathbf{k}$.

Compute.

8. $\displaystyle\int_1^2 \mathbf{f}(t)\,dt$ for $\mathbf{f}(t) = \mathbf{i} + 2t\mathbf{j}$.

*9. $\displaystyle\int_0^\pi \mathbf{r}(t)\,dt$ for $\mathbf{r}(t) = \sin t\,\mathbf{i} + \cos t\,\mathbf{j} + t\,\mathbf{k}$.

10. $\displaystyle\int_0^1 \mathbf{g}(t)\,dt$ for $\mathbf{g}(t) = e^t\mathbf{i} + \sqrt{2}\,\mathbf{j} + e^{-t}\mathbf{k}$.

*11. $\displaystyle\int_0^1 \mathbf{h}(t)\,dt$ for $\mathbf{h}(t) = e^{-t}[t^2\mathbf{i} + \sqrt{2}t\,\mathbf{j} + \mathbf{k}]$.

Find $\mathbf{f}(t)$ given that

*12. $\mathbf{f}'(t) = \mathbf{i} + t^2\mathbf{j}$ and $\mathbf{f}(0) = \mathbf{j} - \mathbf{k}$.

13. $\mathbf{f}'(t) = t\mathbf{i} + \dfrac{t}{\sqrt{1+t^2}}\,\mathbf{j} + te^t\mathbf{k}$ and $\mathbf{f}(0) = \mathbf{i} + 2\mathbf{j} + 3\mathbf{k}$.

*14. $\mathbf{f}'(t) = 2\mathbf{f}(t)$ and $\mathbf{f}(0) = \mathbf{i} - \mathbf{k}$.

15. Show that, if $\mathbf{f}'(t) = \mathbf{0}$ for all t in some interval I, then $\mathbf{f}$ is constant on I. (That is, there exists a vector $\mathbf{c}$ such that $\mathbf{f}(t) = \mathbf{c}$ for all $t \in I$.)

16. Show that, if $\mathbf{f}'(t) = \mathbf{g}'(t)$ for all t in some interval I, then $\mathbf{f}$ and $\mathbf{g}$ differ by a constant on I.

17. Prove (15.1.9).

18. (*Important*) Show that

(15.1.11)

$$\lim_{t \to t_0} \mathbf{f}(t) = \mathbf{L} \quad \text{iff} \quad \lim_{t \to t_0} \|\mathbf{f}(t) - \mathbf{L}\| = 0.$$

15.2 Differentiation Formulas

Vector functions with a common domain can be combined in many ways to form new functions. From $\mathbf{f}$ and $\mathbf{g}$ we can form the sum $\mathbf{f} + \mathbf{g}$:

$$(\mathbf{f} + \mathbf{g})(t) = \mathbf{f}(t) + \mathbf{g}(t).$$

We can form scalar multiples $\alpha\mathbf{f}$ and thus linear combinations $\alpha\mathbf{f} + \beta\mathbf{g}$:

$$(\alpha\mathbf{f})(t) = \alpha\mathbf{f}(t), \qquad (\alpha\mathbf{f} + \beta\mathbf{g})(t) = \alpha\mathbf{f}(t) + \beta\mathbf{g}(t).$$

We can form the dot product $\mathbf{f} \cdot \mathbf{g}$:

$$(\mathbf{f} \cdot \mathbf{g})(t) = \mathbf{f}(t) \cdot \mathbf{g}(t).$$

We can also form the cross product $\mathbf{f} \times \mathbf{g}$:

$$(\mathbf{f} \times \mathbf{g})(t) = \mathbf{f}(t) \times \mathbf{g}(t).$$

There are two ways to bring scalar functions (real-valued functions) into play. If the scalar function u has the same domain as $\mathbf{f}$, we can form the product $u\mathbf{f}$:

$$(u\mathbf{f})(t) = u(t)\mathbf{f}(t).$$

If for each t in some interval $u(t)$ is in the domain of $\mathbf{f}$, then we can form the composition $\mathbf{f} \circ u$:

$$(\mathbf{f} \circ u)(t) = \mathbf{f}(u(t)).$$

For example, with

$$u(t) = \sin t \quad \text{and} \quad \mathbf{f}(t) = e^t\mathbf{i} + e^{-t}\mathbf{j}$$

we have

$$(u\mathbf{f})(t) = u(t)\mathbf{f}(t) = \sin t(e^t\mathbf{i} + e^{-t}\mathbf{j}) \quad \text{and} \quad (\mathbf{f} \circ u)(t) = \mathbf{f}(u(t)) = e^{\sin t}\mathbf{i} + e^{-\sin t}\mathbf{j}.$$

It's not hard to see that if the original functions are continuous, then so are the newly constructed ones. Moreover, if the original functions are differentiable, then so are the new ones. The derivatives can be computed with the aid of the following rules:

(15.2.1)

> (1) $(\mathbf{f} + \mathbf{g})'(t) = \mathbf{f}'(t) + \mathbf{g}'(t).$
> (2) $(\alpha\mathbf{f})'(t) = \alpha\mathbf{f}'(t).$
> (3) $(u\mathbf{f})'(t) = u(t)\mathbf{f}'(t) + u'(t)\mathbf{f}(t).$
> (4) $(\mathbf{f} \cdot \mathbf{g})'(t) = [\mathbf{f}(t) \cdot \mathbf{g}'(t)] + [\mathbf{f}'(t) \cdot \mathbf{g}(t)],$
> (5) $(\mathbf{f} \times \mathbf{g})'(t) = [\mathbf{f}(t) \times \mathbf{g}'(t)] + [\mathbf{f}'(t) \times \mathbf{g}(t)].$
> (6) $(\mathbf{f} \circ u)'(t) = \mathbf{f}'(u(t))u'(t) = u'(t)\mathbf{f}'(u(t)).$ (the chain rule)

We'll take up the derivation of these rules later. First a few comments. Rules (1) and (2) look exactly like those we used for differentiating ordinary functions. Rule (2) is just a special case of Rule (3): set $u(t) = \alpha$ for all t. Rules (3), (4), and (5) are all "product" rules and should remind you of the rule for differentiating the product of ordinary functions. Keep in mind, however, that $\times$ is not commutative and therefore the order in Rule (5) is important. In Rule (6) we first wrote the scalar part $u'(t)$ on the right so that the formula would look like the chain rule for ordinary functions. In general, $\mathbf{a}\alpha$ has the same meaning as $\alpha\mathbf{a}$.

Examples. Taking

$$\mathbf{f}(t) = 2t^2\mathbf{i} - 3\mathbf{j}, \quad \mathbf{g}(t) = \mathbf{i} + t\mathbf{j} + t^2\mathbf{k}, \quad u(t) = \tfrac{1}{3}t^3$$

we have

$$\mathbf{f}'(t) = 4t\mathbf{i}, \quad \mathbf{g}'(t) = \mathbf{j} + 2t\mathbf{k}, \quad u'(t) = t^2.$$

Thus

(a) $(\mathbf{f} + \mathbf{g})'(t) = \mathbf{f}'(t) + \mathbf{g}'(t)$

$$= 4t\mathbf{i} + (\mathbf{j} + 2t\mathbf{k}) = 4t\mathbf{i} + \mathbf{j} + 2t\mathbf{k};$$

(b) $(u\mathbf{f})'(t) = u(t)\mathbf{f}'(t) + u'(t)\mathbf{f}(t)$

$$= \tfrac{1}{3}t^3(4t\mathbf{i}) + t^2(2t^2\mathbf{i} - 3\mathbf{j}) = \tfrac{10}{3}t^4\mathbf{i} - 3t^2\mathbf{j};$$

(c) $(\mathbf{f} \cdot \mathbf{g})'(t) = [\mathbf{f}(t) \cdot \mathbf{g}'(t)] + [\mathbf{f}'(t) \cdot \mathbf{g}(t)]$

$$= [(2t^2\mathbf{i} - 3\mathbf{j}) \cdot (\mathbf{j} + 2t\mathbf{k})] + [4t\mathbf{i} \cdot (\mathbf{i} + t\mathbf{j} + t^2\mathbf{k})] = -3 + 4t;$$

(d) $(\mathbf{f} \times \mathbf{g})'(t) = [\mathbf{f}(t) \times \mathbf{g}'(t)] + [\mathbf{f}'(t) \times \mathbf{g}(t)]$

$$= [(2t^2\mathbf{i} - 3\mathbf{j}) \times (\mathbf{j} + 2t\mathbf{k})] + [4t\mathbf{i} \times (\mathbf{i} + t\mathbf{j} + t^2\mathbf{k})]$$

$$= (2t^2\mathbf{k} - 4t^3\mathbf{j} - 6t\mathbf{i}) + (4t^2\mathbf{k} - 4t^3\mathbf{j}) = -6t\mathbf{i} - 8t^3\mathbf{j} + 6t^2\mathbf{k}$$

while

$(\mathbf{g} \times \mathbf{f})'(t) = [\mathbf{g}(t) \times \mathbf{f}'(t)] + [\mathbf{g}'(t) \times \mathbf{f}(t)]$

$$= [(\mathbf{i} + t\mathbf{j} + t^2\mathbf{k}) \times 4t\mathbf{i}] + [(\mathbf{j} + 2t\mathbf{k}) \times (2t^2\mathbf{i} - 3\mathbf{j})]$$

$$= (-4t^2\mathbf{k} + 4t^3\mathbf{j}) + (-2t^2\mathbf{k} + 4t^3\mathbf{j} + 6t\mathbf{i})$$

$$= 6t\mathbf{i} + 8t^3\mathbf{j} - 6t^2\mathbf{k} = -(\mathbf{f} \times \mathbf{g})'(t);$$

(f) $(\mathbf{f} \circ u)'(t) = \mathbf{f}'(u(t))u'(t)$

$$= [4u(t)\mathbf{i}]u'(t) = [4(\tfrac{1}{3}t^3)\mathbf{i}]\, t^2 = \tfrac{4}{3}t^5\mathbf{i}. \quad \square$$

These differentiation formulas can all be derived by differentiating the functions in question component by component. Take for example formula (3):

$$(u\mathbf{f})'(t) = u(t)\mathbf{f}'(t) + u'(t)\mathbf{f}(t).$$

DERIVATION. Set

$$\mathbf{f}(t) = f_1(t)\mathbf{i} + f_2(t)\mathbf{j} + f_3(t)\mathbf{k}.$$

Then

$$(u\mathbf{f})(t) = u(t)\mathbf{f}(t) = u(t)f_1(t)\mathbf{i} + u(t)f_2(t)\mathbf{j} + u(t)f_3(t)\mathbf{k}$$

and

$$(u\mathbf{f})'(t) = [u(t)f_1'(t) + u'(t)f_1(t)]\mathbf{i} + [u(t)f_2'(t) + u'(t)f_2(t)]\mathbf{j}$$

$$+ [u(t)f_3'(t) + u'(t)f_3(t)]\mathbf{k}$$

$$= u(t)[f_1'(t)\mathbf{i} + f_2'(t)\mathbf{j} + f_3'(t)\mathbf{k}] + u'(t)[f_1(t)\mathbf{i} + f_2(t)\mathbf{j} + f_3(t)\mathbf{k}]$$

$$= u(t)\mathbf{f}'(t) + u'(t)\mathbf{f}(t). \quad \square$$

Component-by-component derivation of the other formulas is left to you as an exercise.

Let's take another look at Formula (3), this time from a different point of view. Instead of casting everything in terms of components, we will work with the vectors directly.

To evaluate $(u\mathbf{f})'(t)$ is to compute the limit as $h \to 0$ of the difference quotient

(∗)
$$\frac{1}{h}[u(t + h)\mathbf{f}(t + h) - u(t)\mathbf{f}(t)].$$

By adding and subtracting $u(t + h)\mathbf{f}(t)$ inside the brackets and then factoring, we can rewrite this difference quotient as

$$u(t + h)\frac{1}{h}[\mathbf{f}(t + h) - \mathbf{f}(t)] + \frac{1}{h}[u(t + h) - u(t)]\mathbf{f}(t).$$

It is clear that as $h \to 0$

$$u(t + h) \to u(t), \quad \frac{1}{h}[\mathbf{f}(t + h) - \mathbf{f}(t)] \to \mathbf{f}'(t), \quad \frac{1}{h}[u(t + h) - u(t)] \to u'(t).$$

It can now be argued that $(*)$ tends to

$$u(t)\mathbf{f}'(t) + u'(t)\mathbf{f}(t). \quad \square$$

It's instructive to apply the same point of view to Formula (4):

$$(\mathbf{f} \cdot \mathbf{g})'(t) = [\mathbf{f}(t) \cdot \mathbf{g}'(t)] + [\mathbf{f}'(t) \cdot \mathbf{g}(t)].$$

On the left we have the limit as $h \to 0$ of the difference quotient

$$\frac{1}{h}[\mathbf{f}(t + h) \cdot \mathbf{g}(t + h) - \mathbf{f}(t) \cdot \mathbf{g}(t)].$$

Adding and subtracting $\mathbf{f}(t + h) \cdot \mathbf{g}(t)$ inside the brackets and rearranging terms, we obtain

$$\mathbf{f}(t + h) \cdot \frac{1}{h}[\mathbf{g}(t + h) - \mathbf{g}(t)] + \frac{1}{h}[\mathbf{f}(t + h) - \mathbf{f}(t)] \cdot \mathbf{g}(t).$$

It's not hard to see now that as h tends to 0 the entire expression tends to

$$[\mathbf{f}(t) \cdot \mathbf{g}'(t)] + [\mathbf{f}'(t) \cdot \mathbf{g}(t)]. \quad \square$$

All these differentiation formulas can be derived in a component-free manner.

Often Leibniz's notation is used. Then the formulas take the following form:

(15.2.2)

$$(1) \quad \frac{d}{dt}[\mathbf{f}(t) + \mathbf{g}(t)] = \frac{d}{dt}[\mathbf{f}(t)] + \frac{d}{dt}[\mathbf{g}(t)].$$

$$(2) \quad \frac{d}{dt}[\alpha\mathbf{f}(t)] = \alpha\frac{d}{dt}[\mathbf{f}(t)].$$

$$(3) \quad \frac{d}{dt}[u(t)\mathbf{f}(t)] = u(t)\frac{d}{dt}[\mathbf{f}(t)] + \frac{d}{dt}[u(t)]\mathbf{f}(t).$$

$$(4) \quad \frac{d}{dt}[\mathbf{f}(t) \cdot \mathbf{g}(t)] = \left(\mathbf{f}(t) \cdot \frac{d}{dt}[\mathbf{g}(t)]\right) + \left(\frac{d}{dt}[\mathbf{f}(t)] \cdot \mathbf{g}(t)\right).$$

$$(5) \quad \frac{d}{dt}[\mathbf{f}(t) \times \mathbf{g}(t)] = \left(\mathbf{f}(t) \times \frac{d}{dt}[\mathbf{g}(t)]\right) + \left(\frac{d}{dt}[\mathbf{f}(t)] \times \mathbf{g}(t)\right).$$

$$(6) \quad \frac{d}{dt}[\mathbf{f}(u(t))] = \frac{d\mathbf{f}}{du}[u(t)]\frac{du}{dt}(t).$$

Exercises

Compute $\mathbf{f}'(t)$ and $\mathbf{f}''(t)$ for each of the following functions.

*1. $\mathbf{f}(t) = \mathbf{a} + t\mathbf{b}$.

2. $\mathbf{f}(t) = \mathbf{a} + t\mathbf{b} + t^2\mathbf{c}$.

*3. $\mathbf{f}(t) = e^{2t}\mathbf{i} - \sin t\,\mathbf{j}$.

4. $\mathbf{f}(t) = [(t^2\mathbf{i} - \mathbf{j}) \cdot (\mathbf{i} - t^2\mathbf{j})]\mathbf{i}$.

*5. $\mathbf{f}(t) = \mathbf{g}(t^2)$.

6. $\mathbf{f}(t) = t\mathbf{g}(t^2)$.

*7. $\mathbf{f}(t) = t\mathbf{g}(\sqrt{t})$.

8. $\mathbf{f}(t) = (\cos t\,\mathbf{i} - \mathbf{j}) \times (t\,\mathbf{j} + \sin t\,\mathbf{k})$.

Find

*9. $\dfrac{d}{dt}\,[e^{\cos t}\,\mathbf{i} + e^{\sin t}\,\mathbf{j}]$.

10. $\dfrac{d^2}{dt^2}\,[e^t \cos t\,\mathbf{i} + e^t \sin t\,\mathbf{j}]$.

*11. $\dfrac{d^3}{dt^3}\,[e^t\,\mathbf{i} - e^{-t}\,\mathbf{j}]$.

12. $\dfrac{d}{dt}\,[(\mathbf{a} + t\mathbf{b}) \times (\mathbf{c} + t\mathbf{d})]$.

*13. $\dfrac{d^2}{dt^2}\,[(\mathbf{a} + t\mathbf{b}) \times (\mathbf{c} + t\mathbf{d})]$.

14. $\dfrac{d}{dt}\,[(\mathbf{a} + t\mathbf{b}) \cdot (\mathbf{a} + t\mathbf{b} + t^2\mathbf{c})]$.

15. Show that

$$\frac{d}{dt}\,[u_1(t)\mathbf{r}_1(t) \times u_2(t)\mathbf{r}_2(t)]$$

can be written

$$u_1(t)u_2(t)\frac{d}{dt}\,[\mathbf{r}_1(t) \times \mathbf{r}_2(t)] + [\mathbf{r}_1(t) \times \mathbf{r}_2(t)]\frac{d}{dt}\,[u_1(t)u_2(t)].$$

Find $\mathbf{f}(t)$ given that:

16. $\mathbf{f}'(t) = \mathbf{b}$ for all real t, $\mathbf{f}(0) = \mathbf{a}$.

*17. $\mathbf{f}''(t) = \mathbf{c}$ for all real t, $\mathbf{f}'(0) = \mathbf{b}$, $\mathbf{f}(0) = \mathbf{a}$.

18. $\mathbf{f}''(t) = \mathbf{a} + t\mathbf{b}$ for all real t, $\mathbf{f}'(0) = \mathbf{c}$, $\mathbf{f}(0) = \mathbf{d}$.

19. Show that, if $\mathbf{f}$ is differentiable at t, then $\mathbf{f}$ is continuous at t.

20. Show that

$$(\mathbf{g} \times \mathbf{f})'(t) = -(\mathbf{f} \times \mathbf{g})'(t).$$

21. Show that

$$\frac{d}{dt}\,[\mathbf{f}(t) \times \mathbf{f}'(t)] = \mathbf{f}(t) \times \mathbf{f}''(t).$$

22. Show that, if $\mathbf{f}(t) = \cos t\,\mathbf{i} + \sin t\,\mathbf{j}$, then $\mathbf{f}(t) \cdot \mathbf{f}'(t) = 0$.

23. Given that $\mathbf{f}$ is differentiable, show that

$$\|\mathbf{f}(t)\| \text{ is constant} \qquad \text{iff} \qquad \mathbf{f}(t) \cdot \mathbf{f}'(t) = 0 \quad \text{identically.}$$

24. Show that, if $\mathbf{g}(t) = e^{kt}\,\mathbf{i} + e^{-kt}\,\mathbf{j}$, then $\mathbf{g}(t)$ and $\mathbf{g}''(t)$ have the same direction.

25. Give an informal derivation of the formula

$$(\mathbf{f} \times \mathbf{g})'(t) = [\mathbf{f}(t) \times \mathbf{g}'(t)] + [\mathbf{f}'(t) \times \mathbf{g}(t)]$$

without appealing to components.

26. Give a detailed proof of the chain rule for vector functions:

$$(\mathbf{f} \circ u)'(t) = \mathbf{f}(u(t))u'(t).$$

15.3   Curves and Tangents

A linear vector function

$$\mathbf{r}(t) = \mathbf{r}_0 + t\mathbf{d}$$

serves as a parametrization for a line l in the sense that, as t ranges over the set of real numbers, the tip of the position vector $\mathbf{r}(t)$ traces out the line l. The vector function

$$\mathbf{r}(t) = \cos t \, \mathbf{i} + \sin t \, \mathbf{j}, \qquad t \in [0, 2\pi]$$

serves as a parametrization for the unit circle in the xy-plane in the sense that, as t ranges over $[0, 2\pi]$, the tip of the vector $\mathbf{r}(t)$ traces out the unit circle in the xy-plane.

In general a vector function

$$\mathbf{r}(t) = x(t)\mathbf{i} + y(t)\mathbf{j} + z(t)\mathbf{k}$$

differentiable on some interval I serves as a *parametrization for a curve C* in the sense that, as t ranges over I, the tip of the vector $\mathbf{r}(t)$ traces out the curve C. (See Figure 15.3.1.)

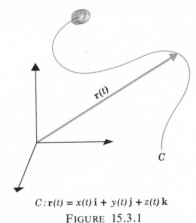

$$C : \mathbf{r}(t) = x(t)\,\mathbf{i} + y(t)\,\mathbf{j} + z(t)\,\mathbf{k}$$
FIGURE 15.3.1

It is time we gave a geometric interpretation to the derivative

$$\mathbf{r}'(t) = x'(t)\mathbf{i} + y'(t)\mathbf{j} + z'(t)\mathbf{k}.$$

Definition 15.3.1

Let

$$C: \mathbf{r}(t) = x(t)\mathbf{i} + y(t)\mathbf{j} + z(t)\mathbf{k}$$

be a differentiable curve. The vector $\mathbf{r}'(t)$, if not $\mathbf{0}$, is said to be *tangent* to C at the tip of $\mathbf{r}(t)$.

· To see why, recall that

$$\mathbf{r}'(t) = \lim_{h \to 0} \frac{\mathbf{r}(t + h) - \mathbf{r}(t)}{h}.$$

With $\mathbf{r}'(t) \neq \mathbf{0}$, we can be sure that for $t + h$ close enough to t, the vector

$$\mathbf{r}(t + h) - \mathbf{r}(t)$$

will not be $\mathbf{0}$. Consequently, we can think of the vector $\mathbf{r}(t + h) - \mathbf{r}(t)$ as pictured in Figure 15.3.2.

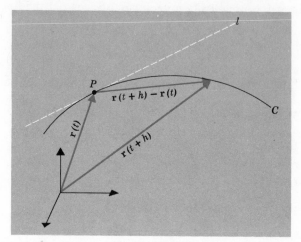

FIGURE 15.3.2

Let's agree that the line marked l in the figure corresponds to our intuitive notion of tangent. We know that l passes through the tip of $\mathbf{r}(t)$. To determine l, we need only find a direction vector for l. As h tends to zero, the vector

$$\mathbf{r}(t + h) - \mathbf{r}(t)$$

comes increasingly closer to serving as a direction vector for l. It may be tempting therefore to take the limiting case

$$\lim_{h \to 0} \left[\mathbf{r}(t + h) - \mathbf{r}(t) \right]$$

and call that a direction vector for l. The trouble here is that this limit vector is $\mathbf{0}$ and $\mathbf{0}$ has no direction. We can circumvent this difficulty by looking instead at the vector

$$\frac{1}{h} \left[\mathbf{r}(t + h) - \mathbf{r}(t) \right].$$

For each value of h, this vector is parallel to $\mathbf{r}(t + h) - \mathbf{r}(t)$, and therefore its limit,

$$\mathbf{r}'(t) = \lim_{h \to 0} \frac{1}{h} \left[\mathbf{r}(t + h) - \mathbf{r}(t) \right],$$

which by assumption is not $\mathbf{0}$, can be taken as a direction vector for l. This in a nutshell is why we call the vector $\mathbf{r}'(t)$ tangent to the curve C at the tip of $\mathbf{r}(t)$.

If $\mathbf{r}'(t_0) \neq \mathbf{0}$, then $\mathbf{r}'(t_0)$ is the tangent vector at the tip of $\mathbf{r}_0 = \mathbf{r}(t_0)$. The *tangent line* at this point can be parametrized by setting

(15.3.2) $\boxed{\mathbf{R}(u) = \mathbf{r}_0 + u\mathbf{r}'(t_0).}$

Problem.   Show that in the case of the circle

$$C: \mathbf{r}(t) = \rho \cos t \,\mathbf{i} + \rho \sin t \,\mathbf{j},$$

the tangent vector is perpendicular to the radius vector.

SOLUTION.   As a tangent vector we can take

$$\mathbf{r}'(t) = -\rho \sin t \,\mathbf{i} + \rho \cos t \,\mathbf{j}.$$

Since

$$\mathbf{r}(t) \cdot \mathbf{r}'(t) = (\rho \cos t \,\mathbf{i} + \rho \sin t \,\mathbf{j}) \cdot (-\rho \sin t \,\mathbf{i} + \rho \cos t \,\mathbf{j})$$
$$= -\rho^2 \cos t \sin t + \rho^2 \cos t \sin t = 0,$$

we have

$$\mathbf{r}(t) \perp \mathbf{r}'(t). \quad \square$$

the twisted cubic, $t > 0$
FIGURE 15.3.3

Problem.   The curve

$$\mathbf{r}(t) = t\mathbf{i} + t^2\mathbf{j} + t^3\mathbf{k} \qquad \text{(Figure 15.3.3)}$$

is called the "twisted" cubic. Find a vector tangent to the twisted cubic at the point $P(2, 4, 8)$ and find the tangent line.

SOLUTION

$$\mathbf{r}'(t) = \mathbf{i} + 2t\mathbf{j} + 3t^2\mathbf{k}.$$

Since $P(2, 4, 8)$ is the tip of $\mathbf{r}(2)$, a suitable tangent vector is

$$\mathbf{r}'(2) = \mathbf{i} + 4\mathbf{j} + 12\mathbf{k}.$$

The tangent line is given by

$$\mathbf{R}(u) = (2\mathbf{i} + 4\mathbf{j} + 8\mathbf{k}) + u(\mathbf{i} + 4\mathbf{j} + 12\mathbf{k}). \quad \square$$

Problem.   Find the point P on the curve

$$\mathbf{r}(t) = (1 - 2t)\mathbf{i} + t^2\mathbf{j} + 2e^{2(t-1)}\mathbf{k}$$

at which the tangent vector $\mathbf{r}'(t)$ is parallel to $\mathbf{r}(t)$.

SOLUTION

$$\mathbf{r}'(t) = -2\mathbf{i} + 2t\mathbf{j} + 4e^{2(t-1)}\mathbf{k}.$$

For $\mathbf{r}'(t)$ to be parallel to $\mathbf{r}(t)$ there must exist a scalar α such that

$$\mathbf{r}(t) = \alpha\mathbf{r}'(t).$$

This implies the scalar equations

$$1 - 2t = -2\alpha, \qquad t^2 = 2\alpha t, \qquad 2e^{2(t-1)} = 4\alpha e^{2(t-1)}.$$

The last equation requires $\alpha = \frac{1}{2}$. This in turn tells us that $t = 1$. The point in question is $P(-1, 1, 2)$.   □

Two curves

$$C_1: \mathbf{r}_1(t) = x_1(t)\mathbf{i} + y_1(t)\mathbf{j} + z_1(t)\mathbf{k} \quad \text{and} \quad C_2: \mathbf{r}_2(u) = x_2(u)\mathbf{i} + y_2(u)\mathbf{j} + z_2(u)\mathbf{k}$$

intersect iff there are numbers t and u such that

$$\mathbf{r}_1(t) = \mathbf{r}_2(u).$$

The angle of intersection is the angle between their tangent lines.

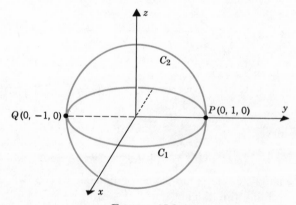

FIGURE 15.3.4

Problem.   Show that the circles

$C_1: \mathbf{r}_1(t) = \cos t\,\mathbf{i} + \sin t\,\mathbf{j}$ and $C_2: \mathbf{r}_2(u) = \cos u\,\mathbf{j} + \sin u\,\mathbf{k}$ (Figure 15.3.4)

meet at right angles at the points $P(0, 1, 0)$ and $Q(0, -1, 0)$.

SOLUTION.   Since

$$\mathbf{r}_1(\tfrac{1}{2}\pi) = \mathbf{j} = \mathbf{r}_2(0),$$

the curves meet at the tip of $\mathbf{j}$, which is $P(0, 1, 0)$. Since

$$\mathbf{r}_1(\tfrac{3}{2}\pi) = -\mathbf{j} = \mathbf{r}_2(\pi),$$

the curves meet at the tip of $-\mathbf{j}$, which is $P(0, -1, 0)$. Obviously

$$\mathbf{r}_1'(t) = -\sin t\,\mathbf{i} + \cos t\,\mathbf{j} \quad \text{and} \quad \mathbf{r}_2'(u) = -\sin u\,\mathbf{j} + \cos u\,\mathbf{k}.$$

Since

$$\mathbf{r}_1'(\tfrac{1}{2}\pi) = -\mathbf{i} \quad \text{and} \quad \mathbf{r}_2'(0) = \mathbf{k},$$

we have

$$\mathbf{r}_1'(\tfrac{1}{2}\pi) \cdot \mathbf{r}_2'(0) = 0.$$

This tells us that the curves are perpendicular at $P(0, 1, 0)$. Since

$$\mathbf{r}_1'(\tfrac{3}{2}\pi) = \mathbf{i} \quad \text{and} \quad \mathbf{r}_2'(\pi) = -\mathbf{k},$$

we have

$$\mathbf{r}_1'(\tfrac{3}{2}\pi) \cdot \mathbf{r}_2'(\pi) = 0.$$

This tells us that the curves are perpendicular at $P(0, -1, 0)$. $\quad\square$

Exercises

1. Find the tangent vector and the tangent line at the indicated point.
   - *(a) $\mathbf{r}(t) = \cos \pi t\,\mathbf{i} + \sin \pi t\,\mathbf{j} + t\,\mathbf{k}$ at $t = 2$.
   - (b) $\mathbf{r}(t) = e^t\mathbf{i} + e^{-t}\mathbf{j} - \log t\,\mathbf{k}$ at $t = 1$.
   - *(c) $\mathbf{r}(t) = \mathbf{a} + t\mathbf{b} + t^2\mathbf{c}$ at $t = -1$.
   - (d) $\mathbf{r}(t) = (t + 1)\mathbf{i} + (t^2 + 1)\mathbf{j} + (t^3 + 1)\mathbf{k}$ at $P(1, 1, 1)$.
   - *(e) $\mathbf{r}(t) = 2t^2\mathbf{i} + (1 - t)\mathbf{j} + (3 + 2t^2)\mathbf{k}$ at $P(2, 0, 5)$.
2. Show that $\mathbf{r}(t) = a \cos \omega t\,\mathbf{i} + b \sin \omega t\,\mathbf{j}$ traces out an ellipse. Find an equation for the ellipse in x and y.
*3. Show that $\mathbf{r}(t) = at\mathbf{i} + bt^2\mathbf{j}$ traces out a parabola. Find an equation for the parabola in x and y.
4. Show that $\mathbf{r}(t) = \tfrac{1}{2}a(e^{\omega t} + e^{-\omega t})\mathbf{i} + \tfrac{1}{2}a(e^{\omega t} - e^{-\omega t})\mathbf{j}$ traces out the hyperbola $x^2 - y^2 = a^2$.
*5. Find (a) the points on the plane curve $\mathbf{r}(t) = t\mathbf{i} + (1 + t^2)\mathbf{j}$ at which $\mathbf{r}(t)$ and $\mathbf{r}'(t)$ are perpendicular; (b) the points at which they have the same direction; (c) the points at which they have opposite directions.
*6. Find the curve given that $\mathbf{r}'(t) = \alpha\mathbf{r}(t)$ for all real t and $\mathbf{r}(0) = \mathbf{i} + 2\mathbf{j} + 3\mathbf{k}$.
7. The curves

$$\mathbf{r}_1(t) = (e^t - 1)\,\mathbf{i} + 2 \sin t\,\mathbf{j} + \log (t + 1)\,\mathbf{k},$$
$$\mathbf{r}_2(u) = (u + 1)\,\mathbf{i} + (u^2 - 1)\,\mathbf{j} + (u^3 + 1)\,\mathbf{k}$$

intersect at the origin. Find the angle of intersection.
8. Show that, if $\mathbf{r}'(t)$ and $\mathbf{r}(t)$ are parallel, then the tangent line passes through the origin.

*9. Find the point at which the curves

$$\mathbf{r}_1(t) = e^t\mathbf{i} + 2 \sin (t + \tfrac{1}{2}\pi)\,\mathbf{j} + (t^2 - 2)\mathbf{k}, \quad \mathbf{r}_2(u) = u\mathbf{i} + 2\mathbf{j} + (u^2 - 3)\mathbf{k}$$

intersect and find the angle of intersection.

10. The curve

$$\mathbf{r}(t) = a \cos \omega t\,\mathbf{i} + a \sin \omega t\,\mathbf{j} + b\omega t\,\mathbf{k} \qquad\qquad (a > 0, \quad b > 0, \quad \omega > 0)$$

is called a *circular helix*.

(a) Show that the circular helix climbs along the cylinder

$$x^2 + y^2 = a^2, \quad z \text{ unrestricted.} \qquad\text{(Figure 15.4.2)}$$

(b) Show that it intersects lines parallel to the z-axis at a constant angle.

15.4 Velocity and Acceleration

A curve

$$C\colon \mathbf{r}(t) = x(t)\mathbf{i} + y(t)\mathbf{j} + z(t)\mathbf{k}, \qquad t \in I$$

can be thought of as the path traced out by a moving object. We can interpret I as a time interval and use $\mathbf{r}(t)$ to indicate the position of the object at time t. Let's assume that $\mathbf{r}$ is twice differentiable. In this context the derivatives $\mathbf{r}'(t)$ and $\mathbf{r}''(t)$ have special names and special significance.

Definition 15.4.1

The vector $\mathbf{r}'(t)$ is called the *velocity* of the object at time t, and $\mathbf{r}''(t)$ is called the *acceleration*.

In symbols, we have

(15.4.2) $$\boxed{\mathbf{r}'(t) = \mathbf{v}(t) \quad \text{and} \quad \mathbf{r}''(t) = \mathbf{v}'(t) = \mathbf{a}(t).}$$

There should be nothing surprising about this. As before, we have velocity as the time rate of change of position and acceleration as the time rate of change of velocity. Since $\mathbf{v}(t) = \mathbf{r}'(t)$, we know from the last section that the velocity vector, when not $\mathbf{0}$, is tangent to the path of the motion. The direction of the velocity vector gives us the direction of the motion. (Figure 15.4.1.)

FIGURE 15.4.1

The magnitude of the velocity vector tells us how fast the object is moving. The details will be made clear in the next section. At this point consider it true by definition; namely, for an object moving with velocity $\mathbf{v}(t)$ we define

(15.4.3) $\boxed{\|\mathbf{v}(t)\| = \text{speed at time } t.}$

Problem.   A particle moves along a circular helix

$$\mathbf{r}(t) = a \cos \omega t\,\mathbf{i} + a \sin \omega t\,\mathbf{j} + b\omega t\,\mathbf{k}. \qquad (a > 0,\ b > 0,\ \omega > 0)$$

For each time t, find

(a) the velocity.   (b) the speed.   (c) the acceleration.
(d) the magnitude of the acceleration vector.
(e) the angle between the velocity vector and the acceleration vector.

circular helix

FIGURE 15.4.2

SOLUTION

(a) $\mathbf{v}(t) = \mathbf{r}'(t) = -a\omega \sin \omega t\,\mathbf{i} + a\omega \cos \omega t\,\mathbf{j} + b\omega\,\mathbf{k}.$

(b) speed at time t: $\|\mathbf{v}(t)\| = \sqrt{a^2\omega^2 \sin^2 \omega t + a^2\omega^2 \cos^2 \omega t + b^2\omega^2}$
$$= \sqrt{a^2\omega^2 + b^2\omega^2} = \omega\sqrt{a^2 + b^2}.$$

The speed is thus constant.

(c) $\mathbf{a}(t) = \mathbf{v}'(t) = -a\omega^2 \cos \omega t\,\mathbf{i} - a\omega^2 \sin \omega t\,\mathbf{j} = -a\omega^2(\cos \omega t\,\mathbf{i} + \sin \omega t\,\mathbf{j}).$

Since the speed is constant, the acceleration comes from the change in direction of the velocity vector.

(d) the magnitude of the acceleration vector: $\|\mathbf{a}(t)\| = a\omega^2.$

(e) $\cos \theta = \dfrac{\mathbf{v}(t) \cdot \mathbf{a}(t)}{\|\mathbf{v}(t)\| \, \|\mathbf{a}(t)\|}$

$= \dfrac{[-a\omega \sin \omega t \, \mathbf{i} + a\omega \cos \omega t \, \mathbf{j} + b\omega \mathbf{k}] \cdot [-a\omega^2 (\cos \omega t \, \mathbf{i} + \sin \omega t \, \mathbf{j})]}{a\omega^3 \sqrt{a^2 + b^2}}$

$= \dfrac{a(\sin \omega t \cos \omega t - \cos \omega t \sin \omega t)}{\sqrt{a^2 + b^2}} = 0.$

The angle is therefore $\tfrac{1}{2}\pi$. □

In classical mechanics the *momentum* $\mathbf{p}$ of an object is defined by the equation

$$\mathbf{p} = m\mathbf{v}$$

where m is the mass of the object and $\mathbf{v}$ is its velocity. To indicate the time dependence, we write

$$\mathbf{p}(t) = m\mathbf{v}(t).$$

The acceleration of an object is related to the *force* acting on it by Newton's second law of motion:

$$\mathbf{F}(t) = m\mathbf{a}(t). \qquad (\textit{force equals mass times acceleration})$$

Differentiating the momentum equation, we find that

$$\mathbf{p}'(t) = m\mathbf{v}'(t) = m\mathbf{a}(t).$$

Obviously then

$$\mathbf{F}(t) = \mathbf{p}'(t).$$

In other words, force is simply the time rate of change of momentum.

Problem.   An object of mass m follows an elliptical path in such a manner that at each time t it has position

$$\mathbf{r}(t) = a \cos \omega t \, \mathbf{i} + b \sin \omega t \, \mathbf{j}.$$

Find the force acting on the object.

SOLUTION

$$\mathbf{v}(t) = \mathbf{r}'(t) = -a\omega \sin \omega t \, \mathbf{i} + b\omega \cos \omega t \, \mathbf{j}.$$
$$\mathbf{a}(t) = \mathbf{v}'(t) = \mathbf{r}''(t) = -a\omega^2 \cos \omega t \, \mathbf{i} - b\omega^2 \sin \omega t \, \mathbf{j} = -\omega^2 \mathbf{r}(t).$$
$$\mathbf{F}(t) = -m\omega^2 \mathbf{r}(t). \quad \square$$

This force is called a *central* force because it acts along the line of the radius vector. It is called an *attractive* force because its direction is opposite to that of the radius vector. In magnitude this force is proportional to the distance of the object from the origin. The force is illustrated in Figure 15.4.3.

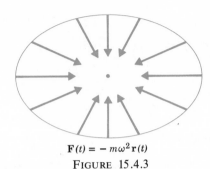

$$\mathbf{F}(t) = -m\omega^2 \mathbf{r}(t)$$

FIGURE 15.4.3

Problem.   A force of magnitude α directed upward from the xy-plane is continually applied to an object of mass m. Given that the object starts at $P(0, y_0, z_0)$ with initial velocity $2\mathbf{j}$ find:

(a) the velocity of the object t seconds later.
(b) the speed of the object t seconds later.
(c) the momentum of the object t seconds later.
(d) the path followed by the object, both in vector form and in Cartesian coordinates.

SOLUTION.   We begin with the force equation

$$\mathbf{F}(t) = \alpha\mathbf{k}.$$

In general, $\mathbf{F}(t) = m\,\mathbf{a}(t)$, so that here

$$\mathbf{a}(t) = \frac{\alpha}{m}\,\mathbf{k}.$$

Integration gives

$$\mathbf{v}(t) = C_1\mathbf{i} + C_2\mathbf{j} + \left(\frac{\alpha}{m}t + C_3\right)\mathbf{k}.$$

Since $\mathbf{v}(0) = 2\mathbf{j}$, you can see that

$$C_1 = 0, \quad C_2 = 2, \quad C_3 = 0.$$

Consequently

$$\mathbf{v}(t) = 2\mathbf{j} + \frac{\alpha}{m}t\mathbf{k}.$$

Another integration gives

$$\mathbf{r}(t) = D_1\mathbf{i} + (2t + D_2)\mathbf{j} + \left(\frac{\alpha}{2m}t^2 + D_3\right)\mathbf{k}.$$

Since $\mathbf{r}(0) = y_0\mathbf{j} + z_0\mathbf{k}$, we have

$$D_1 = 0, \quad D_2 = y_0, \quad D_3 = z_0$$

and thus

$$\mathbf{r}(t) = (2t + y_0)\mathbf{j} + \left(\frac{\alpha}{2m}\,t^2 + z_0\right)\mathbf{k}, \quad t \geq 0.$$

This path is a parabolic arc in the yz-plane. (See Figure 15.4.4.)

FIGURE 15.4.4

To get an equation for this arc in Cartesian coordinates, we write out the components

$$x(t) = 0, \quad y(t) = 2t + y_0, \quad z(t) = \frac{\alpha}{2m}\,t^2 + z_0. \qquad (t \geq 0)$$

From the second equation we have

$$t = \tfrac{1}{2}[y(t) - y_0]. \qquad (y(t) \geq y_0)$$

Substituting this into the third equation, we get

$$z(t) = \frac{\alpha}{8m}\,[y(t) - y_0]^2 + z_0. \qquad (y(t) \geq y_0)$$

Eliminating t altogether, we have

$$z = \frac{\alpha}{8m}\,(y - y_0)^2 + z_0. \qquad (y \geq y_0)$$

Of course we must bear in mind the extra condition $x = 0$.

Answers to (a) through (d):

(a) velocity: $\mathbf{v}(t) = 2\mathbf{j} + \dfrac{\alpha}{m}\,t\mathbf{k}.$

(b) speed: $\|\mathbf{v}(t)\| = \dfrac{1}{m}\sqrt{4m^2 + \alpha^2 t^2}.$

(c) momentum: $m\mathbf{v}(t) = 2m\mathbf{j} + \alpha t\mathbf{k}.$

(d) path in vector form: $\mathbf{r}(t) = (2t + y_0)\mathbf{j} + \left(\dfrac{\alpha}{2m}\,t^2 + z_0\right)\mathbf{k}, \quad t \geq 0.$

path in Cartesian coordinates: $z = \dfrac{\alpha}{8m}\,(y - y_0)^2 + z_0,\ y \geq y_0,\ x = 0. \quad \square$

If an object of mass m moves with velocity $\mathbf{v}(t)$ subject to a force $\mathbf{F}(t)$, the scalar product

(15.4.4)
$$\mathbf{F}(t) \cdot \mathbf{v}(t)$$

is called the *power* (expended by the force) and the number

(15.4.5)
$$\tfrac{1}{2}m\|\mathbf{v}(t)\|^2$$

is called the *kinetic energy* of the object. In the exercises you are asked to show that power is the time rate of change of kinetic energy:

(15.4.6)
$$\mathbf{F}(t) \cdot \mathbf{v}(t) = \frac{d}{dt}\left(\tfrac{1}{2}m\|\mathbf{v}(t)\|^2\right).$$

We will return to these notions later. Now let's talk briefly about the dynamics of rotations. The central notions here are *torque*, usually denoted by τ (the Greek letter tau), and *angular momentum,* usually denoted by $\mathbf{L}$. *Torque,* from the Latin word for "twist," is defined by the equation

(15.4.7)
$$\tau(t) = \mathbf{r}(t) \times \mathbf{F}(t).$$

More precisely, this is the torque about the origin produced by the force $\mathbf{F}(t)$. *Angular momentum* about the origin is defined by setting

(15.4.8)
$$\mathbf{L}(t) = \mathbf{r}(t) \times m\mathbf{v}(t).$$

Just as force is the time rate of change of momentum, torque is the time rate of change of angular momentum:

$$\begin{aligned}
\mathbf{L}'(t) &= [\mathbf{r}(t) \times m\mathbf{v}'(t)] + [\mathbf{r}'(t) \times m\mathbf{v}(t)] \\
&= [\mathbf{r}(t) \times m\mathbf{a}(t)] + [\mathbf{v}(t) \times m\mathbf{v}(t)] \\
&= [\mathbf{r}(t) \times m\mathbf{a}(t)] + \mathbf{0} \\
&= \tau(t). \quad \square
\end{aligned}$$

Central forces, forces which act along the line of the radius vector, produce no torque about the origin. To verify this note that every central force has the form

$$\mathbf{F}(t) = u(t)\mathbf{r}(t).$$

Obviously then

$$\tau(t) = \mathbf{r}(t) \times \mathbf{F}(t) = \mathbf{r}(t) \times u(t)\mathbf{r}(t) = \mathbf{0}. \quad \square$$

Using these ideas, we can demonstrate Johann Kepler's marvelous equal-time equal-area law for planetary motion.†

(15.4.9)

> **Kepler's Second Law**
>
> As a planet moves around the sun, the radius vector sweeps out equal areas in equal times.

---

† The second of three major discoveries made by the great German astronomer Johann Kepler (1571–1630).

A DERIVATION.   For convenience picture the sun at the origin of our coordinate system as in Figure 15.4.5.

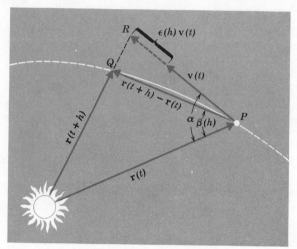

FIGURE 15.4.5

The force involved is the gravitational attraction of the sun. Since this is a central force, we know that the torque is zero and thus the angular momentum of the planet

$$\mathbf{L}(t) = \mathbf{r}(t) \times m\mathbf{v}(t)$$

is constant. In particular, its norm is constant, and thus

$$\|\mathbf{r}(t)\| \, \|\mathbf{v}(t)\| \, |\sin \alpha|$$

is constant.

We will use $A(t)$ to indicate the area swept out by the radius vector from some fixed time up to time t. Our task is to show that $A'(t)$ is constant. This we will do by showing that

$$A'(t) = \tfrac{1}{2}\|\mathbf{r}(t)\| \, \|\mathbf{v}(t)\| \, |\sin \alpha|.$$

The area swept out during the time interval $[t, t + h]$ is simply

$$A(t + h) - A(t).$$

From obvious geometric considerations,

area of $\triangle OPQ \leq A(t + h) - A(t) \leq$ area of $\triangle OPQ$ + area of $\triangle PQR$†

and thus

$$(*) \quad \frac{\text{area of } \triangle OPQ}{h} \leq \frac{A(t + h) - A(t)}{h} \leq \frac{\text{area of } \triangle OPQ}{h} + \frac{\text{area of } \triangle PQR}{h}.$$

---

† Here we are assuming that the path curves toward the sun and not away from it. This is physically obvious: gravitational force is an attracting force, not a repelling force.

Note that

$$\frac{\text{area of } \triangle OPQ}{h} = \tfrac{1}{2} \|\mathbf{r}(t)\| \left\| \frac{\mathbf{r}(t+h) - \mathbf{r}(t)}{h} \right\| |\sin \beta(h)|.$$

As $h \downarrow 0$, $\beta(h) \to \alpha$ and $|\sin \beta(h)| \to |\sin \alpha|$. Since

$$\left\| \frac{\mathbf{r}(t+h) - \mathbf{r}(t)}{h} \right\| \to \|\mathbf{v}(t)\|,$$

we have

$$\lim_{h \downarrow 0} \frac{\text{area of } \triangle OPQ}{h} = \tfrac{1}{2} \|\mathbf{r}(t)\| \|\mathbf{v}(t)\| |\sin \alpha|.$$

We turn our attention now to the other triangle. Here

$$\frac{\text{area of } \triangle PQR}{h} = \tfrac{1}{2} \left\| \frac{\mathbf{r}(t+h) - \mathbf{r}(t)}{h} \right\| \|\mathbf{v}(t) + \epsilon(h)\mathbf{v}(t)\| |\sin [\alpha - \beta(h)]|.$$

As $h \downarrow 0$, $\alpha - \beta(h) \to 0$, and thus $|\sin [\alpha - \beta(h)]| \to 0$. Since

$$\left\| \frac{\mathbf{r}(t+h) - \mathbf{r}(t)}{h} \right\| \to \|\mathbf{v}(t)\|$$

and $\|\mathbf{v}(t) + \epsilon(h)\mathbf{v}(t)\|$ certainly remains bounded, you can see that

$$\lim_{h \downarrow 0} \frac{\text{area of } \triangle PQR}{h} = 0.$$

It follows now from (∗) that

$$\lim_{h \downarrow 0} \frac{A(t+h) - A(t)}{h} = \tfrac{1}{2} \|\mathbf{r}(t)\| \|\mathbf{v}(t)\| |\sin \alpha|.$$

Since a similar argument applies to the case $h < 0$, we can conclude that

$$A'(t) = \tfrac{1}{2} \|\mathbf{r}(t)\| \|\mathbf{v}(t)\| |\sin \alpha|$$

as asserted.   □

Exercises

1. Show that for a point P in circular motion

$$\mathbf{r}(t) = a \cos \theta(t) \, \mathbf{i} + a \sin \theta(t) \, \mathbf{j}$$

the speed of P is the *angular speed* $|\theta'(t)|$ multiplied by the length of the radius vector.

2. An object moves so that $\mathbf{r}(t) = (a \cos t + b)\mathbf{i} + (a \sin t + c)\mathbf{j}$. (a) Show that the speed remains constant. (b) What is the acceleration at time 0?

3. A moving point traces out the curve $\mathbf{r}(t) = at\mathbf{i} + b \sin at \, \mathbf{j}$. Show that the magnitude of its acceleration remains proportional to its distance from the x-axis.

*4. A point traces out the curve $\mathbf{r}(t) = 2\mathbf{i} + t^2\mathbf{j} + (t-1)^2\mathbf{k}$. At what time is the speed a minimum?

5. The path of a plane motion is the curve $\mathbf{r}(t) = 2 \cos 2t\,\mathbf{i} + 3 \cos t\,\mathbf{j}$. *(a) Show that the object oscillates on an arc of the parabola $4y^2 - 9x = 18$. (b) Draw the path. *(c) What are the acceleration vectors at the points of zero velocity? (d) Draw these vectors at the points in question.

6. An object of mass m moves along the path

$$\mathbf{r}(t) = \tfrac{1}{2}a(e^{\omega t} + e^{-\omega t})\mathbf{i} + \tfrac{1}{2}b(e^{\omega t} - e^{-\omega t})\mathbf{j}.$$

(a) What is the velocity at $t = 0$? (b) Show that the acceleration vector is a constant positive multiple of the position vector. (c) What does (b) imply about the angular momentum and the torque? Verify your answers by direct calculation.

7. Show that a particle subject to no outside forces is either stationary or moves with constant speed along a straight line.

8. Show that power is the time rate of change of kinetic energy.

9. Show that for an object of constant velocity the angular momentum is constant.

*10. Find the force required to propel a particle of mass m so that $\mathbf{r}(t) = t\mathbf{j} + t^2\mathbf{k}$.

*11. At each point $P(x(t), y(t), z(t))$ of its motion an object of mass m is subject to a force

$$\mathbf{F}(t) = m\pi^2[a \cos \pi t\,\mathbf{i} + b \sin \pi t\,\mathbf{j}]. \qquad (a > 0, b > 0)$$

Given that $\mathbf{v}(0) = -\pi b\mathbf{j} + \mathbf{k}$ and $\mathbf{r}(0) = b\mathbf{j}$ find the following at time $t = 1$:
(a) the velocity. (b) the speed. (c) the acceleration.
(d) the momentum. (e) the power. (f) the kinetic energy.
(g) the angular momentum. (h) the torque.

12. (*Elliptic harmonic motion.*) Show that, if the acceleration of a particle is given by an equation of the form

$$\mathbf{a}(t) = -\beta^2\mathbf{r}(t),$$

then the path of the particle may be written

$$\mathbf{r}(t) = \cos \beta t\,\mathbf{A} + \sin \beta t\,\mathbf{B}$$

with $\mathbf{A}$ and $\mathbf{B}$ arbitrary constant vectors. Explain the significance of $\mathbf{A}$ and $\mathbf{B}$. HINT: Recall (Exercise 9, Section 7.5) that every solution of the differential equation

$$x''(t) = -\beta^2 x(t)$$

may be written in the form

$$x(t) = A \cos \beta t + B \sin \beta t.$$

15.5 Arc Length and Speed

Arc Length

In Figure 15.5.1 we have sketched a curve C and a polygonal approximation to it. The length of C is by definition the least upper bound of the lengths of such polygonal approximations.

a space curve and a polygonal approximation

FIGURE 15.5.1

Earlier we considered plane curves and used the arc length formula

$$l(C) = \int_a^b \sqrt{[x'(t)]^2 + [y'(t)]^2}\, dt.$$

For space curves the formula becomes

$$l(C) = \int_a^b \sqrt{[x'(t)]^2 + [y'(t)]^2 + [z'(t)]^2}\, dt.$$

Both of these results can be cast in the form

$$l(C) = \int_a^b \|\mathbf{r}'(t)\|\, dt.$$

For plane curves

$$\|\mathbf{r}'(t)\| = \|x'(t)\mathbf{i} + y'(t)\mathbf{j}\| = \sqrt{[x'(t)]^2 + [y'(t)]^2},$$

and for space curves

$$\|\mathbf{r}'(t)\| = \|x'(t)\mathbf{i} + y'(t)\mathbf{j} + z'(t)\mathbf{k}\| = \sqrt{[x'(t)]^2 + [y'(t)]^2 + [z'(t)]^2}.$$

Theorem 15.5.1

The length of a continuously differentiable curve

$$C\colon \mathbf{r}(t) = x(t)\mathbf{i} + y(t)\mathbf{j} + z(t)\mathbf{k}, \qquad t \in [a, b]$$

is given by the formula

$$l(C) = \int_a^b \|\mathbf{r}'(t)\|\, dt.$$

PROOF.   First we show that

$$l(C) \leq \int_a^b \|\mathbf{r}'(t)\| \, dt.$$

To do this, we begin with an arbitrary partition P of $[a, b]$

$$P = \{a = t_0, \ldots, t_{i-1}, t_i, \ldots, t_n = b\}.$$

Such a partition gives rise to a finite number of points of C

$$\mathbf{r}(a) = \mathbf{r}(t_0), \ldots, \mathbf{r}(t_{i-1}), \mathbf{r}(t_i), \ldots, \mathbf{r}(t_n) = \mathbf{r}(b)$$

and thus to a polygonal path of total length

$$l_P = \sum_{i=1}^n \|\mathbf{r}(t_i) - \mathbf{r}(t_{i-1})\|. \qquad \text{(Figure 15.5.2)}$$

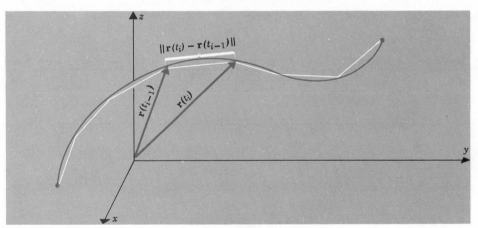

FIGURE 15.5.2

For each i,

$$\mathbf{r}(t_i) - \mathbf{r}(t_{i-1}) = \int_{t_{i-1}}^{t_i} \mathbf{r}'(t) \, dt.$$

This gives

$$\|\mathbf{r}(t_i) - \mathbf{r}(t_{i-1})\| = \left\| \int_{t_{i-1}}^{t_i} \mathbf{r}'(t) \, dt \right\| \leq \int_{t_{i-1}}^{t_i} \|\mathbf{r}'(t)\| \, dt$$

and thus

$$l_P = \sum_{i=1}^n \|\mathbf{r}(t_i) - \mathbf{r}(t_{i-1})\| \leq \sum_{i=1}^n \int_{t_{i-1}}^{t_i} \|\mathbf{r}'(t)\| \, dt = \int_a^b \|\mathbf{r}'(t)\| \, dt.$$

From the arbitrariness of P we know that the inequality

$$l_P \leq \int_a^b \|\mathbf{r}'(t)\| \, dt$$

must hold for all the l_P. This makes the integral on the right an upper bound for all the l_P. Since $l(C)$ is the *least* upper bound of all the l_P, we can conclude right now that

$$l(C) \leq \int_a^b \|\mathbf{r}'(t)\| \, dt.$$

To show equality, we need to know that arc length is additive; we need to know (see Figure 15.5.3) that the length of the curve from P to Q, plus the length of the curve from Q to R, is exactly equal to the length of the curve from P to R.† A proof of this is given in the supplement to this section.

$$l(\widehat{PQ}) + l(\widehat{QR}) = l(\widehat{PR})$$

FIGURE 15.5.3

Let's go on. In Figure 15.5.4 we display the initial vector $\mathbf{r}(a)$, a general position vector $\mathbf{r}(t)$, and a nearby vector $\mathbf{r}(t + h)$. We now set

$$s(t) = \text{length of the curve from } \mathbf{r}(a) \text{ to } \mathbf{r}(t),$$
$$s(t + h) = \text{length of the curve from } \mathbf{r}(a) \text{ to } \mathbf{r}(t + h).$$

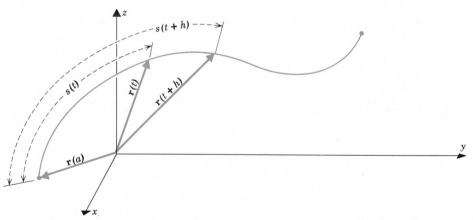

FIGURE 15.5.4

---

† It is obvious that arc length should have this property. What we have to verify is that arc length, *as we have defined it,* has this property.

Assuming the additivity of arc length, we get

$$s(t + h) - s(t) = \text{length of the curve from } \mathbf{r}(t) \text{ to } \mathbf{r}(t + h).$$

From what we have shown already, you can see that

$$\|\mathbf{r}(t + h) - \mathbf{r}(t)\| \le s(t + h) - s(t) \le \int_t^{t+h} \|\mathbf{r}'(u)\|\, du. \quad \text{(Exercise 10)}$$

Dividing this inequality by h (which we are taking as positive), we get

$$\left\| \frac{\mathbf{r}(t + h) - \mathbf{r}(t)}{h} \right\| \le \frac{s(t + h) - s(t)}{h} \le \frac{1}{h} \int_t^{t+h} \|\mathbf{r}'(u)\|\, du.$$

As $h \downarrow 0$, both extremes of the inequality tend to $\|\mathbf{r}'(t)\|$ and consequently so does the middle term.† In short,

$$\lim_{h \downarrow 0} \frac{s(t + h) - s(t)}{h} = \|\mathbf{r}'(t)\|.$$

By taking $h < 0$ and proceeding in a similar manner, one can show that

$$\lim_{h \uparrow 0} \frac{s(t + h) - s(t)}{h} = \|\mathbf{r}'(t)\|.$$

It follows that

$$s'(t) = \lim_{h \to 0} \frac{s(t + h) - s(t)}{h} = \|\mathbf{r}'(t)\|.$$

Recall now that $s(t) = $ length of the curve from $\mathbf{r}(a)$ to $\mathbf{r}(t)$. Obviously then $s(a) = 0$. This, together with the fact that $s'(t) = \|\mathbf{r}'(t)\|$, implies that

$$s(t) = \text{length of the curve from } \mathbf{r}(a) \text{ to } \mathbf{r}(t) = \int_a^t \|\mathbf{r}'(u)\|\, du.$$

The total length of C is therefore

$$s(b) = \int_a^b \|\mathbf{r}'(t)\|\, dt. \quad \square$$

Problem.   Find the length of the curve

$$C: \mathbf{r}(t) = 2 \cos t\, \mathbf{i} + 2 \sin t\, \mathbf{j} + t^2 \mathbf{k} \quad \text{from } t = 0 \text{ to } t = 1.$$

SOLUTION

$$\mathbf{r}'(t) = -2 \sin t\, \mathbf{i} + 2 \cos t\, \mathbf{j} + 2t\, \mathbf{k}.$$
$$\|\mathbf{r}'(t)\| = \sqrt{4 \sin^2 t + 4 \cos^2 t + 4t^2} = 2\sqrt{\sin^2 t + \cos^2 t + t^2} = 2\sqrt{1 + t^2}.$$
$$l(C) = \int_0^1 \|\mathbf{r}'(t)\|\, dt = \int_0^1 2\sqrt{1 + t^2}\, dt = \left[t\sqrt{1 + t^2} + \log\,(t + \sqrt{1 + t^2}) \right]_0^1$$
$$= \sqrt{2} + \log\,(1 + \sqrt{2}). \quad \square$$

---

† One way to show that the right-hand extreme of the inequality tends to $\|\mathbf{r}'(t)\|$ is to use The First Mean-Value Theorem For Integrals (Theorem 10.1.5).

Speed

In the last section we considered motions parametrized by time and, somewhat arbitrarily, called $\|\mathbf{r}'(t)\|$ the speed of the motion. We are now in a position to justify this. Imagine that an object moves so that at time t it has position

$$\mathbf{r}(t) = x(t)\mathbf{i} + y(t)\mathbf{j} + z(t)\mathbf{k}.$$

During a time interval $[0, t]$, the object moves along its path from $\mathbf{r}(0)$ to $\mathbf{r}(t)$ for a total distance which is equal to the length of the curve from $\mathbf{r}(0)$ to $\mathbf{r}(t)$. If, as before, we denote this distance by $s(t)$, then we have

$$s(t) = \int_0^t \|\mathbf{r}'(u)\|\, du \quad \text{and} \quad s'(t) = \|\mathbf{r}'(t)\|.$$

In other words, $\|\mathbf{r}'(t)\|$ is the *rate of change of distance with respect to time*. That is why we call it the *speed*.

In this setting, speed is commonly abbreviated by the greek letter nu. In symbols, we have

$$(15.5.2) \qquad \boxed{\; s'(t) = \nu(t) \quad \text{or, in the Leibniz notation,} \quad \frac{ds}{dt} = \nu. \;}$$

While integration of the velocity vector gives the position vector

$$\mathbf{r}(t) = \mathbf{r}(0) + \int_0^t \mathbf{v}(u)\, du,$$

integration of the speed gives the distance traveled:

$$s(t) = s(0) + \int_0^t \nu(u)\, du.$$

Exercises

Find the length of the curve.

*1. $\mathbf{r}(t) = a \cos t\, \mathbf{i} + a \sin t\, \mathbf{j} + bt\mathbf{k}$ from $t = 0$ to $t = 2\pi$.

2. $\mathbf{r}(t) = t\mathbf{i} + \frac{2}{3}\sqrt{2}\, t^{3/2}\mathbf{j} + \frac{1}{2}t^2\mathbf{k}$ from $t = 0$ to $t = 2$.

*3. $\mathbf{r}(t) = t\mathbf{i} + \log(\sec t)\, \mathbf{j} + 3\mathbf{k}$ from $t = 0$ to $t = \frac{1}{4}\pi$.

4. $\mathbf{r}(t) = t^3\mathbf{i} + t^2\mathbf{j}$ from $t = 0$ to $t = 1$.

*5. $\mathbf{r}(t) = \arctan t\, \mathbf{i} + \frac{1}{2}\log(1 + t^2)\, \mathbf{j}$ from $t = 0$ to $t = 1$.

6. $\mathbf{r}(t) = t\mathbf{i} + \mathbf{j} + (\frac{1}{6}t^3 + \frac{1}{2}t^{-1})\mathbf{k}$ from $t = 1$ to $t = 3$.

*7. $\mathbf{r}(t) = e^t[\cos t\, \mathbf{i} + \sin t\, \mathbf{j}]$ from $t = 0$ to $t = \pi$.

8. $\mathbf{r}(t) = 3t \cos t\, \mathbf{i} + 3t \sin t\, \mathbf{j} + 4t\mathbf{k}$ from $t = 0$ to $t = 4$.

*9. $\mathbf{r}(t) = 2t\mathbf{i} + (t^2 - 2)\mathbf{j} + (1 - t^2)\mathbf{k}$ from $t = 0$ to $t = 2$.

10. Justify in detail the statement

$$\|\mathbf{r}(t + h) - \mathbf{r}(t)\| \le s(t + h) - s(t) \le \int_t^{t+h} \|\mathbf{r}'(u)\|\, du$$

made in the proof of Theorem 15.5.1.

Optional | Estimate within 0.01 the length of the curve from $t = 0$ to $t = \frac{1}{2}$. Evaluate the integral by expanding the integrand in a power series.
*11. $\mathbf{r}(t) = t\mathbf{i} + \frac{1}{3}t^3\mathbf{j}$.   12. $\mathbf{r}(t) = \frac{2}{5}t^{5/2}\mathbf{j} + t\mathbf{k}$.

Supplement

The Additivity of Arc Length

With P, Q, and R as in Figure 15.5.3 we wish to show that

$$l(\widehat{PQ}) + l(\widehat{QR}) = l(\widehat{PR}).$$

If L_1 is a polygonal path inscribed in $\widehat{PQ}$ and L_2 is a polygonal path inscribed in $\widehat{QR}$, then $L_1 \cup L_2$ is inscribed in $\widehat{PR}$. Since

$$l(L_1) + l(L_2) = l(L_1 \cup L_2) \quad \text{and} \quad l(L_1 \cup L_2) \le l(\widehat{PR}),$$

we see that

$$l(L_1) + l(L_2) \le l(\widehat{PR}) \quad \text{and thus} \quad l(L_1) \le l(\widehat{PR}) - l(L_2).$$

Since L_1 is an arbitrary polygonal path inscribed in $\widehat{PQ}$, we conclude that

$$l(\widehat{PQ}) \le l(\widehat{PR}) - l(L_2).$$

It follows that

$$l(L_2) \le l(\widehat{PR}) - l(\widehat{PQ}).$$

From the arbitrariness of L_2 we conclude that

$$l(\widehat{QR}) \le l(\widehat{PR}) - l(\widehat{PQ}),$$

which we rewrite as

$$l(\widehat{PQ}) + l(\widehat{QR}) \le l(\widehat{PR}).$$

We now set out to prove that

$$l(\widehat{PR}) \le l(\widehat{PQ}) + l(\widehat{QR}).$$

To do this, we need only take

$$L = \overline{T_0 T_1} \cup \cdots \cup \overline{T_{n-1} T_n}$$

as an arbitrary polygonal path inscribed in $\widehat{PR}$ and show that

$$l(L) \le l(\widehat{PQ}) + l(QR).$$

If Q is one of the T_i, say $Q = T_k$, then

$$L_1 = \overline{T_0 T_1} \cup \cdots \cup \overline{T_{k-1} T_k} \quad \text{is inscribed in } \widehat{PQ}$$

and

$$L_2 = \overline{T_k T_{k+1}} \cup \cdots \cup \overline{T_{n-1} T_n} \quad \text{is inscribed in } \widehat{QR}.$$

Moreover, $l(L) = l(L_1) + l(L_2)$, so that

$$l(L) \le l(\widehat{PQ}) + l(\widehat{QR}).$$

If Q is none of the T_i, then Q lies between two consecutive points T_k and T_{k+1}. Set

$$L' = \overline{T_0 T_1} \cup \cdots \cup \overline{T_k Q} \cup \overline{Q T_{k+1}} \cup \cdots \cup \overline{T_{n-1} T_n}.$$

Since

$$d(T_k, T_{k+1}) \le d(T_k, Q) + d(Q, T_{k+1}),$$

we have

$$l(L) \le l(L').$$

Proceed as before and you'll see that

$$l(L') \le l(\widehat{PQ}) + l(\widehat{QR}),$$

and once again

$$l(L) \le l(\widehat{PQ}) + l(\widehat{QR}). \quad \square$$

15.6   The Curvature of a Plane Curve

The simplest approach to curvature is based on the use of arc length as a parameter. Suppose that we have a differentiable plane curve C of length l:

$$C \colon \mathbf{r}(s) = x(s)\mathbf{i} + y(s)\mathbf{j}, \qquad s \in I.$$

C is said to be *parametrized by arc length* iff $I = [0, l]$ and, for each s in that interval, $\mathbf{r}(s)$ is that point of C which is at an arc distance s from the initial point $\mathbf{r}(0)$. The idea is illustrated in Figure 15.6.1.

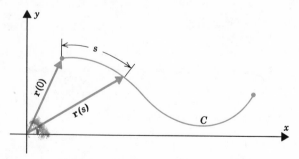

a curve parametrized by arc length

FIGURE 15.6.1

There is a simple way to determine whether or not a given curve is being parametrized by arc length. In general

$$\int_0^s \|\mathbf{r}'(u)\| \, du = \text{the arc distance from } \mathbf{r}(0) \text{ to } \mathbf{r}(s).$$

The parametrization is with respect to arc length iff for each s

$$\int_0^s \|\mathbf{r}'(u)\| \, du = s$$

Optional | which, by differentiation, happens iff for each s

$$\|\mathbf{r}'(s)\| = 1.$$

(15.6.1) | Thus a curve is parametrized by arc length iff $\|\mathbf{r}'(s)\|$, the norm of the tangent vector, is constantly 1.

Examples.   We can parametrize the unit circle by arc length by writing

$$\mathbf{r}(s) = \cos s \, \mathbf{i} + \sin s \, \mathbf{j}, \qquad s \in [0, 2\pi].$$

Here

$$\mathbf{r}'(s) = -\sin s \, \mathbf{i} + \cos s \, \mathbf{j} \quad \text{and} \quad \|\mathbf{r}'(s)\| = \sqrt{\sin^2 s + \cos^2 s} = 1.$$

We can generalize to a circle of radius ρ by setting

$$\mathbf{r}(s) = \rho[\cos (s/\rho) \, \mathbf{i} + \sin (s/\rho) \, \mathbf{j}], \qquad s \in [0, 2\pi\rho].$$

This parametrization is also with respect to arc length. In this case

$$\mathbf{r}'(s) = -\sin (s/\rho) \, \mathbf{i} + \cos (s/\rho) \, \mathbf{j},$$
$$\|\mathbf{r}'(s)\| = \sqrt{\sin^2 (s/\rho) + \cos^2 (s/\rho)} = 1. \quad \square$$

Consider now a twice differentiable plane curve parametrized by arc length. Although the length of the tangent vector $\mathbf{r}'(s)$ remains constantly 1, the direction of this vector can change as s changes. The vector $\mathbf{r}''(s)$ signals this change in direction and is called the *curvature vector*.

The first thing to note is that at each point

(15.6.2) | the curvature vector is perpendicular to the tangent vector.

PROOF.   From the identity

$$\|\mathbf{r}'(s)\| = 1$$

you can see that

$$\mathbf{r}'(s) \cdot \mathbf{r}'(s) = 1,$$
$$\mathbf{r}'(s) \cdot \mathbf{r}''(s) + \mathbf{r}''(s) \cdot \mathbf{r}'(s) = 0,$$
$$2[\mathbf{r}'(s) \cdot \mathbf{r}''(s)] = 0,$$

and therefore

$$\mathbf{r}'(s) \perp \mathbf{r}''(s). \quad \square$$

The magnitude of the curvature vector

$$k(s) = \|\mathbf{r}''(s)\|$$

Optional is called the *curvature*. The curvature gives the magnitude of the change in direction per unit of arc length.

Theorem 15.6.3

On a circle of radius ρ the curvature is constantly $1/\rho$.

PROOF.   We can parametrize a circle of radius ρ in terms of arc length by setting

$$\mathbf{r}(s) = \rho[\cos\,(s/\rho)\,\mathbf{i} + \sin\,(s/\rho)\,\mathbf{j}], \qquad s \in [0, 2\pi\rho].$$

This gives

$$\mathbf{r}'(s) = -\sin\,(s/\rho)\,\mathbf{i} + \cos\,(s/\rho)\,\mathbf{j},$$
$$\mathbf{r}''(s) = -(1/\rho)[\cos\,(s/\rho)\,\mathbf{i} + \sin\,(s/\rho)\,\mathbf{j}].$$

It follows then that

$$k(s) = \|\mathbf{r}''(s)\| = 1/\rho$$

as asserted.   □

In every circle $\mathbf{r}''(s)$ is a negative multiple of $\mathbf{r}(s)$. (See the last proof.) This says that the direction of the curvature vector is always *opposite* to that of the radius vector. See Figure 15.6.2.

FIGURE 15.6.2

That the curvature of a circle of radius ρ should be $1/\rho$ is intuitively obvious. It is geometrically and physically evident that in a circular path the change in direction takes place at a constant rate. Since a complete revolution entails a change in direction of 2π radians and this change is carried out on a path of length $2\pi\rho$, the change in direction per unit arc length must be $1/\rho$.

In general a curve C has curvature $1/\rho$ at the point P iff its rate of turning at that point is the rate of turning of a circle of radius ρ. The idea is illustrated in Figure 15.6.3.

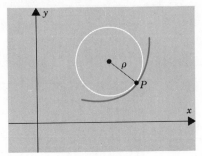

FIGURE 15.6.3

Not all curves are given to us parametrized by arc length. It is therefore useful to know how to compute curvature in terms of an arbitrary parameter t.

Theorem 15.6.4

Let

$$C: \mathbf{r}(t) = x(t)\mathbf{i} + y(t)\mathbf{j}, \qquad t \in [a, b]$$

be a twice differentiable curve. If $\mathbf{r}'(t) \neq \mathbf{0}$, then the curvature is given by the formula

$$k(t) = \frac{|x'(t)y''(t) - y'(t)x''(t)|}{([x'(t)]^2 + [y'(t)]^2)^{3/2}}.$$

PROOF.   Assume that $\mathbf{r}'(t) \neq \mathbf{0}$ and take

$$\mathbf{R}(s) = X(s)\mathbf{i} + Y(s)\mathbf{j}$$

as the arc length parametrization of C which begins at $\mathbf{r}(a)$. Then

$$\mathbf{R}(s(t)) = \mathbf{r}(t) \quad \text{where} \quad s(t) = \int_a^t \|\mathbf{r}'(u)\| \, du:$$

both $\mathbf{R}(s(t))$ and $\mathbf{r}(t)$ give that point of C which is at an arc distance $s(t)$ from the initial point $\mathbf{r}(a)$. See Figure 15.6.4.

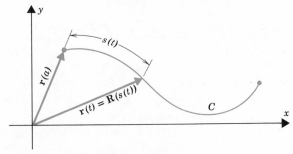

FIGURE 15.6.4

For the unit tangent vector we have

$$\mathbf{T}(t) = \mathbf{R}'(s(t)).$$

Differentiating again, we get

$$\mathbf{T}'(t) = \mathbf{R}''(s(t))s'(t) = \mathbf{R}''(s(t))\|\mathbf{r}'(t)\|$$

$$s'(t) = \|\mathbf{r}'(t)\|$$

and thus

$$k(t) = \|\mathbf{R}''(s(t))\| = \frac{\|\mathbf{T}'(t)\|}{\|\mathbf{r}'(t)\|}.$$

A straightforward but lengthy computation that you can carry out yourself shows that

$$\frac{\|\mathbf{T}'(t)\|}{\|\mathbf{r}'(t)\|} = \frac{|x'(t)y''(t) - y'(t)x''(t)|}{([x'(t)]^2 + [y'(t)]^2)^{3/2}}. \quad \square$$

Problem. Find the curvature of

$$\mathbf{r}(t) = t\mathbf{i} + \tfrac{1}{2}t^2\mathbf{j} \qquad \text{at the point } P(-1, \tfrac{1}{2}).$$

SOLUTION

$$x(t) = t, \quad x'(t) = 1, \quad x''(t) = 0,$$
$$y(t) = \tfrac{1}{2}t^2, \quad y'(t) = t, \quad y''(t) = 1.$$

Therefore

$$k(t) = \frac{|x'(t)y''(t) - y'(t)x''(t)|}{([x'(t)]^2 + [y'(t)]^2)^{3/2}} = \frac{|(1)(1) - (t)(0)|}{(1 + t^2)^{3/2}} = \frac{1}{(1 + t^2)^{3/2}}.$$

At $t = -1$, the curvature is $1/2^{3/2} = \tfrac{1}{4}\sqrt{2}$. $\square$

Problem. Find the curvature of

$$\mathbf{r}(t) = a\cos^3 t\,\mathbf{i} + a\sin^3 t\,\mathbf{j} \qquad \text{at } t = \tfrac{1}{4}\pi.$$

SOLUTION. Here

$$x'(t) = -3a\sin t\cos^2 t, \quad x''(t) = 6a\sin^2 t\cos t - 3a\cos^3 t,$$
$$y'(t) = 3a\sin^2 t\cos t, \quad y''(t) = -3a\sin^3 t + 6a\sin t\cos^2 t,$$

so that

$$x'(\tfrac{1}{4}\pi) = -\tfrac{3}{4}\sqrt{2}a, \quad x''(\tfrac{1}{4}\pi) = \tfrac{3}{4}\sqrt{2}a,$$
$$y'(\tfrac{1}{4}\pi) = \tfrac{3}{4}\sqrt{2}a, \quad y''(\tfrac{1}{4}\pi) = \tfrac{3}{4}\sqrt{2}a.$$

As you can verify,

$$k(\tfrac{1}{4}\pi) = \frac{2}{3|a|}. \quad \square$$

Optional   For the graph of a function $y = f(x)$ the curvature formula reduces to

(15.6.5)
$$k(x) = \frac{|f''(x)|}{(1 + [f'(x)]^2)^{3/2}}.$$

PROOF.   We can parametrize the graph of f by setting

$$\mathbf{r}(t) = t\mathbf{i} + f(t)\mathbf{j}.$$

Here

$$x(t) = t, \qquad x'(t) = 1, \qquad x''(t) = 0$$
$$y(t) = f(t), \quad y'(t) = f'(t), \quad y''(t) = f''(t),$$

and the curvature formula reduces to

$$k(t) = \frac{|f''(t)|}{(1 + [f'(t)]^2)^{3/2}}. \qquad \text{(check this)}$$

Now replace t by x. $\square$

Problem.   Find the curvature of the graph of f at the point where

$$f(x) = 3x - x^3$$

attains its maximum value.

SOLUTION.   Here

$$f'(x) = 3 - 3x^2, \qquad f''(x) = -6x.$$

The tangent is horizontal both at $x = -1$ and at $x = 1$, but by the second derivative test only $x = 1$ gives rise to a maximum. The curvature at $x = 1$ is

$$k(1) = \frac{|f''(1)|}{(1 + [f'(1)]^2)^{3/2}} = 6. \quad \square$$

Problem.   Find the curvature of the ellipse

$$x^2 + 4y^2 = 8 \qquad \text{at the point } P(2, 1).$$

SOLUTION.   We will use Leibniz's notation. In that notation

$$k = \frac{|d^2y/dx^2|}{[1 + (dy/dx)^2]^{3/2}}.$$

We begin by differentiating implicitly:

$$2x + 8y\frac{dy}{dx} = 0 \quad \text{so that} \quad x + 4y\frac{dy}{dx} = 0.$$

A second differentiation gives

Optional

$$1 + 4y \frac{d^2y}{dx^2} + 4 \left(\frac{dy}{dx}\right)^2 = 0.$$

From these last two equations you can see that when $x = 2$ and $y = 1$

$$\frac{dy}{dx} = -\frac{1}{2} \quad \text{and} \quad \frac{d^2y}{dx^2} = -\frac{1}{2}.$$

The curvature is then $\frac{4}{25}\sqrt{5}$:

$$k = \frac{\left|-\frac{1}{2}\right|}{\left(\frac{5}{4}\right)^{3/2}} = \frac{4}{25}\sqrt{5}. \quad \square$$

Exercises

Determine the curvature.

*1. $y = x - x^2$.   2. $y = e^{-x}$.   *3. $y = x^3$.   4. $y = \log(\sec x)$.

*5. Find the point of maximal curvature on the curve $y = \log x$.
6. Find the curvature of the parabola $x^2 = 4cy$ at the vertex.
*7. Find the curvature of the ellipse $x^2/a^2 + y^2/b^2 = 1$ at the vertices.
8. Find the curvature of the hyperbola $x^2 - 4y^2 = 9$ at the point $P(5, 2)$.
*9. Find the curvature of $x^2 + 4xy - 2y^2 = 10$ at the point $P(2, 1)$.
10. Give an arc length parametrization for the ray that begins at $P_0(x_0, y_0, z_0)$ and passes through $P_1(x_1, y_1, z_1)$.
11. The quicker the turning, the larger the curvature. The slower the turning, the smaller the curvature. Along a straight line there is no turning. Show that the curvature of a straight line is 0.

Find the curvature at the point indicated.

12. $\mathbf{r}(t) = t^2\mathbf{i} + (t^4 - t^5)\mathbf{j}$ at $t = 0$.
*13. $\mathbf{r}(t) = 2abt\mathbf{i} + b^2t^2\mathbf{j}$ at $t = 1$.
14. $\mathbf{r}(t) = (t - \sin t)\mathbf{i} + (1 - \cos t)\mathbf{j}$ at $t = \pi$.
*15. $\mathbf{r}(t) = e^t\mathbf{i} + e^{-t}\mathbf{j}$ at $t = 0$.

*16. Find a formula for the curvature of a polar curve $r = f(\theta)$.
17. Find the curvature of the logarithmic spiral $r = e^{a\theta}$.
*18. Find the curvature of the spiral of Archimedes $r = a\theta$.
19. Find the curvature of the cardioid $r = a(1 - \cos \theta)$ in terms of r.
*20. Find the curvature at the highest point of an arch of the cycloid

$$x(t) = r(t - \sin t), \quad y(t) = r(1 - \cos t).$$

Functions of
Several Variables

16.1 What are They?

First we consider functions of two variables. To create such a function we choose a nonempty subset of the *xy*-plane and call it *D*. To each point (x, y) in *D* we assign a real number $f(x, y)$. The set of all these assignments is called a *function of two variables*. We call the function f and the set *D* its *domain*.

Example.   Take *D* as the entire *xy*-plane and to each point (x, y) assign the number

$$f(x, y) = xy. \quad \square$$

Example.   Take *D* as the set of all points (x, y) such that $y \neq 0$. To each such point (x, y) assign the number

$$f(x, y) = \text{arc tan}\left(\frac{x}{y}\right). \quad \square$$

Example.   Take *D* as the *open unit disc*:

$$D = \{(x, y): x^2 + y^2 < 1\}.$$

This set consists of all points inside the unit circle; the circle itself is omitted.† To each point (x, y) in *D* assign the number

$$f(x, y) = \frac{1}{\sqrt{1 - (x^2 + y^2)}}. \quad \square$$

———————

† The *closed unit disc*

$$D = \{(x, y): x^2 + y^2 \leq 1\}$$

consists of all points on or inside the unit circle; the points on the circular boundary are included.

To create a function of three variables we choose a nonempty subset of space and call it D. To each point (x, y, z) in D we assign a real number $f(x, y, z)$. The set of all these assignments is called a *function of three variables*. We call the function f and the set D its *domain*.

Example.　Take D as all of three space and to each point (x, y, z) assign the number

$$g(x, y, z) = xyz. \quad \square$$

Example.　Take D as the set of all (x, y, z) such that $z \neq x + y$ and to each such point assign the number

$$h(x, y, z) = \text{arc tan} \left(\frac{1}{x + y - z} \right). \quad \square$$

Example.　Take D as the *open unit ball:*

$$D = \{(x, y, z): x^2 + y^2 + z^2 < 1\}.$$

This set consists of all the points inside a spherical shell of radius 1 centered at the origin; the shell itself is omitted.† To each point (x, y, z) in D assign the number

$$f(x, y, z) = \sqrt{1 - (x^2 + y^2 + z^2)}. \quad \square$$

Functions of several variables arise naturally in very elementary settings. Here are some examples. For all x and y positive, $f(x, y) = xy$ gives the area of a rectangle of width x and length y, $f(x, y) = \pi x^2 y$ gives the volume of a cylinder with base radius x and height y, and $f(x, y) = 2\pi x(x + y)$ gives its surface area. For all $x, y,$ and z positive, $f(x, y, z) = xyz$ gives the volume of a rectangular solid of dimensions x, y, z. For all x and y, $f(x, y) = \sqrt{x^2 + y^2}$ gives the norm of the vector $(x, y) = x\mathbf{i} + y\mathbf{j}$, and for all $x, y,$ and z, $f(x, y, z) = \sqrt{x^2 + y^2 + z^2}$ gives the norm of the vector $(x, y, z) = x\mathbf{i} + y\mathbf{j} + z\mathbf{k}$. $\quad \square$

If the domain of a function of several variables is not explicitly given, it is to be understood that the domain is the maximal set for which the definition makes sense. Thus in the case of

$$f(x, y) = \frac{1}{x - y}$$

the domain is understood to be all the pairs (x, y) with $y \neq x$. That is, all points of the plane not on the line $y = x$. In the case of

$$g(x, y, z) = \text{arc sin } xyz$$

———————

† The *closed unit ball*

$$D = \{(x, y, z): x^2 + y^2 + z^2 \leq 1\}$$

consists of all points on or inside a spherical shell of radius 1 centered at the origin; the points of the shell are included.

the domain is understood to be all the triples (x, y, z) such that $-1 \leq xyz \leq 1$.   □
   The *range* of a function of several variables is the set of values that the function takes on. To say that a function of several variables is *bounded* is to say that its range is bounded. Since the function

$$f(x, y) = \frac{1}{x - y}$$

takes on all values other than 0, its range is $(-\infty, 0) \cup (0, \infty)$. The function is unbounded. In the case of

$$g(x, y, z) = \arc \sin xyz$$

the range is the closed interval $[-\tfrac{1}{2}\pi, \tfrac{1}{2}\pi]$. This function is bounded, below by $-\tfrac{1}{2}\pi$ and above by $\tfrac{1}{2}\pi$.

Problem.   Find the domain and range of

$$f(x, y) = \frac{1}{\sqrt{4x^2 - y^2}}.$$

SOLUTION.   A pair (x, y) is in the domain of f iff $4x^2 - y^2 > 0$. This occurs iff $y^2 < 4x^2$ and thus iff $-2|x| < y < 2|x|$. (check this) The domain is shown in Figure 16.1.1. It consists of all points of the xy-plane which lie strictly between $y = -2|x|$ and $y = 2|x|$. On this set $\sqrt{4x^2 - y^2}$ takes on all positive values, and so does its reciprocal $f(x, y)$. The range of f is $(0, \infty)$.   □

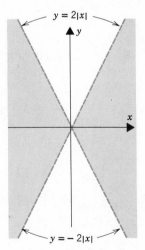

FIGURE 16.1.1

Problem.   Find the domain and range of

$$f(x, y, z) = \frac{\sqrt{a^2 - z^2}}{b^2 + \sqrt{r^2 - (x^2 + y^2)}}. \qquad (a > 0)$$

SOLUTION.   Here we must have

$$x^2 + y^2 \le r^2 \quad \text{and} \quad z^2 \le a^2.$$

The domain is the solid circular cylinder of radius r and height $2a$ given in Figure 16.1.2.

Since $\sqrt{a^2 - z^2}$ takes on all values from 0 to a and $b^2 + \sqrt{r^2 - (x^2 + y^2)}$ takes on all values from b^2 to $b^2 + r$, the quotient

$$f(x, y, z) = \frac{\sqrt{a^2 - z^2}}{b^2 + \sqrt{r^2 - (x^2 + y^2)}}$$

takes on all values from 0 to a/b^2. The range is the closed interval $[0, a/b^2]$. $\square$

FIGURE 16.1.2

Exercises

Find the domain and range of each of the following functions.

*1. $f(x, y) = \sqrt{xy}$.

  2. $f(x, y) = \sqrt{1 - xy}$.

*3. $f(x, y) = \dfrac{1}{x + y}$.

  4. $f(x, y) = \dfrac{1}{x^2 + y^2}$.

*5. $f(x, y) = \dfrac{e^x - e^y}{e^x + e^y}$.

  6. $f(x, y) = \dfrac{x^2}{x^2 + y^2}$.

*7. $f(x, y) = \log (xy)$.

  8. $f(x, y) = \sqrt{9 - x^2} - \sqrt{4 - y^2}$.

*9. $f(x, y) = \dfrac{1}{\sqrt{y - x^2}}$.

10. $f(x, y) = \dfrac{\sqrt{9 - x^2}}{1 + \sqrt{1 - y^2}}$.

*11. $f(x, y, z) = \sqrt{x^2 + y^2 + z^2}$.

12. $f(x, y, z) = \cos x + \cos y + \cos z$.

*13. $f(x, y, z) = \dfrac{x + y + z}{|x + y + z|}$.

14. $f(x, y, z) = \dfrac{z^2}{x^2 - y^2}$.

*15. $f(x, y, z) = -\dfrac{z^2}{\sqrt{x^2 - y^2}}$.

16. $f(x, y, z) = \dfrac{z}{x - y}$.

*17. $f(x, y) = \dfrac{1}{\sqrt{1 - (x^2 + y^2)}}$.

18. $f(x, y, z) = \dfrac{\sqrt{1 - x^2} + \sqrt{4 - y^2}}{1 + \sqrt{9 - z^2}}$.

16.2   A Brief Catalog of the Quadric Surfaces

We represented functions of one variable by curves in the xy-plane. We will represent functions of two variables by surfaces in three-dimensional space.

The simplest curved surfaces are those given by quadratic equations:

$$Ax^2 + By^2 + Cz^2 + Dxy + Exz + Fyz + Hx + Iy + Jz + K = 0.$$

Such surfaces are called *quadric surfaces.*

By suitable translations and rotations of the coordinate axes we can simplify such equations and thereby show that the nondegenerate† quadrics fall into nine distinct types:

1. the ellipsoid
2. the hyperboloid of one sheet
3. the hyperboloid of two sheets
4. the quadric cone
5. the elliptic paraboloid
6. the hyperbolic paraboloid
7. the parabolic cylinder
8. the elliptic cylinder
9. the hyperbolic cylinder.

As you go on with calculus, you'll encounter these surfaces time and time again. Here we give you a picture of each one, together with its equation in standard form and some information about its special properties. These are some of the things to look for:

(a) the *intercepts* (the points at which the surface intersects the coordinate axes)
(b) the *traces* (the intersections with the coordinate planes)
(c) the *sections* (the intersections with planes in general)
(d) the *center* (some quadrics have a center; some do not)
(e) *symmetry*
(f) *boundedness, unboundedness.*

1.   The Ellipsoid

$$\frac{x^2}{a^2} + \frac{y^2}{b^2} + \frac{z^2}{c^2} = 1. \qquad \text{(Figure 16.2.1)}$$

The ellipsoid is centered at the origin and is symmetric about the three coordinate planes. It intersects the coordinate axes at six points: $(\pm a, 0, 0)$, $(0, \pm b, 0)$, $(0, 0, \pm c)$. These points are called the *vertices.* The surface is bounded, being contained in the rectangular solid: $|x| \le a$, $|y| \le b$, $|z| \le c$.†† All three traces are

---

† We are excluding such degenerate quadrics as

$$1 + x^2 + y^2 = -z^2 \quad \text{and} \quad x^2 + y^2 = -z^2.$$

The first one has no points and the second consists of only one point, the origin.

†† Throughout this section we take a, b, c as positive.

ellipses; thus, for example, the trace in the xy-plane (set $z = 0$) is the ellipse

$$\frac{x^2}{a^2} + \frac{y^2}{b^2} = 1.$$

All sections parallel to the coordinate planes are also ellipses; for example, taking $y = y_0$ we have

$$\frac{x^2}{a^2} + \frac{z^2}{c^2} = 1 - \frac{y_0^2}{b^2}.$$

This ellipse is the intersection of the ellipsoid with the plane $y = y_0$. The numbers a, b, c are called the *semiaxes* of the ellipsoid. If two of the semiaxes are equal, then we have an *ellipsoid of revolution*. If the three semiaxes are equal, the surface is a *sphere*. □

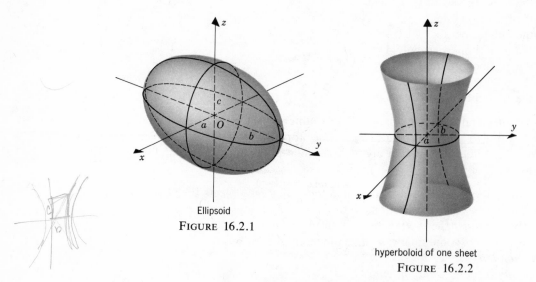

Ellipsoid
FIGURE 16.2.1

hyperboloid of one sheet
FIGURE 16.2.2

2. The Hyperboloid of One Sheet

$$\frac{x^2}{a^2} + \frac{y^2}{b^2} - \frac{z^2}{c^2} = 1. \qquad \text{(Figure 16.2.2)}$$

The surface is unbounded. It is centered at the origin and is symmetric about the three coordinate planes. The surface intersects the coordinate axes at four points: $(\pm a, 0, 0)$, $(0, \pm b, 0)$. The trace in the xy-plane (set $z = 0$) is the ellipse

$$\frac{x^2}{a^2} + \frac{y^2}{b^2} = 1.$$

All sections parallel to the xy-plane are ellipses. The trace in the xz-plane (set $y = 0$) is the hyperbola

$$\frac{x^2}{a^2} - \frac{z^2}{c^2} = 1,$$

and the trace in the yz-plane (set $x = 0$) is the hyperbola

$$\frac{y^2}{b^2} - \frac{z^2}{c^2} = 1.$$

All sections parallel to the xz-plane or yz-plane are hyperbolas. If $a = b$, we have a *hyperboloid of revolution.* □

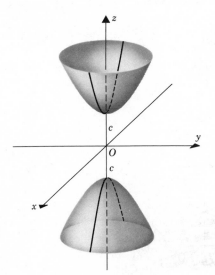

hyperboloid of two sheets

FIGURE 16.2.3

3. The Hyperboloid of Two Sheets

$$\frac{x^2}{a^2} + \frac{y^2}{b^2} - \frac{z^2}{c^2} = -1. \qquad \text{(Figure 16.2.3)}$$

The surface intersects the coordinate axes only at the two vertices $(0, 0, \pm c)$. The surface consists of two parts: one for which $z \geq c$, another for which $z \leq -c$. Each of these parts is unbounded. All sections parallel to the xy-plane are ellipses: for $z = z_0$ we have

$$\frac{x^2}{a^2} + \frac{y^2}{b^2} = \frac{z_0^2}{c^2} - 1.$$

Sections parallel to the other coordinate planes are hyperbolas; for example, for $y = y_0$ we have

$$\frac{z^2}{c^2} - \frac{x^2}{a^2} = 1 + \frac{y_0^2}{b^2}.$$

The surface is symmetric about all three coordinate planes and is centered at the origin. □

quadric cone
FIGURE 16.2.4

elliptic paraboloid
FIGURE 16.2.5

4. The Quadric Cone

$$\frac{x^2}{a^2} + \frac{y^2}{b^2} = z^2. \qquad \text{(Figure 16.2.4)}$$

The surface intersects the coordinate axes only at the origin. The surface is unbounded. Once again there is symmetry about the three coordinate planes. The trace in the xz-plane is a pair of intersecting lines: $z = \pm x/a$. The trace in the yz-plane is also a pair of intersecting lines: $z = \pm y/b$. The trace in the xy-plane is just the origin. Sections parallel to the xy-plane are ellipses. If $a = b$, we have the usual *circular cone*. ☐

5. The Elliptic Paraboloid

$$\frac{x^2}{a^2} + \frac{y^2}{b^2} = z. \qquad \text{(Figure 16.2.5)}$$

The surface does not extend below the xy-plane. Otherwise it is unbounded. The origin is called the *vertex*. Sections parallel to the xy-plane are ellipses; sections parallel to the other coordinate planes are parabolas; hence the name. The surface is symmetric about the xz-plane and about the yz-plane. It is also symmetric about the z-axis. If $a = b$, then the surface is a *paraboloid of revolution*. ☐

6. The Hyperbolic Paraboloid

$$\frac{x^2}{a^2} - \frac{y^2}{b^2} = z. \qquad \text{(Figure 16.2.6)}$$

Here there is symmetry about the xz-plane and yz-plane. Sections parallel to the xy-plane are hyperbolas. Sections parallel to the other coordinate planes are parabolas. The origin is a minimum point for the trace in the xz-plane but a maximum point for the trace in the yz-plane. The origin is called a *minimax* or *saddle point* of the surface. ☐

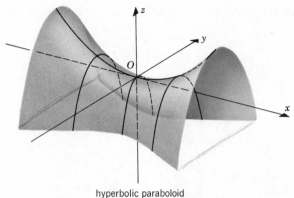

hyperbolic paraboloid
FIGURE 16.2.6

The remaining quadric surfaces are called *cylinders*. In the standard equation of a cylinder one of the three coordinates is missing from the equation. The missing coordinate is unrestricted. In the examples that follow we leave the z-coordinate unrestricted.

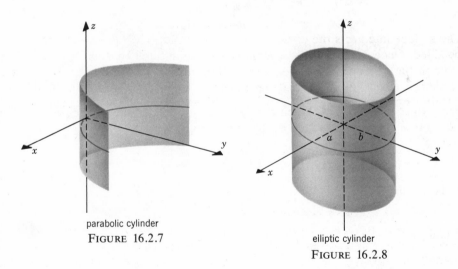

parabolic cylinder
FIGURE 16.2.7

elliptic cylinder
FIGURE 16.2.8

7. The Parabolic Cylinder

$$x^2 = 4cy.$$ (Figure 16.2.7)

This surface is generated by a line parallel to the z-axis which moves so as to trace the parabola $x^2 = 4cy$ in the xy-plane. □

8. The Elliptic Cylinder

$$\frac{x^2}{a^2} + \frac{y^2}{b^2} = 1.$$ (Figure 16.2.8)

The surface is generated by a line parallel to the z-axis which moves so as to trace the ellipse

$$\frac{x^2}{a^2} + \frac{y^2}{b^2} = 1$$

in the xy-plane. If $a = b$, we have the usual *circular cylinder*.　□

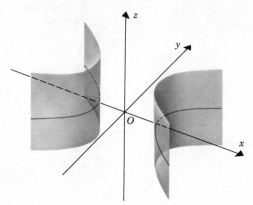

hyperbolic cylinder

FIGURE 16.2.9

9.　The Hyperbolic Cylinder

$$\frac{x^2}{a^2} - \frac{y^2}{b^2} = 1. \qquad \text{(Figure 16.2.9)}$$

The surface has two parts. Each is generated by a line parallel to the z-axis which moves so as to trace a branch of the hyperbola

$$\frac{x^2}{a^2} - \frac{y^2}{b^2} = 1$$

in the xy-plane.　□

Exercises

Identify the surface.

*1. $x^2 + 4y^2 - 16z^2 = 0.$

2. $x^2 + 4y^2 + 16z^2 - 12 = 0.$

*3. $x - 4y^2 = 0.$

4. $x^2 - 4y^2 - 2z = 0.$

*5. $5x^2 + 2y^2 - 6z^2 - 10 = 0.$

6. $2x^2 + 4y^2 - 1 = 0.$

*7. $x^2 + y^2 + z^2 - 4 = 0.$

8. $5x^2 + 2y^2 - 6z^2 + 10 = 0.$

*9. $x^2 + 2y^2 - 4z = 0.$

10. $2x^2 - 3y^2 - 6 = 0.$

*11. $x - y^2 + 2z^2 = 0.$

12. $x - y^2 - 6z^2 = 0.$

Sketch the following cylinders.

13. $x^2 - y + 1 = 0.$

14. $25x^2 + 4y^2 - 100 = 0.$

15. $y^2 - z = 0.$

16. $25y^2 + 4z^2 - 100 = 0.$

17. $y^2 + z = 0.$

18. $25x^2 - 9y^2 - 225 = 0.$

Identify the surface and find the traces. Then sketch the surface.

*19. $9x^2 + 4y^2 - 36z = 0$.   20. $9x^2 + 4y^2 + 36z^2 - 36 = 0$.
*21. $9x^2 + 4y^2 - 36z^2 = 0$.   22. $9x^2 + 4y^2 - 36z^2 - 36 = 0$.
*23. $9x^2 + 4y^2 - 36z^2 + 36 = 0$.   24. $9x^2 - 4y^2 - 36z = 0$.

25. Find the planes of symmetry of the cylinder $x - 4y^2 = 0$.
*26. Write an equation for the surface obtained by revolving the parabola $4z - y^2 = 0$ about the z-axis.
27. Write an equation for the surface obtained by revolving the hyperbola $c^2y^2 - b^2z^2 - b^2c^2 = 0$ about the z-axis.
*28. Determine where the sphere $x^2 + y^2 + z^2 = 1$ intersects the hyperboloid $x^2 + y^2 - z^2 = 1$.
29. The paraboloids $x^2 + y^2 + z - 4 = 0$ and $x^2 + 3y^2 - z = 0$ intersect in a space curve C. Determine the set of points in the xy-plane that lie directly below C. This set of points is called the *projection of C onto the xy-plane*.
*30. The parabolic cylinder $y^2 + z - 4 = 0$ intersects the elliptic paraboloid $x^2 + 3y^2 - z = 0$ in a curve C. Determine the projection of C onto the xy-plane.

16.3   Graphs; Level Curves and Level Surfaces

We begin with a function f of two variables defined on a subset D of the xy-plane. By the *graph of f* we mean the graph of the equation

$$z = f(x, y), \qquad (x, y) \in D.$$

Example.   In the case of

$$f(x, y) = x^2 + y^2$$

the domain is the entire plane. The graph of f is a paraboloid of revolution:

$$z = x^2 + y^2.$$

This surface can be generated by revolving the parabola $z = x^2$ about the z-axis. See Figure 16.3.1.   □

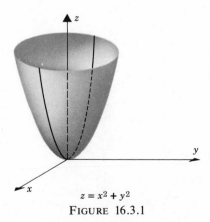

$z = x^2 + y^2$

FIGURE 16.3.1

Example.   The domain of

$$g(x, y) = ax + by + c$$

is also the entire xy-plane. The graph of g is the plane

$$z = ax + by + c$$

with intercepts $x = -c/a, y = -c/b, z = c$. See Figure 16.3.2. $\square$

$z = ax + by + c$
FIGURE 16.3.2

$z = \sqrt{r^2 - (x^2 + y^2)}$
FIGURE 16.3.3

Example.   As another function with a simple graph we can take

$$f(x, y) = \sqrt{r^2 - (x^2 + y^2)}.$$

Here the domain is the disc of radius r centered at the origin:

$$\{(x, y): x^2 + y^2 \leq r^2\}.$$

The graph is the surface

$$z = \sqrt{r^2 - (x^2 + y^2)}, \qquad\qquad \text{(Figure 16.3.3)}$$

the upper half of a sphere of radius r centered at the origin. $\square$

Example.   The function

$$f(x, y) = xy$$

is simple enough, but its graph, the surface

$$z = xy,$$

is quite difficult to draw. It is a "saddle-shaped" surface, a hyperbolic paraboloid: rotate the x and y axes by $\frac{1}{4}\pi$ radians and the equation takes the form

$$\frac{X^2}{2} - \frac{Y^2}{2} = z. \dagger \quad \square$$

$\dagger$ To see this, apply formula 9.7.2 setting $\alpha = \frac{1}{4}\pi$.

Level Curves

In practice the graph of a function of two variables is usually too difficult to draw. Moreover, if drawn, it is often difficult to interpret. Fortunately there is an easier way to convey the same information.

We take our cue from the art of the map maker. In mapping mountainous terrain it is a common practice to sketch curves joining points of constant elevation. A collection of such curves, properly labeled, gives a good idea of the altitude variations in a region. (See Figure 16.3.4.)

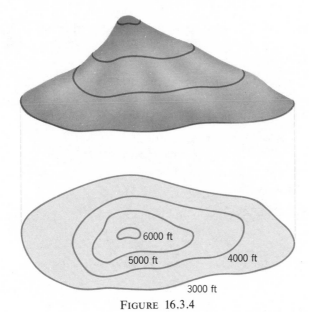

6000 ft

5000 ft   4000 ft

3000 ft

FIGURE 16.3.4

We can do the same thing to portray functions of two variables. Suppose that f is defined on some portion of the xy-plane. If c is some value taken on by f, then we can sketch the curve

$$f(x, y) = c.$$

Such a curve is called a *level curve* for f. It lies in the domain of f, and on it f is constantly c. A collection of such curves, properly drawn and labeled, can give a good idea of the behavior of f.

Example.   We begin with the function

$$f(x, y) = x^2 + y^2,$$

the graph of which is sketched in Figure 16.3.1. The level curves of f are circles centered at the origin:

$$x^2 + y^2 = c, \quad c > 0. \qquad \text{(Figure 16.3.5)}$$

The function takes on the value c on the circle of radius $\sqrt{c}$. At the origin the function is zero. □

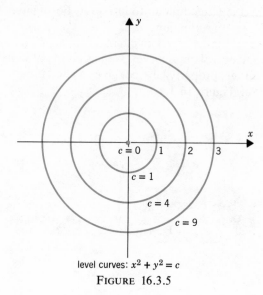

level curves: $x^2 + y^2 = c$

FIGURE 16.3.5

Example. In the case of

$$g(x, y) = x + y + 1$$

the level curves are parallel lines $x + y + 1 = c$. □   (Figure 16.3.6)

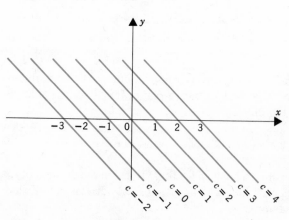

level curves: $x + y + 1 = c$

FIGURE 16.3.6

Example. Now consider the function

$$h(x, y) = \begin{cases} \sqrt{x^2 + y^2}, & x \ge 0 \\ |y|, & x < 0 \end{cases}.$$

For $x \geq 0$, $h(x, y)$ is the distance from (x, y) to the origin. For $x < 0$, $h(x, y)$ is the distance from (x, y) to the x-axis. The level curves are pictured in Figure 16.3.7. The zero level curve is the nonpositive x-axis. The other level curves are horseshoe-shaped: pairs of horizontal rays capped on the right by semicircles.   □

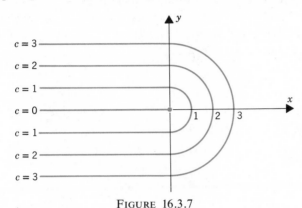

FIGURE 16.3.7

Example.   Let's return to the function

$$f(x, y) = xy.$$

Earlier we noted that the graph of f is a "saddle-shaped" surface that is difficult to draw. The surface is readily understood, however, in terms of its level curves. The 0-level curve consists of the two coordinate axes. The other level curves are simply hyperbolas: $xy = c$. Some are displayed in Figure 16.3.8.

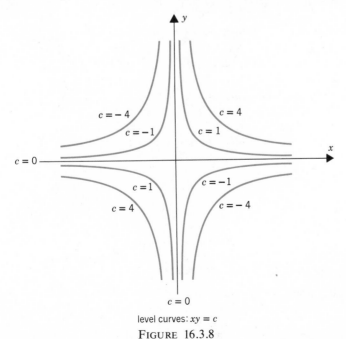

level curves: $xy = c$

FIGURE 16.3.8

Level Surfaces

Drawing graphs for functions of two variables is usually quite difficult. Drawing graphs for functions of three variables is actually impossible. To draw such things we would need four dimensions at our disposal. One way to try to visualize the behavior of a function of three variables is to examine its *level surfaces*. These are the sets with equations of the form

$$f(x, y, z) = c.$$

They are the counterparts of the level curves in the two-variable case. Unlike level curves, level surfaces are usually hard to draw. Nevertheless, a knowledge of what they are is often helpful. We shall restrict ourselves to a few very simple examples.

Example.   In the case of

$$f(x, y, z) = Ax + By + Cz$$

the level surfaces are parallel planes

$$Ax + By + Cz = c. \quad \square$$

Example.   For

$$g(x, y, z) = \sqrt{x^2 + y^2 + z^2}$$

the level surfaces are concentric spheres

$$x^2 + y^2 + z^2 = c^2. \quad \square$$

Example.   As our final example we take the function

$$f(x, y, z) = \frac{|z|}{x^2 + y^2}.$$

We extend this function to the origin by defining it to be zero there. At other points of the z-axis we leave f undefined. In the first place we note that f takes on only nonnegative values. Since it is zero only when $z = 0$, you can see that the 0-level surface is the xy-plane. To find the other level surfaces, we take $c > 0$ and set

$$f(x, y, z) = c.$$

This gives

$$\frac{|z|}{x^2 + y^2} = c \quad \text{and thus} \quad |z| = c(x^2 + y^2). \qquad \text{(Figure 16.3.9)}$$

Each of these surfaces is a double-paraboloid.† $\square$

---

† Surface 5 of the last section together with its mirror image below the xy-plane.

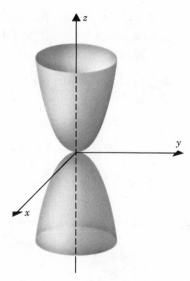

level surface: $|z| = c(x^2 + y^2)$

FIGURE 16.3.9

Exercises

Represent the following functions by level curves.

1. $f(x, y) = x - y.$ 2. $f(x, y) = x^2 - y.$ 3. $f(x, y) = 2x - y.$

4. $f(x, y) = x(y - 1).$ 5. $f(x, y) = \dfrac{x}{x + y}.$ 6. $f(x, y) = \dfrac{y}{x^2}.$

7. $f(x, y) = x^3 - y.$ 8. $f(x, y) = e^{xy}.$ 9. $f(x, y) = x^2 - y^2.$

10. $f(x, y) = x^2.$ 11. $f(x, y) = \dfrac{x^2}{x^2 + y^2}.$ 12. $f(x, y) = x^2y^2.$

Identify the c-level surface and sketch it.

*13. $f(x, y, z) = x + 2y + 3z, \quad c = 0.$
14. $f(x, y, z) = x^2 + y^2, \quad c = 4.$
*15. $f(x, y, z) = z(x^2 + y^2)^{-1/2}, \quad c = 1.$
16. $f(x, y, z) = x^2/4 + y^2/6 + z^2/9, \quad c = 1.$
*17. $f(x, y, z) = z^2 - 36x^2 - 9y^2, \quad c = 1.$
18. $f(x, y, z) = 4x^2 + 9y^2 - 72z, \quad c = 0.$

16.4 Partial Derivatives

Two Variables

The idea of partial differentiation is not hard to understand. Start with a function f of two variables, say

$$f(x, y) = 3x^2y - 5x \cos \pi y.$$

The *partial derivative of f with respect to x* is the function f_x obtained by differentiating f with respect to x, treating y as a constant; in this case

$$f_x(x, y) = 6xy - 5 \cos \pi y.$$

The *partial derivative of f with respect to y* is the function f_y obtained by differentiating f with respect to y, treating x as a constant; in this case

$$f_y(x, y) = 3x^2 + 5\pi x \sin \pi y.$$

The limit definitions of these partial derivatives are as follows:

(16.4.1)

$$f_x(x, y) = \lim_{h \to 0} \frac{f(x + h, y) - f(x, y)}{h},$$

$$f_y(x, y) = \lim_{k \to 0} \frac{f(x, y + k) - f(x, y)}{k}.$$

Example. For

$$g(x, y) = x \text{ arc tan } xy$$

we have

$$g_x(x, y) = x \frac{y}{1 + (xy)^2} + \text{arc tan } xy = \frac{xy}{1 + x^2 y^2} + \text{arc tan } xy$$

and

$$g_y(x, y) = x \frac{x}{1 + (xy)^2} = \frac{x^2}{1 + x^2 y^2}. \quad \square$$

In the one-variable case the derivative gives the rate of change of the function. In the two-variable case, f_x *gives the rate of change of f with respect to x* and f_y *gives the rate of change of f with respect to y.*

Example. In the case of

$$f(x, y) = x \tan y - y \tan x$$

the rate of change of f with respect to x is

$$f_x(x, y) = \tan y - y \sec^2 x$$

and the rate of change of f with respect to y is

$$f_y(x, y) = x \sec^2 y - \tan x.$$

At the point $(0, \tfrac{1}{4}\pi)$ we have

$$f_x(0, \tfrac{1}{4}\pi) = \tan \tfrac{1}{4}\pi - \tfrac{1}{4}\pi \sec^2 0 = 1 - \tfrac{1}{4}\pi$$

and

$$f_y(0, \tfrac{1}{4}\pi) = 0 \sec^2 \tfrac{1}{4}\pi - \tan 0 = 0. \quad \square$$

A Geometric Interpretation

In Figure 16.4.1 we have sketched a surface $z = f(x, y)$. Through this surface we have passed a plane $y = y_0$ perpendicular to the y-axis. This plane cuts the xy-plane in a line PQ and the surface above it in a curve C.

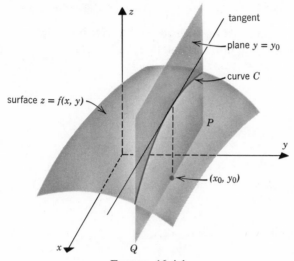

FIGURE 16.4.1

At each point (x, y_0) of PQ, $f(x, y_0)$ gives the height of C above (x, y_0). As x changes, $f(x, y_0)$ changes. The partial derivative $f_x(x_0, y_0)$ gives the rate of change when $x = x_0$.

The other partial derivative f_y can be given a similar interpretation. In Figure 16.4.2 you see the same surface $z = f(x, y)$, this time sliced by a plane $x = x_0$ perpendicular to the x-axis. On the line RS, x remains x_0 and only y changes. As y changes, $f(x_0, y)$ changes. The partial derivative $f_y(x_0, y_0)$ gives the rate of this change when $y = y_0$.

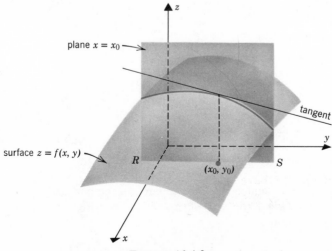

FIGURE 16.4.2

Three Variables

In the case of a function of three variables we can look for three partial derivatives: the partial with respect to x, the partial with respect to y, and also the partial with respect to z. These partials

$$f_x(x, y, z), \quad f_y(x, y, z), \quad f_z(x, y, z)$$

are obtained by differentiating with respect to the subscript variable keeping the other two variables constant. The function f_x gives *the rate of change with respect to x*, f_y gives *the rate of change with respect to y*, and f_z gives *the rate of change with respect to z.*

Example.   For

$$f(x, y, z) = xy^2z^3$$

the rates of change are

$$f_x(x, y, z) = y^2z^3, \quad f_y(x, y, z) = 2xyz^3, \quad f_z(x, y, z) = 3xy^2z^2.$$

In particular

$$f_x(1, -2, -1) = -4, \quad f_y(1, -2, -1) = 4, \quad f_z(1, -2, -1) = 12. \quad \Box$$

Example.   For

$$g(x, y, z) = x^2e^{y/z}$$

we have

$$g_x(x, y, z) = 2xe^{y/z}, \quad g_y(x, y, z) = \frac{x^2}{z} e^{y/z}, \quad g_z(x, y, z) = - \frac{x^2y}{z^2} e^{y/z}. \quad \Box$$

Example.   If $T(x, y, z)$ gives the temperature at the point (x, y, z), then T_x gives the rate of change of T with respect to x, T_y the rate of change of T with respect to y, and T_z the rate of change of T with respect to z. $\quad \Box$

Example.   For a function of the form

$$f(x, y, z) = F(x, y)G(y, z)$$

the partial derivatives are

$$f_x(x, y, z) = F_x(x, y)G(y, z),$$
$$f_y(x, y, z) = F(x, y)G_y(y, z) + F_y(x, y)G(y, z),$$
$$f_z(x, y, z) = F(x, y)G_z(y, z). \quad \Box$$

The geometric significance that we attached to the partials of a function of two variables has no analog in the case of three variables. Figures 16.4.1 and 16.4.2 do not generalize to three variables.

Other Notations

The idea of partial derivative is not tied to the letters x, y, z. In some cases it is more convenient to use other letters.

Problem.   Find the rate of change of the volume of a cylinder first with respect to its radius and then with respect to its height.

SOLUTION.   We could use the letter x for radius and the letter y for height, but it seems more natural to use r for radius and h for height. The volume then becomes

$$V(r, h) = \pi r^2 h.$$

The rate of change of V with respect to r is

$$V_r(r, h) = 2\pi r h.$$

The rate of change of V with respect to h is

$$V_h(r, h) = \pi r^2. \quad \square$$

Problem.   Let V be the volume of the frustum of a circular cone of lower base radius R, upper base radius r, and height h. See Figure 16.4.3. Find the rate of change of V with respect to each of these variables.

frustum of a cone
FIGURE 16.4.3

SOLUTION.   The volume is given by the formula

$$V = \tfrac{1}{3}\pi h(R^2 + Rr + r^2).$$

To emphasize the dependence of V on R, r, and h we write

$$V(R, r, h) = \tfrac{1}{3}\pi h(R^2 + Rr + r^2).$$

We find the rates of change by partial differentiation:

$$V_R(R, r, h) = \tfrac{1}{3}\pi h(2R + r),$$
$$V_r(R, r, h) = \tfrac{1}{3}\pi h(R + 2r),$$
$$V_h(R, r, h) = \tfrac{1}{3}\pi(R^2 + Rr + r^2). \quad \square$$

The subscript notation is not the only one used for partial differentiation. A variant of Leibniz's double-d notation is also commonly used. In this notation the partials f_x, f_y, f_z are denoted by

$$\frac{\partial f}{\partial x}, \quad \frac{\partial f}{\partial y}, \quad \frac{\partial f}{\partial z}.$$

Thus for

$$f(x, y, z) = x^3 y^2 z + \sin xy$$

we have

$$\frac{\partial f}{\partial x}(x, y, z) = 3x^2 y^2 z + y \cos xy,$$

$$\frac{\partial f}{\partial y}(x, y, z) = 2x^3 yz + x \cos xy,$$

$$\frac{\partial f}{\partial z}(x, y, z) = x^3 y^2,$$

or more simply,

$$\frac{\partial f}{\partial x} = 3x^2 y^2 z + y \cos xy,$$

$$\frac{\partial f}{\partial y} = 2x^3 yz + x \cos xy,$$

$$\frac{\partial f}{\partial z} = x^3 y^2.$$

We can also write

$$\frac{\partial}{\partial x}(x^3 y^2 z + \sin xy) = 3x^2 y^2 z + y \cos xy,$$

$$\frac{\partial}{\partial y}(x^3 y^2 z + \sin xy) = 2x^3 yz + x \cos xy,$$

$$\frac{\partial}{\partial z}(x^3 y^2 z + \sin xy) = x^3 y^2.$$

Finally we remark that the "double-decker" notation is not restricted to the letters x, y, z. For instance,

$$\frac{\partial}{\partial r}(r^2 \cos \theta + e^{\theta r}) = 2r \cos \theta + \theta e^{\theta r},$$

$$\frac{\partial}{\partial \theta}(r^2 \cos \theta + e^{\theta r}) = -r^2 \sin \theta + r e^{\theta r}.$$

For

$$\rho = \sin 2\theta \cos 3\phi$$

we have

$$\frac{\partial \rho}{\partial \theta} = 2 \cos 2\theta \cos 3\phi \quad \text{and} \quad \frac{\partial \rho}{\partial \phi} = -3 \sin 2\theta \sin 3\phi.$$

Exercises

Find the partial derivatives.

*1. $f(x, y) = 3x^2 - xy + y$.

*3. $z = Ax^2 + Bxy + Cy^2$.

*5. $f(x, y) = e^{x-y} - x^2y$.

*7. $g(x, y) = \dfrac{Ax + By}{Cx + Dy}$.

*9. $u = xy + yz + zx$.

*11. $\rho = (\theta - \frac{1}{2}\pi) \sin (\phi - \frac{1}{2}\pi)$.

*13. $\rho = e^{\theta+\phi} \cos (\theta - \phi)$.

2. $g(x, y) = x^2e^{-y}$.

4. $\rho = \sin^2 (\theta - \phi)$.

6. $z = \sqrt{x^2 - 3y}$.

8. $u = \dfrac{e^z}{xy^2}$.

10. $f(x, y, z) = z \sin (x - y)$.

12. $g(u, v, w) = \log (u^2 + vw - w^2)$.

14. $f(x, y) = (x + y) \sin (x - y)$.

*15. For $f(x, y) = e^x \log y$, find $f_x(0, e)$ and $f_y(0, e)$.

16. For $f(x, y) = \dfrac{x}{x + y}$, find $f_x(1, 2)$ and $f_y(1, 2)$.

*17. For $g(x, y) = e^{-x} \sin (x + 2y)$, find $g_x(0, \frac{1}{4}\pi)$ and $g_y(0, \frac{1}{4}\pi)$.

18. Show that

$$u = Ax^4 + 2Bx^2y^2 + Cy^4 \quad \text{satisfies the equation} \quad x \frac{\partial u}{\partial x} + y \frac{\partial u}{\partial y} = 4u. \dagger$$

19. Show that

$$u = \frac{x^2 y^2}{x + y} \quad \text{satisfies the equation} \quad x \frac{\partial u}{\partial x} + y \frac{\partial u}{\partial y} = 3u.$$

20. Show that

$$u = x^2y + y^2z + z^2x \quad \text{satisfies the equation} \quad \frac{\partial u}{\partial x} + \frac{\partial u}{\partial y} + \frac{\partial u}{\partial z} = (x + y + z)^2.$$

*21. The area of a triangle is given by the formula

$$K = \tfrac{1}{2}bc \sin A.$$

Suppose that $b = 10$ inches, $c = 20$ inches, and $A = \frac{1}{3}\pi$ radians.

(a) Find the area of the triangle.

(b) Find the rate of change of the area with respect to the side b if c and A remain constant.

(c) Find the rate of change of the area with respect to the angle A if b and c remain constant.

(d) Using the rate found in (c), calculate (by differentials) the approximate change in area if the angle is increased by one degree.

(e) Find the rate of change of c with respect to b if the area and the angle remain constant.

---

† Equations which involve partial derivatives are called *partial differential equations*. Such equations play a prominent role in mathematical physics.

22. The law of cosines for a triangle reads

$$a^2 = b^2 + c^2 - 2bc \cos A.$$

Suppose that $b = 10$ inches, $c = 15$ inches, and $A = \frac{1}{3}\pi$ radians.

(a) Find a.
(b) Find the rate of change of a with respect to b if c and A remain constant.
(c) Using the rate found in (b), calculate (by differentials) the approximate change in a if b is decreased by 1 inch.
(d) Find the rate of change of a with respect to A if b and c remain constant.
(e) Find the rate of change of c with respect to A if a and b remain constant.

16.5   Open Sets and Closed Sets

The domain of a function of several variables can be thought of as a set of points (or vectors) $\mathbf{x}$. In the two-variable case the points are points of the plane:

$$\mathbf{x} = (x, y).$$

In the three-variable case the points are in space:

$$\mathbf{x} = (x, y, z).$$

In this section we study some ways of describing such sets of points.
    There are five simple notions that we wish to establish here:

(1) neighborhood of a point
(2) interior of a set
(3) boundary of a set
(4) open set
(5) closed set.

From our point of view the fundamental notion here is the notion of neighborhood. The other four notions can be derived from it.

Definition 16.5.1   A Neighborhood of a Point

A *neighborhood* of a point $\mathbf{x}_0$ is a set of the form

$$\{\mathbf{x}: \|\mathbf{x} - \mathbf{x}_0\| < \delta\}$$

where δ is some number greater than zero.

In the plane a neighborhood of $\mathbf{x}_0 = (x_0, y_0)$ is the interior of a disc centered at (x_0, y_0). In three-space a neighborhood of $\mathbf{x}_0 = (x_0, y_0, z_0)$ is the interior of a ball centered at (x_0, y_0, z_0). (Figure 16.5.1)

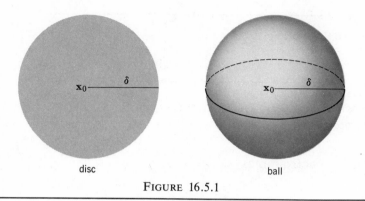

disc   ball

FIGURE 16.5.1

Definition 16.5.2   The Interior of a Set

A point $\mathbf{x}_0$ is said to be in the *interior* of the set S iff S contains some neighborhood of $\mathbf{x}_0$.

Example.   The points marked $\mathbf{x}_1$ and $\mathbf{x}_2$ in Figure 16.5.2 are interior points of the plane set marked Ω, but $\mathbf{x}_3$ and $\mathbf{x}_4$ are not interior points.   □

FIGURE 16.5.2

Definition 16.5.3   The Boundary of a Set

A point $\mathbf{x}_0$ is said to be on the *boundary* of the set S iff every neighborhood of $\mathbf{x}_0$ contains points which are in S and points which are not in S.

Example.   The points marked $\mathbf{x}_3$ and $\mathbf{x}_4$ in Figure 16.5.2 are boundary points. Each neighborhood of $\mathbf{x}_3$ and of $\mathbf{x}_4$ contains points in the set and points not in the set.   □

It is now easy to phrase what we mean by open set and closed set.

Definition 16.5.4   Open Set

A set S is said to be *open* iff it contains a neighborhood of each of its points (iff each of its points is an interior point) (iff it contains no boundary points).

Definition 16.5.5   Closed Set

A set S is said to be *closed* iff it contains its boundary.

Here are some examples of sets which are open, sets which are closed, and sets which are neither open nor closed:

Two-Dimensional Examples

The sets

$$S_1 = \{(x, y): 1 < x < 2, \quad 1 < y < 2\},$$
$$S_2 = \{(x, y): 3 \le x \le 4, \quad 1 \le y \le 2\},$$
$$S_3 = \{(x, y): 5 \le x \le 6, \quad 1 < y < 2\}$$

are displayed in Figure 16.5.3. S_1 is the inside of the first square. S_1 is open because it contains a neighborhood of each of its points. S_2 is the inside of the second square together with the four bounding line segments. S_2 is closed because it contains its entire boundary. S_3 is the inside of the last square together with the two vertical bounding line segments. S_3 is not open because it contains part of its boundary, and it is not closed because it does not contain all of its boundary.   □

FIGURE 16.5.3

Three-Dimensional Examples

We now examine

$$S_1 = \{(x, y, z): z > x^2 + y^2\},$$
$$S_2 = \{(x, y, z): z \ge x^2 + y^2\},$$
$$S_3 = \left\{(x, y, z): 1 \ge \frac{x^2 + y^2}{z}\right\}.$$

Each of these sets has as its boundary the paraboloid

$$z = x^2 + y^2. \qquad \text{(Figure 16.3.1)}$$

The first set consists of all points above this surface. This is an open set because, if a point is above this surface, then all points sufficiently close to it will also be above this surface. The second set is closed because it contains all of its boundary. The third set is neither open nor closed. It is not open because it contains some boundary points; for example, it contains the point $(\frac{1}{2}, \frac{1}{2}, \frac{1}{2})$. It is not closed because it fails to contain the boundary point $(0, 0, 0)$. $\square$

Exercises

Specify the interior and the boundary of the set. State whether the set is open, closed, or neither. Sketch the set.

*1. $\{(x, y): 2 \leq x \leq 4, \quad 1 \leq y \leq 3\}$.   2. $\{(x, y): 2 < x < 4, \quad 1 < y < 3\}$.
*3. $\{(x, y): 1 < x^2 + y^2 < 4\}$.   4. $\{(x, y): 1 \leq x^2 \leq 4\}$.
*5. $\{(x, y): 1 < x^2 \leq 4\}$.   6. $\{(x, y): y < x^2\}$.   *7. $\{(x, y): y \leq x^2\}$.
  8. $\{(x, y, z): 1 \leq x \leq 2, \quad 1 \leq y \leq 2, \quad 1 \leq z \leq 2\}$.
*9. $\{(x, y, z): (x - 1)^2 + (y - 1)^2 + (z - 1)^2 < \frac{1}{4}\}$.
10. $\{(x, y, z): x^2 + y^2 \leq 1, \quad 0 \leq z \leq 4\}$.

16.6 Limits and Continuity; Equality of Mixed Partials

The Basic Notions

The limit process used in taking partial derivatives involved nothing new because in each instance all but one of the variables remained fixed. In this section we take up limits of the form

$$\lim_{(x,y)\to(x_0,y_0)} f(x, y) \quad \text{and} \quad \lim_{(x,y,z)\to(x_0,y_0,z_0)} f(x, y, z).$$

To avoid having to treat the two- and three-variable cases separately, we will write instead

$$\lim_{\mathbf{x}\to\mathbf{x}_0} f(\mathbf{x}).$$

This gives us both the two-variable case [set $\mathbf{x} = (x, y)$ and $\mathbf{x}_0 = (x_0, y_0)$] and the three-variable case [set $\mathbf{x} = (x, y, z)$ and $\mathbf{x}_0 = (x_0, y_0, z_0)$].

Let's begin with a function f defined at least at all points of a neighborhood of $\mathbf{x}_0$, except possibly at $\mathbf{x}_0$ itself. To say that

$$\lim_{\mathbf{x}\to\mathbf{x}_0} f(\mathbf{x}) = l$$

is to say that for $\mathbf{x}$ sufficiently close to $\mathbf{x}_0$ but different from $\mathbf{x}_0$, the number $f(\mathbf{x})$ is close to l; or, to put it another way, as $\|\mathbf{x} - \mathbf{x}_0\|$ tends to zero but remains different from zero, $|f(\mathbf{x}) - l|$ tends to zero. In terms of ϵ, δ we have the following definition:

Definition 16.6.1   The Limit of a Function of Several Variables

$$\lim_{\mathbf{x}\to\mathbf{x}_0} f(\mathbf{x}) = l \quad \text{iff} \quad \begin{cases} \text{for each } \epsilon > 0 \text{ there exists } \delta > 0 \text{ such that,} \\ \text{if} \quad 0 < \|\mathbf{x} - \mathbf{x}_0\| < \delta, \quad \text{then} \quad |f(\mathbf{x}) - l| < \epsilon. \end{cases}$$

As in the one-variable case, the limit (if it exists) is unique. Moreover, if

$$\lim_{\mathbf{x}\to\mathbf{x}_0} f(\mathbf{x}) = l \quad \text{and} \quad \lim_{\mathbf{x}\to\mathbf{x}_0} g(\mathbf{x}) = m,$$

then

$$\lim_{\mathbf{x}\to\mathbf{x}_0} [f(\mathbf{x}) + g(\mathbf{x})] = l + m, \quad \lim_{\mathbf{x}\to\mathbf{x}_0} [\alpha f(\mathbf{x})] = \alpha l, \quad \lim_{\mathbf{x}\to\mathbf{x}_0} [f(\mathbf{x})g(\mathbf{x})] = lm,$$

and

$$\lim_{\mathbf{x}\to\mathbf{x}_0} [f(\mathbf{x})/g(\mathbf{x})] = l/m \quad \text{provided } m \neq 0.$$

These results are not hard to derive. You can do it by simply imitating the comparable arguments in the one-variable case.

Suppose now that $\mathbf{x}_0$ is an interior point of the domain of f. To say that f is *continuous* at $\mathbf{x}_0$ is to say that

(16.6.2)
$$\boxed{\lim_{\mathbf{x}\to\mathbf{x}_0} f(\mathbf{x}) = f(\mathbf{x}_0)}$$

or equivalently that

(16.6.3)
$$\boxed{\lim_{\mathbf{h}\to\mathbf{0}} f(\mathbf{x}_0 + \mathbf{h}) = f(\mathbf{x}_0).}$$

For two variables we can write

$$\lim_{(x,y)\to(x_0,y_0)} f(x, y) = f(x_0, y_0)$$

and for three variables

$$\lim_{(x,y,z)\to(x_0,y_0,z_0)} f(x, y, z) = f(x_0, y_0, z_0).$$

To say that f is *continuous on an open set S* is to say that f is continuous at all points of S.

Some Examples of Continuous Functions

The simplest functions which are everywhere continuous are the polynomials. These are the functions that can be written as finite linear combinations of finite products of x, y, and z: for example,

$$P(x, y) = x^2 y + 3x^3 y^4 - x + 2y, \quad Q(x, y, z) = 6x^3 z - z^3 y + 2xyz.$$

Rational functions, by which we mean quotients of polynomials, are continuous everywhere except where the denominator is zero. Thus

$$f(x, y) = \frac{2x - y}{x^2 + y^2}$$

is continuous at each point of the xy-plane except at the origin $(0, 0)$;

$$g(x, y) = \frac{x^4}{x - y}$$

is continuous except on the line $y = x$;

$$h(x, y) = \frac{1}{x^2 - y}$$

is continuous except on the parabola $y = x^2$;

$$F(x, y, z) = \frac{2x}{x^2 + y^2 + z^2}$$

is continuous at each point of space except at the origin $(0, 0, 0)$;

$$G(x, y, z) = \frac{x^5 - y}{ax + by + cz}$$

is continuous except on the plane $ax + by + cz = 0$. $\square$

You can construct more elaborate continuous functions by forming composites:

$$\arctan \left(\frac{xz^2}{x + y} \right), \qquad \sqrt{x^2 + y^4 + z^6}, \qquad \sin xyz.$$

The first function is continuous except along the vertical plane $x + y = 0$. The last two functions are continuous at each point of space. The continuity of such composites follows from a simple theorem which we state and prove below.

Theorem 16.6.4   The Continuity of Composite Functions

If g is continuous at the point $\mathbf{x}_0$ and f is continuous at the number $g(\mathbf{x}_0)$, then the composition $f \circ g$ is continuous at the point $\mathbf{x}_0$.

PROOF.   We begin with $\epsilon > 0$. We must show that there exists $\delta > 0$ such that

$$\text{if} \quad \|\mathbf{x} - \mathbf{x}_0\| < \delta \quad \text{then} \quad |f(g(\mathbf{x})) - f(g(\mathbf{x}_0))| < \epsilon.$$

From the continuity of f at $g(\mathbf{x}_0)$ we know that there exists $\delta_1 > 0$ such that

$$\text{if} \quad |y - g(\mathbf{x}_0)| < \delta_1 \quad \text{then} \quad |f(y) - f(g(\mathbf{x}_0))| < \epsilon.$$

From the continuity of g at $\mathbf{x}_0$ we know that there exists $\delta > 0$ such that

$$\text{if} \quad \|\mathbf{x} - \mathbf{x}_0\| < \delta \quad \text{then} \quad |g(\mathbf{x}) - g(\mathbf{x}_0)| < \delta_1.$$

This last δ obviously works; namely,

$$\text{if} \quad \|\mathbf{x} - \mathbf{x}_0\| < \delta \quad \text{then} \quad |g(\mathbf{x}) - g(\mathbf{x}_0)| < \delta_1 \quad \text{and} \quad |f(g(\mathbf{x})) - f(g(\mathbf{x}_0))| < \epsilon. \quad \square$$

Continuity in Each Variable Separately

A continuous function of several variables is continuous in each of its variables separately. In the two-variable case this means that, if

$$\lim_{(x,y)\to(x_0,y_0)} f(x, y) = f(x_0, y_0),$$

then

$$\lim_{x\to x_0} f(x, y_0) = f(x_0, y_0) \quad \text{and} \quad \lim_{y\to y_0} f(x_0, y) = f(x_0, y_0).$$

(This is not hard to prove.) The converse is false. *It is possible for a function to be continuous in each variable separately and yet fail to be continuous as a function of several variables.* You can see this kind of behavior in the next example.

Example

$$f(x, y) = \begin{cases} \dfrac{2xy}{x^2 + y^2}, & (x, y) \neq (0, 0) \\ 0, & (x, y) = (0, 0) \end{cases}.$$

Since

$$f(x, 0) = 0 \quad \text{for all } x \quad \text{and} \quad f(0, y) = 0 \quad \text{for all } y,$$

you can see that

$$\lim_{x\to 0} f(x, 0) = f(0, 0) \quad \text{and} \quad \lim_{y\to 0} f(0, y) = f(0, 0).$$

Namely, at the point $(0, 0)$, f is continuous in x and continuous in y. However, as a function of two variables, f is not continuous at $(0, 0)$. An easy way to see this is to note that we can approach $(0, 0)$ as closely as we want by points of the form (t, t) with $t \neq 0$. At such points f takes on the value 1:

$$f(t, t) = \frac{2tt}{t^2 + t^2} = 1.$$

Hence f cannot tend to $f(0, 0) = 0$ as required. $\square$

Continuity and Partial Differentiability

For functions of one variable the existence of the derivative guarantees continuity. (3.1.2) *For functions of several variables the existence of partial derivatives fails to guarantee continuity.*

To show this, we can use the same example:

$$f(x, y) = \begin{cases} \dfrac{2xy}{x^2 + y^2}, & (x, y) \neq (0, 0) \\ 0, & (x, y) = (0, 0) \end{cases}.$$

Since both $f(x, 0)$ and $f(0, y)$ are constantly zero, both partials exist (and are zero) at $(0, 0)$, and yet, as you saw, the function is discontinuous at $(0, 0)$. $\square$

It's not hard to understand why a function can have partial derivatives and yet fail to be continuous. The existence of $\partial f/\partial x$ at (x_0, y_0) depends on the behavior of f only at points of the form $(x_0 + h, y_0)$. Similarly, the existence of $\partial f/\partial y$ at (x_0, y_0) depends on the behavior of f only at points of the form $(x_0, y_0 + k)$. On the other hand, continuity at (x_0, y_0) depends on the behavior of f at points of the more general form $(x_0 + h, y_0 + k)$. More briefly, we can put it this way: *the existence of a partial derivative depends on the behavior of the function along one linear path, but continuity depends on the behavior of the function in all directions.*

Equality of Mixed Partials

There is nevertheless an interesting tie-in between partial differentiation and continuity. To describe it, we begin with a function f of x and y with first partials

$$\frac{\partial f}{\partial x} \quad \text{and} \quad \frac{\partial f}{\partial y}.$$

These are again functions of x and y and may themselves possess partial derivatives:

$$\frac{\partial}{\partial x}\left(\frac{\partial f}{\partial x}\right) = \frac{\partial^2 f}{\partial x^2}, \quad \frac{\partial}{\partial y}\left(\frac{\partial f}{\partial x}\right) = \frac{\partial^2 f}{\partial y \partial x},$$

$$\frac{\partial}{\partial x}\left(\frac{\partial f}{\partial y}\right) = \frac{\partial^2 f}{\partial x \partial y}, \quad \frac{\partial}{\partial y}\left(\frac{\partial f}{\partial y}\right) = \frac{\partial^2 f}{\partial y^2}.$$

These last functions are called the *second-order partials*. Note that there are two "mixed" partials

$$\frac{\partial^2 f}{\partial y \partial x} \quad \text{and} \quad \frac{\partial^2 f}{\partial x \partial y}.$$

The first of these is obtained by differentiating first with respect to x and then with respect to y. The second is obtained by differentiating first with respect to y and then with respect to x.

Example. For

$$f(x, y) = \sin x^2 y$$

we have

$$\frac{\partial f}{\partial x} = 2xy \cos x^2 y \quad \text{and} \quad \frac{\partial f}{\partial y} = x^2 \cos x^2 y.$$

The second-order partials are

$$\frac{\partial^2 f}{\partial x^2} = -4x^2 y^2 \sin x^2 y + 2y \cos x^2 y,$$

$$\frac{\partial^2 f}{\partial y \partial x} = -2x^3 y \sin x^2 y + 2x \cos x^2 y,$$

$$\frac{\partial^2 f}{\partial x \partial y} = -2x^3 y \sin x^2 y + 2x \cos x^2 y,$$

$$\frac{\partial^2 f}{\partial y^2} = -x^4 \sin x^2 y. \quad \square$$

Example.   For

$$f(x, y) = \log (x^2 + y^3)$$

we have

$$\frac{\partial f}{\partial x} = \frac{2x}{x^2 + y^3} \quad \text{and} \quad \frac{\partial f}{\partial y} = \frac{3y^2}{x^2 + y^3}.$$

The second-order partials are

$$\frac{\partial^2 f}{\partial x^2} = \frac{(x^2 + y^3)2 - 2x(2x)}{(x^2 + y^3)^2} = \frac{2(y^3 - x^2)}{(x^2 + y^3)^2},$$

$$\frac{\partial^2 f}{\partial y\,\partial x} = \frac{-2x(3y^2)}{(x^2 + y^3)^2} = -\frac{6xy^2}{(x^2 + y^3)^2},$$

$$\frac{\partial^2 f}{\partial x\,\partial y} = \frac{-3y^2(2x)}{(x^2 + y^3)^2} = -\frac{6xy^2}{(x^2 + y^3)^2},$$

$$\frac{\partial^2 f}{\partial y^2} = \frac{(x^2 + y^3)6y - 3y^2(3y^2)}{(x^2 + y^3)^2} = \frac{3y(2x^2 - y^3)}{(x^2 + y^3)^2}. \quad \square$$

You may have noticed that in both examples we had

$$\frac{\partial^2 f}{\partial y\,\partial x} = \frac{\partial^2 f}{\partial x\,\partial y}.$$

Since in neither case is f symmetric in x and y, this equality of the mixed partials cannot be due to symmetry. Actually it is due to continuity. It can be proved that

(16.6.5)

$$\frac{\partial^2 f}{\partial y\,\partial x} = \frac{\partial^2 f}{\partial x\,\partial y}$$

on every open set U on which f and its partials

$$\frac{\partial f}{\partial x}, \quad \frac{\partial f}{\partial y}, \quad \frac{\partial^2 f}{\partial y\,\partial x}, \quad \frac{\partial^2 f}{\partial x\,\partial y}$$

are all continuous.†

In the case of a function of three variables you can look for three first partials

$$\frac{\partial f}{\partial x}, \quad \frac{\partial f}{\partial y}, \quad \frac{\partial f}{\partial z}$$

and nine second partials

$$\frac{\partial^2 f}{\partial x^2}, \quad \frac{\partial^2 f}{\partial x\,\partial y}, \quad \frac{\partial^2 f}{\partial x\,\partial z}$$

$$\frac{\partial^2 f}{\partial y\,\partial x}, \quad \frac{\partial^2 f}{\partial y^2}, \quad \frac{\partial^2 f}{\partial y\,\partial z}$$

$$\frac{\partial^2 f}{\partial z\,\partial x}, \quad \frac{\partial^2 f}{\partial z\,\partial y}, \quad \frac{\partial^2 f}{\partial z^2}.$$

---

† For a proof consult a text on advanced calculus.

Here again, there is equality of the mixed partials

$$\frac{\partial^2 f}{\partial y \partial x} = \frac{\partial^2 f}{\partial x \partial y}, \qquad \frac{\partial^2 f}{\partial z \partial x} = \frac{\partial^2 f}{\partial x \partial z}, \qquad \frac{\partial^2 f}{\partial y \partial z} = \frac{\partial^2 f}{\partial z \partial y}$$

provided that all the functions involved are continuous.

Example.   For

$$f(x, y, z) = xe^y \sin \pi z$$

we have

$$\frac{\partial f}{\partial x} = e^y \sin \pi z, \qquad \frac{\partial f}{\partial y} = xe^y \sin \pi z, \qquad \frac{\partial f}{\partial z} = \pi xe^y \cos \pi z,$$

$$\frac{\partial^2 f}{\partial x^2} = 0, \qquad \frac{\partial^2 f}{\partial y^2} = xe^y \sin \pi z, \qquad \frac{\partial^2 f}{\partial z^2} = -\pi^2 xe^y \sin \pi z,$$

$$\frac{\partial^2 f}{\partial y \partial x} = e^y \sin \pi z = \frac{\partial^2 f}{\partial x \partial y},$$

$$\frac{\partial^2 f}{\partial z \partial x} = \pi e^y \cos \pi z = \frac{\partial^2 f}{\partial x \partial z},$$

$$\frac{\partial^2 f}{\partial y \partial z} = \pi xe^y \cos \pi z = \frac{\partial^2 f}{\partial z \partial y}. \quad \square$$

Exercises

Find the second partials.

*1. $f(x, y) = Ax^2 + 2Bxy + Cy^2$.

2. $f(x, y) = Ax^3 + Bx^2 y + Cxy^2$.

*3. $f(x, y) = Ax + By + Ce^{xy}$.

4. $f(x, y) = x^2 \cos y + y^2 \sin x$.

*5. $f(x, y) = \sqrt{x + y^2}$.

6. $f(x, y, z) = (x + y^2 + z^3)^2$.

*7. $f(x, y) = \log \left(\dfrac{x}{x + y} \right)$.

8. $f(x, y) = \dfrac{Ax + By}{Cx + Dy}$.

*9. $f(x, y, z) = (x + y)(y + z)(z + x)$.

10. $f(x, y, z) = \arctan xyz$.

11. Show that

$$\text{if} \quad u = \frac{xy}{x + y} \quad \text{then} \quad x^2 \frac{\partial^2 u}{\partial x^2} + 2xy \frac{\partial^2 u}{\partial x \partial y} + y^2 \frac{\partial^2 u}{\partial y^2} = 0.$$

12. Show that

$$\text{if} \quad u = \log \sqrt{x^2 + y^2} \quad \text{then} \quad \frac{\partial^2 u}{\partial x^2} + \frac{\partial^2 u}{\partial y^2} = 0.$$

13. Show that

$$\text{if} \quad u = \frac{1}{\sqrt{x^2 + y^2 + z^2}} \quad \text{then} \quad \frac{\partial^2 u}{\partial x^2} + \frac{\partial^2 u}{\partial y^2} + \frac{\partial^2 u}{\partial z^2} = 0.$$

Gradients;
Extreme Values;
Differentials

17.1 Differentiability and Gradient

The Notion of Differentiability

Our object here is to generalize the notion of differentiability from functions of one variable to functions of several variables. Partial derivatives alone do not fulfill this role because they reflect behavior only along particular lines.

In the one-variable case we formed the difference quotient

$$\frac{f(x + h) - f(x)}{h}$$

and called f differentiable at x provided that this quotient had a limit as h tended to zero. In the multivariable case we can still form the difference

$$f(\mathbf{x} + \mathbf{h}) - f(\mathbf{x}),$$

but the "quotient"

$$\frac{f(\mathbf{x} + \mathbf{h}) - f(\mathbf{x})}{\mathbf{h}}$$

makes no sense because to divide by a vector makes no sense.

The notion of differentiability in the multivariable case is inspired not by difference quotients (we want to avoid $\mathbf{h}$'s in the denominator) but rather by the characterization of $f'(x)$ as the unique number such that

$$f(x + h) - f(x) = f'(x)h + o(h).$$

By $o(h)$, you will recall, we mean any function $g(h)$ such that

$$\lim_{h \to 0} \frac{g(h)}{h} = 0, \quad \text{or equivalently,} \quad \lim_{h \to 0} \frac{g(h)}{|h|} = 0.$$

To set things up for the multivariable case, let us agree to call $o(\mathbf{h})$ any function $g(\mathbf{h})$ such that

$$\lim_{\mathbf{h} \to \mathbf{0}} \frac{g(\mathbf{h})}{\|\mathbf{h}\|} = 0.$$

We now consider a function f of several variables *defined at least in a neighborhood of* $\mathbf{x}$. In the three-variable case $\mathbf{x} = (x, y, z)$, and in the two-variable case $\mathbf{x} = (x, y)$.

Definition 17.1.1 Differentiability

The function f is said to be *differentiable* at $\mathbf{x}$ iff there exists a vector $\nabla f(\mathbf{x})$ such that

$$f(\mathbf{x} + \mathbf{h}) - f(\mathbf{x}) = \nabla f(\mathbf{x}) \cdot \mathbf{h} + o(\mathbf{h}).$$

The vector $\nabla f(\mathbf{x})$ is called the *gradient* of f at $\mathbf{x}$.

The similarities between the one-variable case,

$$f(x + h) - f(x) = f'(x)h + o(h),$$

and the multivariable case,

$$f(\mathbf{x} + \mathbf{h}) - f(\mathbf{x}) = \nabla f(\mathbf{x}) \cdot \mathbf{h} + o(\mathbf{h}),$$

are obvious. We point to the differences. There are essentially two of them:

(1) While the derivative $f'(x)$ is a number, the gradient $\nabla f(\mathbf{x})$ is a vector.
(2) While $f'(x)h$ is the ordinary product of two real numbers, $\nabla f(\mathbf{x}) \cdot \mathbf{h}$ is the dot product of two vectors.

Computing Gradients

First we compute some gradients by applying the definition directly. Then we state and prove a theorem which makes such computations much easier. Finally, we compute gradients with the aid of the theorem. As for notation, in the two-variable case we will write

$$\nabla f(\mathbf{x}) = \nabla f(x, y) \quad \text{and} \quad \mathbf{h} = (h_1, h_2);$$

in the three-variable case

$$\nabla f(\mathbf{x}) = \nabla f(x, y, z) \quad \text{and} \quad \mathbf{h} = (h_1, h_2, h_3).$$

Example. For

$$f(x, y, z) = ax + by + cz$$

we have

$$f(\mathbf{x} + \mathbf{h}) - f(\mathbf{x}) = f(x + h_1, y + h_2, z + h_3) - f(x, y, z)$$
$$= [a(x + h_1) + b(y + h_2) + c(z + h_3)] - [ax + by + cz]$$
$$= ah_1 + bh_2 + ch_3$$
$$= [a\mathbf{i} + b\mathbf{j} + c\mathbf{k}] \cdot \mathbf{h}.$$

Here the remainder is identically zero and thus certainly $o(\mathbf{h})$. Therefore

$$\nabla f(\mathbf{x}) = a\mathbf{i} + b\mathbf{j} + c\mathbf{k}. \quad \square$$

Example.   For

$$f(x, y) = x^2 + y^2$$

we have

$$f(\mathbf{x} + \mathbf{h}) - f(\mathbf{x}) = f(x + h_1, y + h_2) - f(x, y)$$
$$= [(x + h_1)^2 + (y + h_2)^2] - [x^2 + y^2]$$
$$= [2xh_1 + 2yh_2] + [h_1^2 + h_2^2]$$
$$= [2x\mathbf{i} + 2y\mathbf{j}] \cdot \mathbf{h} + \|\mathbf{h}\|^2.$$

Since the remainder $\|\mathbf{h}\|^2$ is $o(\mathbf{h})$,†

$$\nabla f(\mathbf{x}) = \nabla f(x, y) = 2x\mathbf{i} + 2y\mathbf{j}. \quad \square$$

Theorem 17.1.2

If f has continuous first partials in a neighborhood of $\mathbf{x}$, then f is differentiable at $\mathbf{x}$ and

$$\nabla f(\mathbf{x}) = \frac{\partial f}{\partial x}(\mathbf{x})\,\mathbf{i} + \frac{\partial f}{\partial y}(\mathbf{x})\,\mathbf{j} + \frac{\partial f}{\partial z}(\mathbf{x})\,\mathbf{k}.$$

In two variables,

$$\nabla f(\mathbf{x}) = \frac{\partial f}{\partial x}(\mathbf{x})\,\mathbf{i} + \frac{\partial f}{\partial y}(\mathbf{x})\,\mathbf{j}.$$

PROOF.   We prove the theorem in the two-variable case. The same ideas work for a proof in the three-variable case, but the details are more burdensome.
   In the first place,

$$f(\mathbf{x} + \mathbf{h}) - f(\mathbf{x}) = f(x + h_1, y + h_2) - f(x, y).$$

---

† $\dfrac{\|\mathbf{h}\|^2}{\|\mathbf{h}\|} = \|\mathbf{h}\| \to 0$ as $\mathbf{h} \to \mathbf{0}$.

We can rewrite this as

(1) $f(\mathbf{x} + \mathbf{h}) - f(\mathbf{x}) = [f(x + h_1, y + h_2) - f(x, y + h_2)] + [f(x, y + h_2) - f(x, y)].$

By the mean-value theorem for functions of one variable, we know that there are numbers

$$0 < \theta_1 < 1 \quad \text{and} \quad 0 < \theta_2 < 1$$

such that

$$f(x + h_1, y + h_2) - f(x, y + h_2) = \frac{\partial f}{\partial x}(x + \theta_1 h_1, y + h_2) h_1$$

and

$$f(x, y + h_2) - f(x, y) = \frac{\partial f}{\partial y}(x, y + \theta_2 h_2) h_2.$$

By the continuity of $\partial f/\partial x$,

$$\frac{\partial f}{\partial x}(x + \theta_1 h_1, y + h_2) = \frac{\partial f}{\partial x}(x, y) + \epsilon_1(\mathbf{h})$$

where

$$\epsilon_1(\mathbf{h}) \to 0 \quad \text{as} \quad \mathbf{h} \to \mathbf{0}.†$$

By the continuity of $\partial f/\partial y$,

$$\frac{\partial f}{\partial y}(x, y + \theta_2 h_2) = \frac{\partial f}{\partial y}(x, y) + \epsilon_2(\mathbf{h})$$

where

$$\epsilon_2(\mathbf{h}) \to 0 \quad \text{as} \quad \mathbf{h} \to \mathbf{0}.$$

Substituting these expressions in Equation (1), we find that

$$f(\mathbf{x} + \mathbf{h}) - f(\mathbf{x}) = \left[\frac{\partial f}{\partial x}(x, y) + \epsilon_1(\mathbf{h})\right] h_1 + \left[\frac{\partial f}{\partial y}(x, y) + \epsilon_2(\mathbf{h})\right] h_2$$

and therefore

$$f(\mathbf{x} + \mathbf{h}) - f(\mathbf{x}) = \left[\frac{\partial f}{\partial x}(x, y)\mathbf{i} + \frac{\partial f}{\partial y}(x, y)\mathbf{j}\right] \cdot \mathbf{h} + \left[\epsilon_1(\mathbf{h})\mathbf{i} + \epsilon_2(\mathbf{h})\mathbf{j}\right] \cdot \mathbf{h}.$$

To complete the proof of the theorem we need only show that

(2) $$[\epsilon_1(\mathbf{h})\mathbf{i} + \epsilon_2(\mathbf{h})\mathbf{j}] \cdot \mathbf{h} = o(\mathbf{h}).$$

---

† $$\epsilon(\mathbf{h}) = \frac{\partial f}{\partial x}(x + \theta_1 h_1, y + h_2) - \frac{\partial f}{\partial x}(x, y) \to 0$$

since, by the continuity of $\partial f/\partial x$,

$$\frac{\partial f}{\partial x}(x + \theta_1 h_1, y + h_2) \to \frac{\partial f}{\partial x}(x, y).$$

From Schwarz's inequality, $|\mathbf{a} \cdot \mathbf{b}| \leq \|\mathbf{a}\| \|\mathbf{b}\|$, we know that

$$|[\epsilon_1(\mathbf{h})\mathbf{i} + \epsilon_2(\mathbf{h})\mathbf{j}] \cdot \mathbf{h}| \leq \|\epsilon_1(\mathbf{h})\mathbf{i} + \epsilon_2(\mathbf{h})\mathbf{j}\| \|\mathbf{h}\|.$$

It follows that

$$\frac{|[\epsilon_1(\mathbf{h})\mathbf{i} + \epsilon_2(\mathbf{h})\mathbf{j}] \cdot \mathbf{h}|}{\|\mathbf{h}\|} \leq \|\epsilon_1(\mathbf{h})\mathbf{i} + \epsilon_2(\mathbf{h})\mathbf{j}\| \leq \|\epsilon_1(\mathbf{h})\mathbf{i}\| + \|\epsilon_2(\mathbf{h})\mathbf{j}\| = |\epsilon_1(\mathbf{h})| + |\epsilon_2(\mathbf{h})|.$$

As $\mathbf{h} \to \mathbf{0}$, the expression on the right tends to 0. This shows (2) and completes the proof of the theorem. ☐

Example. For

$$f(x, y) = xe^y - ye^x$$

we have

$$\frac{\partial f}{\partial x}(x, y) = e^y - ye^x, \qquad \frac{\partial f}{\partial y}(x, y) = xe^y - e^x$$

and therefore

$$\nabla f(x, y) = (e^y - ye^x)\mathbf{i} + (xe^y - e^x)\mathbf{j}. \quad ☐$$

When there is no reason to emphasize the point of evaluation, we don't write

$$\nabla f(\mathbf{x}) \quad \text{or} \quad \nabla f(x, y) \quad \text{or} \quad \nabla f(x, y, z)$$

but simply ∇f. In this simpler notation, the partial derivatives of

$$f(x, y) = xe^y - ye^x$$

are written

$$\frac{\partial f}{\partial x} = e^y - ye^x, \quad \frac{\partial f}{\partial y} = xe^y - e^x$$

and the gradient

$$\nabla f = (e^y - ye^x)\mathbf{i} + (xe^y - e^x)\mathbf{j}. \quad ☐$$

Example. For

$$f(x, y, z) = \sin xy^2z^3$$

we have

$$\frac{\partial f}{\partial x} = y^2z^3 \cos xy^2z^3, \quad \frac{\partial f}{\partial y} = 2xyz^3 \cos xy^2z^3, \quad \frac{\partial f}{\partial z} = 3xy^2z^2 \cos xy^2z^3,$$

and

$$\nabla f = (\cos xy^2z^3)[y^2z^3\mathbf{i} + 2xyz^3\mathbf{j} + 3xy^2z^2\mathbf{k}]. \quad ☐$$

Example. As a numerical example we take

$$f(x, y, z) = x \sin \pi y + y \cos \pi z$$

and evaluate ∇f at $(0, 1, 2)$.

Here

$$\frac{\partial f}{\partial x} = \sin \pi y, \quad \frac{\partial f}{\partial y} = \pi x \cos \pi y + \cos \pi z, \quad \frac{\partial f}{\partial z} = -\pi y \sin \pi z.$$

At $(0, 1, 2)$

$$\frac{\partial f}{\partial x} = 0, \quad \frac{\partial f}{\partial y} = 1, \quad \frac{\partial f}{\partial z} = 0 \quad \text{and therefore} \quad \nabla f = \mathbf{j}. \quad \square$$

Uniqueness

We defined the gradient at $\mathbf{x}$ as *the* vector $\nabla f(\mathbf{x})$ such that

$$f(\mathbf{x} + \mathbf{h}) - f(\mathbf{x}) = \nabla f(\mathbf{x}) \cdot \mathbf{h} + o(\mathbf{h}).$$

To be sure that this definition makes sense, we must make sure that no two vectors can satisfy that one condition; namely, we must make sure that, if

$$f(\mathbf{x} + \mathbf{h}) - f(\mathbf{x}) = \mathbf{y} \cdot \mathbf{h} + o(\mathbf{h}) \quad \text{and} \quad f(\mathbf{x} + \mathbf{h}) - f(\mathbf{x}) = \mathbf{z} \cdot \mathbf{h} + o(\mathbf{h}),$$

then

$$\mathbf{y} = \mathbf{z}.$$

You are asked to prove this in Exercise 21.

Exercises

Find the gradient.

*1. $f(x, y) = xe^{xy}$.   2. $f(x, y) = x^2 + y^2$.

*3. $f(x, y) = 3x^2 - xy + y$.   4. $f(x, y) = Ax^2 + Bxy + Cy^2$.

*5. $f(x, y) = e^{x-y} - x^2y$.   6. $f(x, y) = x^2y^{-2}$.

*7. $f(x, y, z) = xy^2e^{-z}$.   8. $f(x, y, z) = z \sin (x - y)$.

*9. $f(x, y, z) = xy + yz + zx$.   10. $f(x, y, z) = \sqrt{x^2 + y^4 + z^6}$.

*11. $f(x, y) = (x + y)e^{x-y}$.   12. $f(x, y) = (x + y) \sin (x - y)$.

*13. $f(x, y) = e^x \log y$.   14. $f(x, y, z) = x^2y + y^2z + z^2x$.

*15. $f(x, y) = \dfrac{Ax + By}{Cx + Dy}$.   16. $f(x, y) = \dfrac{x - y}{x^2 + y^2}$.

Find the gradient vector at the indicated point.

*17. $f(x, y) = 2x^2 - 3xy + 4y^2$ at $(2, 3)$.

18. $f(x, y) = \dfrac{2x}{x - y}$ at $(3, 1)$.

*19. $f(x, y) = e^{-x} \sin (x + 2y)$ at $(0, \frac{1}{4}\pi)$.

20. $f(x, y, z) = (x - y) \cos \pi z$ at $(1, 0, \frac{1}{2})$.

21. Show that, if

$$f(\mathbf{x} + \mathbf{h}) - f(\mathbf{x}) = \mathbf{y} \cdot \mathbf{h} + o(\mathbf{h}) \quad \text{and} \quad f(\mathbf{x} + \mathbf{h}) - f(\mathbf{x}) = \mathbf{z} \cdot \mathbf{h} + o(\mathbf{h}),$$

then

$$\mathbf{y} = \mathbf{z}.$$

17.2　Some Simple Properties of Gradients

Much of the calculus of functions of several variables is centered around gradients. Throughout the rest of this chapter we shall study them. In this section we begin.

Differentiability Implies Continuity

As in the one-variable case, differentiability implies continuity; namely, *if f is differentiable at* $\mathbf{x}$, *then f is continuous at* $\mathbf{x}$. To see this, we need only write

$$f(\mathbf{x} + \mathbf{h}) - f(\mathbf{x}) = \nabla f(\mathbf{x}) \cdot \mathbf{h} + o(\mathbf{h}).$$

As $\mathbf{h} \to \mathbf{0}$,

$$|\nabla f(\mathbf{x}) \cdot \mathbf{h}| \le \|\nabla f(\mathbf{x})\| \|\mathbf{h}\| \to 0 \quad \text{and} \quad o(\mathbf{h}) \to 0.$$

It follows that

$$f(\mathbf{x} + \mathbf{h}) - f(\mathbf{x}) \to 0 \quad \text{so that} \quad f(\mathbf{x} + \mathbf{h}) \to f(\mathbf{x}),$$

and thus f is continuous at $\mathbf{x}$. □

Some Formulas

In many respects gradients behave just like derivatives do in the one-variable case. In particular, if $\nabla f(\mathbf{x})$ and $\nabla g(\mathbf{x})$ exist, then so do

$$\nabla[f(\mathbf{x}) + g(\mathbf{x})], \quad \nabla[\alpha f(\mathbf{x})], \quad \nabla[f(\mathbf{x})g(\mathbf{x})].$$

Moreover

(17.2.1)
$$\nabla[f(\mathbf{x}) + g(\mathbf{x})] = \nabla f(\mathbf{x}) + \nabla g(\mathbf{x}),$$
$$\nabla[\alpha f(\mathbf{x})] = \alpha \nabla f(\mathbf{x}),$$
$$\nabla[f(\mathbf{x})g(\mathbf{x})] = f(\mathbf{x}) \nabla g(\mathbf{x}) + g(\mathbf{x}) \nabla f(\mathbf{x}).$$

The first two formulas are easy to derive. To derive the third formula, let's assume that $\nabla f(\mathbf{x})$ and $\nabla g(\mathbf{x})$ both exist. Our task is to show that

$$f(\mathbf{x} + \mathbf{h})g(\mathbf{x} + \mathbf{h}) - f(\mathbf{x})g(\mathbf{x}) = [f(\mathbf{x}) \nabla g(\mathbf{x}) + g(\mathbf{x}) \nabla f(\mathbf{x})] \cdot \mathbf{h} + o(\mathbf{h}).$$

A sketch of how this can be done is given below. We leave it to you to justify each step. To have enough room on the page we set

$$D(\mathbf{x}, \mathbf{h}) = f(\mathbf{x} + \mathbf{h})g(\mathbf{x} + \mathbf{h}) - f(\mathbf{x})g(\mathbf{x}).$$

Then

$$
\begin{aligned}
D(\mathbf{x}, \mathbf{h}) &= [f(\mathbf{x} + \mathbf{h})g(\mathbf{x} + \mathbf{h}) - f(\mathbf{x})g(\mathbf{x} + \mathbf{h})] + [f(\mathbf{x})g(\mathbf{x} + \mathbf{h}) - f(\mathbf{x})g(\mathbf{x})] \\
&= [f(\mathbf{x} + \mathbf{h}) - f(\mathbf{x})]g(\mathbf{x} + \mathbf{h}) + f(\mathbf{x})[g(\mathbf{x} + \mathbf{h}) - g(\mathbf{x})] \\
&= [\nabla f(\mathbf{x}) \cdot \mathbf{h} + o(\mathbf{h})]g(\mathbf{x} + \mathbf{h}) + f(\mathbf{x})[\nabla g(\mathbf{x}) \cdot \mathbf{h} + o(\mathbf{h})] \\
&= g(\mathbf{x} + \mathbf{h})\nabla f(\mathbf{x}) \cdot \mathbf{h} + f(\mathbf{x})\nabla g(\mathbf{x}) \cdot \mathbf{h} + o(\mathbf{h}) \qquad \text{(Exercise 16)} \\
&= g(\mathbf{x})\nabla f(\mathbf{x}) \cdot \mathbf{h} + f(\mathbf{x})\nabla g(\mathbf{x}) \cdot \mathbf{h} + [g(\mathbf{x} + \mathbf{h}) - g(\mathbf{x})]\nabla f(\mathbf{x}) \cdot \mathbf{h} + o(\mathbf{h}) \\
&= [g(\mathbf{x})\nabla f(\mathbf{x}) + f(\mathbf{x})\nabla g(\mathbf{x}) \cdot \mathbf{h} + o(\mathbf{h}). \quad \square \qquad\qquad \text{(Exercise 17)}
\end{aligned}
$$

Directional Derivatives

Here we take up an idea which generalizes the notion of partial derivative. Its connection to gradients will be made clear as we go on. The partial derivatives

$$
\frac{\partial f}{\partial x}(x, y, z) = \lim_{h \to 0} \frac{f(x + h, y, z) - f(x, y, z)}{h},
$$

$$
\frac{\partial f}{\partial y}(x, y, z) = \lim_{h \to 0} \frac{f(x, y + h, z) - f(x, y, z)}{h},
$$

$$
\frac{\partial f}{\partial z}(x, y, z) = \lim_{h \to 0} \frac{f(x, y, z + h) - f(x, y, z)}{h}
$$

can be expressed more elegantly in the form

$$
\frac{\partial f}{\partial x}(\mathbf{x}) = \lim_{h \to 0} \frac{f(\mathbf{x} + h\mathbf{i}) - f(\mathbf{x})}{h},
$$

$$
\frac{\partial f}{\partial y}(\mathbf{x}) = \lim_{h \to 0} \frac{f(\mathbf{x} + h\mathbf{j}) - f(\mathbf{x})}{h},
$$

$$
\frac{\partial f}{\partial z}(\mathbf{x}) = \lim_{h \to 0} \frac{f(\mathbf{x} + h\mathbf{k}) - f(\mathbf{x})}{h}.
$$

Each of the partials is therefore the limit of a quotient of the form

$$
\frac{f(\mathbf{x} + h\mathbf{u}) - f(\mathbf{x})}{h},
$$

where $\mathbf{u}$ is one of the unit coordinate vectors $\mathbf{i}$, $\mathbf{j}$, or $\mathbf{k}$. There is no reason to be so restrictive on $\mathbf{u}$. If f is defined in a neighborhood of $\mathbf{x}$, then, for small h, the difference quotient

$$
\frac{f(\mathbf{x} + h\mathbf{u}) - f(\mathbf{x})}{h}
$$

makes sense for any unit vector $\mathbf{u}$.

Definition 17.2.2   Directional Derivative

For any *unit vector* **u**, the limit

$$f_{\mathbf{u}}'(\mathbf{x}) = \lim_{h \to 0} \frac{f(\mathbf{x} + h\mathbf{u}) - f(\mathbf{x})}{h},$$

if it exists, is called the *directional derivative* of f at **x** in the direction **u**.

The partials are of course themselves directional derivatives:

$$(17.2.3) \qquad \boxed{\frac{\partial f}{\partial x}(\mathbf{x}) = f_{\mathbf{i}}'(\mathbf{x}), \quad \frac{\partial f}{\partial y}(\mathbf{x}) = f_{\mathbf{j}}'(\mathbf{x}), \quad \frac{\partial f}{\partial z}(\mathbf{x}) = f_{\mathbf{k}}'(\mathbf{x}).}$$

As you know, the partials of f give the rates of change of f in the **i**, **j**, **k** directions. The directional derivative $f_{\mathbf{u}}'$ gives *the rate of change* of f in the direction **u**.

There is an important link between the gradient at **x** and the directional derivatives at **x**.

Theorem 17.2.4

If f is differentiable at **x**, then f has a directional derivative at **x** in every direction and

$$f_{\mathbf{u}}'(\mathbf{x}) = \nabla f(\mathbf{x}) \cdot \mathbf{u}.$$

PROOF.   We take **u** as a unit vector and assume that f is differentiable at **x**. The differentiability at **x** tells us that $\nabla f(\mathbf{x})$ exists and

$$f(\mathbf{x} + h\mathbf{u}) - f(\mathbf{x}) = \nabla f(\mathbf{x}) \cdot h\mathbf{u} + o(h\mathbf{u}).$$

Division by h gives

$$\frac{f(\mathbf{x} + h\mathbf{u}) - f(\mathbf{x})}{h} = \nabla f(\mathbf{x}) \cdot \mathbf{u} + \frac{o(h\mathbf{u})}{h}.$$

It's not hard to see that the remainder term tends to zero with h and thus

$$\frac{f(\mathbf{x} + h\mathbf{u}) - f(\mathbf{x})}{h} \to \nabla f(\mathbf{x}) \cdot \mathbf{u}. \quad \square$$

Earlier (in Theorem 17.1.2) we proved that, if f has *continuous first partials in a neighborhood of* **x**, *then f is differentiable at* **x** *and*

$$\nabla f(\mathbf{x}) = \frac{\partial f}{\partial x}(\mathbf{x})\, \mathbf{i} + \frac{\partial f}{\partial y}(\mathbf{x})\, \mathbf{j} + \frac{\partial f}{\partial z}(\mathbf{x})\, \mathbf{k}.$$

Theorem 17.2.5 goes the other way.

Theorem 17.2.5

If f is differentiable at $\mathbf{x}$, then all the first partials exist at $\mathbf{x}$ and

$$\nabla f(\mathbf{x}) = \frac{\partial f}{\partial x}(\mathbf{x})\,\mathbf{i} + \frac{\partial f}{\partial y}(\mathbf{x})\,\mathbf{j} + \frac{\partial f}{\partial z}(\mathbf{x})\,\mathbf{k}.$$

PROOF.   Assume that f is differentiable at $\mathbf{x}$. The existence of the first partials follows from Theorem 17.2.4 together with the observation that the first partials are directional derivatives:

$$\frac{\partial f}{\partial x}(\mathbf{x}) = f_{\mathbf{i}}'(\mathbf{x}), \qquad \frac{\partial f}{\partial y}(\mathbf{x}) = f_{\mathbf{j}}'(\mathbf{x}), \qquad \frac{\partial f}{\partial z}(\mathbf{x}) = f_{\mathbf{k}}'(\mathbf{x}).$$

Also from Theorem 17.2.4 we have

$$\frac{\partial f}{\partial x}(\mathbf{x}) = \nabla f(\mathbf{x}) \cdot \mathbf{i}, \qquad \frac{\partial f}{\partial y}(\mathbf{x}) = \nabla f(\mathbf{x}) \cdot \mathbf{j}, \qquad \frac{\partial f}{\partial z}(\mathbf{x}) = \nabla f(\mathbf{x}) \cdot \mathbf{k}.$$

Since

$$\nabla f(\mathbf{x}) = [\nabla f(\mathbf{x}) \cdot \mathbf{i}]\,\mathbf{i} + [\nabla f(\mathbf{x}) \cdot \mathbf{j}]\,\mathbf{j} + [\nabla f(\mathbf{x}) \cdot \mathbf{k}]\,\mathbf{k},$$

it follows that

$$\nabla f(\mathbf{x}) = \frac{\partial f}{\partial x}(\mathbf{x})\,\mathbf{i} + \frac{\partial f}{\partial y}(\mathbf{x})\,\mathbf{j} + \frac{\partial f}{\partial z}(\mathbf{x})\,\mathbf{k},$$

as asserted.   □

For an arbitrary unit vector $\mathbf{u}$, we have

$$f_{\mathbf{u}}'(\mathbf{x}) = \nabla f(\mathbf{x}) \cdot \mathbf{u} = \frac{\nabla f(\mathbf{x}) \cdot \mathbf{u}}{\|\mathbf{u}\|} = \mathrm{comp}_{\mathbf{u}}\,\nabla f(\mathbf{x}).$$

Namely, the *directional derivative is the component of the gradient vector in that direction*. We illustrate this in Figure 17.2.1.

FIGURE 17.2.1

With

$$\nabla f(\mathbf{x}) = \frac{\partial f}{\partial x}(\mathbf{x})\,\mathbf{i} + \frac{\partial f}{\partial y}(\mathbf{x})\,\mathbf{j} + \frac{\partial f}{\partial z}(\mathbf{x})\,\mathbf{k},$$

the relation $f_{\mathbf{u}}'(\mathbf{x}) = \nabla f(\mathbf{x}) \cdot \mathbf{u}$ gives

$$f_{\mathbf{u}}'(\mathbf{x}) = \frac{\partial f}{\partial x}(\mathbf{x})\, u_1 + \frac{\partial f}{\partial y}(\mathbf{x})\, u_2 + \frac{\partial f}{\partial z}(\mathbf{x})\, u_3.$$

Setting $\mathbf{x} = (x, y, z)$, we have

(17.2.6) $\quad \boxed{f_{\mathbf{u}}'(x, y, z) = \dfrac{\partial f}{\partial x}(x, y, z)\, u_1 + \dfrac{\partial f}{\partial y}(x, y, z)\, u_2 + \dfrac{\partial f}{\partial z}(x, y, z)\, u_3.}$

In the two-variable case, the z-term drops out and we have

(17.2.7) $\quad \boxed{f_{\mathbf{u}}'(x, y) = \dfrac{\partial f}{\partial x}(x, y)\, u_1 + \dfrac{\partial f}{\partial y}(x, y)\, u_2.}$

It's time again for some computations.

Problem. Find the directional derivative of

$$f(x, y) = x^2 + y^2$$

at $(1, 2)$ in the direction of $2\mathbf{i} - 3\mathbf{j}$.

SOLUTION. In the first place, $2\mathbf{i} - 3\mathbf{j}$ is not a unit vector. Its norm is $\sqrt{13}$. The unit vector we want is

$$\mathbf{u} = \tfrac{1}{13}\sqrt{13}\,[2\mathbf{i} - 3\mathbf{j}].$$

Since we are in the two-variable case,

$$f_{\mathbf{u}}'(x, y) = \frac{\partial f}{\partial x}(x, y)\, u_1 + \frac{\partial f}{\partial y}(x, y)\, u_2.$$

From

$$\frac{\partial f}{\partial x}(x, y) = 2x \quad \text{and} \quad \frac{\partial f}{\partial y}(x, y) = 2y,$$

we get

$$\frac{\partial f}{\partial x}(1, 2) = 2 \quad \text{and} \quad \frac{\partial f}{\partial y}(1, 2) = 4.$$

With $u_1 = \tfrac{2}{13}\sqrt{13}$ and $u_2 = -\tfrac{3}{13}\sqrt{13}$, we have

$$f_{\mathbf{u}}'(1, 2) = \tfrac{4}{13}\sqrt{13} - \tfrac{12}{13}\sqrt{13} = -\tfrac{8}{13}\sqrt{13}. \quad \square$$

Problem. Find the directional derivative of

$$f(x, y, z) = x \cos y \sin z$$

at $(1, \pi, \tfrac{1}{4}\pi)$ in the direction of $2\mathbf{i} - \mathbf{j} + 4\mathbf{k}$.

SOLUTION. The unit vector we want is

$$\mathbf{u} = \tfrac{1}{21}\sqrt{21}\,[2\mathbf{i} - \mathbf{j} + 4\mathbf{k}].$$

Here we are in the three-variable case:

$$f_{\mathbf{u}}'(x, y, z) = \frac{\partial f}{\partial x}(x, y, z)\, u_1 + \frac{\partial f}{\partial y}(x, y, z)\, u_2 + \frac{\partial f}{\partial z}(x, y, z)\, u_3.$$

Note that

$$\frac{\partial f}{\partial x}(x, y, z) = \cos y \sin z, \qquad \frac{\partial f}{\partial y}(x, y, z) = -x \sin y \sin z,$$

$$\frac{\partial f}{\partial z}(x, y, z) = x \cos y \cos z$$

so that

$$\frac{\partial f}{\partial x}(1, \pi, \tfrac{1}{4}\pi) = -\tfrac{1}{2}\sqrt{2}, \quad \frac{\partial f}{\partial y}(1, \pi, \tfrac{1}{4}\pi) = 0, \quad \frac{\partial f}{\partial z}(1, \pi, \tfrac{1}{4}\pi) = -\tfrac{1}{2}\sqrt{2}.$$

With $u_1 = \tfrac{2}{21}\sqrt{21}$, $u_2 = -\tfrac{1}{21}\sqrt{21}$, $u_3 = \tfrac{4}{21}\sqrt{21}$, we have

$$f_{\mathbf{u}}'(1, \pi, \tfrac{1}{4}\pi) = -\tfrac{1}{21}\sqrt{42} + 0 - \tfrac{2}{21}\sqrt{42} = -\tfrac{3}{21}\sqrt{42}. \quad \Box$$

Theorem 17.2.8. If f is differentiable at $\mathbf{x}$, then

$$f_{\mathbf{u}}'(\mathbf{x}) = \|\nabla f(\mathbf{x})\| \cos\theta$$

where θ is the angle between $\nabla f(\mathbf{x})$ and the unit vector $\mathbf{u}$.

PROOF. The relation

$$f_{\mathbf{u}}'(\mathbf{x}) = \nabla f(\mathbf{x}) \cdot \mathbf{u}$$

gives

$$f_{\mathbf{u}}'(\mathbf{x}) = \|\nabla f(\mathbf{x})\|\,\|\mathbf{u}\| \cos\theta \tag{14.4.4}$$

where θ is the angle between $\nabla f(\mathbf{x})$ and $\mathbf{u}$. The result follows since $\|\mathbf{u}\| = 1$. $\Box$

Theorem 17.2.9. If f is differentiable at $\mathbf{x}$, then

$$-\|\nabla f(\mathbf{x})\| \le f_{\mathbf{u}}'(\mathbf{x}) \le \|\nabla f(\mathbf{x})\| \qquad \text{for all directions } \mathbf{u}.$$

Moreover,

$$f_{\mathbf{u}}'(\mathbf{x}) = \|\nabla f(\mathbf{x})\| \qquad \text{if } \mathbf{u} \text{ points in the direction of } \nabla f(\mathbf{x})$$

and

$$f_{\mathbf{u}}'(\mathbf{x}) = -\|\nabla f(\mathbf{x})\| \qquad \text{if } \mathbf{u} \text{ points in the direction of } -\nabla f(\mathbf{x}).$$

PROOF. You have just seen that

$$f_{\mathbf{u}}'(\mathbf{x}) = \|\nabla f(\mathbf{x})\| \cos\theta$$

where θ is the angle between $\nabla f(\mathbf{x})$ and $\mathbf{u}$. The relation

$$-\|\nabla f(\mathbf{x})\| \le f_{\mathbf{u}}'(\mathbf{x}) \le \|\nabla f(\mathbf{x})\|$$

follows directly from the observation that $-1 \le \cos\theta \le 1$.

If $\mathbf{u}$ points in the direction of $\nabla f(\mathbf{x})$, then $\theta = 0$, $\cos\theta = 1$ and

$$f_{\mathbf{u}}'(\mathbf{x}) = \|\nabla f(\mathbf{x})\| \cdot 1 = \|\nabla f(\mathbf{x})\|.$$

If $\mathbf{u}$ points in the direction of $-\nabla f(\mathbf{x})$, then $\theta = \pi$, $\cos\theta = -1$ and

$$f_{\mathbf{u}}'(\mathbf{x}) = \|\nabla f(\mathbf{x})\| \cdot (-1) = -\|\nabla f(\mathbf{x})\|. \quad \square$$

Since the directional derivative of a function gives the rate of change of the function in that direction, you can see from the last theorem that

(17.2.10)

> a differentiable function f increases most rapidly in the direction of the gradient (the rate of change is then $\|\nabla f(\mathbf{x})\|$) and it decreases most rapidly in the opposite direction (the rate of change is then $-\|\nabla f(\mathbf{x})\|$).

Example. In the case of

$$f(x, y) = \sqrt{1 - (x^2 + y^2)}$$

the graph is the upper hemisphere of radius one centered at the origin. The gradient

$$\nabla f(x, y) = \frac{-x}{\sqrt{1 - (x^2 + y^2)}}\,\mathbf{i} + \frac{-y}{\sqrt{1 - (x^2 + y^2)}}\,\mathbf{j}$$

points from (x, y) along the line of the radius vector toward the origin:

$$\nabla f(x, y) = -\frac{1}{\sqrt{1 - (x^2 + y^2)}}\,(x\mathbf{i} + y\mathbf{j}) = -\frac{1}{\sqrt{1 - (x^2 + y^2)}}\,\mathbf{r}. \quad \text{(Figure 17.2.2)}$$

This means that along the surface the path of steepest ascent from the point $P(x, y, f(x, y))$ is "the great circle route through the north pole". $\quad \square$

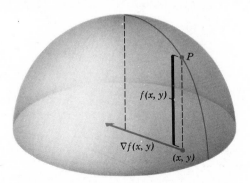

$$f(x, y) = \sqrt{1 - (x^2 + y^2)}$$

FIGURE 17.2.2

Problem.   The temperature at each point of a metal plate is given by the function

$$T(x, y) = e^x \cos y + e^y \cos x.$$

(a) In what direction does the temperature increase most rapidly at the point (0, 0)? What is this rate of increase?

(b) In what direction does the temperature decrease most rapidly at (0, 0)?

SOLUTION

$$\nabla T(x, y) = \frac{\partial T}{\partial x} (x, y) \mathbf{i} + \frac{\partial T}{\partial y} (x, y) \mathbf{j}$$

$$= (e^x \cos y - e^y \sin x) \mathbf{i} + (e^y \cos x - e^x \sin y) \mathbf{j}.$$

(a) At (0, 0) the temperature increases most rapidly in the direction of the gradient

$$\nabla T(0, 0) = \mathbf{i} + \mathbf{j}.$$

This rate of increase is

$$\|\nabla T(0, 0)\| = \|\mathbf{i} + \mathbf{j}\| = \sqrt{2}.$$

(b) The temperature decreases most rapidly in the direction of

$$-\nabla T(0, 0) = -\mathbf{i} - \mathbf{j}. \quad \square$$

Problem.   The density (mass per unit volume) of a metal ball centered at the origin is given by the function

$$\rho(x, y, z) = ke^{-(x^2+y^2+z^2)}, \qquad k \text{ a positive constant.}$$

(a) In what direction does the density increase most rapidly at the point (x, y, z)? What is this rate of density increase?

(b) In what direction does the density decrease most rapidly?

(c) What are the rates of density change at (x, y, z) in the $\mathbf{i}, \mathbf{j}, \mathbf{k}$ directions?

SOLUTION.   The gradient

$$\nabla \rho(x, y, z) = \frac{\partial \rho}{\partial x} (x, y, z) \mathbf{i} + \frac{\partial \rho}{\partial y} (x, y, z) \mathbf{j} + \frac{\partial \rho}{\partial z} (x, y, z) \mathbf{k}$$

$$= -2ke^{-(x^2+y^2+z^2)}[x\mathbf{i} + y\mathbf{j} + z\mathbf{k}]$$

$$= -2\rho(x, y, z)\mathbf{r}$$

points from (x, y, z) along the line of the radius vector toward the origin.

(a) The density increases most rapidly toward the origin. The rate of increase is

$$\|\nabla \rho(x, y, z)\| = 2\rho(x, y, z)\|\mathbf{r}\| = 2\rho(x, y, z)\sqrt{x^2 + y^2 + z^2}.$$

(b) The density decreases most rapidly directly away from the origin.

(c) The rates of density change in the $\mathbf{i}, \mathbf{j}, \mathbf{k}$ directions are given by the directional derivatives

$$\rho_i'(x, y, z) = \nabla\rho(x, y, z) \cdot \mathbf{i} = -2x\rho(x, y, z),$$
$$\rho_j'(x, y, z) = \nabla\rho(x, y, z) \cdot \mathbf{j} = -2y\rho(x, y, z),$$
$$\rho_k'(x, y, z) = \nabla\rho(x, y, z) \cdot \mathbf{k} = -2z\rho(x, y, z).$$

These are just the first partials of ρ. ☐

Problem.　The temperature at each point of a metal plate is given by the function

$$T(x, y) = A + x^2 - y^2.$$

Find the path followed by a heat-seeking particle which starts at the point $(-2, 1)$.

SOLUTION.　The particle moves in the direction of the gradient

$$\nabla T = 2x\mathbf{i} - 2y\mathbf{j}.$$

We want the curve

$$C: \mathbf{r}(t) = x(t)\mathbf{i} + y(t)\mathbf{j}$$

which begins at $(-2, 1)$ and at each point has its tangent vector in the direction of ∇T. We can satisfy the first condition by setting

$$x(0) = -2, \quad y(0) = 1.$$

We can satisfy the second condition by setting

$$x'(t) = 2x(t), \quad y'(t) = -2y(t). \qquad \text{(explain)}$$

These differential equations, together with initial conditions at $t = 0$, imply that

$$x(t) = -2e^{2t}, \quad y(t) = e^{-2t}.$$

We can eliminate the parameter t by noting that

$$x(t)y(t) = (-2e^{2t})(e^{-2t}) = -2.$$

In terms of just x and y we have

$$xy = -2.$$

The particle moves along the hyperbola $xy = -2$ in the direction of decreasing x and y. ☐

Exercises

Find the directional derivative at the given point in the direction indicated.

*1. $f(x, y) = x^2 + 3y^2$　at $(1, 1)$ in the direction of $\mathbf{i} - \mathbf{j}$.
2. $f(x, y) = x + \sin(x + y)$　at $(0, 0)$ in the direction of $2\mathbf{i} + \mathbf{j}$.
*3. $f(x, y) = xe^y - ye^x$　at $(1, 0)$ in the direction of $3\mathbf{i} + 4\mathbf{j}$.

4. $f(x, y) = \dfrac{2x}{x - y}$ at $(1, 0)$ in the direction of $\mathbf{i} - \sqrt{3}\mathbf{j}$.

*5. $f(x, y) = \dfrac{ax + by}{x + y}$ at $(1, 1)$ in the direction of $\mathbf{i} - \mathbf{j}$.

6. $f(x, y) = \dfrac{x + y}{cx + dy}$ at $(1, 1)$ in the direction of $c\mathbf{i} - d\mathbf{j}$.

*7. $f(x, y, z) = xy + yz + zx$ at $(1, -1, 1)$ in the direction of $\mathbf{i} + 2\mathbf{j} + \mathbf{k}$.

8. $f(x, y, z) = x^2y + y^2z + z^2x$ at $(1, 0, 1)$ in the direction of $3\mathbf{j} - \mathbf{k}$.

*9. $f(x, y, z) = (x + y^2 + z^3)^2$ at $(1, -1, 1)$ in the direction of $\mathbf{i} + \mathbf{j}$.

10. $f(x, y, z) = Ax^2 + Bxyz + Cy^2$ at $(1, 2, 1)$ in the direction of $A\mathbf{i} + B\mathbf{j} + C\mathbf{k}$.

*11. Find the directional derivative of $f(x, y) = \log \sqrt{x^2 + y^{2i}}$ at $(x, y) \neq (0, 0)$ toward the origin.

12. Find the directional derivative of $f(x, y) = (x - 1)y^2e^{xy}$ at $(0, 1)$ toward the point $(-1, 3)$.

*13. Find the directional derivative of $f(x, y) = Ax^2 + 2Bxy + Cy^2$ at (a, b) toward (b, a) (a) if $a > b$.   (b) if $a < b$.

14. Find the directional derivative of $f(x, y, z) = xy^2 + y^2z^3 + z^3x$ at $(2, -1, 1)$ in the direction of the most rapid increase of f.

*15. Find the directional derivative of $f(x, y, z) = xe^{y^2 - z^2}$ at $(1, 2, -2)$ in the direction of the path $\mathbf{r}(t) = t\mathbf{i} + 2\cos(t - 1)\mathbf{j} - 2e^{t-1}\mathbf{k}$.

16. Verify that, if g is continuous at $\mathbf{x}$, then $g(\mathbf{x} + \mathbf{h})o(\mathbf{h}) = o(\mathbf{h})$.

17. Verify that, if g is continuous at $\mathbf{x}$, then $[g(\mathbf{x} + \mathbf{h}) - g(\mathbf{x})]\nabla f(\mathbf{x}) \cdot \mathbf{h} = o(\mathbf{h})$.

18. Given the density function $\rho(x, y) = 48 - \frac{4}{3}x^2 - 3y^2$ find the rate of density change (a) at $(1, -1)$ in the direction of the most rapid density decrease. (b) at $(1, 2)$ in the $\mathbf{i}$ direction. (c) at $(2, 2)$ away from the origin.

*19. Determine the path of steepest descent along the surface $z = x^2 + 3y^2$ from the point $(1, 1, 4)$.

20. Same as 19 but take $(1, -2, 13)$ as the starting point.

*21. Determine the path of steepest descent along the surface $z = a^2x^2 + b^2y^2$ from the point $(a^2, b^2, a^4 + b^4)$.

Optional | 22. Determine the path of steepest descent along the surface $z = A + x + 2y - x^2 - 3y^2$ from the point $(0, 0, A)$. HINT: The differential equations can be solved by the methods of Section 6.8.

17.3 The Mean-Value Theorem and Chain Rule

You are already familiar with results bearing these names in the one-variable case. Here we take up the multivariable case. By **ab** we mean the line segment that joins **a** to **b**.

Theorem 17.3.1 The Mean-Value Theorem

If f is differentiable at each point of the line segment **ab**, then there exists on that line segment a point **c** between **a** and **b** such that

$$f(\mathbf{b}) - f(\mathbf{a}) = \nabla f(\mathbf{c}) \cdot (\mathbf{b} - \mathbf{a}).$$

PROOF.   As t ranges from 0 to 1, $\mathbf{a} + t(\mathbf{b} - \mathbf{a})$ traces out the line segment **ab**. The idea of the proof is to apply the one-variable mean-value theorem to the function

$$g(t) = f(\mathbf{a} + t[\mathbf{b} - \mathbf{a}]), \qquad t \in [0, 1].$$

To get the differentiability of g, we take $t \in (0, 1)$ and form

$$
\begin{aligned}
g(t + h) - g(t) &= f(\mathbf{a} + (t + h)[\mathbf{b} - \mathbf{a}]) - f(\mathbf{a} + t[\mathbf{b} - \mathbf{a}]) \\
&= f(\mathbf{a} + t[\mathbf{b} - \mathbf{a}] + h[\mathbf{b} - \mathbf{a}]) - f(\mathbf{a} + t[\mathbf{b} - \mathbf{a}]) \\
&= \nabla f(\mathbf{a} + t[\mathbf{b} - \mathbf{a}]) \cdot h[\mathbf{b} - \mathbf{a}] + o(h[\mathbf{b} - \mathbf{a}]).
\end{aligned}
$$

Since

$$\nabla f(\mathbf{a} + t[\mathbf{b} - \mathbf{a}]) \cdot h(\mathbf{b} - \mathbf{a}) = \left[\nabla f(\mathbf{a} + t[\mathbf{b} - \mathbf{a}]) \cdot (\mathbf{b} - \mathbf{a}) \right] h$$

and the $o(h(\mathbf{b} - \mathbf{a}))$ term is obviously $o(h)$, we can write

$$g(t + h) - g(t) = \left[\nabla f(\mathbf{a} + t[\mathbf{b} - \mathbf{a}]) \cdot (\mathbf{b} - \mathbf{a}) \right] h + o(h).$$

Dividing both sides by h, we see that g is differentiable and

$$g'(t) = \nabla f(\mathbf{a} + t[\mathbf{b} - \mathbf{a}]) \cdot (\mathbf{b} - \mathbf{a}).$$

The function g is clearly continuous at 0 and at 1. Applying the one-variable mean-value theorem to g, we can conclude that there exists a number t_0 between 0 and 1 such that

$$(1) \qquad\qquad g(1) - g(0) = g'(t_0)(1 - 0).$$

Since

$$g(1) = f(\mathbf{b}), \quad g(0) = f(\mathbf{a}), \quad \text{and} \quad g'(t_0) = \nabla f(\mathbf{a} + t_0[\mathbf{b} - \mathbf{a}]) \cdot (\mathbf{b} - \mathbf{a}),$$

condition (1) gives us

$$f(\mathbf{b}) - f(\mathbf{a}) = \nabla f(\mathbf{a} + t_0[\mathbf{b} - \mathbf{a}]) \cdot (\mathbf{b} - \mathbf{a}).$$

Setting $\mathbf{c} = \mathbf{a} + t_0[\mathbf{b} - \mathbf{a}]$, we have

$$f(\mathbf{b}) - f(\mathbf{a}) = \nabla f(\mathbf{c}) \cdot (\mathbf{b} - \mathbf{a}). \quad \square \qquad \text{(Figure 17.3.1)}$$

there exists a point $\mathbf{c}$ between $\mathbf{a}$ and $\mathbf{b}$ such that
$$f(\mathbf{b}) - f(\mathbf{a}) = \nabla f(\mathbf{c}) \cdot [\mathbf{b} - \mathbf{a}]$$

FIGURE 17.3.1

Before stating the chain rule we go over some terminology. A vector-valued function is said to be *continuous* provided that its components are continuous. If $f = f(x, y, z)$ is a scalar-valued function (a real-valued function), then its gradient ∇f is a vector-valued function. We shall say that f is *continuously differentiable* on U iff f is differentiable on U and ∇f is continuous on U. If a curve $\mathbf{r}$ lies in the domain of f, then we can form the composition

$$(f \circ \mathbf{r})(t) = f(\mathbf{r}(t)).$$

The numbers $f(\mathbf{r}(t))$ are the values taken on by f *along the curve* $\mathbf{r}$.

Theorem 17.3.2 Chain Rule

If f is continuously differentiable on an open set U and $\mathbf{r}$ is a differentiable curve in U, then the composition $f \circ \mathbf{r}$ is differentiable and

$$\frac{d}{dt}[f(\mathbf{r}(t))] = \nabla f(\mathbf{r}(t)) \cdot \mathbf{r}'(t).$$

PROOF.   We will show that

$$\lim_{h \to 0} \frac{f(\mathbf{r}(t + h)) - f(\mathbf{r}(t))}{h} = \nabla f(\mathbf{r}(t)) \cdot \mathbf{r}'(t).$$

For $h \neq 0$ and sufficiently small, the line segment joining $\mathbf{r}(t)$ to $\mathbf{r}(t + h)$ lies entirely in U. This we know because U is open and $\mathbf{r}$ is continuous. (See Figure 17.3.2.) For such h, the mean-value theorem just proved tells us that there exists a point $\mathbf{c}(h)$ between $\mathbf{r}(t)$ and $\mathbf{r}(t + h)$ such that

$$f(\mathbf{r}(t + h)) - f(\mathbf{r}(t)) = \nabla f(\mathbf{c}(h)) \cdot [\mathbf{r}(t + h) - \mathbf{r}(t)].$$

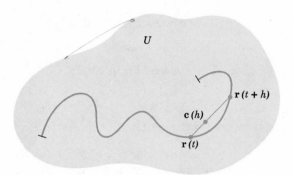

FIGURE 17.3.2

Dividing both sides by h,

$$\frac{f(\mathbf{r}(t + h)) - f(\mathbf{r}(t))}{h} = \nabla f(\mathbf{c}(h)) \cdot \left[\frac{\mathbf{r}(t + h) - \mathbf{r}(t)}{h}\right].$$

As h tends to zero, $\mathbf{c}(h)$ tends to $\mathbf{r}(t)$ and by the continuity of ∇f,

$$\nabla f(\mathbf{c}(h)) \to \nabla f(\mathbf{r}(t)).$$

Since

$$\frac{\mathbf{r}(t + h) - \mathbf{r}(t)}{h} \to \mathbf{r}'(t),$$

the result follows. $\square$

Problem.   Find the rate of change of

$$f(x, y) = \tfrac{1}{3}(x^3 + y^3)$$

with respect to t along the curve $\mathbf{r}(t) = a \cos t\, \mathbf{i} + b \sin t\, \mathbf{j}$.

SOLUTION.   The rate of change of f with respect to t along the curve $\mathbf{r}$ is the derivative

$$\frac{d}{dt} [f(\mathbf{r}(t))].$$

By the chain rule

$$\frac{d}{dt} [f(\mathbf{r}(t))] = \nabla f(\mathbf{r}(t)) \cdot \mathbf{r}'(t).$$

Here

$$\nabla f = x^2 \mathbf{i} + y^2 \mathbf{j}.$$

With $x(t) = a \cos t$ and $y(t) = b \sin t$, we have

$$\nabla f(\mathbf{r}(t)) = a^2 \cos^2 t \, \mathbf{i} + b^2 \sin^2 t \, \mathbf{j}.$$

Since

$$\mathbf{r}'(t) = -a \sin t \, \mathbf{i} + b \cos t \, \mathbf{j},$$

you can see that

$$\frac{d}{dt}[f(\mathbf{r}(t))] = \nabla f(\mathbf{r}(t)) \cdot \mathbf{r}'(t)$$
$$= -a^3 \sin t \cos^2 t + b^3 \sin^2 t \cos t$$
$$= \sin t \cos t \, (b^3 \sin t - a^3 \cos t). \quad \square$$

Problem.   Find the rate of change of

$$f(x, y, z) = x^2 y + z \cos x$$

with respect to t along the twisted cubic $\mathbf{r}(t) = t\mathbf{i} + t^2\mathbf{j} + t^3\mathbf{k}$.

SOLUTION.   Once again we want to find

$$\frac{d}{dt}[f(\mathbf{r}(t))] = \nabla f(\mathbf{r}(t)) \cdot \mathbf{r}'(t).$$

This time

$$\nabla f = (2xy - z \sin x)\mathbf{i} + x^2\mathbf{j} + \cos x \, \mathbf{k}.$$

With $x(t) = t$, $y(t) = t^2$, $z(t) = t^3$, we have

$$\nabla f(\mathbf{r}(t)) = (2t^3 - t^3 \sin t)\mathbf{i} + t^2\mathbf{j} + \cos t \, \mathbf{k}.$$

Since

$$\mathbf{r}'(t) = \mathbf{i} + 2t\mathbf{j} + 3t^2\mathbf{k},$$

we have

$$\frac{d}{dt}[f(\mathbf{r}(t))] = \nabla f(\mathbf{r}(t)) \cdot \mathbf{r}'(t)$$
$$= 2t^3 - t^3 \sin t + 2t^3 + 3t^2 \cos t = 4t^3 - t^3 \sin t + 3t^2 \cos t. \quad \square$$

Another View of the Chain Rule

In the one-variable case we can write the chain rule in two ways. We can write

$$(1) \qquad \frac{d}{dt}[f(g(t))] = f'(g(t))g'(t)$$

or, by setting $u = f(x)$, $x = g(t)$, we can write

(2) $$\frac{du}{dt} = \frac{du}{dx}\frac{dx}{dt}.$$

You can view the formula

$$\frac{d}{dt}[f(\mathbf{r}(t))] = \nabla f(\mathbf{r}(t)) \cdot \mathbf{r}'(t)$$

as the counterpart of (1). The counterpart of (2) can be obtained by setting

$$u = f(x, y, z); \qquad x = g(t), \quad y = h(t), \quad z = k(t).$$

The chain rule then reads

(17.3.3) $$\boxed{\frac{du}{dt} = \frac{\partial u}{\partial x}\frac{dx}{dt} + \frac{\partial u}{\partial y}\frac{dy}{dt} + \frac{\partial u}{\partial z}\frac{dz}{dt}.}$$

In the two-variable case the z-term drops out and we have

(17.3.4) $$\boxed{\frac{du}{dt} = \frac{\partial u}{\partial x}\frac{dx}{dt} + \frac{\partial u}{\partial y}\frac{dy}{dt}.}$$

Problem. Find du/dt if

$$u = x^2 - y^2 \quad \text{and} \quad x = t^2 - 1, \quad y = 3\sin \pi t.$$

Solution.   Here we are in the two-variable case

$$\frac{du}{dt} = \frac{\partial u}{\partial x}\frac{dx}{dt} + \frac{\partial u}{\partial y}\frac{dy}{dt}.$$

Since

$$\frac{\partial u}{\partial x} = 2x, \quad \frac{\partial u}{\partial y} = -2y, \quad \frac{dx}{dt} = 2t, \quad \frac{dy}{dt} = 3\pi \cos \pi t,$$

we have

$$\frac{du}{dt} = (2x)(2t) + (-2y)(3\pi \cos \pi t)$$

$$= 2(t^2 - 1)(2t) + (-2)(3\sin \pi t)(3\pi \cos \pi t)$$

$$= 4t^3 - 4t - 18\sin \pi t \cos \pi t. \quad \square$$

Other forms of the chain rule are left to the exercises. They can all be deduced from Theorem 17.3.2.

Problem.   The upper radius of the frustum of a cone is 10 inches, the lower radius is 12 inches, and the height is 18 inches. At what rate is the volume changing if the upper radius decreases at the rate of 2 inches per minute, the lower radius increases at the rate of 3 inches per minute, and the height decreases at the rate of 4 inches per minute?

SOLUTION.   Let x be the upper radius, y the lower radius, and z the height. Then

$$V = \frac{1}{3}\pi z(x^2 + xy + y^2)$$

so that

$$\frac{\partial V}{\partial x} = \frac{1}{3}\pi z(2x + y), \quad \frac{\partial V}{\partial y} = \frac{1}{3}\pi z(x + 2y), \quad \frac{\partial V}{\partial z} = \frac{1}{3}\pi(x^2 + xy + y^2).$$

Since

$$\frac{dV}{dt} = \frac{\partial V}{\partial x}\frac{dx}{dt} + \frac{\partial V}{\partial y}\frac{dy}{dt} + \frac{\partial V}{\partial z}\frac{dz}{dt},$$

we have here

$$\frac{dV}{dt} = \frac{1}{3}\pi z(2x + y)\frac{dx}{dt} + \frac{1}{3}\pi z(x + 2y)\frac{dy}{dt} + \frac{1}{3}\pi(x^2 + xy + y^2)\frac{dz}{dt}.$$

Set

$$x = 10, \quad y = 12, \quad z = 18, \quad \frac{dx}{dt} = -2, \quad \frac{dy}{dt} = 3, \quad \frac{dz}{dt} = -4,$$

and you'll find that

$$\frac{dV}{dt} = -\frac{772}{3}\pi.$$

The volume decreases at the rate of $\frac{772}{3}\pi$ cubic inches per minute.   □

In Chapter 4 you saw that, if $f'(x) = 0$ for all x in an open interval I, then f is constant on I. We have a similar result for functions of several variables.

Theorem 17.3.5

Let U be an open set with the property that any two points of U can be joined by a differentiable curve which lies entirely in U:

if $\nabla f(\mathbf{x}) = \mathbf{0}$ for all $\mathbf{x}$ in U, then f is constant on U.

PROOF.   Let $\mathbf{a}$ and $\mathbf{b}$ be two points of U. We join these two points by a differentiable curve $\mathbf{r}(t)$ with $\mathbf{r}(a) = \mathbf{a}$ and $\mathbf{r}(b) = \mathbf{b}$. (See Figure 17.3.3.) Let f be the function in question. Since

$$\nabla f(\mathbf{r}(t)) = \mathbf{0} \qquad \text{for all } t \in [a, b],$$

we have

$$\int_a^b [\nabla f(\mathbf{r}(t)) \cdot \mathbf{r}'(t)] \, dt = 0.$$

On the other hand,

$$\int_a^b [\nabla f(\mathbf{r}(t)) \cdot \mathbf{r}'(t)] \, dt = \int_a^b \frac{d}{dt} [f(\mathbf{r}(t))] \, dt = f(\mathbf{r}(b)) - f(\mathbf{r}(a)) = f(\mathbf{b}) - f(\mathbf{a}).$$

This tells us that $f(\mathbf{b}) = f(\mathbf{a})$. Since $\mathbf{a}$ and $\mathbf{b}$ are arbitrary points of U, f must be constant on U. $\square$

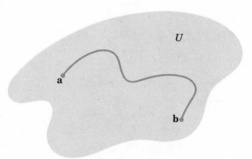

any two points of U can be joined by a
differentiable curve which lies entirely in U

FIGURE 17.3.3

Theorem 17.3.6

If $\nabla f(\mathbf{x}) = \nabla g(\mathbf{x})$ for all $\mathbf{x}$ in U, then f and g differ by a constant on U.

PROOF.   If

$$\nabla f(\mathbf{x}) = \nabla g(\mathbf{x}) \qquad \text{for all } \mathbf{x} \text{ in } U,$$

then

$$\nabla[f(\mathbf{x}) - g(\mathbf{x})] = \nabla f(\mathbf{x}) - \nabla g(\mathbf{x}) = \mathbf{0} \qquad \text{for all } \mathbf{x} \text{ in } U$$

and, by Theorem 17.3.5, $f - g$ must be constant on U. $\square$

CAUTION.   The last two theorems depend strongly on our being able to join any two points of U by a differentiable curve *which lies entirely in U*. If U is "disconnected," as in Figure 17.3.4, the results do not hold.

$U = A \cup B$
$\mathbf{a}$ and $\mathbf{b}$ cannot be joined by a
curve lying entirely in U

FIGURE 17.3.4

Exercises

Find the rate of change of f with respect to t along the given curve.

*1. $f(x, y) = x^2 y$, $\quad \mathbf{r}(t) = e^t \mathbf{i} + e^{-t} \mathbf{j}$.
 2. $f(x, y) = x - y$, $\quad \mathbf{r}(t) = at \mathbf{i} + b \cos at \, \mathbf{j}$.
*3. $f(x, y, z) = xy - yz$, $\quad \mathbf{r}(t) = t \mathbf{i} + t^2 \mathbf{j} + t^3 \mathbf{k}$.
 4. $f(x, y, z) = x^2 + y^2$, $\quad \mathbf{r}(t) = a \cos \omega t \mathbf{i} + b \sin \omega t \mathbf{j} + b \omega t \mathbf{k}$.
*5. $f(x, y, z) = x^2 + y^2 + z$, $\quad \mathbf{r}(t) = a \cos \omega t \mathbf{i} + b \sin \omega t \mathbf{j} + b \omega t \mathbf{k}$.
 6. $f(x, y, z) = y^2 \sin (x + z)$, $\quad \mathbf{r}(t) = 2t \mathbf{i} + \cos t \mathbf{j} + t^3 \mathbf{k}$.

Find du/dt by the method of this section.

*7. $u = x^2 - 3xy + 2y^2$; $\qquad x = \cos t$, $\quad y = \sin t$.
 8. $u = x + 4\sqrt{xy} - 3y$; $\qquad x = t^3$, $\quad y = t^{-1}$. $\qquad (t > 0)$
*9. $u = e^x \sin y + e^y \sin x$; $\qquad x = \tfrac{1}{2}t$, $\quad y = 2t$.
 10. $u = 2x^2 - xy + y^2$; $\qquad x = \cos 2t$, $\quad y = \sin t$.
*11. $u = xy + yz + zx$; $\qquad x = t^2$, $\quad y = t(1 - t)$, $\quad z = (1 - t)^2$.
 12. $u = x \sin \pi y - z \cos \pi x$; $\qquad x = t^2$, $\quad y = 1 - t$, $\quad z = 1 - t^2$.

*13. The height of a cone is 20 inches and the radius is 14 inches. Use the methods of this section to find the rate at which the volume is changing if the height decreases at the rate of 2 inches per second and the radius increases at the rate of 3 inches per second.

 14. Find du/dt given that

$$u = f(x, y); \quad x = g(s), y = h(s); \quad s = k(t).$$

 15. Verify that, if

$$u = f(x, y) \quad \text{where} \quad x = g(s, t) \quad \text{and} \quad y = h(s, t),$$

then

(17.3.7) $\qquad \dfrac{\partial u}{\partial s} = \dfrac{\partial u}{\partial x} \dfrac{\partial x}{\partial s} + \dfrac{\partial u}{\partial y} \dfrac{\partial y}{\partial s} \quad \text{and} \quad \dfrac{\partial u}{\partial t} = \dfrac{\partial u}{\partial x} \dfrac{\partial x}{\partial t} + \dfrac{\partial u}{\partial y} \dfrac{\partial y}{\partial t}.$

*16. Use (17.3.7) to find $\partial u/\partial s$ and $\partial u/\partial t$ given that

$$u = x^2 - xy; \qquad x = s \cos t, \quad y = t \sin s.$$

*17. Find $\partial u/\partial s$ and $\partial u/\partial t$ given that

$$u = f(x, y, z); \qquad x = g(s, t), \quad y = h(s, t), \quad z = k(s, t).$$

*18. Find $\partial u/\partial s$ and $\partial u/\partial t$ given that

$$u = x^2 - xy + z^2; \qquad x = s \cos t, \quad y = \sin (t - s), \quad z = t \sin s.$$

 19. Let $u = f(x, y)$. Show that the change of variables

$$x = r \cos \theta, \quad y = r \sin \theta$$

gives

$$\left(\dfrac{\partial u}{\partial x} \right)^2 + \left(\dfrac{\partial u}{\partial y} \right)^2 = \left(\dfrac{\partial u}{\partial r} \right)^2 + \dfrac{1}{r^2} \left(\dfrac{\partial u}{\partial \theta} \right)^2.$$

20. Let $u = f(r)$, $r = \sqrt{x^2 + y^2}$. Show that

$$\frac{\partial^2 u}{\partial x^2} + \frac{\partial^2 u}{\partial y^2} = \frac{d^2 u}{dr^2} + \frac{1}{r} \frac{du}{dr}.$$

*21. Find f, given that

$$\nabla f(x, y, z) = a_1 \mathbf{i} + a_2 \mathbf{j} + a_3 \mathbf{k} \qquad \text{for all } (x, y, z).$$

22. What can you conclude about f and g given that

$$\nabla f(x, y, z) - \nabla g(x, y, z) = a_1 \mathbf{i} + a_2 \mathbf{j} + a_3 \mathbf{k} \qquad \text{for all } (x, y, z)?$$

23. An object moves so that at time t it has position

$$\mathbf{r}(t) = x(t)\mathbf{i} + y(t)\mathbf{j} + z(t)\mathbf{k}.$$

Show that $d/dt[f(\mathbf{r}(t))]$ can be viewed as the directional derivative of f in the direction of the motion multiplied by the speed of the object.

24. Let $u = f(x, y)$ be a differentiable function. Assume that the equation $f(x, y) = 0$ gives y implicitly as a differentiable function of x.
   (a) Show that

$$\text{if} \quad \frac{\partial f}{\partial y} \neq 0 \quad \text{then} \quad \frac{dy}{dx} = -\frac{\partial f/\partial x}{\partial f/\partial y}.$$

   (b) Use this formula to find dy/dx if $x^2 y^4 + \sin y = 0$.

25. Show that, if $\|\nabla f(\mathbf{x})\| \leq M$ for all $\mathbf{x}$ in some neighborhood of the origin, then

$$|f(\mathbf{x}) - f(\mathbf{y})| \leq M\|\mathbf{x} - \mathbf{y}\| \qquad \text{for all } \mathbf{x} \text{ and } \mathbf{y} \text{ in that neighborhood.}$$

26. (*A chain rule for vector-valued functions.*) Suppose that

$$\mathbf{u} = u_1 \mathbf{i} + u_2 \mathbf{j}$$

where

$$u_1 = f_1(x, y), \quad u_2 = f_2(x, y) \quad \text{and} \quad x = g(t), \quad y = h(t).$$

Show that

(17.3.8)
$$\boxed{\frac{d\mathbf{u}}{dt} = \frac{\partial \mathbf{u}}{\partial x} \frac{dx}{dt} + \frac{\partial \mathbf{u}}{\partial y} \frac{dy}{dt}}$$

where

$$\frac{\partial \mathbf{u}}{\partial x} = \frac{\partial u_1}{\partial x} \mathbf{i} + \frac{\partial u_2}{\partial x} \mathbf{j} \quad \text{and} \quad \frac{\partial \mathbf{u}}{\partial y} = \frac{\partial u_1}{\partial y} \mathbf{i} + \frac{\partial u_2}{\partial y} \mathbf{j}.$$

*27. Take $\mathbf{u}$ as in Exercise 26. Setting $x = g(s, t)$ and $y = h(s, t)$, find $\partial \mathbf{u}/\partial s$ and $\partial \mathbf{u}/\partial t$.

17.4 The Gradient as a Normal; Tangent Lines and Tangent Planes

Functions of Two Variables

You have seen that the gradient of a function points in the direction of the most rapid increase of the function. Here we show that

(17.4.1) | at each point the gradient is perpendicular to the level curve which passes through that point.

PROOF. We begin with a function f and a point (x_0, y_0) in its domain. We suppose that $f(x_0, y_0) = c$ and parametrize the level curve $f(x, y) = c$ by setting

$$\mathbf{r}(t) = x(t)\mathbf{i} + y(t)\mathbf{j}, \qquad t \in I.$$

We take t_0 so that

$$\mathbf{r}(t_0) = x_0\mathbf{i} + y_0\mathbf{j} = (x_0, y_0)$$

and assume that the tangent vector $\mathbf{r}'(t_0)$ exists and is not $\mathbf{0}$. We must show that

$$\nabla f(\mathbf{r}(t_0)) \perp \mathbf{r}'(t_0).$$

Since f is constantly c on that level curve, we have

$$f(\mathbf{r}(t)) = c \qquad \text{for all } t \in I.$$

For such t

$$\frac{d}{dt}[f(\mathbf{r}(t))] = \nabla f(\mathbf{r}(t)) \cdot \mathbf{r}'(t) = 0.$$

In particular

$$\nabla f(\mathbf{r}(t_0)) \cdot \mathbf{r}'(t_0) = 0,$$

which means that

$$\nabla f(\mathbf{r}(t_0)) \perp \mathbf{r}'(t_0). \quad \square$$

Figure 17.4.1 illustrates the result when $\nabla f(x_0, y_0) \neq \mathbf{0}$.

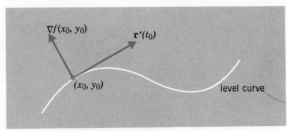

the gradient vector $\nabla f(x_0, y_0)$ is perpendicular
to the level curve at (x_0, y_0)

FIGURE 17.4.1

Example.   For

$$f(x, y) = x^2 + y^2$$

the level curves are concentric circles

$$x^2 + y^2 = c.$$

At each point $(x, y) \neq (0, 0)$ the gradient

$$\nabla f(x, y) = 2x\mathbf{i} + 2y\mathbf{j} = 2\mathbf{r}$$

points away from the origin along the line of the radius vector and is thus perpendicular to the circle in question. At the origin the level curve is reduced to a point and the gradient is simply $\mathbf{0}$. $\square$

Normal Lines and Tangent Lines

Now we consider a curve γ defined implicitly by an equation

$$f(x, y) = c.$$

We can think of this curve as the c-level curve of a function f. If f is differentiable, then the gradient $\nabla f(x_0, y_0)$ is perpendicular to the curve at (x_0, y_0). If $\nabla f(x_0, y_0) = \mathbf{0}$, there is not much more to say. Suppose therefore that $\nabla f(x_0, y_0) \neq \mathbf{0}$. In this case we can use $\nabla f(x_0, y_0)$ as a normal vector and from it derive an equation for the *normal line* at $\mathbf{x}_0 = (x_0, y_0)$. Let's proceed. Using $\nabla f(x_0, y_0)$ as a direction vector, we can parametrize the normal line by setting

$$\mathbf{R}(t) = \mathbf{x}_0 + t\nabla f(\mathbf{x}_0) = \left[x_0 + t\frac{\partial f}{\partial x}(x_0, y_0) \right]\mathbf{i} + \left[y_0 + t\frac{\partial f}{\partial y}(x_0, y_0) \right]\mathbf{j}.$$

In terms of components, we have

$$x(t) = x_0 + t\frac{\partial f}{\partial x}(x_0, y_0) \quad \text{and} \quad y(t) = y_0 + t\frac{\partial f}{\partial y}(x_0, y_0).$$

This gives

$$(x(t) - x_0)\frac{\partial f}{\partial y}(x_0, y_0) = (y(t) - y_0)\frac{\partial f}{\partial x}(x_0, y_0). \qquad \text{(explain)}$$

Eliminating the t altogether, we get

$$(x - x_0)\frac{\partial f}{\partial y}(x_0, y_0) = (y - y_0)\frac{\partial f}{\partial x}(x_0, y_0).$$

Usually the right-hand side is transferred to the left and the equation for the normal line is written

(17.4.2)
$$\boxed{(x - x_0)\frac{\partial f}{\partial y}(x_0, y_0) - (y - y_0)\frac{\partial f}{\partial x}(x_0, y_0) = 0.}$$

The tangent line at (x_0, y_0) is perpendicular to the normal, and therefore its equation can be written *lagr*

(17.4.3) $$\boxed{(x - x_0) \frac{\partial f}{\partial x} (x_0, y_0) + (y - y_0) \frac{\partial f}{\partial y} (x_0, y_0) = 0.}$$ (check this out)

Problem.   Find equations for the lines normal and tangent to the ellipse

$$b^2 x^2 + a^2 y^2 = a^2 b^2 \qquad \text{at the point } (x_0, y_0).$$

SOLUTION.   We can represent the ellipse by setting

$$f(x, y) = b^2 x^2 + a^2 y^2 - a^2 b^2 = 0.$$

Note that

$$\frac{\partial f}{\partial x} (x, y) = 2b^2 x, \qquad \frac{\partial f}{\partial y} (x, y) = 2a^2 y,$$

so that

$$\frac{\partial f}{\partial x} (x_0, y_0) = 2b^2 x_0, \qquad \frac{\partial f}{\partial y} (x_0, y_0) = 2a^2 y_0.$$

The equation for the normal line can be written

$$(x - x_0)2a^2 y_0 - (y - y_0)2b^2 x_0 = 0.$$

This simplifies to

$$(a^2 y_0)x - (b^2 x_0)y = (a^2 - b^2)x_0 y_0.$$

The equation for the tangent line takes the form

$$(x - x_0)2b^2 x_0 + (y - y_0)2a^2 y_0 = 0.$$

Dividing by 2 and noting that

$$b^2 x_0^2 + a^2 y_0^2 = a^2 b^2, \qquad \text{(explain)}$$

we simplify the equation to read

$$(b^2 x_0)x + (a^2 y_0)y = a^2 b^2. \quad \square$$

Functions of Three Variables

Here, instead of level curves, we have level surfaces, but the results are similar:

(17.4.4)   at each point the gradient is perpendicular to the level surface which passes through that point.

PROOF.   We take a differentiable function f, of three variables this time, and fix our attention on a point $\mathbf{x}_0 = (x_0, y_0, z_0)$ of its domain. If $f(x_0, y_0, z_0) = c$, then the level surface through $\mathbf{x}_0 = (x_0, y_0, z_0)$ has equation

$$f(x, y, z) = c.$$

We suppose now that

$$\mathbf{r}(t) = x(t)\mathbf{i} + y(t)\mathbf{j} + z(t)\mathbf{k}, \qquad t \in I$$

is a differentiable curve which lies on the given surface and passes through the point $\mathbf{x}_0 = (x_0, y_0, z_0)$. Moreover, we suppose that

$$\mathbf{r}(t_0) = \mathbf{x}_0 = (x_0, y_0, z_0) \quad \text{and} \quad \mathbf{r}'(t_0) \neq \mathbf{0}.$$

Since $\mathbf{r}(t)$ lies on the surface,

$$f(\mathbf{r}(t)) = c \qquad \text{for every } t.$$

Consequently,

$$\frac{d}{dt}[f(\mathbf{r}(t))] = \nabla f(\mathbf{r}(t)) \cdot \mathbf{r}'(t) = 0.$$

In particular,

$$\nabla f(\mathbf{r}(t_0)) \cdot \mathbf{r}'(t_0) = 0.$$

This tells us that the gradient

$$\nabla f(\mathbf{r}(t_0)) = \nabla f(\mathbf{x}_0) = \nabla f(x_0, y_0, z_0)$$

is perpendicular to the curve in question. This same argument applies to every differentiable curve which lies on the level surface and passes through the point $\mathbf{x}_0 = (x_0, y_0, z_0)$. (See Figure 17.4.2.) Consequently, we view $\nabla f(\mathbf{x}_0)$ as perpendicular to the surface itself. $\square$

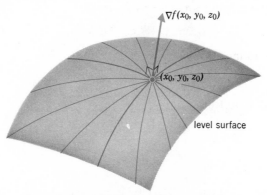

the gradient vector $\nabla f(x_0, y_0, z_0)$ is perpendicular
to the level surface at (x_0, y_0, z_0)[†]
FIGURE 17.4.2

Tangent Planes

You have seen that the gradient vector ∇f is perpendicular to the surface $f(x, y, z) = c$. If the gradient vector ∇f is not zero, we can use it as a normal

---

[†] Here we are assuming that $\nabla f(x_0, y_0, z_0)$ is not zero.

vector. We can then define the *tangent plane at* $\mathbf{x}_0 = (x_0, y_0, z_0)$ *as the plane through* $\mathbf{x}_0$ *with normal* $\nabla f(\mathbf{x}_0)$. (See Figure 17.4.3.) As a vector equation for this tangent plane, we have

(17.4.5)
$$\boxed{\nabla f(\mathbf{x}_0) \cdot (\mathbf{x} - \mathbf{x}_0) = 0.}$$

In Cartesian coordinates, this equation becomes

$$(x - x_0) \frac{\partial f}{\partial x}(x_0, y_0, z_0) + (y - y_0) \frac{\partial f}{\partial y}(x_0, y_0, z_0) + (z - z_0) \frac{\partial f}{\partial z}(x_0, y_0, z_0) = 0.$$

Geometrically, the tangent plane at a point $\mathbf{x}_0$ is the plane that best approximates the surface in a neighborhood of $\mathbf{x}_0$. We'll return to this idea later.

tangent plane to a surface $f(x, y, z) = c$

FIGURE 17.4.3

Problem.   Find an equation for the plane tangent to the surface

$$xy + yz + zx = 11 \qquad \text{at the point } (1, 2, 3).$$

SOLUTION.   Here

$$f(x, y, z) = xy + yz + zx,$$

so that

$$\frac{\partial f}{\partial x}(x, y, z) = y + z, \quad \frac{\partial f}{\partial y}(x, y, z) = x + z, \quad \frac{\partial f}{\partial z}(x, y, z) = x + y.$$

At the point $(1, 2, 3)$

$$\frac{\partial f}{\partial x} = 5, \quad \frac{\partial f}{\partial y} = 4, \quad \frac{\partial f}{\partial z} = 3.$$

The equation for the tangent plane can be written

$$5(x - 1) + 4(y - 2) + 3(z - 3) = 0.$$

This simplifies to

$$5x + 4y + 3z - 22 = 0. \quad \square$$

Problem.   The curve

$$\mathbf{r}(t) = \tfrac{1}{2}t^2\mathbf{i} + 4t^{-1}\mathbf{j} + (\tfrac{1}{2}t - t^2)\mathbf{k}$$

cuts the hyperbolic paraboloid $x^2 - 4y^2 - 4z = 0$ at the point $(2, 2, -3)$. What is the angle of intersection?

SOLUTION.   First we need the value of t which corresponds to the point $(2, 2, -3)$. To find this, we set

$$\tfrac{1}{2}t^2 = 2, \qquad 4t^{-1} = 2, \qquad \tfrac{1}{2}t - t^2 = -3$$

and solve for t. An easy calculation shows that $t = 2$. Since

$$\mathbf{r}'(t) = t\mathbf{i} - 4t^{-2}\mathbf{j} + (\tfrac{1}{2} - 2t)\mathbf{k},$$

we have

$$\mathbf{r}'(2) = 2\mathbf{i} - \mathbf{j} - \tfrac{7}{2}\mathbf{k}.$$

This is the tangent vector of the curve at the point of intersection.
   To represent the surface we will use the gradient. For

$$f(x, y, z) = x^2 - 4y^2 - 4z$$

we have

$$\nabla f = 2x\mathbf{i} - 8y\mathbf{j} - 4\mathbf{k}.$$

At the point $(2, 2, -3)$

$$\nabla f = 4\mathbf{i} - 16\mathbf{j} - 4\mathbf{k}.$$

Since

$$\frac{\mathbf{r}'(2) \cdot \nabla f}{\|\mathbf{r}'(2)\| \|\nabla f\|} = \frac{19}{414}\sqrt{138}, \qquad \text{(check this)}$$

the angle between the curve and the gradient is arc cos $(\tfrac{19}{414}\sqrt{138})$. Since the gradient is perpendicular to the surface, the angle between the curve and the surface itself is

$$\tfrac{1}{2}\pi - \text{arc cos } (\tfrac{19}{414}\sqrt{138}) \cong 1.57 - \text{arc cos } 0.539 \cong 0.57 \text{ radians.} \quad \square$$

   To find the tangent plane for a surface of the form

$$z = g(x, y)$$

we set

$$f(x, y, z) = g(x, y) - z = 0.$$

If g is differentiable, then so is f. Moreover,

$$\frac{\partial f}{\partial x}(x, y, z) = \frac{\partial g}{\partial x}(x, y), \quad \frac{\partial f}{\partial y}(x, y, z) = \frac{\partial g}{\partial y}(x, y), \quad \frac{\partial f}{\partial z}(x, y, z) = -1.$$

The tangent plane at (x_0, y_0, z_0) has equation

$$(x - x_0) \frac{\partial g}{\partial x}(x_0, y_0) + (y - y_0)\frac{\partial g}{\partial y}(x_0, y_0) + (z - z_0)(-1) = 0,$$

which we can rewrite as

(17.4.6) $$(x - x_0)\frac{\partial g}{\partial x}(x_0, y_0) + (y - y_0)\frac{\partial g}{\partial y}(x_0, y_0) = z - z_0.$$

If $\nabla g(x_0, y_0) = \mathbf{0}$, then both partials of g are zero at (x_0, y_0) and the equation reduces to

(17.4.7) $$z = z_0.$$

In this case the tangent plane is *horizontal*.

Problem.   Find an equation for the plane tangent to the surface

$$z = \frac{x^2}{2p} + \frac{y^2}{2q} \qquad \text{at the point } (x_0, y_0, z_0).$$

SOLUTION.   We set

$$g(x, y) = \frac{x^2}{2p} + \frac{y^2}{2q}$$

and note that

$$\frac{\partial g}{\partial x}(x, y) = \frac{x}{p} \quad \text{and} \quad \frac{\partial g}{\partial y}(x, y) = \frac{y}{q}.$$

As an equation for the tangent plane we have

$$\frac{x_0}{p}(x - x_0) + \frac{y_0}{q}(y - y_0) = z - z_0. \quad \square$$

Problem.   At what points of the surface

$$z = 3xy - x^3 - y^3$$

is the tangent plane horizontal?

SOLUTION.   We set

$$g(x, y) = 3xy - x^3 - y^3$$

and seek the pairs (x, y) for which

$$\frac{\partial g}{\partial x}(x, y) = 0 = \frac{\partial g}{\partial y}(x, y).$$

Since

$$\frac{\partial g}{\partial x}(x, y) = 3y - 3x^2 \quad \text{and} \quad \frac{\partial g}{\partial y}(x, y) = 3x - 3y^2,$$

we must solve the equations $3y - 3x^2 = 0$ and $3x - 3y^2 = 0$ simultaneously. This gives $x = 0, y = 0$ and $x = 1, y = 1$. Each pair gives rise to a point at which the tangent plane is horizontal. These points are $(0, 0, 0)$ and $(1, 1, 1)$. $\square$

Exercises

Find equations for the tangent and normal lines at the point indicated.

*1. $x^2 + xy + y^2 = 3$; $(-1, -1)$.
2. $(y - x)^2 = 2x$; $(2, 4)$.
*3. $(x^2 + y^2)^2 = 9(x^2 - y^2)$; $(\sqrt{2}, 1)$.
*4. $x^3 + y^3 = 9$; $(1, 2)$.
5. $xy^2 - 2x^2 + y + 5x = 6$; $(4, 2)$.
*6. $x^5 + y^5 = 2x^3$; $(1, 1)$.
*7. $2x^3 - x^2y^2 - 3x + y + 7 = 0$; $(1, -2)$.

8. Find an equation for the tangent to the curve $x^3 + y^2 + 2x = 6$ at a point where $y = 3$.
*9. Find the slope of the tangent to the curve $x^2y = a^2(a - y)$ at the point $(0, a)$.

Find an equation for the tangent plane at the point indicated.

10. $z = (x^2 + y^2)^2$; $(1, 1, 4)$.
*11. $x^3 + y^3 = 3xyz$; $(1, 2, \frac{3}{2})$.
12. $xy^2 + 2z^2 = 12$; $(1, 2, 2)$.
*13. $z = axy$; $(1, 1/a, 1)$.
14. $\sqrt{x} + \sqrt{y} + \sqrt{z} = 4$; $(1, 4, 1)$.
*15. $z = \sin x + \sin y + \sin(x + y)$; $(0, 0, 0)$.
16. $z = x^2 + xy + y^2 - 6x + 2$; $(4, -2, -10)$.
*17. $b^2c^2x^2 - a^2c^2y^2 - a^2b^2z^2 = a^2b^2c^2$; (x_0, y_0, z_0).

Find the point(s) of the surface at which the tangent plane is horizontal.

18. $z = 4x + 2y - x^2 + xy - y^2$.
*19. $xy + a^3/x + b^3/y - z = 0$.
20. $z - 2x^2 - 2xy + y^2 + 5x - 3y + 2 = 0$.
*21. $z = xy$.
22. $x + y + z + xy - x^2 - y^2 = 0$.

*23. Given that the surfaces $F(x, y, z) = 0$ and $G(x, y, z) = 0$ intersect at right angles in a curve γ, what condition must be satisfied by the partial derivatives of F and G on γ?
24. Show that in the case of a surface of the form $z = xf(x/y)$ with f differentiable all the tangent planes share a point in common.
*25. Show that all pyramids formed by the coordinate planes and a plane tangent to the surface $xyz = a^3$ have the same volume. What is this volume?
26. Show that, for all planes tangent to the surface $\sqrt{x} + \sqrt{y} + \sqrt{z} = \sqrt{a}$, the sum of the intercepts is the same.
27. Show that, for all planes tangent to the surface $x^{2/3} + y^{2/3} + z^{2/3} = a^{2/3}$, the sum of the squares of the intercepts is the same.

*28. The curve $\mathbf{r}(t) = 2t\mathbf{i} + 3t^{-1}\mathbf{j} - 2t^2\mathbf{k}$ and the ellipsoid $x^2 + y^2 + 3z^2 = 25$ intersect at $(2, 3, -2)$. What is the angle of intersection?

29. Show that the curve $\mathbf{r}(t) = \frac{3}{2}(t^2 + 1)\mathbf{i} + (t^4 + 1)\mathbf{j} + t^3\mathbf{k}$ intersects the ellipsoid $x^2 + 2y^2 + 3z^2 = 20$ at right angles at the point $(3, 2, 1)$.

*30. The surfaces $x^2y^2 + 2x + z^3 = 16$ and $3x^2 + y^2 - 2z = 9$ intersect in a curve which passes through the point $(2, 1, 2)$. What are the equations of the respective tangent planes to the two surfaces at this point?

31. Show that the sphere $x^2 + y^2 + z^2 - 8x - 8y - 6z + 24 = 0$ is tangent to the ellipsoid $x^2 + 3y^2 + 2z^2 = 9$ at the point $(2, 1, 1)$.

32. Show that the sphere $x^2 + y^2 + z^2 - 4y - 2z + 2 = 0$ intersects the paraboloid $3x^2 + 2y^2 - 2z = 1$ at right angles at the point $(1, 1, 2)$.

33. Show that the surfaces

$$xy = az^2, \quad x^2 + y^2 + z^2 = b, \quad z^2 + 2x^2 = c(z^2 + 2y^2)$$

are mutually perpendicular.

17.5 Maximum and Minimum Values

In Chapter 4 we talked about the local extreme values of a function of one variable. Here we take up the subject for functions of several variables. The ideas are much the same.

Definition 17.5.1   Local Maximum and Local Minimum

A function f of several variables is said to have a *local maximum* at $\mathbf{x}_0$ iff

$$f(\mathbf{x}_0) \geq f(\mathbf{x}) \qquad \text{for all } \mathbf{x} \text{ in some neighborhood of } \mathbf{x}_0.$$

It is said to have a *local minimum* at $\mathbf{x}_0$ iff

$$f(\mathbf{x}_0) \leq f(\mathbf{x}) \qquad \text{for all } \mathbf{x} \text{ in some neighborhood of } \mathbf{x}_0.$$

As in the one-variable case, the local maxima and local minima together comprise the *local extreme values*.

In the one-variable case we know that, if f is differentiable at x_0 and has a local extreme value at x_0, then

$$f'(x_0) = 0.$$

We have a similar result for functions of several variables.

Theorem 17.5.2

If f is differentiable at $\mathbf{x}_0$ and has a local extreme value at $\mathbf{x}_0$, then

$$\nabla f(\mathbf{x}_0) = \mathbf{0}.$$

PROOF. Let's take f as a function of two variables and assume that f is differentiable at $\mathbf{x}_0 = (x_0, y_0)$. (The three-variable case is similar.) If f has a local extreme value at $\mathbf{x}_0 = (x_0, y_0)$, then $g(x) = f(x, y_0)$ must have a local extreme value at x_0 and, being differentiable there, must satisfy the relation

$$g'(x_0) = \frac{\partial f}{\partial x}(x_0, y_0) = 0.$$

Similarly, $h(y) = f(x_0, y)$ must have a local extreme value at y_0 and, being differentiable there, must satisfy the relation

$$h'(y_0) = \frac{\partial f}{\partial y}(x_0, y_0) = 0.$$

The gradient is $\mathbf{0}$ since both partials are 0. (Theorem 17.2.5) □

Although the ideas introduced so far are completely general, their application to functions of more than two variables is generally laborious. We restrict ourselves mostly to functions of two variables. Not only are the computations less formidable, but also we can make use of our geometric intuition.

Two Variables

A continuously differentiable function of two variables can be viewed geometrically as a surface

$$z = f(x, y).$$

Where f has a local maximum, the surface has a high point. Where f has a local minimum, the surface has a low point. Where f has either a local maximum or a local minimum, the gradient is $\mathbf{0}$ and therefore the tangent plane is horizontal. See Figure 17.5.1.

FIGURE 17.5.1

A zero gradient signals the possibility of an extreme value. It does not guarantee it. For example, in the case of the saddle-shaped surface of Figure 17.5.2, there is a horizontal tangent plane at the origin and therefore the gradient is zero there, but the origin produces neither a local maximum nor a local minimum.

saddle-shaped surface

FIGURE 17.5.2

Points at which the gradient is zero are called *stationary points*. Those stationary points that give rise to extreme values are called *extreme points*. Those stationary points that do not give rise to extreme values are called *saddle points*.

Below we examine some differentiable functions for extreme values. In each case our first step is to seek out the stationary points.

Example.   For

$$f(x, y) = 2x^2 - xy + y^2 - 7y,$$

we have

$$\nabla f(x, y) = (4x - y)\mathbf{i} + (2y - x - 7)\mathbf{j}.$$

To find the stationary points, we set

$$\nabla f(x, y) = \mathbf{0}.$$

This gives

$$4x - y = 0 \quad \text{and} \quad 2y - x - 7 = 0.$$

The only simultaneous solution to these equations is given by $x = 1, y = 4$. The point $(1, 4)$ is therefore the only stationary point.

Now we compare the value of f at $(1, 4)$ with the values of f at nearby points $(1 + h, 4 + k)$:

$$f(1, 4) = 2 - 4 + 16 - 28 = -14,$$
$$f(1 + h, 4 + k) = 2(1 + h)^2 - (1 + h)(4 + k) + (4 + k)^2 - 7(4 + k)$$
$$= 2 + 4h + 2h^2 - 4 - 4h - k - hk + 16 + 8k + k^2 - 28 - 7k$$
$$= 2h^2 + k^2 - hk - 14.$$

The difference

$$f(1 + h, 4 + k) - f(1, 4) = 2h^2 + k^2 - hk = h^2 + (h^2 - hk + k^2)$$

is nonnegative for all small h and k (in fact for all real h and k).† If follows that f has a minimum value at $(1, 4)$. This minimum value is -14.   □

Example.   In the case of

$$f(x, y) = y^2 - xy + 2x + y + 1$$

we have

$$\nabla f(x, y) = (2 - y)\mathbf{i} + (2y - x + 1)\mathbf{j}.$$

The gradient is $\mathbf{0}$ where

$$2 - y = 0 \quad \text{and} \quad 2y - x + 1 = 0.$$

The only simultaneous solution to these equations is $x = 5, y = 2$. The point $(5, 2)$ is the only stationary point. To determine its nature, we compare $f(5, 2)$ with $f(5 + h, 2 + k)$ for small h and k:

$$f(5, 2) = 4 - 10 + 10 + 2 + 1 = 7,$$
$$f(5 + h, 2 + k) = (2 + k)^2 - (5 + h)(2 + k) + 2(5 + h) + (2 + k) + 1$$
$$= 4 + 4k + k^2 - 10 - 2h - 5k - hk + 10 + 2h + 2 + k + 1$$
$$= k^2 - hk + 7.$$

The difference

$$f(5 + h, 2 + k) - f(5, 2) = k^2 - hk = k(k - h)$$

does not keep a constant sign for small h and k. It follows that $(5, 2)$ is a saddle point.   □

Example.   For

$$f(x, y) = x^3 + y^3 - 3xy$$

we have

$$\nabla f(x, y) = (3x^2 - 3y)\mathbf{i} + (3y^2 - 3x)\mathbf{j}.$$

Setting $\nabla f(x, y) = \mathbf{0}$ gives

$$x^2 - y = 0 \quad \text{and} \quad y^2 - x = 0.$$

The only simultaneous solutions to these equations are given by $x = 0, y = 0$ and $x = 1, y = 1$. The points $(0, 0)$ and $(1, 1)$ are the only stationary points. To investigate $(0, 0)$ we first note that $f(0, 0) = 0$. For small $h > 0$ and $k = 0$,

$$f(h, k) = h^3 > 0.$$

For small $h < 0$ and $k = 0$,

$$f(h, k) = h^3 < 0.$$

---

† $h^2 - hk + k^2 \geq h^2 - 2|h||k| + k^2 + \frac{1}{2}|h||k| = (|h| - |k|)^2 + \frac{1}{2}|h||k| \geq 0.$

It follows that $f(0, 0) = 0$ is neither a local maximum nor a local minimum. The origin $(0, 0)$ is therefore a saddle point.

Now we investigate the function at $(1, 1)$ and at nearby points $(1 + h, 1 + k)$:

$$f(1, 1) = 1 + 1 - 3 = -1,$$

$$\begin{aligned}
f(1 + h, 1 + k) &= (1 + h)^3 + (1 + k)^3 - 3(1 + h)(1 + k) \\
&= 1 + 3h + 3h^2 + h^3 + 1 + 3k + 3k^2 + k^3 - 3 \\
&\quad - 3h - 3k - 3hk \\
&= -1 + 3h^2 + 3k^2 - 3hk + h^3 + k^3,
\end{aligned}$$

so that

$$f(1 + h, 1 + k) - f(1, 1) = 3h^2 + 3k^2 - 3hk + h^3 + k^3.$$

The idea now is to play around with the expression on the right-hand side until we can determine whether or not it keeps a constant sign for small h and k. Our problems are over once we realize that we can rewrite the expression as

$$(h^2 + h^3) + (k^2 + k^3) + (h^2 - hk + k^2) + (h^2 - 2hk + k^2)$$

and in turn as

$$h^2(1 + h) + k^2(1 + k) + (h^2 - hk + k^2) + (h^2 - 2hk + k^2).$$

For small h and k all four terms are nonnegative. It follows that

$$f(1 + h, 1 + k) - f(1, 1) \geq 0.$$

The point $(1, 1)$ gives rise to a local minimum. This minimum is -1.   □

The next example is more complicated. The function there has an infinite number of stationary points.

Example.   For all points (x, y) such that $x + y \neq 0$ set

$$f(x, y) = \frac{x^2 + 2y^2}{(x + y)^2}.$$

Then

$$\nabla f(x, y) = 2y \frac{x - 2y}{(x + y)^3} \mathbf{i} + 2x \frac{2y - x}{(x + y)^3} \mathbf{j}.$$

The gradient is $\mathbf{0}$ where

$$2y \frac{x - 2y}{(x + y)^3} = 0 \quad \text{and} \quad 2x \frac{2y - x}{(x + y)^3} = 0.$$

Solving these two equations simultaneously, we get

$$x = 2y.$$

The stationary points are all the points of the line $x = 2y$, except of course for the point $(0, 0)$, which is not in the domain of f. At each stationary point (x_0, y_0) the function takes on the value $\frac{2}{3}$:

$$f(x_0, y_0) = f(2y_0, y_0) = \frac{(2y_0)^2 + 2y_0^2}{(2y_0 + y_0)^2} = \frac{6y_0^2}{9y_0^2} = \frac{2}{3}.$$

In a moment we will show that for all (x, y) in the domain

$$f(x, y) = \frac{x^2 + 2y^2}{(x + y)^2} \geq \frac{2}{3}.$$

It will then follow that every stationary point is a minimum point and the number $\frac{2}{3}$ is a minimum value.

The inequality

$$\frac{x^2 + 2y^2}{(x + y)^2} \geq \frac{2}{3}$$

can be justified by the following sequence of equivalent inequalities, the last of which is obvious:

$$3(x^2 + 2y^2) \geq 2(x + y)^2,$$
$$3x^2 + 6y^2 \geq 2x^2 + 4xy + 2y^2,$$
$$x^2 - 4xy + 4y^2 \geq 0,$$
$$(x - 2y)^2 \geq 0. \quad \square$$

So far we have considered only functions which have gradients on their entire domain. A local extreme value can also occur at a point where the gradient does not exist.

Example. Consider the function

$$f(x, y) = \sqrt{x^2 + y^2}.$$

Its graph is the upper half of a circular cone. (Figure 16.2.4 with $a = b = 1$.) Obviously f has a minimum at $(0, 0)$, but since the partials

$$\frac{\partial f}{\partial x} = \frac{x}{\sqrt{x^2 + y^2}}, \qquad \frac{\partial f}{\partial y} = \frac{y}{\sqrt{x^2 + y^2}}$$

are not defined at $(0, 0)$, the gradient is not defined at $(0, 0)$. (Theorem 17.2.5) The surface comes to a sharp point at $(0, 0)$ and there is no tangent plane. $\quad \square$

Exercises

Classify the stationary points and find the local extreme values.

*1. $2x - x^2 - y^2$.
*3. $2x + 2y - x^2 + y^2 + 5$.
*5. $x^2 + xy + y^2 + 3x + 1$.
*7. $x^3 - 3x + y$.
*9. $x^2 + xy + y^2 - 3x - 3y$.
*11. $|x| + |y|$.
*13. $x^2 + xy + 2x + 2y + 1$.

2. $x^2 + 2y^2 - 4y$.
4. $x^2 - xy + y^2 + 1$.
6. $x^3 - y^2 + 2y$.
8. $x^5 - 3x^2 - 2y$.
10. $x^2 - xy + y^2 + 2x + 2y - 6$.
12. $5xy - 7x^2 - y^2 + 3x - 6y$.
14. $x \sin y, \ -\pi < y < \pi$.

*15. Find the point with the property that the sum of the squares of its distances from $P_1(x_1, y_1)$, $P_2(x_2, y_2)$, $P_3(x_3, y_3)$ is a minimum.

16. A manufacturer finds that he can sell in two distinct markets. If he sells Q_1

units in the first market and Q_2 units in the second market, the revenue functions are

$$R_1 = A_1Q_1 - B_1Q_1^2, \qquad R_2 = A_2Q_2 - B_2Q_2^2$$

and the total cost function is

$$C = A_3 + B_3(Q_1 + Q_2).$$

Given that the A's and B's are known positive constants, find the quantities Q_1 and Q_2 that will maximize profit.

*17. Given that $0 < a < b$, find the two numbers x and y between a and b that maximize the expression

$$\frac{xy}{(a + x)(x + y)(b + y)}.$$

17.6 Second-Partials Test

We start with a function g of one variable. Suppose now that $g'(x_0) = 0$. The second derivative test for extreme values can be phrased this way:

> If $g''(x_0) > 0$, then g has a minimum at x_0.
> If $g''(x_0) < 0$, then g has a maximum at x_0.
> If $g''(x_0) = 0$, then anything can happen.

For functions of two variables we have a similar test. As you might expect, it is a little more complicated to state and definitely more difficult to prove.

Theorem 17.6.1   The Second-Partials Test

Suppose that f has continuous second-order partials in a neighborhood of (x_0, y_0) and that $\nabla f(x_0, y_0) = \mathbf{0}$. Set

$$A = \frac{\partial^2 f}{\partial x^2}(x_0, y_0), \quad B = \frac{\partial^2 f}{\partial y \partial x}(x_0, y_0), \quad C = \frac{\partial^2 f}{\partial y^2}(x_0, y_0)$$

and form the discriminant

$$D = B^2 - AC.$$

1. If $D > 0$, (x_0, y_0) is a saddle point.
2. If $D < 0$, there is a maximum at (x_0, y_0) if $A < 0$ and a minimum if $A > 0$.
3. If $D = 0$, then anything can happen.

A detailed justification of this test would be extremely burdensome at this point.[†] We will not attempt to give one. Instead we move on directly to some examples.

[†] You can find one in most texts on advanced calculus. See for example Section 2-15 in Wilfred Kaplan's *Advanced Calculus* (Addison-Wesley, Reading, Mass., 1952).

Example. For

$$f(x, y) = x^3 - y^3 + 3xy$$

we have

$$\frac{\partial f}{\partial x} = 3x^2 + 3y \quad \text{and} \quad \frac{\partial f}{\partial y} = -3y^2 + 3x.$$

Setting both partials equal to zero, we have

$$x^2 + y = 0 \quad \text{and} \quad -y^2 + x = 0.$$

The only simultaneous solutions to these equations are given by $x = 0, y = 0$ and $x = 1, y = -1$. The points $(0, 0)$ and $(1, -1)$ are the only stationary points. The second partials are

$$\frac{\partial^2 f}{\partial x^2} = 6x, \quad \frac{\partial^2 f}{\partial y \, \partial x} = 3, \quad \frac{\partial^2 f}{\partial y^2} = -6y.$$

At $(0, 0)$ we have $A = 0, B = 3, C = 0$ and thus

$$D = B^2 - AC = 9 > 0.$$

The test tells us that $(0, 0)$ is a saddle point. At $(1, -1)$ we have $A = 6, B = 3, C = 6$ and thus

$$D = B^2 - AC = 9 - 36 = -27 < 0.$$

From $D < 0$ and $A > 0$ we can conclude that $(1, -1)$ gives rise to a local minimum. This minimum is

$$f(1, -1) = 1 + 1 - 3 = -1. \quad \square$$

Example. For

$$f(x, y) = x^2 + xy + y^2 - 3x - 3y$$

we have

$$\frac{\partial f}{\partial x} = 2x + y - 3 \quad \text{and} \quad \frac{\partial f}{\partial y} = x + 2y - 3.$$

Setting both of these partials equal to zero, we have

$$2x + y - 3 = 0 \quad \text{and} \quad x + 2y - 3 = 0.$$

The only simultaneous solution to these equations is $x = 1, y = 1$. The point $(1, 1)$ is the only stationary point. Since at this point

$$\frac{\partial^2 f}{\partial x^2} = 2, \quad \frac{\partial^2 f}{\partial y \, \partial x} = 1, \quad \frac{\partial^2 f}{\partial y^2} = 2,$$

we have $A = 2, B = 1, C = 2$ and thus

$$D = B^2 - AC = 1 - 4 = -3 < 0.$$

With $D < 0$ and $A > 0$ we can conclude that $(1, 1)$ gives a local minimum. This minimum is

$$f(1, 1) = 1 + 1 + 1 - 3 - 3 = -3. \quad \square$$

Example.   We examine the function

$$f(x, y) = \sin x + \sin y + \sin (x + y)$$

on the open square

$$0 < x < \pi, \quad 0 < y < \pi.$$

In the first place

$$\frac{\partial f}{\partial x} = \cos x + \cos (x + y), \qquad \frac{\partial f}{\partial y} = \cos y + \cos (x + y).$$

Setting both of these partials equal to zero, we have

$$\cos x + \cos (x + y) = 0, \qquad \cos y + \cos (x + y) = 0.$$

Together these equations imply that

$$\cos x = \cos y.$$

Since x and y must lie between 0 and π, we conclude that

$$x = y.$$

The condition

$$\cos x + \cos (x + y) = 0$$

now leads to

$$\cos x + \cos 2x = 0.$$

Into this last equation we substitute the identity $\cos 2x = 2 \cos^2 x - 1$ and thereby obtain

$$2 \cos^2 x + \cos x - 1 = 0.$$

This is a quadratic in $\cos x$ with solutions

$$\cos x = \tfrac{1}{4}(-1 \pm \sqrt{1 + 8}) = -1 \text{ or } \tfrac{1}{2}.$$

We throw out the solution -1 because x has to lie between 0 and π. The other possibility, $\cos x = \tfrac{1}{2}$, leads to $x = \tfrac{1}{3}\pi$. Since $x = y$, we also have $y = \tfrac{1}{3}\pi$. This shows that $(\tfrac{1}{3}\pi, \tfrac{1}{3}\pi)$ is the only stationary point.

$$\frac{\partial^2 f}{\partial x^2} = -\sin x - \sin (x + y),$$

$$\frac{\partial^2 f}{\partial y \, \partial x} = -\sin (x + y),$$

$$\frac{\partial^2 f}{\partial y^2} = -\sin y - \sin (x + y).$$

At $(\tfrac{1}{3}\pi, \tfrac{1}{3}\pi)$ we have

$$A = -\sin\tfrac{1}{3}\pi - \sin\tfrac{2}{3}\pi = -\tfrac{1}{2}\sqrt{3} - \tfrac{1}{2}\sqrt{3} = -\sqrt{3},$$
$$B = -\sin\tfrac{2}{3}\pi = -\tfrac{1}{2}\sqrt{3},$$
$$C = -\sin\tfrac{1}{3}\pi - \sin\tfrac{2}{3}\pi = -\tfrac{1}{2}\sqrt{3} - \tfrac{1}{2}\sqrt{3} = -\sqrt{3},$$

and thus
$$D = B^2 - AC = \tfrac{3}{4} - 3 < 0.$$

From $D < 0$ and $A < 0$ we can conclude that $(\tfrac{1}{3}\pi, \tfrac{1}{3}\pi)$ gives a local maximum. This maximum is

$$f(\tfrac{1}{3}\pi, \tfrac{1}{3}\pi) = \sin\tfrac{1}{3}\pi + \sin\tfrac{1}{3}\pi + \sin\tfrac{2}{3}\pi = \tfrac{1}{2}\sqrt{3} + \tfrac{1}{2}\sqrt{3} + \tfrac{1}{2}\sqrt{3} = \tfrac{3}{2}\sqrt{3}. \quad \square$$

Example. That anything can happen when $D = 0$ is illustrated by the behavior of

$$f(x, y) = x^4 + y^4, \qquad g(x, y) = -(x^4 + y^4), \qquad h(x, y) = 2x^2 - 3xy^2 + y^4.$$

As you can readily verify, for each of these functions the point $(0, 0)$ is a stationary point at which $D = 0$, and yet

(1) for f, $(0, 0)$ gives a minimum;
(2) for g, $(0, 0)$ gives a maximum;
(3) for h, $(0, 0)$ is a saddle point.

Statements (1) and (2) are obvious. To confirm (3), note that h can be written in the form
$$h(x, y) = (y^2 - x)(y^2 - 2x).$$

The parabolas
$$y^2 = x \quad \text{and} \quad y^2 = 2x$$

divide the plane as in Figure 17.6.1. The function is zero on the parabolas, negative between the parabolas and positive elsewhere. That the origin is a saddle point follows from the fact that, at the origin, the function is zero, but, in every neighborhood of the origin, the function takes on both positive and negative values. $\square$

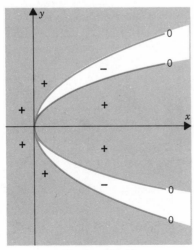

FIGURE 17.6.1

Exercises

Find the local extreme values.

*1. $x^2 + xy + y^2 - 6x + 2$.

2. $x^2 + 2xy + 3y^2 + 2x + 10y + 1$.

*3. $4x + 2y - x^2 + xy - y^2$.

4. $x^2 - xy + y^2 + 2x + 2y - 1$.

*5. $x^3 - 6xy + y^3$.

6. $3x^2 + xy - y^2 + 5x - 5y + 4$.

*7. $x^3 + y^2 - 6xy + 6x + 3y - 2$.

8. $x^2 + xy + y^2 - 6y + 5$.

*9. $e^x \cos y$.

10. $x^2 - 2xy + 2y^2 - 3x + 5y$.

*11. $x \sin y$.

12. $y + x \sin y$.

*13. $(x + y)(xy + 1)$.

14. $x/y - y/x$.

*15. $xy + 1/x + 8/y$.

16. $x^2 - 2xy - y^2 + 1$.

*17. $xy + x^{-1} + y^{-1}$.

18. $(x - y)(xy - 1)$.

*19. Find the maximum volume for a rectangular solid in the first octant† with one vertex at the origin and opposite vertex on the plane $x + y + z = 1$.

20. Describe the behavior of the function $f(x, y) = 2 \cos (x + y) + e^{xy}$ at the origin.

*21. Find the maximum volume for a rectangular solid inscribed in the ellipsoid

$$\frac{x^2}{a^2} + \frac{y^2}{b^2} + \frac{z^2}{c^2} = 1.$$

*22. A pentagon is composed of a rectangle surmounted by an isosceles triangle. (Figure 17.6.2) Given that the perimeter of the pentagon has a fixed value P, find the dimensions for maximum area.

FIGURE 17.6.2

23. Find the distance between the lines $x = \frac{1}{2}y = \frac{1}{3}z$ and $x = y - 2 = z$.

24. A dairy produces two types of cheese at constant average costs of 50 and 60¢ per pound, respectively. If the selling price of the first type is x¢ per pound and of the second y¢ per pound, the number of pounds which can be sold each week is given by the formulas

$$N_1 = 250(y - x), \qquad N_2 = 32,000 + 250(x - 2y).$$

Show that for maximum profit the selling prices should be fixed at 89 and 94¢ per pound, respectively.

---

† $x \geq 0, y \geq 0, z \geq 0$.

*25. A manufacturer produces razors and blades at a constant average cost of 80¢ per razor and 60¢ per dozen blades. If the razors are sold at x¢ each and the blades at y¢ per dozen, the demand of the market each week is

$$\frac{4,000,000}{xy} \text{ razors} \quad \text{and} \quad \frac{8,000,000}{xy} \text{ dozen blades.}$$

Find the selling prices for maximum profit.

*26. Find the maximum value of

$$f(x, y) = \frac{(ax + by + c)^2}{x^2 + y^2 + 1}.$$

17.7   Maxima and Minima with Side Conditions

When we ask for the distance from a point $P(x_0, y_0)$ to a line l: $Ax + By + C = 0$, we are asking for the minimum value of

$$f(x, y) = \sqrt{(x - x_0)^2 + (y - y_0)^2}$$

with (x, y) subject to the side condition $Ax + By + C = 0$. When we ask for the distance from a point $P(x_0, y_0, z_0)$ to a plane p: $Ax + By + Cz + D = 0$, we are asking for the minimum value of

$$f(x, y, z) = \sqrt{(x - x_0)^2 + (y - y_0)^2 + (z - z_0)^2}$$

with (x, y, z) subject to the side condition $Ax + By + Cz + D = 0$.

We have already treated these particular problems by special techniques. Our interest here is to present techniques for handling problems of this sort in general. In the two-variable case, the problems will take the form of maximizing (or minimizing) some function $f(x, y)$ subject to a side condition $g(x, y) = 0$. In the three-variable case, we will seek to maximize (or minimize) some function $f(x, y, z)$ subject to a side condition $g(x, y, z) = 0$. We begin with two simple problems.

Problem.   Maximize xy subject to the side condition $x + y - 1 = 0$.

SOLUTION.   The condition $x + y - 1 = 0$ gives $y = 1 - x$. Our initial problem can therefore be solved simply by maximizing the product

$$h(x) = x(1 - x).$$

The derivative

$$h'(x) = 1 - 2x$$

is 0 at $x = \frac{1}{2}$. Since $h''(x) = -2 < 0$, we know from the second-derivative test that $h(\frac{1}{2}) = \frac{1}{2}(1 - \frac{1}{2}) = \frac{1}{4}$ is the desired maximum. $\square$

Problem.   Find the maximum volume of a rectangular solid given that the sum of the lengths of its edges is $12a$.

SOLUTION.   We denote the dimensions of the solid by x, y, z. The volume then becomes

$$V = xyz.$$

The stipulation on the edges of the solid requires that

$$4(x + y + z) = 12a.$$

Solving this last equation for z, we get

$$z = 3a - (x + y).$$

If we now substitute this expression for z in the volume formula, we get

$$V = xy[3a - (x + y)].$$

Our problem is to maximize this function.

The partials are

$$\frac{\partial V}{\partial x} = xy(-1) + y[3a - (x + y)] = 3ay - 2xy - y^2,$$

$$\frac{\partial V}{\partial y} = xy(-1) + x[3a - (x + y)] = 3ax - x^2 - 2xy.$$

Setting both partials equal to zero, we get

$$(3a - 2x - y)y = 0 \quad \text{and} \quad (3a - x - 2y)x = 0.$$

If either $x = 0$ or $y = 0$, the volume is 0 and cannot be a maximum. The only alternative is that

$$3a - 2x - y = 0 \quad \text{and} \quad 3a - x - 2y = 0.$$

Solving these equations simultaneously, we find that $x = y = a$. The corresponding value of V is a^3. The conditions of the problem make it clear that this is a maximum. □

The last two problems were extremely easy. They were easy in part because the side conditions were such that we could solve for one of the variables in terms of the other(s). In general this is not possible and a more sophisticated approach is required.

The Method of Lagrange †

We begin with what looks like a detour. To avoid having to make separate statements for the two- and three-variable cases, we will use vector notation.

Throughout the discussion, f will be a function of two or three variables continuously differentiable on some open set U. We take

$$C: \mathbf{r}(t), \quad t \in I$$

---

† Another contribution of Joseph Louis Lagrange, the French mathematician cited previously in connection with the remainder of a Taylor polynomial. (13.6.2)

as a curve which lies entirely in U and has at each point a nonzero tangent vector $\mathbf{r}'(t)$. The basic result is this:

(17.7.1)

> if $\mathbf{x}_0$ maximizes (or minimizes) $f(\mathbf{x})$ on C,
> then $\nabla f(\mathbf{x}_0)$ is perpendicular to C at $\mathbf{x}_0$.

PROOF.   Choose t_0 so that

$$\mathbf{r}(t_0) = \mathbf{x}_0.$$

The composition $f(\mathbf{r}(t))$ has a maximum (or minimum) at t_0. Consequently its derivative,

$$\frac{d}{dt}\left[f(\mathbf{r}(t))\right] = \nabla f(\mathbf{r}(t)) \cdot \mathbf{r}'(t),$$

must be zero at t_0:

$$0 = \nabla f(\mathbf{r}(t_0)) \cdot \mathbf{r}'(t_0) = \nabla f(\mathbf{x}_0) \cdot \mathbf{r}'(t_0).$$

This shows that

$$\nabla f(\mathbf{x}_0) \perp \mathbf{r}'(t_0).$$

Since $\mathbf{r}'(t_0)$ is tangent to C at $\mathbf{x}_0$, $\nabla f(\mathbf{x}_0)$ is perpendicular to C at $\mathbf{x}_0$. $\square$

Figure 17.7.1 illustrates the result when $\nabla f(\mathbf{x}_0) \neq \mathbf{0}$.

two variables   three variables

if $\mathbf{x}_0$ maximizes $f(\mathbf{x})$ on C, then $\nabla f(\mathbf{x}_0) \perp C$

FIGURE 17.7.1

We are now ready for the side condition problems. Suppose that g is a continuously differentiable function of two or three variables defined on a subset of the domain of f. Suppose furthermore that the gradient ∇g is never $\mathbf{0}$. Lagrange made the following observation:

> **(17.7.2)** | if $\mathbf{x}_0$ maximizes (or minimizes) $f(\mathbf{x})$ subject to the side condition $g(\mathbf{x}) = 0$, then $\nabla f(\mathbf{x}_0)$ and $\nabla g(\mathbf{x}_0)$ are collinear. Consequently there exists a scalar λ such that
> $$\nabla f(\mathbf{x}_0) = \lambda \nabla g(\mathbf{x}_0).$$

Such a scalar λ is now called a *Lagrange multiplier.*

A Geometric Argument for (17.7.2)

Suppose that $\mathbf{x}_0$ maximizes (or minimizes) $f(\mathbf{x})$ subject to the side condition $g(\mathbf{x}) = 0$. In the two-variable case we have

$$\mathbf{x}_0 = (x_0, y_0) \quad \text{and} \quad \text{the equation } g(x, y) = 0.$$

This equation defines a curve C, which, we assume, has a nonzero tangent vector. Since (x_0, y_0) maximizes (or minimizes) $f(x, y)$ on C, we know that $\nabla f(x_0, y_0)$ is perpendicular to C at (x_0, y_0). Since $\nabla g(x_0, y_0)$ is also perpendicular to C at (x_0, y_0) (Section 17.4), the two gradients must be collinear.

In the three-variable case we have

$$\mathbf{x}_0 = (x_0, y_0, z_0) \quad \text{and} \quad \text{the equation } g(x, y, z) = 0.$$

This equation defines a surface Γ that lies in the domain of f. Now let C be a curve which lies in Γ and passes through (x_0, y_0, z_0). As usual we assume that C has a nonzero tangent vector. We know that (x_0, y_0, z_0) maximizes (or minimizes) $f(x, y, z)$ on C. Consequently, $\nabla f(x_0, y_0, z_0)$ is perpendicular to C at (x_0, y_0, z_0). Since this is true for each such curve C, $\nabla f(x_0, y_0, z_0)$ must be perpendicular to Γ itself. But $\nabla g(x_0, y_0, z_0)$ is also perpendicular to Γ at (x_0, y_0, z_0). (Section 17.4) It follows that $\nabla f(x_0, y_0, z_0)$ and $\nabla g(x_0, y_0, z_0)$ are collinear. $\square$

It is time to apply Lagrange's idea to some problems.

Problem.   Maximize and minimize

$$f(x, y) = xy \quad \text{on the unit circle} \quad x^2 + y^2 = 1.$$

SOLUTION.   Set

$$g(x, y) = x^2 + y^2 - 1.$$

We want to maximize and minimize

$$f(x, y) = xy \quad \text{subject to the side condition} \quad g(x, y) = 0.$$

First we look for those points (x, y) which satisfy a Lagrange condition

$$\nabla f(x, y) = \lambda \nabla g(x, y).$$

The gradients are

$$\nabla f(x, y) = y\mathbf{i} + x\mathbf{j}, \qquad \nabla g(x, y) = 2x\mathbf{i} + 2y\mathbf{j}.$$

Setting

$$\nabla f(x, y) = \lambda \nabla g(x, y),$$

we obtain

$$y = 2\lambda x, \qquad x = 2\lambda y.$$

Multiplying the first equation by y and the second equation by x, we find that

$$y^2 = 2\lambda xy, \qquad x^2 = 2\lambda xy$$

and thus

$$y^2 = x^2.$$

The side condition $x^2 + y^2 = 1$ now implies that $2x^2 = 1$ and therefore that $x = \pm\frac{1}{2}\sqrt{2}$. The points under consideration are

$$(\tfrac{1}{2}\sqrt{2}, \tfrac{1}{2}\sqrt{2}), (\tfrac{1}{2}\sqrt{2}, -\tfrac{1}{2}\sqrt{2}), (-\tfrac{1}{2}\sqrt{2}, \tfrac{1}{2}\sqrt{2}), (-\tfrac{1}{2}\sqrt{2}, -\tfrac{1}{2}\sqrt{2}).$$

At the first and fourth points f takes on the value $\frac{1}{2}$. At the second and third points f takes on the value $-\frac{1}{2}$. The first is a maximum and the second a minimum. $\square$

Problem.   Find the minimum value taken on by the function

$$f(x, y) = x^2 + (y - 2)^2 \quad \text{on the hyperbola} \quad x^2 - y^2 = 1.$$

SOLUTION.   We set

$$g(x, y) = x^2 - y^2 - 1.$$

Our problem is to minimize $f(x, y)$ subject to the side condition

$$g(x, y) = x^2 - y^2 - 1 = 0.$$

The points of interest must satisfy a Lagrange condition

$$\nabla f(x, y) = \lambda \nabla g(x, y) \quad \text{together with} \quad g(x, y) = 0.$$

Here

$$\nabla f(x, y) = 2x\mathbf{i} + 2(y - 2)\mathbf{j}, \qquad \nabla g(x, y) = 2x\mathbf{i} - 2y\mathbf{j}.$$

The condition

$$\nabla f(x, y) = \lambda \nabla g(x, y)$$

gives

$$2x = 2\lambda x, \qquad 2(y - 2) = -2\lambda y,$$

which we can simplify to

$$x = \lambda x, \qquad y - 2 = -\lambda y.$$

The points of interest must therefore satisfy the following three equations:

$$x = \lambda x, \qquad y - 2 = -\lambda y, \qquad x^2 - y^2 - 1 = 0.$$

The last equation shows that x cannot be zero. Dividing $x = \lambda x$ by x, we get $\lambda = 1$. This means that $y - 2 = -y$ and therefore $y = 1$. With $y = 1$, the last equation gives $x = \pm\sqrt{2}$. The points to be checked are therefore $(-\sqrt{2}, 1)$ and $(\sqrt{2}, 1)$. At each of these points f takes on the value 3. This is the desired minimum. $\square$

Problem. Maximize

$$f(x, y, z) = xyz \quad \text{subject to the side condition} \quad x^3 + y^3 + z^3 = 1$$

for $x \geq 0, y \geq 0, z \geq 0$.

SOLUTION.   We set

$$g(x, y, z) = x^3 + y^3 + z^3 - 1$$

so that the side condition becomes $g(x, y, z) = 0$. We seek those triples (x, y, z) which simultaneously satisfy a Lagrange condition

$$\nabla f(x, y, z) = \lambda \, \nabla g(x, y, z) \quad \text{and the side condition} \quad g(x, y, z) = 0.$$

The gradients are

$$\nabla f(x, y, z) = yz\mathbf{i} + xz\mathbf{j} + xy\mathbf{k}, \qquad \nabla g(x, y, z) = 3x^2\mathbf{i} + 3y^2\mathbf{j} + 3z^2\mathbf{k}.$$

Setting

$$\nabla f(x, y, z) = \lambda \nabla g(x, y, z),$$

we find that

$$yz = \lambda 3x^2, \qquad xz = \lambda 3y^2, \qquad xy = \lambda 3z^2.$$

Multiplying the first equation by x, the second by y, and the third by z, we get

$$xyz = \lambda 3x^3, \qquad xyz = \lambda 3y^3, \qquad xyz = \lambda 3z^3$$

and consequently

$$\lambda x^3 = \lambda y^3 = \lambda z^3.$$

We can exclude $\lambda = 0$ because, if $\lambda = 0$, then at least one of the other variables would have to be zero. That would force xyz to be 0, which is obviously not a maximum. Having excluded $\lambda = 0$, we can divide by λ and get

$$x^3 = y^3 = z^3 \quad \text{and thus} \quad x = y = z.$$

The side condition $x^3 + y^3 + z^3 = 1$ now gives

$$x = (\tfrac{1}{3})^{1/3}, \quad y = (\tfrac{1}{3})^{1/3}, \quad z = (\tfrac{1}{3})^{1/3}.$$

The desired maximum is $\tfrac{1}{3}$. $\square$

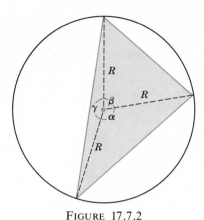

FIGURE 17.7.2

Problem.   Show that, of all the triangles inscribed in a fixed circle, the equilateral triangle has the largest perimeter.

SOLUTION.   In Figure 17.7.2 we have marked the angles α, β, γ. The length of the side subtended by α is $2R \sin \frac{1}{2}\alpha$. The perimeter is therefore given by the function

$$f(\alpha, \beta, \gamma) = 2R(\sin \tfrac{1}{2}\alpha + \sin \tfrac{1}{2}\beta + \sin \tfrac{1}{2}\gamma).$$

As a side condition we have

$$g(\alpha, \beta, \gamma) = \alpha + \beta + \gamma - 2\pi = 0.$$

To maximize the perimeter we form the gradients

$$\nabla f(\alpha, \beta, \gamma) = R[\cos \tfrac{1}{2}\alpha \, \mathbf{i} + \cos \tfrac{1}{2}\beta \, \mathbf{j} + \cos \tfrac{1}{2}\gamma \, \mathbf{k}], \quad \nabla g(\alpha, \beta, \gamma) = \mathbf{i} + \mathbf{j} + \mathbf{k}.$$

The Lagrange condition

$$\nabla f(\alpha, \beta, \gamma) = \lambda \, \nabla g(\alpha, \beta, \gamma)$$

gives

$$\lambda = R \cos \tfrac{1}{2}\alpha, \quad \lambda = R \cos \tfrac{1}{2}\beta, \quad \lambda = R \cos \tfrac{1}{2}\gamma.$$

Under the conditions of the problem this means that

$$\alpha = \beta = \gamma.$$

Since the central angles are equal, the sides are equal. The triangle is therefore equilateral. $\square$

Exercises

*1. Minimize $x^2 + y^2$ on the hyperbola $xy = 1$.

2. Maximize xy on the ellipse $b^2x^2 + a^2y^2 = a^2b^2$.

*3. Minimize xy on the ellipse $b^2x^2 + a^2y^2 = a^2b^2$.

4. Minimize xy^2 on the unit circle $x^2 + y^2 = 1$.

*5. Maximize xy^2 on the ellipse $b^2x^2 + a^2y^2 = a^2b^2$.

6. Maximize $x + y$ on the curve $x^4 + y^4 = 1$.

*7. Minimize xyz on the unit sphere $x^2 + y^2 + z^2 = 1$.

8. Maximize xyz on the ellipsoid $x^2/a^2 + y^2/b^2 + z^2/c^2 = 1$.

*9. Maximize xy^2z^3 on the plane $x + y + z = 1$.

10. Minimize $x + 2y + 4z$ on the sphere $x^2 + y^2 + z^2 = 7$.

*11. Maximize $2x + 3y + 5z$ on the sphere $x^2 + y^2 + z^2 = 19$.

12. Minimize $x^4 + y^4 + z^4$ on the plane $x + y + z = 1$.

*13. Maximize the volume of a rectangular solid in the first octant with one vertex at the origin and opposite vertex on the plane $x/a + y/b + z/c = 1$. (Take $a > 0$, $b > 0$, $c > 0$.)

14. Show that the square has the largest area of all the rectangles with a given perimeter.

*15. Find the distance from the point $(0, 1)$ to the parabola $x^2 = 4y$.

16. Find the distance from the point $(p, 4p)$ to the parabola $y^2 = 2px$.

*17. Maximize the volume of a rectangular solid given that the sum of the areas of the six faces is $6a^2$.

18. Use the method of Lagrange multipliers to find the distance from the origin to the plane $Ax + By + Cz + D = 0$.

*19. Within a triangle there is a point P such that the sum of the squares of its distances from the sides of the triangle is a minimum. Find this minimum.

20. Show that of all the triangles inscribed in a fixed circle the equilateral one has the largest (a) product of the lengths of the sides, (b) sum of the squares of the lengths of the sides.

21. A plane passes through the point (a, b, c). Find its intercepts with the coordinate axes if the volume of the solid bounded by the plane and the coordinate planes is to be a minimum.

22. Show that, of all the triangles with a given perimeter, the equilateral triangle has the largest area. HINT: area $= \sqrt{s(s - a)(s - b)(s - c)}$, where s is the semiperimeter $s = \frac{1}{2}(a + b + c)$.

*23. A manufacturer can produce three distinct products in quantities Q_1, Q_2, Q_3, respectively, and thereby derive a profit

$$P(Q_1, Q_2, Q_3) = 2Q_1 + 8Q_2 + 24Q_3.$$

Find the values of Q_1, Q_2, Q_3 that maximize profit if production is subject to the constraint

$$Q_1^2 + 2Q_2^2 + 4Q_3^2 = 4,500,000,000.$$

17.8 Differentials

We begin by reviewing the one-variable case. If f is differentiable at x, then

$$f(x + h) - f(x) = f'(x)h + o(h).$$

This tells us that, for small $h, f(x + h) - f(x)$ and $f'(x)h$ are approximately equal, and consequently we can use the differential

$$df = f'(x)h$$

to approximate the increment

$$\Delta f = f(x + h) - f(x).$$

In symbols

$$\Delta f \cong df.$$

The differential of a function of several variables is defined in a similar manner and plays a similar approximating role. If f, now a function of several variables, is differentiable at $\mathbf{x}$, then

$$f(\mathbf{x} + \mathbf{h}) - f(\mathbf{x}) = \nabla f(\mathbf{x}) \cdot \mathbf{h} + o(\mathbf{h}).$$

This says that, for small $\mathbf{h}, f(\mathbf{x} + \mathbf{h}) - f(\mathbf{x})$ and $\nabla f(\mathbf{x}) \cdot \mathbf{h}$ are approximately equal. In this setting, we call

(17.8.1)
$$\boxed{\Delta f = f(\mathbf{x} + \mathbf{h}) - f(\mathbf{x})}$$

the *increment* of f and

(17.8.2)
$$\boxed{df = \nabla f(\mathbf{x}) \cdot \mathbf{h}}$$

the *differential* (more formally, the *total differential*) at $\mathbf{x}$ with increment $\mathbf{h}$. The key point is that, as before,

(17.8.3)
$$\boxed{\Delta f \cong df.}$$

In the two-variable case we set

$$\mathbf{x} = (x, y) \quad \text{and} \quad \mathbf{h} = (\Delta x, \Delta y).$$

The increment $\Delta f = f(\mathbf{x} + \mathbf{h}) - f(\mathbf{x})$ then becomes

$$\Delta f = f(x + \Delta x, y + \Delta y) - f(x, y),$$

and the differential $df = \nabla f(\mathbf{x}) \cdot \mathbf{h}$ becomes

$$df = \frac{\partial f}{\partial x}(x, y)\, \Delta x + \frac{\partial f}{\partial y}(x, y)\, \Delta y.$$

By suppressing the point of evaluation, we can write

(17.8.4)
$$\boxed{df = \frac{\partial f}{\partial x}\, \Delta x + \frac{\partial f}{\partial y}\, \Delta y.}$$

The approximation $\Delta f \cong df$ is illustrated in Figure 17.8.1. There we have represented f as a surface $z = f(x, y)$. Through a point $P(x_0, y_0, f(x_0, y_0))$ we have drawn the tangent plane. *The difference*

$$df - \Delta f$$

is the vertical separation between this tangent plane and the surface as measured above the point $(x_0 + \Delta x, y_0 + \Delta y)$.

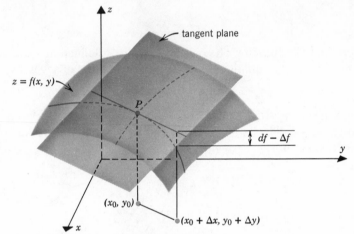

FIGURE 17.8.1

PROOF.   The tangent plane at P has equation

$$(x - x_0) \frac{\partial f}{\partial x}(x_0, y_0) + (y - y_0) \frac{\partial f}{\partial y}(x_0, y_0) = z - f(x_0, y_0).$$

The z-coordinate of this plane over the point $(x_0 + \Delta x, y_0 + \Delta y)$ is

$$f(x_0, y_0) + \frac{\partial f}{\partial x}(x_0, y_0)\,\Delta x + \frac{\partial f}{\partial y}(x_0, y_0)\,\Delta y. \qquad \text{(check this)}$$

The z-coordinate of the surface over this same point is

$$f(x_0 + \Delta x, y_0 + \Delta y).$$

The difference

$$\left[f(x_0, y_0) + \frac{\partial f}{\partial x}(x_0, y_0)\,\Delta x + \frac{\partial f}{\partial y}(x_0, y_0)\,\Delta y \right] - \left[f(x_0 + \Delta x, y_0 + \Delta y) \right]$$

can be written as

$$\left[\frac{\partial f}{\partial x}(x_0, y_0)\,\Delta x + \frac{\partial f}{\partial y}(x_0, y_0)\,\Delta y \right] - \left[f(x_0 + \Delta x, y_0 + \Delta y) - f(x_0, y_0) \right].$$

This is just $df - \Delta f$. $\square$

For the three-variable case we set

$$\mathbf{x} = (x, y, z) \quad \text{and} \quad \mathbf{h} = (\Delta x, \Delta y, \Delta z).$$

The increment becomes

$$\Delta f = f(x + \Delta x, y + \Delta y, z + \Delta z) - f(x, y, z),$$

and the approximating differential becomes

$$df = \frac{\partial f}{\partial x}(x, y, z)\,\Delta x + \frac{\partial f}{\partial y}(x, y, z)\,\Delta y + \frac{\partial f}{\partial z}(x, y, z)\,\Delta z.$$

Suppressing the point of evaluation we have

(17.8.5)
$$df = \frac{\partial f}{\partial x}\,\Delta x + \frac{\partial f}{\partial y}\,\Delta y + \frac{\partial f}{\partial z}\,\Delta z.$$

To illustrate the use of differentials we begin with a rectangle of sides x and y. The area is given by

$$A(x, y) = xy.$$

An increase in the dimensions of the rectangle to $x + \Delta x$ and $y + \Delta y$ produces a change in area

$$
\begin{aligned}
\Delta A &= (x + \Delta x)(y + \Delta y) - xy \\
&= (xy + x\,\Delta y + y\,\Delta x + \Delta x\,\Delta y) - xy \\
&= x\,\Delta y + y\,\Delta x + \Delta x\,\Delta y.
\end{aligned}
$$

The differential estimate for this change in area is

$$dA = \frac{\partial A}{\partial x}\,\Delta x + \frac{\partial A}{\partial y}\,\Delta y = y\,\Delta x + x\,\Delta y. \qquad \text{(Figure 17.8.2)}$$

The error of our estimate, the difference between the actual change and the estimated change, is the difference $\Delta A - dA = \Delta x\,\Delta y$. $\square$

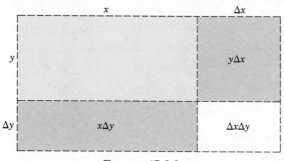

FIGURE 17.8.2

Problem. Given that

$$f(x, y) = yx^{2/5} + x\sqrt{y},$$

estimate by a differential the change in f from $(32, 16)$ to $(35, 18)$.

SOLUTION. Since

$$\frac{\partial f}{\partial x} = \frac{2}{5}\left(\frac{1}{x}\right)^{3/5} y + \sqrt{y} \qquad \text{and} \qquad \frac{\partial f}{\partial y} = x^{2/5} + \frac{x}{2\sqrt{y}},$$

we have

$$df = \left[\frac{2}{5}\left(\frac{1}{x}\right)^{3/5} y + \sqrt{y}\right]\Delta x + \left[x^{2/5} + \frac{x}{2\sqrt{y}}\right]\Delta y.$$

At $x = 32$, $y = 16$, $\Delta x = 3$, $\Delta y = 2$,

$$df = \left[\frac{2}{5}\left(\frac{1}{32}\right)^{3/5} 16 + \sqrt{16}\right]3 + \left[32^{2/5} + \frac{32}{2\sqrt{16}}\right]2 = 30.4.$$

The change increases the value of f by approximately 30.4.   □

Problem.   Use differentials to estimate $\sqrt{27}\sqrt[3]{1021}$.

SOLUTION.   We know $\sqrt{25}$ and $\sqrt[3]{1000}$. What we need is an estimate for the increase of

$$f(x, y) = \sqrt{x}\sqrt[3]{y} = x^{1/2}y^{1/3}$$

from $x = 25$, $y = 1000$ to $x = 27$, $y = 1021$. The differential is

$$df = \tfrac{1}{2}x^{-1/2}y^{1/3}\,\Delta x + \tfrac{1}{3}x^{1/2}y^{-2/3}\,\Delta y.$$

With $x = 25$, $y = 1000$, $\Delta x = 2$, $\Delta y = 21$, df becomes

$$(\tfrac{1}{2}\cdot 25^{-1/2}\cdot 1000^{1/3})2 + (\tfrac{1}{3}\cdot 25^{1/2}\cdot 1000^{-2/3})21 = 2.35.$$

The change increases the value of the function by about 2.35. It follows that

$$\sqrt{27}\sqrt[3]{1021} \cong \sqrt{25}\sqrt[3]{1000} + 2.35 = 52.35. \quad □$$

Problem.   Estimate the change in the volume of the frustum of a cone if the upper radius r is decreased from 3 to 2.7 centimeters, the base radius R is increased from 8 to 8.1 centimeters, and the height h is increased from 6 to 6.3 centimeters.

SOLUTION

$$V(r, R, h) = \tfrac{1}{3}\pi h(R^2 + Rr + r^2),$$

$$dV = \tfrac{1}{3}\pi h(R + 2r)\,\Delta r + \tfrac{1}{3}\pi h(2R + r)\,\Delta R + \tfrac{1}{3}\pi(R^2 + Rr + r^2)\,\Delta h.$$

At $r = 3$, $R = 8$, $h = 6$, $\Delta r = -0.3$, $\Delta R = 0.1$, and $\Delta h = 0.3$,

$$dV = (28\pi)(-0.3) + (38\pi)(0.1) + \tfrac{1}{3}(97\pi)(0.3) = 5.1\pi.$$

The volume increases by about 5.1π cubic centimeters.   □

Exercises

Find the differential df.

*1. $f(x, y) = x^3 y - x^2 y^2$.   2. $f(x, y) = xy + yz + xz$.
*3. $f(x, y) = x\cos y - y\cos x$.   4. $f(x, y, z) = x^2 y e^{2z}$.
*5. $f(x, y, z) = x - y\tan z$.   6. $f(x, y) = (x - y)\log(x + y)$.

*7. Compute Δu and du for $u = x^2 - 3xy + 2y^2$ at $x = 2$, $y = -3$, $\Delta x = -0.3$, $\Delta y = 0.2$.

*8. Compute du for $u = (x + y)\sqrt{x - y}$ at $x = 6$, $y = 2$, $\Delta x = \frac{1}{4}$, $\Delta y = -\frac{1}{2}$.

Use differentials to find the approximate value.

*9. $\sqrt{125}\,\sqrt[4]{17}$.

10. $(1 - \sqrt{10})(1 + \sqrt{24})$.

*11. $\sin \frac{6}{7}\pi \cos \frac{1}{5}\pi$.

12. $\sqrt{8}\,\tan \frac{5}{16}\pi$.

*13. Given that $z = (x - y)(x + y)^{-1}$, use dz to find the approximate change in z if x is increased from 4 to $4\frac{1}{10}$ and y is increased from 2 to $2\frac{1}{10}$. What is the exact change?

*14. Estimate by a differential the change in the volume of a right circular cylinder if the height is increased from 12 to 12.2 inches and the radius is decreased from 8 to 7.7 inches.

15. Estimate the change in total surface area for the cylinder of exercise 14.

16. The radius of a circular cylinder of height h is increased from r to $r + \Delta r$.
*(a) Determine the exact change in h that will keep the volume constant. Then estimate this change in h by using a differential.
(b) Determine the exact change in h that will keep the total surface area constant. Then estimate this change in h by using a differential.

*17. Use a differential to estimate the change in $T = x^2 \cos \pi z - y^2 \sin \pi z$ from $x = 2$, $y = 2$, $z = 2$ to $x = 2.1$, $y = 1.9$, $z = 2.2$.

18. Give bounding estimates for the volume and the lateral area of a right circular cone given that the radius is between 4.8 and 5.2 inches and the slant height is between 11.7 and 12.3 inches.

*19. The specific gravity of a solid is given by the formula $s = A(A - W)^{-1}$ where A is the weight in air and W the weight in water. Estimate s if A is measured to be 9 pounds (within a tolerance of 0.01 pounds) and W is measured to be 5 pounds (within a tolerance of 0.02 pounds).

17.9 Reconstructing a Function from its Gradient

This section has essentially three parts. In the first part we show how to find $f(x, y)$, given its gradient

$$\nabla f(x, y) = \frac{\partial f}{\partial x}(x, y)\,\mathbf{i} + \frac{\partial f}{\partial y}(x, y)\mathbf{j}.$$

In the second part we show that, although all gradients $\nabla f(x, y)$ are expressions of the form

$$P(x, y)\mathbf{i} + Q(x, y)\mathbf{j},$$

(set $P = \partial f/\partial x$ and $Q = \partial f/\partial y$), not all such expressions are gradients. In the third part we tackle the problem of recognizing which expressions $P(x, y)\mathbf{i} + Q(x, y)\mathbf{j}$ are actually gradients.

Part 1.

Problem. Find f given that

$$\nabla f(x, y) = y^2 \mathbf{i} + (2xy - 1)\mathbf{j}.$$

SOLUTION.   Since

$$\nabla f(x, y) = \frac{\partial f}{\partial x}(x, y)\,\mathbf{i} + \frac{\partial f}{\partial y}(x, y)\,\mathbf{j},$$

we must have

$$\frac{\partial f}{\partial x}(x, y) = y^2 \quad \text{and} \quad \frac{\partial f}{\partial y}(x, y) = 2xy - 1.$$

Integrating the first relation with respect to x, we have

$$f(x, y) = xy^2 + \phi(y)$$

where $\phi(y)$ is independent of x but may depend on y. Integrating the second relation with respect to y, we have

$$f(x, y) = xy^2 - y + \psi(x)$$

where $\psi(x)$ is independent of y but may depend on x. We can reconcile these two characterizations of $f(x, y)$ only by having

$$xy^2 + \phi(y) = xy^2 - y + \psi(x).$$

This forces $\phi(y) = -y + \psi(x)$ and thus

$$\phi(y) + y = \psi(x).$$

The left-hand side can depend at most on y, and the right-hand side at most on x. This means that both sides are equal to some constant C. This gives

$$\phi(y) + y = C \quad \text{and} \quad \psi(x) = C.$$

The relation $f(x, y) = xy^2 + \phi(y)$ now gives

$$f(x, y) = xy^2 - y + C.$$

The relation $f(x, y) = xy^2 - y + \psi(x)$ also gives

$$f(x, y) = xy^2 - y + C. \quad \square$$

REMARK.   Once we had derived

$$f(x, y) = xy^2 - y + C$$

from the relation $f(x, y) = xy^2 + \phi(y)$, it was unnecessary to derive the same result from $f(x, y) = xy^2 - y + \psi(x)$. We did it simply to check our computations. It's reassuring to get the same answer from both relations.

Problem. Find f given that

$$\nabla f(x, y) = (y^{1/2} - \tfrac{1}{2}yx^{-1/2} + 2x)\mathbf{i} + (\tfrac{1}{2}xy^{-1/2} - x^{1/2} + 1)\mathbf{j}.$$

SOLUTION.   We begin by observing that

$$\frac{\partial f}{\partial x}(x, y) = y^{1/2} - \tfrac{1}{2}yx^{-1/2} + 2x \quad \text{and} \quad \frac{\partial f}{\partial y}(x, y) = \tfrac{1}{2}xy^{-1/2} - x^{1/2} + 1.$$

Integration of the first relation with respect to x, gives

$$f(x, y) = y^{1/2}x - x^{1/2}y + x^2 + \phi(y)$$

with $\phi(y)$ independent of x. Integration of the second relation with respect to y, gives

$$f(x, y) = xy^{1/2} - x^{1/2}y + y + \psi(x),$$

with $\psi(x)$ independent of y. These two characterizations of $f(x, y)$ can be reconciled only by having

$$x^2 + \phi(y) = y + \psi(x).$$

This in turn gives

$$\phi(y) - y = \psi(x) - x^2.$$

The left-hand side can depend at most on y, and the right-hand side at most on x. Both sides must therefore be constant. The relations

$$\phi(y) - y = C \quad \text{and} \quad \psi(x) - x^2 = C$$

lead to

$$\phi(y) = y + C \quad \text{and} \quad \psi(x) = x^2 + C.$$

Each of these relations gives

$$f(x, y) = xy^{1/2} - x^{1/2}y + x^2 + y + C. \quad \square$$

The function

$$f(x, y) = xy^{1/2} - x^{1/2}y + x^2 + y + C$$

that we just found is in its most general form in the sense that it contains an arbitrary constant C. If we want f to take on a particular value z_0 at some particular point (x_0, y_0), we need only adjust the constant C accordingly. Let's suppose, for example, that we require

$$f(1, 4) = 5.$$

Note that

$$f(1, 4) = 1\sqrt{4} - \sqrt{1} \cdot 4 + 1^2 + 4 + C = 3 + C.$$

To get $f(1, 4) = 5$ we must have $3 + C = 5$ and thus $C = 2$. $\quad \square$

Part 2.

The next problem shows that not all linear combinations $P(x, y)\mathbf{i} + Q(x, y)\mathbf{j}$ are actually gradients.

Problem.   Show that there is no function f such that

$$\nabla f(x, y) = y\mathbf{i} - x\mathbf{j}.$$

SOLUTION.   Suppose on the contrary that there is such a function. Then of course we must have

$$\frac{\partial f}{\partial x}(x, y) = y \quad \text{and} \quad \frac{\partial f}{\partial y}(x, y) = -x.$$

Integrating the first equation with respect to x, we get

$$f(x, y) = xy + \phi(y),$$

where $\phi(y)$ is independent of x. Integrating the second equation with respect to y, we get

$$f(x, y) = -xy + \psi(x),$$

where $\psi(x)$ is independent of y. To reconcile these two characterizations of f, we must have $xy + \phi(y) = -xy + \psi(x)$ and consequently

$$\psi(x) = 2xy + \phi(y).$$

Differentiation with respect to y gives

$$\frac{\partial}{\partial y}[\psi(x)] = 2x + \phi'(y), \quad 0 = 2x + \phi'(y), \quad \text{and thus} \quad -2x = \phi'(y).$$

Differentiation of this last equation with respect to x gives

$$-2 = 0.$$

The assumption that $y\mathbf{i} - x\mathbf{j}$ is a gradient has led to an absurdity.   □

Part 3.

We come now to the problem of recognizing which linear combinations

$$P(x, y)\mathbf{i} + Q(x, y)\mathbf{j}$$

are gradients.

Theorem 17.9.1

Let P and Q be functions of two variables each continuously differentiable in an open rectangle S, which we take to have sides parallel to the coordinate axes. The linear combination $P(x, y)\mathbf{i} + Q(x, y)\mathbf{j}$ is a gradient in S iff

$$\frac{\partial P}{\partial y}(x, y) = \frac{\partial Q}{\partial x}(x, y) \qquad \text{for all } (x, y) \in S.$$

PROOF.   Suppose that $P(x, y)\mathbf{i} + Q(x, y)\mathbf{j}$ is a gradient on S, say

$$\nabla f(x, y) = P(x, y)\mathbf{i} + Q(x, y)\mathbf{j}.$$

Since

$$\nabla f(x, y) = \frac{\partial f}{\partial x}(x, y)\, \mathbf{i} + \frac{\partial f}{\partial y}(x, y)\, \mathbf{j},$$

we have

$$P = \frac{\partial f}{\partial x} \quad \text{and} \quad Q = \frac{\partial f}{\partial y}.$$

Since P and Q have continuous first partials, f has continuous second partials. Consequently, its mixed partials are equal (16.6.5), and

$$\frac{\partial P}{\partial y} = \frac{\partial^2 f}{\partial y\, \partial x} = \frac{\partial^2 f}{\partial x\, \partial y} = \frac{\partial Q}{\partial x}.$$

Conversely, suppose that

$$\frac{\partial P}{\partial y}(x, y) = \frac{\partial Q}{\partial x}(x, y) \qquad \text{for all } (x, y) \in S.$$

We must show that $P(x, y)\mathbf{i} + Q(x, y)\mathbf{j}$ is a gradient in S. To do this, we choose a point (x_0, y_0) and form the function

$$f(x, y) = C + \int_{x_0}^{x} P(u, y_0)\, du + \int_{y_0}^{y} Q(x, v)\, dv, \qquad (x, y) \in S.$$

[If you want to visualize f, you can refer to Figure 17.9.1. The function P is being integrated along the horizontal line segment joining (x_0, y_0) to (x, y_0), and Q is being integrated along the vertical line segment joining (x, y_0) to (x, y). Our assumptions on S guarantee that these line segments remain in S. C is simply an arbitrary constant.]

FIGURE 17.9.1

The first integral is independent of y. Hence

$$\frac{\partial f}{\partial y}(x, y) = \frac{\partial}{\partial x}\left(\int_{y_0}^{y} Q(x, v)\, dv \right) = Q(x, y).$$

The last equality holds because we are differentiating an integral with respect to its upper limit.† The differentiation of f with respect to x is more complicated:

$$\frac{\partial f}{\partial x}(x, y) = \frac{\partial}{\partial x}\left(\int_{x_0}^{x} P(u, y_0)\, du \right) + \frac{\partial}{\partial x}\left(\int_{y_0}^{y} Q(x, v)\, dv \right).$$

---

† This is a fundamental notion. If you are unsure about it, see Theorem 5.3.5.

In the first term the variable x appears as an upper limit. In the second term the variable x appears under the sign of integration. Differentiation under the sign of integration is justified in more advanced texts.† Anticipating this result, we get

$$\frac{\partial f}{\partial x}(x, y) = P(x, y_0) + \int_{y_0}^{y} \frac{\partial Q}{\partial x}(x, v)\, dv$$

$$= P(x, y_0) + \int_{y_0}^{y} \frac{\partial P}{\partial y}(x, v)\, dv$$

$$= P(x, y_0) + P(x, y) - P(x, y_0) = P(x, y).$$

We have shown that

$$P(x, y) = \frac{\partial f}{\partial x}(x, y) \quad \text{and} \quad Q(x, y) = \frac{\partial f}{\partial y}(x, y).$$

It follows that $P(x, y)\mathbf{i} + Q(x, y)\mathbf{j}$ is the gradient of f. □

REMARK.   The result we just proved for a rectangle can be shown to hold for any *simply connected region* (for any open connected plane set with a connected boundary), but for this a much more complicated proof is required.

Examples

(1) To determine whether $xy\mathbf{i} + \frac{1}{2}(x + 1)^2y^2\mathbf{j}$ is a gradient, we set

$$P(x, y) = xy \quad \text{and} \quad Q(x, y) = \tfrac{1}{2}(x + 1)^2y^2.$$

Since

$$\frac{\partial P}{\partial y}(x, y) = x \quad \text{and} \quad \frac{\partial Q}{\partial x}(x, y) = (x + 1)y^2,$$

we have

$$\frac{\partial P}{\partial y}(x, y) \neq \frac{\partial Q}{\partial x}(x, y).$$

The expression is not a gradient.

(2) In the case of $2x \sin y\, \mathbf{i} + x^2 \cos y\, \mathbf{j}$ we have

$$P(x, y) = 2x \sin y, \qquad Q(x, y) = x^2 \cos y,$$

$$\frac{\partial P}{\partial y}(x, y) = 2x \cos y, \qquad \frac{\partial Q}{\partial x}(x, y) = 2x \cos y.$$

Since the two partials are equal, the expression is a gradient. □

---

† See, for example, Einar Hille, *Analysis,* Vol. 2 (Blaisdell, Waltham, Mass., 1966).

Exercises

Determine whether or not the expression is a gradient $\nabla f(x, y)$. If so, find $f(x, y)$.

*1. $xy^2\mathbf{i} + x^2y\mathbf{j}$.

2. $x\mathbf{i} + y\mathbf{j}$.

*3. $y\mathbf{i} + x\mathbf{j}$.

4. $(x^2 + y)\mathbf{i} + (y^3 + x)\mathbf{j}$.

*5. $(y^3 + x)\mathbf{i} + (x^2 + y)\mathbf{j}$.

6. $(y^2e^x - y)\mathbf{i} + (2ye^x - x)\mathbf{j}$.

*7. $(\cos x - y \sin x)\mathbf{i} + \cos x\mathbf{j}$.

8. $(1 + e^y)\mathbf{i} + (xe^y + y^2)\mathbf{j}$.

*9. $ye^x(1 + x)\mathbf{i} + (xe^x - e^{-y})\mathbf{j}$.

10. $(x \tan y + \sec^2 x)\mathbf{i} + (\tfrac{1}{2}x^2 \sec^2 y + \pi y)\mathbf{j}$.

*11. $(x^2 \text{ arc sin } y)\mathbf{i} + \left(\dfrac{x^3}{3\sqrt{1 - y^2}} - \log y\right)\mathbf{j}$.

12. $\left(\dfrac{\text{arc tan } y}{\sqrt{1 - x^2}} + \dfrac{x}{y}\right)\mathbf{i} + \left(\dfrac{\text{arc sin } x}{1 + y^2} - \dfrac{x^2}{2y^2} + 1\right)\mathbf{j}$.

Optional | *13. $(\cosh x + \sinh y)\mathbf{i} + (x \cosh y)\mathbf{j}$.

14. $(\cosh x - x \sinh y)\mathbf{i} + (2y \cosh x)\mathbf{j}$.

Double and
Triple Integrals

18

We began with ordinary integrals

$$\int_a^b f(x)\, dx.$$

Here we'll study double integrals

$$\iint\limits_{\Omega} f(x, y)\, dxdy$$

and a little later triple integrals

$$\iiint\limits_{T} f(x, y, z)\, dxdydz.$$

Our first step is to introduce some new notation.

18.1 Multiple-Sigma Notation

In an ordinary sequence each term a_i is indexed by a single integer i. The sum of all the a_i from $i = 1$ to $i = m$ is then denoted by

$$\sum_{i=1}^{m} a_i.$$

When two indices are involved, say

$$a_{ij} = 2^i 5^j, \qquad a_{ij} = i/j, \qquad \text{or} \qquad a_{ij} = (1 + i)^j,$$

then we use double-sigma notation. By

(18.1.1)
$$\sum_{i=1}^{m} \sum_{j=1}^{n} a_{ij}$$

we mean *the sum of all the a_{ij} where i ranges from 1 to m and j ranges from 1 to n.*

Since addition is associative and commutative, we can add the terms of (18.1.1) in any order we choose. Usually we set

(18.1.2)
$$\sum_{i=1}^{m} \sum_{j=1}^{n} a_{ij} = \sum_{i=1}^{m} \left(\sum_{j=1}^{n} a_{ij} \right).$$

We can expand the expression on the right by expanding first with respect to i and then with respect to j:

$$\sum_{i=1}^{m} \left(\sum_{j=1}^{n} a_{ij} \right) = \sum_{j=1}^{n} a_{1j} + \sum_{j=1}^{n} a_{2j} + \cdots + \sum_{j=1}^{n} a_{mj}$$

$$= (a_{11} + a_{12} + \cdots + a_{1n}) + (a_{21} + a_{22} + \cdots + a_{2n})$$

$$+ \cdots + (a_{m1} + a_{m2} + \cdots + a_{mn})$$

or we can expand first with respect to j and then with respect to i:

$$\sum_{i=1}^{m} \left(\sum_{j=1}^{n} a_{ij} \right) = \sum_{i=1}^{m} (a_{i1} + a_{i2} + \cdots + a_{in})$$

$$= (a_{11} + a_{12} + \cdots + a_{1n}) + (a_{21} + a_{22} + \cdots + a_{2n})$$

$$+ \cdots + (a_{m1} + a_{m2} + \cdots + a_{mn}).$$

The results are the same.

For example, we can write

$$\sum_{i=1}^{3} \left(\sum_{j=1}^{2} a_{ij} \right) = \sum_{j=1}^{2} a_{1j} + \sum_{j=1}^{2} a_{2j} + \sum_{j=1}^{2} a_{3j} = (a_{11} + a_{12}) + (a_{21} + a_{22}) + (a_{31} + a_{32}),$$

or we can write

$$\sum_{i=1}^{3} \left(\sum_{j=1}^{2} a_{ij} \right) = \sum_{i=1}^{3} (a_{i1} + a_{i2}) = (a_{11} + a_{12}) + (a_{21} + a_{22}) + (a_{31} + a_{32}).$$

Since constants can be factored through single sums, they can also be factored through double sums; namely,

(18.1.3)
$$\sum_{i=1}^{m} \sum_{j=1}^{n} \alpha\, a_{ij} = \alpha \sum_{i=1}^{m} \sum_{j=1}^{n} a_{ij}.$$

Moreover

(18.1.4)
$$\sum_{i=1}^{m} \sum_{j=1}^{n} (a_{ij} + b_{ij}) = \sum_{i=1}^{m} \sum_{j=1}^{n} a_{ij} + \sum_{i=1}^{m} \sum_{j=1}^{n} b_{ij}.$$

The easiest double sums to handle are those where each term a_{ij} appears as a product $b_i c_j$ in which each of the factors bears only one index. In that case we can express the double sum as the product of two single sums:

(18.1.5)
$$\sum_{i=1}^{m} \sum_{j=1}^{n} b_i c_j = \left(\sum_{i=1}^{m} b_i \right) \left(\sum_{j=1}^{n} c_j \right).$$

PROOF.   Set

$$B = \sum_{i=1}^{m} b_i, \qquad C = \sum_{j=1}^{n} c_j.$$

Then

$$\sum_{i=1}^{m} \sum_{j=1}^{n} b_i c_j = \sum_{i=1}^{m} \left(\sum_{j=1}^{n} b_i c_j \right) \overset{\dagger}{=} \sum_{i=1}^{m} b_i \left(\sum_{j=1}^{n} c_j \right) = \sum_{i=1}^{m} b_i C = C \sum_{i=1}^{m} b_i = CB = BC. \quad \square$$

For example,

$$\sum_{i=1}^{4} \sum_{j=1}^{2} 2^i 5^j = \left(\sum_{i=1}^{4} 2^i \right) \left(\sum_{j=1}^{2} 5^j \right) = (2 + 2^2 + 2^3 + 2^4)(5 + 5^2) = (30)(30) = 900.$$

Triple-sigma notation is used when three indices are involved. The sum of all the a_{ijk} where i ranges from 1 to m, j from 1 to n, and k from 1 to q can be written

(18.1.6)
$$\sum_{i=1}^{m} \sum_{j=1}^{n} \sum_{k=1}^{q} a_{ijk}.$$

Multiple sums will appear in the following sections. We introduced them here so as to avoid lengthy asides later.

Exercises

Let

$$P_1 = \{x_0, x_1, \cdots, x_m\} \text{ be a partition of } [a_1, a_2],$$
$$P_2 = \{y_0, y_1, \cdots, y_n\} \text{ be a partition of } [b_1, b_2],$$
$$P_3 = \{z_0, z_1, \cdots, z_q\} \text{ be a partition of } [c_1, c_2].$$

As usual take

$$\Delta x_i = x_i - x_{i-1}, \qquad \Delta y_j = y_j - y_{j-1}, \qquad \Delta z_k = z_k - z_{k-1}.$$

Now evaluate the following sums.

*1. $\displaystyle\sum_{i=1}^{m} \Delta x_i$.

*2. $\displaystyle\sum_{j=1}^{n} \Delta y_j$.

---

† Since b_i is independent of j, it can be factored through the j-summation.

*3. $\displaystyle\sum_{i=1}^{m} \sum_{j=1}^{n} \Delta x_i \Delta y_j.$

*4. $\displaystyle\sum_{j=1}^{n} \sum_{k=1}^{q} \Delta y_j \Delta z_k.$

*5. $\displaystyle\sum_{i=1}^{m} (x_i + x_{i-1}) \Delta x_i$

*6. $\displaystyle\sum_{j=1}^{n} \tfrac{1}{2}(y_j^2 + y_j y_{j-1} + y_{j-1}^2) \Delta y_j.$

*7. $\displaystyle\sum_{i=1}^{m} \sum_{j=1}^{n} (x_i + x_{i-1}) \Delta x_i \Delta y_j.$

*8. $\displaystyle\sum_{i=1}^{m} \sum_{j=1}^{n} (y_j + y_{j-1}) \Delta x_i \Delta y_j.$

*9. $\displaystyle\sum_{i=1}^{m} \sum_{j=1}^{n} (2\Delta x_i - 3\Delta y_j)$

*10. $\displaystyle\sum_{i=1}^{m} \sum_{j=1}^{n} (3\Delta x_i - 2\Delta y_j).$

*11. $\displaystyle\sum_{i=1}^{m} \sum_{j=1}^{n} \sum_{k=1}^{q} \Delta x_i \Delta y_j \Delta z_k.$

*12. $\displaystyle\sum_{i=1}^{m} \sum_{j=1}^{n} \sum_{k=1}^{q} (x_i + x_{i-1}) \Delta x_i \Delta y_j \Delta z_k.$

18.2 The Double Integral Over a Rectangle

We start with a function f continuous on a rectangle

$$R = \{(x, y): a \le x \le b, \quad c \le y \le d\}.†$$

See Figure 18.2.1.

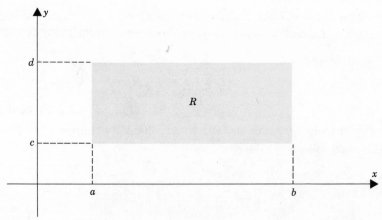

FIGURE 18.2.1

Our object is to define the double integral

$$\iint_R f(x, y)\, dx dy.$$

---

† Until now we've considered continuity for functions of several variables only at interior points of the domain. Continuity at a boundary point $\mathbf{x}_0$ is defined by requiring that for each $\epsilon > 0$ there exist $\delta > 0$ such that $|f(\mathbf{x}) - f(\mathbf{x}_0)| < \epsilon$ for all x *in the domain of f* such that $\|\mathbf{x} - \mathbf{x}_0\| < \delta$. This is in complete analogy with one-sided continuity at the endpoints of a closed interval $[a, b]$ as defined in Section 2.6.

To define

$$\int_a^b f(x)\, dx$$

we first introduced some auxiliary notions: partition P, upper sum $U_f(P)$, and lower sum $L_f(P)$. We were then able to define

$$\int_a^b f(x)\, dx$$

as the unique number I which satisfies the inequality

$$L_f(P) \le I \le U_f(P) \qquad \text{for all partitions } P \text{ of } [a, b].\dagger$$

We will follow exactly the same procedure to define the double integral

$$\int\int_R f(x, y)\, dxdy.$$

First we explain what we mean by a partition of the rectangle R. To do this, we begin with a partition

$$P_1 = \{x_0, x_1, \ldots, x_m\} \quad \text{of} \quad [a, b]$$

and a partition

$$P_2 = \{y_0, y_1, \ldots, y_n\} \quad \text{of} \quad [c, d].$$

A set of the form

$$P = P_1 \times P_2 = \{(x_i, y_j): x_i \in P_1, y_j \in P_2\}\dagger\dagger$$

is called a *partition of* R. The idea is illustrated in Figure 18.2.2. P consists of all the grid points (x_i, y_j).

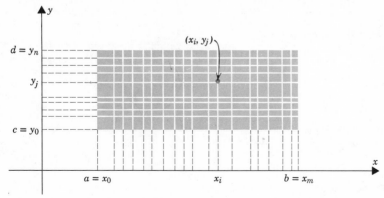

FIGURE 18.2.2

---

† Section 5.2.

†† $P_1 \times P_2$ is a Cartesian product. Cartesian products are discussed briefly in Appendix A.1.

rectangle R_{ij}

FIGURE 18.2.3

The partition P breaks up R into $m \times n$ non-overlapping rectangles

$$R_{ij} = \{(x, y): x_{i-1} \leq x \leq x_i, \quad y_{j-1} \leq y \leq y_j\}. \quad \text{(Figure 18.2.3)}$$

On each rectangle R_{ij}, the function f takes on a maximum value M_{ij} and a minimum value m_{ij}.† The sum of all the products

$$M_{ij}(\text{area of } R_{ij}) = M_{ij}(x_i - x_{i-1})(y_j - y_{j-1}) = M_{ij} \Delta x_i \Delta y_j$$

is called the *P upper sum* for f:

(18.2.1) $$U_f(P) = \sum_{i=1}^{m} \sum_{j=1}^{n} M_{ij} (\text{area of } R_{ij}) = \sum_{i=1}^{m} \sum_{j=1}^{n} M_{ij} \Delta x_i \Delta y_j.$$

The sum of all the products

$$m_{ij}(\text{area of } R_{ij}) = m_{ij}(x_i - x_{i-1})(y_j - y_{j-1}) = m_{ij} \Delta x_i \Delta y_j$$

is called the *P lower sum* for f:

(18.2.2) $$L_f(P) = \sum_{i=1}^{m} \sum_{j=1}^{n} m_{ij}(\text{area of } R_{ij}) = \sum_{i=1}^{m} \sum_{j=1}^{n} m_{ij} \Delta x_i \Delta y_j.$$

Example. Consider the function

$$f(x, y) = x + y - 2$$

on the rectangle

$$R = \{(x, y): 1 \leq x \leq 4, \ 1 \leq y \leq 3\}.$$

---

† It can be shown that every continuous function takes on both a maximum and a minimum on every closed bounded set. Since R_{ij} is closed and bounded, we can be sure that f takes on both a maximum and a minimum there.

As a partition of $[1, 4]$ take

$$P_1 = \{1, 2, 3, 4\},$$

and as a partition of $[1, 3]$ take

$$P_2 = \{1, \tfrac{3}{2}, 3\}.$$

The partition

$$P = P_1 \times P_2$$

then breaks up the initial rectangle into the six rectangles marked in Figure 18.2.4.

FIGURE 18.2.4

On each rectangle R_{ij}, the function f takes on its maximum value M_{ij} at the corner furthest away from the origin, (x_i, y_j); namely,

$$M_{ij} = f(x_i, y_j) = x_i + y_j - 2.$$

$$2 + \tfrac{3}{2} - 2$$

As the upper sum we have

$$U_f(P) = M_{11}(\text{area of } R_{11}) + M_{12}(\text{area of } R_{12}) + M_{21}(\text{area of } R_{21}) + M_{22}(\text{area of } R_{22})$$
$$+ M_{31}(\text{area of } R_{31}) + M_{32}(\text{area of } R_{32})$$
$$= \tfrac{3}{2}(\tfrac{1}{2}) + 3(\tfrac{3}{2}) + \tfrac{5}{2}(\tfrac{1}{2}) + 4(\tfrac{3}{2}) + \tfrac{7}{2}(\tfrac{1}{2}) + 5(\tfrac{3}{2}) = \tfrac{87}{4}.$$

On each rectangle R_{ij}, f takes on its minimum value m_{ij} at the corner closest to the origin, (x_{i-1}, y_{j-1}); namely,

$$m_{ij} = f(x_{i-1}, y_{j-1}) = x_{i-1} + y_{j-1} - 2.$$

The lower sum is given by

$$L_f(P) = m_{11}(\text{area of } R_{11}) + m_{12}(\text{area of } R_{12}) + m_{21}(\text{area of } R_{21}) + m_{22}(\text{area of } R_{22})$$
$$+ m_{31}(\text{area of } R_{31}) + m_{32}(\text{area of } R_{32})$$
$$= 0(\tfrac{1}{2}) + \tfrac{1}{2}(\tfrac{3}{2}) + 1(\tfrac{1}{2}) + \tfrac{3}{2}(\tfrac{3}{2}) + 2(\tfrac{1}{2}) + \tfrac{5}{2}(\tfrac{3}{2}) = \tfrac{33}{4}. \quad \square$$

We return now to the general situation. As in the one-variable case, it can be shown that, if f is continuous, then there exists one and only one number I which satisfies the inequality

$$L_f(P) \le I \le U_f(P) \qquad \text{for all partitions } P \text{ of } R.$$

Definition 18.2.3   The Double Integral over a Rectangle R

The unique number I which satisfies the inequality

$$L_f(P) \le I \le U_f(P) \qquad \text{for all partitions } P \text{ of } R$$

is called the *double integral* of f over R and is denoted by

$$\iint_R f(x, y)\, dxdy.$$

The Double Integral as a Volume

If f is nonnegative on the rectangle R, the equation

$$z = f(x, y)$$

represents a surface that lies over R. The double integral

$$\iint_R f(x, y)\, dxdy$$

then gives the volume of the solid T which is bounded below by R and above by the surface $z = f(x, y)$. See Figure 18.2.5.

$z = f(x, y)$

R

R_{ij}

FIGURE 18.2.5

To see this, consider a partition P of R. P breaks up R into subrectangles R_{ij} and thus the entire solid T into parts T_{ij}. Since T_{ij} contains a rectangular solid with base R_{ij} and height m_{ij} (Figure 18.2.6), we must have

$$m_{ij}(\text{area of } R_{ij}) \leq \text{volume of } T_{ij}.$$

Since T_{ij} is contained in a rectangular solid with base R_{ij} and height M_{ij} (Figure 18.2.7), we must have

$$\text{volume of } T_{ij} \leq M_{ij}(\text{area of } R_{ij}).$$

FIGURE 18.2.6   FIGURE 18.2.7

In short, for each pair of indices i and j, we must have

$$m_{ij}(\text{area of } R_{ij}) \leq \text{volume of } T_{ij} \leq M_{ij}(\text{area of } R_{ij}).$$

Adding up these inequalities, we are forced to conclude that

$$L_f(P) \leq \text{volume of } T \leq U_f(P).$$

Since P is arbitrary, the volume of T must be the double integral:

(18.2.4)
$$\text{volume of } T = \iint_R f(x, y) \, dxdy.$$

The double integral

$$\iint_R 1 \, dxdy = \iint_R dxdy$$

gives the volume of a solid of constant height 1 erected over R. In square units this is the area of R:

(18.2.5)
$$\text{area of } R = \iint_R dxdy.$$

Some Computations

Double integrals are generally computed by techniques that we will take up later. It is possible, however, to evaluate simple double integrals directly from the definition.

Problem. Evaluate

$$\iint_R \alpha \, dxdy$$

where

$$R = \{(x, y): a \leq x \leq b, c \leq y \leq d\}.$$

SOLUTION. Here

$$f(x, y) = \alpha \qquad \text{for all } (x, y) \in R.$$

We begin with $P_1 = \{x_0, x_1, \ldots, x_m\}$ as an arbitrary partition of $[a, b]$ and $P_2 = \{y_0, y_1, \ldots, y_n\}$ as an arbitrary partition of $[c, d]$. This gives

$$P = P_1 \times P_2 = \{(x_i, y_j): x_i \in P_1, y_j \in P_2\}$$

as an arbitrary partition of R. On each rectangle R_{ij}, f has constant value α. This gives $M_{ij} = \alpha$ and $m_{ij} = \alpha$ throughout. Thus

$$U_f(P) = \sum_{i=1}^m \sum_{j=1}^n \alpha \Delta x_i \Delta y_j = \alpha \left(\sum_{i=1}^m \Delta x_i \right) \left(\sum_{j=1}^n \Delta y_j \right) = \alpha(b - a)(d - c).$$

Similarly

$$L_f(P) = \alpha(b - a)(d - c).$$

Since

$$L_f(P) \leq \alpha(b - a)(d - c) \leq U_f(P)$$

and P was chosen arbitrarily, the inequality must hold for all partitions P of R. This means that

$$\iint_R f(x, y) \, dxdy = \alpha(b - a)(d - c). \quad \square$$

REMARK. If $\alpha > 0$,

$$\iint_R \alpha \, dxdy = \alpha(b - a)(d - c)$$

gives the volume of the rectangular solid of constant height α erected over the rectangle R.

Problem. Evaluate

$$\iint_R (x + y) \, dxdy$$

where again R is the rectangle

$$\{(x, y): a \leq x \leq b, c \leq y \leq d\}.$$

SOLUTION. We take $P_1 = \{x_0, x_1, \ldots, x_m\}$ as an arbitrary partition of $[a, b]$ and $P_2 = \{y_0, y_1, \ldots, y_n\}$ as an arbitrary partition of $[c, d]$. This gives

$$P = P_1 \times P_2 = \{(x_i, y_j): x_i \in P_1, y_j \in P_2\}$$

as an arbitrary partition of R. On each rectangle

$$R_{ij} = \{(x, y): x_{i-1} \leq x \leq x_i, \; y_{j-1} \leq y \leq y_j\}$$

the function

$$f(x, y) = x + y$$

has

a maximum $M_{ij} = x_i + y_j$ and   a minimum $m_{ij} = x_{i-1} + y_{j-1}$.

Thus

$$U_f(P) = \sum_{i=1}^{m} \sum_{j=1}^{n} (x_i + y_j)\,\Delta x_i \Delta y_j \quad \text{and} \quad L_f(P) = \sum_{i=1}^{m} \sum_{j=1}^{n} (x_{i-1} + y_{j-1})\,\Delta x_i \Delta y_j.$$

For each pair of indices i and j,

$$x_{i-1} + y_{j-1} \leq \tfrac{1}{2}(x_i + x_{i-1}) + \tfrac{1}{2}(y_j + y_{j-1}) \leq x_i + y_j. \qquad \text{(explain)}$$

Consequently,

$$L_f(P) \leq \sum_{i=1}^{m} \sum_{j=1}^{n} [\tfrac{1}{2}(x_i + x_{i-1}) + \tfrac{1}{2}(y_j + y_{j-1})]\,\Delta x_i \Delta y_j \leq U_f(P).$$

The middle term can be written

$$\sum_{i=1}^{m} \sum_{j=1}^{n} \tfrac{1}{2}(x_i + x_{i-1})\,\Delta x_i \Delta y_j + \sum_{i=1}^{m} \sum_{j=1}^{n} \tfrac{1}{2}(y_j + y_{j-1})\,\Delta x_i \Delta y_j.$$

The first double sum reduces to

$$\sum_{i=1}^{m} \sum_{j=1}^{n} \tfrac{1}{2}(x_i^2 - x_{i-1}^2)\,\Delta y_j = \tfrac{1}{2}\left(\sum_{i=1}^{m} (x_i^2 - x_{i-1}^2)\right)\left(\sum_{j=1}^{n} \Delta y_j\right) = \tfrac{1}{2}(b^2 - a^2)(d - c).$$

$$\underset{(18.1.5)}{\underline{\qquad\qquad}}$$

The second sum reduces to

$$\sum_{i=1}^{m} \sum_{j=1}^{n} \tfrac{1}{2}\Delta x_i(y_j^2 - y_{j-1}^2) = \tfrac{1}{2}\left(\sum_{i=1}^{m} \Delta x_i\right)\left(\sum_{j=1}^{n} (y_j^2 - y_{j-1}^2)\right) = \tfrac{1}{2}(b - a)(d^2 - c^2).$$

The sum of these two numbers,

$$I = \tfrac{1}{2}(b^2 - a^2)(d - c) + \tfrac{1}{2}(b - a)(d^2 - c^2) = \tfrac{1}{2}(a + b + c + d)(b - a)(d - c),$$

satisfies the inequality

$$L_f(P) \leq I \leq U_f(P).$$

Since P is arbitrary, we have

$$I = \tfrac{1}{2}(a + b + c + d)(b - a)(d - c)$$

as the desired integral:

$$\iint_R (x + y) \, dxdy = \tfrac{1}{2}(a + b + c + d)(b - a)(d - c). \quad \square$$

REMARK. If $x + y$ is positive on the rectangle

$$R = \{(x, y): a \leq x \leq b, c \leq y \leq d\},$$

then the double integral

$$\iint_R (x + y) \, dxdy = \tfrac{1}{2}(a + b + c + d)(b - a)(d - c)$$

gives the volume of the prism bounded above by the plane $z = x + y$ and below by R.

Problem. Evaluate

$$\iint_R (y - 2x) \, dxdy$$

where

$$R = \{(x, y): 1 \leq x \leq 2, 3 \leq y \leq 5\}.$$

SOLUTION. With

$$P_1 = \{x_0, x_1, \ldots, x_m\} \text{ as an arbitrary partition of } [1, 2]$$

and

$$P_2 = \{y_0, y_1, \ldots, y_n\} \text{ as an arbitrary partition of } [3, 5]$$

we have

$$P = P_1 \times P_2 = \{(x_i, y_j): x_i \in P_1, y_j \in P_2\}$$

as an arbitrary partition of R. On each rectangle

$$R_{ij} = \{(x, y): x_{i-1} \leq x \leq x_i, \quad y_{j-1} \leq y \leq y_j\}$$

the function

$$f(x, y) = y - 2x$$

has a maximum

$$M_{ij} = y_j - 2x_{i-1} \quad (y \text{ is maximized and } x \text{ is minimized})$$

and a minimum

$$m_{ij} = y_{j-1} - 2x_i. \quad (y \text{ is minimized and } x \text{ is maximized})$$

Thus

$$U_f(P) = \sum_{i=1}^{m}\sum_{j=1}^{n} (y_j - 2x_{i-1})\,\Delta x_i \Delta y_j \quad \text{and} \quad L_f(P) = \sum_{i=1}^{m}\sum_{j=1}^{n} (y_{j-1} - 2x_i)\,\Delta x_i \Delta y_j.$$

For each pair of indices i and j,

$$y_{j-1} - 2x_i \le \tfrac{1}{2}(y_j + y_{j-1}) - (x_i + x_{i-1}) \le y_j - 2x_{i-1}.$$

This means that for arbitrary P we have

$$L_f(P) \le \sum_{i=1}^{m}\sum_{j=1}^{n} [\tfrac{1}{2}(y_j + y_{j-1}) - (x_i + x_{i-1})]\,\Delta x_i \Delta y_j \le U_f(P).$$

The middle term can be written

$$\sum_{i=1}^{m}\sum_{j=1}^{n} \tfrac{1}{2}(y_j + y_{j-1})\,\Delta x_i \Delta y_j + \sum_{i=1}^{m}\sum_{j=1}^{n} - (x_i + x_{i-1})\,\Delta x_i \Delta y_j.$$

The first double sum reduces to

$$\sum_{i=1}^{m}\sum_{j=1}^{n} \tfrac{1}{2}\Delta x_i(y_j^2 - y_{j-1}^2) = \tfrac{1}{2}\left(\sum_{i=1}^{m} \Delta x_i\right)\left(\sum_{j=1}^{n} (y_j^2 - y_{j-1}^2)\right) = \tfrac{1}{2}(2 - 1)(25 - 9) = 8.$$

The second double sum reduces to

$$-\sum_{i=1}^{m}\sum_{j=1}^{n} (x_i^2 - x_{i-1}^2)\,\Delta y_j = -\left(\sum_{i=1}^{m} (x_i^2 - x_{i-1}^2)\right)\left(\sum_{j=1}^{n} \Delta y_j\right) = -(4 - 1)(5 - 3) = -6.$$

The sum of these two numbers

$$I = 2$$

satisfies the inequality

$$L_f(P) \le 2 \le U_f(P) \qquad \text{for arbitrary } P.$$

The integral is therefore 2:

$$\iint\limits_{R} (y - 2x)\,dx\,dy = 2. \quad \square$$

REMARK.   This last integral should not be interpreted as a volume. The expression $y - 2x$ does not keep a constant sign on

$$R = \{(x, y): 1 \le x \le 2,\, 3 \le y \le 5\}. \quad \square$$

Exercises

For Exercises 1–9 take

$$f(x, y) = x + 2y \quad \text{on} \quad R = \{(x, y): 0 \le x \le 2,\, 0 \le y \le 1\}$$

and P as the partition $P = P_1 \times P_2$.

*1. Find $L_f(P)$ if $P_1 = \{0, 1, 2\}$ and $P_2 = \{0, 1\}$.

*2. Find $U_f(P)$ with P_1 and P_2 as in Exercise 1.

3. Find $L_f(P)$ if $P_1 = \{0, 1, 2\}$ and $P_2 = \{0, \frac{1}{2}, 1\}$.

4. Find $U_f(P)$ with P_1 and P_2 as in Exercise 3.

*5. Find $L_f(P)$ if $P_1 = \{0, 1, \frac{3}{2}, 2\}$ and $P_2 = \{0, \frac{1}{2}, 1\}$.

*6. Find $U_f(P)$ with P_1 and P_2 as in Exercise 5.

7. Find $L_f(P)$ if $P_1 = \{x_0, x_1, \ldots, x_m\}$ is an arbitrary partition of $[0, 2]$ and $P_2 = \{y_0, y_1, \ldots, y_n\}$ is an arbitrary partition of $[0, 1]$.

8. Find $U_f(P)$ with P_1 and P_2 as in Exercise 7.

Optional   *9. Using Exercises 7 and 8, evaluate the double integral

$$\iint_R (x + 2y) \, dxdy$$

and give a geometric interpretation to your answer.

For Exercises 10–13 take

$$f(x, y) = x - y \qquad \text{on} \quad R = \{(x, y): 0 \le x \le 1, 0 \le y \le 1\}$$

and P as the partition $P = P_1 \times P_2$.

*10. Find $L_f(P)$ if $P_1 = \{0, \frac{1}{2}, \frac{3}{4}, 1\}$ and $P_2 = \{0, \frac{1}{2}, 1\}$.

*11. Find $U_f(P)$ with P_1 and P_2 as in Exercise 10.

12. Find $L_f(P)$ if $P_1 = \{x_0, x_1, \ldots, x_m\}$ and $P_2 = \{y_0, y_1, \ldots, y_n\}$ are both arbitrary partitions of $[0, 1]$.

13. Find $U_f(P)$ with P_1 and P_2 as in Exercise 12.

Optional   *14. Using Exercises 12 and 13, evaluate the double integral

$$\iint_R (x - y) \, dxdy.$$

*15. Evaluate the double integral

$$\iint_R (x^2 + y) \, dxdy \qquad \text{with} \quad R = \{(x, y): 0 \le x \le 1, 0 \le y \le 2\}$$

by any method at your disposal. HINT: Think in terms of volume. You can use the method of this section and/or the method of parallel cross sections.

18.3 The Double Integral Over More General Regions

We start with a region Ω that we depict in Figure 18.3.1. We suppose that f is continuous on Ω. Our object is to define the double integral

$$\iint_\Omega f(x, y) \, dxdy.$$

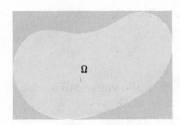

FIGURE 18.3.1   FIGURE 18.3.2

To do this, we surround Ω by a rectangle R in the manner of Figure 18.3.2. We now extend f to all of R by setting f equal to 0 outside of Ω. This extended function f is now continuous on all of R, except possibly at the boundary of Ω. If this boundary is not too wild (the details here are outside the scope of this course), the double integral

$$\iint_R f(x, y)\, dxdy$$

will still exist. We can then define the double integral over Ω by setting

(18.3.1)
$$\boxed{\iint_\Omega f(x, y)\, dxdy = \iint_R f(x, y)\, dxdy.}$$

If f is nonnegative over Ω, the extended f is nonnegative on all of R. The double integral gives the volume of the solid trapped between the surface $z = f(x, y)$ and the rectangle R. But since the surface has height 0 outside of Ω, the volume outside of Ω is 0. It follows therefore that

$$\iint_\Omega f(x, y)\, dxdy$$

gives the volume of the solid T bounded above by $z = f(x, y)$ and below by Ω:

(18.3.2)
$$\boxed{\text{volume of } T = \iint_\Omega f(x, y)\, dxdy.}$$ (Figure 18.3.3)

$z = f(x, y)$

Ω

FIGURE 18.3.3

The double integral

$$\iint\limits_{\Omega} 1 \, dxdy = \iint\limits_{\Omega} dxdy$$

gives the volume of a solid of constant height 1 over Ω. In square units this is the area of Ω:

(18.3.3)
$$\boxed{\text{area of } \Omega = \iint\limits_{\Omega} dxdy.}$$

Some Properties of the Double Integral

Like the ordinary integral, the double integral is linear:

$$\iint\limits_{\Omega} [\alpha f(x, y) + \beta g(x, y)] \, dxdy = \alpha \iint\limits_{\Omega} f(x, y) \, dxdy + \beta \iint\limits_{\Omega} g(x, y) \, dxdy.$$

It preserves order:

if $f(x, y) \geq 0$ for all $(x, y) \in \Omega$, then

$$\iint\limits_{\Omega} f(x, y) \, dxdy \geq 0;$$

if $f(x, y) \leq g(x, y)$ for all $(x, y) \in \Omega$, then

$$\iint\limits_{\Omega} f(x, y) \, dxdy \leq \iint\limits_{\Omega} g(x, y) \, dxdy.$$

It is additive:

if Ω is broken up into two non-overlapping sets Ω_1, Ω_2 over which f is still integrable, then

$$\iint\limits_{\Omega_1} f(x, y) \, dxdy + \iint\limits_{\Omega_2} f(x, y) \, dxdy = \iint\limits_{\Omega} f(x, y) \, dxdy.$$

18.4 The Evaluation of Double Integrals by Repeated Integrals

The Reduction Formulas

Now we discuss the problem of evaluating the double integral

$$\iint\limits_{\Omega} f(x, y) \, dxdy.$$

For simplicity we restrict our attention to sets Ω of the form depicted in Figure 18.4.1. In both cases we assume that the bounding functions ϕ_1 and ϕ_2 are continuous. The fundamental idea is that double integrals over sets of this form can be reduced to "repeated" ordinary integrals.

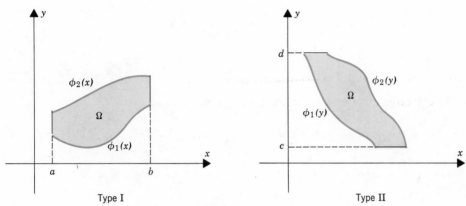

Type I   Type II

FIGURE 18.4.1

If Ω is of type I,

$$\Omega = \{(x, y): a \le x \le b,\ \phi_1(x) \le y \le \phi_2(x)\},$$

then

(18.4.1)

$$\iint_\Omega f(x, y)\, dxdy = \int_a^b \left(\int_{\phi_1(x)}^{\phi_2(x)} f(x, y)\, dy \right) dx.$$

Here we integrate first with respect to y, keeping x fixed; then we integrate with respect to x.

If Ω is of type II,

$$\Omega = \{(x, y): c \le y \le d,\ \phi_1(y) \le x \le \phi_2(y)\},$$

then

(18.4.2)

$$\iint_\Omega f(x, y)\, dxdy = \int_c^d \left(\int_{\phi_1(y)}^{\phi_2(y)} f(x, y)\, dx \right) dy.$$

This time we integrate first with respect to x, keeping y fixed; then we integrate with respect to y.

These formulas are easy to understand geometrically.

The Reduction Formulas Viewed Geometrically

Let's take f as nonnegative and Ω of type I. The double integral over Ω gives the volume of the solid T which is trapped between Ω and the surface $z = f(x, y)$:

$$(1) \qquad\qquad \iint\limits_{\Omega} f(x, y)\, dxdy = \text{volume of } T. \qquad\qquad \text{(Figure 18.4.2)}$$

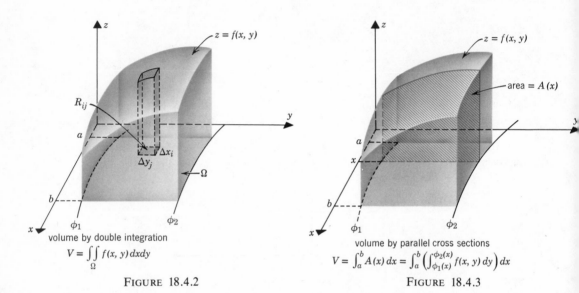

volume by double integration
$$V = \iint\limits_{\Omega} f(x, y)\, dxdy$$

FIGURE 18.4.2

volume by parallel cross sections
$$V = \int_a^b A(x)\, dx = \int_a^b \left(\int_{\phi_1(x)}^{\phi_2(x)} f(x, y)\, dy \right) dx$$

FIGURE 18.4.3

We can also compute the volume of T by the method of parallel cross sections. As in Figure 18.4.3, let $A(x)$ be the area of that cross section of T which has first coordinate x. Then by (10.2.1)

$$\int_a^b A(x)\, dx = \text{volume of } T.$$

Since

$$A(x) = \int_{\phi_1(x)}^{\phi_2(x)} f(x, y)\, dy,$$

we have

$$(2) \qquad\qquad \int_a^b \left(\int_{\phi_1(x)}^{\phi_2(x)} f(x, y)\, dy \right) dx = \text{volume of } T.$$

Combining (1) with (2), we have the first reduction formula

$$\iint\limits_{\Omega} f(x, y)\, dxdy = \int_a^b \left(\int_{\phi_1(x)}^{\phi_2(x)} f(x, y)\, dy \right) dx.$$

The other reduction formula can be obtained in a similar manner.

Note that our argument is a very loose one and certainly not a proof. How do we know, for example, that the "volume" obtained by double integration is the same as the "volume" obtained by the method of parallel cross sections? Intuitively it seems evident, but actually it's quite difficult to prove. We go on directly to some computations.

Computations

Problem. Evaluate

$$\iint_{\Omega} (x^2 - y) \, dx \, dy$$

with Ω as in Figure 18.4.4.

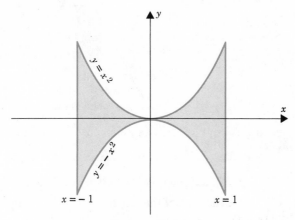

FIGURE 18.4.4

SOLUTION. For each x from -1 to 1, y lies between $-x^2$ and x^2:

$$\Omega = \{(x, y): -1 \le x \le 1, \ -x^2 \le y \le x^2\}.$$

We integrate last between the constant limits:

$$\iint_{\Omega} (x^2 - y) \, dx \, dy = \int_{-1}^{1} \left(\int_{-x^2}^{x^2} (x^2 - y) \, dy \right) dx$$

$$= \int_{-1}^{1} \left[x^2 y - \tfrac{1}{2} y^2 \right]_{-x^2}^{x^2} dx$$

$$= \int_{-1}^{1} [(x^4 - \tfrac{1}{2} x^4) - (-x^4 - \tfrac{1}{2} x^4)] \, dx$$

$$= \int_{-1}^{1} 2x^4 \, dx = \left[\tfrac{2}{5} x^5 \right]_{-1}^{1} = \tfrac{4}{5}. \quad \square$$

Problem. Evaluate

$$\iint_\Omega (xy - y^3)\,dxdy$$

with Ω as in Figure 18.4.5.

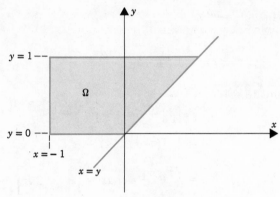

FIGURE 18.4.5

SOLUTION.   For each y from 0 to 1, x lies between -1 and y:

$$\Omega: \{(x, y): 0 \le y \le 1, -1 \le x \le y\}.$$

Again we integrate last between the constant limits:

$$\iint_\Omega (xy - y^3)\,dxdy = \int_0^1 \left(\int_{-1}^y (xy - y^3)\,dx \right) dy$$

$$= \int_0^1 \left[\tfrac{1}{2}x^2 y - xy^3 \right]_{-1}^y dy$$

$$= \int_0^1 (-\tfrac{1}{2}y - \tfrac{1}{2}y^3 - y^4)\,dy$$

$$= \left[-\tfrac{1}{4}y^2 - \tfrac{1}{8}y^4 - \tfrac{1}{5}y^5 \right]_0^1 = -\tfrac{23}{40}. \quad \square$$

Repeated integrals

$$\int_a^b \left(\int_{\phi_1(x)}^{\phi_2(x)} f(x, y)\,dy \right) dx \quad \text{and} \quad \int_c^d \left(\int_{\phi_1(y)}^{\phi_2(y)} f(x, y)\,dx \right) dy$$

can be written in more compact form by omitting the large parentheses. From now on we will simply write

$$\int_a^b \int_{\phi_1(x)}^{\phi_2(x)} f(x, y)\,dy\,dx \quad \text{and} \quad \int_c^d \int_{\phi_1(y)}^{\phi_2(y)} f(x, y)\,dx\,dy.$$

Problem. Evaluate

$$\iint\limits_{R} (y - 2x)\, dxdy$$

where R is the rectangle

$$\{(x, y): 1 \le x \le 2,\ 3 \le y \le 5\}.$$

SOLUTION. Here both x and y range over constant limits and we can integrate in either order. We can integrate first with respect to y:

$$\iint\limits_{R} (y - 2x)\, dxdy = \int_{1}^{2} \int_{3}^{5} (y - 2x)\, dy\, dx$$

$$= \int_{1}^{2} \left[\tfrac{1}{2} y^2 - 2xy \right]_{3}^{5} dx$$

$$= \int_{1}^{2} [(\tfrac{25}{2} - 10x) - (\tfrac{9}{2} - 6x)]\, dx$$

$$= \int_{1}^{2} (8 - 4x)\, dx = \left[8x - 2x^2 \right]_{1}^{2} = 2.$$

We can get the same result by integrating first with respect to x:

$$\iint\limits_{R} (y - 2x)\, dxdy = \int_{3}^{5} \int_{1}^{2} (y - 2x)\, dx\, dy$$

$$= \int_{3}^{5} \left[yx - x^2 \right]_{1}^{2} dy$$

$$= \int_{3}^{5} [(2y - 4) - (y - 1)]\, dy$$

$$= \int_{3}^{5} (y - 3)\, dy = \left[\tfrac{1}{2} y^2 - 3y \right]_{3}^{5} = 2. \quad \square$$

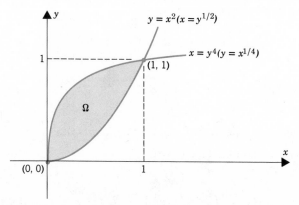

FIGURE 18.4.6

Problem.   Evaluate

$$\iint\limits_{\Omega} (x^{1/2} - y^2) \, dxdy$$

with Ω as in Figure 18.4.6.

SOLUTION.   Here also we have a choice in the order of integration. We can take the point of view that for each x from 0 to 1, y lies between x^2 and $x^{1/4}$. This gives

$$\iint\limits_{\Omega} (x^{1/2} - y^2) \, dxdy = \int_0^1 \int_{x^2}^{x^{1/4}} (x^{1/2} - y^2) \, dy \, dx$$

$$= \int_0^1 \left[x^{1/2}y - \tfrac{1}{3}y^3 \right]_{x^2}^{x^{1/4}} dx$$

$$= \int_0^1 (\tfrac{2}{3}x^{3/4} - x^{5/2} + \tfrac{1}{3}x^6) \, dx$$

$$= \left[\tfrac{8}{21}x^{7/4} - \tfrac{2}{7}x^{7/2} + \tfrac{1}{21}x^7 \right]_0^1 = \tfrac{8}{21} - \tfrac{2}{7} + \tfrac{1}{21} = \tfrac{1}{7}.$$

On the other hand, we can take the point of view that for each y from 0 to 1, x lies between y^4 and $y^{1/2}$. This gives

$$\iint\limits_{\Omega} (x^{1/2} - y^2) \, dxdy = \int_0^1 \int_{y^4}^{y^{1/2}} (x^{1/2} - y^2) \, dx \, dy$$

$$= \int_0^1 \left[\tfrac{2}{3}x^{3/2} - y^2 x \right]_{y^4}^{y^{1/2}} dy$$

$$= \int_0^1 (\tfrac{2}{3}y^{3/4} - y^{5/2} + \tfrac{1}{3}y^6) \, dy$$

$$= \left[\tfrac{8}{21}y^{7/4} - \tfrac{2}{7}y^{7/2} + \tfrac{1}{21}y^7 \right]_0^1 = \tfrac{8}{21} - \tfrac{2}{7} + \tfrac{1}{21} = \tfrac{1}{7}. \quad \Box$$

Problem.   Calculate by double integration the area of the region Ω which lies between $\sqrt{x} + \sqrt{y} = \sqrt{a}$ and $x + y = a$.

SOLUTION.   The region Ω is pictured in Figure 18.4.7. Its area is given by the double integral

$$\iint\limits_{\Omega} dxdy.$$

Here again we can integrate in either order. We can write the boundaries as functions of x,

$$y = (\sqrt{a} - \sqrt{x})^2 \quad \text{and} \quad y = a - x,$$

and note that, for each x from 0 to a, y lies between $(\sqrt{a} - \sqrt{x})^2$ and $a - x$. This gives

$$\iint\limits_{\Omega} dxdy = \int_0^a \int_{(\sqrt{a}-\sqrt{x})^2}^{a-x} dy \, dx$$

$$= \int_0^a [(a-x) - (\sqrt{a} - \sqrt{x})^2] \, dx$$

$$= \int_0^a (-2x + 2\sqrt{a} \, \sqrt{x}) \, dx$$

$$= \left[-x^2 + \tfrac{4}{3}\sqrt{ax} \, \sqrt{x}\right]_0^a = -a^2 + \tfrac{4}{3}a^2 = \tfrac{1}{3}a^2.$$

We can also take the point of view that, for each y from 0 to a, x is between $(\sqrt{a} - \sqrt{y})^2$ and $a - y$. This gives

$$\iint\limits_{\Omega} dxdy = \int_0^a \int_{(\sqrt{a}-\sqrt{y})^2}^{a-y} dx \, dy,$$

which is also $\tfrac{1}{3}a^2$. $\square$

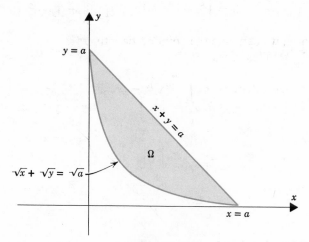

FIGURE 18.4.7

Problem. Calculate the volume within the cylinder $x^2 + y^2 = b^2$ between the planes $y + z = a^2$ and $z = 0$.

SOLUTION. See Figure 18.4.8. The solid in question is bounded below by the disc

$$\Omega = \{(x, y): x^2 + y^2 \le b^2\}$$

and above by the plane

$$z = a^2 - y.$$

The volume is given by the double integral

$$\iint\limits_{\Omega} (a^2 - y) \, dxdy.$$

FIGURE 18.4.8

Once again we have a choice in the order of integration. Taking the point of view that

$$\Omega = \{(x, y): -b \le x \le b, \ -\sqrt{b^2 - x^2} \le y \le \sqrt{b^2 - x^2}\},$$

we have

$$V = \iint_{\Omega} (a^2 - y) \, dxdy = \int_{-b}^{b} \int_{-\sqrt{b^2-x^2}}^{\sqrt{b^2-x^2}} (a^2 - y) \, dy \, dx$$

$$= \int_{-b}^{b} \left[a^2 y - \tfrac{1}{2} y^2 \right]_{-\sqrt{b^2-x^2}}^{\sqrt{b^2-x^2}} dx$$

$$= \int_{-b}^{b} 2a^2 \sqrt{b^2 - x^2} \, dx = \pi a^2 b^2. \quad \square$$

FIGURE 18.4.9

When two orders of integration are possible, one order may be easy to carry out, while the other may present serious difficulties. Thus, for example, with Ω as in Figure 18.4.9,

$$\iint_{\Omega} \cos \left(\tfrac{1}{2} \pi x^2 \right) dxdy$$

can be written

$$\int_0^1 \int_0^x \cos\left(\tfrac{1}{2}\pi x^2\right) dy\, dx \qquad \text{or} \qquad \int_0^1 \int_y^1 \cos\left(\tfrac{1}{2}\pi x^2\right) dx\, dy.$$

The first expression is easy to calculate, but the second is not. The only practical way to handle the second expression is to expand $\cos\left(\tfrac{1}{2}\pi x^2\right)$ as a series in x and then integrate term by term.

Exercises

Evaluate.

*1. $\displaystyle\iint_\Omega x^2\, dxdy$ with $\Omega = \{(x, y): -1 \le x \le 1,\ 0 \le y \le 3\}$.

2. $\displaystyle\iint_\Omega e^{x+y}\, dxdy$ with Ω as in Exercise 1.

*3. $\displaystyle\iint_\Omega \cos \tfrac{1}{2}\pi x \cos \tfrac{1}{2}\pi y\, dxdy$ with Ω as in Exercise 1.

4. $\displaystyle\iint_\Omega x^3 y\, dxdy$ with $\Omega = \{(x, y): 0 \le x \le 1,\ 0 \le y \le x\}$.

*5. $\displaystyle\iint_\Omega (y + \sin \pi x^2)\, dxdy$ with Ω as in Exercise 4.

6. $\displaystyle\iint_\Omega \cos (x + y)\, dxdy$ with $\Omega = \{(x, y): 0 \le x \le \tfrac{1}{2}\pi,\ 0 \le y \le \tfrac{1}{2}\pi\}$.

*7. $\displaystyle\iint_\Omega \frac{x}{1 + xy}\, dxdy$ with $\Omega = \{(x, y): 0 \le x \le 1,\ 0 \le y \le 1\}$.

8. $\displaystyle\iint_\Omega (x^2 + 3y^2)\, dxdy$ with $\Omega = \{(x, y): x^2 + y^2 \le 1\}$.

*9. $\displaystyle\iint_\Omega \sqrt{xy}\, dxdy$ with $\Omega = \{(x, y): 0 \le y \le 1,\ y^2 \le x \le y\}$.

10. $\displaystyle\iint_\Omega ye^x\, dxdy$ with $\Omega = \{(x, y): 0 \le y \le 1,\ 0 \le x \le y^2\}$.

*11. $\displaystyle\iint_\Omega (4 - y^2)\, dxdy$ with Ω the region between $y^2 = 2x$ and $y^2 = 8 - 2x$.

12. $\iint\limits_{\Omega} e^{x^2}\, dxdy$ with Ω the triangle formed by $2y = x$, $x = 2$, and the x-axis.

*13. $\iint\limits_{\Omega} e^{-y^2/2}\, dxdy$ with Ω the triangle formed by $2y = x$, $y = 1$, and the y-axis.

Calculate by double integration the area of the bounded region determined by the following pairs of curves.

14. $x^2 = 4y$, $2y - x - 4 = 0$.   *15. $y = x$, $x = 4y - y^2$.
16. $y = x$, $4y^3 = x^2$.   *17. $x + y = 5$, $xy = 6$.

18. Find the volume of the solid bounded above by $z = x + y$ and below by the triangle with vertices $(0, 0)$, $(0, 1)$, $(1, 0)$.
*19. Find the volume of the solid bounded by $\frac{1}{2}x + \frac{1}{3}y + \frac{1}{4}z = 1$ and the coordinate planes.
20. Find the volume of the solid bounded above by the plane $z = 2x + 3y$ and below by the unit square $\{(x, y): 0 \le x \le 1, 0 \le y \le 1\}$.
*21. Find the volume of the solid bounded above by $z = x^3 y$ and below by the triangle with vertices $(0, 0)$, $(2, 0)$, $(0, 1)$.
22. Find the volume of the solid bounded above by $z = 2x + 1$ and below by $\{(x, y): (x - 1)^2 + y^2 \le 1\}$.
*23. Find the volume of the solid which lies between the paraboloid $z = x^2 + y^2$ and the unit disc $\{(x, y): x^2 + y^2 \le 1\}$.
24. Find the volume of the solid bounded above by $z = 4 - y^2 - \frac{1}{4}x^2$ and below by the disc $\{(x, y): (y - 1)^2 + x^2 \le 1\}$.
25. Show that, if $R = \{(x, y): a \le x \le b, c \le y \le d\}$, then

$$\iint\limits_{R} f(x)g(y)\, dxdy = \left[\int_a^b f(x)\, dx\right] \cdot \left[\int_c^d g(y)\, dy\right].$$

Optional | 26. Let Ω be the triangle with vertices $(0, 0)$, $(0, 1)$, $(1, 1)$. Show that

$$\text{if}\quad \int_0^1 f(x)\, dx = 0 \qquad \text{then}\quad \iint\limits_{\Omega} f(x)f(y)\, dxdy = 0.$$

18.5   Double Integrals in Polar Coordinates

In this section we will integrate over regions given in polar coordinates (r, θ). For simplicity we will take $r \ge 0$ and $0 \le \theta \le 2\pi$. We start with a region R given in polar coordinates by a pair of inequalities

$$r^* \le r \le r^{**}, \qquad \theta^* \le \theta \le \theta^{**}. \qquad\qquad \text{(Figure 18.5.1)}$$

Such a set will be called a "polar rectangle." (It does not look like a rectangle, but analytically it has the structure of a rectangle.)

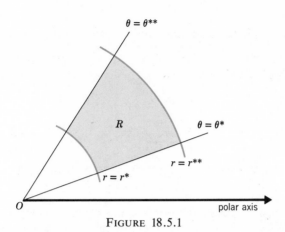

FIGURE 18.5.1

Over R we place a surface $z = f(r, \theta)$. (Figure 18.5.2)

$$V = \iint_R f(r, \theta)\, r\, dr d\theta$$

FIGURE 18.5.2

The solid which lies between R and the surface $z = f(r, \theta)$ has volume

(18.5.1)
$$\boxed{V = \iint_R f(r, \theta)\, r\, dr d\theta.}$$
(note the extra r present)

To show how this comes about, we begin by partitioning the intervals $[r^*, r^{**}]$ and $[\theta^*, \theta^{**}]$:

$$r^* = r_0 < \cdots < r_{i-1} < r_i < \cdots < r_m = r^{**},$$
$$\theta^* = \theta_0 < \cdots < \theta_{j-1} < \theta_j < \cdots < \theta_n = \theta^{**}.$$

These partitions break up R into $m \times n$ "polar rectangles"

$$R_{ij}: \ r_{i-1} \le r \le r_i, \ \theta_{j-1} \le \theta \le \theta_j. \qquad \text{(Figure 18.5.3)}$$

FIGURE 18.5.3

On each such "polar rectangle" f has a maximum value M_{ij} and a minimum value m_{ij}. (We are assuming as usual that f is continuous.) The volume V that we seek is the only number that can be counted upon to satisfy the inequality

$$\sum_{i=1}^{m} \sum_{j=1}^{n} m_{ij}(\text{area of } R_{ij}) \le V \le \sum_{i=1}^{m} \sum_{j=1}^{n} M_{ij}(\text{area of } R_{ij}). \qquad \text{(explain)}$$

We now argue that the number

$$V = \iint_{R} f(r, \theta) \, r \, dr d\theta$$

has this property. In the first place, it can be shown that the integral is additive over the R_{ij} and therefore

$$\iint_{R} f(r, \theta) \, r \, dr d\theta = \sum_{i=1}^{m} \sum_{j=1}^{n} \iint_{R_{ij}} f(r, \theta) \, r \, dr d\theta.$$

All we have to show then is that

$$m_{ij}(\text{area of } R_{ij}) \le \iint_{R_{ij}} f(r, \theta) \, r \, dr d\theta \le M_{ij}(\text{area of } R_{ij}).$$

Observe now that

$$\iint_{R_{ij}} r \, dr d\theta = \int_{\theta_{j-1}}^{\theta_j} \int_{r_{i-1}}^{r_i} r \, dr d\theta = \tfrac{1}{2}(r_i^2 - r_{i-1}^2)(\theta_j - \theta_{j-1}) = \text{area of } R_{ij}.$$

from last section   from Section 11.4

From this it follows that

$$\iint_{R_{ij}} f(r, \theta) \, r \, dr d\theta \le \iint_{R_{ij}} M_{ij} r \, dr d\theta = M_{ij} \iint_{R_{ij}} r \, dr d\theta = M_{ij}(\text{area of } R_{ij})$$

and

$$\iint_{R_{ij}} f(r, \theta) \, r \, dr d\theta \ge \iint_{R_{ij}} m_{ij} r \, dr d\theta = m_{ij} \iint_{R_{ij}} r \, dr d\theta = m_{ij}(\text{area of } R_{ij}). \qquad \square$$

The volume formula that we just verified for "polar rectangles" holds for more general regions. In particular, it holds for regions Ω of the type depicted in Figure 18.5.4.

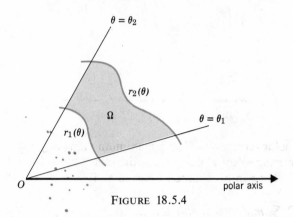

FIGURE 18.5.4

For such Ω we have

(18.5.2)
$$V = \iint_{\Omega} f(r, \theta) \, r \, drd\theta = \int_{\theta_1}^{\theta_2} \int_{r_1(\theta)}^{r_2(\theta)} f(r, \theta) \, r \, dr \, d\theta.$$

If $f(r, \theta) = 1$ throughout, the double integral takes the form

$$\iint_{\Omega} r \, drd\theta.$$

In cubic units, it gives the volume of a solid with base Ω and constant height 1. In square units, it gives the area of Ω:

(18.5.3)
$$\text{area of } \Omega = \iint_{\Omega} r \, drd\theta$$

Problem.   Calculate the volume of a sphere by the methods of this section.

SOLUTION.   We take the sphere to have radius R and center at the origin. The upper spherical surface is then given by the equation

$$z = \sqrt{R^2 - r^2}. \qquad \text{(check this out)}$$

As a base region we can take

$$\Omega: 0 \le r \le R, 0 \le \theta \le 2\pi.$$

Consequently,

$$V = 2 \iint\limits_{\Omega} \sqrt{R^2 - r^2}\, r\, dr d\theta$$

$$= \int_0^{2\pi} \int_0^R \sqrt{R^2 - r^2}\, (2r)\, dr\, d\theta$$

$$= \int_0^{2\pi} \left[-\tfrac{2}{3}(R^2 - r^2)^{3/2} \right]_0^R d\theta = \int_0^{2\pi} \tfrac{2}{3}R^3\, d\theta = \tfrac{4}{3}\pi R^3. \quad \square$$

Problem.   Calculate by double integration the area of the region Ω enclosed by the cardioid

$$r = 1 - \cos \theta.$$

SOLUTION.   We can express Ω by the inequalities

$$0 \le r \le 1 - \cos \theta, \quad 0 \le \theta \le 2\pi.$$

Consequently,

$$\text{area of } \Omega = \iint\limits_{\Omega} r\, dr d\theta$$

$$= \int_0^{2\pi} \int_0^{1 - \cos \theta} r\, dr\, d\theta$$

$$= \int_0^{2\pi} \left[\tfrac{1}{2} r^2 \right]_0^{1 - \cos \theta} d\theta$$

$$= \int_0^{2\pi} \tfrac{1}{2}(1 - \cos \theta)^2\, d\theta$$

$$= \tfrac{1}{2} \int_0^{2\pi} (1 - 2 \cos \theta + \cos^2 \theta)\, d\theta.$$

Since

$$\tfrac{1}{2} \int_0^{2\pi} (1 - 2 \cos \theta)\, d\theta = \pi$$

and

$$\tfrac{1}{2} \int_0^{2\pi} \cos^2 \theta\, d\theta = \tfrac{1}{4} \int_0^{2\pi} (1 + \cos 2\theta)\, d\theta = \tfrac{1}{2}\pi,$$

we have

$$\text{area of } \Omega = \tfrac{3}{2}\pi. \quad \square$$

Problem.   Calculate the volume of the solid bounded above by $z = 2 - r$ and below by the plane region

$$\Omega: 0 \le r \le 2 \cos \theta, \ -\tfrac{1}{2}\pi \le \theta \le \tfrac{1}{2}\pi.$$

SOLUTION

$$V = \iint\limits_{\Omega} (2 - r)r \, drd\theta$$

$$= \int_{-\pi/2}^{\pi/2} \int_0^{2 \cos \theta} (2r - r^2) \, dr \, d\theta$$

$$= \int_{-\pi/2}^{\pi/2} \left[r^2 - \tfrac{1}{3}r^3 \right]_0^{2 \cos \theta} d\theta$$

$$= \int_{-\pi/2}^{\pi/2} (4 \cos^2 \theta - \tfrac{8}{3} \cos^3 \theta) \, d\theta$$

$$= 2 \int_0^{\pi/2} [2(1 + \cos 2\theta) - \tfrac{8}{3}(1 - \sin^2 \theta) \cos \theta] \, d\theta$$

$$= 2 \left[2\theta + \sin 2\theta - \tfrac{8}{3} \sin \theta + \tfrac{8}{9} \sin^3 \theta \right]_0^{\pi/2} = 2(\pi - \tfrac{8}{3} + \tfrac{8}{9}) = \tfrac{2}{9}(9\pi - 16). \quad \square$$

Change of Variables Formula

Some double integrals

$$\iint\limits_{\Omega} g(x, y) \, dxdy$$

are easier to evaluate when expressed in polar coordinates. The change to polar coordinates can be effected as follows:

(18.5.4)
$$\iint\limits_{\Omega} g(x, y) \, dxdy = \iint\limits_{\Omega} g(r \cos \theta, r \sin \theta) \, r \, drd\theta.$$

We will verify this formula under the assumption that g is nonnegative.
VERIFICATION. The surface $z = g(x, y)$ is given in polar coordinates by

$$z = g(r \cos \theta, r \sin \theta) = f(r, \theta).$$

The double integral

$$\iint\limits_{\Omega} g(x, y) \, dxdy$$

gives the volume of the solid trapped between Ω and the given surface. By (18.5.2), the double integral

$$\iint\limits_{\Omega} g(r \cos \theta, r \sin \theta) \, r \, drd\theta = \iint\limits_{\Omega} f(r, \theta) \, r \, drd\theta$$

gives the volume of this same solid. $\square$

The change from rectangular coordinates to polar coordinates is particularly helpful when the original integrand $g(x, y)$ is expressed as a function of $r = \sqrt{x^2 + y^2}$. For then in polar coordinates the integrand becomes a function of r alone.

Problem.   Evaluate the double integral

$$\iint_\Omega e^{-(x^2+y^2)} \, dxdy$$

where Ω is the unit disc

$$\{(x, y)\colon x^2 + y^2 \leq 1\}.$$

SOLUTION.   If you try to compute this integral by the usual methods, you'll run into integrals that cannot be evaluated in closed form. We can sidestep these difficulties by switching to polar coordinates. In polar coordinates we have

$$\Omega\colon 0 \leq r \leq 1, \ 0 \leq \theta \leq 2\pi.$$

The change of variables formula gives

$$\iint_\Omega e^{-(x^2+y^2)} \, dxdy = \iint_\Omega e^{-r^2} r \, drd\theta$$

$$= \int_0^{2\pi} \int_0^1 e^{-r^2} r \, dr \, d\theta$$

$$= \int_0^{2\pi} \left[-\tfrac{1}{2} e^{-r^2} \right]_0^1 d\theta = \int_0^{2\pi} \tfrac{1}{2} \left(1 - \frac{1}{e} \right) d\theta = \pi \left(1 - \frac{1}{e} \right). \quad \square$$

Problem.   Find the volume of the solid bounded below by the xy-plane, above by the ellipsoid of revolution $b^2x^2 + b^2y^2 + a^2z^2 = a^2b^2$, and on the sides by the cylinder $x^2 + y^2 - ay = 0$.

SOLUTION.   Setting

$$x = r \cos \theta, \quad y = r \sin \theta$$

we can rewrite the upper boundary as

$$z = \frac{b}{a} \sqrt{a^2 - r^2}.$$

The boundary of the base region becomes

$$r = a \sin \theta. \qquad\qquad \text{(Figure 18.5.5)}$$

The base region itself can be written

$$\Omega\colon 0 \leq r \leq a \sin \theta, \ 0 \leq \theta \leq \pi.$$

The volume can now be computed as follows:

$$V = \iint\limits_{\Omega} \frac{b}{a}\sqrt{a^2 - r^2}\, r\, drd\theta = \frac{b}{a}\int_0^{\pi}\int_0^{a\sin\theta} r\sqrt{a^2 - r^2}\, dr\, d\theta = \tfrac{2}{9}a^2b(3\pi - 4). \quad \square$$

FIGURE 18.5.5

Problem. Evaluate

$$I = \iint\limits_{\Omega} \frac{1}{(1 + x^2 + y^2)^{3/2}}\, dxdy$$

where Ω is the triangle of Figure 18.5.6.

FIGURE 18.5.6   FIGURE 18.5.7

SOLUTION. By the change of variables formula,

$$\iint\limits_{\Omega} \frac{1}{(1 + x^2 + y^2)^{3/2}}\, dxdy = \iint\limits_{\Omega} \frac{r}{(1 + r^2)^{3/2}}\, drd\theta.$$

In terms of polar coordinates, Ω is given by

$$0 \le \theta \le \tfrac{1}{4}\pi, \quad 0 \le r \le \sec\theta. \qquad \text{(Figure 18.5.7)}$$

Therefore

$$I = \int_0^{\pi/4} \int_0^{\sec\theta} \frac{r}{(1 + r^2)^{3/2}} \, dr \, d\theta$$

$$= - \int_0^{\pi/4} \left[(1 + r^2)^{-1/2} \right]_0^{\sec\theta} d\theta$$

$$= \int_0^{\pi/4} [1 - (1 + \sec^2\theta)^{-1/2}] \, d\theta$$

$$= \int_0^{\pi/4} \left(1 - \frac{\cos\theta}{\sqrt{2 - \sin^2\theta}} \right) d\theta$$

$$= \left[\theta - \arcsin\left(\tfrac{1}{2}\sqrt{2} \sin\theta\right) \right]_0^{\pi/4} = \tfrac{1}{4}\pi - \tfrac{1}{6}\pi = \tfrac{1}{12}\pi. \quad \square$$

Exercises

*1. Find the area of the region which lies inside the circle $r = \frac{3}{2}$ but to the right of the line $4r \cos\theta = 3$.

2. Find the area of the region which lies inside the circle $r = 3\cos\theta$ but outside the circle $r = \frac{3}{2}$.

*3. Find the area of the region which lies inside the circle $r = 3\cos\theta$ but outside the circle $r = \cos\theta$.

4. Find the area of the region which lies inside the cardioid $r = 1 + \cos\theta$ but to the right of the line $4r \cos\theta = 3$.

*5. Find the area of the region which lies inside the cardioid $r = 1 + \cos\theta$ but outside the circle $r = 1$.

6. Find the area of the region which lies inside the circle $r = 1$ but outside the cardioid $r = 1 + \cos\theta$.

*7. Find the area of the region which lies inside the circle $r = 1$ but outside the parabola $r(1 + \cos\theta) = 1$.

8. Integrate $g(x, y) = \cos(x^2 + y^2)$ over the unit disc.

*9. Integrate $g(x, y) = \sin(\sqrt{x^2 + y^2})$ over the unit disc.

10. Find the volume of the solid bounded below by the xy-plane and above by the paraboloid $z = 1 - (x^2 + y^2)$.

*11. Find the volume of the solid bounded above by the plane $z = y + b$, below by the xy-plane, and on the sides by the circular cylinder $x^2 + y^2 = b^2$.

12. Find the volume of the solid bounded above by $z = 1 - (x^2 + y^2)$, below by the xy-plane, and on the sides by the cylinder $x^2 + y^2 - x = 0$.

*13. Find the volume of the solid bounded below by the xy-plane, above by the spherical surface $x^2 + y^2 + z^2 = 4$, and on the sides by the cylinder $x^2 + y^2 = 1$.

14. Find the volume of the solid bounded above by the cone $z^2 = x^2 + y^2$ and below by the region Ω which lies inside the curve $x^2 + y^2 = 2ax$.

18.6 Triple Integrals

Now that you are familiar with double integrals

$$\iint_{\Omega} f(x, y) \, dxdy,$$

you will find it easy to understand triple integrals

$$\iiint_{T} f(x, y, z) \, dxdydz.$$

Basically the only difference is this: instead of dealing with functions of two variables continuous on a plane region Ω, we will be dealing with functions of three variables continuous on some portion T of space.

Instead of beginning with a rectangle

$$R = \{(x, y): a_1 \le x \le a_2, \ b_1 \le y \le b_2\},$$

we begin with a *box* (a rectangular solid)

$$\Pi = \{(x, y, z): a_1 \le x \le a_2, b_1 \le y \le b_2, c_1 \le z \le c_2\}.$$

FIGURE 18.6.1

To partition this box, we first partition the edges. Taking

$$P_1 = \{x_0, \ \ldots \ , x_m\} \quad \text{as a partition of } [a_1, a_2],$$
$$P_2 = \{y_0, \ \ldots \ , y_n\} \quad \text{as a partition of } [b_1, b_2],$$
$$P_3 = \{z_0, \ \ldots \ , z_q\} \quad \text{as a partition of } [c_1, c_2],$$

we form the set

$$P = P_1 \times P_2 \times P_3 = \{(x_i, y_j, z_k): x_i \in P_1, \quad y_j \in P_2, \quad z_k \in P_3\}\dagger$$

and call this a *partition of* Π. P breaks up Π into $m \times n \times q$ nonoverlapping boxes

---

$\dagger$ $P_1 \times P_2 \times P_3$ is a Cartesian product. Cartesian products are discussed briefly in Appendix A.1.

$$\Pi_{ijk} = \{(x, y, z): x_{i-1} \le x \le x_i,\ y_{j-1} \le y \le y_j,\ z_{k-1} \le z \le z_k\}.$$

A typical such box is pictured in Figure 18.6.2.

FIGURE 18.6.2

Taking

$$M_{ijk} \text{ as the maximum value of } f \text{ on } \Pi_{ijk}$$

and

$$m_{ijk} \text{ as the minimum value of } f \text{ on } \Pi_{ijk},$$

we form the *upper sum*

$$U_f(P) = \sum_{i=1}^{m} \sum_{j=1}^{n} \sum_{k=1}^{q} M_{ijk}(\text{volume of } \Pi_{ijk}) = \sum_{i=1}^{m} \sum_{j=1}^{n} \sum_{k=1}^{q} M_{ijk}\, \Delta x_i\, \Delta y_j\, \Delta z_k$$

and the *lower sum*

$$L_f(P) = \sum_{i=1}^{m} \sum_{j=1}^{n} \sum_{k=1}^{q} m_{ijk}(\text{volume of } \Pi_{ijk}) = \sum_{i=1}^{m} \sum_{j=1}^{n} \sum_{k=1}^{q} m_{ijk}\, \Delta x_i\, \Delta y_j\, \Delta z_k.$$

As in the case of functions of one and two variables, it turns out that, with f continuous on Π, there is one and only one number I which satisfies the inequality

$$L_f(P) \le I \le U_f(P) \qquad \text{for all partitions } P \text{ of } \Pi.$$

Definition 18.6.1   The Triple Integral Over a Box Π

The unique number I which satisfies the inequality

$$L_f(P) \le I \le U_f(P) \qquad \text{for all partitions } P \text{ of } \Pi$$

is called the *triple integral* of f over Π and is denoted by

$$\iiint\limits_{\Pi} f(x, y, z)\, dx\, dy\, dz.$$

To obtain the integral over a more general solid T we first encase T in a rectangular box Π. We then extend f to all of Π by defining f to be 0 outside of T.[†] This extended function f is now continuous on all of Π except possibly at the boundary of T. If this boundary is not too wild (the details here are outside the scope of this course), the triple integral

$$\iiint_\Pi f(x, y, z)\, dxdydz$$

will still exist. The triple integral over T is then defined as the triple integral over Π:

(18.6.2)
$$\boxed{\iiint_T f(x, y, z)\, dxdydz = \iiint_\Pi f(x, y, z)\, dxdydz.}$$

Interpretation of the Triple Integral as a Mass

The density of a solid T is by definition its mass per unit volume. If the density ρ is constant throughout T, then we can get the total mass of T simply by multiplying the density by the volume V:

(18.6.3)
$$\boxed{M = \rho V.}$$

Suppose now that the density varies continuously from point to point; say

$$\rho = f(x, y, z).$$

What is the total mass of T then? The answer is simple:

(18.6.4)
$$\boxed{\text{mass of } T = \iiint_T f(x, y, z)\, dxdydz.}$$

To justify this assertion, we first encase T in a box Π assigning density 0 to points outside of T. The mass of T will then be the mass of Π. We now take P as a partition of Π and use it to break up Π into little sub-boxes Π_{ijk} in the manner previously discussed. For each little box

$$\left(\begin{array}{c}\text{min density} \\ \text{on } \Pi_{ijk}\end{array}\right) \cdot (\text{volume of } \Pi_{ijk}) \leq \text{mass of } \Pi_{ijk} \leq \left(\begin{array}{c}\text{max density} \\ \text{on } \Pi_{ijk}\end{array}\right) \cdot (\text{volume of } \Pi_{ijk}).$$

min mass *max mass*

In our previous notation,

$$m_{ijk}(\text{volume of } \Pi_{ijk}) \leq \text{mass of } \Pi_{ijk} \leq M_{ijk}(\text{volume of } \Pi_{ijk}).$$

Adding up these inequalities, we see that the total mass of Π must satisfy the condition

$$L_f(P) \leq \text{mass of } \Pi \leq U_f(P).$$

[†] Does all this sound familiar?

Since P was chosen arbitrarily, this inequality must hold for all partitions P; consequently, we must have

$$\text{mass of } \Pi = \iiint\limits_{\Pi} f(x, y, z)\, dxdydz.$$

Since the density outside of T was defined to be 0, we have

$$\iiint\limits_{T} f(x, y, z)\, dxdydz = \iiint\limits_{\Pi} f(x, y, z)\, dxdydz,$$

and, since mass of T = mass of Π, we have

$$\text{mass of } T = \iiint\limits_{T} f(x, y, z)\, dxdydz. \quad \Box$$

The triple integral

$$\iiint\limits_{T} 1\, dxdydz = \iiint\limits_{T} dxdydz$$

gives the mass of T when T has constant density 1. In cubic units this is the volume of T:

(18.6.5)
$$\boxed{\text{volume of } T = \iiint\limits_{T} dxdydz.}$$

18.7 Reduction to Repeated Integrals

As with double integrals, we usually evaluate triple integrals by first reducing them to repeated integrals. If T is given by inequalities of the form

$$a_1 \leq x \leq a_2, \quad \phi_1(x) \leq y \leq \phi_2(x), \quad \psi_1(x, y) \leq z \leq \psi_2(x, y)$$

(see Figure 18.7.1), then

(18.7.1)
$$\boxed{\iiint\limits_{T} f(x, y, z)\, dxdydz = \int_{a_1}^{a_2} \left[\int_{\phi_1(x)}^{\phi_2(x)} \left(\int_{\psi_1(x,y)}^{\psi_2(x,y)} f(x, y, z)\, dz \right) dy \right] dx,}$$

which, by omitting the parentheses and brackets, can be written

(18.7.2)
$$\boxed{\int_{a_1}^{a_2} \int_{\phi_1(x)}^{\phi_2(x)} \int_{\psi_1(x,y)}^{\psi_2(x,y)} f(x, y, z)\, dz\, dy\, dx.}†$$

---

† In this section we give no proofs. All the functions in question, f, ϕ_1, ϕ_2, ψ_1, ψ_2 are assumed to be continuous.

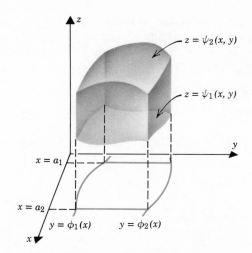

$$a_1 \le x \le a_2, \quad \phi_1(x) \le y \le \phi_2(x), \quad \psi_1(x, y) \le z \le \psi_2(x, y)$$

FIGURE 18.7.1

Here we integrate first with respect to z, then with respect to y, and then with respect to x. Other orders of integration are possible. If, for instance, S is of the form

$$b_1 \le y \le b_2, \quad \phi_1(y) \le z \le \phi_2(y), \quad \psi_1(y, z) \le x \le \psi_2(y, z),$$

then the triple integral over T leads to the repeated integral

$$\int_{b_1}^{b_2} \left[\int_{\phi_1(y)}^{\phi_2(y)} \left(\int_{\psi_1(y,z)}^{\psi_2(y,z)} f(x, y, z) \, dx \right) dz \right] dy = \int_{b_1}^{b_2} \int_{\phi_1(y)}^{\phi_2(y)} \int_{\psi_1(y,z)}^{\psi_2(y,z)} f(x, y, z) \, dx \, dz \, dy.$$

In this case we integrate first with respect to x, then with respect to z, and then with respect to y. Still other orders of integration are of course possible.

We illustrate the technique in some problems.

Problem 1. Use triple integration to find the volume of the tetrahedron T of Figure 18.7.2.

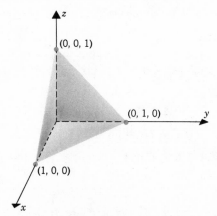

FIGURE 18.7.2

SOLUTION.   We begin by referring to Figure 18.7.3. The inclined face of T has equation $x + y + z = 1$. It intersects the xy-plane in a line which has equation $x + y = 1$. As x ranges from 0 to 1 (see Figure 18.7.4), y ranges from 0 to $1 - x$, and z ranges from 0 to $1 - x - y$. In short, T is given by the inequalities

$$0 \le x \le 1, \quad 0 \le y \le 1 - x, \quad 0 \le z \le 1 - x - y.$$

FIGURE 18.7.3

FIGURE 18.7.4

The volume of T is therefore

$$V = \iiint_T dx\,dy\,dz$$

$$= \int_0^1 \int_0^{1-x} \int_0^{1-x-y} dz\,dy\,dx$$

$$= \int_0^1 \int_0^{1-x} (1 - x - y)\,dy\,dx$$

$$= \int_0^1 \left[(1 - x)y - \tfrac{1}{2}y^2 \right]_0^{1-x} dx = \int_0^1 \tfrac{1}{2}(1 - x)^2\,dx = \left[-\tfrac{1}{6}(1 - x)^3 \right]_0^1 = \tfrac{1}{6}. \quad \square$$

REMARK.   In the last problem we integrated first with respect to z, then with respect to y, and then with respect to x. This came from expressing the tetrahedron by the inequalities

$$0 \le x \le 1, \quad 0 \le y \le 1 - x, \quad 0 \le z \le 1 - x - y.$$

Other characterizations are also possible. If we use the inequalities

$$0 \le z \le 1, \quad 0 \le x \le 1 - z, \quad 0 \le y \le 1 - z - x,$$

then we obtain

$$V = \int_0^1 \int_0^{1-z} \int_0^{1-z-x} dy\,dx\,dz.$$

This formulation of the problem calls for us to integrate first with respect to y, then with respect to x, and then with respect to z. As you can check, this procedure leads to the same result.

Problem 2.   Find the total mass of the tetrahedron of Problem 1 if the density at each point (x, y, z) is xy.

SOLUTION.   Using the first characterization of T, we have

$$\text{mass} = \iiint_T xy \, dx \, dy \, dz = \int_0^1 \int_0^{1-x} \int_0^{1-x-y} xy \, dz \, dy \, dx.$$

Since

$$\int_0^{1-x-y} xy \, dz = xy(1 - x - y) = x(1 - x)y - xy^2,$$

we have

$$\int_0^{1-x} \int_0^{1-x-y} xy \, dz \, dy = \int_0^{1-x} [x(1 - x)y - xy^2] \, dy$$

$$= \left[\tfrac{1}{2}x(1 - x)y^2 - \tfrac{1}{3}xy^3 \right]_0^{1-x}$$

$$= \tfrac{1}{6}x(1 - x)^3 = \tfrac{1}{6}(x - 3x^2 + 3x^3 - x^4).$$

Consequently,

$$M = \int_0^1 \tfrac{1}{6}(x - 3x^2 + 3x^3 - x^4) \, dx$$

$$= \tfrac{1}{6}\left[\tfrac{1}{2}x^2 - x^3 + \tfrac{3}{4}x^4 - \tfrac{1}{5}x^5 \right]_0^1 = \tfrac{1}{6}(\tfrac{1}{2} - 1 + \tfrac{3}{4} - \tfrac{1}{5}) = \tfrac{1}{120}. \quad \square$$

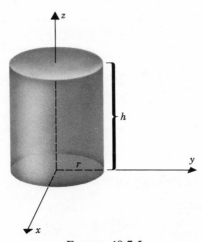

FIGURE 18.7.5

Problem 3.   Find the mass of the cylindrical solid of Figure 18.7.5 if the density at each point is proportional to its distance from the lower base.

SOLUTION.   The solid, call it T, can be characterized by the following inequalities:

$$-r \le x \le r, \quad -\sqrt{r^2 - x^2} \le y \le \sqrt{r^2 - x^2}, \quad 0 \le z \le h.$$

Taking k as the constant of proportionality, we have

$$M = \iiint_T kz \, dxdydz$$

$$= \int_{-r}^{r} \int_{-\sqrt{r^2-x^2}}^{\sqrt{r^2-x^2}} \int_0^h kz \, dz \, dy \, dx$$

$$= 4k \int_0^r \int_0^{\sqrt{r^2-x^2}} \int_0^h z \, dz \, dy \, dx \qquad \text{(explain)}$$

$$= 4k \int_0^r \int_0^{\sqrt{r^2-x^2}} \tfrac{1}{2} h^2 \, dy \, dx$$

$$= 2kh^2 \int_0^r \sqrt{r^2 - x^2} \, dx = 2kh^2 \cdot \tfrac{1}{4}\pi r^2 = \tfrac{1}{2} kh^2 r^2 \pi. \quad \square$$

Problem 4.   Integrate

$$f(x, y, z) = xy + yz + zx$$

over that part of the first octant $x \ge 0, y \ge 0, z \ge 0$ which is cut off by the ellipsoid

$$\frac{x^2}{a^2} + \frac{y^2}{b^2} + \frac{z^2}{c^2} = 1.$$

SOLUTION.   See Figure 18.7.6. Call the solid T. The upper boundary of T has equation

$$z = \psi(x, y) = c \left(1 - \frac{x^2}{a^2} - \frac{y^2}{b^2} \right)^{1/2}.$$

This surface intersects the xy-plane in the curve

$$y = \phi(x) = b \left(1 - \frac{x^2}{a^2} \right)^{1/2}.$$

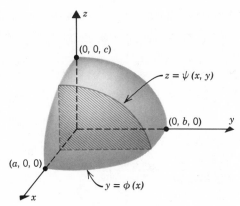

FIGURE 18.7.6

As x ranges from 0 to a, y ranges from 0 to $\phi(x)$, and z ranges from 0 to $\psi(x, y)$. We are therefore led to the repeated integral

$$I = \int_0^a \int_0^{\phi(x)} \int_0^{\psi(x,y)} (xy + yz + zx) \, dz \, dy \, dx.$$

A straightforward computation shows that $I = \frac{1}{15} abc(ab + bc + ca)$. $\square$

Problem 5. Use triple integration to find the volume of the solid T bounded above by the parabolic cylinder $z = 4 - y^2$ and bounded below by the elliptic paraboloid $z = x^2 + 3y^2$.

SOLUTION. Solving the two equations simultaneously we have

$$4 - y^2 = x^2 + 3y^2 \quad \text{and thus} \quad x^2 + 4y^2 = 4.$$

This tells us that the two surfaces intersect in a space curve which lies along the elliptic cylinder $x^2 + 4y^2 = 4$. The projection of this intersection onto the xy-plane is the ellipse $x^2 + 4y^2 = 4$. This is illustrated in Figure 18.7.7.

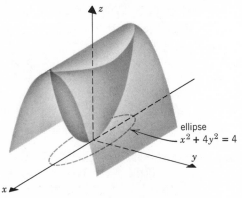

FIGURE 18.7.7

The projection of T onto the xy-plane is the set of all (x, y) such that

$$-2 \le x \le 2, \quad -\tfrac{1}{2}\sqrt{4 - x^2} \le y \le \tfrac{1}{2}\sqrt{4 - x^2}.$$

The solid T is then the set of all (x, y, z) such that

$$-2 \le x \le 2, \quad -\tfrac{1}{2}\sqrt{4 - x^2} \le y \le \tfrac{1}{2}\sqrt{4 - x^2}, \quad x^2 + 3y^2 \le z \le 4 - y^2.$$

Its volume is given by

$$V = \int_{-2}^{2} \int_{-(\sqrt{4-x^2})/2}^{(\sqrt{4-x^2})/2} \int_{x^2+3y^2}^{4-y^2} dz\, dy\, dx = 4 \int_{0}^{2} \int_{0}^{(\sqrt{4-x^2})/2} \int_{x^2+3y^2}^{4-y^2} dz\, dy\, dx = 4\pi. \quad \square$$

Exercises

Evaluate.

*1. $\displaystyle\int_{0}^{a} \int_{0}^{b} \int_{0}^{c} dx\, dy\, dz.$

2. $\displaystyle\int_{0}^{1} \int_{0}^{x} \int_{0}^{y} y\, dz\, dy\, dx.$

*3. $\displaystyle\int_{0}^{1} \int_{1}^{2y} \int_{0}^{x} (x + 2z)\, dz\, dx\, dy.$

4. $\displaystyle\int_{0}^{1} \int_{1-x}^{1+x} \int_{0}^{xy} 4z\, dz\, dy\, dx.$

Express 5–10 by repeated integrals. Do not evaluate.

*5. The mass of a ball $\{(x, y, z): x^2 + y^2 + z^2 \le r^2\}$ if the density at each point is proportional to its distance from the outer shell.

6. The mass of a solid cone bounded above by $z = 1$ and bounded below by $z = \sqrt{x^2 + y^2}$ if the density at each point is proportional to its distance from the origin.

*7. The volume of the solid bounded above by the parabolic cylinder $z = 1 - y^2$, below by the plane $2x + 3y + z + 10 = 0$, and on the sides by the circular cylinder $x^2 + y^2 - x = 0$.

8. The volume of the solid bounded above by the paraboloid $z = 4 - x^2 - y^2$ and bounded below by the parabolic cylinder $z = 2 + y^2$.

*9. The mass of the solid bounded by the elliptic paraboloids $z = 4 - x^2 - \tfrac{1}{4}y^2$ and $z = 3x^2 + \tfrac{1}{4}y^2$ if the density at each point is proportional to its vertical distance from the lower surface.

10. The mass of the solid bounded by the paraboloid $z = x^2 + 2y^2$ and the parabolic cylinder $z = 4 - x^2$ if the density at each point is proportional to its distance from the x-axis.

*11. Find the volume of the tetrahedron of Figure 18.7.8.

12. Find the mass of the tetrahedron of Exercise 11 if $a = 1, b = 1, c = 1$, and the density at (x, y, z) is (a) $x^2 + y^2$. (b) $\tfrac{1}{2}(x + y + z)$.

FIGURE 18.7.8

*13. Find the mass of a block in the shape of a unit cube if its density at each point is proportional (a) to its distance from one of the faces. (b) to the square of its distance from one of the vertices.

14. Let T be the solid bounded above by the plane $z = y$, below by the xy-plane, and on the sides by the planes $x = 0, x = 1, y = 0, y = 1$. Find the mass of T given that the density at each point is proportional to the square of its distance from the origin.

*15. Find the volume of the solid bounded above by the cylinder $y^2 + z = 4$, below by the plane $y + z = 2$, and on the sides by the planes $x = 0$ and $x = 2$.

16. Find the mass of the solid of Exercise 15 if the density at (x, y, z) is
    (a) x.   *(b) y^2.   (c) $x^2 y^2$.

17. Find the mass of the solid discussed in Problem 5, p. 799, if the density function is $f(x, y, z) = |x|$.

*18. Use triple integration to find the volume of the solid bounded above by the plane $z - y = 0$ and bounded below by the paraboloid $z = x^2 + y^2$.

18.8 Averages and Centroids

In Chapter 10 we defined the average of a continuous function by setting

$$\text{average of } f \text{ on } [a, b] = \frac{1}{b - a} \int_a^b f(x)\, dx = \frac{1}{\text{length of } [a, b]} \int_a^b f(x)\, dx.$$

Averages for functions of several variables are defined in a similar manner. If f is continuous on a plane region Ω, its average on Ω is defined by setting

(18.8.1)

$$\text{average of } f \text{ on } \Omega = \frac{1}{\text{area of } \Omega} \iint\limits_{\Omega} f(x, y) \, dxdy.$$

If f is continuous on a portion T of space, its average on T is defined by setting

(18.8.2)

$$\text{average of } f \text{ on } T = \frac{1}{\text{volume of } T} \iiint\limits_{T} f(x, y, z) \, dxdydz.$$

Problem. Calculate the average height (average z-coordinate) of the plane $x + \frac{1}{3}y + \frac{1}{4}z = 1$ over the triangular region Ω with vertices at $(0, 0)$, $(1, 0)$, $(0, 1)$.

SOLUTION.   The equation for the plane can be written

$$z = 4(1 - x - \tfrac{1}{3}y).$$

The triangular region is pictured in Figure 18.8.1.

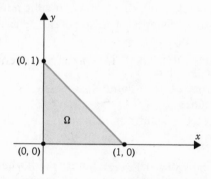

FIGURE 18.8.1

Since the line through $(1, 0)$ and $(0, 1)$ has equation $y = 1 - x$, Ω can be characterized as the set of all points (x, y) such that

$$0 \le x \le 1, \quad 0 \le y \le 1 - x.$$

As you can verify,

$$\iint\limits_{\Omega} 4(1 - x - \tfrac{1}{3}y) \, dxdy = 4 \int_0^1 \int_0^{1-x} (1 - x - \tfrac{1}{3}y) \, dy \, dx = \tfrac{10}{9}.$$

Since the area of Ω is $\frac{1}{2}$, the average height is $\frac{20}{9}$.   □

Problem. Find the average density of the cylindrical solid

$$T: -r \le x \le r, \quad -\sqrt{r^2 - x^2} \le y \le \sqrt{r^2 - x^2}, \quad 0 \le z \le h$$

if the density at each point is given by the function $\rho(x, y, z) = y^2\sqrt{r^2 - x^2}$.

SOLUTION. The average we want is the quotient

$$\frac{\text{mass of } T}{\text{volume of } T} = \frac{1}{V} \iiint_T \rho(x, y, z) \, dxdydz.$$

In the first place,

$$\iiint_T \rho(x, y, z) \, dxdydz = \int_{-r}^r \int_{-\sqrt{r^2-x^2}}^{\sqrt{r^2-x^2}} \int_0^h y^2 \sqrt{r^2 - x^2} \, dz \, dy \, dx$$

$$= 4 \int_0^r \int_0^{\sqrt{r^2-x^2}} \int_0^h y^2 \sqrt{r^2 - x^2} \, dz \, dy \, dx$$

$$= 4 \int_0^r \int_0^{\sqrt{r^2-x^2}} hy^2 \sqrt{r^2 - x^2} \, dy \, dx$$

$$= 4h \int_0^r \tfrac{1}{3}(r^2 - x^2)^2 \, dx$$

$$= \tfrac{4}{3}h \int_0^r (r^4 - 2r^2x^2 + x^4) \, dx$$

$$= \tfrac{4}{3}h \left[r^4 x - \tfrac{2}{3}r^2 x^3 + \tfrac{1}{5}x^5 \right]_0^r = \tfrac{32}{45}hr^5.$$

Since the volume of the cylindrical solid is $\pi r^2 h$, the average density is

$$\frac{1}{V} M = \frac{1}{\pi r^2 h} \cdot \frac{32}{45} hr^5 = \frac{32}{45\pi} r^3. \quad \square$$

Centroids

The *centroid* of a plane region Ω is the point $(\bar{x}, \bar{y})$ where $\bar{x}$ is the average x-coordinate of points in Ω and $\bar{y}$ is the average y-coordinate of points in Ω:

(18.8.3)

$$\boxed{\begin{aligned} \bar{x} &= \frac{1}{\text{area of } \Omega} \iint_\Omega x \, dxdy, \\[2ex] \bar{y} &= \frac{1}{\text{area of } \Omega} \iint_\Omega y \, dxdy. \end{aligned}}$$

If you think of Ω as a thin distribution of matter of constant density, then the centroid of Ω becomes the center of gravity, the balance point.

For regions that have a geometric center, the center and the centroid coincide. For example, in the case of a rectangle

$$R: a \le x \le b, \, c \le y \le d$$

we have

$$(\bar{x}, \bar{y}) = (\tfrac{1}{2}[a + b], \tfrac{1}{2}[c + d]).$$

This follows from observing that

$$\iint_R x\, dxdy = \int_a^b \int_c^d x\, dy\, dx = \tfrac{1}{2}(b^2 - a^2)(d - c)$$

$$= \tfrac{1}{2}(a + b)(b - a)(d - c) = \tfrac{1}{2}(a + b)(\text{area of } R)$$

and

$$\iint_R y\, dxdy = \int_a^b \int_c^d y\, dy\, dx = \tfrac{1}{2}(d^2 - c^2)(b - a)$$

$$= \tfrac{1}{2}(c + d)(b - a)(d - c) = \tfrac{1}{2}(c + d)(\text{area of } R). \quad \square$$

A region may fail to have a center and yet be symmetric with respect to some line l. The centroid will then lie on l.

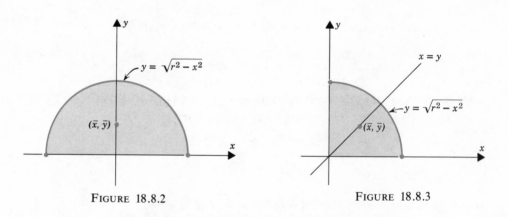

FIGURE 18.8.2   FIGURE 18.8.3

Problem.   Find the centroid of the semicircular region in Figure 18.8.2.

SOLUTION.   Here

$$\Omega: -r \le x \le r,\ 0 \le y \le \sqrt{r^2 - x^2}.$$

Since Ω is symmetric with respect to the y-axis, we can expect $\bar{x} = 0$. You can verify this by showing that

$$\iint_\Omega x\, dxdy = 0.$$

Since

$$\iint_\Omega y\, dxdy = \int_{-r}^r \int_0^{\sqrt{r^2-x^2}} y\, dy\, dx = \int_{-r}^r \tfrac{1}{2}(r^2 - x^2)\ dx = \left[\tfrac{1}{2}(r^2 x - \tfrac{1}{3}x^3)\right]_{-r}^r = \tfrac{2}{3}r^3$$

and the area of $\Omega = \frac{1}{2}\pi r^2$, we have

$$\bar{y} = \frac{1}{\text{area of } \Omega} \iint\limits_{\Omega} y \, dxdy = \frac{4r}{3\pi}. \quad \square$$

Problem.   Find the centroid of the quarter disc depicted in Figure 18.8.3.

SOLUTION.   Here we have symmetry with respect to the line $x = y$. We can therefore expect $\bar{x} = \bar{y}$. As in the last problem

$$\bar{y} = \frac{1}{\text{area of } \Omega} \iint\limits_{\Omega} y \, dxdy = \frac{4r}{3\pi}. \quad\quad\quad \text{(check this)}$$

The centroid is the point

$$(\bar{x}, \bar{y}) = \left(\frac{4r}{3\pi}, \frac{4r}{3\pi} \right). \quad \square$$

Problem.   Find the centroid of the region bounded above by the line $y = 2x$ and below by the parabola $y = x^2$.

SOLUTION.   The line intersects the parabola at the origin and at the point $(2, 4)$. The region Ω can be characterized by the inequalities

$$0 \le x \le 2, \quad x^2 \le y \le 2x.$$

Here

$$\text{area of } \Omega = \iint\limits_{\Omega} dxdy = \int_0^2 \int_{x^2}^{2x} dy \, dx = \int_0^2 (2x - x^2) \, dx = \left[x^2 - \tfrac{1}{3}x^3 \right]_0^2 = \tfrac{4}{3},$$

$$\iint\limits_{\Omega} x \, dxdy = \int_0^2 \int_{x^2}^{2x} x \, dy \, dx = \int_0^2 (2x^2 - x^3) \, dx = \left[\tfrac{2}{3}x^3 - \tfrac{1}{4}x^4 \right]_0^2 = \tfrac{4}{3},$$

$$\iint\limits_{\Omega} y \, dxdy = \int_0^2 \int_{x^2}^{2x} y \, dy \, dx = \tfrac{1}{2} \int_0^2 (4x^2 - x^4) \, dx = \tfrac{1}{2} \left[\tfrac{4}{3}x^3 - \tfrac{1}{5}x^5 \right]_0^2 = \tfrac{32}{15},$$

$$\bar{x} = \frac{1}{\text{area of } \Omega} \iint\limits_{\Omega} x \, dxdy = 1, \quad \bar{y} = \frac{1}{\text{area of } \Omega} \iint\limits_{\Omega} y \, dxdy = \tfrac{8}{5}.$$

The centroid is the point $(1, \tfrac{8}{5})$. $\square$

There is a simple relation between centroids and volumes of revolution first pointed out by Pappus of Alexandria (circa 300 A.D.).

Theorem 18.8.4   (Pappus)†

When a plane region Ω is revolved about an axis which lies in its plane but does not cross it, the volume of the resulting solid of revolution is the area of Ω multiplied by the circumference of the circle described by the centroid of Ω.

To confirm the theorem, we take Ω as in Figure 18.8.4 and see what happens when we revolve Ω about each of the coordinate axes.

$$\Omega: \quad a \leq x \leq b, \quad \phi_1(x) \leq y \leq \phi_2(x)$$

FIGURE 18.8.4

The coordinates of the centroid are

$$\bar{x} = \frac{1}{\text{area of } \Omega} \iint_\Omega x \, dxdy, \quad \bar{y} = \frac{1}{\text{area of } \Omega} \iint_\Omega y \, dxdy.$$

When Ω is revolved about the x-axis, the resulting solid has volume

$$V_x \underset{\underset{\text{by } 10.2.3}{\uparrow}}{=} \int_a^b \pi[\phi_2(x)]^2 \, dx - \int_a^b \pi[\phi_1(x)]^2 \, dx = \pi \int_a^b ([\phi_2(x)]^2 - [\phi_1(x)]^2) \, dx.$$

When Ω is revolved about the y-axis, the resulting solid has volume

$$V_y \underset{\underset{\text{by } 10.3.3}{\uparrow}}{=} \int_a^b 2\pi x \phi_2(x) \, dx - \int_a^b 2\pi x \phi_1(x) \, dx = 2\pi \int_a^b x[\phi_2(x) - \phi_1(x)] \, dx.$$

According to Pappus we must have

$$V_x = 2\pi\bar{y} \, (\text{area of } \Omega) \quad \text{and} \quad V_y = 2\pi\bar{x} \, (\text{area of } \Omega).$$

---

† This theorem is found in Book VII of Pappus's "Mathematical Collection," largely a survey of ancient geometry to which Pappus made many original contributions (among them this theorem). Much of what we know today of Greek geometry we owe to Pappus.

Since

$$2\pi \bar{y} \cdot \text{area of } \Omega = 2\pi \iint_{\Omega} y \, dxdy$$

$$= 2\pi \int_a^b \int_{\phi_1(x)}^{\phi_2(x)} y \, dy \, dx$$

$$= 2\pi \int_a^b \left[\tfrac{1}{2} y^2 \right]_{\phi_1(x)}^{\phi_2(x)} dx$$

$$= 2\pi \int_a^b \tfrac{1}{2}([\phi_2(x)]^2 - [\phi_1(x)]^2) \, dx$$

$$= \pi \int_a^b ([\phi_2(x)]^2 - [\phi_1(x)]^2) \, dx = V_x,$$

the first relation is confirmed. We now confirm the second relation:

$$2\pi \bar{x} \cdot \text{area of } \Omega = 2\pi \iint_{\Omega} x \, dxdy$$

$$= 2\pi \int_a^b \int_{\phi_1(x)}^{\phi_2(x)} x \, dy \, dx$$

$$= 2\pi \int_a^b x[\phi_2(x) - \phi_1(x)] \, dx = V_y. \quad \square$$

The *centroid* of a solid T is the point $(\bar{x}, \bar{y}, \bar{z})$ where $\bar{x}$ is the average x-coordinate of points in T, $\bar{y}$ the average y-coordinate, and $\bar{z}$ the average z-coordinate. The defining formulas are:

(18.8.5)

$$\bar{x} = \frac{1}{\text{volume of } T} \iiint_T x \, dxdydz,$$

$$\bar{y} = \frac{1}{\text{volume of } T} \iiint_T y \, dxdydz,$$

$$\bar{z} = \frac{1}{\text{volume of } T} \iiint_T z \, dxdydz.$$

For a solid of constant density the centroid is *the center of mass*. Centers of mass are important in physics. The reason is roughly this. When a complicated object (or collection of objects) T of total mass M is subjected to an outside force $\mathbf{F}$, not every part of T reacts with the same acceleration. What happens then to Newton's famous $\mathbf{F} = M\mathbf{a}$? It still holds, not for every part of T, but for the center of mass. In short, Newton's $\mathbf{F} = M\mathbf{a}$ should be read

$$\begin{pmatrix} \text{total outside force} \\ \text{applied to } T \end{pmatrix} = \begin{pmatrix} \text{total mass} \\ \text{of } T \end{pmatrix} \cdot \begin{pmatrix} \text{acceleration of the} \\ \text{center of mass of } T \end{pmatrix}.$$

As in the two-dimensional case, if T has a geometric center, the center and centroid coincide. If T fails to have a center but has an axis or plane of symmetry, then the centroid will lie on that axis or on that plane.

Problem.   Find the centroid of the tetrahedron with vertices $(0, 0, 0)$, $(1, 0, 0)$, $(0, 1, 0)$, $(0, 0, 1)$.

SOLUTION.   Call the tetrahedron T. T is characterized by the inequalities

$$0 \le x \le 1, \quad 0 \le y \le 1 - x, \quad 0 \le z \le 1 - x - y. \quad \text{(Figure 18.7.4)}$$

As you have seen before, $V = \frac{1}{6}$. Consequently,

$$\bar{x} = \frac{1}{V} \iiint_T x \, dxdydz = 6 \int_0^1 \int_0^{1-x} \int_0^{1-x-y} x \, dz \, dy \, dx$$

$$= 6 \int_0^1 \int_0^{1-x} [x(1 - x) - xy] \, dy \, dx$$

$$= 6 \int_0^1 \left[x(1 - x)y - \tfrac{1}{2}xy^2 \right]_0^{1-x} dx$$

$$= 6 \int_0^1 \tfrac{1}{2}x(1 - x)^2 \, dx$$

$$= 3 \int_0^1 (x - 2x^2 + x^3) \, dx$$

$$= 3 \left[\tfrac{1}{2}x^2 - \tfrac{2}{3}x^3 + \tfrac{1}{4}x^4 \right]_0^1 = \tfrac{1}{4}.$$

In a similar manner (or by symmetry), you can verify that $\bar{y}$ and $\bar{z}$ are also $\frac{1}{4}$. The centroid is the point $(\frac{1}{4}, \frac{1}{4}, \frac{1}{4})$. $\square$

Exercises

Find the centroid of the bounded region determined by the following curves.

*1. $y = 6x - x^2, \quad y = x$.
2. $y = 4x - x^2, \quad y = 2x - 3$.
*3. $x^2 = 4y, \quad x - 2y + 4 = 0$.
4. $y = x^2, \quad 2x - y + 3 = 0$.
*5. $y^3 = x^2, \quad 2y = x$.
6. $y^2 = 2x, \quad y = x - x^2$.
*7. $y = x^2 - 2x, \quad y = 6x - x^2$.
8. $y = 6x - x^2, \quad x + y = 6$.
*9. $x + 1 = 0, \quad x + y^2 = 0$.
10. $\sqrt{x} + \sqrt{y} = \sqrt{a}, \quad x = 0, \quad y = 0$.

*11. Use the theorem of Pappus to find the volume of the torus generated by revolving the circle

$$(x - b)^2 + y^2 = a^2, \quad b > a$$

about the y-axis.

12. What is the centroid of the solid bounded above by the plane $z = 1 + x + y$, below by the plane $z = -2$, and on the sides by the planes $x = 1$, $x = 2$, $y = 1$, $y = 2$?

*13. Taking a, b, c as positive, find the centroid of the tetrahedron with vertices $(0, 0, 0)$, $(a, 0, 0)$, $(0, b, 0)$, and $(0, 0, c)$.

14. Find the centroid of the solid bounded above by the cylindrical surface $x^2 + z = 4$, below by the plane $x + z = 2$, and on the sides by the planes $y = 0$ and $y = 3$.

*15. The rectangle of Figure 18.8.5 is revolved about the line marked l. Find the volume of the resulting solid.

FIGURE 18.8.5

For a solid of varying density, $\rho = f(x, y, z)$, *the center of mass* is given by the equations

$$x_M = \frac{1}{M} \iiint\limits_T xf(x, y, z)\, dxdydz, \quad y_M = \frac{1}{M} \iiint\limits_T yf(x, y, z)\, dxdydz$$

$$z_M = \frac{1}{M} \iiint\limits_T zf(x, y, z)\, dxdydz.$$

*16. Find the center of mass of a unit cube $0 \le x \le 1, 0 \le y \le 1, 0 \le z \le 1$ if the density is proportional to (a) the square of the distance from the origin. (b) the distance from the xy-plane. (c) the square of the distance from the diagonal which joins $(0, 0, 0)$ to $(1, 1, 1)$.

17. Find the center of mass of the solid described in Exercise 14, Section 18.7.

18. Find the center of mass of the tetrahedron of Exercise 13 if the density at each point (x, y, z) is *(a) yz. (b) xyz.

19. Find the center of mass of the solid of Exercise 14 if the density at each point (x, y, z) is $1 + y$.

20. Show that in polar coordinates the formulas for the centroid become

$$\bar{x} = \frac{1}{\text{area of } \Omega} \iint\limits_{\Omega} r^2 \cos\theta\, drd\theta, \quad \bar{y} = \frac{1}{\text{area of } \Omega} \iint\limits_{\Omega} r^2 \sin\theta\, drd\theta.$$

*21. Find the centroid of the region enclosed by the cardioid $r = a(1 + \cos\theta)$.

18.9   Integration in Cylindrical Coordinates

The cylindrical coordinates (r, θ, z) of a point P in space are illustrated in Figure 18.9.1. The first two coordinates, r and θ, are the usual polar coordinates. The third coordinate is the third rectangular coordinate z. For simplicity we will take $r \geq 0$ and $0 \leq \theta \leq 2\pi$.

cylindrical coordinates
FIGURE 18.9.1

In rectangular coordinates the coordinate surfaces

$$x = x_0, \quad y = y_0, \quad z = z_0$$

are three mutually perpendicular planes. In cylindrical coordinates the coordinate surfaces take the form

$$r = r_0, \quad \theta = \theta_0, \quad z = z_0.$$

FIGURE 18.9.2

These surfaces are drawn in Figure 18.9.2. The first surface is a circular cylinder of radius r_0. The central axis of the cylinder is the z-axis. The surface $\theta = \theta_0$ is a vertical half-plane hinged at the z-axis. The plane lies at an angle of θ_0 radians from the positive x-axis. The last coordinate surface is the plane $z = z_0$ of rectangular coordinates.

The easiest solids to describe in terms of the coordinate surfaces are sets of the form

$$\Pi: a_1 \le r \le a_2, \ b_1 \le \theta \le b_2, \ c_1 \le z \le c_2.$$

Such sets are called *cylindrical wedges*. You can see what one looks like in Figure 18.9.3.

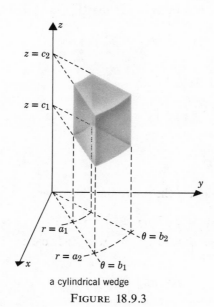

a cylindrical wedge

FIGURE 18.9.3

We introduced triple integrals in (x, y, z) coordinates by working first with rectangular solids, which we called "boxes." Here we work first with cylindrical wedges.

Suppose that Π is a cylindrical wedge. If the density $f(r, \theta, z)$ varies continuously on Π, then

(18.9.1)

$$\text{the mass of } \Pi = \iiint\limits_{\Pi} f(r, \theta, z) \, r \, dr d\theta dz.$$

PROOF.   Break up Π in an obvious manner into little cylindrical wedges

$$\Pi_{ijk}: r_{i-1} \le r \le r_i, \ \theta_{j-1} \le \theta \le \theta_j, \ z_{k-1} \le z \le z_k.$$

On each Π_{ijk} the density function takes on a maximum value, M_{ijk}, and a minimum value, m_{ijk}. Clearly

$$m_{ijk}(\text{volume of } \Pi_{ijk}) \leq \text{mass of } \Pi_{ijk} \leq M_{ijk}(\text{volume of } \Pi_{ijk}).$$

Adding up all these inequalities, we have an estimate for the total mass:

$$\Sigma\Sigma\Sigma \, m_{ijk}(\text{volume of } \Pi_{ijk}) \leq \text{mass of } \Pi \leq \Sigma\Sigma\Sigma \, M_{ijk}(\text{volume of } \Pi_{ijk}).$$

We now argue that

$$\text{mass of } \Pi = \iiint\limits_{\Pi} f(r, \theta, z) \, r \, dr d\theta dz$$

by arguing that this integral satisfies this last inequality. In the first place it can be shown that the integral is additive over the Π_{ijk}, and thus

$$\iiint\limits_{\Pi} f(r, \theta, z) \, r \, dr d\theta \, dz = \Sigma\Sigma\Sigma \iiint\limits_{\Pi_{ijk}} f(r, \theta, z) \, r \, dr d\theta dz.$$

All we have to show therefore is that

$$m_{ijk}(\text{volume of } \Pi_{ijk}) \leq \iiint\limits_{\Pi_{ijk}} f(r, \theta, z) \, r \, dr d\theta dz \leq M_{ijk}(\text{volume of } \Pi_{ijk}).$$

Observe now that

$$\iiint\limits_{\Pi_{ijk}} r \, dr d\theta \, dz = \int_{z_{k-1}}^{z_k} \int_{\theta_{j-1}}^{\theta_j} \int_{r_{i-1}}^{r_i} r \, dr d\theta \, dz$$

$$= \tfrac{1}{2}(r_i^2 - r_{i-1}^2)(\theta_j - \theta_{j-1})(z_k - z_{k-1}) = \text{volume of } \Pi_{ijk}.^{\dagger}$$

The rest is straightforward:

$$\iiint\limits_{\Pi_{ijk}} f(r, \theta, z) \, r \, dr d\theta dz \leq \iiint\limits_{\Pi_{ijk}} M_{ijk} \, r \, dr d\theta dz$$

$$= M_{ijk} \iiint\limits_{\Pi_{ijk}} r \, dr d\theta dz = M_{ijk}(\text{volume of } \Pi_{ijk}),$$

$$\iiint\limits_{\Pi_{ijk}} f(r, \theta, z) \, r \, dr d\theta dz \geq \iiint\limits_{\Pi_{ijk}} m_{ijk} \, r \, dr d\theta dz$$

$$= m_{ijk} \iiint\limits_{\Pi_{ijk}} r \, dr d\theta dz = m_{ijk}(\text{volume of } \Pi_{ijk}).$$

---

† The base is a "polar rectangle" of area $\tfrac{1}{2}(r_i^2 - r_{i-1}^2)(\theta_j - \theta_{j-1})$, and the thickness is $z_i - z_{i-1}$.

Our argument that

$$\text{mass of } \Pi = \iint\limits_{\Pi_{ijk}} \int f(r, \theta, z)\, r\, dr d\theta dz$$

is now complete. □

To find the mass of a more general solid T we first encase T in a cylindrical wedge Π. We then extend the density function to all of Π by defining f to be zero outside of T. Obviously then

$$\text{mass of } T = \text{mass of } \Pi.$$

If the boundary of T is not too wild, the triple integral over T will still exist and be equal to the triple integral over Π. Consequently we can set

(18.9.2)
$$\boxed{\text{mass of } T = \iint\limits_{T} \int f(r, \theta, z) r\, dr d\theta dz.}$$

The triple integral

$$\iint\limits_{T} \int 1r\, dr d\theta dz = \iint\limits_{T} \int r\, dr d\theta dz$$

gives the mass of T when T has constant density 1. In cubic units this is the volume of T:

(18.9.3)
$$\boxed{\text{volume of } T = \iint\limits_{T} \int r\, dr d\theta dz.}$$

Cylindrical coordinates are often convenient for physical applications when an axis of symmetry is present. The axis of symmetry is then taken as the z-axis.

Problem.   Find the mass of a cylindrical solid of radius R and height h if the density at each point is proportional to its distance from the axis of the cylinder.

SOLUTION.   We call the solid T and express it by the inequalities

$$0 \le r \le R, \quad 0 \le \theta \le 2\pi, \quad 0 \le z \le h.$$

The density function is of the form

$$f(r, \theta, z) = kr.$$

According to our formula,

$$\text{mass of } T = \iiint_T kr^2 \, dr d\theta dz$$

$$= \int_0^h \int_0^{2\pi} \int_0^R kr^2 \, dr \, d\theta \, dz$$

$$= \int_0^h \int_0^{2\pi} \tfrac{1}{3}kR^3 \, d\theta \, dz = \int_0^h \tfrac{2}{3}\pi kR^3 \, dz = \tfrac{2}{3}\pi kR^3 h. \quad \square$$

Problem. Use cylindrical coordinates to find the volume of the solid T bounded above by the plane $z = y$ and below by the paraboloid $z = x^2 + y^2$.

SOLUTION. Solving the two equations simultaneously we have $r = \sin \theta$. This tells us that the two surfaces intersect in a space curve which lies along the circular cylinder $r = \sin \theta$. The projection of this intersection onto the xy-plane is the circle $r = \sin \theta$. (See Figure 18.9.4.)

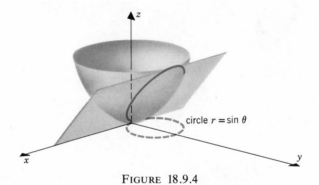

FIGURE 18.9.4

The projection of T onto the xy-plane is the disc

$$0 \le \theta \le \pi, \quad 0 \le r \le \sin \theta.$$

The solid T is then the set of all (r, θ, z) such that

$$0 \le \theta \le \pi, \quad 0 \le r \le \sin \theta, \quad r^2 \le z \le r \sin \theta.$$

The volume is then

$$V = \iiint_T r \, dr d\theta dz$$

$$= \int_0^\pi \int_0^{\sin \theta} \int_{r^2}^{r \sin \theta} r \, dz \, dr \, d\theta$$

$$= \int_0^\pi \int_0^{\sin \theta} (r^2 \sin \theta - r^3) \, dr \, d\theta$$

$$= \int_0^\pi \left[\tfrac{1}{3}r^3 \sin \theta - \tfrac{1}{4}r^4 \right]_0^{\sin \theta} d\theta$$

$$= \tfrac{1}{12} \int_0^\pi \sin^4 \theta \ d\theta = \tfrac{1}{12}(\tfrac{3}{8}\pi) = \tfrac{1}{32}\pi. \quad \square$$

by Exercise 5, Section 8.3

Problem. Find the mass of the solid of the last problem if the density at each point is proportional to the square of its distance from the z-axis.

SOLUTION. Here the density function is of the form

$$f(r, \theta, z) = kr^2.$$

According to the mass formula,

$$\text{mass of } T = \int \int \int_T kr^3 \ dr d\theta dz$$

$$= \int_0^\pi \int_0^{\sin \theta} \int_{r^2}^{r \sin \theta} kr^3 \ dz \ dr \ d\theta$$

$$= \int_0^\pi \int_0^{\sin \theta} k(r^4 \sin \theta - r^5) \ dr \ d\theta$$

$$= k \int_0^\pi \left[\tfrac{1}{5}r^5 \sin \theta - \tfrac{1}{6}r^6 \right]_0^{\sin \theta} d\theta$$

$$= k \int_0^\pi \tfrac{1}{30} \sin^6 \theta \ d\theta = \tfrac{1}{96}k\pi. \quad \square$$

Exercises

*1. Find the mass of a cylindrical solid of radius R and height h if the density at each point is proportional to (a) the distance from one of the bases. (b) the square root of the distance from the axis of the cylinder. (c) the square of the distance from the curved boundary.

2. Use cylindrical coordinates to show that a cone of base radius R and height h has volume $V = \tfrac{1}{3}\pi R^2 h$.

3. Find the mass of the cone of Exercise 2 if the density is proportional to the distance (a) from the axis. (b) from the base.

*4. Use cylindrical coordinates to find the volume of the solid S which is bounded above by the paraboloid $z = 1 - (x^2 + y^2)$ and below by the xy-plane.

5. Find the mass of the solid of Exercise 4 if the density is proportional to (a) the distance from the xy-plane. (b) the square of the distance from the origin.

*6. Find the volume of the solid bounded by the paraboloid of revolution $x^2 + y^2 = az$, the xy-plane, and the cylinder $x^2 + y^2 = 2ax$.

*7. Find the volume of the solid bounded above by the cone $z^2 = x^2 + y^2$, below by the xy-plane, and on the sides by the cylinder $x^2 + y^2 = 2ax$.

*8. Find the volume of the solid bounded above by the plane $2z = 4 + x$, below by the xy-plane, and on the sides by the cylinder $x^2 + y^2 = 2x$.

*9. Find the volume of the solid bounded above by the cone $z = a - \sqrt{x^2 + y^2}$, below by the xy-plane, and on the sides by the cylinder $x^2 + y^2 = ax$.

10. Find the volume of the solid bounded above by $x^2 + y^2 + z^2 = 25$ and below by $z = \sqrt{x^2 + y^2} + 1$.

*11. Find the volume of the solid bounded by the plane $z = x$ and the paraboloid $z = x^2 + y^2$.

12. Find the volume of the solid bounded by the hyperboloid $z^2 = a^2 + x^2 + y^2$ and the upper nappe of the cone $z^2 = 2(x^2 + y^2)$.

18.10 Integration in Spherical Coordinates

The spherical coordinates (ρ, θ, ϕ) of a point P in space are illustrated in Figure 18.10.1. The coordinate surfaces

$$\rho = \rho_0, \quad \theta = \theta_0, \quad \phi = \phi_0$$

are described in the answers to Exercises 4, 5, and 6.

Rectangular coordinates (x, y, z) are related to spherical coordinates by the equations

$$x = \rho \sin \phi \cos \theta, \quad y = \rho \sin \phi \sin \theta, \quad z = \rho \cos \phi.$$

You can verify these formulas by referring to Figure 18.10.1.

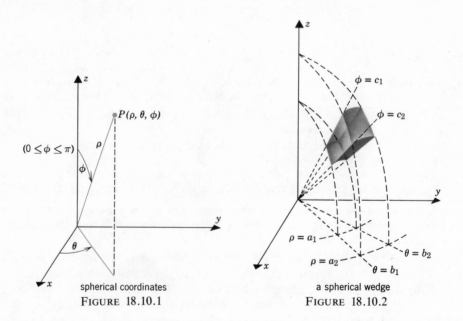

spherical coordinates
FIGURE 18.10.1

a spherical wedge
FIGURE 18.10.2

In spherical coordinates the fundamental solids are sets of the form

$$\Pi : a_1 \le r \le a_2, \ b_1 \le \theta \le b_2, \ c_1 \le \phi \le c_2.$$

Such solids are called *spherical wedges*. A typical one is pictured in Figure 18.10.2.

The volume of a spherical wedge Π is given by the formula

(18.10.1)
$$V = \iiint_{\Pi} \rho^2 \sin \phi \, d\rho \, d\theta \, d\phi.$$

PROOF.   Note first that Π is a solid of revolution. One way to obtain Π is to rotate the face

$$\Omega: a_1 \leq \rho \leq a_2, \; \theta = b_1, \; c_1 \leq \phi \leq c_2$$

about the z-axis for $b_2 - b_1$ radians. (See Figure 18.10.2.)

As you can verify (see Exercise 20 of Section 18.8), the centroid of Ω is at a distance

$$\frac{1}{\text{area of } \Omega} \iint_{\Omega} \rho^2 \sin \phi \, d\rho \, d\phi$$

from the z-axis. As the face Ω is rotated from $\theta = b_1$ to $\theta = b_2$, the centroid travels through a circular arc of length

$$s = (b_2 - b_1) \cdot \frac{1}{\text{area of } \Omega} \iint_{\Omega} \rho^2 \sin \phi \, d\rho \, d\phi.$$

From the theorem of Pappus we know that the volume of a solid of revolution is the length of the path traveled by the centroid multiplied by the area of Ω. In this case we have

$$\text{volume of } \Pi = s \, (\text{area of } \Omega) = (b_2 - b_1) \iint_{\Omega} \rho^2 \sin \phi \, d\rho \, d\phi.$$

Since

$$(b_2 - b_1) \iint_{\Omega} \rho^2 \sin \phi \, d\rho \, d\phi = (b_2 - b_1) \int_{a_1}^{a_2} \int_{c_1}^{c_2} \rho^2 \sin \phi \, d\phi \, d\rho$$

$$= \int_{b_1}^{b_2} \int_{a_1}^{a_2} \int_{c_1}^{c_2} \rho^2 \sin \phi \, d\phi \, d\rho \, d\theta$$

$$= \iiint_{\Pi} \rho^2 \sin \phi \, d\rho \, d\theta \, d\phi,$$

we have the desired formula

$$V = \iiint_{\Pi} \rho^2 \sin \phi \, d\rho \, d\theta \, d\phi. \quad \square$$

From the fact that a spherical wedge Π has volume

$$\iiint_{\Pi} \rho^2 \sin \phi \, d\rho \, d\theta \, d\phi,$$

one can argue that the mass of Π is given by the integral

$$\iiint_{\Pi} f(\rho, \theta, \phi) \, \rho^2 \sin \phi \, d\rho \, d\theta \, d\phi$$

where, as usual, f is the density function. This formula for mass is then extended to a more general solid T in the usual manner. Encase T in a spherical wedge Π and define the density outside of T to be 0. Obviously then

$$\text{mass of } \Pi = \text{mass of } T$$

and, since

$$\iiint_{\Pi} f(\rho, \theta, \phi) \, \rho^2 \sin \phi \, d\rho \, d\theta \, d\phi = \iiint_{T} f(\rho, \theta, \phi) \, \rho^2 \sin \phi \, d\rho \, d\theta \, d\phi,$$

we have

(18.10.2) $\qquad \boxed{\text{mass of } T = \iiint_{T} f(\rho, \theta, \phi) \, \rho^2 \sin \phi \, d\rho \, d\theta \, d\phi.}$

If the density is constantly 1, the integral on the right becomes

$$\iiint_{T} 1 \, \rho^2 \sin \phi \, d\rho \, d\theta \, d\phi = \iiint_{T} \rho^2 \sin \phi \, d\rho \, d\theta \, d\phi.$$

This is the volume of T:

(18.10.3) $\qquad \boxed{\text{volume of } T = \iiint_{T} \rho^2 \sin \phi \, d\rho \, d\theta \, d\phi.}$

Spherical coordinates are commonly used in applications where there is a center of symmetry. The center of symmetry is then taken as the origin.

Problem. Find the mass of a solid ball of radius 1 if the density at each point d units from the center is

$$\frac{1}{1 + d^2}.$$

SOLUTION.   In spherical coordinates the unit ball is the solid

$$T: 0 \leq \rho \leq 1, \ 0 \leq \theta \leq 2\pi, \ 0 \leq \phi \leq \pi.$$

The density function is

$$f(\rho, \theta, \phi) = \frac{1}{1 + \rho^2}.$$

We therefore have

$$\text{mass of } T = \iiint\limits_T \frac{\rho^2}{1 + \rho^2} \sin \phi \, d\rho \, d\theta \, d\phi$$

$$= \int_0^\pi \int_0^{2\pi} \int_0^1 \frac{\rho^2}{1 + \rho^2} \sin \phi \, d\rho \, d\theta \, d\phi$$

$$= \int_0^\pi \int_0^{2\pi} \int_0^1 \left(1 - \frac{1}{1 + \rho^2}\right) \sin \phi \, d\rho \, d\theta \, d\phi$$

$$= \int_0^\pi \int_0^{2\pi} (1 - \tfrac{1}{4}\pi) \sin \phi \, d\theta \, d\phi$$

$$= (1 - \tfrac{1}{4}\pi) \int_0^\pi 2\pi \sin \phi \, d\phi = 4\pi - \pi^2. \quad \square$$

Problem.   Find the volume of the solid T enclosed by the surface

$$(x^2 + y^2 + z^2)^2 = 2z(x^2 + y^2).$$

SOLUTION.   We begin by expressing the surface in spherical coordinates. Since z remains nonnegative, ϕ must remain less than or equal to $\frac{1}{2}\pi$.† With $x = \rho \sin \phi \cos \theta, y = \rho \sin \phi \sin \theta, z = \rho \cos \phi$, we find that the bounding surface becomes

$$\rho = 2 \sin^2 \phi \cos \phi. \qquad \text{(check this out)}$$

We can express the solid inside this surface by the following inequalities:

$$0 \le \theta \le 2\pi, \quad 0 \le \phi \le \tfrac{1}{2}\pi, \quad 0 \le \rho \le 2 \sin^2 \phi \cos \phi.$$

Consequently,

$$V = \iiint\limits_T \rho^2 \sin \phi \, d\rho \, d\theta \, d\phi$$

$$= \int_0^{2\pi} \int_0^{\pi/2} \int_0^{2 \sin^2 \phi \cos \phi} \rho^2 \sin \phi \, d\rho \, d\phi \, d\theta$$

$$= \int_0^{2\pi} \int_0^{\pi/2} \tfrac{8}{3} \sin^7 \phi \cos^3 \phi \, d\phi \, d\theta$$

$$= \tfrac{8}{3} \int_0^{2\pi} \int_0^{\pi/2} (\sin^7 \phi \cos \phi - \sin^9 \phi \cos \phi) \, d\phi \, d\theta$$

$$= \tfrac{8}{3} \int_0^{2\pi} \tfrac{1}{40} \, d\theta = \tfrac{2}{15}\pi. \quad \square$$

---

† See Figure 18.10.1.

Exercises

*1. Find the spherical coordinates (ρ, θ, ϕ) of the point with rectangular coordinates $(1, 1, 1)$.

2. Find the rectangular coordinates of the point with spherical coordinates $(2, \frac{1}{6}\pi, \frac{1}{4}\pi)$.

*3. Find the spherical coordinates of the point with cylindrical coordinates $(2, \frac{2}{3}\pi, 6)$.

The following equations are given in spherical coordinates. Interpret each one geometrically.

*4. $\rho = \rho_0$.   *5. $\theta = \theta_0$.   *6. $\phi = \phi_0$.

7. $\tan \theta = 1$.   *8. $\rho \cos \phi = 1$.   9. $\rho = \cos \phi$.

10. Derive the formula for the volume of a sphere of radius a using spherical coordinates.

*11. A wedge is cut from a ball of radius a by two planes which meet at a diameter. Find the volume of the wedge if the angle between the planes is α radians.

12. Find the mass of a ball of radius a if the density at each point is proportional to its distance from the boundary.

*13. Use spherical coordinates to derive the formula for the volume of a circular cone of base radius r and height h.

*14. Find the mass of a circular cone of base radius r and height h if the density at each point is proportional to its distance from the vertex.

*15. Find the volume of the solid common to the sphere $\rho = a$ and the cone $\phi = \alpha$. Take $\alpha \in (0, \frac{1}{2}\pi)$.

*16. Find an equation for the sphere $x^2 + y^2 + (z - b)^2 = b^2$ in spherical coordinates.

17. Find the mass of the ball $\rho \leq 2b \cos \phi$ if the density at each point (ρ, θ, ϕ) is
(a) ρ.   *(b) $\rho \sin \phi$.   (c) $\rho \cos^2 \theta \sin \phi$.

*18. Find the volume of the solid common to the two spheres $\rho = 2\sqrt{2} \cos \phi$ and $\rho = 2$.

19. Find the volume of the solid enclosed by the surface $\rho = 1 - \cos \phi$.

Line Integrals and Surface Integrals

19

19.1 Work and Line Integrals

The work done by a constant force $\mathbf{F}$ on an object that moves along a straight line is, by definition, the component of $\mathbf{F}$ in the direction of the displacement multiplied by the length of the displacement vector $\mathbf{d}$:

$$W = (\text{comp}_{\mathbf{d}}\,\mathbf{F}) \times \|\mathbf{d}\|.$$

We can write this more briefly as a dot product:

(19.1.1)
$$\boxed{W = \mathbf{F} \cdot \mathbf{d}.}$$

This elementary notion of work is useful, but it is not sufficient. Consider, for example, an object that moves through a magnetic field or a gravitational field. The path of the motion is then usually not a straight line but a curve, and the force, rather than remain constant, tends to vary from point to point. What we want now is a notion of work that applies to this more general situation.

Let's suppose that at time t the object in question has position $\mathbf{r}(t)$ and is subject to a force $\mathbf{F}(\mathbf{r}(t))$. As the object moves through the force field from time a to time b, it traces out a curve

$$C: \mathbf{r}(t) = x(t)\mathbf{i} + y(t)\mathbf{j} + z(t)\mathbf{k}, \qquad t \in [a, b].$$

We will assume that this curve is *smooth*: that the tangent vector $\mathbf{r}'$ is continuous on $[a, b]$ and never $\mathbf{0}$. What we want to do here is define *the total work done by* $\mathbf{F}$ *along the curve C*.

To decide how to do this, we begin by focusing our attention on what happens during a short time interval $[t, t + h]$. As an estimate for the work done during this short time interval, we can use the dot product

$$\mathbf{F}(\mathbf{r}(t)) \cdot [\mathbf{r}(t + h) - \mathbf{r}(t)].$$

In making this estimate (see Figure 19.1.1), we are using the force vector at time t and we are replacing the curved path from $\mathbf{r}(t)$ to $\mathbf{r}(t + h)$ by the corresponding line segment.

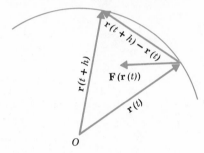

FIGURE 19.1.1

If we set

$$W(t) = \text{total work done up to time } t$$

and

$$W(t + h) = \text{total work done up to time } t + h,$$

then the work done from time t to time $t + h$ must be the difference

$$W(t + h) - W(t).$$

Bringing our estimate into play, we are led to the approximate equation

$$W(t + h) - W(t) \cong \mathbf{F}(\mathbf{r}(t)) \cdot [\mathbf{r}(t + h) - \mathbf{r}(t)],$$

which, upon division by h, becomes

$$\frac{W(t + h) - W(t)}{h} \cong \mathbf{F}(\mathbf{r}(t)) \cdot \frac{[\mathbf{r}(t + h) - \mathbf{r}(t)]}{h}.$$

The quotients here are average rates of change, and the equation is only an approximate one. The notion of work is made precise by requiring that both sides have exactly the same limit as h tends to zero; in other words, by requiring that

$$W'(t) = \mathbf{F}(\mathbf{r}(t)) \cdot \mathbf{r}'(t).$$

The rest is now determined. Since

$$W(a) = 0 \quad \text{and} \quad W(b) = \text{total work done},$$

we have

$$\text{total work done} = W(b) - W(a) = \int_a^b W'(t) \, dt = \int_a^b [\mathbf{F}(\mathbf{r}(t)) \cdot \mathbf{r}'(t)] \, dt.$$

In short, we have arrived at the following definition of work:

(19.1.2)
$$W = \int_a^b [\mathbf{F}(\mathbf{r}(t)) \cdot \mathbf{r}'(t)] \, dt.$$

The integral on the right is called a *line integral*. More precisely, it is called *the line integral of* $\mathbf{F}$ *over* C. Usually it's written

(19.1.3)
$$\int_C \mathbf{F}(\mathbf{r}) \cdot \mathbf{dr}.$$

In this notation we have

(19.1.4)
$$W = \int_C \mathbf{F}(\mathbf{r}) \cdot \mathbf{dr} = \int_a^b [\mathbf{F}(\mathbf{r}(t)) \cdot \mathbf{r}'(t)] \, dt.$$

Problem.   Find the total work done on an object that traverses the parabolic arc

$$C: \mathbf{r}(t) = t\mathbf{i} + t^2\mathbf{j}, \qquad t \in [0, 1]$$

subject to a force

$$\mathbf{F}(x, y) = xy\mathbf{i} + y^2\mathbf{j}.$$

SOLUTION.   The curve gives

$$x(t) = t, \quad y(t) = t^2.$$

Therefore

$$\mathbf{F}(\mathbf{r}(t)) = \mathbf{F}(x(t), y(t)) = x(t)y(t)\mathbf{i} + [y(t)]^2\mathbf{j} = t^3\mathbf{i} + t^4\mathbf{j}.$$

Since

$$\mathbf{r}'(t) = \mathbf{i} + 2t\mathbf{j},$$

we have

$$\mathbf{F}(\mathbf{r}(t)) \cdot \mathbf{r}'(t) = t^3 + 2t^5$$

and

$$W = \int_C \mathbf{F}(\mathbf{r}) \cdot \mathbf{dr} = \int_0^1 (t^3 + 2t^5) \, dt = \tfrac{7}{12}. \quad \square$$

Problem.   Find the total work done on an object that traverses the twisted cubic

$$C: \mathbf{r}(t) = t\mathbf{i} + t^2\mathbf{j} + t^3\mathbf{k}, \qquad t \in [-1, 1]$$

subject to a force

$$\mathbf{F}(x, y, z) = xy\mathbf{i} + yz\mathbf{j} + xz\mathbf{k}.$$

SOLUTION.   Here

$$x(t) = t, \quad y(t) = t^2, \quad z(t) = t^3.$$

Thus

$$\mathbf{F}(\mathbf{r}(t)) = \mathbf{F}(x(t), y(t), z(t)) = x(t)y(t)\mathbf{i} + y(t)z(t)\mathbf{j} + x(t)z(t)\mathbf{k} = t^3\mathbf{i} + t^5\mathbf{j} + t^4\mathbf{k}.$$

Since

$$\mathbf{r}'(t) = \mathbf{i} + 2t\mathbf{j} + 3t^2\mathbf{k},$$

you can see that

$$\mathbf{F}(\mathbf{r}(t)) \cdot \mathbf{r}'(t) = t^3 + 2t^6 + 3t^6 = t^3 + 5t^6.$$

It follows that

$$W = \int_C \mathbf{F}(\mathbf{r}) \cdot d\mathbf{r} = \int_{-1}^{1} [t^3 + 5t^6]\, dt = \tfrac{10}{7}. \quad \square$$

If a curve C is not smooth but is made up of a finite number of adjoining smooth pieces $C_1, C_2, \ldots, C_n$, then we define the integral over C (and therefore the work done) as the sum of the integrals over the C_i:

(19.1.5)
$$\boxed{\int_C = \int_{C_1} + \int_{C_2} + \cdots + \int_{C_n}.}$$

A curve of this type is said to be *piecewise smooth*. See Figure 19.1.2.

three piecewise smooth curves

FIGURE 19.1.2

Triangles are piecewise smooth curves. In the next problem we integrate over a triangle. We do this by integrating over each of the sides and then adding up the results. Keep in mind that the directed line segment which begins at $\mathbf{a} = (a_1, a_2)$ and ends at $\mathbf{b} = (b_1, b_2)$ can be parametrized by setting

$$\mathbf{r}(t) = (1 - t)\mathbf{a} + t\mathbf{b}, \qquad t \in [0, 1].$$

Problem.　Evaluate the line integral

$$\int_C \mathbf{F}(\mathbf{r}) \cdot d\mathbf{r}$$

if $\mathbf{F}(x, y) = e^y \mathbf{i} - \sin \pi x \, \mathbf{j}$ and C is the triangle with vertices $(1, 0), (0, 1), (-1, 0)$ traversed counterclockwise.

SOLUTION.　The path C is made up of three line segments:

$$C_1\colon \mathbf{r}(t) = (1 - t)\mathbf{i} + t\mathbf{j}, \qquad t \in [0, 1],$$
$$C_2\colon \mathbf{r}(t) = (1 - t)\mathbf{j} + t(-\mathbf{i}) = -t\mathbf{i} + (1 - t)\mathbf{j}, \qquad t \in [0, 1],$$
$$C_3\colon \mathbf{r}(t) = (1 - t)(-\mathbf{i}) + t\mathbf{i} = (2t - 1)\mathbf{i}, \qquad t \in [0, 1].$$

See Figure 19.1.3.

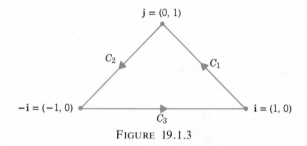

FIGURE 19.1.3

As you can verify,

$$\int_{C_1} \mathbf{F}(\mathbf{r}) \cdot d\mathbf{r} = \int_0^1 \left[e^{y(t)} x'(t) - \sin\left[\pi x(t)\right] y'(t) \right] dt$$

$$= \int_0^1 \left[-e^t - \sin\left[\pi(1 - t)\right] \right] dt$$

$$= \left[-e^t - \frac{1}{\pi} \cos\left[\pi(1 - t)\right] \right]_0^1 = 1 - e - \frac{2}{\pi};$$

$$\int_{C_2} \mathbf{F}(\mathbf{r}) \cdot d\mathbf{r} = \int_0^1 \left[e^{y(t)} x'(t) - \sin\left[\pi x(t)\right] y'(t) \right] dt$$

$$= \int_0^1 \left[-e^{1-t} + \sin\left(-\pi t\right) \right] dt$$

$$= \left[e^{1-t} + \frac{1}{\pi} \cos\left(-\pi t\right) \right]_0^1 = 1 - e - \frac{2}{\pi};$$

$$\int_{C_3} \mathbf{F}(\mathbf{r}) \cdot d\mathbf{r} = \int_0^1 \left[e^{y(t)} x'(t) - \sin\left[\pi x(t)\right] y'(t) \right] dt = \int_0^1 2 \, dt = 2.$$

As the integral over the entire triangle, we have

$$\int_C \mathbf{F}(\mathbf{r}) \cdot d\mathbf{r} = \left(1 - e - \frac{2}{\pi} \right) + \left(1 - e - \frac{2}{\pi} \right) + 2 = 4 - 2e - \frac{4}{\pi}. \quad \square$$

When we integrate over a parametrized curve, we integrate in the direction determined by the parametrization. If we integrate in the opposite direction, our answer is altered by a factor of -1. To see how this comes about (and incidentally clarify this idea), let C be a piecewise smooth curve and let $-C$ denote the same path traversed in the *opposite direction*. See Figure 19.1.4.

FIGURE 19.1.4

If C is parametrized by a vector function $\mathbf{r}$ defined on $[a, b]$, then $-C$ can be parametrized by setting

$$\mathbf{R}(u) = \mathbf{r}(a + b - u), \qquad u \in [a, b].†$$

Both $\mathbf{r}$ and $\mathbf{R}$ generate exactly the same points, but while C begins at $\mathbf{a}\,[\mathbf{r}(a) = \mathbf{a}]$ and ends at $\mathbf{b}\,[\mathbf{r}(b) = \mathbf{b}]$, the curve $-C$ begins at $\mathbf{b}\,[\mathbf{R}(a) = \mathbf{r}(b) = \mathbf{b}]$ and ends at $\mathbf{a}$ $[\mathbf{R}(b) = \mathbf{r}(a) = \mathbf{a}]$. Our assertion is that

(19.1.6)
$$\int_{-C} \mathbf{F}(\mathbf{R}) \cdot d\mathbf{R} = -\int_{C} \mathbf{F}(\mathbf{r}) \cdot d\mathbf{r}$$

or more briefly that

(19.1.7)
$$\int_{-C} = -\int_{C}.$$

PROOF.   We prove the result first in the case that C is smooth. Setting

$$t(u) = a + b - u \qquad \text{with } u \in [a, b],$$

we have

$$t(b) = a \quad \text{and} \quad t(a) = b.$$

Moreover,

$$\mathbf{R}(u) = \mathbf{r}(t(u)) \quad \text{and} \quad \mathbf{R}'(u) = \mathbf{r}'(t(u))t'(u).$$

---

† In a more formal treatment this is how $-C$ would be defined.

It follows that

$$\int_C \mathbf{F}(\mathbf{r}) \cdot d\mathbf{r} = \int_{a=t(b)}^{b=t(a)} [\mathbf{F}(\mathbf{r}(t)) \cdot \mathbf{r}'(t)]\, dt$$

$$= \int_b^a [\mathbf{F}(\mathbf{r}(t(u))) \cdot \mathbf{r}'(t(u))]t'(u)\, du \,^\dagger$$

$$= \int_b^a [\mathbf{F}(\mathbf{R}(u)) \cdot \mathbf{R}'(u)]\, du$$

$$= -\int_a^b [\mathbf{F}(\mathbf{R}(u)) \cdot \mathbf{R}'(u)]\, du$$

$$= -\int_{-C} \mathbf{F}(\mathbf{R}) \cdot d\mathbf{R}.$$

If C is only piecewise smooth, the result still holds because it holds for each of the smooth pieces. □

We'll come back to these ideas in the next section.

Exercises

Compute the line integral of $\mathbf{F}$ along the indicated curve.

*1. $\mathbf{F}(x, y) = y\mathbf{i} + x\mathbf{j}$; $\mathbf{r}(t) = t\mathbf{i} + t^2\mathbf{j}$, $t \in [0, 1]$.

2. $\mathbf{F}(x, y) = (x^2 - y)\mathbf{i} + x\mathbf{j}$; $\mathbf{r}(t) = (t - 2)\mathbf{i}$, $t \in [0, 2]$.

*3. $\mathbf{F}(x, y) = y\mathbf{i} + x\mathbf{j}$;   the unit circle traversed clockwise.

4. $\mathbf{F}(x, y) = y^2x\mathbf{i} + 2\mathbf{j}$; $\mathbf{r}(t) = e^t\mathbf{i} + e^{-t}\mathbf{j}$, $t \in [0, 1]$.

*5. $\mathbf{F}(x, y) = (x - y)\mathbf{i} + xy\mathbf{j}$;   the line segment from $(2, 3)$ to $(1, 2)$.

6. $\mathbf{F}(x, y) = (x - y)\mathbf{i} + xy\mathbf{j}$;   the line segment from $(1, 2)$ to $(2, 3)$.

*7. $\mathbf{F}(x, y) = (\cos x - y \sin x)\mathbf{i} + \cos x\,\mathbf{j}$;   the triangle with vertices $(0, 2)$, $(2, 3)$, $(1, 4)$ traversed counterclockwise.

8. $\mathbf{F}(x, y) = e^{x-y}\mathbf{i} + e^{x+y}\mathbf{j}$;   the line segment from $(-1, 1)$ to $(1, 2)$.

*9. $\mathbf{F}(x, y) = \dfrac{1}{xy^2}\mathbf{i} + \dfrac{1}{x^2y}\mathbf{j}$; $\mathbf{r}(t) = \sqrt{t}\,\mathbf{i} + \sqrt{1+t}\,\mathbf{j}$, $t \in [1, 4]$.

10. $\mathbf{F}(x, y) = \dfrac{1}{xy^2}\mathbf{i} + \dfrac{1}{x^2y}\mathbf{j}$;   the curve of Exercise 9 traversed in the opposite sense.

*11. $\mathbf{F}(x, y) = (x + y)\mathbf{i} + (y^2 - x)\mathbf{j}$;   the closed curve which begins at $(-1, 0)$, proceeds along the x-axis to $(1, 0)$, and returns to $(-1, 0)$ by the upper part of the unit circle.

12. $\mathbf{F}(x, y, z) = x\mathbf{i} + xy\mathbf{j} + xyz\mathbf{k}$;   the line segment from $(0, 1, 4)$ to $(1, 0, -4)$.

*13. $\mathbf{F}(x, y, z) = yz\mathbf{i} + x^2\mathbf{j} + xz\mathbf{k}$; $\mathbf{r}(t) = t\mathbf{i} + t^2\mathbf{j} + t^3\mathbf{k}$, $t \in [0, 1]$.

14. $\mathbf{F}(x, y, z) = e^{\sqrt{y}}\mathbf{i} + yz\mathbf{j} + e^{xy}\mathbf{k}$;   the curve of Exercise 13.

$\dagger$ Here we applied the change of variables formula

$$\int_{t(b)}^{t(a)} f(t)\, dt = \int_b^a f(t(u))t'(u)\, du \quad \text{to} \quad f(t) = \mathbf{F}(\mathbf{r}(t)) \cdot \mathbf{r}'(t).$$

*15. An object orbits an ellipse in standard position subject to a force

$$\mathbf{F}(x, y) = -\tfrac{1}{2}[y\mathbf{i} - x\mathbf{j}].$$

Find a relation between the work done and the area of the ellipse.

16. Show that, if C is a smooth curve which begins and ends at the same point, then

$$\int_C \nabla f(\mathbf{r}) \cdot d\mathbf{r} = 0.$$

19.2   The Fundamental Theorem for Line Integrals

There are many kinds of energy recognized in physics. Among the most important of these is *kinetic energy*, the energy possessed by an object because of its motion. The kinetic energy of an object depends on its mass m and on its speed v:

(19.2.1)
$$\boxed{\text{K.E.} = \tfrac{1}{2}mv^2.}$$

There is an important relation between work and kinetic energy which we now derive.

Suppose that an object of mass m moves along a curve

$$C: \mathbf{r}(t), \quad t \in [a, b]$$

subject to a force $\mathbf{F}$. The total work done on the object along the curve is given by the formula

$$W = \int_C \mathbf{F}(\mathbf{r}) \cdot d\mathbf{r} = \int_a^b [\mathbf{F}(\mathbf{r}(t)) \cdot \mathbf{r}'(t)]\, dt.$$

From Newton's second law of motion, we know that at time t,

$$\mathbf{F}(\mathbf{r}(t)) = m\,\mathbf{a}(t) = m\,\mathbf{r}''(t).$$

It follows then that

$$\mathbf{F}(\mathbf{r}(t)) \cdot \mathbf{r}'(t) = m\,\mathbf{r}''(t) \cdot \mathbf{r}'(t)$$
$$= \frac{d}{dt}\left[\frac{1}{2}m(\mathbf{r}'(t) \cdot \mathbf{r}'(t))\right] = \frac{d}{dt}\left[\frac{1}{2}m\|\mathbf{r}'(t)\|^2\right] = \frac{d}{dt}\left[\frac{1}{2}m[v(t)]^2\right].$$

Substituting this last expression into the work integral, we find that

$$W = \int_a^b \frac{d}{dt}\left(\frac{1}{2}m[v(t)]^2\right) dt = \frac{1}{2}m[v(b)]^2 - \frac{1}{2}m[v(a)]^2.$$

This last equation says that the work done on an object along a curve C is exactly equal to the difference in the kinetic energy of the object as measured at the endpoints of C. In short, we have the following result:

(19.2.2)
$$\boxed{\text{work done} = \text{change in kinetic energy.}}$$

In general, if an object moves from one point to another, the work done depends on the path of the motion. There is, however, an important exception. If the force field is a *gradient field*,

$$\mathbf{F} = \nabla f,$$

then the work done depends only on the endpoints of the path and not on the path itself. Returning to pure mathematics for a moment, we have the following theorem.

Theorem 19.2.3   The Fundamental Theorem for Line Integrals

Let C: $\mathbf{r}(t)$, $t \in [a, b]$ be a piecewise smooth curve which begins at $\mathbf{a}$ $[\mathbf{r}(a) = \mathbf{a}]$ and ends at $\mathbf{b}$ $[\mathbf{r}(b) = \mathbf{b}]$. If f is continuously differentiable on an open set that contains C, then

$$\int_C \nabla f(\mathbf{r}) \cdot d\mathbf{r} = f(\mathbf{b}) - f(\mathbf{a}). \qquad \text{(Figure 19.2.1)}$$

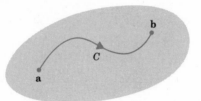

FIGURE 19.2.1

PROOF. If C is smooth,

$$\int_C \nabla f(\mathbf{r}) \cdot d\mathbf{r} = \int_a^b [\nabla f(\mathbf{r}(t)) \cdot \mathbf{r}'(t)] \, dt$$

$$= \int_a^b \frac{d}{dt} [f(\mathbf{r}(t))] \, dt = f(\mathbf{r}(b)) - f(\mathbf{r}(a)) = f(\mathbf{b}) - f(\mathbf{a}).$$

If C is not smooth but only piecewise smooth, then we break up C into smooth pieces

$$C = C_1 \cup C_2 \cup \cdots \cup C_n.$$

With obvious notation,

$$\int_C \nabla f(\mathbf{r}) \cdot d\mathbf{r} = \int_{C_1} \nabla f(\mathbf{r}) \cdot d\mathbf{r} + \int_{C_2} \nabla f(\mathbf{r}) \cdot d\mathbf{r} + \cdots + \int_{C_n} \nabla f(\mathbf{r}) \cdot d\mathbf{r}$$

$$= [f(\mathbf{a}_1) - f(\mathbf{a}_0)] + [f(\mathbf{a}_2) - f(\mathbf{a}_1)] + \cdots + [f(\mathbf{a}_n) - f(\mathbf{a}_{n-1})]$$

$$= f(\mathbf{a}_n) - f(\mathbf{a}_0) = f(\mathbf{b}) - f(\mathbf{a}). \quad \square$$

If the curve C is *closed* (that is, if $\mathbf{a} = \mathbf{b}$), then of course $f(\mathbf{b}) = f(\mathbf{a})$ and

$$\int_C \nabla f(\mathbf{r}) \cdot d\mathbf{r} = 0.$$

If a force field is a gradient field, then the work done along any closed path is 0. *An object which starts at a point with a given kinetic energy returns to that same point with exactly the same kinetic energy.*

We can use these ideas in computation.

Problem.   Find the work done on an object that traverses the helix

$$C: \mathbf{r}(t) = \cos t \, \mathbf{i} + \sin t \, \mathbf{j} + t \, \mathbf{k}, \qquad t \in [0, 2\pi]$$

subject to a force

$$\mathbf{F}(\mathbf{r}) = \frac{K}{\|\mathbf{r}\|^2} \, \mathbf{r}.$$

SOLUTION.   First we try to determine whether $\mathbf{F}$ is a gradient. In terms of x, y, z we have

$$\mathbf{F}(x, y, z) = \frac{K}{x^2 + y^2 + z^2} \, [x\mathbf{i} + y\mathbf{j} + z\mathbf{k}].$$

It's not hard to see that

$$\mathbf{F}(x, y, z) = \boldsymbol{\nabla} f(x, y, z) \qquad \text{where} \quad f(x, y, z) = \tfrac{1}{2} K \log (x^2 + y^2 + z^2).$$

In terms of $\mathbf{r}$, we have

$$f(\mathbf{r}) = \tfrac{1}{2} K \log \|\mathbf{r}\|^2 = K \log \|\mathbf{r}\|.$$

To compute the work integral, we need only evaluate f at the endpoints of C. Since

$$\mathbf{r}(0) = \mathbf{i} \quad \text{and} \quad \mathbf{r}(2\pi) = \mathbf{i} + 2\pi\mathbf{k},$$

we have

$$\begin{aligned}
W = \int_C \mathbf{F}(\mathbf{r}) \cdot d\mathbf{r} &= f(\mathbf{i} + 2\pi\mathbf{k}) - f(\mathbf{i}) \\
&= K \log \|\mathbf{i} + 2\pi\mathbf{k}\| - K \log \|\mathbf{i}\| \\
&= K \log \sqrt{1 + 4\pi^2} - K \log 1 = \tfrac{1}{2} K \log (1 + 4\pi^2). \quad \square
\end{aligned}$$

Problem.   Find the work done on an object that traverses the circular arc

$$C: \mathbf{r}(t) = \cos t \, \mathbf{i} + \sin t \, \mathbf{j}, \qquad t \in [0, \tfrac{1}{2}\pi]$$

subject to a force

$$\mathbf{F}(x, y) = y^2\mathbf{i} + (2xy - e^y)\mathbf{j}.$$

SOLUTION. $\mathbf{F}(x, y)$ has the form $P(x, y)\mathbf{i} + Q(x, y)\mathbf{j}$ with

$$P(x, y) = y^2 \quad \text{and} \quad Q(x, y) = 2xy - e^y.$$

Since P and Q are continuously differentiable everywhere and

$$\frac{\partial P}{\partial y} (x, y) = 2y = \frac{\partial Q}{\partial x} (x, y),$$

we can conclude that $\mathbf{F}(x, y)$ is a gradient (Theorem 17.9.1). Since the work integral depends then only on the endpoints of C and not on C itself, we can simplify the computations by integrating over the line segment C' that joins these same endpoints. See Figure 19.2.2.

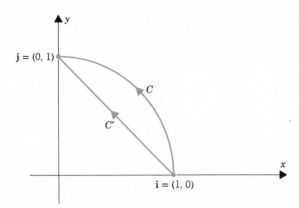

FIGURE 19.2.2

We parametrize C' by setting
$$\mathbf{r}(t) = (1 - t)\mathbf{i} + t\mathbf{j}, \qquad t \in [0, 1].$$
Then we have

$$\int_C \mathbf{F}(\mathbf{r}) \cdot d\mathbf{r} = \int_{C'} \mathbf{F}(\mathbf{r}) \cdot d\mathbf{r}$$

$$= \int_0^1 [\mathbf{F}(\mathbf{r}(t)) \cdot \mathbf{r}'(t)]\, dt$$

$$= \int_0^1 \left[[y(t)]^2 x'(t) + [2x(t)y(t) - e^{y(t)}]y'(t) \right] dt$$

$$= \int_0^1 [2t - 3t^2 - e^t]\, dt = \left[t^2 - t^3 - e^t \right]_0^1 = 1 - e. \quad \square$$

ALTERNATIVE SOLUTION.   Once we recognize that $\mathbf{F}(x, y) = y^2\mathbf{i} + (2xy - e^y)\mathbf{j}$ is a gradient $\nabla f(x, y)$, we can determine $f(x, y)$ by the methods of Section 17.9. Since

$$\frac{\partial f}{\partial x}(x, y) = y^2 \quad \text{and} \quad \frac{\partial f}{\partial y}(x, y) = 2xy - e^y,$$

we have

$$f(x, y) = xy^2 + \phi(y) \quad \text{and} \quad f(x, y) = xy^2 - e^y + \psi(x).$$

Equating these two expressions for $f(x, y)$, we have

$$\phi(y) = -e^y + \psi(x) \quad \text{and therefore} \quad \phi(y) + e^y = \psi(x).$$

It is obvious now that $\psi(x)$ is constant, so that

$$f(x, y) = xy^2 - e^y + c.$$

Since the curve C begins at $(1, 0)$ and ends at $(0, 1)$, we have

$$\int_C \mathbf{F(r)} \cdot \mathbf{dr} = f(0, 1) - f(1, 0) = -e + 1 = 1 - e. \quad \square$$

Problem.   Evaluate the line integral

$$\int_C \mathbf{F(r)} \cdot \mathbf{dr}$$

where C is the unit circle

$$\mathbf{r}(t) = \cos t \, \mathbf{i} + \sin t \, \mathbf{j}, \qquad t \in [0, 2\pi]$$

and

$$\mathbf{F}(x, y, z) = (y^2 + y)\mathbf{i} + (2xy - e^y)\mathbf{j}.$$

SOLUTION.   Although $(y^2 + y)\mathbf{i} + (2xy - e^y)\mathbf{j}$ is not a gradient, part of it,

$$y^2\mathbf{i} + (2xy - e^y)\mathbf{j},$$

is a gradient. (See the previous problem.) Since we are integrating over a closed curve, the contribution of the gradient part is 0. The contribution of the remaining part is

$$\int_C \mathbf{G(r)} \cdot \mathbf{dr} \quad \text{with} \quad \mathbf{G}(x, y) = y\mathbf{i}.$$

This last integral is easy to evaluate

$$\int_C \mathbf{G(r)} \cdot \mathbf{dr} = \int_0^{2\pi} [\mathbf{G}(\mathbf{r}(t)) \cdot \mathbf{r}'(t)] \, dt = \int_0^{2\pi} y(t)x'(t) \, dt = \int_0^{2\pi} -\sin^2 t \, dt = -\pi. \quad \square$$

Exercises

Compute the line integral of $\mathbf{F}$ along the indicated curve.

  *1. $\mathbf{F}(x, y) = x\mathbf{i} + y\mathbf{j}$; $\mathbf{r}(t) = a \cos t \, \mathbf{i} + b \sin t \, \mathbf{j}$, $t \in [0, 2\pi]$.
  2. $\mathbf{F}(x, y) = (x + y)\mathbf{i} + y\mathbf{j}$;   the curve of Exercise 1.
  *3. $\mathbf{F}(x, y) = \cos \pi y \, \mathbf{i} - \pi x \sin \pi y \, \mathbf{j}$; $\mathbf{r}(t) = t^2 \mathbf{i} - t^3 \mathbf{j}$, $t \in [0, 1]$.
  4. $\mathbf{F}(x, y) = (x^2 + \cos \pi y)\mathbf{i} + (x - \pi x \sin \pi y)\mathbf{j}$;   the curve of Exercise 1.
  *5. $\mathbf{F}(x, y) = xy^2\mathbf{i} + x^2 y\mathbf{j}$; $\mathbf{r}(t) = t \sin \pi t \, \mathbf{i} + \cos \pi t^2 \, \mathbf{j}$, $t \in [0, 1]$.
  6. $\mathbf{F}(x, y) = (1 + e^y)\mathbf{i} + (xe^y - x)\mathbf{j}$;   the square with vertices $(-1, -1), (1, -1),$
     $(1, 1), (-1, 1)$ traversed counterclockwise.
  *7. $\mathbf{F}(x, y) = (1 + e^y)\mathbf{i} + (xe^y - 4)\mathbf{j}$; $\mathbf{r}(t) = \cos t \, \mathbf{i} + \sin t \, \mathbf{j}$, $t \in [0, \pi]$.
Optional   8. $\mathbf{F}(x, y) = 2xy \sinh x^2 y\mathbf{i} + x^2 \sinh x^2 y\mathbf{j}$;   the curve of Exercise 1.
        *9. $\mathbf{F}(x, y) = [2x^2 y \sinh x^2 + y \cosh x^2]\mathbf{i} + [x(1 + \cosh x^2)]\mathbf{j}$;   the
        square of Exercise 6.

10. Show that a central force field

$$\mathbf{F(r)} = \frac{K}{\|\mathbf{r}\|^n} \mathbf{r}, \qquad n \text{ a positive integer}$$

is a gradient field. Find f such that $\nabla f(\mathbf{r}) = \mathbf{F}(\mathbf{r})$ if *(a) $n = 2$, *(b) $n \neq 2$.

11. Set

$$P(x, y) = \frac{y}{x^2 + y^2} \quad \text{and} \quad Q(x, y) = -\frac{x}{x^2 + y^2}$$

on the *punctured unit disc* S: $0 < x^2 + y^2 < 1$.

(a) Verify that P and Q are continuously differentiable on S and that

$$\frac{\partial P}{\partial y}(x, y) = \frac{\partial Q}{\partial x}(x, y) \qquad \text{for all } (x, y) \in S.$$

(b) Verify that, in spite of (a), the vector function $P(x, y)\mathbf{i} + Q(x, y)\mathbf{j}$ is not a gradient in S. HINT: Integrate the vector function over a circle of radius less than one centered at the origin.

(c) Explain how (b) does not contradict our remark after the proof of Theorem 17.9.1.

12. (*Important*) Let f be a real-valued function continuous on a piecewise smooth curve C: $\mathbf{r}(t) = x(t)\mathbf{i} + y(t)\mathbf{j} + z(t)\mathbf{k}$, $t \in [a, b]$. The line integral of f over C *with respect to arc length* is defined by setting

(19.2.4)
$$\int_C f(\mathbf{r}) \, ds = \int_a^b f(\mathbf{r}(t)) s'(t) \, dt.$$

If C represents a thin wire of varying density $\rho(\mathbf{r}) = \rho(x, y, z)$, then the total mass of the wire is given by

$$M = \int_C \rho(r) \, ds$$

and the center of mass is determined by the equations

$$x_M = \frac{1}{M} \int_C x \rho(x, y, z) ds, \quad y_M = \frac{1}{M} \int_C y \rho(x, y, z) ds, \quad z_M = \frac{1}{M} \int_C z \rho(x, y, z) ds.$$

*(a) Find the total mass of a wire in the form of a circular arc

$$C: \mathbf{r}(t) = a(\cos t \, \mathbf{i} + \sin t \, \mathbf{j}), \qquad t \in [0, \tfrac{1}{2}\pi]$$

if $\rho(x, y) = |x| + |y|$. Where is the center of mass?

(b) Find the total mass of a wire in the form

$$C: \mathbf{r}(t) = t\mathbf{i} + t^2\mathbf{j} + \tfrac{2}{3}t^3\mathbf{k}, \qquad t \in [-1, 1]$$

if $\rho(x, y, z) = x^2 + y^2 + z^2$. Where is the center of mass?

*(c) Show that

(19.2.5)
$$\int_C \mathbf{F}(\mathbf{r}) \cdot d\mathbf{r} = \int_C [\mathbf{F}(\mathbf{r}) \cdot \mathbf{T}(\mathbf{r})] \, ds$$

where $\mathbf{T}$ is the unit tangent vector.

19.3 Green's Theorem

A line integral taken over a simple closed plane curve can be expressed as a double integral taken over the region bounded by the curve. This result, known as Green's theorem, was discovered by George Green, an Englishman. He published the result in 1828.

Theorem 19.3.1   Green's Theorem

Let Ω be a closed region totally bounded by a piecewise smooth simple closed curve C. If

$$\mathbf{F}(x, y) = P(x, y)\mathbf{i} + Q(x, y)\mathbf{j}$$

is continuously differentiable on an open set that contains Ω, then

$$\int_C \mathbf{F(r)} \cdot \mathbf{dr} = \iint_\Omega \left[\frac{\partial Q}{\partial x}(x, y) - \frac{\partial P}{\partial y}(x, y) \right] dxdy$$

where the line integral is taken along C in the counterclockwise direction.

PROOF. We will prove the theorem only for regions Ω which are both of type I and type II in the sense of Figure 19.3.1.

FIGURE 19.3.1

Taking Ω to be of type I, we can show that

$$(1) \qquad \int_C P(\mathbf{r})\mathbf{i} \cdot \mathbf{dr} = \iint_\Omega - \frac{\partial P}{\partial y}(x, y)\, dxdy.$$

In the first place

$$\iint_\Omega - \frac{\partial P}{\partial y}(x, y)\, dxdy = - \int_a^b \int_{\phi_1(x)}^{\phi_2(x)} \frac{\partial P}{\partial y}(x, y)\, dy\, dx$$

$$= -\int_a^b \{P[x, \phi_2(x)] - P[x, \phi_1(x)]\}\, dx$$

$$(*) \qquad = \int_a^b P[x, \phi_1(x)]\, dx - \int_a^b P[x, \phi_2(x)]\, dx.$$

With

$$C_1: \mathbf{r}_1(t) = t\mathbf{i} + \phi_1(t)\mathbf{j}, \quad t \in [a, b] \qquad \text{and} \qquad C_2: \mathbf{r}_2(t) = t\mathbf{i} + \phi_2(t)\mathbf{j}, \quad t \in [a, b],$$

we have

$$\int_C P(\mathbf{r})\mathbf{i} \cdot d\mathbf{r} = \int_{C_1} P(\mathbf{r})\mathbf{i} \cdot d\mathbf{r} - \int_{C_2} P(\mathbf{r})\mathbf{i} \cdot d\mathbf{r}$$

$$= \int_a^b P[t, \phi_1(t)]\, dt - \int_a^b P[t, \phi_2(t)]\, dt.$$

Comparison with (∗) proves (1).

We leave it to you to verify that

$$(2) \qquad \int_C Q(\mathbf{r})\mathbf{j} \cdot d\mathbf{r} = \iint_\Omega \frac{\partial Q}{\partial x}(x, y)\, dxdy$$

by taking Ω to be of type II. This will complete the proof of the theorem. $\quad\square$

It is customary to set

$$\int_C P(\mathbf{r})\mathbf{i} \cdot d\mathbf{r} = \int_C P(x, y)\, dx \quad \text{and} \quad \int_C Q(\mathbf{r})\mathbf{j} \cdot d\mathbf{r} = \int_C Q(x, y)\, dy.$$

Green's equation is then written

$$(19.3.2) \qquad \boxed{\int_C P(x, y)\, dx + Q(x, y)\, dy = \iint_\Omega \left[\frac{\partial Q}{\partial x}(x, y) - \frac{\partial P}{\partial y}(x, y)\right] dxdy.}$$

Problem. Use Green's theorem to evaluate

$$\int_C xy^2\, dx + x^2 y\, dy$$

where C is the circle $x^2 + y^2 = a^2$ traversed counterclockwise.

SOLUTION. Let Ω be the disc $0 \le x^2 + y^2 \le a$. Then

$$\int_C xy^2\, dx + x^2 y\, dy = \iint_\Omega (2xy - 2xy)\, dxdy = 0. \quad \square$$

Problem. Use Green's theorem to evaluate

$$\int_C (1 + 10xy + y^2)\, dx + (6xy + 5x^2)\, dy$$

where C is the square with vertices $(0, 0)$, $(a, 0)$, (a, a), $(0, a)$ traversed counterclockwise.

SOLUTION.   Here

$$\frac{\partial Q}{\partial x}(x, y) = 6y + 10x \quad \text{and} \quad \frac{\partial P}{\partial y}(x, y) = 10x + 2y$$

so that the line integral equals

$$\iint_\Omega 4y \, dxdy = 4 \iint_\Omega y \, dxdy = 4\bar{y} \text{ (area of } \Omega) = 4a^2\bar{y}$$

$$\text{————(18.8.3)}$$

where $\bar{y}$ is the y-coordinate of the centroid of the square. Obviously $\bar{y} = \frac{1}{2}a$ and the integral equals $2a^3$. $\square$

Problem.   Use Green's theorem to evaluate

$$\int_C (4xy + y^2) \, dx - (xy + 3x^2) \, dy$$

where C is the circle $(x + 1)^2 + (y - 2)^2 = 9$ traversed counterclockwise.

SOLUTION.   In this case

$$\frac{\partial Q}{\partial x}(x, y) = -y - 6x \quad \text{and} \quad \frac{\partial P}{\partial y}(x, y) = 4x + 2y.$$

The line integral therefore equals

$$\iint_\Omega (-10x - 3y) \, dxdy = (-10\bar{x} - 3\bar{y})(\text{area of } \Omega) = (10 - 6)9\pi = 36\pi.   \square$$

Exercises

Verify Green's theorem in these two cases.

1. $\mathbf{F}(x, y) = x\mathbf{i} + y\mathbf{j}$; $\Omega: 0 \leq x^2 + y^2 \leq a^2$.
2. $\mathbf{F}(x, y) = -y\mathbf{i} + x\mathbf{j}$; $\Omega: 0 \leq x^2 + y^2 \leq a^2$.

Evaluate by Green's theorem integrating over C counterclockwise.

*3. $\int_C 3y \, dx + 5x \, dy$; $C: x^2 + y^2 = 1$.

4. $\int_C 5x \, dx + 3y \, dy$; $C: (x - 1)^2 + (y + 1)^2 = 1$.

*5. $\int_C x^2 \, dy$; C the rectangle with vertices $(0, 0)$, $(a, 0)$, (a, b), $(0, b)$.

6. $\int_C y^2 \, dx$; C the rectangle with vertices $(0, 0)$, $(a, 0)$, (a, b), $(0, b)$.

*7. $\int_C (3xy + y^2) \, dx + (2xy + 5x^2) \, dy$; $C: (x - 1)^2 + (y + 2)^2 = 1$.

8. $\int_C (xy + 3y^2) \, dx + (5xy + 2x^2) \, dy$; $C: (x - 1)^2 + (y + 2)^2 = 1$.

*9. $\int_C (2x^2 + xy - y^2) \, dx + (3x^2 - xy + 2y^2) \, dy$; $C: (x - a)^2 + y^2 = r^2$.

10. $\int_C (x^2 - 2xy + 3y^2) \, dx + (5x + 1) \, dy$; $C: x^2 + (y - b)^2 = r^2$.

*11. $\int_C e^x \sin y \, dx + e^x \cos y \, dy$; $C: (x - a)^2 + (y - b)^2 = r^2$.

12. $\int_C e^x \cos y \, dx + e^x \sin y \, dy$; C the rectangle with vertices $(0, 0)$, $(1, 0)$, $(1, \pi)$, $(0, \pi)$.

13. Show that, if C is a piecewise smooth simple closed curve traversed counter-clockwise, then

$$\int_C x \, dy, \qquad \int_C - y \, dx, \qquad \tfrac{1}{2} \int_C x \, dy - y \, dx$$

all give the area enclosed by C.

14. Use Exercise 13 to calculate the area enclosed by
    (a) the ellipse $b^2 x^2 + a^2 y^2 = a^2 b^2$.   *(b) the astroid $x^{2/3} + y^{2/3} = a^{2/3}$.
    HINT: Use trigonometric parametrizations.

*15. Determine the area under one arch of the cycloid:

$$x(t) = r(t - \sin t), \quad y(t) = r(1 - \cos t); \qquad t \in [0, 2\pi].$$

16. Determine the simple closed curve C that maximizes the line integral

$$\int_C y^3 \, dx + (3x - x^3) \, dy.$$

17. Complete the proof of Green's theorem by showing that

$$\int_C Q(\mathbf{r})\mathbf{j} \cdot \mathbf{dr} = \iint_\Omega \frac{\partial Q}{\partial x} (x, y) \, dxdy.$$

19.4   Multiple Riemann Sums; Surface Area

Multiple Riemann Sums

Earlier in this text we described two methods of obtaining the definite integral

$$I = \int_a^b f(x) \, dx.$$

The Darboux method: I is the unique number which satisfies the inequality

$$L_f(P) \leq I \leq U_f(P) \qquad \text{for all partitions } P \text{ of } [a, b]$$

and the Riemann method:

$$I = \lim_{\|P\| \mapsto 0} S^*(P) = \lim_{\|P\| \mapsto 0} \sum_{i=1}^{m} f(x_i^*) \Delta x_i.$$

So far our work with multiple integrals has been based entirely on the ideas of Darboux. It is time to describe the method of Riemann. For simplicity we begin with the two variable case.

Consider a function f continuous on a rectangle

$$R: a \leq x \leq b, \ c \leq y \leq d.$$

Following Darboux, we defined the double integral

$$\iint_R f(x, y) \, dx dy$$

as the unique number I such that

$$L_f(P) \leq I \leq U_f(P) \qquad \text{for all partitions } P \text{ of } R.$$

We can obtain this same integral by the method of Riemann.

Starting with

$$P_1 = \{x_0, \ \ldots, x_m\} \quad \text{as a partition of } [a, b]$$

and

$$P_2 = \{y_0, \ \ldots, y_n\} \quad \text{as a partition of } [c, d]$$

we form

$$P = P_1 \times P_2 = \{(x_i, y_j): x_i \in P_1, \ y_j \in P_2\}.$$

This is a partition of R which breaks up R into $m \times n$ nonoverlapping rectangles

$$R_{ij} = \{(x, y): x_{i-1} \leq x \leq x_i, \ y_{j-1} \leq y \leq y_j\}.$$

We define $\|P\|$, the *norm* of P, by setting

$$\|P\| = \|P_1\| + \|P_2\| = \max \Delta x_i + \max \Delta y_j.$$

If we were to follow Darboux, we would form the usual lower and upper sums:

$$L_f(P) = \sum_{i=1}^{m} \sum_{j=1}^{n} m_{ij} \Delta x_i \Delta y_j \quad \text{and} \quad U_f(P) = \sum_{i=1}^{m} \sum_{j=1}^{n} M_{ij} \Delta x_i \Delta y_j.$$

Instead of doing that, we choose an arbitrary point (x_i^*, y_j^*) from each rectangle R_{ij} and form the sum

$$S^*(P) = \sum_{i=1}^{m} \sum_{j=1}^{n} f(x_i^*, y_j^*) \Delta x_i \Delta y_j.$$

Such a sum is called a *Riemann sum*. The double integral

$$\iint_R f(x, y) \, dxdy$$

can be obtained as the limit of such Riemann sums in the sense that, given any $\epsilon > 0$, there exists a $\delta > 0$ such that

$$\text{if} \quad \|P\| < \delta \quad \text{then} \quad \left| S^*(P) - \iint_R f(x, y) \, dxdy \right| < \epsilon$$

no matter how the $(x_i{}^*, y_j{}^*)$ are chosen within the R_{ij}. In symbols we write

$$(19.4.1) \quad \iint_R f(x, y) \, dxdy = \lim_{\|P\| \to 0} S^*(P) = \lim_{\|P\| \to 0} \sum_{i=1}^m \sum_{j=1}^n f(x_i^*, y_j^*) \, \Delta x_i \Delta y_j.$$

For functions of three variables we have

$$(19.4.2) \quad \iiint_\Pi f(x, y, z) \, dxdydz = \lim_{\|P\| \to 0} \sum_{i=1}^m \sum_{j=1}^n \sum_{k=1}^q f(x_i^*, y_j^*, z_k^*) \Delta x_i \Delta y_j \Delta z_k$$

where this time $(x_i{}^*, y_j{}^*, z_k{}^*)$ is chosen arbitrarily from the box Π_{ijk} and the norm of P is defined by setting

$$\|P\| = \|P_1\| + \|P_2\| + \|P_3\| = \max \Delta x_i + \max \Delta y_j + \max \Delta z_k.$$

Surface Area

Our discussion of surface area in Section 11.8 was limited to surfaces of revolution. Here we explain how to calculate the area of a surface $z = f(x, y)$ defined over a region Ω of the xy-plane.

We begin with a simple case: a plane surface

$$z = Ax + By + C$$

defined over a rectangle R. See Figure 19.4.1.

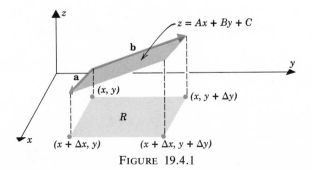

FIGURE 19.4.1

As suggested by the figure, this surface is a parallelogram. What we want here is the area of the parallelogram.

The vector marked **a** in the figure begins at the point $(x, y, Ax + By + C)$ and ends at the point $(x + \Delta x, y, A(x + \Delta x) + By + C)$. It's easy to see then that

$$\mathbf{a} = \Delta x \mathbf{i} + A \Delta x \mathbf{k} = \Delta x(\mathbf{i} + A\mathbf{k}).$$

The vector marked **b** begins at the point $(x, y, Ax + By + C)$ and ends at the point $(x, y + \Delta y, Ax + B(y + \Delta y) + C)$. This gives

$$\mathbf{b} = \Delta y \mathbf{j} + B \Delta y \mathbf{k} = \Delta y(\mathbf{j} + B\mathbf{k}).$$

The area of the parallelogram is the norm of the cross product of these two vectors:

$$\|\mathbf{a} \times \mathbf{b}\| = \|\Delta x(\mathbf{i} + A\mathbf{k}) \times \Delta y(\mathbf{j} + B\mathbf{k})\|$$
$$= \|(\mathbf{i} + A\mathbf{k}) \times (\mathbf{j} + B\mathbf{k})\| \, \Delta x \, \Delta y$$
$$= \|-A\mathbf{i} - B\mathbf{j} + \mathbf{k}\| \, \Delta x \, \Delta y$$
$$(*) = \sqrt{A^2 + B^2 + 1} \, \Delta x \, \Delta y.$$

This result will help us determine the area of a more general surface.

FIGURE 19.4.2

Figure 19.4.2 shows a surface S which projects onto a region Ω of the xy-plane. If S is given by a continuously differentiable function $z = f(x, y)$, then

(19.4.3)   the surface area of $S = \displaystyle\iint\limits_{\Omega} \sqrt{\left[\frac{\partial f}{\partial x}(x, y)\right]^2 + \left[\frac{\partial f}{\partial y}(x, y)\right]^2 + 1} \; dxdy.$

We will derive the formula under the simplifying assumption that the base region is a rectangle $R: a \le x \le b, c \le y \le d$ as in Figure 19.4.3. A partition of R breaks up R into $m \times n$ little rectangles

$$R_{ij} = \{(x, y): x_{i-1} \le x \le x_i, \; y_{j-1} \le y \le y_j\}.$$

FIGURE 19.4.3   FIGURE 19.4.4

Figure 19.4.3 singles out such a rectangle and the portion of the surface directly above it. As in Figure 19.4.4, we take (x_i^*, y_j^*) as the center of R_{ij} and replace the surface over R_{ij} by the tangent plane at $(x_i^*, y_j^*, f(x_i^*, y_j^*))$. The equation of this tangent plane can be written

$$z = \frac{\partial f}{\partial x} (x_i^*, y_j^*) x + \frac{\partial f}{\partial y} (x_i^*, y_j^*) y + C. \tag{17.4.6}$$

By $(*)$ its area over R_{ij} is given by the expression

$$\sqrt{\left[\frac{\partial f}{\partial x} (x_i^*, y_j^*)\right]^2 + \left[\frac{\partial f}{\partial y} (x_i^*, y_j^*)\right]^2 + 1} \; \Delta x_i \Delta y_j.$$

This expression gives an estimate for the area of the surface over R_{ij}. Adding up all these area estimates, we have an estimate for the total area of the surface:

$$\sum_{i=1}^{m} \sum_{j=1}^{n} \sqrt{\left[\frac{\partial f}{\partial x} (x_i^*, y_j^*)\right]^2 + \left[\frac{\partial f}{\partial y} (x_i^*, y_j^*)\right]^2 + 1} \; \Delta x_i \Delta y_j.$$

This double sum is a Riemann sum. As the norm of the partition tends to zero, this Riemann sum tends to the double integral

$$\iint\limits_{R} \sqrt{\left[\frac{\partial f}{\partial x} (x, y)\right]^2 + \left[\frac{\partial f}{\partial y} (x, y)\right]^2 + 1} \; dxdy. \quad \square$$

Problem.   Find the surface area of that part of the parabolic cylinder $z = y^2$ which lies over the triangle with vertices $(0, 0)$, $(1, 0)$, $(1, 1)$ in the xy-plane.

SOLUTION. Here $f(x, y) = y^2$ so that

$$\frac{\partial f}{\partial x}(x, y) = 0, \qquad \frac{\partial f}{\partial y}(x, y) = 2y.$$

The base triangle can be expressed by writing

$$\Omega: 0 \le y \le 1, \ 0 \le x \le y.$$

The surface has area

$$A[S] = \iint_\Omega \sqrt{\left[\frac{\partial f}{\partial x}(x, y)\right]^2 + \left[\frac{\partial f}{\partial y}(x, y)\right]^2 + 1} \, dxdy$$

$$= \int_0^1 \int_0^y \sqrt{4y^2 + 1} \, dxdy$$

$$= \int_0^1 y\sqrt{4y^2 + 1} \, dy$$

$$= \left[\tfrac{1}{12}(4y^2 + 1)^{3/2}\right]_0^1 = \tfrac{1}{12}(5\sqrt{5} - 1). \quad \square$$

Problem. Find the surface area of that part of the hyperbolic paraboloid $z = xy$ which lies inside the cylinder $x^2 + y^2 = a^2$.

SOLUTION. Here $f(x, y) = xy$ so that

$$\frac{\partial f}{\partial x}(x, y) = y, \quad \frac{\partial f}{\partial y}(x, y) = x.$$

The formula gives

$$A[S] = \iint_\Omega \sqrt{x^2 + y^2 + 1} \, dxdy.$$

In polar coordinates the base region takes the form

$$0 \le r \le a, \quad 0 \le \theta \le 2\pi.$$

Thus we have

$$A[S] = \iint_\Omega \sqrt{r^2 + 1} \, r \, drd\theta = \int_0^{2\pi} \int_0^a \sqrt{r^2 + 1} \, r \, drd\theta$$

$$= \tfrac{2}{3}\pi[(a^2 + 1)\sqrt{a^2 + 1} - 1]. \quad \square$$

The surface area formula

$$A[S] = \iint_\Omega \sqrt{\left[\frac{\partial f}{\partial x}(x, y)\right]^2 + \left[\frac{\partial f}{\partial y}(x, y)\right]^2 + 1} \, dxdy$$

cannot be applied directly to a hemispherical surface

$$f(x, y) = \sqrt{a^2 - (x^2 + y^2)}, \qquad 0 \le x^2 + y^2 \le a^2$$

because the partial derivatives

$$\frac{\partial f}{\partial x}(x, y) = \frac{-x}{\sqrt{a^2 - (x^2 + y^2)}}, \qquad \frac{\partial f}{\partial y}(x, y) = \frac{-y}{\sqrt{a^2 - (x^2 + y^2)}}$$

are both infinite on the boundary of the base region Ω. You can overcome this difficulty by starting with a slightly smaller base region

$$\Omega_\epsilon : 0 \le x^2 + y^2 \le (a - \epsilon)^2,$$

computing

$$\iint\limits_{\Omega_\epsilon} \sqrt{\left[\frac{\partial f}{\partial x}(x, y)\right]^2 + \left[\frac{\partial f}{\partial y}(x, y)\right]^2 + 1}\; dxdy$$

and then taking the limit as $\epsilon \downarrow 0$. Carry out this procedure and you'll see that

$$A[S] = \lim_{\epsilon \downarrow 0} \iint\limits_{\Omega_\epsilon} \sqrt{\left[\frac{\partial f}{\partial x}(x, y)\right]^2 + \left[\frac{\partial f}{\partial y}(x, y)\right]^2 + 1}\; dxdy$$

$$= \lim_{\epsilon \downarrow 0} \iint\limits_{\Omega_\epsilon} \frac{a}{\sqrt{a^2 - (x^2 + y^2)}}\; dxdy$$

$$= \lim_{\epsilon \downarrow 0} \int_0^{2\pi} \int_0^{a-\epsilon} \frac{ar}{\sqrt{a^2 - r^2}}\; drd\theta = 2\pi a^2. \quad \square$$

Exercises

Find the area of the surface.

*1. $bcx + acy + abz = abc$ with $x \ge 0$, $y \ge 0$, $z \ge 0$.

2. $z^2 = x^2 + y^2$ with $0 \le x^2 + y^2 \le 1$, $z \ge 0$.

*3. $2z = a^2(x^2 + y^2)$ over the sector $r_1 \le r \le r_2$, $\theta_1 \le \theta \le \theta_2$.

4. $z^2 = 2xy$ with $0 \le x \le a$, $0 \le y \le b$, $z \ge 0$.

*5. $z = a^2 - (x^2 + y^2)$ with $\frac{1}{4}a^2 \le x^2 + y^2 \le a^2$.

6. $3z^2 = (x + y)^3$ with $x + y \le 2$, $x \ge 0$, $y \ge 0$.

*7. $3z = x^{3/2} + y^{3/2}$ with $0 \le x \le 1$, $0 \le y \le x$.

8. $x^2 + y^2 + z^2 - 2az = 0$ with $0 \le x^2 + y^2 \le bz$. Assume $a > b > 0$.

*9. $x^2 + y^2 + z^2 - 4z = 0$ with $0 \le 3(x^2 + y^2) \le z^2$, $z \ge 2$.

10. (*Important*) A surface S is said to be *smooth* iff it has a unit normal **n** which is continuous throughout S. Show that a surface of the form

$$S: z = f(x, y), \quad (x, y) \in \Omega$$

is smooth if f is continuously differentiable on Ω.

11. (*Important*) Let S be a smooth surface projectable onto a region Ω of the xy-plane:

$$S: z = f(x, y), (x, y) \in \Omega \qquad \text{with } f \text{ continuously differentiable.}$$

Show that

(19.4.4)
$$\sqrt{\left[\frac{\partial f}{\partial x}(x, y)\right]^2 + \left[\frac{\partial f}{\partial y}(x, y)\right]^2 + 1} = \sec\left[\gamma(x, y)\right]$$

where $\gamma(x, y)$ is the angle between the upper normal at $(x, y, f(x, y))$ and the positive z-axis. The surface area formula can thus be written

(19.4.5)
$$A[S] = \int\int_\Omega \sec\left[\gamma(x, y)\right] dx dy.$$

*12. (a) Find a formula for the area of a smooth surface S which is projectable onto a region Ω of the yz-plane; say

$$S: x = g(y, z), \quad (y, z) \in \Omega.$$

(b) Find a formula for the area of a smooth surface S which is projectable onto a region Ω of the xz-plane; say

$$S: y = h(x, z), \quad (x, z) \in \Omega.$$

19.5 The Mass of a Material Surface; Surface Integrals

The Mass of a Material Surface

Imagine a material surface, say a curved thin sheet of metal. If the density of the metal is constant throughout the surface, then the total mass of the surface is the density times the surface area. If, on the other hand, the density varies from point to point, then the total mass of the surface must be calculated by integration.

Figure 19.5.1 depicts a material surface S which projects onto a rectangle R of the xy-plane. Let's suppose that the density varies continuously from point to point, say $\rho = \rho(x, y, z)$, and let's suppose that S itself is given by a continuously differentiable function $z = f(x, y), (x, y) \in R$. To estimate the mass of S we take a partition P of R. This breaks up R into $m \times n$ little rectangles

$$R_{ij} = \{(x, y): x_{i-1} \leq x \leq x_i, \ y_{j-1} \leq y \leq y_j\}$$

and breaks up S into $m \times n$ little pieces

$$S_{ij}: z = f(x, y), \quad (x, y) \in R_{ij}.$$

One such little piece S_{ij} is singled out in Figure 19.5.2. If (x_i^*, y_j^*) is the center of R_{ij}, then the density at the corresponding point of S_{ij} is $\rho[x_i^*, y_j^*, f(x_i^*, y_j^*)]$. If R_{ij} is small enough, then the density is approximately the same throughout S_{ij} (remember f and ρ are continuous) and therefore

$$\text{mass of } S_{ij} \cong \rho[x_i^*, y_j^*, f(x_i^*, y_j^*)] \times (\text{surface area of } S_{ij}).$$

FIGURE 19.5.1

FIGURE 19.5.2

Taking

$$\text{surface area of } S_{ij} \cong \sqrt{\left[\frac{\partial f}{\partial x}(x_i^*, y_j^*)\right]^2 + \left[\frac{\partial f}{\partial y}(x_i^*, y_j^*)\right]^2 + 1}\; \Delta x_i \Delta y_j$$

$$= \sec\left[\gamma(x_i^*, y_j^*)\right] \Delta x_i \Delta y_j,$$

(19.4.4)

we have

$$\text{mass of } S_{ij} \cong \rho[x_i^*, y_j^*, f(x_i^*, y_j^*)] \sec\left[\gamma(x_i^*, y_j^*)\right] \Delta x_i \Delta y_j.$$

Adding up these estimates you can see that

$$\text{the total mass of } S \cong \sum_{i=1}^{n} \sum_{j=1}^{m} \rho[x_i^*, y_j^*, f(x_i^*, y_j^*)] \sec\left[\gamma(x_i^*, y_j^*)\right] \Delta x_i \Delta y_j.$$

The expression on the right is a Riemann sum which, as an estimate for the total mass of S, becomes increasingly more accurate as the norm of the partition tends to zero. We conclude therefore that

$$\text{the total mass of } S = \lim_{\|P\|\to 0} \sum_{i=1}^{n} \sum_{j=1}^{m} \rho[x_i^*, y_j^*, f(x_i^*, y_j^*)] \sec\left[\gamma(x_i^*, y_j^*)\right] \Delta x_i \Delta y_j$$

$$= \iint_{R} \rho[x, y, f(x, y)] \sec\left[\gamma(x, y)\right] dx dy.$$

If S projects, not onto a rectangle R, but onto a region Ω, then the double integral on the right must be taken over the region Ω. For a material surface of the form $S: z = f(x, y)$, $(x, y) \in \Omega$ of varying density $\rho = \rho(x, y, z)$

(19.5.1) $\quad$ the total mass of $S = \displaystyle\iint_{\Omega} \rho[x, y, f(x, y)] \sec\left[\gamma(x, y)\right] dx dy.$

Surface Integrals

The integral on the right in (19.5.1) makes sense if we replace the density ρ by any continuous function H.

Definition 19.5.2   Surface Integral

For any function H continuous on a smooth surface $S: z = f(x, y), (x, y) \in \Omega$ we define

$$\iint_S H(x, y, z) \, d\sigma = \iint_\Omega H[x, y, f(x, y)] \sec [\gamma(x, y)] \, dxdy$$

where $\gamma(x, y)$ is the angle between the upper normal at $(x, y, f(x, y))$ and the positive z-axis.

If $H(x, y, z)$ is constantly 1, the surface integral takes the form

$$\iint_S d\sigma.$$

This just gives the *surface area of S*. In general

$$\iint_S H(x, y, z) \, d\sigma$$

can be interpreted as *the average value of H on S times the surface area of S*. If M is the total mass of a material surface, then

$$M = \iint_S \rho(x, y, z) \, d\sigma = \text{(the average density of } S) \times \text{(the surface area of } S).$$

Problem.   Find the total mass of a material surface

$$S: z = xy, \quad 0 \le x^2 + y^2 \le 1$$

if the density at each point is proportional to its distance from the z-axis.

SOLUTION

$$z = f(x, y) = xy; \qquad \frac{\partial f}{\partial x} = y, \quad \frac{\partial f}{\partial y} = x.$$

$$\sec[\gamma(x, y)] = \sqrt{\left(\frac{\partial f}{\partial x}\right)^2 + \left(\frac{\partial f}{\partial y}\right)^2 + 1} = \sqrt{y^2 + x^2 + 1} = \sqrt{x^2 + y^2 + 1}.$$

Therefore

$$M = \iint_S k \sqrt{x^2 + y^2} \, d\sigma$$

$$= \iint_\Omega k \sqrt{x^2 + y^2} \sec[\gamma(x, y)] \, dxdy = \iint_\Omega k \sqrt{x^2 + y^2} \sqrt{x^2 + y^2 + 1} \ \ dxdy$$

where Ω is the unit disc. Changing to polar coordinates, we have

$$M = \iint_\Omega kr\sqrt{r^2 + 1} \, r \, dr d\theta$$

$$= k \int_0^{2\pi} \int_0^1 r^2 \sqrt{r^2 + 1} \, dr d\theta$$

$$= 2\pi k \int_0^1 r^2 \sqrt{r^2 + 1} \, dr$$

set $r = \tan\phi$

$$= 2\pi k \int_0^{\pi/4} \tan^2\phi \sec^3\phi \, d\phi = 2\pi k \int_0^{\pi/4} [\sec^5\phi - \sec^3\phi] \, d\phi$$

$$= \tfrac{1}{4}\pi k[3\sqrt{2} - \log(\sqrt{2} + 1)]. \quad \square$$

In physical applications vector-valued functions $\mathbf{v} = \mathbf{v}(x, y, z)$ are usually called *vector fields*. By contrast real-valued functions $f = f(x, y, z)$ are called *scalar fields*.

Surface integrals are commonly used in connection with vector fields. Suppose that S is a smooth surface on which we have chosen a continuous unit normal $\mathbf{n} = \mathbf{n}(x, y, z)$. For vector field $\mathbf{v} = \mathbf{v}(x, y, z)$ continuous on S, the surface integral

$$\iint_S (\mathbf{v} \cdot \mathbf{n}) \, d\sigma = \iint_S [\mathbf{v}(x, y, z) \cdot \mathbf{n}(x, y, z)] \, d\sigma$$

is called *the flux of* $\mathbf{v}$ *across* S.

This flux integral definitely depends on the choice of unit normal. If $-\mathbf{n}$ is chosen instead of $\mathbf{n}$, the sign of the flux is reversed:

$$\iint_S (\mathbf{v} \cdot -\mathbf{n}) \, d\sigma = \iint_S -(\mathbf{v} \cdot \mathbf{n}) \, d\sigma = -\iint_S (\mathbf{v} \cdot \mathbf{n}) \, d\sigma.$$

Flux integrals play a prominent role in the study of fluid motion. Imagine a surface S within a fluid and choose a unit normal $\mathbf{n}$. Think of $\mathbf{v}$ as the velocity of the fluid that passes through S. You can expect $\mathbf{v}$ to vary from point to point, but let's stipulate that $\mathbf{v}$ does not change with time. Such a time-independent flow is said to be *steady-state*. The flux of $\mathbf{v}$ across S is the average component of $\mathbf{v}$ in the direction of $\mathbf{n}$ times the area of the surface S. This is just the volume of fluid that passes through S in unit time, from the $-\mathbf{n}$ side of S to the $\mathbf{n}$ side of S. If S is a closed surface and $\mathbf{n}$ is chosen as the *outer* unit normal, the flux across S gives the volume of fluid that flows *out* through S in unit time; if $\mathbf{n}$ is chosen as the *inner* unit normal, then the flux gives the volume of liquid that flows *in* through S in unit time.

Theorem 19.5.3

Let $S: z = f(x, y)$, $(x, y) \in \Omega$ be a smooth surface and let $\mathbf{n}$ be the upper unit normal. For each vector field $\mathbf{v}$ continuous on S,

$$\iint_S (\mathbf{v} \cdot \mathbf{n}) \, d\sigma = \iint_\Omega \left(-v_1 \frac{\partial f}{\partial x} - v_2 \frac{\partial f}{\partial y} + v_3 \right) dx\,dy.$$

PROOF.   The surface is given implicitly by the equation

$$G(x, y, z) = 0 \qquad \text{where} \quad G(x, y, z) = z - f(x, y).$$

Since

$$\nabla G = \frac{\partial G_1}{\partial x} \mathbf{i} + \frac{\partial G_2}{\partial y} \mathbf{j} + \frac{\partial G_3}{\partial z} \mathbf{k} = -\frac{\partial f}{\partial x} \mathbf{i} - \frac{\partial f}{\partial y} \mathbf{j} + \mathbf{k}$$

is perpendicular to the surface at each point, the upper unit normal can be written

$$\mathbf{n} = \frac{1}{\sqrt{(\partial f/\partial x)^2 + (\partial f/\partial y)^2 + 1}} \left(-\frac{\partial f}{\partial x} \mathbf{i} - \frac{\partial f}{\partial y} \mathbf{j} + \mathbf{k} \right)$$

$$= \frac{1}{\sec \gamma} \left(-\frac{\partial f}{\partial x} \mathbf{i} - \frac{\partial f}{\partial y} \mathbf{j} + \mathbf{k} \right).$$

It follows that

$$\iint_S (\mathbf{v} \cdot \mathbf{n}) \, d\sigma = \iint_\Omega (\mathbf{v} \cdot \mathbf{n}) \sec \gamma \, dx\,dy = \iint_\Omega \left(-v_1 \frac{\partial f}{\partial x} - v_2 \frac{\partial f}{\partial y} + v_3 \right) dx\,dy. \quad \square$$

Problem.   Evaluate the flux

$$\iint_S (\mathbf{v} \cdot \mathbf{n}) \, d\sigma$$

for the vector field $\mathbf{v}(x, y, z) = x\mathbf{i} + y\mathbf{j} + z\mathbf{k}$ over the surface $z = 1 - (x^2 + y^2)$, $z \geq 0$ if $\mathbf{n}$ is chosen as the upper unit normal.

SOLUTION

$$f(x, y) = 1 - (x^2 + y^2); \qquad \frac{\partial f}{\partial x} = -2x, \quad \frac{\partial f}{\partial y} = -2y.$$

$$\iint_S (\mathbf{v} \cdot \mathbf{n}) \, d\sigma = \iint_\Omega \left(-v_1 \frac{\partial f}{\partial x} - v_2 \frac{\partial f}{\partial y} + v_3 \right) dx\,dy$$

$$= \iint_\Omega [-x(-2x) - y(-2y) + 1 - (x^2 + y^2)] \, dxdy$$

$$= \iint_\Omega (1 + x^2 + y^2) \, dxdy$$

in polar coordinates ——↑ $\quad = \iint_\Omega (1 + r^2)r \, drd\theta = \int_0^{2\pi} \int_0^1 (r + r^3) \, drd\theta = \tfrac{3}{2}\pi. \quad \square$

We have been evaluating surface integrals over smooth surfaces which are projectable onto the xy-plane. If S is projectable onto the xz-plane, say

$$S: y = g(x, z) \quad \text{with} \quad (x, z) \in \Omega,$$

then

$$\iint_S H \, d\sigma = \iint_\Omega H \sec \beta \, dxdz$$

where β is the angle between the unit normal with the positive $\mathbf{j}$ component and the positive y-axis. If S is projectable onto the yz-plane, say

$$S: x = h(y, z) \quad \text{with} \quad (y, z) \in \Omega,$$

then

$$\iint_S H \, d\sigma = \iint_\Omega H \sec \alpha \, dydz$$

where α is the angle between the unit normal with positive $\mathbf{i}$ component and the positive x-axis.

Problem.　Compute the total flux of

$$\mathbf{v}(x, y, z) = xy\mathbf{i} + 4yz^2\mathbf{j} + yz\mathbf{k}$$

out of the unit cube: $0 \le x \le 1, 0 \le y \le 1, 0 \le z \le 1$.

SOLUTION.　The total flux will be the sum of the fluxes across the faces of the cube:

face	$\mathbf{n}$	$\mathbf{v} \cdot \mathbf{n}$	flux
$x = 0$	$-\mathbf{i}$	$-xy = 0$	0
$x = 1$	$\mathbf{i}$	$xy = y$	$\tfrac{1}{2}$
$y = 0$	$-\mathbf{j}$	$-4yz^2 = 0$	0
$y = 1$	$\mathbf{j}$	$4yz^2 = 4z^2$	$\tfrac{4}{3}$
$z = 0$	$-\mathbf{k}$	$-yz = 0$	0
$z = 1$	$\mathbf{k}$	$yz = y$	$\tfrac{1}{2}$

The total flux is $\tfrac{1}{2} + \tfrac{4}{3} + \tfrac{1}{2} = \tfrac{7}{3}$. $\quad \square$

Exercises

1. Evaluate these integrals over the surface $S: z = \frac{1}{2}y^2, 0 \le x \le 1, 0 \le y \le 1$.

*(a) $\displaystyle\iint_S d\sigma$.

(b) $\displaystyle\iint_S x^2 \, d\sigma$.

*(c) $\displaystyle\iint_S 3y \, d\sigma$.

*(d) $\displaystyle\iint_S (x - y) \, d\sigma$.

(e) $\displaystyle\iint_S \sqrt{2z} \, d\sigma$.

*(f) $\displaystyle\iint_S \sqrt{1 + y^2} \, d\sigma$.

2. Find the mass of a material surface in the shape of a triangle $(a, 0, 0)$, $(0, a, 0)$, $(0, 0, a)$ with the given density function.

   *(a) $\rho(x, y, z) = k$.   (b) $\rho(x, y, z) = k(x + y)$.   *(c) $\rho(x, y, z) = kx^2$.

3. Find the mass of the material surface $S: z = 1 - \frac{1}{2}(x^2 + y^2)$ with $0 \le x \le 1$, $0 \le y \le 1$ if the density at each point (x, y, z) is proportional to xy.

4. Determine the flux

$$\iint_S (\mathbf{v} \cdot \mathbf{n}) \, d\sigma$$

across the triangle $(a, 0, 0)$, $(0, a, 0)$, $(0, 0, a)$ for each of the following vector functions. Take $\mathbf{n}$ as the upper unit normal.

   *(a) $\mathbf{v} = x\mathbf{i} + y\mathbf{j} + z\mathbf{k}$.   (b) $\mathbf{v} = x^2\mathbf{i} - y^2\mathbf{j}$.   *(c) $\mathbf{v} = (x + z)\mathbf{k}$.

5. Determine the flux across $S: z = xy$ with $0 \le x \le 1, 0 \le y \le 2$ for each of the following vector fields. Take $\mathbf{n}$ as the upper unit normal.

   *(a) $\mathbf{v} = x^2y\mathbf{i} + z^2\mathbf{k}$.   (b) $\mathbf{v} = -xy^2\mathbf{i} + z\mathbf{j}$.   *(c) $\mathbf{v} = xz\mathbf{j} - xy\mathbf{k}$.

6. Determine the flux across $S: z = \frac{2}{3}(x^{3/2} + y^{3/2})$ with $0 \le x \le 1, 0 \le y \le 1 - x$ for each of the following vector fields. Take $\mathbf{n}$ as the upper unit normal.

   *(a) $\mathbf{v} = x\mathbf{i} - y\mathbf{j} + \frac{3}{2}z\mathbf{k}$.   (b) $\mathbf{v} = x^2\mathbf{i}$.

   *(c) $\mathbf{v} = y^2\mathbf{j}$.   (d) $\mathbf{v} = y\mathbf{i} - \sqrt{xy}\,\mathbf{j}$.

The *centroid* of a surface S is the point $(\bar{x}, \bar{y}, \bar{z})$ where

$$\bar{x} = \frac{1}{\text{area of } S} \iint_S x \, d\sigma, \quad \bar{y} = \frac{1}{\text{area of } S} \iint_S y \, d\sigma, \quad \bar{z} = \frac{1}{\text{area of } S} \iint_S z \, d\sigma.$$

*7. Find the centroid of the triangle $(1, 0, 0)$, $(0, 1, 0)$, $(0, 0, 1)$.

8. Find the centroid of the surface $z = \frac{1}{2}(x^2 + y^2), 0 \le z \le 4$.

*9. Find the centroid of a hemisphere $x^2 + y^2 + z^2 = a^2, z \ge 0$.

10. Find the total flux of $\mathbf{v} = xz\mathbf{i} + 4xyz^2\mathbf{j} + 2z\mathbf{k}$ out of the unit cube: $0 \le x \le 1$, $0 \le y \le 1, 0 \le z \le 1$.

19.6 The Divergence Theorem

The Divergence Theorem

The *divergence* of a vector field $\mathbf{v} = \mathbf{v}(x, y, z)$ is defined by setting

(19.6.1)
$$\operatorname{div} \mathbf{v} = \frac{\partial v_1}{\partial x} + \frac{\partial v_2}{\partial y} + \frac{\partial v_3}{\partial z},$$

provided, of course, that these partials exist. For plane vector fields $\mathbf{v} = v_1\mathbf{i} + v_2\mathbf{j}$ we have

(19.6.2)
$$\operatorname{div} \mathbf{v} = \frac{\partial v_1}{\partial x} + \frac{\partial v_2}{\partial y}.$$

Green's theorem enables us to express a line integral over a simple closed plane curve as a double integral over the region enclosed by the curve. The relation

$$\int_C \mathbf{F}(\mathbf{r}) \cdot d\mathbf{r} = \iint_\Omega \left(\frac{\partial Q}{\partial x} - \frac{\partial P}{\partial y} \right) dx\, dy$$

can be written

(19.6.3)
$$\int_C (\mathbf{v} \cdot \mathbf{n})\, ds = \iint_\Omega \operatorname{div} \mathbf{v}\, dx\, dy,$$

where $\mathbf{n}$ is taken as the outer unit normal.

PROOF. Begin with $\mathbf{F} = P\mathbf{i} + Q\mathbf{j}$ and set $\mathbf{v} = Q\mathbf{i} - P\mathbf{j}$. Obviously

$$\iint_\Omega \operatorname{div} \mathbf{v}\, dx\, dy = \iint_\Omega \left(\frac{\partial v_1}{\partial x} + \frac{\partial v_2}{\partial y} \right) dx\, dy = \iint_\Omega \left(\frac{\partial Q}{\partial x} - \frac{\partial P}{\partial y} \right) dx\, dy.$$

Remember that in Green's theorem the curve C is traversed counterclockwise. From this it follows that $\mathbf{n} = \mathbf{T} \times \mathbf{k}$, where $\mathbf{T}$ is the unit tangent vector. (Draw a figure.) Thus

$$\mathbf{v} \cdot \mathbf{n} = \mathbf{v} \cdot (\mathbf{T} \times \mathbf{k}) = (-\mathbf{v}) \cdot (\mathbf{k} \times \mathbf{T}) = (-\mathbf{v} \times \mathbf{k}) \cdot \mathbf{T}.$$

in general $\mathbf{a} \cdot (\mathbf{b} \times \mathbf{c}) = (\mathbf{a} \times \mathbf{b}) \cdot \mathbf{c}$

Since

$$-\mathbf{v} \times \mathbf{k} = (P\mathbf{j} - Q\mathbf{i}) \times \mathbf{k} = P\mathbf{i} + Q\mathbf{j} = \mathbf{F},$$

we have

$$\int_C (\mathbf{v} \cdot \mathbf{n})\, ds = \int_C (\mathbf{F} \cdot \mathbf{T})\, ds = \int_C \mathbf{F}(\mathbf{r}) \cdot d\mathbf{r}. \quad \square$$

19.2.5

Green's theorem as expressed in (19.6.3) has a higher dimensional analog which is known as the divergence theorem.†

Theorem 19.6.4   The Divergence Theorem

Let T be a solid totally bounded by a closed surface S which, if not smooth, consists of at most a finite number of smooth pieces. If the components of $\mathbf{v} = \mathbf{v}(x, y, z)$ are continuously differentiable on T, then

$$\iint_S (\mathbf{v} \cdot \mathbf{n}) \, d\sigma = \iiint_T \operatorname{div} \mathbf{v} \, dxdydz$$

where $\mathbf{n}$ is the outer unit normal.

PROOF.   We will carry out the proof under the assumption that S is smooth and that no line parallel to one of the coordinate axes intersects S at more than two points.

Taking $\mathbf{n} = \cos \alpha_1 \, \mathbf{i} + \cos \alpha_2 \, \mathbf{j} + \cos \alpha_3 \, \mathbf{k}$, we have

$$\mathbf{v} \cdot \mathbf{n} = v_1 \cos \alpha_1 + v_2 \cos \alpha_2 + v_3 \cos \alpha_3.$$

The equation to be proved is then equivalent to the following three equations:

$$\iint_S v_1 \cos \alpha_1 \, d\sigma = \iiint_T \frac{\partial v_1}{\partial x} \, dxdydz,$$

$$\iint_S v_2 \cos \alpha_2 \, d\sigma = \iiint_T \frac{\partial v_2}{\partial y} \, dxdydz,$$

$$\iint_S v_3 \cos \alpha_3 \, d\sigma = \iiint_T \frac{\partial v_3}{\partial z} \, dxdydz.$$

All three equations can be verified in much the same manner. We will carry out the details for the third equation. Let's denote by Ω the projection of T onto the xy-plane. Vertical lines through the points of Ω intersect S in at most two points, an upper point P^+ and a lower point P^-. The set of all upper points forms a surface

$$S^+: z = f^+(x, y), \quad (x, y) \in \Omega$$

and the set of all lower points forms a surface

$$S^-: z = f^-(x, y), \quad (x, y) \in \Omega.$$

---

† Sometimes called Gauss's theorem after the German mathematician Carl Friedrich Gauss (1777–1855). Often referred to as "The Prince of Mathematicians," Gauss is regarded by many as one of the greatest geniuses of all time.

Together these surfaces comprise the total surface S:

$$S = S^+ \cup S^-. \qquad\qquad \text{(Figure 19.6.1)}$$

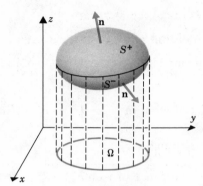

FIGURE 19.6.1

The solid T can thus be characterized as the set of all points (x, y, z) such that

$$f^-(x, y) \leq z \leq f^+(x, y), \quad (x, y) \in \Omega.$$

Observe now that

$$\iiint_T \frac{\partial v_3}{\partial z}\, dx\, dy\, dz = \iint_\Omega \left(\int_{f^-(x,y)}^{f^+(x,y)} \frac{\partial v_3}{\partial z}\, dz \right) dx\, dy$$

$$= \iint_\Omega \{v_3[x, y, f^+(x, y)] - v_3[x, y, f^-(x, y)]\}\, dx\, dy$$

$$(*) \qquad = \iint_\Omega v_3[x, y, f^+(x, y)]\, dx\, dy - \iint_\Omega v_3[x, y, f^-(x, y)]\, dx\, dy.$$

On S^+ the outer unit normal $\mathbf{n}$ is the upper unit normal, and on S^- it is the lower unit normal. Thus $\cos \alpha_3 \sec \gamma$ is 1 on S^+ but -1 on S^-. It follows that

$$\iint_{S^+} v_3 \cos \alpha_3\, d\sigma = \iint_\Omega v_3[x, y, f^+(x, y)]\, dx\, dy,$$

$$\iint_{S^-} v_3 \cos \alpha_3\, d\sigma = - \iint_\Omega v_3[x, y, f^-(x, y)]\, dx\, dy$$

and thus

$$\iint_S v_3 \cos \alpha_3\, d\sigma = \iint_{S^+} v_3 \cos \alpha_3\, d\sigma + \iint_{S^-} v_3 \cos \alpha_3\, d\sigma$$

$$= \iint_\Omega v_3[x, y, f^+(x, y)]\, dx\, dy - \iint_\Omega v_3[x, y, f^-(x, y)]\, dx\, dy.$$

Combining this with (∗) we have the third equation

$$\iint_S v_3 \cos \alpha_3 \, d\sigma = \iiint_T \frac{\partial v_3}{\partial z} \, dx\,dy\,dz. \quad \square$$

Div v as Flux per Unit Volume

Choose a point P and surround it by a closed ball N_ϵ of radius ϵ. According to the divergence theorem

$$\text{the flux of } \mathbf{v} \text{ out of } N_\epsilon = \iiint_N \text{div } \mathbf{v} \, dx\,dy\,dz.$$

In other words the flux of $\mathbf{v}$ out of N_ϵ is

(the average divergence of $\mathbf{v}$ on N_ϵ) $\times$ (the volume of N_ϵ).

Division by the volume factor gives

$$\text{the average divergence of } \mathbf{v} \text{ on } N_\epsilon = \frac{\text{the flux of } \mathbf{v} \text{ out of } N_\epsilon}{\text{the volume of } N_\epsilon}.$$

Take the limit of both sides as ϵ shrinks to 0, and you will see that

$$\textbf{(19.6.5)} \qquad \text{the divergence of } \mathbf{v} \text{ at } P = \lim_{\epsilon \downarrow 0} \frac{\text{the flux of } \mathbf{v} \text{ out of } N_\epsilon}{\text{the volume of } N_\epsilon}.$$

In this sense divergence is *flux per unit volume*.

Think of $\mathbf{v}$ for the moment as the velocity of a fluid in steady-state motion. A positive divergence at a point signals a movement of liquid away from that point:

$$\text{div } \mathbf{v} > 0 \text{ at } P \Rightarrow \text{flux out of } N_\epsilon > 0 \Rightarrow \text{net flow out of } N_\epsilon.$$

A negative divergence at a point signals a movement of liquid toward that point:

$$\text{div } \mathbf{v} < 0 \text{ at } P \Rightarrow \text{flux out of } N_\epsilon < 0 \Rightarrow \text{net flow into } N_\epsilon.$$

Points at which the divergence is positive are called *sources*; points at which the divergence is negative are called *sinks*. If the divergence of $\mathbf{v}$ is 0 throughout, then the flow has no sources and no sinks and $\mathbf{v}$ is called *solenoidal*.

Exercises

Find the divergence.

*1. $\mathbf{v}(x, y) = x\mathbf{i} + y\mathbf{j}$.   2. $\mathbf{v}(x, y) = -y\mathbf{i} + x\mathbf{j}$.

*3. $\mathbf{v}(x, y) = \dfrac{x}{x^2 + y^2} \mathbf{i} + \dfrac{y}{x^2 + y^2} \mathbf{j}$.

*4. $\mathbf{v}(x, y, z) = xyz\mathbf{i} + xz\mathbf{j} + z\mathbf{k}$.   5. $\mathbf{v}(x, y, z) = x^2 y\mathbf{i} + y^2 z\mathbf{j} + z^2 x\mathbf{k}$.

*6. $\mathbf{v}(x, y, z) = \dfrac{x}{x^2 + y^2 + z^2} \mathbf{i} + \dfrac{y}{x^2 + y^2 + z^2} \mathbf{j} + \dfrac{z}{x^2 + y^2 + z^2} \mathbf{k}$.

7. Verify that

$$\text{div } (\mathbf{u} + \mathbf{v}) = \text{div } \mathbf{u} + \text{div } \mathbf{v} \qquad \text{and} \qquad \text{div } (\alpha \mathbf{v}) = \alpha \text{ div } \mathbf{v}.$$

8. Show that

$$\text{div } (f\mathbf{v}) = f \text{ div } \mathbf{v} + \nabla f \cdot \mathbf{v}.$$

9. The divergence of a gradient is the *Laplacian*

$$\nabla^2 f = \frac{\partial^2 f}{\partial x^2} + \frac{\partial^2 f}{\partial y^2} + \frac{\partial^2 f}{\partial z^2}.$$

Show that

$$\text{div } (f \, \nabla f) = f \, \nabla^2 f + \|\nabla f\|^2.$$

*10. Compute div $(\nabla f \times \nabla g)$ if f and g have continuous second partials.

Verify the divergence theorem on the unit ball $x^2 + y^2 + z^2 \leq 1$ for the following vector fields.

*11. $\mathbf{v}(x, y, z) = x\mathbf{i} + y\mathbf{j} + z\mathbf{k}$.
12. $\mathbf{v}(x, y, z) = (1 - x)\mathbf{i} + (2 - y)\mathbf{j} + (3 - z)\mathbf{k}$.
*13. $\mathbf{v}(x, y, z) = x^2\mathbf{i} + y^2\mathbf{j} + z^2\mathbf{k}$.
14. $\mathbf{v}(x, y, z) = (1 - x^2)\mathbf{i} - y^2\mathbf{j} + z\mathbf{k}$.

Verify the divergence theorem on the unit cube $0 \leq x \leq 1, 0 \leq y \leq 1, 0 \leq z \leq 1$ for the following vector fields.

*15. $\mathbf{v}(x, y, z) = x\mathbf{i} + y\mathbf{j} + z\mathbf{k}$.   16. $\mathbf{v}(x, y, z) = xy\mathbf{i} + yz\mathbf{j} + xz\mathbf{k}$.
*17. $\mathbf{v}(x, y, z) = x^2\mathbf{i} - xz\mathbf{j} + z^2\mathbf{k}$.   18. $\mathbf{v}(x, y, z) = x\mathbf{i} + xy\mathbf{j} + xyz\mathbf{k}$.

Use the divergence theorem to find the total flux out of the given solid.

*19. $\mathbf{v}(x, y, z) = x\mathbf{i} + 2y^2\mathbf{j} + 3z^2\mathbf{k};$ $0 \leq x^2 + y^2 \leq 9,$ $0 \leq z \leq 1.$
20. $\mathbf{v}(x, y, z) = xy\mathbf{i} + yz\mathbf{j} + xz\mathbf{k};$ $0 \leq x \leq 1,$ $0 \leq y \leq 1 - x,$ $0 \leq z \leq 1 - x - y.$

19.7 Stokes's Theorem

The *curl* of a vector field $\mathbf{v} = \mathbf{v}(x, y, z)$ is defined by the equation

(19.6.6)
$$\text{curl } \mathbf{v} = \left(\frac{\partial v_3}{\partial y} - \frac{\partial v_2}{\partial z} \right) \mathbf{i} + \left(\frac{\partial v_1}{\partial z} - \frac{\partial v_3}{\partial x} \right) \mathbf{j} + \left(\frac{\partial v_2}{\partial x} - \frac{\partial v_1}{\partial y} \right) \mathbf{k},$$

provided that these partials exist. If you are familiar with determinants, you can write

(19.7.1)
$$\text{curl } \mathbf{v} = \begin{vmatrix} \mathbf{i} & \mathbf{j} & \mathbf{k} \\ \dfrac{\partial}{\partial x} & \dfrac{\partial}{\partial y} & \dfrac{\partial}{\partial z} \\ v_1 & v_2 & v_3 \end{vmatrix}$$

as a mnemonic device and then expand by elements of the first row.

For plane vector fields $\mathbf{v} = v_1\mathbf{i} + v_2\mathbf{j}$, the definition reduces to

(19.7.2)
$$\text{curl } \mathbf{v} = \left(\frac{\partial v_2}{\partial x} - \frac{\partial v_1}{\partial y}\right)\mathbf{k}$$

and therefore Green's theorem can be written

(19.7.3)
$$\int_C \mathbf{v(r)} \cdot d\mathbf{r} = \int\int_\Omega (\text{curl } \mathbf{v} \cdot \mathbf{k})\, dx dy.$$

This formulation of Green's theorem is generalized by Stokes's theorem.†

Theorem 19.7.4   Stokes's Theorem.

Let $S: z = f(x, y)$, $(x, y) \in \Omega$ be a smooth surface with upper unit normal $\mathbf{n}$. Assume that f has continuous second partials and that the boundary of S is a piecewise smooth curve C which we take to be directed counterclockwise with respect to $\mathbf{n}$. If $\mathbf{v} = \mathbf{v}(x, y, z)$ has continuously differentiable components on S, then

$$\int_C \mathbf{v(r)} \cdot d\mathbf{r} = \int\int_S (\text{curl } \mathbf{v} \cdot \mathbf{n})\, d\sigma.$$

PROOF.   Set $\mathbf{n} = \cos \alpha_1 \mathbf{i} + \cos \alpha_2 \mathbf{j} + \cos \alpha_3 \mathbf{k}$. The surface integral can be written

$$\int\int_S \left[\left(\frac{\partial v_3}{\partial y} - \frac{\partial v_2}{\partial z}\right)\cos \alpha_1 + \left(\frac{\partial v_1}{\partial z} - \frac{\partial v_3}{\partial x}\right)\cos \alpha_2 + \left(\frac{\partial v_2}{\partial x} - \frac{\partial v_1}{\partial y}\right)\cos \alpha_3\right] d\sigma.$$

Recall that

$$\mathbf{n} = \left(-\frac{\partial f}{\partial x}\mathbf{i} - \frac{\partial f}{\partial y}\mathbf{j} + \mathbf{k}\right)\cos \gamma$$

where γ is the angle between the upper normal and the positive z-axis. Thus

$$\cos \alpha_1 = -\frac{\partial f}{\partial x}\cos \gamma, \quad \cos \alpha_2 = -\frac{\partial f}{\partial y}\cos \gamma, \quad \cos \alpha_3 = \cos \gamma.$$

We will prove the theorem by showing that

$$\int_C \mathbf{v(r)} \cdot d\mathbf{r} = \int\int_\Omega \left[\left(\frac{\partial v_2}{\partial z} - \frac{\partial v_3}{\partial y}\right)\frac{\partial f}{\partial x} + \left(\frac{\partial v_3}{\partial x} - \frac{\partial v_1}{\partial z}\right)\frac{\partial f}{\partial y} + \left(\frac{\partial v_2}{\partial x} - \frac{\partial v_1}{\partial y}\right)\right] dx dy.$$

---

† The result was formulated by George Gabriel Stokes (1819–1903), who, like Green, was a Cambridge don. What we give here is a very restricted statement of Stokes's theorem. More general versions are outside the scope of this book.

Our approach will be to verify that

(1)
$$\int_C v_1 \, dx = -\iint_\Omega \left(\frac{\partial v_1}{\partial y} + \frac{\partial v_1}{\partial z} \frac{\partial f}{\partial y} \right) dxdy,$$

(2)
$$\int_C v_2 \, dy = \iint_\Omega \left(\frac{\partial v_2}{\partial x} + \frac{\partial v_2}{\partial z} \frac{\partial f}{\partial x} \right) dxdy,$$

(3)
$$\int_C v_3 \, dz = \iint_\Omega \left(\frac{\partial v_3}{\partial x} \frac{\partial f}{\partial y} - \frac{\partial v_3}{\partial y} \frac{\partial f}{\partial x} \right) dxdy.$$

Take

$$C: \mathbf{r}(t) = x(t)\mathbf{i} + y(t)\mathbf{j} + z(t)\mathbf{k}, \qquad t \in [0, 1]$$

and let C_{xy} be the projection of C onto the xy-plane:

$$C_{xy}: \mathbf{r}(t) = x(t)\mathbf{i} + y(t)\mathbf{j}, \qquad t \in [0, 1].$$

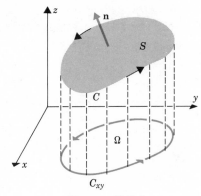

FIGURE 19.7.1

In the first place

$$\int_C v_1 \, dx = \int_0^1 v_1[x(t), y(t), z(t)]x'(t) \, dt$$

$$= \int_0^1 v_1[x(t), y(t), f(x(t), y(t))]x'(t) \, dt$$

$$\underset{\substack{\uparrow \\ \text{set } H(x,y) = v_1[x, y, f(x, y)]}}{=} \int_0^1 H[x(t), y(t)]x'(t)$$

$$= \int_{C_{xy}} H \, dx$$

$$\underset{\substack{\uparrow \\ \text{by Green's theorem}}}{=} -\iint_\Omega \frac{\partial H}{\partial y} \, dxdy$$

$$\underset{\substack{\uparrow \\ \text{by the chain rule}}}{=} -\iint_\Omega \left(\frac{\partial v_1}{\partial y} + \frac{\partial v_1}{\partial z} \frac{\partial f}{\partial y} \right) dxdy.$$

This verifies equation (1). We leave it to you to verify equation (2) and go on to the last equation.

$$\int_C v_3 \, dz = \int_0^1 v_3[x(t), y(t), z(t)]z'(t) \, dt$$

$$= \int_0^1 v_3[x(t), y(t), f(x(t), y(t))] \frac{d}{dt} [f(x(t), y(t))] \, dt.$$

Setting $H(x, y) = v_3[x, y, f(x, y)]$ and noting that

$$\frac{d}{dt} [f(x(t), y(t))] = \frac{\partial f}{\partial x} (x(t), y(t))x'(t) + \frac{\partial f}{\partial y} (x(t), y(t))y'(t),$$

we have

$$\int_C v_3 \, dz = \int_0^1 \left[H(x(t), y(t)) \frac{\partial f}{\partial x} (x(t), y(t))x'(t) \right.$$

$$\left. + H(x(t), y(t)) \frac{\partial f}{\partial y} (x(t), y(t))y'(t) \right] dt$$

$$= \int_{C_{xy}} H \frac{\partial f}{\partial x} \, dx + H \frac{\partial f}{\partial y} \, dy$$

$$\underset{\text{by Green's theorem}}{=} \iint_\Omega \left(H \frac{\partial^2 f}{\partial x \, \partial y} + \frac{\partial H}{\partial x} \frac{\partial f}{\partial y} - H \frac{\partial^2 f}{\partial y \, \partial x} - \frac{\partial H}{\partial y} \frac{\partial f}{\partial x} \right) dx dy$$

$$\underset{\text{since } \frac{\partial^2 f}{\partial x \partial y} = \frac{\partial^2 f}{\partial y \partial x}}{=} \iint_\Omega \left(\frac{\partial H}{\partial x} \frac{\partial f}{\partial y} - \frac{\partial H}{\partial y} \frac{\partial f}{\partial x} \right) dx dy$$

$$\underset{\text{by the chain rule}}{=} \iint_\Omega \left[\left(\frac{\partial v_3}{\partial x} + \frac{\partial v_3}{\partial z} \frac{\partial f}{\partial x} \right) \frac{\partial f}{\partial y} - \left(\frac{\partial v_3}{\partial y} + \frac{\partial v_3}{\partial z} \frac{\partial f}{\partial y} \right) \frac{\partial f}{\partial x} \right] dx dy$$

$$= \iint_\Omega \left(\frac{\partial v_3}{\partial x} \frac{\partial f}{\partial y} - \frac{\partial v_3}{\partial y} \frac{\partial f}{\partial x} \right) dx dy.$$

This proves equation (3) and thus the theorem. $\square$

The Normal Component of Curl **v** as Circulation per Unit Area

Take $\mathbf{v} = \mathbf{v}(x, y, z)$ as the velocity of a fluid flow. Let C be a circle within the flow and let $\mathbf{T}$ be the unit tangent vector. The line integral

$$\int_C \mathbf{v}(\mathbf{r}) \cdot \mathbf{dr} = \int_C (\mathbf{v} \cdot \mathbf{T}) \, ds$$

is called the *circulation of* **v** *around* C.

For velocities **v** of a given magnitude the circulation measures the extent to which **v** keeps the direction of **T**, which is to say, the extent to which the flow is a rotation along C.

Now take a point P within the flow and at that point choose a unit vector **n**. Let D_ϵ be the ϵ-disc which is centered at P and is perpendicular to **n**. Let C_ϵ be the boundary of D_ϵ directed counterclockwise with respect to **n**. See Figure 19.7.2.

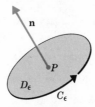

FIGURE 19.7.2

By Stokes's theorem the circulation of **v** around C_ϵ equals

(the average **n**-component of curl **v** on D_ϵ) × (the area of D_ϵ).

It follows that

$$\text{the average } \textbf{n}\text{-component of curl } \textbf{v} \text{ on } D_\epsilon = \frac{\text{the circulation of } \textbf{v} \text{ around } C_\epsilon}{\text{the area of } D_\epsilon}.$$

Taking the limit as ϵ shrinks to 0, you can see that

$$\text{the } \textbf{n}\text{-component of curl } \textbf{v} \text{ at } P = \lim_{\epsilon \downarrow 0} \frac{\text{the circulation of } \textbf{v} \text{ around } C_\epsilon}{\text{the area of } D_\epsilon}.$$

*At each point P the component of curl **v** in any direction **n** is the circulation of **v** per unit area in the plane normal to **n**.*

Curl **v** and the Angular Velocity Vector

To specify a rotation about a point P we must specify an axis of rotation, a sense of rotation, and an angular speed. All this can be done simply and elegantly by a single vector $\boldsymbol{\omega}$, called the *angular velocity vector*: the axis of rotation is the line determined by the vector $\boldsymbol{\omega}$ placed so as to emanate from P; the sense of rotation is set so that the direction appears counterclockwise to an observer at the tip of $\boldsymbol{\omega}$; the angular speed is taken as $\|\boldsymbol{\omega}\|$. See Figure 19.7.3.

angular speed = $\|\boldsymbol{\omega}\|$

FIGURE 19.7.3

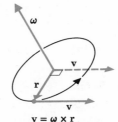

$\textbf{v} = \boldsymbol{\omega} \times \textbf{r}$

FIGURE 19.7.4

It's not hard to show that

(19.7.5)

if **v** is the velocity vector of a rotation, then

$$\mathbf{v} = \boldsymbol{\omega} \times \mathbf{r}$$

where $\boldsymbol{\omega}$ is the angular velocity vector and **r** is the radius vector.

PROOF. In the first place $\mathbf{v} \perp \boldsymbol{\omega}$ and $\mathbf{v} \perp \mathbf{r}$. Moreover, as illustrated in Figure 19.7.4, the vectors $\boldsymbol{\omega}, \mathbf{r}, \mathbf{v}$ form a right-handed system. All we need to verify is that

$$\|\mathbf{v}\| = \|\boldsymbol{\omega}\| \|\mathbf{r}\| \sin \theta \qquad \text{where } \theta \text{ is the angle between } \boldsymbol{\omega} \text{ and } \mathbf{r}.$$

This follows from observing that $\sin \theta = 1$ and recalling that in any circular motion the speed $\|\mathbf{v}\|$ is the angular speed times the length of the radius vector. (Exercise 1, Section 15.4.) ☐

Theorem 19.7.6 The Curl of a Rotation

If **v** is the velocity vector of a rotation, then

$$\text{curl } \mathbf{v} = 2\boldsymbol{\omega}$$

where $\boldsymbol{\omega}$ is the angular velocity vector.

PROOF. Taking (x_0, y_0, z_0) as the center of rotation, we have

$$\begin{aligned}
\mathbf{v} = \boldsymbol{\omega} \times \mathbf{r} &= [\omega_1 \mathbf{i} + \omega_2 \mathbf{j} + \omega_3 \mathbf{k}] \times [(x - x_0)\mathbf{i} + (y - y_0)\mathbf{j} + (z - z_0)\mathbf{k}] \\
&= [\omega_2(z - z_0) - \omega_3(y - y_0)]\mathbf{i} + [\omega_3(x - x_0) - \omega_1(z - z_0)]\mathbf{j} \\
&\qquad + [\omega_1(y - y_0) - \omega_2(x - x_0)]\mathbf{k}.
\end{aligned}$$

Straightforward computation shows that

$$\text{curl } \mathbf{v} = 2\omega_1 \mathbf{i} + 2\omega_2 \mathbf{j} + 2\omega_3 \mathbf{k} = 2\boldsymbol{\omega}. \quad ☐$$

Curl measures rotational tendency. If curl $\mathbf{v} = \mathbf{0}$ throughout the domain of **v**, then we say that the flow is *irrotational*.

Exercises

Find the curl.

*1. $\mathbf{v}(x, y) = x\mathbf{i} + y\mathbf{j}.$ 2. $\mathbf{v}(x, y) = -y\mathbf{i} + x\mathbf{j}.$

*3. $\mathbf{v}(x, y) = \dfrac{x}{x^2 + y^2}\mathbf{i} + \dfrac{y}{x^2 + y^2}\mathbf{j}.$

*4. $\mathbf{v}(x, y, z) = x\mathbf{i} + y\mathbf{j} + z\mathbf{k}.$ 5. $\mathbf{v}(x, y, z) = yz\mathbf{i} + xz\mathbf{j} + xy\mathbf{k}.$

*6. $\mathbf{v}(x, y, z) = xyz\mathbf{i} + xz\mathbf{j} + z\mathbf{k}.$ 7. $\mathbf{v}(x, y, z) = x^2 y\mathbf{i} + y^2 z\mathbf{j} + xy^2\mathbf{k}.$

*8. $\mathbf{v}(x, y, z) = \dfrac{x}{x^2 + y^2 + z^2}\mathbf{i} + \dfrac{y}{x^2 + y^2 + z^2}\mathbf{j} + \dfrac{z}{x^2 + y^2 + z^2}\mathbf{k}.$

9. Verify that

$$\text{curl}(\mathbf{u} + \mathbf{v}) = \text{curl } \mathbf{u} + \text{curl } \mathbf{v} \qquad \text{and} \qquad \text{curl}(\alpha\mathbf{v}) = \alpha \text{ curl } \mathbf{v}.$$

10. Show that

$$\text{curl}(f\mathbf{v}) = f \text{ curl } \mathbf{v} + \nabla f \times \mathbf{v}.$$

11. Show that gradients are irrotational:

$$\text{curl}(\nabla f) = \mathbf{0}$$

and curls are solenoidal:

$$\text{div}(\text{curl } \mathbf{v}) = 0.\dagger$$

*12. Express $\text{div}(\mathbf{u} \times \mathbf{v})$ in terms of $\mathbf{u}$, $\mathbf{v}$, curl $\mathbf{u}$, curl $\mathbf{v}$.

13. Show that

$$\text{curl}(\text{curl } \mathbf{v}) = \nabla(\text{div } \mathbf{v}) - \nabla^2\mathbf{v} \qquad \text{where} \quad \nabla^2\mathbf{v} = \nabla^2 v_1\mathbf{i} + \nabla^2 v_2\mathbf{j} + \nabla^2 v_3\mathbf{k}.\dagger\dagger$$

Verify Stokes's theorem on the upper half of the unit sphere $x^2 + y^2 + z^2 = 1$.

*14. $\mathbf{v}(x, y, z) = x\mathbf{i} + y\mathbf{j} + z\mathbf{k}$.
15. $\mathbf{v}(x, y, z) = y\mathbf{i} - x\mathbf{j} + z\mathbf{k}$.
*16. $\mathbf{v}(x, y, z) = z^2\mathbf{i} + 2x\mathbf{j} - y^3\mathbf{k}$.
17. $\mathbf{v}(x, y, z) = 6xz\mathbf{i} - x^2\mathbf{j} - 3y^2\mathbf{k}$.

Verify Stokes's theorem on the triangle $(2, 0, 0)$, $(0, 2, 0)$, $(0, 0, 2)$.

*18. $\mathbf{v}(x, y, z) = 2z\mathbf{i} - y\mathbf{j} + x\mathbf{k}$.
19. $\mathbf{v}(x, y, z) = (x^2 + y^2)\mathbf{i} + y^2\mathbf{j} + (x^2 + z^2)\mathbf{k}$.
*20. $\mathbf{v}(x, y, z) = x^4\mathbf{i} + xy\mathbf{j} + z^4\mathbf{k}$,

---

$\dagger$ Subject to some technical restrictions, gradients are the only irrotational fields:

$$\text{curl } \mathbf{v} = \mathbf{0} \Rightarrow \mathbf{v} = \nabla f \quad \text{for some scalar field } f$$

and curls are the only solenoidal fields:

$$\text{div } \mathbf{v} = 0 \Rightarrow \mathbf{v} = \text{curl } \mathbf{u} \quad \text{for some vector field } \mathbf{u}.$$

$\dagger\dagger$ For the ∇^2 notation see Exercise 9 of Section 19.6.

Appendix A
Some Elementary Topics

A.1 Sets

A *set* is a collection of objects. The objects in a set are called the *elements* (or *members*) of the set.

We might, for example, consider the set of capital letters appearing on this page, or the set of motorcycles licensed in Idaho, or the set of rational numbers. Suppose, however, that we wanted to find the set of rational people. Everybody might have a different collection. Which would be the right one? To avoid such problems we insist that sets be unambiguously defined. Collections based on highly subjective judgments—such as "all good football players" or "all likeable children"—are not sets.

Notation

To indicate that an object x is in the set A, we write

$$x \in A.$$

To indicate that x is not in A, we write

$$x \notin A.$$

Thus

$$\sqrt{2} \in \text{the set of real numbers} \quad \text{but} \quad \sqrt{2} \notin \text{the set of rational numbers.}$$

Sets are often denoted by braces. The set consisting of a alone is written $\{a\}$; that consisting of a, b is written $\{a, b\}$; that consisting of $a, b, c, \{a, b, c\}$; and so on. Thus

$$0 \in \{0, 1, 2\}, \quad 1 \in \{0, 1, 2\}, \quad 2 \in \{0, 1, 2\}, \quad \text{but} \quad 3 \notin \{0, 1, 2\}.$$

We can also use braces for infinite sets:

$\{1, 2, 3, \ldots\}$ is the set of positive integers,

$\{-1, -2, -3, \ldots\}$ is the set of negative integers,

$\{1, 2, 2^2, 2^3, \ldots\}$ is the set of powers of 2.

Sets are often defined by a property. We write $\{x\colon P\}$ to indicate *the set of all x for which property P holds.* Thus

$\{x\colon x > 2\}$ is the set of all numbers greater than 2;

$\{x\colon x^2 > 9\}$ is the set of all numbers whose squares are greater than 9;

$\{p/q\colon p, q \text{ integers}, q \neq 0\}$ is the set of all rational numbers.

If A is a set, then $\{x\colon x \in A\}$ is A itself.

Containment and Equality

If A and B are sets, then A is said to be *contained* in B, in symbols $A \subseteq B$, iff† every element of A is also an element of B. For example,

the set of equilateral triangles $\subseteq$ the set of all triangles,

the set of all college freshmen $\subseteq$ the set of all college students,

the set of rational numbers $\subseteq$ the set of real numbers.

If A is contained in B, then A is called a *subset* of B. Thus

the set of equilateral triangles is a subset of the set of all triangles,

the set of college freshmen is a subset of the set of all college students,

the set of rational numbers is a subset of the set of real numbers.

Two sets are said to be *equal* iff they have exactly the same membership. In symbols,

(A.1.1)
$$\boxed{A = B \quad \text{iff} \quad A \subseteq B \ \text{ and } \ B \subseteq A.}$$

Examples

$$\{x\colon x^2 = 4\} = \{-2, 2\},$$
$$\{x\colon x^2 < 4\} = \{x\colon -2 < x < 2\},$$
$$\{x\colon x^2 > 4\} = \{x\colon x < -2 \text{ or } x > 2\}. \quad \square$$

The Intersection of Two Sets

The set of elements common to two sets A and B is called the *intersection* of A and

---

† By "iff" we mean "if and only if." This expression is used so often in mathematics that it is convenient to have an abbreviation for it.

B and is denoted by $A \cap B$. The idea is illustrated in Figure A.1.1. In symbols,

(A.1.2)

$$x \in A \cap B \quad \text{iff} \quad x \in A \quad \text{and} \quad x \in B.$$

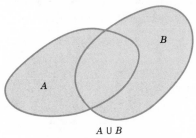

FIGURE A.1.1

Examples

1. If A is the set of all nonnegative numbers, and B is the set of all nonpositive numbers, then $A \cap B = \{0\}$.
2. If A is the set of all multiples of 3, and B is the set of all multiples of 4, then $A \cap B$ = the set of all multiples of 12.
3. If $A = \{a, b, c, d, e\}$, and $B = \{c, d, e, f\}$, then $A \cap B = \{c, d, e\}$.
4. If $A = \{x : x > 1\}$ and $B = \{x : x < 4\}$, then $A \cap B = \{x : 1 < x < 4\}$. $\square$

The Union of Two Sets

The *union* of two sets A and B, written $A \cup B$, is the set of elements which are either in A or in B. This does not exclude objects which are elements of both A and B. (See Figure A.1.2.) In symbols,

(A.1.3)

$$x \in A \cup B \quad \text{iff} \quad x \in A \quad \text{or} \quad x \in B.$$

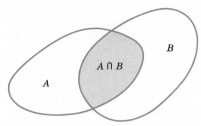

$A \cup B$
FIGURE A.1.2

Examples

1. If A is the set of all nonnegative numbers and B is the set of all nonpositive numbers, then $A \cup B$ = the set of all real numbers.
2. If $A = \{a, b, c, d, e\}$ and $B = \{c, d, e, f\}$, then $A \cup B = \{a, b, c, d, e, f\}$.

3. If $A = \{x: 0 < x < 1\}$ and $B = \{0, 1\}$, then $A \cup B = \{x: 0 \le x \le 1\}$.
4. If $A = \{x: x > 1\}$ and $B = \{x: x > 2\}$, then $A \cup B = \{x: x > 1\}$.   □

The Empty Set

If the sets A and B have no elements in common, we say that A and B are *disjoint* and write $A \cap B = \varnothing$. We regard $\varnothing$ as a set with no elements and refer to it as the *empty set*.

Examples

1. If A is the set of all positive numbers, and B is the set of all negative numbers, then $A \cap B = \varnothing$.
2. If $A = \{0, 1, 2, 3\}$ and $B = \{4, 5, 6, 7, 8\}$, then $A \cap B = \varnothing$.
3. The set of all irrational rational numbers is empty; so is the set of all even odd integers; so is the set of real numbers whose squares are negative.   □

The empty set $\varnothing$ plays a role in the theory of sets which is strikingly similar to the role played by 0 in the arithmetic of numbers. Without pursuing the matter very far, note that for numbers,

$$a + 0 = 0 + a = a, \qquad a \cdot 0 = 0 \cdot a = 0,$$

and for sets,

$$A \cup \varnothing = \varnothing \cup A = A, \qquad A \cap \varnothing = \varnothing \cap A = \varnothing.$$

Cartesian Products

If A and B are nonempty sets, then $A \times B$, the *Cartesian product* of A and B, is the set of all ordered pairs (a, b) with $a \in A$ and $b \in B$. In set notation

(A.1.4)
$$A \times B = \{(a, b): a \in A, \ b \in B\}.$$

Examples

1. For $A = \{0, 1\}$, $B = \{1, 2, 3\}$, $A \times B = \{(0, 1), (0, 2), (0, 3), (1, 1), (1, 2), (1, 3)\}$ and $B \times A = \{(1, 0), (1, 1), (2, 0), (2, 1), (3, 0), (3, 1)\}$.
2. If A is the set of rational numbers and B is the set of irrational numbers, then $A \times B$ is the set of all pairs (a, b) with a rational and b irrational.   □

The Cartesian product $A \times B \times C$ consists of all ordered triples (a, b, c) with $a \in A, b \in B, c \in C$:

(A.1.5)
$$A \times B \times C = \{(a, b, c): a \in A, \ b \in B, \ c \in C.\}$$

Exercises†

For Exercises 1–12 take

$$A = \{0, 2\}, \quad B = \{-1, 0, 1\}, \quad C = \{1, 2, 3, 4\},$$
$$D = \{2, 4, 6, 8, \ldots\}, \quad E = \{-2, -4, -6, -8, \ldots\},$$

and determine the following sets:

*1. $A \cup B$.	2. $A \cap B$.	*3. $B \cup C$.
*4. $B \cap C$.	5. $A \cup D$.	*6. $A \cap D$.
*7. $B \cap D$.	8. $D \cup E$.	*9. $C \cap D$.
*10. $A \times B$.	11. $B \times C$.	*12. $A \times B \times A$.

For Exercises 13–18 take

$$A = \{x : x > 2\}, \quad B = \{x : x \leq 4\}, \quad C = \{x : x > 3\},$$

and determine the following sets:

*13. $A \cup C$.	14. $A \cap C$.	*15. $B \cup C$.
*16. $B \cap C$.	17. $A \cap (B \cap C)$.	*18. $A \cap (B \cup C)$.

19. Given that $A \subseteq B$, find (a) $A \cup B$. (b) $A \cap B$.
*20. List all the nonempty subsets of $\{0, 1, 2\}$.
21. What can you conclude about A and B given that
    (a) $A \cup B = A$?   (b) $A \cap B = A$?   (c) $A \cup B = A$ and $A \cap B = A$?

A.2   Induction

Suppose that you were asked to show that a certain set S contains the set of positive integers. You could start by verifying that $1 \in S$, and $2 \in S$, and $3 \in S$, and so on, but even if each such step took you only one-hundredth of a second, you would still never finish.

To avoid such a bind, mathematicians use a special procedure called *induction*. That induction works is an *assumption* that we make.

Axiom A.2.1   Axiom of Induction

Let S be a set of integers. If

$$(A) \; 1 \in S \quad \text{and} \quad (B) \; k \in S \text{ implies } k + 1 \in S,$$

then all the positive integers are in S.

You can take the axiom of induction as a kind of "domino theory." If the first

---

† The starred exercises have answers at the back of the book.

domino falls (Figure A.2.1), and if each domino that falls causes the next one to fall, then, according to the axiom of induction, all the dominoes will fall.

domino theory

FIGURE A.2.1

While we cannot prove that this axiom is valid (axioms are by their very nature assumptions and therefore not subject to proof), we can argue that it is *plausible*.

Let's assume that we have a set S that satisfies conditions (A) and (B). Now let's choose a positive integer m and "argue" that $m \in S$.

From (A) we know that $1 \in S$. Since $1 \in S$, we know from (B) that $1 + 1 \in S$, and thus that $(1 + 1) + 1 \in S$, and so on. Since m can be obtained from 1 by adding 1 successively $(m - 1)$ times, it *seems clear* that $m \in S$. □

As an example of this procedure, note that

$$5 = \{[(1 + 1) + 1] + 1\} + 1$$

and thus $5 \in S$.

Problem.   Show that

$$\text{if } \quad 0 \leq a < b \quad \text{ then } \quad a^n < b^n \quad \text{ for all positive integers } n.$$

SOLUTION.   Suppose that $0 \leq a < b$, and let S be the set of positive integers n for which $a^n < b^n$.

Obviously $1 \in S$. Let's assume now that $k \in S$. This assures us that $a^k < b^k$. It follows that

$$a^{k+1} = a \cdot a^k < a \cdot b^k < b \cdot b^k = b^{k+1}$$

and thus that $k + 1 \in S$.

We have shown that

$$1 \in S \quad \text{ and that } \quad k \in S \quad \text{ implies } \quad k + 1 \in S.$$

By the axiom of induction, we can conclude that all the positive integers are in S. □

Problem.   Show that, if $x \geq -1$, then

$$(1 + x)^n \geq 1 + nx \quad \text{ for all positive integers } n.$$

SOLUTION.   Take $x \geq -1$ and let S be the set of positive integers n for which

$$(1 + x)^n \geq 1 + nx.$$

Since

$$(1 + x)^1 \geq 1 + 1 \cdot x,$$

you can see that $1 \in S$.

Assume now that $k \in S$. By the definition of S,

$$(1 + x)^k \geq 1 + kx.$$

Since

$$(1 + x)^{k+1} = (1 + x)^k(1 + x) \geq (1 + kx)(1 + x) \qquad \text{(explain)}$$

and

$$(1 + kx)(1 + x) = 1 + (k + 1)x + kx^2 \geq 1 + (k + 1)x,$$

it follows that

$$(1 + x)^{k+1} \geq 1 + (k + 1)x$$

and thus that $k + 1 \in S$.

We have shown that

$$1 \in S \qquad \text{and that} \qquad k \in S \quad \text{implies} \quad k + 1 \in S.$$

By the axiom of induction, all the positive integers are in S. $\square$

Exercises

Show that the following statements hold for all positive integers n:

*1. $n(n + 1)$ is divisible by 2. [HINT: $(k + 1)(k + 2) = k(k + 1) + 2(k + 1)$.]

2. $n(n + 1)(n + 2)$ is divisible by 6.

3. $1 + 2 + 3 + \cdots + n = \frac{1}{2}n(n + 1)$.

*4. $1 + 3 + 5 + \cdots + (2n - 1) = n^2$.

5. $1^2 + 2^2 + 3^2 + \cdots + n^2 = \frac{1}{6}n(n + 1)(2n + 1)$.

6. $1^3 + 2^3 + 3^3 + \cdots + n^3 = (1 + 2 + 3 + \cdots + n)^2$. [HINT: Use Exercise 3.]

*7. $3^{2n+1} + 2^{n+2}$ is divisible by 7.

8. For what integers n is $9^n - 8n - 1$ divisible by 64? Prove that your answer is correct.

9. Prove that all sets with n elements have 2^n subsets. Count the empty set $\varnothing$ as a subset.

10. Show that for all positive integers n

$$1^3 + 2^3 + \cdots + (n - 1)^3 < \tfrac{1}{4}n^4 < 1^3 + 2^3 + \cdots + n^3.$$

11. Find a simplifying expression for the product

$$\left(1 - \frac{1}{2}\right)\left(1 - \frac{1}{3}\right) \cdots \left(1 - \frac{1}{n}\right)$$

and verify its validity for all integers $n \geq 2$.

Appendix B
Some Additional
Proofs

In this appendix we present some proofs that many would consider too advanced for the main body of the text. Some details are omitted. These are left to you.

The arguments presented in B.1, B.2, and B.4 require some familiarity with the *least upper bound axiom*. This is discussed in Section 11.7. In addition D.4 requires some understanding of *sequences*, for which we refer you to Sections 12.1 and 12.2.

B.1 The Intermediate-Value Theorem

Lemma B.1.1

Let f be continuous on $[a, b]$. If $f(a) < 0 < f(b)$ or $f(b) < 0 < f(a)$, then there is a number c between a and b such that $f(c) = 0$.

PROOF. Suppose that $f(a) < 0 < f(b)$. (The other case can be treated in a similar manner.) Since $f(a) < 0$, we know from the continuity of f that there exists a number ξ such that f is negative on $[a, \xi)$. Let

$$c = \text{lub } \{\xi : f \text{ is negative on } [a, \xi)\}.$$

Clearly, $c \le b$. We cannot have $f(c) > 0$, for then f would be positive on some interval extending to the left of c, and we know that, to the left of c, f is negative. Incidentally this argument excludes the possibility $c = b$ and means that $c < b$. We cannot have $f(c) < 0$, for then there would be an interval $[a, t)$, with $t > c$, on which f is negative, and this would contradict the definition of c. It follows that $f(c) = 0$. $\square$

871

Theorem B.1.2   The Intermediate-Value Theorem

If f is continuous on $[a, b]$ and C is a number between $f(a)$ and $f(b)$, then there is at least one number c between a and b such that $f(c) = C$.

PROOF.   Suppose for example that

$$f(a) < C < f(b).$$

(The other possibility can be handled in a similar manner.) The function

$$g(x) = f(x) - C$$

is continuous on $[a, b]$. Since

$$g(a) = f(a) - C < 0 \qquad \text{and} \qquad g(b) = f(b) - C > 0,$$

we know from the lemma that there is a number c between a and b such that $g(c) = 0$. Obviously then $f(c) = C$. $\square$

B.2   The Maximum-Minimum Theorem

Lemma B.2.1

If f is continuous on $[a, b]$, then f is bounded on $[a, b]$.

PROOF.   Consider

$$\{x: x \in [a, b] \text{ and } f \text{ is bounded on } [a, x]\}.$$

It is easy to see that this set is nonempty and bounded above by b. Thus we can set

$$c = \text{lub } \{x: f \text{ is bounded on } [a, x]\}.$$

Now we argue that $c = b$. To do so, we suppose that $c < b$. From the continuity of f at c, it is easy to see that f is bounded on $[c - \epsilon, c + \epsilon]$ for some $\epsilon > 0$. Being bounded on $[a, c - \epsilon]$ and on $[c - \epsilon, c + \epsilon]$, it is obviously bounded on $[a, c + \epsilon]$. This contradicts our choice of c. We can therefore conclude that $c = b$. This tells us that f is bounded on $[a, x]$ for all $x < b$. We are now almost through. From the continuity of f, we know that f is bounded on some interval of the form $[b - \epsilon, b]$. Since $b - \epsilon < b$, we know from what we just proved that f is bounded on $[a, b - \epsilon]$. Being bounded on $[a, b - \epsilon]$ and on $[b - \epsilon, b]$, it is bounded on $[a, b]$. $\square$

Theorem B.2.2   The Maximum-Minimum Theorem

If f is continuous on $[a, b]$, then f takes on both a maximum value M and a minimum value m on $[a, b]$.

PROOF.   By the lemma, f is bounded on $[a, b]$. Set

$$M = \text{lub } \{f(x): x \in [a, b]\}.$$

We must show that there exists c in $[a, b]$ such that $f(c) = M$. To do this, we set

$$g(x) = \frac{1}{M - f(x)}.$$

If f does not take on the value M, then g is continuous on $[a, b]$ and thus, by the lemma, bounded on $[a, b]$. A look at the definition of g makes it clear that g cannot be bounded on $[a, b]$. The assumption that f does not take on the value M has led to a contradiction. (That f takes on a minimum value m can be proved in a similar manner.) $\square$

B.3   Inverses

Theorem B.3.1   Continuity of the Inverse

Let f be a one-to-one function defined on an interval (a, b). If f is continuous, then its inverse f^{-1} is also continuous.

PROOF.   If f is continuous, then, being one-to-one, f either increases throughout (a, b) or it decreases throughout (a, b). The proof of this assertion we leave to you.

Let's suppose now that f increases throughout (a, b). Let's take c in the domain of f^{-1} and show that f^{-1} is continuous at c.

We first observe that $f^{-1}(c)$ lies in (a, b) and choose $\epsilon > 0$ sufficiently small so that $f^{-1}(c) - \epsilon$ and $f^{-1}(c) + \epsilon$ also lie in (a, b). We seek $\delta > 0$ such that

$$\text{if }\ \ c - \delta < x < c + \delta \ \ \ \ \text{ then }\ \ f^{-1}(c) - \epsilon < f^{-1}(x) < f^{-1}(c) + \epsilon.$$

This condition can be met by choosing δ to satisfy

$$f(f^{-1}(c) - \epsilon) < c - \delta \ \ \text{ and }\ \ c + \delta < f(f^{-1}(c) + \epsilon)$$

for then, if $c - \delta < x < c + \delta$, then

$$f(f^{-1}(c) - \epsilon) < x < f(f^{-1}(c) + \epsilon)$$

and, since f^{-1} also increases,

$$f^{-1}(c) - \epsilon < f^{-1}(x) < f^{-1}(c) + \epsilon.$$

The case where f decreases throughout (a, b) can be handled in a similar manner. $\square$

Theorem B.3.2   Differentiability of the Inverse

Let f be a one-to-one function defined on an interval (a, b). If f is differentiable and its derivative does not take on the value 0, then f^{-1} is differentiable and

$$(f^{-1})'(x) = \frac{1}{f'(f^{-1}(x))}.$$

PROOF.   Let x be in the domain of f^{-1}. We take $\epsilon > 0$ and show that there exists $\delta > 0$ such that

$$\text{if }\quad 0 < |t - x| < \delta \quad\quad \text{then}\quad \left| \frac{f^{-1}(t) - f^{-1}(x)}{t - x} - \frac{1}{f'(f^{-1}(x))} \right| < \epsilon.$$

Since f is differentiable at $f^{-1}(x)$ and $f'(f^{-1}(x)) \neq 0$, there exists $\delta_1 > 0$ such that, if

$$0 < |y - f^{-1}(x)| < \delta_1,$$

then

$$\left| \frac{1}{\dfrac{f(y) - f(f^{-1}(x))}{y - f^{-1}(x)}} - \frac{1}{f'(f^{-1}(x))} \right| < \epsilon$$

and therefore

$$\left| \frac{y - f^{-1}(x)}{f(y) - f(f^{-1}(x))} - \frac{1}{f'(f^{-1}(x))} \right| < \epsilon.$$

By the previous theorem, f^{-1} is continuous at x and therefore there exists $\delta > 0$ such that

$$\text{if }\quad 0 < |t - x| < \delta \quad\quad \text{then}\quad 0 < |f^{-1}(t) - f^{-1}(x)| < \delta_1.$$

It follows from the special property of δ_1 that

$$\left| \frac{f^{-1}(t) - f^{-1}(x)}{t - x} - \frac{1}{f'(f^{-1}(x))} \right| < \epsilon. \quad \square$$

B.4   The Integrability of Continuous Functions

The aim here is to prove that, if f is continuous on $[a, b]$, then there is one and only one number I which satisfies the inequality

$$L_f(P) \le I \le U_f(P) \quad\quad \text{for all partitions } P \text{ of } [a, b].$$

Definition B.4.1

A function f is said to be *uniformly continuous* on $[a, b]$ iff for each $\epsilon > 0$ there exists $\delta > 0$ such that,

$$\text{if }\quad x, y \in [a, b] \quad \text{and} \quad |x - y| < \delta, \quad\quad \text{then}\quad |f(x) - f(y)| < \epsilon.$$

For convenience, let's agree to say that *the interval* $[a, b]$ *has the property* P_ϵ iff there exist sequences $\{x_n\}, \{y_n\}$ satisfying

$$x_n, y_n \in [a, b], \quad |x_n - y_n| < 1/n, \quad |f(x_n) - f(y_n)| \geq \epsilon.$$

Lemma B.4.2

If f is not uniformly continuous on $[a, b]$, then $[a, b]$ has the property P_ϵ for some $\epsilon > 0$.

The proof is easy. $\square$

Lemma B.4.3

Let f be continuous on $[a, b]$. If $[a, b]$ has the property P_ϵ, then at least one of the subintervals $[a, \frac{1}{2}(a + b)]$, $[\frac{1}{2}(a + b), b]$ has the property P_ϵ.

PROOF.   Let's suppose that the lemma is false. For convenience, we let

$$c = \tfrac{1}{2}(a + b),$$

so that the halves become $[a, c]$ and $[c, b]$. Since $[a, c]$ fails to have the property P_ϵ, there exists an integer p such that,

if $x, y \in [a, c]$ and $|x - y| < 1/p$,   then $|f(x) - f(y)| < \epsilon$.

Since $[c, b]$ fails to have the property P_ϵ, there exists an integer q such that,

if $x, y \in [c, b]$ and $|x - y| < 1/q$,   then $|f(x) - f(y)| < \epsilon$.

Since f is continuous at c, there exists an integer r such that, if $|x - c| < 1/r$, then $|f(x) - f(c)| < \frac{1}{2}\epsilon$. Set $s = \max\{p, q, r\}$ and suppose that

$$x, y \in [a, b], \quad |x - y| < 1/s.$$

If x, y are both in $[a, c]$ or both in $[c, b]$, then

$$|f(x) - f(y)| < \epsilon.$$

The only other possibility is that $x \in [a, c]$ and $y \in [c, b]$. In this case we have

$$|x - c| < 1/r, \quad |y - c| < 1/r,$$

and thus

$$|f(x) - f(c)| < \tfrac{1}{2}\epsilon, \quad |f(y) - f(c)| < \tfrac{1}{2}\epsilon.$$

By the triangle inequality, we again have

$$|f(x) - f(y)| < \epsilon.$$

In summary, we have obtained the existence of an integer s with the property that

$$x, y \in [a, b], \quad |x - y| < 1/s \quad \text{implies} \quad |f(x) - f(y)| < \epsilon.$$

Hence $[a, b]$ does not have the property P_ϵ. This is a contradiction and proves the lemma. $\square$

Theorem B.4.4

If f is continuous on $[a, b]$, then f is uniformly continuous on $[a, b]$.

PROOF.   We suppose that f is not uniformly continuous on $[a, b]$ and base our argument on a mathematical version of the classical maxim "Divide and Conquer."

By the first lemma of this section, we know that $[a, b]$ has the property P_ϵ for some $\epsilon > 0$. We bisect $[a, b]$ and note by the second lemma that one of the halves, say $[a_1, b_1]$, has the property P_ϵ. We then bisect $[a_1, b_1]$ and note that one of the halves, say $[a_2, b_2]$, has the property P_ϵ. Continuing in this manner, we obtain a sequence of intervals $[a_n, b_n]$, each with the property P_ϵ. For each n, we can choose $x_n, y_n \in [a_n, b_n]$ such that

$$|x_n - y_n| < 1/n \quad \text{and} \quad |f(x_n) - f(y_n)| \geq \epsilon.$$

Since

$$a \leq a_n \leq a_{n+1} < b_{n+1} \leq b_n \leq b,$$

we see that the sequences $\{a_n\}$ and $\{b_n\}$ are both bounded and monotonic. Thus they are convergent. Since $b_n - a_n \to 0$, we see that $\{a_n\}$ and $\{b_n\}$ both converge to the same limit, say l. From the inequality

$$a_n \leq x_n \leq y_n \leq b_n,$$

we conclude that

$$x_n \to l \quad \text{and} \quad y_n \to l.$$

This tells us that

$$|f(x_n) - f(y_n)| \to |f(l) - f(l)| = 0,$$

which contradicts the statement that $|f(x_n) - f(y_n)| \geq \epsilon$ for all n. $\square$

Lemma B.4.5

If P and Q are partitions of $[a, b]$, then $L_f(P) \leq U_f(Q)$.

PROOF. $P \cup Q$ is a partition of $[a, b]$ which contains both P and Q. It's obvious then that

$$L_f(P) \le L_f(P \cup Q) \le U_f(P \cup Q) \le U_f(Q). \quad \square$$

From the last lemma it follows that the set of all lower sums is bounded above and has a least upper bound L. The number L satisfies the inequality

$$L_f(P) \le L \le U_f(P) \qquad \text{for all partitions } P$$

and is clearly the least of such numbers. Similarly, we find that the set of all upper sums is bounded below and has a greatest lower bound U. The number U satisfies the inequality

$$L_f(P) \le U \le U_f(P) \qquad \text{for all partitions } P$$

and is clearly the greatest of such numbers.

We are now ready to prove the basic theorem.

Theorem B.4.6.   The Integrability Theorem

If f is continuous on $[a, b]$, then there exists one and only one number I which satisfies the inequality

$$L_f(P) \le I \le U_f(P) \qquad \text{for all partitions } P \text{ of } [a, b].$$

PROOF.   We know that

$$L_f(P) \le L \le U \le U_f(P) \qquad \text{for all } P,$$

so that existence is no problem. We will have uniqueness if we can prove that

$$L = U.$$

To do this, we let $\epsilon > 0$ and note that f, being continuous on $[a, b]$, is uniformly continuous on $[a, b]$. Thus there exists $\delta > 0$ such that, if

$$x, y \in [a, b] \quad \text{and} \quad |x - y| < \delta, \qquad \text{then} \quad |f(x) - f(y)| < \frac{\epsilon}{b - a}.$$

We now choose a partition $P = \{x_0, x_1, \ldots, x_n\}$ for which max $\Delta x_i < \delta$. For this partition P, we have

$$
\begin{aligned}
U_f(P) - L_f(P) &= \sum_{i=1}^{n} M_i \Delta x_i - \sum_{i=1}^{n} m_i \Delta x_i \\
&= \sum_{i=1}^{n} (M_i - m_i) \Delta x_i \\
&< \sum_{i=1}^{n} \frac{\epsilon}{b - a} \Delta x_i \\
&= \frac{\epsilon}{b - a} \sum_{i=1}^{n} \Delta x_i = \frac{\epsilon}{b - a} (b - a) = \epsilon.
\end{aligned}
$$

Since
$$U_f(P) - L_f(P) < \epsilon \qquad \text{and} \qquad 0 \le U - L \le U_f(P) - L_f(P),$$
you can see that
$$0 \le U - L < \epsilon.$$
Since ϵ was chosen arbitrarily, we must have $U - L = 0$ and $L = U$. $\square$

B.5   The Integral as the Limit of Riemann Sums

For the notation we refer to Section 5.12.

Theorem B.5.1

If f is continuous on $[a, b]$, then
$$\int_a^b f(x)\, dx = \lim_{\|P\| \mapsto 0} S^*(P).$$

PROOF.   Let $\epsilon > 0$. We must show that there exists $\delta > 0$ such that
$$\text{if} \quad \|P\| < \delta \qquad \text{then} \quad \left| S^*(P) - \int_a^b f(x)\, dx \right| < \epsilon.$$

From the proof of Theorem B.4.6 we know that there exists $\delta > 0$ such that
$$\text{if} \quad \|P\| < \delta \qquad \text{then} \quad U_f(P) - L_f(P) < \epsilon.$$

For such P we have
$$U_f(P) - \epsilon < L_f(P) \le S^*(P) \le U_f(P) < L_f(P) + \epsilon.$$

This gives
$$\int_a^b f(x)\, dx - \epsilon < S^*(P) < \int_a^b f(x)\, dx + \epsilon$$

and therefore
$$\left| S^*(P) - \int_a^b f(x)\, dx \right| < \epsilon. \quad \square$$

B.6   A Variant of Bliss's Theorem

For the notation we refer to Section 10.3.

Theorem B.6.1　(Bliss)

If f and g are continuous and nonnegative on $[a, b]$, then the definite integral

$$\int_a^b f(x)g(x)\, dx$$

is the only number I which satisfies the inequality

$$L_{f,g}^*(P) \le I \le U_{f,g}^*(P) \qquad \text{for all partitions } P \text{ of } [a, b].$$

PROOF.　Take f and g as continuous and nonnegative on $[a, b]$ and take P as an arbitrary partition. It's easy to show first of all that

$$L_{f,g}^*(P) \le L_{fg}(P) \quad \text{and} \quad U_{fg}(P) \le U_{f,g}^*(P).$$

Since in general

$$L_{fg}(P) \le \int_a^b f(x)g(x)\, dx \le U_{fg}(P),$$

it follows that

$$L_{f,g}^*(P) \le \int_a^b f(x)g(x)\, dx \le U_{f,g}^*(P).$$

We have shown that the integral lies between $L_{f,g}^*(P)$ and $U_{f,g}^*(P)$ for all partitions P of $[a, b]$. Now we show that the integral is the *only* such number. We assume that there are two such numbers I_1 and I_2 and obtain a contradiction by showing the existence of a partition P for which

$$U_{f,g}^*(P) - L_{f,g}^*(P) < |I_1 - I_2|.$$

Since f and g are continuous on $[a, b]$, they are there uniformly continuous. It follows that there exists $\delta > 0$ such that, if

$$x, y \in [a, b] \quad \text{and} \quad |x - y| < \delta,$$

then

$$|f(x) - f(y)| < \frac{|I_1 - I_2|}{2(b - a)[1 + M(g)]} \quad \text{and} \quad |g(x) - g(y)| < \frac{|I_1 - I_2|}{2(b - a)[1 + M(f)]},$$

where $M(f)$ and $M(g)$ are the maximum values of f and g on all of $[a, b]$. We now choose a partition $P = \{x_0, x_1, \ldots, x_n\}$ for which $\max \Delta x_i < \delta$. For this partition P we have

$$U_{f,g}^*(P) - L_{f,g}^*(P) = \sum_{i=1}^n M_i(f)M_i(g)\Delta x_i - \sum_{i=1}^n m_i(f)m_i(g)\Delta x_i$$

$$= \sum_{i=1}^n [M_i(f)M_i(g) - m_i(f)m_i(g)]\Delta x_i$$

$$= \sum_{i=1}^{n} [M_i(f)M_i(g) - M_i(f)m_i(g) + M_i(f)m_i(g) - m_i(f)m_i(g)]\Delta x_i$$

$$= \sum_{i=1}^{n} \{M_i(f)[M_i(g) - m_i(g)] + m_i(g)[M_i(f) - m_i(f)]\}\Delta x_i$$

$$\leq \sum_{i=1}^{n} \left\{ \frac{M(f)|I_1 - I_2|}{2(b - a)[1 + M(f)]} + \frac{M(g)|I_1 - I_2|}{2(b - a)[1 + M(g)]} \right\} \Delta x_i$$

$$< \sum_{i=1}^{n} \left(\frac{|I_1 - I_2|}{b - a} \right) \Delta x_i$$

$$= \frac{|I_1 - I_2|}{b - a} \sum_{i=1}^{n} \Delta x_i = \frac{|I_1 - I_2|}{b - a} (b - a) = |I_1 - I_2|. \quad \square$$

Appendix C
Tables

TABLE 1. Natural logs

x	$\log x$	x	$\log x$	x	$\log x$
		4.0	1.386	8.0	2.079
0.1	-2.303	4.1	1.411	8.1	2.092
0.2	-1.609	4.2	1.435	8.2	2.104
0.3	-1.204	4.3	1.459	8.3	2.116
0.4	-0.916	4.4	1.482	8.4	2.128
0.5	-0.693	4.5	1.504	8.5	2.140
0.6	-0.511	4.6	1.526	8.6	2.152
0.7	-0.357	4.7	1.548	8.7	2.163
0.8	-0.223	4.8	1.569	8.8	2.175
0.9	-0.105	4.9	1.589	8.9	2.186
1.0	0.000	5.0	1.609	9.0	2.197
1.1	0.095	5.1	1.629	9.1	2.208
1.2	0.182	5.2	1.649	9.2	2.219
1.3	0.262	5.3	1.668	9.3	2.230
1.4	0.336	5.4	1.686	9.4	2.241
1.5	0.405	5.5	1.705	9.5	2.251
1.6	0.470	5.6	1.723	9.6	2.262
1.7	0.531	5.7	1.740	9.7	2.272
1.8	0.588	5.8	1.758	9.8	2.282
1.9	0.642	5.9	1.775	9.9	2.293
2.0	0.693	6.0	1.792	10	2.303
2.1	0.742	6.1	1.808	20	2.996
2.2	0.788	6.2	1.825	30	3.401
2.3	0.833	6.3	1.841	40	3.689
2.4	0.875	6.4	1.856	50	3.912
2.5	0.916	6.5	1.872	60	4.094
2.6	0.956	6.6	1.887	70	4.248
2.7	0.993	6.7	1.902	80	4.382
2.8	1.030	6.8	1.917	90	4.500
2.9	1.065	6.9	1.932	100	4.605
3.0	1.099	7.0	1.946		
3.1	1.131	7.1	1.960		
3.2	1.163	7.2	1.974		
3.3	1.194	7.3	1.988		
3.4	1.224	7.4	2.001		
3.5	1.253	7.5	2.015		
3.6	1.281	7.6	2.028		
3.7	1.308	7.7	2.041		
3.8	1.335	7.8	2.054		
3.9	1.361	7.9	2.067		

TABLE 2.   Exponentials (0.01 to 0.99)

x	e^x	e^{-x}	x	e^x	e^{-x}	x	e^x	e^{-x}
0.01	1.010	0.990	0.34	1.405	0.712	0.67	1.954	0.512
0.02	1.020	0.980	0.35	1.419	0.705	0.68	1.974	0.507
0.03	1.030	0.970	0.36	1.433	0.698	0.69	1.994	0.502
0.04	1.041	0.961	0.37	1.448	0.691	0.70	2.014	0.497
0.05	1.051	0.951	0.38	1.462	0.684	0.71	2.034	0.492
0.06	1.062	0.942	0.39	1.477	0.677	0.72	2.054	0.487
0.07	1.073	0.932	0.40	1.492	0.670	0.73	2.075	0.482
0.08	1.083	0.923	0.41	1.507	0.664	0.74	2.096	0.477
0.09	1.094	0.914	0.42	1.522	0.657	0.75	2.117	0.472
0.10	1.105	0.905	0.43	1.537	0.651	0.76	2.138	0.468
0.11	1.116	0.896	0.44	1.553	0.644	0.77	2.160	0.463
0.12	1.127	0.887	0.45	1.568	0.638	0.78	2.181	0.458
0.13	1.139	0.878	0.46	1.584	0.631	0.79	2.203	0.454
0.14	1.150	0.869	0.47	1.600	0.625	0.80	2.226	0.449
0.15	1.162	0.861	0.48	1.616	0.619	0.81	2.248	0.445
0.16	1.174	0.852	0.49	1.632	0.613	0.82	2.270	0.440
0.17	1.185	0.844	0.50	1.649	0.607	0.83	2.293	0.436
0.18	1.197	0.835	0.51	1.665	0.600	0.84	2.316	0.432
0.19	1.209	0.827	0.52	1.682	0.595	0.85	2.340	0.427
0.20	1.221	0.819	0.53	1.699	0.589	0.86	2.363	0.423
0.21	1.234	0.811	0.54	1.716	0.583	0.87	2.387	0.419
0.22	1.246	0.803	0.55	1.733	0.577	0.88	2.411	0.415
0.23	1.259	0.795	0.56	1.751	0.571	0.89	2.435	0.411
0.24	1.271	0.787	0.57	1.768	0.566	0.90	2.460	0.407
0.25	1.284	0.779	0.58	1.786	0.560	0.91	2.484	0.403
0.26	1.297	0.771	0.59	1.804	0.554	0.92	2.509	0.399
0.27	1.310	0.763	0.60	1.822	0.549	0.93	2.535	0.395
0.28	1.323	0.756	0.61	1.840	0.543	0.94	2.560	0.391
0.29	1.336	0.748	0.62	1.859	0.538	0.95	2.586	0.387
0.30	1.350	0.741	0.63	1.878	0.533	0.96	2.612	0.383
0.31	1.363	0.733	0.64	1.896	0.527	0.97	2.638	0.379
0.32	1.377	0.726	0.65	1.916	0.522	0.98	2.664	0.375
0.33	1.391	0.719	0.66	1.935	0.517	0.99	2.691	0.372

TABLE 3.   Exponentials (1.0 to 4.9)

x	e^x	e^{-x}	x	e^x	e^{-x}
1.0	2.718	0.368	3.0	20.086	0.050
1.1	3.004	0.333	3.1	22.198	0.045
1.2	3.320	0.301	3.2	24.533	0.041
1.3	3.669	0.273	3.3	27.113	0.037
1.4	4.055	0.247	3.4	29.964	0.033
1.5	4.482	0.223	3.5	33.115	0.030
1.6	4.953	0.202	3.6	36.598	0.027
1.7	5.474	0.183	3.7	40.447	0.025
1.8	6.050	0.165	3.8	44.701	0.024
1.9	6.686	0.150	3.9	49.402	0.020
2.0	7.389	0.135	4.0	54.598	0.018
2.1	8.166	0.122	4.1	60.340	0.017
2.2	9.025	0.111	4.2	66.686	0.015
2.3	9.974	0.100	4.3	73.700	0.014
2.4	11.023	0.091	4.4	81.451	0.012
2.5	12.182	0.082	4.5	90.017	0.011
2.6	13.464	0.074	4.6	99.484	0.010
2.7	14.880	0.067	4.7	109.947	0.009
2.8	16.445	0.061	4.8	121.510	0.008
2.9	18.174	0.055	4.9	134.290	0.007

TABLE 4.   Sines, cosines, tangents (radian measure)

x	$\sin x$	$\cos x$	$\tan x$	x	$\sin x$	$\cos x$	$\tan x$
0.00	0.000	1.000	0.000	0.46	0.444	0.896	0.495
0.01	0.010	1.000	0.010	0.47	0.453	0.892	0.508
0.02	0.020	1.000	0.020	0.48	0.462	0.887	0.521
0.03	0.030	1.000	0.030	0.49	0.471	0.882	0.533
0.04	0.040	0.999	0.040	0.50	0.479	0.878	0.546
0.05	0.050	0.999	0.050	0.51	0.488	0.873	0.559
0.06	0.060	0.998	0.060	0.52	0.497	0.868	0.573
0.07	0.070	0.998	0.070	0.53	0.506	0.863	0.586
0.08	0.080	0.997	0.080	0.54	0.514	0.858	0.599
0.09	0.090	0.996	0.090	0.55	0.523	0.853	0.613
0.10	0.100	0.995	0.100	0.56	0.531	0.847	0.627
0.11	0.110	0.994	0.110	0.57	0.540	0.842	0.641
0.12	0.120	0.993	0.121	0.58	0.548	0.836	0.655
0.13	0.130	0.992	0.131	0.59	0.556	0.831	0.670
0.14	0.140	0.990	0.141	0.60	0.565	0.825	0.684
0.15	0.149	0.989	0.151	0.61	0.573	0.820	0.699
0.16	0.159	0.987	0.161	0.62	0.581	0.814	0.714
0.17	0.169	0.986	0.172	0.63	0.589	0.808	0.729
0.18	0.179	0.984	0.182	0.64	0.597	0.802	0.745
0.19	0.189	0.982	0.192	0.65	0.605	0.796	0.760
0.20	0.199	0.980	0.203	0.66	0.613	0.790	0.776
0.21	0.208	0.978	0.213	0.67	0.621	0.784	0.792
0.22	0.218	0.976	0.224	0.68	0.629	0.778	0.809
0.23	0.228	0.974	0.234	0.69	0.637	0.771	0.825
0.24	0.238	0.971	0.245	0.70	0.644	0.765	0.842
0.25	0.247	0.969	0.255	0.71	0.652	0.758	0.860
0.26	0.257	0.966	0.266	0.72	0.659	0.752	0.877
0.27	0.267	0.964	0.277	0.73	0.667	0.745	0.895
0.28	0.276	0.961	0.288	0.74	0.674	0.738	0.913
0.29	0.286	0.958	0.298	0.75	0.682	0.732	0.932
0.30	0.296	0.955	0.309	0.76	0.689	0.725	0.950
0.31	0.305	0.952	0.320	0.77	0.696	0.718	0.970
0.32	0.315	0.949	0.331	0.78	0.703	0.711	0.989
0.33	0.324	0.946	0.343	0.79	0.710	0.704	1.009
0.34	0.333	0.943	0.354	0.80	0.717	0.697	1.030
0.35	0.343	0.939	0.365	0.81	0.724	0.689	1.050
0.36	0.352	0.936	0.376	0.82	0.731	0.682	1.072
0.37	0.362	0.932	0.388	0.83	0.738	0.675	1.093
0.38	0.371	0.929	0.399	0.84	0.745	0.667	1.116
0.39	0.380	0.925	0.411	0.85	0.751	0.660	1.138
0.40	0.389	0.921	0.423	0.86	0.758	0.652	1.162
0.41	0.399	0.917	0.435	0.87	0.764	0.645	1.185
0.42	0.408	0.913	0.447	0.88	0.771	0.637	1.210
0.43	0.417	0.909	0.459	0.89	0.777	0.629	1.235
0.44	0.426	0.905	0.471	0.90	0.783	0.622	1.260
0.45	0.435	0.900	0.483	0.91	0.790	0.614	1.286

TABLE 4   (continued)

x	sin x	cos x	tan x	x	sin x	cos x	tan x
0.92	0.796	0.606	1.313	1.26	0.952	0.306	3.113
0.93	0.802	0.598	1.341	1.27	0.955	0.296	3.224
0.94	0.808	0.590	1.369	1.28	0.958	0.287	3.341
0.95	0.813	0.582	1.398	1.29	0.961	0.277	3.467
0.96	0.819	0.574	1.428	1.30	0.964	0.267	3.602
0.97	0.825	0.565	1.459	1.31	0.966	0.258	3.747
0.98	0.830	0.557	1.491	1.32	0.969	0.248	3.903
0.99	0.836	0.549	1.524	1.33	0.971	0.238	4.072
1.00	0.841	0.540	1.557	1.34	0.973	0.229	4.256
1.01	0.847	0.532	1.592	1.35	0.976	0.219	4.455
1.02	0.852	0.523	1.628	1.36	0.978	0.209	4.673
1.03	0.857	0.515	1.665	1.37	0.980	0.199	4.913
1.04	0.862	0.506	1.704	1.38	0.982	0.190	5.177
1.05	0.867	0.498	1.743	1.39	0.984	0.180	5.471
1.06	0.872	0.489	1.784	1.40	0.985	0.170	5.798
1.07	0.877	0.480	1.827	1.41	0.987	0.160	6.165
1.08	0.882	0.471	1.871	1.42	0.989	0.150	6.581
1.09	0.887	0.462	1.917	1.43	0.990	0.140	7.055
1.10	0.891	0.454	1.965	1.44	0.991	0.130	7.602
1.11	0.896	0.445	2.014	1.45	0.993	0.121	8.238
1.12	0.900	0.436	2.066	1.46	0.994	0.111	8.989
1.13	0.904	0.427	2.120	1.47	0.995	0.101	9.887
1.14	0.909	0.418	2.176	1.48	0.996	0.091	10.983
1.15	0.913	0.408	2.234	1.49	0.997	0.081	12.350
1.16	0.917	0.399	2.296	1.50	0.997	0.071	14.101
1.17	0.921	0.390	2.360	1.51	0.998	0.061	16.428
1.18	0.925	0.381	2.427	1.52	0.999	0.051	19.670
1.19	0.928	0.372	2.498	1.53	0.999	0.041	24.498
1.20	0.932	0.362	2.572	1.54	1.000	0.031	32.461
1.21	0.936	0.353	2.650	1.55	1.000	0.021	48.078
1.22	0.939	0.344	2.733	1.56	1.000	0.011	92.620
1.23	0.942	0.334	2.820	1.57	1.000	0.001	1255.770
1.24	0.946	0.325	2.912	1.58	1.000	−0.009	−108.649
1.25	0.949	0.315	3.010				

TABLE 5.   Sines, cosines, tangents (degree measure)

x	$\sin x$	$\cos x$	$\tan x$	x	$\sin x$	$\cos x$	$\tan x$
0°	0.00	1.00	0.00	45°	0.71	0.71	1.00
1	0.02	1.00	0.02	46	0.72	0.69	1.04
2	0.03	1.00	0.03	47	0.73	0.68	1.07
3	0.05	1.00	0.05	48	0.74	0.67	1.11
4	0.07	1.00	0.07	49	0.75	0.66	1.15
5	0.09	1.00	0.09	50	0.77	0.64	1.19
6	0.10	0.99	0.11	51	0.78	0.63	1.23
7	0.12	0.99	0.12	52	0.79	0.62	1.28
8	0.14	0.99	0.14	53	0.80	0.60	1.33
9	0.16	0.99	0.16	54	0.81	0.59	1.38
10	0.17	0.98	0.18	55	0.82	0.57	1.43
11	0.19	0.98	0.19	56	0.83	0.56	1.48
12	0.21	0.98	0.21	57	0.84	0.54	1.54
13	0.22	0.97	0.23	58	0.85	0.53	1.60
14	0.24	0.97	0.25	59	0.86	0.52	1.66
15	0.26	0.97	0.27	60	0.87	0.50	1.73
16	0.28	0.96	0.29	61	0.87	0.48	1.80
17	0.29	0.96	0.31	62	0.88	0.47	1.88
18	0.31	0.95	0.32	63	0.89	0.45	1.96
19	0.33	0.95	0.34	64	0.90	0.44	2.05
20	0.34	0.94	0.36	65	0.91	0.42	2.14
21	0.36	0.93	0.38	66	0.91	0.41	2.25
22	0.37	0.93	0.40	67	0.92	0.39	2.36
23	0.39	0.92	0.42	68	0.93	0.37	2.48
24	0.41	0.91	0.45	69	0.93	0.36	2.61
25	0.42	0.91	0.47	70	0.94	0.34	2.75
26	0.44	0.90	0.49	71	0.95	0.33	2.90
27	0.45	0.89	0.51	72	0.95	0.31	3.08
28	0.47	0.88	0.53	73	0.96	0.29	3.27
29	0.48	0.87	0.55	74	0.96	0.28	3.49
30	0.50	0.87	0.58	75	0.97	0.26	3.73
31	0.52	0.86	0.60	76	0.97	0.24	4.01
32	0.53	0.85	0.62	77	0.97	0.22	4.33
33	0.54	0.84	0.65	78	0.98	0.21	4.70
34	0.56	0.83	0.67	79	0.98	0.19	5.14
35	0.57	0.82	0.70	80	0.98	0.17	5.67
36	0.59	0.81	0.73	81	0.99	0.16	6.31
37	0.60	0.80	0.75	82	0.99	0.14	7.12
38	0.62	0.79	0.78	83	0.99	0.12	8.14
39	0.63	0.78	0.81	84	0.99	0.10	9.51
40	0.64	0.77	0.84	85	1.00	0.09	11.43
41	0.66	0.75	0.87	86	1.00	0.07	14.30
42	0.67	0.74	0.90	87	1.00	0.05	19.08
43	0.68	0.73	0.93	88	1.00	0.03	28.64
44	0.69	0.72	0.97	89	1.00	0.02	57.29
45	0.71	0.71	1.00	90	1.00	0.00	—

TABLE 6.   Integrals

1. $\int \alpha\, dx = \alpha x + C.$

2. $\int x^n\, dx = \dfrac{x^{n+1}}{n+1} + C, \quad n \neq -1.$

3. $\int \dfrac{dx}{x} = \log|x| + C.$

4. $\int e^x\, dx = e^x + C.$

5. $\int p^x\, dx = \dfrac{p^x}{\log p} + C.$

6. $\int \log x\, dx = x \log x - x + C.$

7. $\int \cos x\, dx = \sin x + C.$

8. $\int \sin x\, dx = -\cos x + C.$

9. $\int \sec^2 x\, dx = \tan x + C.$

10. $\int \mathrm{cosec}^2 x\, dx = -\cot x + C.$

11. $\int \sec x \tan x\, dx = \sec x + C.$

12. $\int \mathrm{cosec}\, x \cot x\, dx = -\mathrm{cosec}\, x + C.$

13. $\int \tan x\, dx = \log|\sec x| + C.$

14. $\int \cot x\, dx = \log|\sin x| + C.$

15. $\int \sec x\, dx = \log|\sec x + \tan x| + C.$

16. $\int \mathrm{cosec}\, x\, dx = \log|\mathrm{cosec}\, x - \cot x| + C.$

17. $\int \mathrm{arc}\sin x\, dx = x\, \mathrm{arc}\sin x + \sqrt{1 - x^2} + C.$

18. $\int \mathrm{arc}\tan x\, dx = x\, \mathrm{arc}\tan x - \tfrac{1}{2}\log(1 + x^2) + C.$

19. $\int \dfrac{dx}{x^2 + a^2} = \dfrac{1}{a}\, \mathrm{arc}\tan\left(\dfrac{x}{a}\right) + C.$ \qquad (take $a > 0$ throughout)

20. $\int \dfrac{dx}{x^2 - a^2} = \dfrac{1}{2a}\log\left|\dfrac{x - a}{x + a}\right| + C.$

21. $\int \dfrac{dx}{\sqrt{x^2 \pm a^2}} = \log\left(x + \sqrt{x^2 \pm a^2}\right) + C.$

22. $\int \dfrac{dx}{\sqrt{a^2 - x^2}} = \mathrm{arc}\sin\left(\dfrac{x}{a}\right) + C.$

23. $\int \sqrt{x^2 \pm a^2}\, dx = \dfrac{x}{2}\sqrt{x^2 \pm a^2} \pm \dfrac{a^2}{2}\log\left(x + \sqrt{x^2 \pm a^2}\right) + C.$

24. $\int \sqrt{a^2 - x^2}\, dx = \dfrac{x}{2}\sqrt{a^2 - x^2} + \dfrac{a^2}{2}\, \mathrm{arc}\sin\left(\dfrac{x}{a}\right) + C.$

TABLE 6.   (continued)

Hyperbolic Functions

25. $\displaystyle\int \cosh x \; dx = \sinh x + C.$

26. $\displaystyle\int \sinh x \; dx = \cosh x + C.$

27. $\displaystyle\int \operatorname{sech}^2 x \; dx = \tanh x + C.$

28. $\displaystyle\int \operatorname{cosech}^2 x \; dx = -\coth x + C.$

29. $\displaystyle\int \operatorname{sech} x \tanh x \; dx = -\operatorname{sech} x + C.$

30. $\displaystyle\int \operatorname{cosech} x \coth x \; dx = -\operatorname{cosech} x + C.$

31. $\displaystyle\int \frac{dx}{\sqrt{x^2 + a^2}} = \sinh^{-1}\left(\frac{x}{a}\right) + C.$ (take $a > 0$ throughout)

32. $\displaystyle\int \frac{dx}{\sqrt{x^2 - a^2}} = \cosh^{-1}\left(\frac{x}{a}\right) + C.$

33. $\displaystyle\int \sqrt{x^2 + a^2} \; dx = \frac{x}{2}\sqrt{x^2 + a^2} + \frac{a^2}{2}\sinh^{-1}\left(\frac{x}{a}\right) + C.$

34. $\displaystyle\int \sqrt{x^2 - a^2} \; dx = \frac{x}{2}\sqrt{x^2 - a^2} - \frac{a^2}{2}\cosh^{-1}\left(\frac{x}{a}\right) + C.$

Answers to Starred Exercises

Section 1.3

1. (a) $\frac{1}{2}$ (c) 0   (e) $-y_0/x_0$
2. (a) slope 1, y-intercept 4   (c) slope undefined, no y-intercept
   (e) slope -1, y-intercept -1 (g) slope $-\frac{7}{4}$, y-intercept -1
   (i) slope 0, y-intercept $-\frac{5}{2}$
3. (a) $y = 5x + 2$ (c) $y = 5x - 2$ 4. (a) $y = 3$ 5. (b) $x = -3$
6. (a) $y = 7$ (c) $x + y - 9 = 0$ (f) $2x + 3y - 25 = 0$
7. (a) $y = -2x + 4$ (c) $x/2 + y/4 = 1$
8. (a) $45°$ (c) $90°$
9. (a) $y = \frac{1}{3}\sqrt{3}x + 2$ (c) $y = -\frac{1}{3}\sqrt{3}x + \sqrt{3}$
10. (a) $(\frac{3}{5}, -\frac{3}{5})$; approx. $23°$ (c) $(-\frac{2}{23}, \frac{38}{23})$; approx. $17°$
11. (a) $90°$, $45°$, $45°$ (c) $16°$-$17°$, $32°$-$33°$, $130°$-$131°$

Section 1.4

1. $(-\infty, 1)$ 3. $(-\infty, -3]$ 5. $[-1, \infty)$
7. $(-1, 1)$ 9. $(-\infty, 1) \cup (2, \infty)$ 11. $(0, 1) \cup (2, \infty)$
13. $[0, \infty)$ 15. $(-\infty, -\frac{1}{5})$
17. $(-5 - 2\sqrt{6}, -5 + 2\sqrt{6})$ 19. $(-\infty, 0) \cup (0, \infty)$
21. $(-1, 0) \cup (1, \infty)$ 23. $(-\infty, 0] \cup (5, \infty)$
25. $(-\infty, -\frac{5}{3}) \cup (5, \infty)$ 27. $(-\infty, -\frac{1}{3}\sqrt{3}] \cup [\frac{1}{3}\sqrt{3}, \infty)$

Section 1.5

1. $(-2, 2)$ 3. $(-\infty, -3) \cup (3, \infty)$ 4. $(0, 2)$
6. $(-\frac{3}{2}, \frac{5}{2})$ 7. $(-\frac{9}{4}, -\frac{7}{4})$ 9. $(-1, 0) \cup (0, 1)$

10. $(-\frac{1}{2}, 0) \cup (0, \frac{1}{2})$ 12. $(-\frac{3}{2}, \frac{1}{2}) \cup (\frac{1}{2}, \frac{5}{2})$ 13. $(-5, 3) \cup (3, 11)$

15. $(1, 2)$ 16. $(\frac{2}{3}, \frac{8}{3})$ 18. $(\frac{1}{2}, \frac{7}{10})$

19. $(-\infty, -4) \cup (-1, \infty)$ 21. $(-\infty, -\frac{8}{5}) \cup (2, \infty)$

22. $(-\infty, -5) \cup (\frac{3}{2}, \infty)$ 24. $(-\sqrt{3}, -1) \cup (1, \sqrt{3})$

25. $(-\sqrt{2}, 0) \cup (0, \sqrt{2})$ 27. $(-\infty, -2) \cup (-\sqrt{2}, \sqrt{2}) \cup (2, \infty)$

28. $(-\infty, -5) \cup (-\sqrt{7}, \sqrt{7}) \cup (5, \infty)$

30. $(-\infty, 0) \cup (2 - \sqrt{2}, 2 + \sqrt{2}) \cup (4, \infty)$

Section 1.6

1. $x = 0$ 3. $x = -2$

5. dom $(f) = (-\infty, \infty)$; ran $(f) = [0, \infty)$

7. dom $(f) = (-\infty, \infty)$; ran $(f) = (-\infty, \infty)$

9. dom $(f) = (-\infty, 0) \cup (0, \infty)$; ran $(f) = (0, \infty)$

11. dom $(f) = (-\infty, 1]$; ran $(f) = [0, \infty)$

13. dom $(f) = (-\infty, 1]$; ran $(f) = [-1, \infty)$

15. dom $(f) = (-\infty, 1)$; ran $(f) = (0, \infty)$

27. dom $(f) = (-\infty, 0) \cup (0, \infty)$; ran $(f) = \{-1, 1\}$; Figure A.1.6.1

29. dom $(f) = [0, \infty)$; ran $(f) = [1, \infty)$; Figure A.1.6.2

FIGURE A.1.6.1

FIGURE A.1.6.2

31. (a) $(6f + 3g)(x) = 9\sqrt{x}, \quad x > 0$ (b) $(fg)(x) = x - 2/x + 1, \quad x > 0$

    (c) $(f/g)(x) = \dfrac{x - 1}{x + 2}, \quad x > 0$

33. (a) $(f + g)(x) = \begin{cases} 1 - x, & x \le 1 \\ 2x - 1, & 1 < x < 2 \\ 2x - 2, & x \ge 2 \end{cases}$

    (b) $(f - g)(x) = \begin{cases} 1 - x, & x \le 1 \\ 2x - 1, & 1 < x < 2 \\ 2x, & x \ge 2 \end{cases}$ (c) $(fg)(x) = \begin{cases} 0, & x < 2 \\ 1 - 2x, & x \ge 2 \end{cases}$

35. Figure A.1.6.3   37. Figure A.1.6.4   39. Figure A.1.6.5

41. Figure A.1.6.6

43. odd   45. neither   46. odd

48. odd   49. even   50. even

51. odd   52. symmetry about the y-axis   53. symmetry about the origin

FIGURE A.1.6.3

FIGURE A.1.6.4

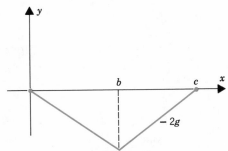

FIGURE A.1.6.5

FIGURE A.1.6.6

Section 1.7

1. $(f \circ g)(x) = 2x^2 + 5$

3. $(f \circ g)(x) = \sqrt{x^2 + 5}$

5. $(f \circ g)(x) = x, \ x \neq 0$

7. $(f \circ g)(x) = 1/x - 1$

9. $(f \circ g)(x) = \dfrac{1}{x^2 + 1}$

11. $(f \circ g \circ h)(x) = 4(x^2 - 1)$

13. $(f \circ g \circ h)(x) = (x^4 - 1)^2$

15. $(f \circ g \circ h)(x) = 2x^2 + 1$

17. (a) $f(x) = 1/x$ (c) $f(x) = \sqrt{a^2 + x}$

18. (b) $g(x) = a^2 x^2, \ x \neq 0$

19. (a) $(f \circ g)(x) = \begin{cases} x^2, & x < 0 \\ 1 + x, & 0 \le x < 1 \\ (1 + x)^2, & x \ge 1 \end{cases}$

(b) $(g \circ f)(x) = \begin{cases} 2 - x, & x \le 0 \\ -x^2, & 0 < x < 1 \\ 1 + x^2, & x \ge 1 \end{cases}$

Section 1.8

1. $f^{-1}(x) = \frac{1}{5}(x - 3)$

3. $f^{-1}(x) = \frac{1}{4}(x + 7)$

5. not one-to-one

7. $f^{-1}(x) = (x - 1)^{1/5}$

9. $f^{-1}(x) = [\frac{1}{3}(x - 1)]^{1/3}$

11. $f^{-1}(x) = 1 - x^{1/3}$

13. $f^{-1}(x) = (x - 2)^{1/3} - 1$

15. $f^{-1}(x) = x^{5/3}$

17. $f^{-1}(x) = \frac{1}{3}(2 - x^{1/3})$

19. $f^{-1}(x) = 1/x$

21. not one-to-one

23. $f^{-1}(x) = (1/x - 1)^{1/3}$

25. f

Section 2.1

1. -1 3. 4   4. does not exist   6. 4
7. does not exist   9. 0   10. does not exist   12. 2
13. 2   15. does not exist   17. does not exist   19. does not exist

Section 2.2

1. $\frac{1}{2}$ 3. does not exist   4. $\frac{4}{3}$ 6. does not exist
7. 4   9. 1   10. δ_1 and δ_2 12. $\delta = \frac{1}{5}\epsilon$
14. $\delta = 5\epsilon$
16. Take $\delta = \frac{1}{3}\epsilon$. If $0 < |x - 2| < \frac{1}{3}\epsilon$, then $|3x - 6| < \epsilon$ and therefore $|(3x - 1) - 5| < \epsilon$.
18. Take $\delta = \frac{1}{5}\epsilon$. If $0 < |x - 0| < \frac{1}{5}\epsilon$, then $|5x| < \epsilon$ and so $|(2 - 5x) - 2| < \epsilon$.
20. Take $\delta = \epsilon$. If $0 < |x - 2| < \epsilon$, then $||x - 2| - 0| < \epsilon$.
22. Take $\delta = $ minimum of 9 and ϵ. If $0 < |x - 9| < \delta$, then $x > 0$ and $|x - 9| < \epsilon$. It follows that $|\sqrt{x} - 3||\sqrt{x} + 3| < \epsilon$ and, since $|\sqrt{x} + 3| > 1$, that $|\sqrt{x} - 3| < \epsilon$.
24. Take $\delta = $ minimum of 4 and ϵ. If $0 < |x - 5| < \delta$, then $x > 1$ and $|(x - 1) - 4| < \epsilon$. It follows that $|\sqrt{x - 1} - 2||\sqrt{x - 1} + 2| < \epsilon$ and, since $|\sqrt{x - 1} + 2| > 1$, that $|\sqrt{x - 1} - 2| < \epsilon$.

Section 2.3

1. (a) $\frac{7}{8}$ (c) 4   (e) does not exist   (g) $-\frac{23}{20}$
   (i) 2   (k) 0   (m) 0   (o) does not exist
   (q) $\frac{8}{9}$ (s) $\frac{5}{8}$ (u) does not exist
2. (a) does not exist   (c) -1 (e) $-\frac{1}{2}$ (g) 4
3. (a) 3   (d) 0   (e) does not exist
4. (a) 27   (c) 0
6. set $f(x) = 1/x$ and $g(x) = -1/x$

Section 2.4

1. (a) 0   (c) 1   (e) 10
2. (a) 9   (c) 0   (d) 0   (f) -1 (g) 0   (i) -1
   (j) -1 (l) 0   (m) does not exist   (o) does not exist
3. (a) (i) $\lim\limits_{x \to 3} \dfrac{1}{x - 1} = \dfrac{1}{2}$ (ii) $\lim\limits_{x \to 3} \left(\dfrac{1}{x - 1} - \dfrac{1}{2} \right) = 0$

   (iii) $\lim\limits_{x \to 3} \left| \dfrac{1}{x - 1} - \dfrac{1}{2} \right| = 0$ (iv) $\lim\limits_{h \to 0} \dfrac{1}{(3 + h) - 1} = \dfrac{1}{2}$
5. (a) does not exist   (b) 1   7. (a) 0
8. (b) $\frac{1}{2}\{[f(x) + g(x)] - |f(x) - g(x)|\}$

Section 2.5

1. (b) (i) -1 (iii) does not exist
2. (b) (i) 1   (iii) 1   (v) 2   (vii) 4   (ix) does not exist
3. (a) 1   (c) -1 (d) 1   (f) 1   (g) 0
   (i) 1   (j) 0   (l) 1   (m) does not exist
4. (b) (i) -1 (iii) -1 (v) 0   (vii) -1 (ix) -1
5. (b) (i) 0   (ii) 1   (iii) 1   (iv) 1   (v) does not exist   (vi) 1
9. (a) 0   (c) 1

Section 2.6

1. continuous
3. essential discontinuity
5. continuous
7. removable discontinuity
9. essential discontinuity
11. continuous
13. essential discontinuity
15. no discontinuities
17. essential discontinuity at 1
19. no discontinuities
21. no discontinuities
23. essential discontinuities at 0 and 2
25. $f(1) = 2$
27. impossible
31. (a) nowhere   (b) at 0

Section 2.7

1. -1
3. does not exist   4. 0
6. does not exist
7. 1
9. -3
10. 1
12. $\frac{1}{2}$
13. does not exist   15. 0
16. does not exist   18. $\frac{1}{2}$
19. 0
21. 1
22. -2
24. does not exist
25. 1
27. 0
28. Take $\delta = \frac{1}{5}\epsilon$. If $0 < |x - 2| < \frac{1}{5}\epsilon$, then $|5x - 10| < \epsilon$ and therefore $|(5x - 4) - 6| < \epsilon$.
30. Take $\delta = \frac{1}{2}\epsilon$. If $0 < |x + 4| < \frac{1}{2}\epsilon$, then $|2x + 8| < \epsilon$. Since

$$||2x + 5| - 3| = ||2x + 5| - |-3|| \le |2x + 5 - (-3)| = |2x + 8|,$$
$$\underset{\text{by 1.5.5}}{\underline{\hspace{3cm}}}$$

it follows that $||2x + 5| - 3| < \epsilon$.
32. (b) (i) 2   (iii) 2   (iv) 1   (vi) does not exist   (vii) 1
   (ix) does not exist
   (c) (i) at -1 (iii) at -1
33. 0

Section 3.1

1. $f'(x) = 0$
3. $f'(x) = 3$
5. $f'(x) = -1/x^2$
7. $f'(x) = x$
9. $f'(x) = 6x^2$
11. $f'(x) = 2(x - 1)$
13. $f'(x) = -\frac{1}{2}x^{-3/2}$
15. $f'(x) = -\frac{1}{2}(x - 1)^{-3/2}$
17. $f'(2) = -\frac{1}{9}$
19. $f'(2) = \frac{1}{4}$
25. $f'(c) = 4$
27. does not exist
29. does not exist

Section 3.2

1. $F'(x) = -1$

3. $F'(x) = 55x^4 - 18x^2$

5. $F'(x) = 2ax + b$

7. $F'(x) = 2/x^3$

9. $F'(x) = \dfrac{bc - ad}{(cx - d)^2}$

11. $G'(x) = \dfrac{3x^2 - 2x^3}{(1 - x)^2}$

13. $G'(x) = 3x^2 - 6x - 1$

15. $G'(x) = 2x - 3$

17. $G'(x) = -\dfrac{2(3x^2 - x + 1)}{x^2(x - 2)^2}$

19. $f'(x) = (9x^8 - 8x^9)\left(1 - \dfrac{1}{x^2}\right) + \left(x + \dfrac{1}{x}\right)(72x^7 - 72x^8)$

$= -80x^9 + 81x^8 - 64x^7 + 63x^6$

21. $f'(0) = -\frac{1}{4}, \quad f'(1) = -1$

23. $f'(0) = 0, \quad f'(1) = -1$

25. $f'(0) = \dfrac{ad - bc}{d^2}, \quad f'(1) = \dfrac{ad - bc}{(c + d)^2}$

27. $f'(0) = 3$

29. $f'(0) = \frac{20}{9}$

Section 3.3

1. $\dfrac{dy}{dx} = 12x^3 - 2x$

3. $\dfrac{dy}{dx} = 1 + \dfrac{1}{x^2}$

5. $\dfrac{dy}{dx} = \dfrac{1 - x^2}{(1 + x^2)^2}$

7. $\dfrac{dy}{dx} = \dfrac{2x - x^2}{(1 - x)^2}$

9. $\dfrac{dy}{dx} = \dfrac{-6x^2}{(x^3 - 1)^2}$

11. 2

13. $21x^2$

15. $\dfrac{-4t}{(t^2 - 1)^2}$

17. $\dfrac{2t^3(t^3 - 2)}{(2t^3 - 1)^2}$

19. $\dfrac{2}{(1 - 2u)^2}$

21. $-\left[\dfrac{1}{(u - 1)^2} + \dfrac{1}{(u + 1)^2}\right] = -2\left[\dfrac{u^2 + 1}{(u^2 - 1)^2}\right]$

23. $2x\left[\dfrac{1}{(1 - x^2)^2} + \dfrac{1}{x^4}\right]$

25. $\dfrac{-2}{(x - 1)^2}$

27. 47

29. $\frac{1}{4}$

Section 3.4

1. 8

3. 4

5. $-\frac{5}{36}$

8. $\dfrac{dA}{dr} = 8\pi r, \quad 8\pi r_0, \quad r_0 = 1/8\pi$

10. $-h/b$

11. (a) $\frac{1}{2}r^2$ (c) $-4Ar^{-3} = -2\theta/r$

12. (b) $2\pi(2r + h)$

14. -4

16. $-\dfrac{r}{2r + h}$

Section 3.5

1. $2(1 - 2x)^{-2}$

3. $20(x^5 - x^{10})^{19}(5x^4 - 10x^9)$

5. $4\left(x - \dfrac{1}{x}\right)^3\left(1 + \dfrac{1}{x^2}\right)$

7. $4(x - x^3 - x^5)^3(1 - 3x^2 - 5x^4)$

9. $200t(t^2 - 1)^{99}$

11. $-4(t^{-1} + t^{-2})^3(t^{-2} + 2t^{-3})$

13. $324x^3\left[\dfrac{1 - x^2}{(x^2 + 1)^5}\right]$

15. $2(x^4 + x^2 + x)(4x^3 + 2x + 1)$

17. $-\left(\dfrac{x^3}{3} + \dfrac{x^2}{2} + \dfrac{x}{1}\right)^{-2}(x^2 + x + 1)$

19. $3\left(\dfrac{1}{x + 2} - \dfrac{1}{x - 2}\right)^2\left[\dfrac{-1}{(x + 2)^2} + \dfrac{1}{(x - 2)^2}\right] = \dfrac{384x}{(x + 2)^4(x - 2)^4}$

21. $\dfrac{dy}{dx} = -\dfrac{2x + 1}{(2x^2 + 2x + 1)^2}$

23. $\dfrac{dy}{dx} = \dfrac{80x(5x^2 + 1)^3}{[1 - 4(5x^2 + 1)^4]^2}$

25. $\dfrac{dy}{dt} = \dfrac{dy}{du}\dfrac{du}{dx}\dfrac{dx}{dt} = \dfrac{7(2t - 5)^4 + 12(2t - 5)^2 - 2}{[(2t - 5)^4 + 2(2t - 5)^2 + 2]^2}[4(2t - 5)]$

27. 1   29. 1   31. 1

33. 2   35. 2   37. 0

39. $2xf'(x^2 + 1)$ 41. $2f(x)f'(x)$

43. (a) $x = 0$ (b) $x < 0$ (c) $x > 0$

45. (a) $x = -1, x = 1$ (b) $-1 < x < 1$ (c) $x < -1, x > 1$

48. $-\dfrac{r}{2\theta}$ 50. $2[(7x - x^{-1})^{-2} + 3x^2]^{-2}[(7x - x^{-1})^{-3}(7 + x^{-2}) - 3x]$

52. $3[(x + x^{-1})^2 - (x^2 + x^{-2})^{-1}]^2[2(x + x^{-1})(1 - x^{-2}) + (x^2 + x^{-2})^{-2}(2x - 2x^{-3})]$

Section 3.6

1. $42x - 120x^3$

3. $\dfrac{4ab^2}{(a - bx)^3}$

5. $6ax + 12bx^{-5}$

7. $\dfrac{2x(x^2 - 3a^2)}{(x^2 + a^2)^3}$

9. 2

11. 48

13. $\dfrac{2c(bc - ad)}{(cx + d)^3}$

15. 0

17. $\dfrac{2(1 - 2x)}{(1 + x)^4}$

21. (a) $x = 0$ (b) $x > 0$ (c) $x < 0$

23. (a) $x = -2, x = 1$ (b) $x < -2, x > 1$ (c) $-2 < x < 1$

26. $b^n n!$

Section 3.7

1. $\dfrac{3}{2}x^2(x^3 + 1)^{-1/2}$

3. $\dfrac{2x^2 + 1}{\sqrt{x^2 + 1}}$

5. $\dfrac{x}{(\sqrt[4]{2x^2 + 1})^3}$

7. $\dfrac{x(2x^2 - 5)}{\sqrt{2 - x^2}\sqrt{3 - x^2}}$

9. $\dfrac{1}{2}\left(\dfrac{1}{\sqrt{x}} - \dfrac{1}{x\sqrt{x}}\right)$

11. $\dfrac{1}{(\sqrt{x^2 + 1})^3}$

12. $-\dfrac{1}{x^2\sqrt{x^2+1}}$

14. $\dfrac{ad-bc}{2(cx+d)^2}\sqrt{\dfrac{cx+d}{ax+b}}$

16. $\dfrac{a^2}{(\sqrt{a^2+x^2})^3}$

18. $\dfrac{x(2x^2-3a^2)}{(\sqrt{a^2-x^2})^3}$

20. $\dfrac{1}{2\sqrt{x}}\,f'(\sqrt{x}+1)$

22. $\dfrac{xf'(x^2+1)}{\sqrt{f(x^2+1)}}$

24. $\dfrac{1}{x}$

26. $\dfrac{1}{\sqrt{1-x^2}}$

27. (a) elastic   (c) unitary   (e) inelastic   for $0<P<20$,   unitary   at $P=20$,   elastic for $20<P<40$

Section 3.8

1. $\dfrac{dy}{dx}=\dfrac{2}{3}x^{-1/3}$

3. $\dfrac{dy}{dx}=-\dfrac{a}{x^2}+c$

4. $\dfrac{dy}{dx}=\dfrac{1}{4\sqrt{x}}+\dfrac{1}{x\sqrt{x}}$

6. $\dfrac{ds}{dt}=-\dfrac{a}{2t\sqrt{t}}+\dfrac{b}{2\sqrt{t}}+\dfrac{3c\sqrt{t}}{2}$

7. $\dfrac{dr}{d\theta}=-\dfrac{1}{\sqrt{1-2\theta}}$

9. $f'(x)=\dfrac{x}{(\sqrt{a^2-x^2})^3}$

10. $\dfrac{dy}{dx}=\dfrac{2b}{x^2}\left(a-\dfrac{b}{x}\right)$

12. $\dfrac{dy}{dx}=\dfrac{2a+3bx}{2\sqrt{a+bx}}$

13. $\dfrac{ds}{dt}=\dfrac{a^2+2t^2}{\sqrt{a^2+t^2}}$

15. $\dfrac{dy}{dx}=\dfrac{4a^2x}{(a^2-x^2)^2}$

16. $\dfrac{dy}{dx}=-\dfrac{a^2}{x^2\sqrt{a^2+x^2}}$

18. $\dfrac{dy}{dx}=\dfrac{a^2}{(\sqrt{a^2-x^2})^3}$

19. $\dfrac{dy}{dx}=\dfrac{2a-bx}{2(\sqrt{a-bx})^3}$

21. $\dfrac{dy}{dx}=-\dfrac{c}{(1+cx)^2}\sqrt{\dfrac{1+cx}{1-cx}}$

22. $f'(x)=\dfrac{2+4x}{(\sqrt[3]{2+3x})^2}$

24. $\dfrac{ds}{dt}=\dfrac{4}{(2-3t)^2}\left(\sqrt[3]{\dfrac{2-3t}{2+3t}}\right)^2$

25. $\dfrac{dr}{d\theta}=-\dfrac{3a+2b\theta}{3\theta^2(\sqrt[3]{a+b\theta})^2}$

27. $\dfrac{dy}{dx}=-\dfrac{bx}{a\sqrt{a^2-x^2}}$

28. $\dfrac{dy}{dx}=-\dfrac{(a^{3/5}-x^{3/5})^{2/3}}{x^{2/5}}$

30. $\dfrac{dy}{dx}=\dfrac{(x+2)(3x^2+2x+4)}{\sqrt{x^2+2}}$

31. 540   33. $\frac{1}{12}$ 35. $\frac{5}{6}$ 37. $\frac{3}{64}$

39. $-\frac{41}{90}$ 41. 0   43. 20

Section 3.9

1. (a) tangent $11x-y+16=0$; normal $x+11y+68=0$
   (c) tangent $4x+y=0$; normal $2x-8y-17=0$

2. (a) $(\frac{1}{4},\frac{1}{2})$ (c) $(2,6)$ 3. (a) $(-2,-10)$ (c) $(0,1)$

4. (a) $(1,\frac{2}{3})$ (c) $(\frac{1}{3},\frac{2}{27}\sqrt{3})$ 7. $\frac{425}{8}$

9. $y=x-3$ 11. $A=-1$, $B=0$, $C=4$

13. $C = \dfrac{1 - 2\sqrt{AB}}{B}$; $\left(\dfrac{\sqrt{AB}}{A}, \dfrac{1 - \sqrt{AB}}{B}\right)$ is the equilibrium point.

Section 3.10

1. $-\dfrac{x}{y}$
3. $-\dfrac{b^2 x}{a^2 y}$
5. $-\left(\dfrac{y}{x}\right)^{1/3}$

7. $-\dfrac{x^2(x + 3y)}{x^3 + y^3}$
9. $-\dfrac{2y + 1}{2x + 1}$

11. tangent $2x + 3y - 5 = 0$; normal $3x - 2y + 12 = 0$
13. tangent $x + 2y + 8 = 0$; normal $2x - y + 1 = 0$
15. at right angles
18. $\dfrac{16}{(x + y)^3}$
21. $-\dfrac{4a^2}{y^3}$

23. $-\dfrac{b^4}{a^2 y^3}$
24. $\dfrac{dy}{dx} = \frac{5}{8}$, $\dfrac{d^2 y}{dx^2} = -\frac{9}{128}$

26. $2x + y - 12 = 0$, $14x + y - 60 = 0$

Section 4.1

1. $c = \frac{3}{2}$
3. $c = \frac{1}{3}\sqrt{39}$
5. $c = \frac{1}{2}\sqrt{2}$
7. $c = 0$
9. f is not differentiable on all of $(-1, 2)$; it fails to be differentiable at $x = \frac{1}{2}$.
12. Set $P(x) = 6x^4 - 7x + 1$. If there existed three numbers $a < b < c$ at which $P(x) = 0$, then by Rolle's theorem $P'(x)$ would have to be zero for some x in (a, b) and also for some x in (b, c). This is not the case: $P'(x) = 24x^3 - 7$ is zero only at $x = \left(\frac{7}{24}\right)^{1/3}$.
14. Set $P(x) = x^3 + 9x^2 + 33x - 8$. Note that $P(0) < 0$ and $P(1) > 0$. Thus by the intermediate-value theorem there exists some number c between 0 and 1 at which $P(x) = 0$. If the equation $P(x) = 0$ had an additional real root, then by Rolle's theorem there would have to be some real number at which $P'(x) = 0$. This is not the case: $P'(x) = 3x^2 + 18x + 33$ is never 0 since the discriminant $b^2 - 4ac = (18)^2 - 12(33) < 0$.

Section 4.2

1. increasing on $(-\infty, -1]$ and $[1, \infty)$, decreasing on $[-1, 1]$
3. increasing on $(-\infty, -1]$ and $[1, \infty)$, decreasing on $[-1, 0)$ and $(0, 1]$
5. increasing on $(-\infty, -1 - \frac{1}{3}\sqrt{3}]$ and $[-1 + \frac{1}{3}\sqrt{3}, \infty)$, decreasing on $[-1 - \frac{1}{3}\sqrt{3}, -1 + \frac{1}{3}\sqrt{3}]$
7. increasing on $(-\infty, 2)$, decreasing on $(2, \infty)$
9. increasing on $(-\infty, -1)$ and $(-1, 0]$, decreasing on $[0, 1)$ and $(1, \infty)$
11. increasing on $[-\sqrt{5}, 0]$ and $[\sqrt{5}, \infty)$, decreasing on $(-\infty, -\sqrt{5}]$ and $[0, \sqrt{5}]$
13. increasing on $(-\infty, -1)$ and $(-1, \infty)$
15. decreasing on $[0, \infty)$
17. increasing on $[0, \infty)$, decreasing on $(-\infty, 0]$
19. $f(x) = \frac{1}{3}x^3 - x + 1$
21. $f(x) = x^2 - 5x + 10$
23. $f(x) = -2x^{-2} + 2$, $x > 0$
25. $f(x) = -\frac{1}{4}x^{-4} - \frac{25}{4}x^{4/5} + \frac{13}{2}$, $x > 0$

Section 4.3

1. no critical pts; no local extreme values
3. critical pts ± 1; local max $f(-1) = -2$, local min $f(1) = 2$
5. critical pts $-1 \pm \frac{1}{3}\sqrt{3}$; local max $f(-1 - \frac{1}{3}\sqrt{3}) = \frac{2}{9}\sqrt{3}$,
   local min $f(-1 + \frac{1}{3}\sqrt{3}) = -\frac{2}{9}\sqrt{3}$
7. no critical pts; no local extreme values
9. no critical pts; no local extreme values
11. critical pts $-4, 0, 4$; local max $f(0) = 16$, local min $f(-4) = f(4) = 0$
13. critical pt 2; no local extreme values
15. critical pts $-1, \frac{1}{2}$; local max $f(\frac{1}{2}) = \frac{27}{16}$
17. critical pt 0; local min $f(0) = 0$
19. critical pts $-2, -\frac{1}{2}, 1$; local max $f(-\frac{1}{2}) = \frac{9}{4}$, local min $f(-2) = f(1) = 0$
21. critical pts $-\frac{3}{2}, 0$; local max $f(-\frac{3}{2}) = -\frac{27}{4}$
23. critical pt $-\frac{3}{2}$; local max $f(-\frac{3}{2}) = -4$

Section 4.4

1. critical pt -2; $f(-2) = 0$ endpt min and absolute min
3. critical pts $0, 2, 3$; $f(0) = 1$ endpt max and absolute max, $f(2) = -3$ local min
   and absolute min, $f(3) = -2$ endpt max
5. critical pt $2^{-1/3}$; $f(2^{-1/3}) = 3 \cdot 2^{-2/3}$ local min
7. critical pts $\frac{1}{10}, 2^{-1/3}, 2$; $f(\frac{1}{10}) = 100\frac{1}{100}$ endpt max and absolute max;
   $f(2^{-1/3}) = 3 \cdot 2^{-2/3}$ local min and absolute min, $f(2) = 4\frac{1}{2}$ endpt max
9. critical pts $0, \frac{3}{2}, 2$; $f(0) = 2$ endpt max and absolute max, $f(\frac{3}{2}) = -\frac{1}{4}$ local min
   and absolute min, $f(2) = 0$ endpt max
11. critical pt 0; $f(0) = \frac{1}{3}$ endpt max
13. critical pts $0, \frac{1}{4}, 1$; $f(0) = 0$ endpt min and absolute min, $f(\frac{1}{4}) = \frac{1}{16}$ local max,
    $f(1) = 0$ local min and absolute min
15. critical pts $2, 3$; $f(2) = 2$ local max and absolute max, $f(3) = 0$ endpt min
17. critical pt 1; no extreme values
19. critical pts $1, 4$; $f(1) = -2$ local min, $f(4) = 1$ local max

Section 4.5

1. 20, 20
3. 20 by 10
5. $2r^2$
7. $x = \frac{1}{2}a, \quad y = \frac{1}{2}b$
9. $\dfrac{\beta - b}{2(a + \alpha)}$ tons
11. $l = \frac{1}{3}\sqrt{3}\,d, \quad w = \frac{1}{3}\sqrt{6}\,d$
13. $\frac{1}{3}p, \frac{1}{3}p, \frac{1}{3}p$
15. -3
17. (1, 1)
19. $s = \frac{1}{6}(a + b - \sqrt{a^2 - ab + b^2})$
23. $(0, \frac{1}{3}\sqrt{3}a)$
26. (a) use it all for the circle   (b) use $9l/(\sqrt{3}\pi + 9)$ for the triangle
27. (a) base radius $\frac{2}{3}r$ and height $\frac{1}{3}h$
29. $2\sqrt{3}r, \quad 2\sqrt{3}r, \quad 2\sqrt{3}r$
31. (b) $r = \frac{1}{2}\sqrt{2}R, \quad h = \sqrt{2}R$
33. $\frac{2}{27}\sqrt{3}\,\pi a^3$
34. 10

Section 4.6

1. concave down on $(-\infty, 0)$, concave up on $(0, \infty)$
3. concave down on $(-\infty, 0]$, concave up on $[0, \infty)$; pt of inflection $(0, 2)$
5. concave up on $(-\infty, -\frac{1}{3}\sqrt{3}]$, concave down on $[-\frac{1}{3}\sqrt{3}, \frac{1}{3}\sqrt{3}]$, concave up on $[\frac{1}{3}\sqrt{3}, \infty)$; pts of inflection $(-\frac{1}{3}\sqrt{3}, -\frac{5}{36})$, $(\frac{1}{3}\sqrt{3}, -\frac{5}{36})$
7. concave down on $(-\infty, -1)$ and on $[0, 1)$, concave up on $(-1, 0]$ and on $(1, \infty)$; pt of inflection $(0, 0)$
9. concave up on $(-\infty, -\frac{1}{3}\sqrt{3}]$, concave down on $[-\frac{1}{3}\sqrt{3}, \frac{1}{3}\sqrt{3}]$, concave up on $[\frac{1}{3}\sqrt{3}, \infty)$; pts of inflection $(-\frac{1}{3}\sqrt{3}, \frac{4}{9})$, $(\frac{1}{3}\sqrt{3}, \frac{4}{9})$
11. concave up on $(0, \infty)$
13. $d = \frac{1}{3}(a + b + c)$ 15. $a = -\frac{1}{2}, \quad b = \frac{1}{2}$

Section 4.7

1. critical pt 2; $f(2) = 0$ local min and absolute min; decreasing on $(-\infty, 2]$, increasing on $[2, \infty)$; concave up everywhere; no pts of inflection
3. critical pts $\frac{1}{3}$, 1; $f(\frac{1}{3}) = 1\frac{4}{27}$ local max, $f(1) = 1$ local min; increasing on $(-\infty, \frac{1}{3}]$, decreasing on $[\frac{1}{3}, 1]$, increasing on $[1, \infty)$; concave down on $(-\infty, \frac{2}{3}]$, concave up on $[\frac{2}{3}, \infty)$; pt of inflection $(\frac{2}{3}, 1\frac{2}{27})$
5. critical pts $-4, 0, 4$; $f(-4) = 32$ endpt max, $f(0) = 0$ local and absolute min, $f(4) = 160$ endpt and absolute max; decreasing on $[-4, 0]$, increasing on $[0, 4]$; concave down on $(-4, -2]$, concave up on $[-2, 4)$; pt of inflection $(-2, 16)$
7. critical pts $-2, \frac{5}{2}$; $f(-2) = 11\frac{2}{3}$ local max, $f(\frac{5}{2}) = -18\frac{17}{24}$ local min; increasing on $(-\infty, -2]$, decreasing on $[-2, \frac{5}{2}]$, increasing on $[\frac{5}{2}, \infty)$; concave down on $(-\infty, \frac{1}{4}]$, concave up on $[\frac{1}{4}, \infty)$; pt of inflection $(\frac{1}{4}, -3\frac{25}{48})$
9. critical pts 0, 1; $f(1) = 0$ local and absolute min; decreasing on $(-\infty, 1]$, increasing on $[1, \infty)$; concave up on $(-\infty, 0]$, concave down on $[0, \frac{2}{3}]$, concave up on $[\frac{2}{3}, \infty)$; pts of inflection $(0, 1)$ and $(\frac{2}{3}, \frac{11}{27})$
11. critical pts 0, 1, 4; $f(0) = 0$ endpt and absolute min, $f(1) = 1$ local and absolute max, $f(4) = 0$ endpt and absolute min; increasing on $[0, 1]$, decreasing on $[1, 4]$; concave down throughout; no pts of inflection
13. critical pt -1; $f(-1) = 2$ local min and absolute min; decreasing on $(-\infty, -1]$, increasing on $[-1, \infty)$; concave up everywhere; no pts of inflection
15. critical pt 0; no extreme values; increasing everywhere; concave down on $(-\infty, 0]$, concave up on $[0, \infty)$; pt of inflection $(0, 0)$
16. critical pts $-1, 0$; $f(-1) = -1$ local and absolute min; decreasing on $(-\infty, -1]$, increasing on $[-1, \infty)$; concave up on $(-\infty, -\frac{2}{3}]$, concave down on $[-\frac{2}{3}, 0]$, concave up on $[0, \infty)$; pts of inflection $(-\frac{2}{3}, -\frac{16}{17})$, $(0, 0)$
18. critical pt 2; no extreme values; increasing everywhere; concave down on $(-\infty, 2]$, concave up on $[2, \infty)$; pt of inflection $(2, 1)$
19. critical pts $-1, -\frac{1}{2}, 0$; $f(-1) = f(0) = 0$ local and absolute min, $f(-\frac{1}{2}) = \frac{1}{16}$ local max; decreasing on $(-\infty, -1]$, increasing on $[-1, -\frac{1}{2}]$, decreasing on $[-\frac{1}{2}, 0]$, increasing on $[0, \infty)$; concave up on $(-\infty, -\frac{1}{2} - \frac{1}{6}\sqrt{3}]$, concave down on $[-\frac{1}{2} - \frac{1}{6}\sqrt{3}, -\frac{1}{2} + \frac{1}{6}\sqrt{3}]$, concave up on $[-\frac{1}{2} + \frac{1}{6}\sqrt{3}, \infty)$; pts of inflection $(-\frac{1}{2} \pm \frac{1}{6}\sqrt{3}, \frac{1}{36})$

21. critical pts $\frac{2}{3}$, 1; $f(\frac{2}{3}) = \frac{2}{9}\sqrt{3}$ local max and absolute max, $f(1) = 0$ endpt min; increasing on $(-\infty, \frac{2}{3}]$, decreasing on $[\frac{2}{3}, 1]$; concave down; no pts of inflection

22. critical pts 0, $\frac{1}{2}$, 1; $f(0) = 0 = f(1)$ endpt and absolute min, $f(\frac{1}{2}) = \frac{1}{2}$ local and absolute max; increasing on $[0, \frac{1}{2}]$, decreasing on $[\frac{1}{2}, 1]$; concave down; no pts of inflection

24. no critical pts; no extreme values; decreasing on $(-\infty, \frac{3}{4})$ and on $(\frac{3}{4}, \infty)$; concave down on $(-\infty, \frac{3}{4})$, concave up on $(\frac{3}{4}, \infty)$; no pts of inflection

25. critical pts $-\frac{2}{3}$, 0; $f(-\frac{2}{3}) = -\frac{4}{9}$ local max, $f(0) = 0$ local min; increasing on $(-\infty, -\frac{2}{3}]$, decreasing on $[-\frac{2}{3}, -\frac{1}{3})$, decreasing on $(-\frac{1}{3}, 0]$, increasing on $[0, \infty)$; concave down on $(-\infty, -\frac{1}{3})$, concave up on $(-\frac{1}{3}, \infty)$; no pts of inflection

27. critical pt $\frac{1}{3}$; $f(\frac{1}{3}) = \frac{1}{12}$ local and absolute max; decreasing on $(-\infty, -\frac{1}{3})$, increasing on $(-\frac{1}{3}, \frac{1}{3}]$, decreasing on $[\frac{1}{3}, \infty)$; concave down on $(-\infty, -\frac{1}{3})$ and on $(-\frac{1}{3}, \frac{2}{3}]$, concave up on $[\frac{2}{3}, \infty)$; pt of inflection $(\frac{2}{3}, \frac{2}{27})$

28. critical pt 1; $f(1) = \frac{1}{2}$ local and absolute max; decreasing on $(-\infty, -1)$, increasing on $(-1, 1]$, decreasing on $[1, \infty)$; concave down on $(-\infty, -1)$ and on $(-1, 2]$, concave up on $[2, \infty)$; pt of inflection $(2, \frac{4}{9})$

30. critical pt 0; $f(0) = 0$ local and absolute min; decreasing on $(-\infty, 0]$, increasing on $[0, \infty)$; concave down on $(-\infty, -\frac{1}{3}\sqrt{6}]$, concave up on $[-\frac{1}{3}\sqrt{6}, \frac{1}{3}\sqrt{6}]$, concave down on $[\frac{1}{3}\sqrt{6}, \infty)$; pt of inflection $(\pm\frac{1}{3}\sqrt{6}, \frac{1}{2})$

Section 4.8

1. (a) -2 units per second   (b) 4 units per second
3. increasing $\frac{8}{3}$ units per second
5. $\frac{1}{2}\sqrt{3}\alpha k$ square centimeters per minute
7. boat A
9. 10 cubic feet per hour
11. (a) $-\frac{5}{4}\sqrt{2}$ (b) $-\frac{7}{16}\sqrt{2}$ (c) $\frac{5}{4}\sqrt{2}$ (d) $\frac{7}{16}\sqrt{2}$
13. $1/(50\pi)$ feet per minute; $\frac{8}{5}$ square feet per minute
15. $\frac{1600}{3}$ feet per minute
17. (a) $\dfrac{1}{\pi}$ inches per minute   (b) $\dfrac{1}{2\sqrt[3]{2}\pi}$ inches per minute
19. (a) 144 feet
    (b) $(9\sqrt{265} - 135)^2$ feet (about 130 feet, taking $\sqrt{265} \cong 16.28$)

Section 4.9

3. $10\frac{1}{30}$ taking $x = 1000$
5. $1\frac{31}{32}$ taking $x = 16$
7. 1.975 taking $x = 32$
9. 8.15 taking $x = 32$
13. $2\pi rht$
14. (b) 38.4π cubic centimeters
15. error ≤ 0.01 feet
16. (a) $x > 2500$
17. about 0.00153 seconds
19. within $\frac{1}{2}\%$
20. (a) within $1/n\%$

Section 4.10

1. (a) $x_0 = 2$ (b) $x_0 = \pm\sqrt{2}$ (c) $x_0 = \frac{1}{4}$
3. 54 inches by 72 inches
4. $A = 1,\ \ B = -3,\ \ C = 5$ 6. $A = 9,\ \ B = 1$
8. $f(-2) = \frac{3}{2}$ local and absolute max, $f(2) = \frac{5}{6}$ local and absolute min
10. $f(\frac{4}{3}) = \frac{1}{8}$ local and absolute max
12. 20 inches   14. (3, 6)   16. 120 feet per minute
17. (a) decreasing 2.8 mph   (b) increasing $\frac{86}{97}\sqrt{97}$ mph (about 8.73 mph)
18. increasing 15 mph or $15\sqrt{3}$ mph
20. (a) about 0.0533 seconds per inch   (b) 0.0106 seconds
21. 984 feet   23. 25 feet
26. base $= 2a\sqrt{3},\ \ $ height $= 3b$ 28. a decrease of 4 inches per minute
29. 30 units per week   32. $\frac{1}{2}(\beta - b)$
33. $\frac{2}{9}\sqrt{3}r^{-2}$ 35. $(\frac{1}{3}, \frac{1}{9}\sqrt{3})$
36. (a) 1   (b) $\frac{4}{3}$ 37. $\frac{1}{10}$ foot per minute
39. $\frac{3}{2}a^2k$ square units per second   40. decreasing 4 units per second

42. \$5400   43. $\dfrac{a^{1/3}s}{a^{1/3} + b^{1/3}}$

Section 5.2

1. $L_f(P) = \frac{5}{8},\ \ U_f(P) = \frac{11}{8}$ 3. $L_f(P) = \frac{9}{64},\ \ U_f(P) = \frac{37}{64}$
5. $L_f(P) = \frac{17}{16},\ \ U_f(P) = \frac{25}{16}$ 7. $L_f(P) = \frac{7}{16},\ \ U_f(P) = \frac{25}{16}$
9. $L_f(P) = \frac{3}{16},\ \ U_f(P) = \frac{43}{32}$
13. (a) $L_f(P) = -3x_1(x_1 - x_0) - 3x_2(x_2 - x_1) - \cdots - 3x_n(x_n - x_{n-1})$
        $U_f(P) = -3x_0(x_1 - x_0) - 3x_1(x_2 - x_1) - \cdots - 3x_{n-1}(x_n - x_{n-1})$
    (b) $-\frac{3}{2}(b^2 - a^2)$

Section 5.3

1. (a) 5   (b) -2 (c) -1 (d) 0   (e) -4 (f) 1
2. (a) Since $P_1 \subseteq P_2,\ U_f(P_2) \le U_f(P_1)$.
3. (b) $F'(x) = x\sqrt{x + 1}$ (d) $F(2) = \displaystyle\int_0^2 t\sqrt{t + 1}\ dt$

9. (a) $\frac{1}{10}$ (c) $\frac{4}{37}$

Section 5.4

1. -2 3. 1   4. $\frac{16}{3}$ 6. $\frac{16}{3}$ 7. $\frac{32}{3}$
9. $\frac{2}{3}$ 10. $-\frac{2}{3}$ 12. $-\frac{16}{21}$ 13. $\frac{1}{18}(2^{18} - 1)$ 15. $\frac{1}{6}a^2$
16. $\frac{26}{3}$ 18. $\frac{1}{4}$ 19. $-\frac{1}{12}$ 21. $\frac{21}{2}$

Section 5.5

1. part of the motion is in the negative direction

3. $\frac{2}{3}$ 5. 2   7. $\frac{27}{4}$ 8. $\frac{1}{3}$ 10. $21\frac{1}{3}$ 12. 36

14. $\frac{1}{3}a + \frac{1}{2}b + c$

16. (a) $285\frac{1}{3}$ units to the left of the origin   (b) $283\frac{2}{3}$ units

18. $(x_0 + \frac{28}{15}, y_0 + \frac{2}{3})$

Section 5.7

1. $-\dfrac{1}{3x^3} + C$

3. $\frac{1}{2}ax^2 + bx + C$

5. $2\sqrt{1 + x} + C$

7. $\frac{2}{3}x^{3/2} - 2x^{1/2} + C$

9. $\frac{1}{3}t^3 - \frac{1}{2}(a + b)t^2 + abt + C$

11. $-\dfrac{1}{g(x)} + C$

13. $\dfrac{d}{dx}\left(\displaystyle\int f(x)\,dx\right) = f(x),\quad \displaystyle\int \dfrac{d}{dx}[f(x)]\,dx = f(x) + C$

14. (a) $x(t) = x_0 + v_0 t + At^2 + Bt^3$

Section 5.8

1. $\dfrac{1}{3(2 - 3x)} + C$

3. $\frac{1}{3}(2x + 1)^{3/2} + C$

5. $\dfrac{4(ax + b)^{7/4}}{7a} + C$

7. $-\dfrac{1}{8(4x^2 + 9)} + C$

9. $\frac{1}{75}(5x^3 + 9)^5 + C$

11. $\frac{4}{15}(1 + x^3)^{5/4} + C$

13. $-\dfrac{1}{4(1 + s^2)^2} + C$

15. $\sqrt{x^2 + 1} + C$

17. $-\dfrac{b^3}{2a^4}\sqrt{1 - a^4 x^4} + C$

19. $\frac{15}{8}$

21. $\frac{31}{2}$

23. 0

25. $\frac{1}{3}|a|^3$

27. $\frac{2}{5}(x + 1)^{5/2} - \frac{2}{3}(x + 1)^{3/2} + C$

29. $\frac{1}{10}(2x - 1)^{5/2} + \frac{1}{6}(2x - 1)^{3/2} + C$

Section 5.9

1. yes; $\displaystyle\int_a^b [f(x) - g(x)]\,dx = \int_a^b f(x)\,dx - \int_a^b g(x)\,dx > 0.$

2. no; take for example the functions

$$f(x) = x \quad \text{and} \quad g(x) = 0 \quad \text{on} \quad [-\tfrac{1}{2}, 1]$$

3. yes; otherwise we would have

$$f(x) \le g(x) \qquad \text{for all } x \in [a, b]$$

and it would follow that

$$\int_a^b f(x)\, dx \le \int_a^b g(x)\, dx.$$

4. no; take for example

$$f(x) = 0 \quad \text{and} \quad g(x) = -1 \quad \text{on} \quad [0, 1].$$

5. yes; $\int_a^b |f(x)|\, dx > \int_a^b f(x)\, dx$ and we are assuming that

$$\int_a^b f(x)\, dx > \int_a^b g(x)\, dx.$$

6. no; take

$$f(x) = 0 \quad \text{and} \quad g(x) = -1 \quad \text{on} \quad [0, 1].$$

16. $\dfrac{2}{x}$ 17. $\dfrac{2x}{\sqrt{2x^2 + 7}}$ 21. $\dfrac{1}{x}$

Section 5.10

1. $10\frac{2}{3}$ 3. $121\frac{1}{2}$ 5. 18 7. $\frac{9}{128}a^4$
9. $10\frac{2}{15}$ 11. 36 13. $5\frac{3}{4}$

Section 5.11

1. $\frac{2}{3}[(x - a)^{3/2} - (x - b)^{3/2}] + C$ 3. $\frac{1}{2}(x^{2/3} - 1)^3 + C$

5. $x + \frac{8}{3}x^{3/2} + 2x^2 + C$ 7. $\dfrac{2(a + b\sqrt{x + 1})^3}{3b} + C$

9. $-\dfrac{1}{2[g(x)]^2} + C$ 11. $\frac{13}{3}$

13. $\frac{2}{9}$

15. (a) $\displaystyle\int_0^{x_0} g(x)\, dx - x_0 y_0$ (b) $x_0 y_0 - \displaystyle\int_0^{x_0} f(x)\, dx$

(c) $\displaystyle\int_0^{x_0} f(x)\, dx + \int_{x_0}^a g(x)\, dx$

(d) $ab + \frac{1}{2}a(c - b) - \displaystyle\int_0^{x_0} g(x)\, dx - \int_{x_0}^a f(x)\, dx$

(e) $\displaystyle\int_{x_0}^a [f(x) - g(x)]\, dx$

16. 2 18. $\frac{9}{2}$ 20. $\frac{5}{12}$ 22. 3.6

24. necessarily holds: $L_g(P) \le \displaystyle\int_a^b g(x)\, dx < \int_a^b f(x)\, dx \le U_f(P)$

26. necessarily holds: $L_g(P) \le \displaystyle\int_a^b g(x)\, dx < \int_a^b f(x)\, dx$

27. need not hold: take $f(x) = \frac{3}{4}$ and $g(x) = x$ on the interval $[0, 1]$ and use $P = \{0, 1\}$

29. need not hold: take the functions suggested for Exercise 27

33. $\dfrac{1}{1 + x^2}$

35. $\dfrac{2x}{1 + x^4} - \dfrac{1}{1 + x^2}$

Section 5.12

1. (a) $\dfrac{1}{n^2}(1 + 2 + \cdots + n) = \dfrac{1}{n^2}\left[\dfrac{n(n+1)}{2}\right] = \dfrac{1}{2} + \dfrac{1}{2n}$

   (b) $S_n^* = \dfrac{1}{2} + \dfrac{1}{2n}, \displaystyle\int_0^1 x\, dx = \left[\dfrac{1}{2}x^2\right]_0^1 = \dfrac{1}{2}$

   $\left| S_n^* - \displaystyle\int_0^1 x\, dx \right| = \dfrac{1}{2n} < \dfrac{1}{n} < \epsilon$ if $n > \dfrac{1}{\epsilon}$

3. (a) $\dfrac{1}{n^4}(1^3 + 2^3 + \cdots + n^3) = \dfrac{1}{n^4}\left[\dfrac{n^2(n+1)^2}{4}\right] = \dfrac{1}{4} + \dfrac{1}{2n} + \dfrac{1}{4n^2}$

   (b) $S_n^* = \dfrac{1}{4} + \dfrac{1}{2n} + \dfrac{1}{4n^2}, \displaystyle\int_0^1 x^3\, dx = \left[\dfrac{1}{4}x^4\right]_0^1 = \dfrac{1}{4}$

   $\left| S_n^* - \displaystyle\int_0^1 x^3\, dx \right| = \dfrac{1}{2n} + \dfrac{1}{4n^2} < \dfrac{1}{n} < \epsilon$ if $n > \dfrac{1}{\epsilon}$

Section 6.2

1. $\log 2 + \log 10 \cong 2.99$

3. $2 \log 4 - \log 10 \cong 0.48$

4. $4 \log 3 \cong 4.40$

6. $2 \log 5 - \log 10 \cong 0.92$

7. $\log 8 + \log 9 - \log 10 \cong 1.98$

9. $\frac{1}{2} \log 2 \cong 0.35$

11. $-\log x = \log 1/x$

13. (a) 1.65 (c) 1.71

14. (b) 2.26

15. $x = e^2$

18. $x = 1$

19. $x = 1$

Section 6.3

1. domain $(0, \infty)$, $f'(x) = \dfrac{1}{x}$

3. domain $(-1, \infty)$, $f'(x) = \dfrac{3x^2}{x^3 + 1}$

5. domain $(-\infty, \infty)$, $f'(x) = \dfrac{x}{1 + x^2}$

7. domain all $x \neq \pm 1$, $f'(x) = \dfrac{4x^3}{x^4 - 1}$

9. domain $(0, 1) \cup (1, \infty)$, $f'(x) = -\dfrac{1}{x(\log x)^2}$

11. domain $(0, \infty)$, $f'(x) = 1 + \log x$

13. domain $(-1, \infty)$, $f'(x) = \dfrac{1 - \log(x + 1)}{(x + 1)^2}$

15. $\log |x + 1| + C$

17. $-\frac{1}{2} \log |3 - x^2| + C$

19. $\dfrac{1}{2(3 - x^2)} + C$

21. $\log \left| \dfrac{x - a}{x - b} \right| + C$

23. $\dfrac{1}{2} \log \left| \dfrac{x^2 - a^2}{x^2 - b^2} \right| + C$ 25. $\frac{2}{3} \log |1 + x\sqrt{x}| + C$

27. 1   29. 1   31. $\frac{1}{2} \log \frac{8}{5}$ 33. $\frac{1}{2}(\log 2)^2$

35. $g'(x) = (x^2 + 1)^2(x - 1)^5 x^3 \left(\dfrac{4x}{x^2 + 1} + \dfrac{5}{x - 1} + \dfrac{3}{x} \right)$

37. $g'(x) = \dfrac{x^4(x - 1)}{x + 2} \left(\dfrac{4}{x} + \dfrac{1}{x - 1} - \dfrac{1}{x + 2} \right)$

39. $g'(x) = \dfrac{1}{2} \sqrt{\dfrac{(x - 1)(x - 2)}{(x - 3)(x - 4)}} \left(\dfrac{1}{x - 1} + \dfrac{1}{x - 2} - \dfrac{1}{x - 3} - \dfrac{1}{x - 4} \right)$

41. (a) $\frac{15}{8} - \log 4$ 42. (a) log 5 feet

45. (a) domain $(0, \infty)$; increasing throughout; no extreme values; concave down throughout; no pts of inflection

(c) domain $(-\infty, -4)$; decreasing throughout; no extreme values; concave up throughout; no points of inflection

(e) domain $(0, \infty)$; decreasing on $(0, 1/e]$, increasing on $[1/e, \infty)$; $f(1/e) = -1/e$ local and absolute min; concave up throughout; no points of inflection

(g) domain $(0, \infty)$; increasing on $(0, 1]$, decreasing on $[1, \infty)$; $f(1) = -\log 2$ local and absolute max; concave down on $(0, \sqrt{2 + \sqrt{5}}]$, concave up on $[\sqrt{2 + \sqrt{5}}, \infty)$; pt of inflection at $x = \sqrt{2 + \sqrt{5}}$

Section 6.4

1. $-2e^{-2x}$ 3. $2xe^{x^2 - 1}$

4. $-8e^{-4x}$ 6. $x^2 e^x + 2xe^x$

7. $-(x^{-1} + x^{-2})e^{-x}$ 9. $\frac{1}{2}(e^x - e^{-x})$

10. $\frac{1}{2}(e^x + e^{-x})$ 12. $-8e^{4x}(1 - e^{4x})$

13. $2(e^{2x} - e^{-2x})$ 15. $4xe^{x^2}(e^{x^2} + 1)$

16. $4(e^{4x} - e^{-4x})$ 18. $(x^2 + 2x)e^x - (2x^2 + 1)e^{x^2}$

19. $\dfrac{2e^x}{(e^x + 1)^2}$ 21. $2(a - b) \dfrac{e^{(a+b)x}}{(e^{ax} + e^{bx})^2}$

22. $\frac{1}{2}e^{2x} + C$ 24. $\dfrac{1}{k} e^{kx} + C$

25. $\dfrac{1}{a} e^{ax+b} + C$ 27. $-\frac{1}{2}e^{-x^2} + C$

28. $-e^{1/x} + C$ 30. $-\frac{1}{2}e^{-2x} + 2e^{-x} + x + C$

31. $-8e^{-x/2} + C$ 33. $2\sqrt{e^x + 1} + C$

34. $3(e^x + 1)^{2/3} + C$ 36. $\dfrac{1}{2a} \log (e^{ax^2} + 1) + C$

37. $e - 1$ 39. $\frac{1}{6}(1 - \pi^{-6})$

40. $\frac{1}{2}(1 - e^{-1})$ 42. $3 - 4e^{-1}$

43. $\log \left(\dfrac{3}{4 - e} \right)$ 45. $4 - 4e^{-1} + \frac{9}{2}e^{-2} - \frac{1}{2}e^{-4}$

46. $e^{-0.4} = \dfrac{1}{e^{0.4}} \cong \dfrac{1}{1.49} \cong 0.67$

48. $e^{2.8} = (e^2)(e^{0.8}) \cong (7.39)(2.23) \cong 16.48$

50. 7.61   52. 23.10   56. $\dfrac{2}{a}(e^{a^2} + e^{-a^2} - 2)$

58. (i) domain $(-\infty, \infty)$ (ii) increasing on $[0, \infty)$, decreasing on $(-\infty, 0]$
    (iii) $f(0) = 1$ local and absolute min   (iv) concave up everywhere
60. (i) domain $(-\infty, \infty)$ (ii) increasing on $[0, \infty)$, decreasing on $(-\infty, 0]$
    (iii) $f(0) = 1$ local and absolute min   (iv) concave up everywhere
62. (i) domain $(-\infty, \infty)$ (ii) increasing on $[-1, \infty)$, decreasing on $(-\infty, -1]$
    (iii) $f(-1) = -1/e$ local and absolute min   (iv) concave up on $[-2, \infty)$, con-
    cave down on $(-\infty, -2]$, point of inflection at $x = -2$
64. (i) domain $(-\infty, 0) \cup (0, \infty)$ (ii) increasing on $(-\infty, 0)$, decreasing on
    $(0, \infty)$ (iii) no extreme values   (iv) concave up on $(-\infty, 0)$ and on $(0, \infty)$

Section 6.5

1. (a) 6   (c) $-\frac{1}{6}$ (d) -2 (f) -1 (g) 3   (i) $\frac{3}{5}$ (j) $-\frac{2}{5}$ (l) $\frac{1}{4}$
3. (a) 0   (c) $e^{\pm\sqrt{(\log 2)(\log 3)}}$ (d) 2   (f) $\frac{1}{2}$
4. (a) $t_1 < \log a < t_2$ (b) $x_1 < e^b < x_2$
6. (a) $\dfrac{3^x}{\log 3} + C$ (c) $-\dfrac{2^{-x}}{\log 2} + C$

    (d) $\dfrac{10^{x^2}}{2 \log 10} + C$ (f) $\dfrac{2^x - 2^{-x}}{\log 2} + C$

    (g) $\log_5 x + C$ (i) $\frac{1}{4}(\log 3)(\log_3 x)^2 - \log x + C$

7. (a) $\dfrac{1}{e \log 3}$ (c) $\dfrac{1}{e}$

9. (a) $(x + 1)^x \left[\dfrac{x}{x + 1} + \log (x + 1) \right]$

    (c) $[(x^2 + 2)^{\log x}] \left[\dfrac{2x \log x}{x^2 + 2} + \dfrac{\log (x^2 + 2)}{x} \right]$

    (d) $-\left(\dfrac{1}{x}\right)^x (1 + \log x)$ (f) $[(\log x)^{\log x}] \left[\dfrac{1 + \log (\log x)}{x} \right]$

11. (a) domain $(-\infty, \infty)$; increasing on $(-\infty, 0]$, decreasing on $[0, \infty)$; $f(0) = 10$
    local and absolute max
    (b) domain all $x \neq \pm 1$; decreasing on $(-\infty, -1)$ and on $(-1, 0]$, increasing on
    $[0, 1)$ and on $(1, \infty)$; $f(0) = 10$ local min
    (c) domain $[-1, 1]$; increasing on $[-1, 0]$, decreasing on $[0, 1]$; $f(0) = 10$
    local and absolute max, $f(-1) = f(1) = 1$ endpt and absolute min

12. (a) $\dfrac{2^{2\pi+1} - 1}{2\pi + 1}$ (c) $\dfrac{3}{\log 4}$ (d) $\dfrac{b(a^2 - 1)}{2 \log a}$

    (f) 2   (g) $(\log 2)(\log 10)$ (i) $\dfrac{10}{\log p}(p^{\sqrt{2}} - p)$

14. (a) $\log_{10} 4 = \dfrac{\log 4}{\log 10} \cong \dfrac{1.39}{2.30} \cong 0.60$

   (c) $\log_{10} 12 = \dfrac{\log 12}{\log 10} = \dfrac{\log 3 + \log 4}{\log 10} \cong \dfrac{1.10 + 1.39}{2.30} \cong 1.08$

Section 6.6

1. (a) $f(x) = 2e^{3x}$ (c) $f(x) = e^{3x-5}$
2. (a) about \$325 ($e^{0.5} \cong 1.65$)   (c) about \$505 ($e^{0.7} \cong 2.01$)
3. (a) about \$820 ($e^{0.2} \cong 1.22$)   (c) about \$671 ($e^{0.4} \cong 1.49$)
4. (a) about 17 years and 3 months   (c) about 8 years and $7\frac{1}{2}$ months
6. (a) about $5\frac{1}{2}\%$ 7. 10,000 log 2 gallons (about 6900 gallons)
9. (a) $15e^{-(1/2)\log 1.5}$ (about 12.3) pounds
   (b) $15e^{-(3/2)\log 1.5}$ (about 8.2) pounds
11. 640 pounds

12. (a) (i) $\dfrac{4 \log 2}{\log 4 - \log 3}$ (about $9\frac{1}{2}$) years

   (b) (i) $\frac{16}{3}$ grams   (iii) $3(\frac{3}{4})^{10}$ grams
14. about $16\frac{4}{5}$ years

Section 6.7

1. $-xe^{-x} - e^{-x} + C$
3. $\frac{1}{2}(x^2 - 1) \log(x + 1) - \frac{1}{4}x^2 + \frac{1}{2}x + C$

4. $\dfrac{x2^x}{\log 2} - \dfrac{2^x}{(\log 2)^2} + C$ 6. $x \log(-x) - x + C$

7. $\frac{2}{3}x^{3/2} \log x - \frac{4}{9}x^{3/2} + C$ 9. $-\frac{1}{2}x^2 e^{-x^2} - \frac{1}{2}e^{-x^2} + C$
10. $e^x(x^2 - 2x + 2) - \frac{1}{3}x^3 + C$
12. $-2x^2(1 - x)^{1/2} - \frac{8}{3}x(1 - x)^{3/2} - \frac{16}{15}(1 - x)^{5/2} + C$
13. $\frac{2}{3}x(x + 1)^{3/2} - \frac{4}{15}(x + 1)^{5/2} + C$
15. $x(\log x)^2 - 2x \log x + 2x + C$
16. $-e^{-x}(x^2 + 2x + 2) + C$
18. $-\frac{1}{13}x(x + 5)^{-13} - \frac{1}{156}(x + 5)^{-12} + C$
19. $\frac{1}{10}x^2(x + 1)^{10} - \frac{1}{55}x(x + 1)^{11} + \frac{1}{660}(x + 1)^{12} + C$
21. $\frac{3}{8}$ 22. $\frac{1}{4}e^2 + \frac{29}{12}$ 24. $\frac{1}{5} + \frac{11}{2}(\log 2)^{-1} - 8(\log 2)^{-2} + 4(\log 2)^{-3}$

Section 6.8

1. $y(x) = -\frac{1}{2} + Ce^{2x}$ 3. $y(x) = \frac{2}{5} + Ce^{-(5/2)x}$
5. $y(x) = x + Ce^{2x}$ 7. $y(x) = nx + Cx^3$
9. $y(x) = \frac{2}{11}(x + 1)^{7/2} + C(x + 1)^{-2}$
11. $y(x) = e^x + Cx^{-1}$ 13. $y(x) = x + 2e^{-x} - 1$
15. $y(x) = e^{-x}[\log(1 + e^x) + e - \log 2]$
17. $y(x) = x^2(e^x - e)$

19. $y(x) = \left(\dfrac{5}{e} - 1\right)\dfrac{1}{x^2} - e^{-x}\left(1 + \dfrac{2}{x} + \dfrac{2}{x^2}\right)$

21. (a) $200(\frac{4}{5})^{t/5}$ liters (b) $200(\frac{4}{5})^{t^2/25}$ liters

23. $20 + 5e^{-0.03t}$ pounds

Section 6.9

1. $2\sqrt{ab}$ 3. $\left(\pm\dfrac{1}{\sqrt{2}}, \dfrac{1}{\sqrt{e}}\right)$ 5. $x = \dfrac{1}{\sqrt{e}}$ 7. 1

9. (a) $y = \frac{1}{2}(e^{\alpha x} - e^{-\alpha x})$ (c) $y = e^{\alpha x}$

11. $a^2 \log 2$ 13. $\dfrac{176}{\log 2}$ feet (about 254 feet)

Section 7.1

1. $3 \cos 3x$ 3. $x \sec^2 x + \tan x$

4. $\cos x - \sin x$ 6. $\dfrac{1}{2\sqrt{x}} \cos \sqrt{x}$

7. $\pi[\cos^2 \pi x - \sin^2 \pi x] = \pi \cos 2\pi x$ 9. $\pi x^2 \cos \pi x + 2x \sin \pi x$

10. $\sin x \cos x$ 12. $6 \tan^2 2x \sec^2 2x$ 13. $e^x \cos x + e^x \sin x$

15. $-\dfrac{x \sin x^2}{\sqrt{\cos x^2}}$ 16. $(\cos \pi x)^x[\log (\cos \pi x) - \pi x \tan \pi x]$

18. $ke^{-kx}[\cos kx - \sin kx]$ 19. $\pi 10^x \sec^2 \pi x + 10^x \log 10 \tan \pi x$

21. $\dfrac{1}{2}\left(\dfrac{\sec^2 x}{\tan x}\right) = \operatorname{cosec} 2x$ 22. $e^x(\frac{1}{2}\pi \cos \frac{1}{2}\pi x + \sin \frac{1}{2}\pi x)$

24. $\pi \tan^3 \frac{1}{4}\pi x \sec^2 \frac{1}{4}\pi x$ 25. $(\cos x)e^{\sin x}$ 27. $2\tan x \sec^2 x [\sec^2 x + \tan^2 x]$

28. $n \sin^{n-1} x [\sin nx \cos x + \sin x \cos nx] = n \sin^{n-1}x \sin [(n + 1)x]$

30. $-\tan 2x$ 31. $2 \tan \theta \sec^2 \theta$ 33. $-6 \operatorname{cosec}^2 3\theta \cot 3\theta$

34. $-k^2 \sin kx$ 36. $-x \cos x - 2 \sin x$ 37. $e^{2t}(3 \cos t - 4 \sin t)$

39. $-e^{-t}(3 \sin 2t + 4 \cos 2t)$ 40. $\dfrac{\sin (x - y)}{\sin (x - y) - 1}$

42. $-\dfrac{1}{1 + (x + y) \sin y}$ 44. $\tan \theta = 2\sqrt{2}, \quad \theta \cong 1.23$ radians

46. (a) 0.868 (c) 0.248 47. (a) 0.884 (c) 0.647

48. 3 50. $\frac{3}{2}$ 51. 0 53. does not exist

54. 4 56. $\frac{3}{2}$ 57. $\frac{9}{2}$ 59. 0

65. increasing on $[0, 2\pi]$; $f(0) = 0$ endpt and absolute min, $f(2\pi) = 2\pi$ endpt and absolute max; concave down on $(0, \pi]$, concave up on $[\pi, 2\pi)$; point of inflection (π, π)

67. increasing on $[0, \frac{1}{4}\pi]$, decreasing on $[\frac{1}{4}\pi, \frac{5}{4}\pi]$, increasing on $[\frac{5}{4}\pi, 2\pi]$; $f(0) = 1$ endpt min, $f(\frac{1}{4}\pi) = \frac{1}{2}\sqrt{2}e^{\pi/4}$ local max, $f(\frac{5}{4}\pi) = -\frac{1}{2}\sqrt{2}e^{5\pi/4}$ local and absolute min, $f(2\pi) = e^{2\pi}$ endpt and absolute max; concave down on $(0, \pi]$, concave up on $[\pi, 2\pi)$; pt of inflection $(\pi, -e^\pi)$

69. decreasing on $[0, \frac{1}{2}\pi]$, increasing on $[\frac{1}{2}\pi, \pi]$; $f(0) = 1$ endpt and absolute max, $f(\frac{1}{2}\pi) = 0$ local and absolute min, $f(\pi) = 1$ endpt and absolute max; concave

down on $(0, \frac{1}{4}\pi]$, concave up on $[\frac{1}{4}\pi, \frac{3}{4}\pi]$, concave down on $[\frac{3}{4}\pi, \pi)$; pts of inflection $(\frac{1}{4}\pi, \frac{1}{2})$ and $(\frac{3}{4}\pi, \frac{1}{2})$.

Section 7.2

1. $\frac{1}{3}\sin(3x - 1) + C$

3. $\sin^3 x + C$

4. $\frac{1}{2}\sec 2x + C$

6. $-\frac{2}{3}(1 + \cos x)^{3/2} + C$

7. $-e^{-\sin x} + C$

9. $-2\cos x^{1/2} + C$

10. $-2\sqrt{1 - \sin x} + C$

12. $-\log(1 + \cos x) + C$

13. $-\dfrac{1}{3\pi}\cos^3 \pi x + C$

15. $\dfrac{1}{\pi}\log|\sin \pi x| + C$

16. $\frac{1}{2}\log|1 + 2\sin x| + C$

18. $-\frac{1}{2}\log(2 - \sin 2x) + C$

19. $\frac{1}{2}\tan^2 x + C$

21. $\frac{2}{3}(1 + \tan x)^{3/2} + C$

22. $-\dfrac{1}{3a^2}\cos(a^3 x^3 + b^3) + C$

23. $\frac{1}{2}(x + \sin x) + C$

25. $\dfrac{1}{2}\left(x + \dfrac{1}{2\pi}\sin 2\pi x\right) + C$

27. $\frac{5}{2}$

29. $x\sin x + \cos x + C$

31. $\frac{1}{2}e^x(\sin x - \cos x) + C$

Section 7.3

1. $\frac{1}{2}\pi$ 3. 0 4. $\frac{1}{2}\pi$ 6. $\frac{1}{4}\pi$

7. $\frac{1}{4}\pi$ 9. $\frac{1}{3}\pi$ 10. $-\frac{1}{6}\pi$ 12. 0

13. $\dfrac{1}{x^2 + 2x + 2}$

15. $\dfrac{2x}{\sqrt{1 - x^4}}$

17. $\dfrac{2x}{\sqrt{1 - 4x^2}} + \arcsin 2x$

19. $\dfrac{2\arcsin x}{\sqrt{1 - x^2}}$

21. $\dfrac{x - (1 + x^2)\arctan x}{x^2(1 + x^2)}$

23. $\dfrac{1}{(1 + 4x^2)\sqrt{\arctan 2x}}$

25. $\dfrac{1}{x[1 + (\log x)^2]}$

27. $-\dfrac{r}{|r|\sqrt{1 - r^2}}$

29. $\dfrac{1}{1 + r^2}$

31. $\sqrt{\dfrac{c - x}{c + x}}$

33. $\dfrac{x^2}{(c^2 - x^2)^{3/2}}$

35. $\dfrac{x}{\sqrt{1 - x^2}}$

37. $\dfrac{\sqrt{1 - x^2}}{x}$

41. $\frac{1}{4}\pi$ 43. $\frac{1}{4}\pi$ 44. $\frac{1}{6}\pi$

46. $\frac{1}{8}\pi$ 47. $\frac{1}{24}\pi$ 49. $\frac{1}{6}\pi$

50.

$$y = \text{arc sec } x$$

$$\sec y = x$$

$$\sec y \tan y \frac{dy}{dx} = 1$$

$$\frac{dy}{dx} = \cos y \cot y = \frac{1}{x\sqrt{x^2 - 1}}$$

51. $\dfrac{d}{dx} (\text{arc cos } x) = -\dfrac{1}{\sqrt{1-x^2}}$

53. $\dfrac{d}{dx} (\text{arc cosec } x) = -\dfrac{1}{x\sqrt{x^2-1}}$

Section 7.4

1. absolute min 0, absolute max π
3. absolute min $\frac{1}{6}\pi - \frac{1}{2}\sqrt{3}$, absolute max $\frac{5}{6}\pi + \frac{1}{2}\sqrt{3}$
4. absolute min -1, absolute max $\frac{7}{12}\pi + \frac{1}{2}\sqrt{3}$
6. absolute min $-5\sqrt{2}e^{-3\pi/4}$, absolute max 10
7. $\sqrt{a^2 + b^2}$
9. (a) $\frac{1}{4}\pi$ (c) $1 - \frac{1}{32}\pi^2$ (e) $-\frac{1}{2}$
10. $\frac{1}{2}\log 2$ 12. $\frac{1}{6}\pi$ 14. $\frac{1}{6}(\sqrt{3}\pi - 3)$ 16. $\frac{1}{2}$
20. decreasing 0.04 radians per minute
22. 5π miles per minute   23. 12 feet
24. $r = 2\sqrt{6}$ inches, $h = 4\sqrt{3}$ inches   26. $\theta = \text{arc tan } m$
27. $\theta = \frac{1}{4}\pi + \frac{1}{2}\alpha$ radians

Section 7.5

1. $x(t) = \sin (8t + \frac{1}{2}\pi);$ $a = 1,$ $f = 4/\pi$
3. $\pm\dfrac{2\pi a}{p}$ 5. $x(t) = (15/\pi) \sin \frac{1}{3}\pi t$
7. $x = \pm\frac{1}{2}\sqrt{3}\, C_2$ 10. $x(t) = x_0 \sin (\sqrt{k/m}\, t + \frac{1}{2}\pi)$

Section 7.6

1. $2x \cosh x^2$ 3. $\dfrac{a \sinh ax}{2\sqrt{\cosh ax}}$
5. $\dfrac{1}{1 - \cosh x}$ 7. $ab(\cosh bx - \sinh ax)$
9. $\dfrac{a \cosh ax}{\sinh ax}$ 16. minimum value of 3
18. no extreme values   21. $y = 2 \cosh 3x + \frac{1}{3} \sinh 3x$

Section 7.7

1. $2 \tanh x \, \text{sech}^2 x$ 3. $\text{sech } x \, \text{cosech } x$ 5. $\dfrac{2e^{2x} \cosh (\text{arc tan } e^{2x})}{1 + e^{4x}}$
7. $\dfrac{-x \, \text{cosech}^2 (\sqrt{x^2 + 1})}{\sqrt{x^2 + 1}}$ 9. $\dfrac{-\text{sech } x \, (\tanh x + 2 \sinh x)}{(1 + \cosh x)^2}$
15. (a) $\frac{3}{5}$ (b) $\frac{5}{3}$ (c) $\frac{4}{3}$ (d) $\frac{5}{4}$ (e) $\frac{3}{4}$

Section 8.1

1. $-e^{2-x} + C$ 3. $\frac{3}{2} \sin \frac{2}{3} x + C$ 4. $\dfrac{\pi}{4c}$ 6. $e - \sqrt{e}$

7. $-\dfrac{1}{5^x \log 5} + C$ 9. $\frac{1}{2}(x^2 - 1) \log (x + 1) - \frac{1}{4}x^2 + \frac{1}{2}x + C$

10. $-\log (2\sqrt{3} - 3)$ 12. 2   13. $\tan (x - 1) + C$

15. $\dfrac{x5^x}{\log 5} - \dfrac{5^x}{(\log 5)^2} + C$ 16. $\frac{1}{4} \log 2$ 18. $\frac{1}{2} \log 3$

19. $-\frac{1}{5}e^{-x} (\sin 2x + 2 \cos 2x) + C$ 21. $\frac{1}{2} \arcsin x^2 + C$

22. $\frac{1}{4}(2x^2 \arcsin 2x^2 + \sqrt{1 - 4x^4}) + C$

24. $\frac{1}{2} \log [(x + \alpha)^2 + a^2] - \dfrac{\alpha}{a} \arctan \left(\dfrac{x + \alpha}{a} \right) + C$

25. $\frac{2}{3}\sqrt{3 \tan \theta + 1} + C$ 27. $\dfrac{\log |ae^x - b|}{a} + C$

28. $\dfrac{a^x e^x}{1 + \log a} + C$ 30. $\arctan (x + 3) + C$

31. $\frac{1}{3} \arctan [\frac{1}{3}(x - 2)] + C$ 33. $\frac{1}{4} \sinh 2x - \frac{1}{2}x + C$

35. $\frac{1}{2}(\cos x \cosh x + \sin x \sinh x) + C$

36. $\log (\cosh x) - \frac{1}{2} \tanh^2 x + C$ 38. $\dfrac{1}{4a} e^{2ax} + \dfrac{x}{2} + C$

Section 8.2

1. $\log \left| \dfrac{x - 2}{x + 5} \right| + C$ 3. $\log \left| \dfrac{x^2 - 1}{x} \right| + C$

4. $5 \log |x - 1| - 3 \log |x| + \dfrac{3}{x} + C$

6. $\dfrac{x^4}{4} + \dfrac{4x^3}{3} + 6x^2 + 32x - \dfrac{32}{x - 2} + 80 \log |x - 2| + C$

7. $5 \log |x - 2| - 4 \log |x - 1| + C$

9. $\dfrac{-1}{2(x - 1)^2} + C$

10. $\dfrac{3}{4} \log |x - 1| - \dfrac{1}{2(x - 1)} + \dfrac{1}{4} \log |x + 1| + C$

12. $\dfrac{1}{3} \left(\dfrac{1}{x + 1} - \dfrac{1}{x - 2} \right) + C$ 13. $\dfrac{1}{32} \log \left| \dfrac{x - 2}{x + 2} \right| - \dfrac{1}{16} \arctan \dfrac{x}{2} + C$

15. $\dfrac{1}{16} \log \left(\dfrac{x^2 + 2x + 2}{x^2 - 2x + 2} \right) + \dfrac{1}{8} [\arctan (x + 1) + \arctan (x - 1)] + C$

16. $\frac{1}{16}\pi$

21. (a) $C(t) = 2A_0 \left(\dfrac{t}{t + 1} \right)$ (b) $C(t) = 4A_0 \dfrac{3^t - 2^t}{2(3^t) - 2^t}$

(c) $C(t) = A_0(m + n) \left(\dfrac{m^t - n^t}{m^{t+1} - n^{t+1}} \right)$

Section 8.3

1. $\frac{1}{3}\cos^3 x - \cos x + C$

3. $-\frac{1}{5}\cos^5 x + \frac{1}{7}\cos^7 x + C$

4. $\frac{2}{3}$

6. $\frac{1}{2}\cos x - \frac{1}{10}\cos 5x + C$

7. $\frac{3}{16}\pi$

9. $\frac{1}{5}\cos^5 x - \frac{1}{3}\cos^3 x + C$

10. $\frac{1}{4}\sin^4 x - \frac{1}{6}\sin^6 x + C$

12. $\frac{8}{15}$

13. $-\frac{1}{3}\cos^3 x + \frac{2}{5}\cos^5 x - \frac{1}{7}\cos^7 x + C$

15. $\frac{1}{16}x - \frac{1}{64}\sin 4x + \frac{1}{48}\sin^3 2x + C$

16. $\frac{3}{8}\arctan x + \dfrac{x}{2(x^2 + 1)} + \dfrac{x(1 - x^2)}{8(x^2 + 1)^2} + C$

18. $\frac{1}{2}\left[\arctan (x + 1) + \dfrac{x + 1}{x^2 + 2x + 2}\right] + C$

20. 1   22. 0   24. 0

Section 8.4

1. $\frac{1}{2}\tan^2 x + \log |\cos x| + C$

3. $\tan (x + 1) + C$

4. $-\log |\sin x| - \frac{1}{2}\cot^2 x + C$

6. $\frac{1}{2}\cot^2 x - \frac{1}{4}\cot^4 x + \log |\sin x| + C$

7. $\frac{1}{7}\tan^7 x + \frac{1}{5}\tan^5 x + C$

9. $\frac{1}{12}\tan^4 3x - \frac{1}{6}\tan^2 3x + \frac{1}{3}\log |\sec 3x| + C$

10. $-\frac{4}{3}\cot^3 \frac{1}{4}x - 4\cot \frac{1}{4}x + C$

12. $-\operatorname{cosec} x + C$

13. $-\frac{1}{3}\cot^3 x + C$

15. $\frac{1}{4}\sec^3 x \tan x + \frac{3}{8}\sec x \tan x + \frac{3}{8}\log |\sec x + \tan x| + C$

Section 8.5

1. $\arcsin \left(\dfrac{x}{a}\right) + C$

3. $\dfrac{x}{5\sqrt{5 - x^2}} + C$

4. $\sqrt{x^2 - 4} + C$

6. $-\sqrt{4 - x^2} + C$

7. $\frac{1}{2}x\sqrt{x^2 - 4} + 2\log |x + \sqrt{x^2 - 4}| + C$

9. $\frac{1}{2}x\sqrt{x^2 + 4} - 2\log (x + \sqrt{x^2 + 4}) + C$

10. $\dfrac{1}{\sqrt{1 - x^2}} + C$

12. $\frac{1}{2}\log (a^2 + x^2) + C$

13. $-\frac{1}{3}(4 - x^2)^{3/2} + C$

15. $\log (\sqrt{8 + x^2} + x) - \dfrac{x}{\sqrt{8 + x^2}} + C$

16. $-\dfrac{(1 - x^2)^{3/2}}{3x^3} + C$

18. $\log (x + \sqrt{x^2 + a^2}) + C$

19. $\log |x + \sqrt{x^2 - a^2}| + C$

21. $\frac{1}{2}x\sqrt{x^2 - a^2} - \frac{1}{2}a^2 \log |x + \sqrt{x^2 - a^2}| + C$

22. $\dfrac{1}{a}\log \left|\dfrac{x}{a + \sqrt{a^2 - x^2}}\right| + C$

24. $-\dfrac{1}{a^2 x}\sqrt{a^2 - x^2} + C$

25. $\dfrac{1}{a^2 x}\sqrt{x^2 - a^2} + C$

27. $\dfrac{1}{a}\arccos \left(\dfrac{a}{x}\right) + C$

Section 8.6

1. $\tan \frac{1}{2}x + C$ 3. $-\cot \frac{1}{2}x + C$ 4. $\frac{2}{5}\sqrt{5}$ arc tan $\frac{1}{5}\sqrt{5}$

6. $\frac{4}{3}\sqrt{3}$ arc tan $[\frac{2}{3}\sqrt{3} \tan \frac{1}{2}x - \frac{1}{3}\sqrt{3}] - x + C$

7. $\frac{2}{3}$ arc tan $(\frac{1}{3}\tan \frac{1}{2}x) + C$ 9. $-\frac{1}{3}x + \frac{5}{6}$ arc tan $(2 \tan \frac{1}{2}x) + C$

10. $-\cot \frac{1}{2}x - x + C$ 12. $\tan \frac{1}{2}x + 2 \log |\sec \frac{1}{2}x| + C$

Section 8.7

1. $-2(\sqrt{x} + \log |1 - \sqrt{x}|) + C$

3. $2 \log (\sqrt{1 + e^x} - 1) - x + 2\sqrt{1 + e^x} + C$

5. $\frac{2}{5}(1 + x)^{5/2} - \frac{2}{3}(1 + x)^{3/2} + C$ 7. $\frac{2}{5}(x - 1)^{5/2} + 2(x - 1)^{3/2} + C$

9. $-\dfrac{1 + 2x^2}{4(1 + x^2)^2} + C$ 11. $x + 2\sqrt{x} + 2 \log |\sqrt{x} - 1| + C$

13. $x + 4\sqrt{x - 1} + 4 \log |\sqrt{x - 1} - 1| + C$

15. $\log \left(\dfrac{\sqrt{e^x + 1} - 1}{\sqrt{e^x + 1} + 1}\right) + C$ 17. $\frac{2}{3}(x - 8)\sqrt{x + 4} + C$

19. $\frac{1}{16}(4x + 1)^{1/2} + \frac{1}{8}(4x + 1)^{-1/2} - \frac{1}{48}(4x + 1)^{-3/2} + C$

21. $\dfrac{4b + 2ax}{a^2\sqrt{ax + b}} + C$

Section 8.8

1. (a) 506   (b) 650   (c) 572   (d) 578   (e) 576

3. (a) 1.39   (b) 0.91   (c) 1.18   (d) 1.15   (e) 1.16

5. (a) $\frac{1}{4}\pi \cong 0.78$ (b) $\frac{1}{4}\pi \cong 0.78$

Section 8.9

1. $\dfrac{10^{nx}}{n \log 10} + C$ 3. $\frac{1}{3}(2x + 1)^{3/2} + C$

4. $\frac{1}{15}(3x - 1)(2x + 1)\sqrt{2x + 1} + C$ 6. $\frac{2}{3}(x + 1)^{3/2} + \frac{2}{3}x^{3/2} + C$

7. $\dfrac{1}{ab}$ arc tan $\left(\dfrac{ax}{b}\right) + C$ 9. $-2e^{-\sqrt{x}} + C$

10. $\frac{1}{2} \log |\sec 2x + \tan 2x| + C$ 12. $\dfrac{x2^x}{\log 2} - \dfrac{2^x}{(\log 2)^2} + C$

13. $\frac{1}{2}x - \dfrac{n}{2\pi} \sin \left(\dfrac{\pi}{n}x\right) \cos \left(\dfrac{\pi}{n}x\right) + C$ 15. $\dfrac{2\sqrt{x}}{a} - \dfrac{2b}{a^2} \log |a\sqrt{x} + b| + C$

16. $x - 2 \log |\sec x + \tan x| + \tan x + C$

18. a arc sin $\left(\dfrac{x}{a}\right) + \sqrt{a^2 - x^2} + C$ 19. $\frac{1}{2} \tan^2 x + C$

21. $\frac{1}{6} \tan^6 x + C$ 22. $\log |\tan x| + C$

24. $-\dfrac{2}{1 + \tan x} - x + C$

25. $-\dfrac{\sqrt{a^2 - x^2}}{x} - \arcsin\left(\dfrac{x}{a}\right) + C$

27. $\log|x + 1| + \dfrac{1}{x + 1} + C$

28. $\frac{1}{2}\sqrt{2}[\log|\tan\frac{1}{2}x + 1 + \sqrt{2}| - \log|\tan\frac{1}{2}x + 1 - \sqrt{2}|] + C$

30. $\dfrac{1}{a}(ax + b)\log(ax + b) - x + C$

31. $\dfrac{x^2}{2}\log(ax + b) - \dfrac{x^2}{4} + \dfrac{bx}{2a} - \dfrac{b^2}{2a^2}\log(ax + b) + C$

33. $-2\cos\sqrt{x} + C$

35. $x - \arctan x + C$

36. $\frac{1}{6}x\sqrt{x^2 - 9} - \frac{3}{2}\log|x + \sqrt{x^2 - 9}| + C$

38. $\dfrac{e^x}{1 + \pi^2}[\sin\pi x - \pi\cos\pi x] + C$

39. $\frac{1}{4}\sqrt{2}[\log|\tan\frac{1}{2}x + 3 - 2\sqrt{2}| - \log|\tan\frac{1}{2}x + 3 + 2\sqrt{2}|]$

41. $\frac{1}{3}(x^2 - 2)\sqrt{1 + x^2} + C$

42. $\frac{1}{2}(x^2 + 1)\log\sqrt{x^2 + 1} - \frac{1}{4}x^2 + C$

44. $-2\sqrt{x} - 4\log|\sqrt{x} - 2| + C$

46. $\arcsin(x - 1) + C$

47. $-\cot x - \frac{3}{2}x - \frac{1}{4}\sin 2x + C$

49. $\sqrt{x^2 + 4} + \log(\sqrt{x^2 + 4} - 2) - \log(\sqrt{x^2 + 4} + 2) + C$

50. $2\sqrt{e^x + 1} + C$

52. $\frac{1}{4}\sqrt{2}\log\left|\dfrac{e^x - \sqrt{2}}{e^x + \sqrt{2}}\right| + C$

53. $-2\cos\frac{1}{2}x + \frac{4}{3}\cos^3\frac{1}{2}x - \frac{2}{5}\cos^5\frac{1}{2}x + C$

55. $2\cos^3 x - \frac{8}{5}\cos^5 x + C$

57. $(x - 1)\log(1 - \sqrt{x}) - \frac{1}{2}x - \sqrt{x} + C$

59. $-\frac{1}{2}\sqrt{2}\arctan[\frac{1}{2}\sqrt{2}(\cos x - 1)] + C$

61. $\frac{1}{4}\sinh 2x - \frac{1}{2}x + C$

63. $\frac{1}{2}x - \frac{1}{4}e^{-2x} + C$

Section 9.2

1. (a) $\frac{2}{13}$ (c) $\frac{2}{13}\sqrt{13}$ (e) $\frac{1}{5}\sqrt{5}$ 3. (a) $\frac{17}{2}$

5. decreasing $\frac{8}{5}b^2\pi$ units per minute

Section 9.3

1. (a) $y^2 = 8x$ (b) $y^2 = -8x$

    (d) $(y - 2)^2 = 8(x - 1)$ (f) $(x - 1)^2 = 4y$

    (h) $y^2 + 4y + 6x - 17 = 0$

2. (a) vertex $(0, 0)$, focus $(\frac{1}{2}, 0)$, axis $y = 0$, directrix $x = -\frac{1}{2}$

    (c) vertex $(1, 0)$, focus $(\frac{3}{2}, 0)$, axis $y = 0$, directrix $x = \frac{1}{2}$

    (e) vertex $(-2, \frac{3}{2})$, focus $(-2, \frac{7}{2})$, axis $x = -2$, directrix $y = -\frac{1}{2}$

    (g) vertex $(-\frac{1}{2}, \frac{3}{4})$, focus $(-\frac{1}{2}, 1)$, axis $x = -\frac{1}{2}$, directrix $y = \frac{1}{2}$

3. (a) $(x - y)^2 = 6x + 10y - 9$ (c) $(x + y)^2 = -12x + 20y + 28$

7. $2y = x^2 - 4x + 7$, $18y = x^2 - 4x + 103$

9. $y = -\dfrac{16}{v_0^2} (\sec^2 \theta) x^2 + (\tan \theta) x$ 11. $\frac{1}{16} v_0^2 \cos \theta \sin \theta$ feet   13. $\frac{1}{4}\pi$

15. (b) vertex $\left(-\dfrac{B}{2A}, \dfrac{4AC - B^2}{4A} \right)$, focus $\left(-\dfrac{B}{2A}, \dfrac{4AC - B^2 + 1}{4A} \right)$,

      directrix $y = \dfrac{4AC - B^2 - 1}{4A}$

Section 9.4

1. foci $(\pm \sqrt{5}, 0)$, major axis has length 6, minor axis has length 4

3. foci $(0, \pm \sqrt{2})$, major axis has length $2\sqrt{6}$, minor axis has length 4

5. foci $(\pm 1, 0)$, major axis has length 4, minor axis has length $2\sqrt{3}$

7. foci $(1, \pm 4\sqrt{3})$, major axis has length 16, minor axis has length 8

9. $\dfrac{x^2}{9} + \dfrac{y^2}{8} = 1$ 11. $\dfrac{(x - 6)^2}{25} + \dfrac{(y - 1)^2}{16} = 1$

13. $\dfrac{(x - 1)^2}{21} + \dfrac{(y - 3)^2}{25} = 1$ 15. $\dfrac{(x - 3)^2}{25} + \dfrac{(y + 1)^2}{9} = 1$

16. πab

18. the ellipse tends to a circle of radius a

20. $\dfrac{x^2}{25} + \dfrac{4y^2}{75} = 1$

Section 9.5

1. $\dfrac{x^2}{9} - \dfrac{y^2}{16} = 1$ 3. $\dfrac{y^2}{25} - \dfrac{x^2}{144} = 1$

5. $\dfrac{x^2}{9} - \dfrac{(y - 1)^2}{16} = 1$ 7. $xy = \frac{1}{2}$

8. transverse axis 2, vertices $(\pm 1, 0)$, foci $(\pm \sqrt{2}, 0)$, asymptotes $y = \pm x$

10. transverse axis 6, vertices $(\pm 3, 0)$, foci $(\pm 5, 0)$, asymptotes $y = \pm\frac{4}{3}x$

12. transverse axis 8, vertices $(0, \pm 4)$, foci $(0, \pm 5)$, asymptotes $y = \pm\frac{4}{3}x$

14. transverse axis 6, vertices $(4, 3)$ and $(-2, 3)$, foci $(6, 3)$ and $(-4, 3)$, asymptotes $y = \pm\frac{4}{3}(x - 1) + 3$

16. transverse axis 2, vertices $(0, 0)$ and $(-2, 0)$, foci $(1, 0)$ and $(-3, 0)$, asymptotes $y = \pm\sqrt{3}(x + 1)$

18. transverse axis $2\sqrt{2}$, vertices $(1, 1)$ and $(-1, -1)$, foci $(\sqrt{2}, \sqrt{2})$ and $(-\sqrt{2}, -\sqrt{2})$, asymptotes $x = 0$ and $y = 0$

Section 9.6

1. parabola with vertex at $(0, -1)$ and focus at the origin

3. hyperbola with foci at $(5 \pm 2\sqrt{5}, 0)$ and transverse axis 4

5. ellipse with foci at $(-3 \pm \frac{1}{3}\sqrt{6}, 0)$ and major axis 2

7. parabola with vertex at $(\frac{3}{2}, -2)$ and focus at $(1, -2)$

9. ellipse with foci at $(\pm\frac{4}{15}, -2)$ and major axis $\frac{2}{3}$

11. hyperbola with foci at $(-1 \pm \frac{6}{7}\sqrt{21}, -4)$ and transverse axis $\frac{6}{7}\sqrt{35}$

13. the union of the parabola $x^2 = 4y$ and the ellipse $\frac{1}{9}x^2 + \frac{1}{4}y^2 = 1$

Section 9.7

1. $\alpha = \frac{1}{4}\pi$, $\frac{1}{2}(X^2 - Y^2) = 1$ 3. $\alpha = \frac{1}{6}\pi$, $4X^2 - Y^2 = 1$

5. $\alpha = \frac{1}{4}\pi$, $2Y^2 + \sqrt{2}X = 0$ 7. $\alpha = -\frac{1}{6}\pi$, $Y^2 + X = 0$

9. $\alpha = \frac{1}{8}\pi$, $\cos \alpha + \frac{1}{2}\sqrt{2 + \sqrt{2}}$, $\sin \alpha = \frac{1}{2}\sqrt{2 - \sqrt{2}}$

Section 10.1

1. $A.V. = \frac{1}{2}mc + b$, $x = \frac{1}{2}c$ 3. $A.V. = 0$, $x = 0$

5. $A.V. = \dfrac{1}{k}(e^k - 1)$ 7. $\frac{1}{4}\pi a$ 9. 0

15. (a) the terminal velocity is twice the average velocity

(b) the average velocity during the first $\frac{1}{2}x$ seconds is one-third of the average velocity during the next $\frac{1}{2}x$ seconds

18. $\frac{1}{2}\pi b$ and $\frac{1}{2}\pi a$

Section 10.2

1. $\frac{1}{5}\pi$ 3. $\frac{1}{2}\pi(e^2 - 1)$ 5. $\frac{12}{7}(2^{1/3})\pi$ 7. $\frac{1}{2}\pi$

9. $\frac{3}{10}\pi$ 11. $\frac{1}{3}\pi^4 - \frac{1}{2}\pi^2$ 13. $\frac{1}{3}\pi h(R^2 + Rr + r^2)$

14. $\frac{1}{4}\pi a^3$

16. (a) $\frac{4}{3}ab^2$ (b) $\frac{16}{3}ab^2$ (c) πab

17. (a) $31\frac{1}{4}\%$ (b) $14\frac{22}{27}\%$

19. (a) $4h$ (b) $\frac{3}{2}\sqrt{3}$ (c) $\frac{3}{2}$

20. (a) 64π (b) $\frac{1024}{35}\pi$ (c) $\frac{704}{5}\pi$ (d) $\frac{512}{7}\pi$

22. $\frac{16}{3}r^3$

Section 10.3

1. $\frac{1}{2}\pi$ 3. 2π 5. $3(4^{1/3})\pi$ 7. 2π

9. $\frac{4}{5}\pi$ 11. $\frac{3}{10}\pi$ 13. $\frac{2}{3}\pi^4 - 2\pi^2$ 15. $\frac{1}{4}\sqrt{3}a^3\pi$

Section 10.4

1. (a) 25 foot-pounds   (b) $\frac{225}{4}$ foot-pounds   3. $\frac{3}{2}$ feet

5. (a) $6480\pi + 8640$ foot-pounds   (b) $15{,}120\pi + 8640$ foot-pounds

6. $6480\pi - 8640$ foot-pounds   9. $GmM\left(\dfrac{1}{r_2} - \dfrac{1}{r_1}\right)$

11. 788 foot-pounds

13. (a) $\frac{1}{2}\sigma l^2$ foot-pounds   (b) $\frac{3}{2}\sigma l^2$ foot-pounds

Section 10.5

1. 2160 pounds   3. $\frac{8000}{3}\sqrt{2}$ pounds   5. 2560 pounds

7. (a) 41,250 pounds   (b) 11 feet

Section 10.6

1. (a) \$3700   (b) \$3425

4. (a) \$1008.40   (b) \$897

5. (a) \$4086   (b) \$3946

Section 10.7

1. $\dfrac{v_0^2 \sin^2 \theta}{3g}$ 3. $\frac{1}{4}\pi a^3$ 5. $\frac{1}{15}\pi a^3$

7. $\frac{32}{35}\pi ab^2$ 9. $\frac{64}{5}\pi$ 11. $\frac{4}{5}\pi a^2 b$

13. (a) $(2\sqrt{2} - 2)h$ (d) $\frac{1}{8}\sqrt{3}(\pi - 2)$

14. (a) disc: $V = \int_0^a \pi[f(x)]^2\, dx$;   shell: $V = \int_0^b 2\pi y[a - f^{-1}(y)]\, dy$

(c) disc: $V = \int_0^b \pi[a - f^{-1}(y)]^2\, dy$;   shell: $V = \int_0^a 2\pi x f(a - x)\, dx$

15. $wh + \frac{1}{2}\sigma h^2$

Section 11.1

9. $(-2, 0)$ 11. $(-\frac{3}{2}, \frac{3}{2}\sqrt{3})$ 13. $(1, 0)$ 15. $(0, 3)$

17. $[1, \frac{1}{2}\pi + 2n\pi]$, $[-1, \frac{3}{2}\pi + 2n\pi]$ 19. $[3, \pi + 2n\pi]$, $[-3, 2n\pi]$

21. $[2\sqrt{2}, \frac{7}{4}\pi + 2n\pi]$, $[-2\sqrt{2}, \frac{3}{4}\pi + 2n\pi]$

23. $[8, \frac{1}{6}\pi + 2n\pi]$, $[-8, \frac{7}{6}\pi + 2n\pi]$ 25. $[2, \frac{2}{3}\pi + 2n\pi]$, $[-2, \frac{5}{3}\pi + 2n\pi]$

27. symmetry about the x-axis

29. no symmetry about axes or the origin

30. symmetry about both axes and the origin

32. symmetry about both axes and the origin

33. $r^2 \sin 2\theta = 1$ 35. $r = 2$

37. $r = a(1 - \cos \theta)$ 39. the horizontal line $y = 4$

41. the line $y = \sqrt{3}\, x$ 42. the lines $y = \pm\sqrt{3}$

44. the circle $x^2 + (y + 2)^2 = 4$

Section 11.2

1. circle of radius 2 centered at the pole

3. line through the pole with 30° inclination

5. spiral (Figure A.11.2.1)   7. circle of radius $\frac{1}{2}$ centered at $(\frac{1}{2}, 0)$

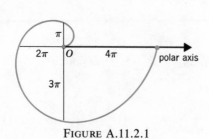

FIGURE A.11.2.1   FIGURE A.11.2.2

9. circle of radius 1 centered at (1, 0)
11. Figure A.11.2.2   13. Figure A.11.2.3   15. Figure A.11.2.4

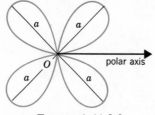

FIGURE A.11.2.3   FIGURE A.11.2.4

Section 11.3

1. yes; $[1, \pi] = [-1, 0]$ and the pair $r = -1$, $\theta = 0$ satisfies the equation
3. yes; the pair $r = \frac{1}{2}$, $\theta = \frac{1}{2}\pi$ satisfies the equation
6. $(-1, 0)$, $(1, 0)$ 8. $(\frac{1}{2}[b^2 - 1], b)$
10. $(0, 0)$, $(\frac{1}{4}, \pm\frac{1}{4}\sqrt{3})$

Section 11.4

1. $\frac{1}{4}\pi a^2$ 3. $\frac{1}{2}a^2$ 5. $\frac{1}{2}\pi a^2$
7. $\frac{1}{4} - \frac{1}{16}\pi$ 9. $\frac{3}{16}\pi + \frac{3}{8}$ 11. $\frac{5}{2}a^2$
13. $\frac{1}{12}(3e^{2\pi} - 3 - 2\pi^3)$ 15. $\frac{1}{4}(e^{2\pi} + 1 - 2e^\pi)$

Section 11.5

1. $(y - 1)^2 = 4x$ 3. $a^3y = (x - b)^3$ 5. $y = e^{(x+1)/2}$
7. $3x - y - 3 = 0$ 9. $y = 1$ 11. $3x + y - 3 = 0$
13. $2x + 2y - \sqrt{2} = 0$ '   15. $2x + y - 8 = 0$ 17. $x - 5y + 4 = 0$
19. $x + 2y + 1 = 0$
23. (a) no horizontal tangents   (b) vertical tangents at $(2, 2)$ and $(-2, 0)$
25. (a) horizontal tangents at $(3, 7)$ and $(3, 1)$
    (b) vertical tangents at $(-1, 4)$ and $(7, 4)$
27. (a) paths intersect at $(2, 10)$ and $(5, 3)$
    (b) particles collide at $(5, 3)$

Section 11.6

4. $x = r \text{ arc cos} \left(\dfrac{r - y}{r} \right) - \sqrt{2ry - y^2}$

7. $2\pi r^2$
9. $\frac{3}{8}\pi a^2$

Section 11.7

1. 2
3. 2
5. $\sqrt{2}$

7. e
8. 0
11. 2

12. $\sqrt{m^2 + 1}(d - c)$
14. $2\sqrt{3}$
16. $2 \log (\sqrt{2} + 1)$

18. $2 - \sqrt{2} - \frac{1}{2} \log 3 + \log (\sqrt{2} + 1)$
20. $\frac{2}{3}[2\sqrt{2} - 1]$

22. $\frac{1}{3}\pi + \frac{1}{2}\sqrt{3}$
24. $\frac{1}{2}[\sqrt{2} + \log (1 + \sqrt{2})]$

26. $\sqrt{2}(e^\pi - 1)$
28. $\sqrt{2}(e^{4\pi} - 1)$

30. $\frac{1}{2}\sqrt{5}(e^{4\pi} - 1)$
32. $2a$

35. $\log (1 + \sqrt{2})$
37. $f(x) = \cosh x = \frac{1}{2}(e^x + e^{-x})$

Section 11.8

1. $2\pi r$
3. $2\pi[\sqrt{2} + \log (1 + \sqrt{2})]$

5. $\frac{1}{9}\pi(17^{3/2} - 1)$
7. $\frac{61}{432}\pi$

9. $3\pi\sqrt{13}$
11. $2\pi r^2$
13. $\frac{6}{5}\pi a^2$

15. $2\pi \displaystyle\int_{\theta_1}^{\theta_2} \rho(\theta) \sin \theta \sqrt{[\rho(\theta)]^2 + [\rho'(\theta)]^2} \, d\theta$

17. $\pi s(R + r)$ where s is the slant height
19. $\pi a^2(20 + \log 3)$

21. (a) $2\pi b^2 + \dfrac{2\pi a b}{e} \text{ arc sin } e$
(b) $2\pi a^2 + \dfrac{\pi b^2}{e} \log \left| \dfrac{1 + e}{1 - e} \right|$,

where e is the eccentricity $\dfrac{\sqrt{a^2 - b^2}}{a}$

Section 11.9

1. Figure A.11.9.1, area = 4
3. Figure A.11.9.2, area = $\sqrt{2} - \frac{1}{8}\pi$

FIGURE A.11.9.1

ray $\theta = \frac{1}{4}\pi$

$r = \sec \theta + \tan \theta$

polar axis

FIGURE A.11.9.2

5. $\frac{19}{27}$
7. $\frac{335}{27}a$
9. $6a$
11. $8r$

13. $\frac{52}{3}\pi p^2$
15. $\frac{47}{16}\pi a^2$
17. $\frac{64}{3}\pi r^2$

Section 12.1

1. decreasing; bounded below by 0 and above by 2
3. increasing; bounded below by 1 but not bounded above
4. not monotonic; bounded below by 0 and above by $\frac{3}{2}$
6. decreasing; bounded below by 0 and above by 0.9
7. increasing; bounded below by $\frac{1}{2}$ but not bounded above
9. increasing; bounded below by $\frac{4}{5}\sqrt{5}$ and above by 2
10. increasing; bounded below by $\frac{2}{51}$ but not bounded above
12. decreasing; bounded below by 0 and above by $\frac{1}{2}(10^{10})$
13. not monotonic; not bounded below and not bounded above
15. increasing; bounded below by 0 and above by log 2
16. not monotonic; bounded below by $-\frac{1}{2}$ and above by $\frac{1}{4}$
18. increasing; bounded below by $\sqrt{3}$ and above by 2
19. increasing; bounded below by $\frac{1}{2}$ and above by 1
21. decreasing; bounded below by 0 and above by $\frac{1}{2}$
22. decreasing; bounded below by 0 and above by $\frac{1}{3}$ log 3
24. decreasing; bounded below by 1 and above by $\sqrt{2}$
25. increasing; bounded below by $\frac{3}{4}$ but not bounded above
27. not monotonic; not bounded below and not bounded above

Section 12.2

1. converges to 0
3. diverges
4. converges to 1
6. converges to 0
7. diverges
9. converges to 4
10. diverges
12. converges to 0
13. diverges
15. converges to log 2
16. converges to 0
18. diverges
19. converges to 1
21. converges to 1
22. converges to $\frac{1}{2}\sqrt{2}$
24. diverges
25. converges to 0
27. converges to 0
28. converges to $\frac{1}{2}$
30. converges to e^2

Section 12.3

1. converges to 1
3. converges to 0
4. converges to 0
6. converges to 0
7. converges to 0
9. converges to 0
10. converges to 0
12. converges to 1
13. converges to $1/e$
15. converges to 1
16. converges to $\frac{1}{2}$
18. converges to 0
19. converges to π
21. converges to 2
22. diverges
24. diverges
27. $\lim\limits_{n\to\infty} 2n \sin(\pi/n) = 2\pi$; the number $2n \sin(\pi/n)$ is the perimeter of a regular polygon of n sides inscribed in the unit circle; as n tends to ∞, the perimeter of the polygon tends to the circumference of the circle
28. $\frac{1}{2}$ 30. $\frac{1}{8}$ 33. converges to $1/e$ 35. converges to 1
37. converges to 0 if $t < 1$, converges to e^x if $t = 1$, diverges if $t > 1$

Section 12.4

1. 0   3. 1   4. 0   6. 1   7. a^2 9. 0
10. 1   12. 0   13. $\frac{1}{2}\pi$ 15. 0   17. $\frac{1}{2}$

19. Figure A.12.4.1; horizontal asymptote x-axis
21. Figure A.12.4.2; horizontal asymptote $y = -2$

FIGURE A.12.4.1

FIGURE A.12.4.2

22. Figure A.12.4.3; horizontal asymptotes $y = \pm\frac{1}{2}$
24. Figure A.12.4.4; horizontal asymptote x-axis

FIGURE A.12.4.3

FIGURE A.12.4.4

Section 12.5

1. 0   3. 1   4. $\dfrac{1}{n}\,a^{1-n}$ 6. 1

7. 0   9. 2   10. $\dfrac{1+\pi}{1-\pi}$ 12. $\log\left(\dfrac{a}{a+1}\right)$

13. $-\frac{1}{2}$ 15. -2 17. $\frac{1}{3}\sqrt{6}$ 18. 0
20. e^2 21. $-\frac{1}{2}$ 23. $\frac{1}{2}$ 25. $-\frac{1}{6}$
27. 1   31. $\frac{1}{3}$

Section 12.6

1. ∞ 3. $-\infty$ 4. -1 6. 0   7. $\frac{1}{5}$
9. 1   10. 0   12. 0   13. 1   15. 1
16. 1   18. $\frac{1}{3}$ 19. 1   21. 1   22. $\frac{1}{2}$
24. 0   26. 0   28. 1   29. 0   31. 1

32. Figure A.12.6.1; $x = 4$ vertical asymptote, x-axis horizontal asymptote
34. Figure A.12.6.2; $x = \pm 2$ vertical asymptotes, x-axis horizontal asymptote

FIGURE A.12.6.1   FIGURE A.12.6.2

35. Figure A.12.6.3; $x = \pm 2$ vertical asymptotes, $y = 1$ horizontal asymptote
37. Figure A.12.6.4; y-axis vertical asymptote, $x = \pm\frac{1}{3}$ horizontal asymptotes

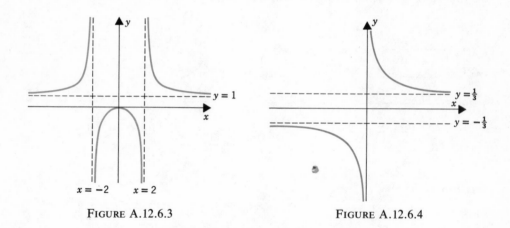

FIGURE A.12.6.3   FIGURE A.12.6.4

38. Figure A.12.6.5; x-axis horizontal asymptote
40. Figure A.12.6.6; y-axis vertical asymptote, x-axis horizontal asymptote

FIGURE A.12.6.5   FIGURE A.12.6.6

Section 12.7

1. 1   3. $\frac{1}{4}\pi$ 4. diverges   6. 2

8. diverges   9. $\frac{1}{2}\pi$ 11. 2   12. $\frac{1}{2}\pi$

14. diverges   15. 1   17. π 18. $\frac{1}{2}\log 3$

20. $\frac{3}{2}$ 21. $\log 2$ 23. $\frac{1}{4}$

24. (b) 1   (c) $\frac{1}{2}\pi$ (d) 2π (e) $\pi[\sqrt{2} + \log(1 + \sqrt{2})]$

26. (b) $\frac{4}{3}$ (c) 2π (d) $\frac{8}{7}\pi$

28. converges by comparison with $\int_0^\infty \dfrac{dx}{x^{3/2}}$

30. diverges since for x large the integrand is greater than $\dfrac{1}{x}$ and $\int_1^\infty \dfrac{1}{x}\,dx$ diverges

31. converges by comparison with $\int_\pi^\infty \dfrac{1}{x^2}\,dx$

33. diverges by comparison with $\int_e^\infty \dfrac{dx}{(x+1)\log(x+1)}$

35. \$2000

Section 13.1

2. (a) 12   (c) -5 (d) -10 (f) $\frac{15}{16}$ (g) $\frac{43}{60}$ (i) $\frac{13}{32}$

5. (a) $\displaystyle\sum_{k=1}^{n} m_k \Delta_k$ (c) $\displaystyle\sum_{k=1}^{n} f(x_k{}^*)\Delta x_k$

6. (a) $\displaystyle\sum_{k=0}^{5} a^{5-k}b^k$ (c) $\displaystyle\sum_{k=0}^{n} a^{n-k}b^k$ (e) $\displaystyle\sum_{k=0}^{n} a_k x^{n-k}$

Section 13.2

1. $\frac{1}{4}$ 3. $\frac{1}{2}$ 4. $\frac{10}{3}$ 6. $\frac{67{,}000}{999}$ 7. $\frac{5}{6}$ 9. $\frac{15}{2}$ 10. $-\frac{3}{2}$

12. $\frac{1}{2}$ 13. $\frac{1}{4}$ 15. $\frac{7}{9}$ 17. $\frac{8}{33}$ 18. $\frac{89}{99}$ 20. $\frac{35}{111}$ 21. $\frac{687}{1100}$

27. (a) $\displaystyle\sum_{k=0}^{\infty} (-1)^k x^{k+1}$ (c) $\displaystyle\sum_{k=0}^{\infty} x^{2k}$

29. $h_0 + \dfrac{2h_0\sigma}{1-\sigma}$ 30. $\displaystyle\sum_{k=1}^{\infty} n_k \left(1 + \dfrac{r}{100}\right)^{-k}$

Section 13.3

1. converges, comparison $\Sigma(1/k^2)$ 3. converges, comparison $\Sigma(1/k^2)$

4. diverges, comparison $\Sigma(1/k)$ 6. converges, comparison $\Sigma(1/k^2)$

7. diverges, comparison $\Sigma(1/3k)$ 9. diverges, comparison $\Sigma(1/k)$

10. converges, integral test   12. diverges, integral test

13. converges, limit comparison $\Sigma(1/k^2)$

15. converges, limit comparison $\Sigma(1/k^4)$

16. diverges, limit comparison $\Sigma(1/k)$

18. converges, limit comparison $\Sigma(1/k^{3/2})$
19. converges, integral test   21. converges, integral test

Section 13.4

1. converges, ratio test   3. converges, root test
4. converges, root test   6. diverges, comparison $\Sigma(1/k)$
7. diverges, limit comparison $\Sigma(1/k)$ 9. converges, root test
10. diverges, comparison $\Sigma(1/k)$ 12. converges, limit comparison $\Sigma(1/k^2)$
13. diverges, ratio test   15. converges, comparison $\Sigma(1/k^{3/2})$
16. converges, ratio test $a_{k+1}/a_k \to 2/e$
18. converges, integral test   19. diverges, integral test
21. diverges, $a_k \not\to 0$ 22. converges, ratio test
24. diverges, $a_k \not\to 0$ 25. converges, root test
27. converges, comparison $\Sigma(1/k^{3/2})$

Section 13.5

1. diverges   3. diverges
5. converges conditionally   7. diverges
8. converges absolutely (terms already positive)
10. converges absolutely   11. diverges
13. converges conditionally
14. converges absolutely (terms already positive)
16. converges conditionally   17. converges absolutely
19. diverges

Section 13.6

1. $1 - x + \dfrac{x^2}{2!} - \dfrac{x^3}{3!} + \dfrac{x^4}{4!} - \dfrac{x^5}{5!}$ 3. $x - \frac{1}{3}x^3 + \frac{1}{5}x^5$

5. $1 - x + x^2 - x^3 + x^4 - x^5$ 7. $x^3 + \frac{1}{3}x^5$
9. $1 + \frac{1}{2}x^2 + \frac{5}{24}x^4$ 11. $x - \frac{1}{2}x^2 + \frac{1}{3}x^3 - \frac{1}{4}x^4$
13. $-\frac{1}{2}x^2 - \frac{1}{12}x^4$ 15. $x + x^2 + \frac{5}{6}x^3 + \frac{1}{2}x^4$
17. Table 2   19. Table 4
21. Table 4   23. Table 3

29. (a) $x + \dfrac{x^3}{3!} + \dfrac{x^5}{5!}$

Section 13.7

1. $6 + 9(x - 1) + 7(x - 1)^2 + 3(x - 1)^3$
3. $-3 + 5(x + 1) - 19(x + 1)^2 + 20(x + 1)^3 - 10(x + 1)^4 + 2(x + 1)^5$

5. $\displaystyle\sum_{k=0}^{\infty} (-1)^k (\tfrac{1}{2})^{k+1}(x - 1)^k$

7. (a) $\dfrac{e^x}{e^a} = e^{x-a} = \displaystyle\sum_{k=0}^{\infty} \dfrac{(x-a)^k}{k!}$, $\quad e^x = e^a \displaystyle\sum_{k=0}^{\infty} \dfrac{(x-a)^k}{k!}$

(b) $R_n(x) = \dfrac{e^c}{n!}(x-a)^n$ with c between a and x

(c) $\dfrac{e^c}{n!}(x-a)^n \to 0$ by (12.3.3)

(d) $e^{a+(x-a)} = e^x = e^a \displaystyle\sum_{k=0}^{\infty} \dfrac{(x-a)^k}{k!}$, $\quad e^{x_1+x_2} = e^{x_1} \displaystyle\sum_{k=0}^{\infty} \dfrac{x_2{}^k}{k!} = e^{x_1}e^{x_2}$.

9. (a) $f^{(n)}(a) = (-1)^{n-1}(n-1)!\,\dfrac{1}{a^n}$, $n \geq 1$

(b) $\log x = \log a + \dfrac{1}{a}(x-a) - \dfrac{1}{2a^2}(x-a)^2 + \dfrac{1}{3a^3}(x-a)^3 - \cdots$

Section 13.8

1. Table 1   3. Table 1   5. Table 4

7. $\log(1+x) = x - \dfrac{x^2}{2} + \dfrac{x^3}{3} - \dfrac{x^4}{4} + \cdots$ for $-1 < x \leq 1$ (13.8.1)

$\log(1-x) = -x - \dfrac{x^2}{2} - \dfrac{x^3}{3} - \dfrac{x^4}{4} + \cdots$ for $-1 \leq x < 1$

$\log\left(\dfrac{1+x}{1-x}\right) = \log(1+x) - \log(1-x) = 2(x + \dfrac{x^3}{3} + \dfrac{x^5}{5} + \cdots)$

for $-1 < x < 1$

9. $\dfrac{1+x}{1-x} = 1.3$ gives $x = \dfrac{3}{23}$

$\log 1.3 \cong 2[\tfrac{3}{23} + \tfrac{1}{3}(\tfrac{3}{23})^3] = 2(\tfrac{1596}{12167}) \cong 0.262$

11. (a) $\log 3 = \log(2+1) \cong \log 2 + 2[\tfrac{1}{5} + \tfrac{1}{3}(\tfrac{1}{5})^3 + \tfrac{1}{5}(\tfrac{1}{5})^5]$

$\cong 0.693 + 2(\tfrac{9503}{46875}) \cong 0.693 + 0.405 = 1.098$

(c) $\log 7 = \log(6+1) \cong \log 6 + 2[\tfrac{1}{13} + \tfrac{1}{3}(\tfrac{1}{13})^3]$

$= \log 3 + \log 2 + 2(\tfrac{508}{6591}) \cong 1.098 + 0.693 + 0.154$

$= 1.945$

13. $\log(a+x) = \log a + \dfrac{x}{a} - \dfrac{x^2}{2a^2} + \dfrac{x^3}{3a^3} - \cdots + (-1)^{n-1}\dfrac{x^n}{na^n} + \cdots$

for $-a < x \leq a$

14. (a) $\tfrac{5}{12}$ (b) $\tfrac{120}{119}$ (c) 1

Section 13.9

1. $(-1, 1)$ 3. $(-\infty, \infty)$ 4. $[-\tfrac{1}{2}, \tfrac{1}{2}]$

6. $(-1, 1]$ 7. $[-2, 2)$ 9. $[-1, 1]$

10. $(-1, 1)$ 12. $[-1, 1)$ 13. $(-1, 1)$

15. $(-10, 10)$ 16. $(-\infty, \infty)$ 18. $(-\infty, \infty)$

20. $(-\tfrac{5}{2}, \tfrac{5}{2})$

Section 13.10

1. (a) $1 + 2x + 3x^2 + \cdots + nx^{n-1} + \cdots$

   (c) $1 + kx + \dfrac{(k+1)k}{2!}x^2 + \cdots + \dfrac{(n+k-1)!}{n!(k-1)!}x^n + \cdots$

   (d) $-x - \dfrac{1}{2}x^2 - \dfrac{1}{3}x^3 - \cdots - \dfrac{1}{n+1}x^{n+1} - \cdots$

   (f) $\log 2 - \dfrac{3}{2}x - \dfrac{1}{2}\left(\dfrac{3}{2}\right)^2 x^2 - \cdots - \dfrac{1}{n+1}\left(\dfrac{3}{2}\right)^{n+1} x^{n+1} - \cdots$

2. (a) $1 + x^2 + \frac{2}{3}x^4 + \frac{17}{45}x^6 + \cdots$
3. (a) -72 4. $0.80 \le I \le 0.81$
6. $0.60 \le I \le 0.61$ 7. $0.29 \le I \le 0.30$
9. $0.60 \le I \le 0.61$ 10. $0.06 \le I \le 0.07$

Section 13.11

6. (a) $1 + \frac{1}{2}x - \frac{1}{8}x^2 + \frac{1}{16}x^3 - \frac{5}{128}x^4$ (c) $1 + \frac{1}{2}x^2 - \frac{1}{8}x^4$
7. (a) 9.90 (c) 4.99 (e) 0.49

Section 13.12

1. 1 3. 0 4. 1 6. 0 7. 0
9. diverges   10. 0 12. 0 13. 0 15. 0
16. $\frac{4}{3}$ 18. $\frac{2}{3}$ 19. 2 21. $\frac{1}{4}$

22. $\dfrac{x}{1-x^5}$, $\;|x| < 1$ 24. $\dfrac{\frac{3}{2}x}{1-x^2}$, $\;|x| < 1$ 25. xe^x, all real x

27. $\frac{1}{4}$ 28. diverges
30. converges absolutely   32. converges conditionally
34. converges absolutely   36. converges conditionally
38. $[\frac{9}{5}, \frac{11}{5})$ 40. $(0, 2)$ 41. $(-\frac{1}{2}, \frac{1}{2})$ 43. $(-9, 9)$
44. converges only at $x = -1$ 46. $(-\infty, \infty)$

47. $1 + (\log a)x + \dfrac{(\log a)^2}{2!}x^2 + \cdots + \dfrac{(\log a)^n}{n!}x^n + \cdots$

49. $x^2 - \frac{1}{2}x^4 + \frac{1}{3}x^6 - \cdots + \dfrac{(-1)^{n-1}}{n}x^{2n} + \cdots$

50. $0.49 \le I \le 0.50$ 52. $3.91 \le \sqrt[3]{60} \le 3.92$

Section 14.1

2. (a) $\sqrt{2}$ (c) 7
3. (a) $x^2 + y^2 + z^2 = r^2$ (b) $(x - a_1)^2 + (y - a_2)^2 + (z - a_3)^2 = r^2$
4. (a) $z = a_3$ (c) $x = a_1$
5. distances to x, y, z axes are $\sqrt{b^2 + c^2}, \sqrt{a^2 + c^2}, \sqrt{a^2 + b^2}$, respectively
7. $x = \frac{1}{3}a_1 + \frac{2}{3}b_1$, $\;\; y = \frac{1}{3}a_2 + \frac{2}{3}b_2$, $\;\; z = \frac{1}{3}a_3 + \frac{2}{3}b_3$
8. (a) $Q(x_0, y_0, -z_0)$ (d) $Q(-x_0, -y_0, -z_0)$

Section 14.3

1. $3\mathbf{i} - 4\mathbf{j} + 6\mathbf{k}$ 3. $-3\mathbf{i} - \mathbf{j} + 8\mathbf{k}$

5. 5   7. 3   9. $\sqrt{6}$

11. (a) $\mathbf{a}, \mathbf{c}, \mathbf{d}$ (b) $\mathbf{a}, \mathbf{c}$ (c) $\mathbf{a}$ and $\mathbf{c}$ both have direction opposite to $\mathbf{d}$

12. (i) $\mathbf{a} + \mathbf{b}$ (ii) $-(\mathbf{a} + \mathbf{b})$ (iii) $\mathbf{a} - \mathbf{b}$ (iv) $\mathbf{b} - \mathbf{a}$

14. (a) $6\mathbf{i} + 3\mathbf{j} + 12\mathbf{k}$ (b) $A = \frac{26}{7}$, $B = -\frac{11}{7}$, $C = \frac{3}{7}$

16. $\alpha = \pm 3$ 19. $\frac{1}{3}\sqrt{6}\,(\mathbf{i} + 2\mathbf{j} - \mathbf{k})$

Section 14.4

1. (a) $\mathbf{a} \cdot \mathbf{b}$ (c) $\|\mathbf{a}\|^2 + \|\mathbf{b}\|^2$

2. (a) $\mathbf{a} \cdot \mathbf{b} = 5$, $\mathbf{a} \cdot \mathbf{c} = 8$, $\mathbf{b} \cdot \mathbf{c} = 18$

   (c) $\mathrm{comp}_b\,\mathbf{a} = \frac{5}{14}\sqrt{14}$, $\mathrm{comp}_c\,\mathbf{a} = \frac{8}{5}$

3. $\frac{1}{2}\mathbf{i} + \frac{1}{2}\sqrt{2}\mathbf{j} - \frac{1}{2}\mathbf{k}$ 5. (a) $\frac{1}{3}\pi$

8. (a) $\mathbf{a} \perp \mathbf{b}$ (b) $\mathbf{a} \perp \mathbf{b}$

9. $\mathbf{a} \cdot \mathbf{b} = \mathbf{a} \cdot \mathbf{c}$ iff $\mathbf{a} \cdot (\mathbf{b} - \mathbf{c}) = 0$ iff $\mathbf{a} \perp \mathbf{b} - \mathbf{c}$

14. (a) $W = \mathbf{F} \cdot \mathbf{r}$ (c) (i) $W_2 = W_1$ (ii) $W_2 = \sqrt{3}W_1$

15. $x = 3$, $x = \frac{5}{2}$

17. $\theta = \mathrm{arc\ cos}\,(\frac{1}{3}\sqrt{3}) \cong \mathrm{arc\ cos}\,0.577 \cong 0.96$ radians

Section 14.5

1. P and Q 2. l_1, l_3, l_4 are parallel

3. all nonzero scalar multiples of $\mathbf{d}$

4. (a) $\mathbf{r}(t) = (3\mathbf{i} + \mathbf{j}) + t\mathbf{k}$

   (d) $\mathbf{r}(t) = x_0\mathbf{i} + y_0\mathbf{j} + z_0\mathbf{k} + t[(x_1 - x_0)\mathbf{i} + (y_1 - y_0)\mathbf{j} + (z_1 - z_0)\mathbf{k}]$

5. (a) $x(t) = 1 + t$, $y(t) = -t$, $z(t) = 3 + t$

   (c) $x(t) = 2$, $y(t) = \phi(t)$, $z(t) = 3$ with ϕ any function with range $(-\infty, \infty)$

7. $\dfrac{x}{x_0} = \dfrac{y}{y_0} = \dfrac{z}{z_0}$ or $\dfrac{x - x_0}{x_0} = \dfrac{y - y_0}{y_0} = \dfrac{z - z_0}{z_0}$ provided that $x_0 y_0 z_0 \neq 0$

9. (b) $P(6, \sqrt{3}, 0), \frac{1}{3}\pi$

   (d) $P(1, 3, 1)$, $\mathrm{arc\ cos}\,(\frac{1}{6}\sqrt{3}) \cong \mathrm{arc\ cos}\,0.289 \cong 1.28$ radians

10. the lines meet at right angles at $P(x_0, y_0, z_0)$

11. the lines are parallel

13. (a) $t_0 = -\dfrac{\mathbf{r}_0 \cdot \mathbf{d}}{\|\mathbf{d}\|^2}$

    (b) $\mathbf{R}(t) = \mathbf{r}(t_0) \pm t\,\dfrac{\mathbf{d}}{\|\mathbf{d}\|}$ where t_0 is as in part (a)

14. (a) $\frac{1}{2}\sqrt{2}$ (b) $\sqrt{2}$

Section 14.6

1. Q
2. R, S
3. $x - 4y + 3z - 2 = 0$
5. $3x - 2y + 5z - 9 = 0$
9. $(-\frac{2}{3}, 0, 0), (0, 2, 0), (0, 0, -\frac{1}{2})$
10. $\pm \frac{1}{62}\sqrt{62}(2\mathbf{i} - 3\mathbf{j} + 7\mathbf{k})$
11. $\frac{1}{2}\pi$
13. $\cos \theta = \frac{2}{21}\sqrt{42} \cong 0.617, \theta \cong 0.91$ radians

15. coplanar
17. not coplanar
19. $\dfrac{|D_2 - D_1|}{\sqrt{A^2 + B^2 + C^2}}$

Section 14.7

1. $-2\mathbf{k}$
3. $\mathbf{i} + \mathbf{j} + \mathbf{k}$
4. $-\mathbf{i} - 2\mathbf{k}$
6. 1
7. -1
9. $\mathbf{0}$
10. $\mathbf{i} + \mathbf{j} - 2\mathbf{k}$
12. -3
14. $5\mathbf{i} - 4\mathbf{j} - \mathbf{k}$
21. (a) $\frac{3}{2}$
22. (a) 1
24. (a) $x + z = 2$
25. (a) $P(-\frac{19}{14}, \frac{15}{7}, \frac{17}{4})$

Section 14.9

1. (a) $3\sqrt{10}$ (c) $(11, -12, 9)$ or $(-1, 9, -6)$
   (e) $x(t) = 6 + 4t, \quad y(t) = 6 - 7t, \quad z(t) = -4 + 5t$
   (g) $x + 27y + 37z - 20 = 0$
2. (a) no
3. $(1, 0, 1)$
4. (a) $2x - z - 1 = 0$ (c) $3x + 2y - z = 0$ (e) $2x - 2y - z - 6 = 0$
6. $x(t) = t - \frac{31}{7}, \quad y(t) = t + \frac{20}{7}, \quad z(t) = t$
8. 1 10. $5x + 2y + 3z - 3 = 0$ 13. 3

Section 15.1

1. $\mathbf{f}'(t) = 2\mathbf{i} - \mathbf{j} + 3\mathbf{k}$

3. $\mathbf{f}'(t) = -\dfrac{1}{2\sqrt{1 - t}}\mathbf{i} + \dfrac{1}{2\sqrt{1 + t}}\mathbf{j} - \dfrac{2}{(1 + t)^2}\mathbf{k}$

5. $\mathbf{f}'(t) = \cos t\mathbf{i} - \sin t\mathbf{j} + \sec^2 t\mathbf{k}$ 7. $\mathbf{f}'(t) = \dfrac{1}{\sqrt{1 - t^2}}\mathbf{i} + \dfrac{2}{1 + 2t}\mathbf{j} + 2t\mathbf{k}$

9. $2\mathbf{i} + \frac{1}{2}\pi^2\mathbf{k}$

11. $\dfrac{1}{e}[(2e - 5)\mathbf{i} + \sqrt{2}(e - 2)\mathbf{j} + (e - 1)\mathbf{k}]$

12. $\mathbf{f}(t) = t\mathbf{i} + (\frac{1}{3}t^3 + 1)\mathbf{j} - \mathbf{k}$ 14. $\mathbf{f}(t) = e^{2t}\mathbf{i} - e^{2t}\mathbf{k}$

Section 15.2

1. $\mathbf{f}'(t) = \mathbf{b}, \quad \mathbf{f}''(t) = \mathbf{0}$
3. $\mathbf{f}'(t) = 2e^{2t}\mathbf{i} - \cos t\mathbf{j}, \quad \mathbf{f}''(t) = 4e^{2t}\mathbf{i} + \sin t\mathbf{j}$
5. $\mathbf{f}'(t) = 2t\mathbf{g}'(t^2), \quad \mathbf{f}''(t) = 4t^2\mathbf{g}''(t^2) + 2\mathbf{g}'(t^2)$

7. $\mathbf{f}'(t) = \frac{1}{2}\sqrt{t}\,\mathbf{g}'(\sqrt{t}) + \mathbf{g}(\sqrt{t}), \quad \mathbf{f}''(t) = \frac{1}{4}\mathbf{g}''(\sqrt{t}) + \dfrac{3}{4\sqrt{t}}\,\mathbf{g}'(\sqrt{t})$

9. $(-\sin t)e^{\cos t}\mathbf{i} + (\cos t)e^{\sin t}\mathbf{j}$ 　　　 11. $e^t\mathbf{i} + e^{-t}\mathbf{j}$

13. $2(\mathbf{b} \times \mathbf{d})$ 　　　　　　　　　　 17. $\mathbf{f}(t) = \mathbf{a} + t\mathbf{b} + \frac{1}{2}t^2\mathbf{c}$

Section 15.3

1. (a) $\pi\mathbf{j} + \mathbf{k}, \quad \mathbf{R}(u) = (\mathbf{i} + 2\mathbf{k}) + u(\pi\mathbf{j} + \mathbf{k})$
   (c) $\mathbf{b} - 2\mathbf{c}, \quad \mathbf{R}(u) = (\mathbf{a} - \mathbf{b} + \mathbf{c}) + u(\mathbf{b} - 2\mathbf{c})$
   (e) $4\mathbf{i} - \mathbf{j} + 4\mathbf{k}, \quad \mathbf{R}(u) = (2\mathbf{i} + 5\mathbf{k}) + u(4\mathbf{i} - \mathbf{j} + 4\mathbf{k})$
3. The scalar components $x(t) = at$ and $y(t) = bt^2$ satisfy

$$a^2 y(t) = a^2 bt^2 = b(at)^2 = b[x(t)]^2$$

and generate the parabola $a^2 y = bx^2$.
5. (a) $P(0, 1)$ 　　 (b) $P(1, 2)$ 　　 (c) $P(-1, 2)$
6. $\mathbf{r}(t) = e^{\alpha t}(\mathbf{i} + 2\mathbf{j} + 3\mathbf{k})$
9. $P(1, 2, -2)$, arc cos $(\frac{1}{5}\sqrt{5}) \cong$ arc cos $0.447 \cong 1.11$ radians

Section 15.4

4. $t = \frac{1}{2}$
5. (a) The scalar components, $x(t) = 2 \cos 2t$ and $y(t) = 3 \cos t$, satisfy

$$
\begin{aligned}
4[y(t)]^2 - 9x(t) &= 4(9 \cos^2 t) - 9(2 \cos 2t) \\
&= 36 \cos^2 t - 18(\cos^2 t - \sin^2 t) \\
&= 18(\cos^2 t + \sin^2 t) = 18.
\end{aligned}
$$

　 (c) $-8\mathbf{i} - 3\mathbf{j}$ at $(2, 3), \quad -8\mathbf{i} + 3\mathbf{j}$ at $(2, -3)$
10. $\mathbf{F}(t) = 2m\mathbf{k}$
11. (a) $\pi b\mathbf{j} + \mathbf{k}$ 　　　　 (b) $\sqrt{\pi^2 b^2 + 1}$ 　　　　　 (c) $-\pi^2 a\mathbf{i}$
　 (d) $m(\pi b\mathbf{j} + \mathbf{k})$ 　　　 (e) 0 　　　　　　　　 (f) $\frac{1}{2}m(\pi^2 b^2 + 1)$
　 (g) $m[b(1 - \pi)\mathbf{i} - 2a\mathbf{j} + 2\pi ab\mathbf{k}]$ 　　　　　 (h) $-m\pi^2 a[\mathbf{j} - b\mathbf{k}]$

Section 15.5

1. $2\pi\sqrt{a^2 + b^2}$ 　　　　　 3. $\log(1 + \sqrt{2})$ 　　　　　 5. $\log(1 + \sqrt{2})$
7. $\sqrt{2}(e^\pi - 1)$ 　　　　 9. $6 + \frac{1}{2}\sqrt{2}\log(2\sqrt{2} + 3)$ 　　 11. 0.50

Section 15.6

1. $\dfrac{\sqrt{2}}{2(1 - 2x + 2x^2)^{3/2}}$ 　 3. $\dfrac{6|x|}{(1 + 9x^4)^{3/2}}$ 　　　 5. $(\frac{1}{2}\sqrt{2}, \frac{1}{2}\log\frac{1}{2})$

7. a/b^2 at $(\pm a, 0)$ and b/a^2 at $(0, \pm b)$ 　　　　　 9. $\frac{3}{10}\sqrt{5}$

13. $\dfrac{|a|}{2(a^2 + b^2)^{3/2}}$ 　 15. $\frac{1}{2}\sqrt{2}$ 　 16. $k(\theta) = \dfrac{|[f(\theta)]^2 + 2[f'(\theta)]^2 - f(\theta)f''(\theta)|}{([f(\theta)]^2 + [f'(\theta)]^2)^{3/2}}$

18. $k(\theta) = \dfrac{\theta^2 + 2}{|a|(\theta^2 + 1)^{3/2}}$ 　　　　 20. $\dfrac{1}{4r}$

Section 16.1

1. dom (f) = the first and third quadrants, including the axes
   ran (f) = $[0, \infty)$
3. dom (f) = the set of all points (x, y) not on the line $y = -x$
   ran (f) = $(-\infty, 0) \cup (0, \infty)$
5. dom (f) = the entire plane
   ran (f) = $(-1, 1)$
7. dom (f) = the first and third quadrants, excluding the axes
   ran (f) = $(-\infty, \infty)$
9. dom (f) = the set of all points (x, y) with $x^2 < y$; in other words, the set of all points of the plane above the parabola $y = x^2$
   ran (f) = $(0, \infty)$
11. dom (f) = all of space
    ran (f) = $[0, \infty)$
13. dom (f) = the set of all points (x, y, z) not on the plane $x + y + z = 0$
    ran (f) = $\{-1, 1\}$
15. dom (f) = the set of all points (x, y, z) with $|y| < |x|$
    ran (f) = $(-\infty, 0]$
17. dom (f) = the set of all points (x, y) such that $x^2 + y^2 < 1$; in other words, the set of all points of the plane inside the unit circle $x^2 + y^2 = 1$
    ran (f) = $[1, \infty)$

Section 16.2

1. a quadric cone   3. a parabolic cylinder
5. a hyperboloid of one sheet   7. sphere of radius 2 centered at the origin
9. an elliptic paraboloid   11. a hyperbolic paraboloid
19. elliptic paraboloid with the following traces: the origin in the xy-plane, the parabola $x^2 = 4z$ in the xz-plane, the parabola $y^2 = 9z$ in the yz-plane
21. quadric cone with the following traces: the origin in the xy-plane, the lines $x = \pm 2z$ in the xz-plane, the lines $y = \pm 3z$ in the yz-plane
23. hyperboloid of two sheets with the following traces: the hyperbola $z^2 - \frac{1}{4}x^2 = 1$ in the xz-plane, the hyperbola $z^2 - \frac{1}{9}y^2 = 1$ in the yz-plane, no trace in the xy-plane
26. $x^2 + y^2 - 4z = 0$ (paraboloid of revolution)
28. along the unit circle $x^2 + y^2 = 1$ in the xy-plane
30. the ellipse $x^2 + 4y^2 - 4 = 0$

Section 16.3

13. $x + 2y + 3z = 0$, plane through the origin
15. $z = \sqrt{x^2 + y^2}$, the upper nappe of the cone $z^2 = x^2 + y^2$ (Figure 16.2.4)
17. the hyperboloid of two sheets

$$\frac{x^2}{(\frac{1}{6})^2} + \frac{y^2}{(\frac{1}{3})^2} - \frac{z^2}{(1)^2} = -1 \quad \text{(Figure 16.2.3)}$$

Section 16.4

1. $\dfrac{\partial f}{\partial x} = 6x - y, \quad \dfrac{\partial f}{\partial y} = 1 - x$

3. $\dfrac{\partial z}{\partial x} = 2Ax + By, \quad \dfrac{\partial z}{\partial y} = Bx + 2Cy$

5. $\dfrac{\partial f}{\partial x} = e^{x-y} - 2xy, \quad \dfrac{\partial f}{\partial y} = -e^{x-y} - x^2$

7. $\dfrac{\partial g}{\partial x} = \dfrac{(AD - BC)y}{(Cx + Dy)^2}, \quad \dfrac{\partial g}{\partial y} = \dfrac{(BC - AD)x}{(Cx + Dy)^2}$

9. $\dfrac{\partial u}{\partial x} = y + z, \quad \dfrac{\partial u}{\partial y} = x + z, \quad \dfrac{\partial u}{\partial z} = x + y$

11. $\dfrac{\partial \rho}{\partial \theta} = \sin(\phi - \tfrac{1}{2}\pi), \quad \dfrac{\partial \rho}{\partial \phi} = (\theta - \tfrac{1}{2}\pi)\cos(\phi - \tfrac{1}{2}\pi)$

13. $\dfrac{\partial \rho}{\partial \theta} = e^{\theta+\phi}[\cos(\theta - \phi) - \sin(\theta - \phi)]$

$\dfrac{\partial \rho}{\partial \phi} = e^{\theta+\phi}[\cos(\theta - \phi) + \sin(\theta - \phi)]$

15. $f_x(0, e) = 1, \quad f_y(0, e) = e^{-1}$

17. $g_x(0, \tfrac{1}{4}\pi) = 0, \quad g_y(0, \tfrac{1}{4}\pi) = 0$

21. (a) $50\sqrt{3}$ square inches   (b) $5\sqrt{3}$ (c) 50
    (d) $\tfrac{5}{18}\pi$ square inches   (e) -2

Section 16.5

1. interior $= \{(x, y): 2 < x < 4, 1 < y < 3\}$ (the inside of the rectangle)
   boundary $=$ the union of the four line segments that bound the rectangle
   set is closed
3. interior $=$ the entire set (region between two concentric circles)
   boundary $= \{(x, y): x^2 + y^2 = 1 \text{ or } x^2 + y^2 = 4\}$ (the two circles)
   set is open
5. interior $= \{(x, y): 1 < x^2 < 4\}$
   $\qquad\quad = \{(x, y): -2 < x < 1\} \cup \{(x, y): 1 < x < 2\}$
   $\qquad\qquad$ (two vertical strips without the boundary lines)
   boundary $= \{(x, y): x = -2, x = -1, x = 1, \text{ or } x = 2\}$
   $\qquad\qquad$ (four vertical lines)
   set is neither open nor closed
7. interior $= \{(x, y): y < x^2\}$ (region below the parabola)
   boundary $= \{(x, y): y = x^2\}$ (the parabola)
   set is closed
9. interior $=$ the entire set [the inside of the ball of radius $\tfrac{1}{2}$ centered at $(1, 1, 1)$]
   boundary $= \{(x, y, z): (x - 1)^2 + (y - 1)^2 + (z - 1)^2 = \tfrac{1}{4}\}$
   $\qquad\qquad$ (the spherical surface)
   set is open

Section 16.6

1. $\dfrac{\partial^2 f}{\partial x^2} = 2A,\quad \dfrac{\partial^2 f}{\partial y^2} = 2C,\quad \dfrac{\partial^2 f}{\partial y\,\partial x} = \dfrac{\partial^2 f}{\partial x\,\partial y} = 2B$

3. $\dfrac{\partial^2 f}{\partial x^2} = Cy^2 e^{xy},\quad \dfrac{\partial^2 f}{\partial y^2} = Cx^2 e^{xy},\quad \dfrac{\partial^2 f}{\partial y\,\partial x} = \dfrac{\partial^2 f}{\partial x\,\partial y} = Ce^{xy}(xy+1)$

5. $\dfrac{\partial^2 f}{\partial x^2} = -\dfrac{1}{4(x+y^2)^{3/2}},\quad \dfrac{\partial^2 f}{\partial y^2} = \dfrac{x}{(x+y^2)^{3/2}},\quad \dfrac{\partial^2 f}{\partial y\,\partial x} = \dfrac{\partial^2 f}{\partial x\,\partial y} = -\dfrac{y}{2(x+y^2)^{3/2}}$

7. $\dfrac{\partial^2 f}{\partial x^2} = \dfrac{1}{(x+y)^2} - \dfrac{1}{x^2},\quad \dfrac{\partial^2 f}{\partial y^2} = \dfrac{1}{(x+y)^2},\quad \dfrac{\partial^2 f}{\partial y\,\partial x} = \dfrac{\partial^2 f}{\partial x\,\partial y} = \dfrac{1}{(x+y)^2}$

9. $\dfrac{\partial^2 f}{\partial x^2} = 2(y+z),\ \dfrac{\partial^2 f}{\partial y^2} = 2(x+z),\ \dfrac{\partial^2 f}{\partial z^2} = 2(x+y)$; the second mixed partials are all $2(x+y+z)$

Section 17.1

1. $\nabla f = e^{xy}[(xy+1)\mathbf{i} + x^2\,\mathbf{j}]$

3. $\nabla f = (6x - y)\mathbf{i} + (1 - x)\mathbf{j}$

5. $\nabla f = (e^{x-y} - 2xy)\mathbf{i} - (e^{x-y} + x^2)\mathbf{j}$

7. $\nabla f = e^{-z}[\,y^2\,\mathbf{i} + 2xy\,\mathbf{j} - xy^2\,\mathbf{k}]$

9. $\nabla f = (y+z)\mathbf{i} + (x+z)\mathbf{j} + (x+y)\mathbf{k}$

11. $\nabla f = e^{x-y}[(1 + x + y)\mathbf{i} + (1 - x - y)\mathbf{j}]$

13. $\nabla f = e^{x}\left[\log y\,\mathbf{i} + \dfrac{1}{y}\,\mathbf{j}\right]$

15. $\nabla f = \dfrac{AD - BC}{(Cx + Dy)^2}\,[\,y\mathbf{i} - x\mathbf{j}]$

17. $\nabla f = -\mathbf{i} + 18\mathbf{j}$

19. $\nabla f = -\mathbf{i}$

Section 17.2

1. $-2\sqrt{2}$

3. $\frac{1}{5}(7 - 4e)$

5. $\frac{1}{4}\sqrt{2}(a - b)$

7. $\frac{2}{3}\sqrt{6}$

9. $-3\sqrt{2}$

11. $-\dfrac{1}{\sqrt{x^2 + y^2}}$

13. (a) $\sqrt{2}[a(B - A) + b(C - B)]$ (b) $\sqrt{2}[a(A - B) + b(B - C)]$

15. $-\frac{7}{5}\sqrt{5}$

19. its projection onto the xy-plane is the curve $y = x^3$ from $(1, 1)$ to $(0, 0)$

21. its projection onto the xy-plane is the curve $(b^2)^{a^2} x^{b^2} = (a^2)^{b^2} y^{a^2}$ from (a^2, b^2) to $(0, 0)$

Section 17.3

1. e^t

3. $3t^2 - 5t^4$

5. $2\omega(b^2 - a^2)\sin \omega t \cos \omega t + b\omega$

7. $\sin 2t - 3\cos 2t$

9. $e^{t/2} \left(\frac{1}{2} \sin 2t + 2 \cos 2t \right) + e^{2t} \left(2 \sin \frac{1}{2}t + \frac{1}{2} \cos \frac{1}{2}t \right)$

11. $1 - 4t + 6t^2 - 4t^3$

13. increasing $\frac{1288}{3}\pi$ cubic inches per second

16. $\dfrac{\partial u}{\partial s} = 2s \cos^2 t - t \sin s \cos t - st \cos s \cos t$

$\dfrac{\partial u}{\partial t} = -2s^2 \sin t \cos t + st \sin s \sin t - s \sin s \cos t$

17. $\dfrac{\partial u}{\partial s} = \dfrac{\partial u}{\partial x} \dfrac{\partial x}{\partial s} + \dfrac{\partial u}{\partial y} \dfrac{\partial y}{\partial s} + \dfrac{\partial u}{\partial z} \dfrac{\partial z}{\partial s}$

$\dfrac{\partial u}{\partial t} = \dfrac{\partial u}{\partial x} \dfrac{\partial x}{\partial t} + \dfrac{\partial u}{\partial y} \dfrac{\partial y}{\partial t} + \dfrac{\partial u}{\partial z} \dfrac{\partial z}{\partial t}$

18. $\dfrac{\partial u}{\partial s} = 2s \cos^2 t - \sin(t - s) \cos t + s \cos t \cos(t - s) + 2t^2 \sin s \cos s$

$\dfrac{\partial u}{\partial t} = -2s^2 \sin t \cos t + s \sin(t - s) \sin t - s \cos t \cos(t - s) + 2t \sin^2 s$

21. $f(x, y, z) = a_1 x + a_2 y + a_3 z + C$

27. $\dfrac{\partial \mathbf{u}}{\partial s} = \dfrac{\partial \mathbf{u}}{\partial x} \dfrac{\partial x}{\partial s} + \dfrac{\partial \mathbf{u}}{\partial y} \dfrac{\partial y}{\partial s}, \quad \dfrac{\partial \mathbf{u}}{\partial t} = \dfrac{\partial \mathbf{u}}{\partial x} \dfrac{\partial x}{\partial t} + \dfrac{\partial \mathbf{u}}{\partial y} \dfrac{\partial y}{\partial t}$

Section 17.4

1. tangent $x + y + 2 = 0$; normal $x - y = 0$

3. tangent $\sqrt{2}x - 5y + 3 = 0$; normal $5x + \sqrt{2}y - 6\sqrt{2} = 0$

4. tangent $x + 4y - 9 = 0$; normal $4x - y - 2 = 0$

6. tangent $x - 5y + 4 = 0$; normal $5x + y - 6 = 0$

7. tangent $x - y - 3 = 0$; normal $x + y + 1 = 0$

9. 0   11. $4x - 5y + 4z = 0$

13. $x + ay - z - 1 = 0$ 15. $2x + 2y - z = 0$

17. $b^2 c^2 x_0 x - a^2 c^2 y_0 y - a^2 b^2 z_0 z - a^2 b^2 c^2 = 0$

19. $(a^2/b, b^2/a, 3ab)$ 21. $(0, 0, 0)$

23. the tangent planes meet at right angles and therefore the normals ∇F and ∇G must meet at right angles:

$$\frac{\partial F}{\partial x} \frac{\partial G}{\partial x} + \frac{\partial F}{\partial y} \frac{\partial G}{\partial y} + \frac{\partial F}{\partial z} \frac{\partial G}{\partial z} = 0$$

25. $\frac{9}{2}a^3 \ (V = \frac{1}{3}Bh)$

28. $\frac{1}{2}\pi - \text{arc cos} \left(\frac{19}{203} \sqrt{29} \right) \cong 1.57 - \text{arc cos } 0.504 \cong 0.53$ radians

30. $3x + 4y + 6z = 22, \quad 6x + y - z = 11$

Section 17.5

1. $(1, 0)$ gives max of 1   3. $(1, -1)$ is a saddle point

5. $(-2, 1)$ gives min of -2

7. no stationary points and no extreme values

9. $(1, 1)$ gives min of -3 11. no stationary points but 0 is min

13. $(-2, 2)$ is a saddle point

15. $(\frac{1}{3}[x_1 + x_2 + x_3], \frac{1}{3}[y_1 + y_2 + y_3])$

17. $x = (a^2b)^{1/3}$, $y = (ab^2)^{1/3}$

Section 17.6

1. $(4, -2)$ gives a local min of -10

3. $(\frac{10}{3}, \frac{8}{3})$ gives a local max of $\frac{28}{3}$

5. $(0, 0)$ is a saddle point; $(2, 2)$ gives a local min of -8

7. $(1, \frac{3}{2})$ is a saddle point; $(5, \frac{27}{2})$ gives a local min of $-\frac{117}{4}$

9. no extreme values

11. no extreme values

13. no extreme values

15. $(\frac{1}{2}, 4)$ gives a local min of 6

17. $(1, 1)$ gives a local min of 3

19. $\frac{1}{27}$

21. $\frac{8}{9}\sqrt{3}\,abc$

22. $\theta = \frac{1}{6}\pi$, $x = (2 - \sqrt{3})P$, $y = \frac{1}{6}(3 - \sqrt{3})P$

25. \$1 per dozen blades, \$2 per razor

26. $a^2 + b^2 + c^2$

Section 17.7

1. 2

3. $-\frac{1}{2}ab$

5. $\frac{2}{9}\sqrt{3}ab^2$

7. $-\frac{1}{9}\sqrt{3}$

9. no maximum exists

11. $19\sqrt{2}$

13. $\frac{1}{27}abc$

15. 1

17. a^3

19. $4A^2(a^2 + b^2 + c^2)^{-1}$ where A is the area of the triangle and a, b, c, are the sides.

23. $Q_1 = 10,000$; $Q_2 = 20,000$; $Q_3 = 30,000$

Section 17.8

1. $df = (3x^2y - 2xy^2)\,\Delta x + (x^3 - 2x^2y)\,\Delta y$

3. $df = (\cos y + y \sin x)\,\Delta x - (x \sin y + \cos x)\,\Delta y$

5. $df = \Delta x - (\tan z)\,\Delta y - (y \sec^2 z)\,\Delta z$

7. $\Delta u = -7.15$, $du = -7.5$

8. $du = 1$

9. $22\frac{249}{352}$ taking $u = x^{1/2}y^{1/4}$, $x = 121$, $y = 16$, $\Delta x = 4$, $\Delta y = 1$

11. $\frac{1}{14}\sqrt{2}\pi$ taking $u = \sin x \cos y$, $x = \pi$, $y = \frac{1}{4}\pi$, $\Delta x = -\frac{1}{7}\pi$, $\Delta y = -\frac{1}{20}\pi$

13. $dz = -\frac{1}{90}$, $\Delta z = -\frac{1}{93}$

14. decreases by about 44.8π cubic inches

16. (a) $\Delta h = -\dfrac{(2r + \Delta r)h}{(r + \Delta r)^2}\,\Delta r$, $\Delta h \cong -\left(\dfrac{2h}{r}\right)\Delta r$

17. decreases about $\frac{4}{5}\pi - \frac{2}{5}$

19. $2.2356 \le s \le 2.2644$

Section 17.9

1. $f(x, y) = \frac{1}{2}x^2y^2 + C$

3. $f(x, y) = xy + C$

5. not a gradient

7. $f(x, y) = \sin x + y \cos x + C$

9. $f(x, y) = xye^x + e^{-y} + C$

11. $f(x, y) = \frac{1}{3}x^3 \text{ arc } \sin y + y - y \log y + C$

13. $f(x, y) = \sinh x + x \sinh y + C$

Section 18.1

1. $a_2 - a_1$
2. $b_2 - b_1$
3. $(a_2 - a_1)(b_2 - b_1)$
4. $(b_2 - b_1)(c_2 - c_1)$
5. $a_2^2 - a_1^2$
6. $\frac{1}{2}(b_2^3 - b_1^3)$
7. $(a_2^2 - a_1^2)(b_2 - b_1)$
8. $(a_2 - a_1)(b_2^2 - b_1^2)$
9. $2n(a_2 - a_1) - 3m(b_2 - b_1)$
10. $3n(a_2 - a_1) - 2m(b_2 - b_1)$
11. $(a_2 - a_1)(b_2 - b_1)(c_2 - c_1)$
12. $(a_2^2 - a_1^2)(b_2 - b_1)(c_2 - c_1)$

Section 18.2

1. $L_f(P) = 1$
2. $U_f(P) = 7$
5. $L_f(P) = 2\frac{1}{4}$
6. $U_f(P) = 5\frac{3}{4}$
9. $I = 4$; the volume of the prism bounded above by the plane $z = x + 2y$ and below by R
10. $L_f(P) = -\frac{7}{16}$
11. $U_f(P) = \frac{7}{16}$
14. $I = 0$
15. $I = 2\frac{2}{3}$

Section 18.4

1. 2
3. $-8/\pi^2$
5. $\frac{1}{6} + 1/\pi$
7. $2 \log 2 - 1$
9. $\frac{2}{27}$
11. $\frac{512}{15}$
13. $2(1 - 1/\sqrt{e})$
15. $\frac{9}{2}$
17. $\frac{5}{2} + 6 \log \frac{2}{3}$
19. 4
21. $\frac{2}{15}$
23. $\frac{1}{2}\pi$

Section 18.5

1. $\frac{3}{4}\pi - \frac{9}{16}\sqrt{3}$
3. 2π
5. $\frac{1}{4}\pi + 2$
7. $\frac{1}{2}\pi - \frac{2}{3}$
9. $2\pi(\sin 1 - \cos 1)$
11. $b^3\pi$
13. $\frac{2}{3}(8 - 3\sqrt{3})\pi$

Section 18.7

1. abc
3. $\frac{2}{3}$
5. $\int_{-r}^{r} \int_{-\sqrt{r^2-x^2}}^{\sqrt{r^2-x^2}} \int_{-\sqrt{r^2-(x^2+y^2)}}^{\sqrt{r^2-(x^2+y^2)}} k(r - \sqrt{x^2 + y^2 + z^2})\, dz\, dy\, dx$
7. $\int_{0}^{1} \int_{-\sqrt{x-x^2}}^{\sqrt{x-x^2}} \int_{-2x-3y-10}^{1-y^2} dz\, dy\, dx$
9. $\int_{-1}^{1} \int_{-2\sqrt{2-2x^2}}^{2\sqrt{2-2x^2}} \int_{3x^2+y^2/4}^{4-x^2-y^2/4} k(z - 3x^2 - \frac{1}{4}y^2)\, dz\, dy\, dx$
11. $\frac{1}{6}abc$
13. (a) $\frac{1}{2}k$ (b) k
15. 9
16. (b) $\frac{63}{10}$
18. $\frac{1}{32}\pi$

Section 18.8

1. $(\frac{5}{2}, 5)$
3. $(1, \frac{8}{5})$
5. $(\frac{10}{3}, \frac{40}{21})$
7. $(2, 4)$
9. $(-\frac{3}{5}, 0)$
11. $V = 2\pi^2 a^2 b$
13. $(\frac{1}{4}a, \frac{1}{4}b, \frac{1}{4}c)$
15. $V = \pi ab(2c + \sqrt{a^2 + b^2})$ (theorem of Pappus)
16. (a) $(\frac{7}{12}, \frac{7}{12}, \frac{7}{12})$ (b) $(\frac{1}{2}, \frac{1}{2}, \frac{2}{3})$ (c) $(\frac{1}{2}, \frac{1}{2}, \frac{1}{2})$
18. (a) $(\frac{1}{6}a, \frac{1}{3}b, \frac{1}{3}c)$
21. $(\frac{5}{6}a, 0)$

Section 18.9

1. (a) $\frac{1}{2}k\pi R^2 h^2$ (b) $\frac{4}{5}k\pi R^{5/2}h$ (c) $\frac{1}{6}k\pi R^4 h$
4. $\frac{1}{2}\pi$ 6. $\frac{3}{2}\pi a^3$ 7. $\frac{32}{9}a^3$ 8. $\frac{5}{2}\pi$ 9. $\frac{1}{36}a^2(9\pi - 16)$ 11. $\frac{1}{32}\pi$

Section 18.10

1. $(\sqrt{3}, \frac{1}{4}\pi, \text{arc cos } \frac{1}{3}\sqrt{3})$ 3. $(2\sqrt{10}, \frac{2}{3}\pi, \text{arc cos } \frac{3}{10}\sqrt{10})$
4. sphere of radius ρ_0 centered at the origin
5. vertical half-plane hinged at the z-axis and making an angle θ_0 with the positive x-axis
6. for $0 < \phi_0 < \frac{1}{2}\pi$ and $\frac{1}{2}\pi < \phi_0 < \pi$, the cone generated by revolving about the z-axis a ray which starts out at the origin and makes an angle of ϕ_0 radians with the positive z-axis; for $\phi_0 = \frac{1}{2}\pi$, the xy-plane; for $\phi_0 = 0$ and $\phi_0 = \pi$, the surface degenerates: in the first case, to the nonnegative z-axis, and in the second case, to the nonpositive z-axis
8. horizontal plane one unit above the xy-plane

11. $V = \displaystyle\int_0^\alpha \int_0^\pi \int_0^a \rho^2 \sin \phi \, d\rho \, d\phi \, d\theta = \frac{2}{3}\alpha a^3$

13. $V = \displaystyle\int_0^{2\pi} \int_0^{\text{arc tan } (r/h)} \int_0^{h \sec \phi} \rho^2 \sin \phi \, d\rho \, d\phi \, d\theta = \frac{1}{3}\pi r^2 h$

14. $M = \displaystyle\int_0^{2\pi} \int_0^{\text{arc tan } (r/h)} \int_0^{h \sec \phi} k\rho^3 \sin \phi \, d\rho \, d\phi \, d\theta = \frac{1}{6}k\pi h[(r^2 + h^2)^{3/2} - h^3]$

15. $V = \displaystyle\int_0^{2\pi} \int_0^\alpha \int_0^a \rho^2 \sin \phi \, d\rho \, d\phi \, d\theta = \frac{2}{3}\pi(1 - \cos \alpha)a^3$

16. $\rho = 2b \cos \phi$

17. (b) $M = \displaystyle\int_0^{2\pi} \int_0^{\pi/2} \int_0^{2b \cos \phi} \rho^3 \sin^2 \phi \, d\rho \, d\phi \, d\theta = \frac{1}{4}\pi^2 b^4$

18. $V = \displaystyle\int_0^{2\pi} \int_0^{\pi/4} \int_0^2 \rho^2 \sin \phi \, d\rho \, d\phi \, d\theta + \int_0^{2\pi} \int_{\pi/4}^{\pi/2} \int_0^{2\sqrt{2} \cos \phi} \rho^2 \sin \phi \, d\rho \, d\phi \, d\theta$
$= \frac{1}{3}(16 - 6\sqrt{2})\pi$

Section 19.1

1. 1   3. 0   5. $-\frac{17}{6}$ 7. 0
9. $\log \frac{8}{5}$ 11. $-\pi$ 13. $\frac{23}{21}$ 15. $|W| = A$

Section 19.2

1. 0   3. -1 5. 0   7. -4 9. 4

10. (a) $f(\mathbf{r}) = K \log \|\mathbf{r}\| + C$ (b) $f(\mathbf{r}) = -\left(\dfrac{K}{n - 2}\right)\dfrac{1}{\|\mathbf{r}\|^{n-2}} + C$

12. (a) $M = 2a^2$, $x_M = y_M = \frac{1}{8}|a|(\pi + 2)$

(c) $\displaystyle\int_C \mathbf{F}(\mathbf{r}) \cdot d\mathbf{r} = \int_a^b [\mathbf{F}(\mathbf{r}(t)) \cdot \mathbf{r}'(t)]\, dt$

$\displaystyle\qquad\qquad = \int_a^b \left[\mathbf{F}(\mathbf{r}(t)) \cdot \frac{\mathbf{r}'(t)}{\|\mathbf{r}'(t)\|} \right] \|\mathbf{r}'(t)\|\, dt$

$\displaystyle\qquad\qquad = \int_a^b [\mathbf{F}(\mathbf{r}(t)) \cdot \mathbf{T}(\mathbf{r}(t))]s'(t)\, dt$

$\displaystyle\qquad\qquad = \int_a^b [\mathbf{F}(\mathbf{r}) \cdot \mathbf{T}(\mathbf{r})]\, ds.$

Section 19.3

3. 2π 5. $a^2 b$ 7. 7π 9. $5a\pi r^2$

11. 0 14. (b) $\frac{3}{8}\pi a^2$ 15. $3\pi r^2$

Section 19.4

1. $\frac{1}{2}\sqrt{a^2 b^2 + a^2 c^2 + b^2 c^2}$

3. $\dfrac{1}{3a^4} (\theta_2 - \theta_1)[(a^4 r_2^2 + 1)^{3/2} - (a^4 r_1^2 + 1)^{3/2}]$

5. $\frac{1}{6}\pi[(4a^2 + 1)^{3/2} - (a^2 + 1)^{3/2}]$

7. $\frac{1}{15}[36\sqrt{6} - 50\sqrt{5} + 32]$ 9. 4π

12. (a) $\displaystyle\iint_\Omega \sqrt{\left[\frac{\partial g}{\partial y}(y, z)\right]^2 + \left[\frac{\partial g}{\partial z}(y, z)\right]^2 + 1}\; dy\,dz = \iint_\Omega \sec\,[\alpha(y, z)]\, dy\,dz$

where α is the angle between the unit normal with positive $\mathbf{i}$ component and the x-axis

(b) $\displaystyle\iint_\Omega \sqrt{\left[\frac{\partial h}{\partial x}(x, z)\right]^2 + \left[\frac{\partial h}{\partial z}(x, z)\right]^2 + 1}\; dx\,dz = \iint_\Omega \sec\,[\beta(x, z)]\, dx\,dz$

where β is the angle between the unit normal with positive $\mathbf{j}$ component and the positive y-axis

Section 19.5

1. (a) $\frac{1}{2}[\sqrt{2} + \log(1 + \sqrt{2})]$ (c) $2\sqrt{2} - 1$

  (d) $\frac{1}{3} - \frac{5}{12}\sqrt{2} + \frac{1}{4}\log(1 + \sqrt{2})$ (f) $\frac{4}{3}$

2. (a) $\frac{1}{2}\sqrt{3}a^2 k$ (c) $\frac{1}{12}\sqrt{3}a^4 k$ 4. (a) $\frac{1}{2}a^3$ (c) $\frac{1}{3}a^3$

5. (a) 0 (c) $-\frac{3}{2}$ 6. (a) $\frac{8}{35}$ (c) $-\frac{4}{63}$

7. $(\frac{1}{3}, \frac{1}{3}, \frac{1}{3})$ 9. $(0, 0, \frac{1}{2}a)$

Section 19.6

1. 2 3. 0 5. $2(xy + yz + xz)$ 6. $(x^2 + y^2 + z^2)^{-1}$ 10. 0

11. Here $\mathbf{n} = x\mathbf{i} + y\mathbf{j} + z\mathbf{k}$ and div $\mathbf{v} = 3$.

$$\iint_S (\mathbf{v} \cdot \mathbf{n}) \, d\sigma = \iint_S (x^2 + y^2 + z^2) \, d\sigma = \iint_S d\sigma = A[S] = 4\pi,$$

$$\iiint_T \text{div } \mathbf{v} \, dxdydz = \iiint_T 3 \, dxdydz = 3 \, V(T) = 4\pi.$$

13. Here $\mathbf{n} = x\mathbf{i} + y\mathbf{j} + z\mathbf{k}$ and div $\mathbf{v} = 2(x + y + z)$.

$$\iint_S (\mathbf{v} \cdot \mathbf{n}) \, d\sigma = \iint_S (x^3 + y^3 + z^3) \, d\sigma,$$

$$\iiint_T \text{div } \mathbf{v} \, dxdydz = \iiint_T 2(x + y + z) \, dxdydz.$$

Both integrals are zero since the integrands have the property

$$f(-x, -y, -z) = -f(x, y, z),$$

and S and T are both symmetric about the origin.

15.

face	$\mathbf{n}$	$\mathbf{v} \cdot \mathbf{n}$	flux	
$x = 0$	$-\mathbf{i}$	0	0	
$x = 1$	$\mathbf{i}$	1	1	
$y = 0$	$-\mathbf{j}$	0	0	
$y = 1$	$\mathbf{j}$	1	1	total flux $= 3$
$z = 0$	$-\mathbf{k}$	0	0	
$z = 1$	$\mathbf{k}$	1	1	

$$\iiint_T \text{div } \mathbf{v} \, dxdydz = \iiint_T 3 \, dxdydz = 3 \, V(T) = 3$$

17.

face	$\mathbf{n}$	$\mathbf{v} \cdot \mathbf{n}$	flux	
$x = 0$	$-\mathbf{i}$	0	0	
$x = 1$	$\mathbf{i}$	1	1	
$y = 0$	$-\mathbf{j}$	xz		fluxes add up to 0 total flux $= 2$
$y = 1$	$\mathbf{j}$	$-xz$		
$z = 0$	$-\mathbf{k}$	0	0	
$z = 1$	$\mathbf{k}$	1	1	

$$\iiint_T \text{div } \mathbf{v} \, dx \, dy \, dz = \iiint_T 2(x + y) \, dx \, dy \, dz$$

$$= 2(\bar{x} + \bar{y}) \, V(T) = 2(\tfrac{1}{2} + \tfrac{1}{2})1 = 2$$

19. flux $= \iiint_T (1 + 4y + 6z) \, dxdydz = (1 + 4\bar{y} + 6\bar{z}) \, V(T)$

$$= (1 + 0 + 3)9\pi = 36\pi$$

Section 19.7

1. **0**   3. **0**   4. **0**   6. $-x\mathbf{i} + xy\mathbf{j} + (1 - x)z\mathbf{k}$

8. **0**   12. $\text{div}(\mathbf{u} \times \mathbf{v}) = \mathbf{v} \cdot \text{curl } \mathbf{u} - \mathbf{u} \cdot \text{curl } \mathbf{v}$

14. Note that curl $\mathbf{v} = \mathbf{0}$ and set C: $\mathbf{r}(t) = \cos t\mathbf{i} + \sin t\mathbf{j}$, $t \in [0, 2\pi]$.

$$\int_C \mathbf{v}(\mathbf{r}) \cdot d\mathbf{r} = \int_C x\,dx + y\,dy = \int_0^{2\pi} 0\,dt = 0,$$

$$\iint_S (\text{curl } \mathbf{v} \cdot \mathbf{n})\,d\sigma = \iint_S 0\,d\sigma = 0.$$

16. C: $\mathbf{r}(t) = \cos t\mathbf{i} + \sin t\mathbf{j}$ with $t \in [0, 2\pi]$,
    curl $\mathbf{v} = -3y^2\mathbf{i} + 2z\mathbf{j} + 2\mathbf{k}$, $\mathbf{n} = x\mathbf{i} + y\mathbf{j} + z\mathbf{k}$.

$$\int_C \mathbf{v}(\mathbf{r}) \cdot d\mathbf{r} = \int_C z^2\,dx + 2x\,dy = \int_0^{2\pi} 2\cos^2 t\,dt = 2\pi,$$

$$\iint_S (\text{curl } \mathbf{v} \cdot \mathbf{n})\,d\sigma = \iint_\Omega 2\,dxdy = 2\pi.$$

18. S: $z = 2 - x - y$, with $0 \le x \le 2, 0 \le y \le 2 - x$;
    $C = C_1 \cup C_2 \cup C_3$ with

$$\mathbf{r}_1(t) = 2(1 - t)\mathbf{i} + 2t\mathbf{j}, \quad t \in [0, 1]$$
$$\mathbf{r}_2(t) = 2(1 - t)\mathbf{j} + 2t\mathbf{k}, \quad t \in [0, 1]$$
$$\mathbf{r}_3(t) = 2(1 - t)\mathbf{k} + 2t\mathbf{i}, \quad t \in [0, 1];$$

$\mathbf{n} = \tfrac{1}{3}\sqrt{3}(\mathbf{i} + \mathbf{j} + \mathbf{k})$.

$$\int_C \mathbf{v}(\mathbf{r}) \cdot d\mathbf{r} = \left(\int_{C_1} + \int_{C_2} + \int_{C_3}\right) \mathbf{v}(\mathbf{r}) \cdot d\mathbf{r} = -2 + 2 + 2 = 2,$$

$$\iint_S (\text{curl } \mathbf{v} \cdot \mathbf{n})\,d\sigma = \iint_S \tfrac{1}{3}\sqrt{3}\,d\sigma = \tfrac{1}{3}\sqrt{3}A[S] = 2.$$

20. $\displaystyle\int_C \mathbf{v}(\mathbf{r}) \cdot d\mathbf{r} = \left(\int_{C_1} + \int_{C_2} + \int_{C_3}\right) \mathbf{v}(\mathbf{r}) \cdot d\mathbf{r} = (\tfrac{4}{3} - \tfrac{32}{5}) + \tfrac{32}{5} + 0 = \tfrac{4}{3}$

$$\iint_S (\text{curl } \mathbf{v} \cdot \mathbf{n})\,d\sigma = \iint_\Omega y\,dxdy = \int_0^2 \int_0^{2-x} y\,dy\,dx = \tfrac{4}{3}$$

Section A.1

1. $\{-1, 0, 1, 2\}$ 3. $\{-1, 0, 1, 2, 3, 4\}$ 4. $\{1\}$

6. $\{2\}$ 7. $\varnothing$ 9. $\{2, 4\}$

10. $\{(0, -1), (0, 0), (0, 1), (2, -1), (2, 0), (2, 1)\}$

12. $\left\{\begin{array}{l}(0, -1, 0), (0, 0, 0), (0, 1, 0), (2, -1, 0), (2, 0, 0), (2, 1, 0) \\ (0, -1, 2), (0, 0, 2), (0, 1, 2), (2, -1, 2), (2, 0, 2), (2, 1, 2)\end{array}\right\}$

13. A 15. the set of real numbers   16. $\{x: 3 < x \le 4\}$ 18. A

20. $\{0, 1, 2\}, \{0, 1\}, \{0, 2\}, \{1, 2\}, \{0\}, \{1\}, \{2\}$

Section A.2

1. Let S be the set of integers for which the statement is true. Since $(1)(2) = 2$ is divisible by 2, $1 \in S$.

    Assume now that $k \in S$. This tells us that $k(k + 1)$ is divisible by 2 and therefore

$$(k + 1)(k + 2) = k(k + 1) + 2(k + 1)$$

    is also divisible by 2. This places $k + 1 \in S$.
    We have shown that

$$1 \in S \quad \text{and that} \quad k \in S \quad \text{implies} \quad k + 1 \in S.$$

    It follows that S contains all the positive integers.

4. Let S be the set of integers for which the statement is true. Since $1 = 1^2$, $1 \in S$.

    Assume now that $k \in S$. This tells us that

$$1 + 3 + 5 + \cdots + (2k - 1) = k^2.$$

    It follows that

$$
\begin{aligned}
1 + 3 + 5 + \cdots + (2k - 1) + [2(k + 1) - 1] &= k^2 + [2(k + 1) - 1] \\
&= k^2 + 2k + 1 \\
&= (k + 1)^2.
\end{aligned}
$$

    This places $k + 1 \in S$.
    We have shown that

$$1 \in S \quad \text{and that} \quad k \in S \quad \text{implies} \quad k + 1 \in S.$$

    It follows that S contains all the positive integers.

7. Let S be the set of integers for which the statement is true. Since

$$3^{2(1)+1} + 2^{1+2} = 27 + 8 = 35$$

    is divisible by 7, $1 \in S$.
    Assume now that $k \in S$. This tells us that

$$3^{2k+1} + 2^{k+2} \text{ is divisible by 7.}$$

    It follows that

$$
\begin{aligned}
3^{2(k+1)+1} + 2^{(k+1)+2} &= 3^2 \cdot 3^{2k+1} + 2 \cdot 2^{k+2} \\
&= 9 \cdot 3^{2k+1} + 2 \cdot 2^{k+2} \\
&= 7 \cdot 3^{2k+1} + 2(3^{2k+1} + 2^{k+2})
\end{aligned}
$$

    is also divisible by 7. This places $k + 1 \in S$.
    We have shown that

$$1 \in S \quad \text{and that} \quad k \in S \quad \text{implies} \quad k + 1 \in S.$$

    It follows that S contains all the positive integers.

Index

NOTES

NOTES

NOTES

NOTES

NOTES

NOTES

NOTES

NOTES

NOTES

NOTES

NOTES

NOTES